Computational Electromagnetics

For a complete listing of the *Artech House Antenna Library*,
turn to the back of this book.

Computational Electromagnetics

Konada Umashankar
Allen Taflove

Artech House
Boston • London

Library of Congress Cataloging-in-Publication Data

Umashankar, Konada
Computational Electromagnetics/Konada Umashankar
Includes bibliographical references and index.
ISBN 0-89006-599-3
1. Electromagnetism. 2. Maxwell equations—Numerical solutions.
3. Maxwell equations—Data processing. 4. Moments method (Statistics).
5. Integro-differential equations—Numerical solutions. I. Taflove, Allen.
II. Title.
QC760.U54 1993 93-12355
537–dc20 CIP

British Library Cataloguing in Publication Data

Umashankar, Konada
Computational Electromagnetics
I. Title. II. Taflove, Allen.
QC760.U54 1993 93-12355
537
ISBN 0-89006-599-3

685 Canton Street
Norwood, MA 02062

International Standard Book Number: 0-89006-599-3
Library of Congress Catalog Card Number: 93-12355

10 9 8 7 6 5 4 3 2 1

Contents

Chapter 1

Introduction

Much of what we call modern technology depends upon the engineering analysis and synthesis of high-frequency electromagnetic systems. In applications as diverse as telecommunication, radar, geophysical studies, medical diagnosis and therapy, and strategic defense, the ability to understand and control the radiation, scattering, and penetration of electromagnetic waves is of great importance. Analysis and synthesis of high-frequency electromagnetic systems ultimately is based upon obtaining accurate solutions of Maxwell's equations for the systems of interest. Computational electromagnetics involves the development and application of analyses and corresponding numerical algorithms for the solution of Maxwell's equations.

This book is concerned with the fundamental principles and rigorous validations of one major contemporary thrust in high-frequency computational electromagnetics – the frequency-domain integro-differential equation approach, with a solution via the *method of moments* (MOM). No work is complete without insight into real-time knowledge of the electromagnetic fields. A sequel to this book will address the fundamental principles and rigorous validations of another major contemporary thrust in high-frequency computational electromagnetics – the time-domain differential-equation approach with a solution via the *finite-difference time-domain* (FD-TD) method. Extensive results reported in this book, based upon the frequency-domain integro-differential equation approach as applied to electromagnetic scattering and interaction, are rigorously validated using the finite-difference time-domain method. Overall, this book provides discussions of basic electrophysics, analysis, algorithm development, relevant computer science issues, and validation studies to illuminate the thrust area and point to research frontiers in computational electromagnetics.

1.1 HISTORICAL SKETCH

The early development of computational electromagnetics was prompted by intellectual curiosity concerning the implications of Maxwell's unified electric and magnetic field theory (1873), especially after Hertz's verification late in the 1880s of the predicted wireless propagation of electromagnetic energy. This early work concentrated on obtaining exact analytical solutions of Maxwell's equations for a variety of diffraction problems. As discussed later, diffraction theory evolved during the next 60 years toward geometrical optics. In the years after World War II, the present geometrical theory of diffraction was formulated and elaborated. During this time, researchers were seeking straightforward techniques to analyze complex electromagnetic interaction problems. As early as 1930s, methods based on integral equation formulation and reduction of the operator type equations to practically manageable matrix formulations were considered in other branches of science. This methodology was introduced to the electromagnetics community by many researchers during and after 1950s and is currently well known as the *method of moments technique*.

1.1.1 HIGH-FREQUENCY DIFFRACTION THEORY

One of the first exact solutions of a diffraction problem was given in 1881 by Rayleigh for the scattering of an electromagnetic wave by a perfectly conducting circular cylinder. The solution was obtained by separation of variables, yielding an infinite series of cylindrical wave functions. An analogous procedure was employed for the sphere by Mie in 1908. In 1891, Kirchoff devised an approximate method for the case of diffraction through a hole in a thin screen.

The separation of variables approach was ultimately applied to the class of scatterer shapes that can be conformably fit into orthonormal coordinate systems. This class is clearly limited in number. Further, only electrically small shapes (spanning a small number of wavelengths) can be treated with this approach since the number of characteristic wave functions required to be summed up to achieve a numerically converged result increases with electrical size.

Attempts to treat electrically large scatterers of a simple shape can be traced back to Sommerfeld's work of 1896, which provided a solution for the scattering of a plane or spherical wave by a semi-infinite, thin screen in terms of Fresnel integrals. Asymptotic evaluation of integrals for surface currents can be followed back to Debye's work of 1908. In the late 1930s and early 1940s, workers including Pauli, Pekeris, and Fock obtained asymptotic expansions for scattering of canonical objects such as the wedge and paraboloid of revolution. In the late 1940s, Luneberg, Friedlander, and Kline devised a general high-frequency theory of diffraction by showing that a field is associated with each point on a geometrical optics ray and that it has an asymptotic expansion in inverse powers of the wave number, k.

In the 1950s, Keller formulated the *geometrical theory of diffraction* (GTD). This was based upon a generalization of Luneberg's asymptotic series to include fields associated with diffracted rays arising from vertices or corners, smooth curved surfaces, discontinuities in curvature, and other features. Keller extended Fermat's principle to derive laws governing these rays. Then, by considering canonical problems, he was able to determine diffraction coefficients of various kinds, decay exponents, and other features. Diffraction theory in the years after Keller's formulation has concentrated on overcoming the defects of GTD. Because GTD yields nonphysical field infinities at focal points, caustics, and shadow boundaries, efforts have been made to represent the field uniformly at these loci in terms of special functions or surface integrals. Because GTD requires diffraction coefficients for each type of edge, vertex structure, and surface that constitutes a scatterer, efforts have been made to employ asymptotic or numerical techniques to obtain the necessary coefficients for a wide variety of potential scatterer features; as an example, dielectric edges and surfaces present extremely challenging problems in this vein. Because it is difficult to apply GTD in cavities, waveguides, and other reentrant regions where rays bounce repeatedly – in fact, a chaotic ray behavior can result – efforts have been made to partially or fully represent the field in terms of modes, rather than rays. Finally, because GTD can employ different numbers and types of rays in different regions of the scatterer, with different expressions for the fields associated with them, efforts have been made to obtain a uniform presentation without greatly increasing the required computation.

1.1.2 WAVE INTERACTIONS AT LOW AND MODERATE FREQUENCIES

Despite continuous progress in radio technology in the 50 years after Hertz's experiments, as exemplified by Marconi and Popov (radio telegraph), Fleming (diode), de Forest (triode amplifier), Armstrong (superheterodyne), Southworth (waveguides), Kraus (directional antennas), and the Varians (magnetron), the first large-scale effort to solve practical electromagnetic engineering problems is generally considered to have been undertaken at the M.I.T Radiation Laboratory during World War II. Computational electromagnetics for arbitrary structures at low and moderate frequencies can perhaps be traced back to the work by Schwinger and others, who applied the variational method to microwave problems, and Marcuvitz, who organized the solutions to many of these problems into the popular handbook. In the 1950s, Rumsey formalized some of the concepts into a more compact notation in his "reaction concept," and other researchers such as Harrington used these variational methods for a number of practical problems.

By the early 1960s, the expansion of industrial and defense needs in electromagnetic engineering, combined with the appearance of readily available computers having substantial speed and storage, prompted a number of researchers to investigate the use of numerical methods to solve the electromagnetic field solutions. For example, Mei and Van Bladel used a subsectional and point-matching method to compute the scattering from rectangular cylinders. Others, including Andreasen and Oshiro, were developing similar methods.

In the mid-1960s Harrington set the agenda for the next 20 years by working out a systematic, functional-space description of electromagnetic interactions which he called the *method of moments*. Here, an integral or integro-differential equation derived from Maxwell's equations for the structure of interest is interpreted as the infinite-dimensional functional equation, $Lf = g$, where L is the linear operator associated with the integral or integro-differential equation, g is a known function related to the excitation, and f is an unknown function such as an induced current distribution that is to be determined. The

MOM approach was to set up a numerical solution by using a projection from the infinite-dimensional functional space onto a finite-dimensional subspace, as rigorously defined by Hilbert and used extensively in quantum mechanics in the 1920s. The projection was accomplished by representing the unknown function, f, by a linear combination of a finite set of basis functions, f_i, in the domain of L. Then, a finite set of weighting functions, w_i, is defined in the range of L. After taking the inner product (usually an integration) of functional expansion with each weighting function, the linearity of the inner product is used to obtain a finite set of equations for the coefficients of the basis functions. This set of equations is then solved to obtain the approximate or exact solution for f (approximate or exact depending upon the choice of the basis and weighting functions).

In the years after Harrington's formulation, most work in the computational electromagnetics at low and moderate frequencies has concentrated on refining the method of moments and applying it to a wide variety of important scattering and penetration problems. For example, MOM has been successfully used to model radiation and scattering by wires and rods, scattering by two-dimensional metal and dielectric cylinders, scattering by metal and dielectric bodies of revolution, radiation and scattering by bodies of translation, scattering by three-dimensional metal and dielectric objects of arbitrary shape, field penetration through apertures in planar conducting screens and subsequent coupling to wires, pulse response of wires and canonical two- and three-dimensional bodies, and high-frequency diffraction coefficients for GTD analyses. The widespread use of MOM for engineering applications is ample testimony to its power and flexibility. This book, in fact, attempts to expose some general ideas behind electromagnetic modeling using the integral equations technique.

1.1.3 HYBRID AND ITERATIVE FORMULATIONS

The two successful approaches, Keller's geometric theory of diffraction and Harrington's method of moments have well served the electromagnetic engineering community for classes of problems that can be fit into the modeling "envelope" of each general approach. However, by the early 1980s, it became clear that rigorous engineering design tools are needed for structures that cannot be well treated by either approach. As a class, such structures are electrically large, but cannot be easily modeled using GTD or its variants because of their complex shape and material composition. For example, the surfaces might have reentrant or cavitylike features that require multiple rays. As noted by Keller, such features may lead to chaotic ray behavior that renders GTD modeling inappropriate. Also, the surfaces might be metal coated with dielectric or permeable media or composed of nonmetallic carbon or ceramic fiber composites. In either case, substantial problems in obtaining the necessary diffraction coefficients may occur, again rendering GTD modeling inappropriate.

In principle, the method of moments can incorporate the physics of reentrant features and nonmetallic composition. Yet, because MOM leads to systems of linear equations having dense, full coefficient matrices, the required computer storage has an N^2 dependence, where N is the number of field unknowns; and the required computer execution time to invert the MOM system matrix has an N^2 to N^3 dependence. With spatial resolution requirements on the order of 0.2 to 0.1 wavelength to avoid aliasing of vital near-field magnitude and phase data, this implies that arbitrary three-dimensional structures spanning more than a very few wavelengths would exhaust computer memories, even if these memories have capacities measured in the billions of bytes. Electrically large structures would also tax the arithmetic speeds of computers, even if these speeds are measured in the billions of floating point operations per second (gigaflops).

The continuing expansion of industrial and defense needs in electromagnetic engineering, combined with perceived fundamental limitations of the pure GTD and MOM approaches, has prompted a number of researchers to formulate hybrid techniques that combine the desirable features of both GTD and MOM. In the mid-1970s, initial work in this area by Thiele and Newhouse and by Burnside, Yu, and Marhefka effectively used MOM to obtain numerical GTD diffraction coefficients of edges and corners and GTD to model the diffraction physics away from the edges and corners. This approach mitigated the intensive computer resource requirements of MOM, since the required locus of MOM application could be confined to the area immediately adjacent to geometric singularities. However, this approach cannot be usefully applied in large multiray regions where the required MOM application zone constitutes the entirety of the multiray region. Additional difficulty arises when the structure surface is coated or not perfectly conducting, again extending the required MOM application zone.

Recognizing limitations of hybrid techniques, a second group of researchers attempted to develop more efficient solution approaches for the pure method of moments. In the early and mid-1980s, initial work in this area by Sarkar and by Mittra concentrated on iterative formulations. Here, the conjugate gradient and fast Fourier transform techniques were used to eliminate the need to store and invert a dense, full, coefficient matrix. This reduces the required MOM computer storage and execution time from an order (N^2) dependence to an order ($N \log N$) or even an order (N) dependence, where N is the number of field unknowns. Such a dimensional reduction in computer resource requirements would permit the detailed modeling capabilities afforded by MOM to be extended to electrically large structures, spanning 10 or more wavelengths in three dimensions, using available computers.

However, iterative formulations for MOM were found to suffer from convergence problems in that progress of the initial guess for the field solution toward the correct answer could slow or even stop during the iterative process, depending upon the nature of the electromagnetic wave interaction problem and the resultant degree of ill-conditioning of the MOM system. As a consequence, research began to focus on developing efficient means of preconditioning the MOM system to ensure uniform, rapid convergence for most problems. Alternative iterative formulation based on spatial decomposition of the scatterer is also being considered.

1.1.4 FINITE-DIFFERENCE FORMULATION

In the mid-1960s, Yee introduced a computationally efficient means of directly solving Maxwell's time-dependent curl equations using finite differences. With this approach, the continuous electromagnetic field in a finite volume of space is sampled at distinct points in a space lattice and at distinct equal-spaced points in time. Wave propagation, scattering, and penetration phenomena are modeled in a self-consistent manner by marching in time, that is, repeatedly implementing the finite-difference analog of the curl equations at each lattice point. This results in a simulation of the continuous actual waves by sampled-data numerical analogs propagating in a data space stored in a computer. Space and time sampling increments are selected to avoid aliasing of the continuous field distribution, and to guarantee stability of the time-marching algorithm. Time marching is completed when the desired late-time or steady-state field behavior is observed. An important example of the latter is the sinusoidal steady state, which eventually results if a continuous sinusoidal incident excitation is assumed.

The Yee formulation, designated as the finite-difference time-domain method, permit in principle the modeling of electromagnetic wave interactions with a level of detail as high as that of the method of moments. Unlike MOM, however, FD-TD does not lead to a system of linear equations defined over the entire problem space; updating each field component requires knowledge of only the immediately adjacent field components computed one-half time step earlier (and thus available in memory). Therefore, overall computer storage and running time requirements for FD-TD are linearly proportional to N, the number of field unknowns in the finite volume of space being modeled. This order (N) dependence of computer resources is, in fact, the goal of the iterative approaches for MOM. It is noted here again, in the 20 years after Harrington's formulation, published MOM models have usually dealt with discretizing the surfaces of structures, so that N for such models is proportional to the surface area, rather than the volume (as required by FD-TD). Demands to model structures of increasing complexity will likely require more volumetric MOM modeling, however.

Despite this potential advantage, the use of Yee's formulation of FD-TD was very limited until the early 1980s because of a number of basic problems. First, Yee's formulation provided for no simulation of the field sampling space extending to infinity. This deficiency caused spurious, nonphysical reflection of the numerical wave analogs at the outermost planes of the space lattice. Second, it provided for no simulation of an incident wave having an arbitrary duration or an arbitrary angle of incidence and angle of polarization. Third, it provided for no means to obtain sinusoidal steady-state magnitude and phase data from the computed transient field response. Fourth, it provided for no means to simulate wave interactions with important structures, such as wires and slots, having dimensions smaller than one lattice cell. Fifth, it provided no means to compute far-field radiation or scattering patterns. And sixth, its required volumetric space discretization caused its computer resource needs to seem prohibitive. By the mid-1980s, the major difficulties with FD-TD were overcome. Extensive novel research publications by Taflove and Umashankar during the 1970s and early 1980s have put the electromagnetic modeling technique based on the FD-TD on a firm credible foundation. The applications of the FD-TD to electromagnetic scattering, penetration, coupling and interaction studies have advanced beyond the capabilities of any other alternative approaches. In fact, an accurate simulation of the extension of the field sampling space to infinity was reported by Mur, who had adapted previous work on numerical wave radiation conditions performed by Engquist and Majda. An accurate simulation of an incident wave having an arbitrary duration and arbitrary angle of incidence and angle of polarization was deduced independently by Mur and by Taflove and Umashankar. This simulation was accomplished by zoning the FD-TD space lattice into a total-field region (in which the structure of interest is embedded) surrounded by a scattered-field region and providing a proper connecting condition between the regions. Taflove and Umashankar also introduced means to obtain sinusoidal steady-state data from the transient response, accurately simulate wave interactions with narrow slots, wires, and wire bundles, and accurately compute far scattered fields and radar cross section.

Finally, the appearance of computers such as those of the CRAY series (supercomputers) provided machines with sufficient central memory and arithmetic speed to contain FD-TD models of the three-dimensional structures spanning 10 to 30 wavelengths. Validations of FD-TD models at all levels for important canonical problems against MOM results and careful experimental procedures indicated that FD-TD and MOM were providing comparable degrees of modeling detail and accuracy, with FD-TD permitting the modeling of structure sizes about one order of magnitude larger than those permitted by the noniterative, pure MOM approach.

1.1.5 FINITE-ELEMENT FORMULATION

In the mid-1970s, Mei, Morgan and Chang introduced a finite-element approach for the Helmholtz equation, suitable for axially symmetric three-dimensional conducting and dielectric structures. With this approach, the continuous electromagnetic field in a finite, axially symmetric volume of space containing the structure is approximated by a set of piecewise-linear (pyramid) functions, U_i, wherein each U_i is defined over a specific triangular spatial element. On defining linear pyramid weighting functions, W_j, wherein each W_j is unity at node j and zero at the surrounding nodes, a Galerkin procedure is used to obtain the coefficients of the basis functions, U_i, and thereby the desired field. Using this procedure, excellent agreement with available analytical and experimental data was obtained for plane wave scattering by a dielectric sphere, a finite dielectric cylinder with a spherical void, and other like geometries, having a maximum dimension of about four wavelengths.

In the early 1980s, Mei and his colleagues shifted their finite-element research to direct solutions of Maxwell's curl equations. By analogy with the Yee finite-difference work, we will call this formulation the *finite-element time-domain method*. Related work in FE-TD was also being conducted during this time by Madsen and his colleagues. Mei's approach involved the use of isoparametric quadrilateral elements, point matching, a uniform rectangular mesh everywhere except at the vicinity of the body contour (where conforming boundary elements are used), and a numerical wave radiation condition to truncate a finite-element mesh. Two-dimensional codes constructed showed excellent results for several convex conducting and dielectric bodies. Three-dimensional codes were also constructed by both Mei's group and Madsen's group. Here, a key consideration is optimizing the efficiency of mesh generation and coordinate storage, because these have a heavy impact on the overall computer resource requirement of any finite-element code.

1.1.6 OVERALL PERSPECTIVE

It should be noted that Maxwell's equations were formulated about 90 years before the appearance of useful electronic computers in an era when solutions were necessarily analytical. Substantial momentum in the analytical area carried through World War II until the early 1960s, when it became clear that emerging computer technology made numerical studies feasible.

In the late 1960s, 1970s, and early 1980s, the appearance of readily available computer systems with speeds on the order of 1 megaflop (1 million floating-point arithmetic operations per second) and semiconductor memory capacities on the order of 1 megabyte (1 million bytes) permitted implementation of high-frequency diffraction theory and the method of moments for practical engineering problems. These two approaches, which essentially limit the wave interaction physics being modeled to the surface of the structure of interest, presented computer resource needs well matched to the existing machines.

By the middle and late 1980s, relentless progress in computer hardware technology had achieved thousandfold advances in speed and memory capacity relative to the machines that had been typically used for electromagnetic engineering. By 1990s, a computer having a processing rate of 10 gigaflops (10 billion floating-point operations per second) and a central memory capacity of 10 gigabytes (10 billion bytes) was feasible. The realization of this capacity could calculate complicated electromagnetic wave interactions from first principles (a direct solution of Maxwell's partial differential equations) and spurred new research activity in the area of finite-difference and finite-element time-domain solutions. Enormously enhanced computer speeds and storage capacities now made possible modeling wave interaction physics within the entirety of the structure of interest. This is essential for understanding multiray regions, penetration and coupling to wires in complicated cavities, and the physics of advanced composite structures. As in the 1960s and 1970s, there was a good match between computer resources needed by the computational electromagnetics algorithms and the resources of the existing machines.

Overall, we note a trend wherein advancing computer hardware technology permits the effective implementation of algorithms based closer to the governing partial differential equations of the physical system of interest. This results in a continuing expansion of the range of problems that can be modeled.

1.2 SYNOPSIS OF CONTEMPORARY APPROACHES

A number of useful alternative analytical and numerical approaches are presently available to study electromagnetic wave interactions with three-dimensional objects. In categorizing these approaches, we find that the shape and material characteristics of the object of interest play important roles in selecting the approach used to model its electromagnetic interaction properties. Further, we find that the need for specific time-domain or frequency-domain modeling data also plays an important role in selecting the modeling approach.

1.2.1 CATEGORIZATION OF APPROACHES

We shall first consider the shape and material characteristics of a structure that influence the choice of numerical modeling approach. Following Table 1.1, a structure can first be classified according to shape as either an open body or a closed body. Open bodies are generally conducting and have apertures (holes) leading to some interior zone, while closed bodies do not and are penetrable only through the possibility of dielectric composition. In the case of open structures, apertures and cavities can play a major role in electromagnetic interactions, providing complex effects that are difficult to analyze, especially when the cavities are loaded by materials. In the case of closed structures, details of the object's composition can assume a key role, especially where possible material inhomogeneities, electrical loss, and anisotropy strongly interact with the incident field energy.

Electromagnetic interaction modeling approaches have also been categorized according to their ability to provide specific time-domain or frequency-domain data. Direct time-domain techniques generally keep the time derivatives in Maxwell's equations and provide the complete solution consisting of the homogeneous part (transient) and the particular solution part that depends on the specific type of excitation function. Direct time-domain approaches have been used to model both sinusoidal and nonsinusoidal excitations, the latter including pulses and tone bursts. Equivalent frequency-domain data have been obtained either by marching the solution to the sinusoidal steady state (for a single-frequency sinusoidal excitation) or by performing a Fourier transform of the transient response (for a suitable impulsive excitation).

Direct frequency-domain techniques generally employ a time-harmonic excitation function; assume that all charges, currents, and fields have a phasor (sinusoidal steady-state) time dependence; and strive to solve for the unknown phasor current or field quantities for the assumed single-frequency excitation. Equivalent time-domain data have been obtained by computing the magnitude and phase of a complete spectrum of the

Table 1.1
Categorization of numerical modeling approaches for electromagnetic interactions with structures

---

A. Shape and Material Characteristics of Structure

1. Open bodies with or without apertures
   - Perfectly conducting objects
   - Electric type
   - Magnetic type

2. Closed bodies
   - Dielectric
   - Inhomogeneous
   - Anisotropic objects

B. Time-Domain or Frequency-Domain Data, or Asymptotic Analysis

1. Frequency domain techniques
   - Method of moments
   - Unimoment method
   - Finite element methods
   - Iterative methods

2. Time-domain techniques
   - Finite-difference time-domain
   - Potential integral approach
   - Singularity expansion method

3. High-frequency methods
   - Geometrical theory of diffraction
   - Physical theory of diffraction
   - Spectral methods

C. Hybrid Methods

1. Combination of preceding methods
2. Method of moments / high-frequency method
3. Finite difference-time domain / method of moments
4. Finite difference-time domain / high-frequency method

---

electric and magnetic field solution in the frequency-domain and then using inverse Fourier transform and convolution techniques to model a specific impulsive excitation.

1.2.2 DIRECT TIME-DOMAIN TECHNIQUES

Direct time-domain techniques fall under two main categories: (a) integral equation formulations; and (b) differential equation formulations. The integral equation form is generally referred to as the *retarded potential integral approach.* This approach sets up an equation for the electromagnetic field boundary conditions at the surface of a wire or a scattering body. Methods have been introduced that permit marching-in-time solution of the integral equation, which is effectively an iterative approach. An important advantage of this approach is that the computation space is generally limited to the surface of the structure. A disadvantage is that back-storage in time is required to perform the retarded integrals. This substantially increases the necessary computer storage and has limited the application of the retarded potential integral approach to wires, metal or dielectric bodies of revolution, and electrically small objects.

The differential equation form is adapted in the *finite-difference time-domain* and *finite-element time-domain* methods. Here, Maxwell's time-dependent curl equations are directly solved in a volume region of space that fully contains the structure being modeled. A separate numerical wave radiation condition is employed to truncate the computation zone, effectively simulating its extension to infinity. Wave propagation, scattering, and penetration phenomena are modeled in a self-consistent manner by marching in time, that is, repeatedly implementing the finite-difference or finite-element analog of the curl equations at each lattice cell or element. This results in a simulation of the continuous actual waves by sampled-data numerical analogs propagating in a data space stored in the computer. An important feature of this approach is that no back-storage in time is required to permit time marching, other than the set of field components computed everywhere in the lattice at the previous time step. A disadvantage is that the computation space is not limited to the surface of a structure, but must include its interior and sufficient exterior space to permit effective implementation of the radiation condition. However, this may not be a disadvantage and, in fact, may be an important advantage, for modeling highly inhomogeneous dielectric structures such as biological tissues, geophysical strata, and reentrant or shielding metal structures having complex interior loading. Published applications include modeling scattering and penetration interactions with structures spanning as many as 30 wavelengths in three dimensions.

1.2.3 DIRECT FREQUENCY-DOMAIN TECHNIQUES

Similar to time-domain techniques, direct frequency-domain techniques can fall under the categories of integral equation formulations and differential equation formulations. Before dealing with these categories, it should be noted that frequency-domain techniques also include the separation of variables approach, wherein analytically exact solutions have been obtained for a small class of simple scattering shapes such as spheres, circular cylinders, and elliptical cylinders. Each such shape has a boundary surface that coincides with a constant-coordinate locus in a separable coordinate systems which permits an eigenfunction expansion of the interior and exterior fields. For an obstacle not deviating greatly from a sphere, namely a prolate spheroid, the perturbation technique has been applied.

The broad category of frequency-domain integral equation formulations treats electromagnetic interactions as a boundary value problem deriving and then solving an integral or integro-differential equation for the electric field and/or magnetic field boundary conditions at the surface of a scatterer. These equations are, in fact, not general and have to be rederived based on the type of scattering geometry and its material characteristics. To obtain meaningful data for the interactions, a numerical technique is employed to solve the applicable equation. As stated earlier, the most widely used contemporary approach for this purpose is the *method of moments*, now augmented by iterative techniques based on preconditioned conjugate gradient methods.

For those objects that have an arbitrary shape, either the volume or surface integral equations can be applied. The volume integral equations is based principally on relating the "polarization current" in terms of the total field. By associating an unknown polarization current coefficient with either each point or cubic cell inside the given region, the operator form of the integral equation is first converted into an equivalent matrix equation. The resulting matrix equation can be easily solved for the unknown coefficients in the low-frequency regime. Since the region of the scatterer is modeled cell by cell, arbitrary inhomogeneity and shape can be easily incorporated into this approach. However, this method leads to very large matrices, which rapidly exhaust computer storage and lead to numerical difficulties such as ill conditioning. To some extent, recent advances in iterative techniques based on fast Fourier transforms and the conjugate gradient approach have extended the applicability of volume integral equation approaches. Scattering results using the volume equivalence principle and method of moments have been reported for dielectric cylinders and for planar slabs, cylinders, and more realistic cross sections of biological tissues. Extensive efforts have been devoted to advance further applications.

The surface integral equation approach is well suited for analyzing conducting, homogeneous dielectric objects or objects made up of distinct layers of homogeneous regions. The usual procedure in this case is to set up coupled integral equations in terms of equivalent electric and magnetic currents on the surfaces of the homogeneous regions. For an object made up of large numbers of layers, the fields induced in any one layer are due to the equivalent electric and magnetic currents on the adjacent layers. This results in a block tridiagonal matrix.

An iterative procedure has been utilized for solving the currents on the outermost layer in terms of the currents on the inner layers. The surface equivalence approach is, in principle, applicable to all layered scatterers, regardless of shape. For simple objects such as dielectric cylinders and bodies of revolution, the surface integral equation method has been applied with good results for surface fields and radar cross section. But, when the desired structure is three dimensional and has an arbitrary shape, modeling of the surface electric and magnetic currents becomes complicated. Recently, an efficient approach utilizing triangular patch modeling of arbitrary three-dimensional conducting and dielectric bodies has been reported and validated.

Some work has been reported in the area of the extended boundary condition approach related to scattering. This approach expresses the fields in terms of integrals over surfaces separating one or more homogeneous regions surrounding a given scatterer. Here, one uses the fact that in all regions complementary to those in which the equivalence is valid, the fields must vanish. In these null-field regions, the integral expressions for the fields are expanded in terms of either the spherical or cylindrical harmonics of the standing wave type for interior three-dimensional or two-dimensional regions. Also, the fields are expanded in terms of outgoing wave modes for the exterior regions. Based on the boundary conditions, a set of equations is obtained for the coefficients of the harmonic functions. This method has been applied to multilayered dielectric scatterers. The limitations of this approach arise because it depends on the object having an interior region in which a circumscribed sphere or circle can be placed (depending upon the problem dimensionality). Hence, the method is better suited for nearly spherical or cylindrical objects than for elongated objects. Detailed modification in the approach is required for other object shapes.

The differential equation formulation is adapted in the unimoment method. In this method, depending on the problem dimensionality, a spherical or a cylindrical region surrounds the given scatterer. The minimum radius of this region is selected to totally enclose the scatterer so that the scattered fields in the exterior region can be expanded in terms of outgoing spherical or cylindrical wave functions. The incident field can also be expanded in terms of incoming wave harmonics. A finite-element approach is applied to solve the Helmholtz equation within the interior region. Appropriate boundary conditions for the tangential fields are enforced at the interface between the interior and exterior zones. This provides a set of equations that determine the coefficients of the unknown scattered fields, as well as the interior penetrating fields. Scattering results using the unimoment method have been reported for three-dimensional dielectric bodies of revolution spanning up to four wavelengths. The method is well suited for such structures and efficiently deals with inhomogeneities. However, it is not clear whether the unimoment approach can be applied efficiently to arbitrary structures that are not bodies of revolution.

1.2.4 HIGH-FREQUENCY DIFFRACTION TECHNIQUES

For electrically large objects, it has been found that high-frequency approximations can yield much information about the far-field scattering response, especially for conducting structures. Here, ray tracing based upon the geometrical theory of diffraction, uniform theory of diffraction, physical theory of diffraction, spectral theory of diffraction, and other variants has been adopted for engineering studies of scattering and radiation properties of aircraft, antennas, and other important structures. Because the effects of small features and material properties cannot be easily handled in the framework of high-frequency diffraction techniques, various hybrid approaches involving the method of moments and the high-frequency techniques have been explored to permit a self-consistent composition of the scattering effects of large as well as small structural features and conducting as well as dielectric material composition.

Additional work has been reported in attempting to apply high-frequency techniques to multiray regions, especially cylindrical cavities, using either hybrid ray/modal approaches or pure ray tracing within the cavities. These studies have substantial practical importance in the defense area, where cavity physics is an important element of penetration and scattering concerns for practical systems.

1.3 ASSESSMENT OF GENERAL APPLICABILITY

Tables 1.2, 1.3 and 1.4 provide an assessment of the general applicability of four of the major types of numerical modeling approaches for electromagnetic wave interactions with structures. The approaches include

1. High-frequency diffraction theory, including GTD and its variants;
2. The frequency-domain integral equation formulation, with solution via the method of moments;
3. Iterative approaches for category 2 employing preconditioned conjugate gradient techniques; and
4. The time-domain differential equation formulation, with solution via either the finite-difference time-domain or the finite-element time-domain methods.

Applicability of the above four major types of numerical modeling approaches is listed in three key areas, assuming that three-dimensional structures are to be modeled

1. Range of structure size, in wavelengths;
2. Generic structural geometry (shape) features; and
3. Structure material composition.

In constructing these tables, it was noted that not all of the major modeling approaches currently have the same level of development in terms of demonstrated working codes and validations. Therefore, table entries represent an assessment of probable capability to be demonstrated by each approach in the late 1980s and beginning of 1990s. Extensive research is still underway to improve numerical modeling capabilities.

Table 1.2

Range of structure size, in wavelengths

Modeling approach	< λ/10	→ λ	→ 10λ	→ 100λ	> 100λ
High-frequency techs.			✓	✓	✓
Method of moments	✓	✓	✍		
MOM / iterative	✓	✓	✓	✍	
FD-TD or FE-TD	✍	✓	✓	✓	✍

Notes: Blank = Not applicable

✍ = Not sure if applicable

✓ = Applicable

Table 1.3

Generic structural geometry (shape) features

Modeling Approach	*Close to Spherical*	*With Edges*	*With Corners*	*With Corner Reflectors*	*With Arbitrary Unloaded Cavities*	*With Arbitrary Loaded Cavities*
High-frequency techs.	✔	✔	✔	✍		
Method of moments	✔	✔	✔	✔	✍	✍
MOM / iterative	✔	✔	✔	✔	✍	✍
FD-TD or FE-TD	✔	✔	✔	✔	✔	✔

Notes: Blank = Not applicable

✍ = Not sure if applicable

✔ = Applicable

Table 1.4
Structure material composition

Modeling Approach	*Perfect Electrical Conductors*	*Perfect Magnetic Conductors*	*Homogeneous Dielectrics*	*Inhomogen Dielectrics*	*Lossy Media*	*Aniostropic Media*	
						Diag	*Genl*
High-frequency techs.	✔	✔	✍				
Method of moments	✔	✔	✔	✔	✔	✔	✍
MOM / iterative	✔	✔	✔	✔	✍	✍	
FD-TD or FE-TD	✔	✔	✔	✔	✔	✔	✍

Notes: Blank = Not applicable

✍ = Not sure if applicable

✔ = Applicable

As pointed out earlier, this book is concerned with the fundamental principles and rigorous validations of one major contemporary thrust – the frequency-domain integro-differential equation approach, with solutions via the method of moments. A sequel to this book will address the fundamental principles and rigorous validations of another major contemporary thrust in high-frequency computational electromagnetics – the time-domain differential-equation approach with solutions via the finite-difference time-domain method.

Extensive numerical results of two-dimensional studies are reported in this book based on the frequency-domain integro-differential equation approach as applied to electromagnetic scattering and interaction. They are rigorously validated using the finite-difference time-domain method. In the MOM numerical technique, only the pulse-type piecewise numerical modeling is extensively considered. In fact, the pulse set is ideal for interaction studies dealing with the complex material and shape of structures. The scope of this book is not intended to cover all other types of piecewise numerical modeling. Overall, this book provides discussions of basic analysis, algorithm development, relevant computer science issues, and validation studies to illuminate the thrust area and point to future research frontiers in computational electromagnetics.

Chapter 2

Maxwell's Equations

Electric currents and charges form the principal sources for simulating electric field and magnetic field distributions in a given region of space. The static electric charges produce a static electric field and the constant electric currents produce a static magnetic field. Time-varying electric currents and charges produce time-varying electric and magnetic fields. But, the time-varying electric charges and the corresponding electric currents are interrelated through the electromagnetic source continuity equation, and thus, the corresponding electric and magnetic fields are also directly coupled to each other.

The study of electromagnetic fields involves understanding the physics of the sources that produce the electric and magnetic fields. It involves the study of physical laws governing the mechanism of the electric and magnetic fields. Familiar electromagnetic laws, such as, Coulomb's law, Gauss's law, Biot-Savart's law, and Ampere's law explain the various relationships between the static sources and their corresponding static electric and magnetic fields. In general, the electric charges and electric currents are time-dependent sources. Correspondingly, the electric and magnetic fields so produced are coupled force fields that are, in fact, vector quantities with both spatial and temporal dependence. The basic static field laws are not sufficient to explain the various physical interactions associated with the time-varying sources and their fields. For explaining the interactions associated with the time-varying fields, Faraday's law and modified Ampere's law are required. These fundamental laws of electromagnetism are referred to as *Maxwell's equations*. These are generally cast in terms of compact vector representation. In general, the time-dependent electric and magnetic fields existing in any a given region of space, to be unique, should satisfy the classical Maxwell's equations.

In the following sections, various types of mathematical modeling for the time-varying electric charge density, electric current density, and the corresponding mathematical relationship between the electric field and the magnetic field quantities are discussed.

2.1 CHARGE DENSITY DISTRIBUTIONS

Let us consider the following macroscopic model for a source charge distribution. The electric charges generally distribute throughout a given region of space depending upon medium characteristics, assuming a mechanism exists to support them. For example, conducting materials can easily support either positive or negative charge distributions.

In a given region of space, the time-dependent charges may be distributed randomly. To account for the charges quantitatively, charge density distributions can be defined on a macroscopic level in an average sense. Such mathematical models are convenient and useful in electromagnetic field studies. Three different practical cases of the electric charge density distributions – such as volume, surface, and line distributions – are considered later.

Let us consider first the volume distribution of electric charges in a very large three-dimensional region that is linear, homogeneous, and isotropic medium. To account for the time-varying charge distribution at each and every point within the source region, consider an arbitrary source point (x', y', z'), which is enclosed within a small incremental volume Δv. It is assumed that there exist in this small volume $+Nv$ number of positive charges and $-Nv$ number of negative charges. Based on these charge numbers (assuming that the charges can be counted), one can calculate the net charge distribution as either positive or negative within the small incremental volume.

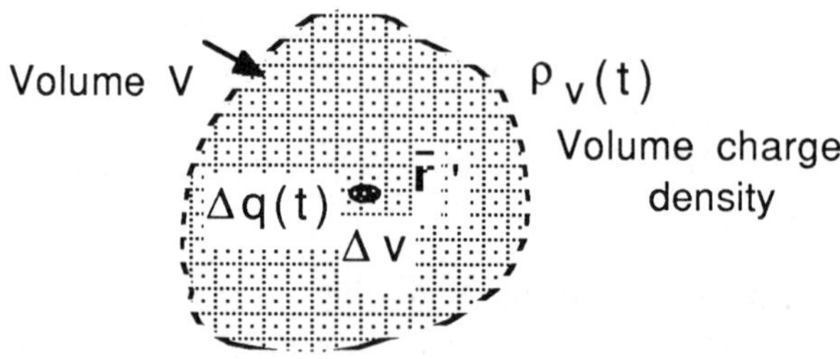

Figure 2.1 Continuous distribution of electric charge density – Volume case.

Referring to Figure 2.1, let

$\Delta q(\bar{r}',t)$: net charge in the incremental volume, in coulombs;
$\Delta v(\bar{r}')$: incremental volume, in cubic meters;
t : time coordinate variable, in second;
$\bar{r}'$: location of the source point under consideration

$$= (x',y',z'). \tag{2.1.1}$$

Considering a ratio of the net electric charge in the incremental volume to its volume, and then taking a limit as the incremental volume tends to zero, gives the volume charge density at the source point:

$$\rho_v(\bar{r}',t) = \lim_{\Delta v \to 0} \frac{\Delta q(\bar{r}',t)}{\Delta v(\bar{r}')}$$

$$= \frac{\partial q(\bar{\mathbf{r}}',t)}{\partial v(\bar{\mathbf{r}}')} \qquad \text{coulombs per cubic meter} \tag{2.1.2}$$

Hence, if a time-dependent volume charge density is known within a given volume, the total charge contained in the volume can be calculated by considering volume integration over the complete volume charge density distribution. In Figure 2.1, using the definition of (2.1.2), the incremental charge contained in the incremental volume is given by

$$dq(\bar{\mathbf{r}}',t) = \rho_V(\bar{\mathbf{r}}',t)\ dv(\bar{\mathbf{r}}') \tag{2.1.3}$$

The total charge, in coulombs, contained in the volume *V* is

$$q(t) = \iiint\limits_V \rho_V(t)\ dv \tag{2.1.4}$$

where the triple integration is performed over the complete volume containing the source charge distribution. Any coordinate systems, either rectangular or cylindrical or spherical, can be selected to simplify (2.1.4). But, if one invokes the symmetry of the charge density distribution (if any) with respect to the specific coordinate variables, then, simplification of the integral expression (2.1.4) is generally straightforward.

The time-dependent surface distribution of charges can be considered in a similar manner. All charges are assumed to be confined to an arbitrary surface. To account for the source charges at each and every point on a given surface, consider an arbitrary source point (x', y', z') which is enclosed within an incremental surface Δs. It is assumed that there exist $+Ns$ number of positive charges and $-Ns$ number of negative charges in the incremental surface. Based on these charge numbers, one can calculate the net charge as either positive or negative within the incremental surface.

Referring to Figure 2.2, using the earlier notation, let

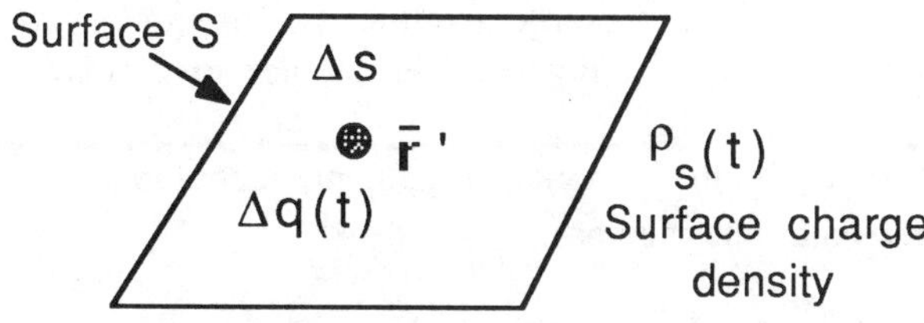

Figure 2.2 Continuous distribution of electric charge density – Surface case.

$\Delta q(\bar{r}',t)$: net charge in the incremental surface, in coulombs;
$\Delta s(\bar{r}')$: incremental surface area, in square meter.

Considering a ratio of the net charge in the incremental surface to its area and then taking a limit as the incremental surface tends to zero, gives the surface charge density at the source point:

$$\rho_S(\bar{r}',t) = \lim_{\Delta s \to 0} \frac{\Delta q(\bar{r}',t)}{\Delta s(\bar{r}')}$$

$$= \frac{\partial q(\bar{r}',t)}{\partial s(\bar{r}')} \qquad \text{coulombs per square meter} \qquad (2.1.5)$$

Hence, if the time-dependent surface charge density on a given surface is known, the total charge contained on the surface can be calculated by surface integration over the complete surface charge density distribution. In Figure 2.2, using the definition of (2.1.5), the incremental charge contained in the incremental surface is given by

$$dq(\bar{r}',t) = \rho_S(\bar{r}',t)\ ds(\bar{r}') \qquad (2.1.6)$$

The total charge, in coulombs, contained on the surface S is

$$q(t) = \iint_S \rho_S(t)\ ds \qquad (2.1.7)$$

The time-dependent line source distribution of electric charges can be modeled in a similar way. It is assumed that all charges are confined to an arbitrary contour path, C. To account for the charge distribution at each and every point along a given contour, consider an arbitrary source point (x', y', z') that is enclosed within an incremental line segment ΔL. Again, it is assumed that there exists in the incremental line segment $+N\ell$ number of positive charges and $-N\ell$ number of negative charges. Based on these charge numbers, one can calculate the net charge as either positive or negative within the incremental line segment. Referring to Figure 2.3, using the earlier notation, let

$\Delta q(\bar{r}',t)$: net charge in the incremental segment, in coulombs;
$\Delta L(\bar{r}')$: incremental line segment, in meter.

Considering a ratio of the net charge within the incremental line segment to its length, and then taking a limit as the incremental length tends to zero, gives the line charge density at the source point.

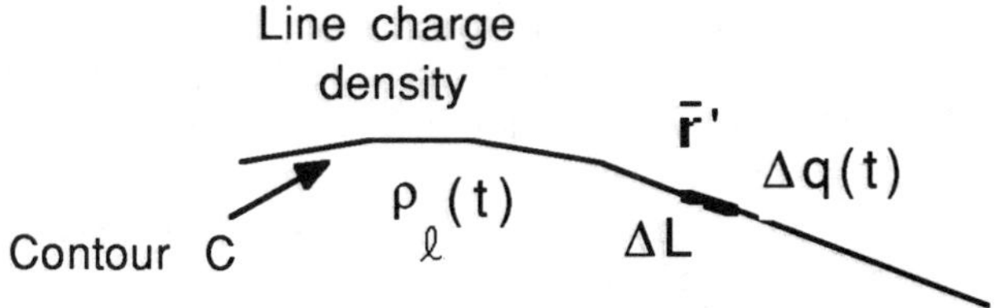

Figure 2.3 Continuous distribution of electric charge density – Line case.

$$\rho_\ell(\bar{r}',t) = \lim_{\Delta L \to 0} \frac{\Delta q(\bar{r}',t)}{\Delta L(\bar{r}')}$$

$$= \frac{\partial q(\bar{r}',t)}{\partial L(\bar{r}')} \qquad \text{coulombs per meter} \tag{2.1.8}$$

Hence, if the time-dependent line charge density is known along a given contour path, the total charge contained along the contour can be calculated by line integration over the line charge density distribution. In Figure 2.3, using the definition of (2.1.8), the incremental charge contained within the incremental line segment is given by

$$dq(\bar{r}',t) = \rho_\ell(\bar{r}',t)\ dL(\bar{r}') \tag{2.1.9}$$

The total charge, in coulombs, contained along contour C is

$$q(t) = \int_C \rho_\ell(t)\ dL \tag{2.1.10}$$

The macroscopic mathematical models of the electric charge density just defined are utilized in later sections to study electromagnetic field distributions.

2.1.1 ELECTRIC CURRENT DENSITY

The electric current is the time rate of change of electric charges. To come up with a suitable mathematical model for time-dependent electric currents in a very large three-dimensional medium, let us consider an arbitrary distribution of electric charges. By applying an external electric force field, the electric charges can be drifted in the direction of the applied electric field. The positive charges are drifted in the same direction, but the negative charges are drifted in a direction opposite to the applied electric field. When the electric charges move in a three-dimensional medium, a volume current density is set up and is directly proportional to the product of volume charge density and drift velocity of the charges. The drift velocity is a vector, and so is the electric current density. The direction of current density is same as that of the drift velocity. The electric current obtained is referred to as the *convection electric current*, which exists principally due to the velocity of the moving charges:

$$\bar{J}_m(t) = \rho_v \bar{\nu}_d \tag{2.1.11a}$$

where

$\bar{J}_m(t)$: electric current density, in amperes per square meter;

ρ_v : volume charge density distribution, in coulombs per cubic meter;

$\bar{\nu}_d$: drift velocity of the electric charges, in meters per second.

The same current density concept can be extended to the flow of charges in a conducting material medium as well. The drift velocity in the material medium is, in fact, directly proportional to the applied electric field and mobility of the net positive and negative charges, which principally depends upon the specific nature of the conducting material. The charge density and the corresponding charge mobility can be combined together to define conductivity property of a given material medium. The electric current inside the conducting medium is generally referred to as the *conduction electric current.* It is principally the mobility of charges inside the conducting material medium, which, in fact, is exerted by an external electric field. Thus,

$$\bar{J}_c(\bar{r},t) = \sigma \bar{E}(\bar{r},t) \tag{2.1.11b}$$

where

$\bar{J}_c(\bar{r},t)$: conduction current density, in amperes per square meter;

σ : conductivity of the medium, in mhos per meter;

$\bar{E}(\bar{r},t)$: applied electric field, in volts per meter.

Expression (2.1.11b) is referred to as the *generalized Ohm's law*. In a conducting material medium, there is always an electric current flowing whose density is directly proportional to the external electric field, and the proportionality constant is the conductivity of conducting material medium. It is interesting to note, the conductivity of a perfectly conducting material is very large and in a limit tends to infinity. In such a case, the conduction electric current correspondingly flows at the boundary surface of the conducting material medium. In the case of an imperfect conducting material medium, the electric current penetrates due to *skin-depth* of the medium.

2.1.2 ELECTRIC CURRENT DENSITY DISTRIBUTIONS

Rarely, in practical situations, does the time-dependent electric currents exist at specific spatial points. Generally, the electric currents are distributed throughout a given region of space, depending on the medium characteristics. To account for the electric current distribution at each and every point within a given source volume, let us consider an arbitrary source point (x', y', z') which is enclosed within an incremental volume with an incremental cross-sectional area Δs.

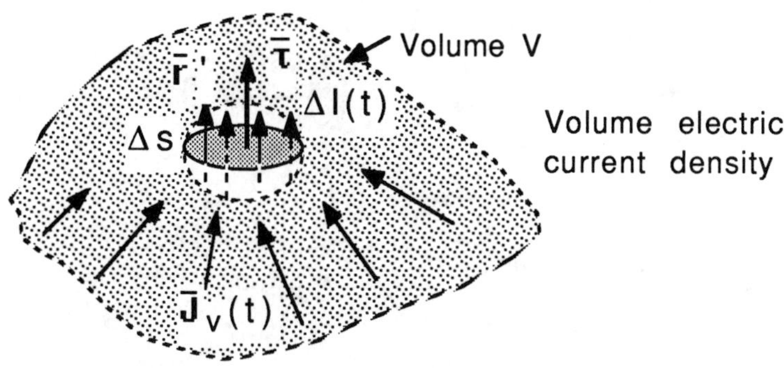

Figure 2.4a Continuous distribution of electric current density – Volume case.

Referring to Figure 2.4a, let

ΔI : net electric current crossing incremental surface, in amperes;
Δs : area of incremental surface, in square meters;
$\hat{\tau}$: unit vector normal to the incremental surface in the direction of the electric current flow.

Considering a ratio of the net electric current crossing the incremental cross-sectional area in the direction perpendicular to the incremental surface area, and taking a limit as the incremental surface shrinks to zero, the time-dependent volume current density at the source point is given by

$$\bar{J}_V(\bar{r}',t) = J_V \hat{\tau} \tag{2.1.12a}$$

where the magnitude of volume current density in amperes per square meter is obtained by the ratio

$$J_V = \lim_{\Delta s \to 0} \frac{\Delta I}{\Delta s} \tag{2.1.12b}$$

Hence, if the volume current density is known at each and every point within a given volume, then the total electric current crossing an arbitrary surface can be calculated by taking the surface integration over the complete current density distribution.

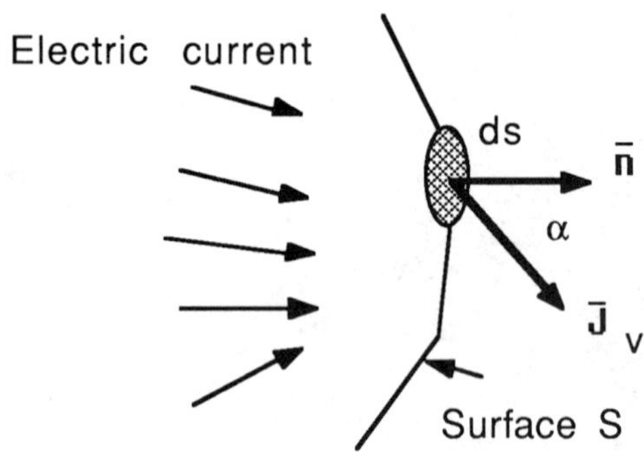

Figure 2.4b Total current crossing a surface.

Referring to Figure 2.4b, the incremental electric current dI crossing the incremental surface ds in the perpendicular direction is given by

$$dI = J_V \, ds \, \cos\alpha \tag{2.1.13a}$$

$$= \bar{J}_V \bullet ds\hat{n} \tag{2.1.13b}$$

$$= \bar{J}_V \bullet d\bar{s} \tag{2.1.13c}$$

The total electric current crossing an arbitrary open surface S is

$$I(t) = \iint_S \bar{J}_V \bullet d\bar{s} \tag{2.1.14}$$

The case of time-dependent electric currents confined to an arbitrary shaped surface can be considered in a similar way. It is assumed that the surface has no thickness. The electric current distribution is tangential at each and every point on a given surface, and practically no electric current exists normal to the surface. Let us consider an arbitrary source point (x', y', z'), which is enclosed within an incremental surface with a corresponding incremental cross sectional length ΔL.

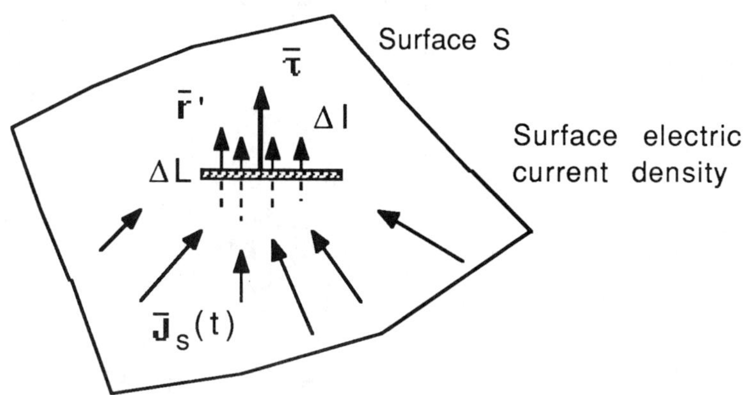

Figure 2.5a Continuous distribution of electric current density – Surface case.

Referring to Figure 2.5a, let

ΔI : net electric current crossing the incremental length, in amperes;
ΔL : length of incremental line segment, in meters;
$\hat{\tau}$: unit vector normal to the incremental line in the direction of the electric current flow.

Considering a ratio of the net electric current crossing the incremental line in the direction perpendicular to the incremental line segment and taking a limit as the incremental line segment shrinks to zero, the time-dependent surface current density at the source point is given by

$$\bar{J}_S(\bar{r}',t) = J_S \hat{\tau} \tag{2.1.15a}$$

where the magnitude of the surface current density in amperes per meter

$$J_S = \lim_{\Delta L \to 0} \frac{\Delta I}{\Delta L} \tag{2.1.15b}$$

Hence, if the tangential surface current density is known on a given surface, then the total electric current crossing a given contour can be calculated by taking the line integration over the complete current density distribution. Assuming the arbitrary surface has no physical thickness, referring to Figure 2.5b, the incremental electric current dI crossing the incremental length dL in the perpendicular direction is given by

$$dI = J_S \, dL \, \cos\alpha \tag{2.1.16a}$$

$$= \bar{J}_S \bullet dL\hat{n} \tag{2.1.16b}$$

$$= \bar{J}_S \bullet d\bar{L} \tag{2.1.16c}$$

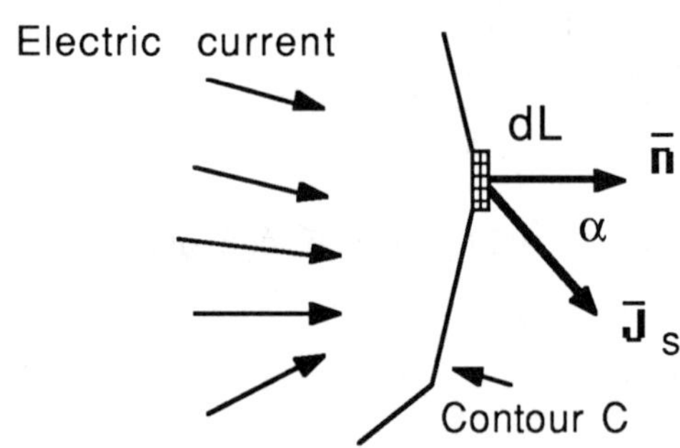

Figure 2.5b Total current crossing a contour.

Therefore, the total electric current crossing the arbitrary contour C is

$$I(t) = \int_C \bar{J}_S \bullet d\bar{L} \tag{2.1.17}$$

where the line integration is to be performed over the complete contour containing the source electric current density distribution. The treatment of the time-dependent line distribution of electric current is straightforward. To account for the electric current distribution at each and every point along a given contour, let us consider an arbitrary source point (x', y', z'), which is enclosed within an incremental length.

Referring to Figure 2.6, let

I : net electric current flowing along incremental length, in amperes;

$\hat{\tau}$: unit tangential vector to contour C in the direction of current flow.

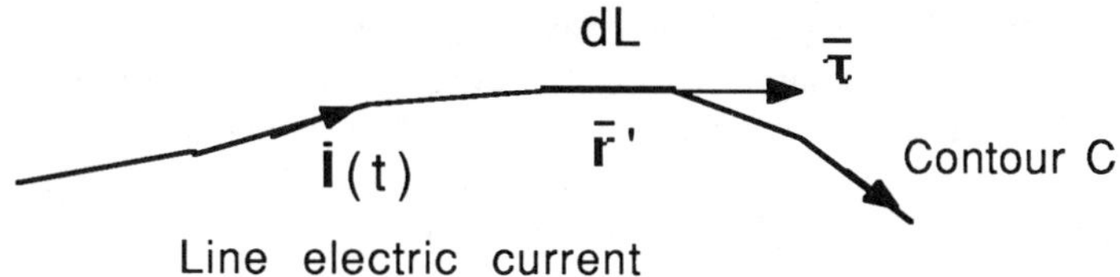

Figure 2.6 Line electric current distribution.

Then, the net electric current flowing along the contour C in the tangential direction at the source point is given by

$$\bar{I}(\bar{r}',t) = I\hat{\tau}\,(x',y',z') \tag{2.1.18}$$

The macroscopic mathematical models adapted previously for various types of source electric current distributions are quite useful in the calculation of electromagnetic field distributions.

2.2 CONTINUITY EQUATION

The electric charge and the electric current sources are directly interrelated through the electromagnetic continuity equation. Referring to Figure 2.7, let us consider a closed surface S bounding a volume V in which there exists an arbitrary distribution of time-dependent electric charges.

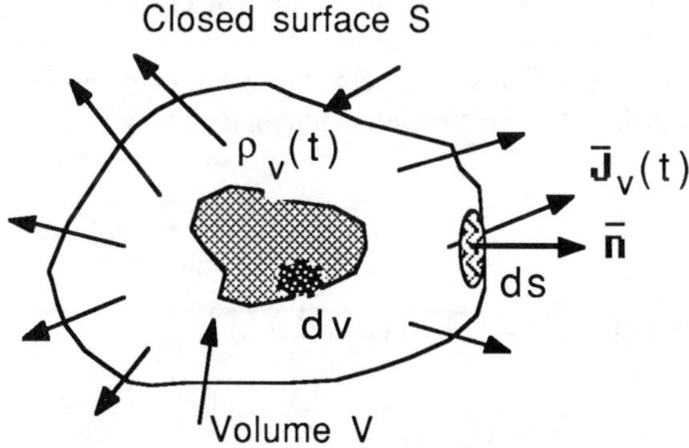

Figure 2.7 Continuity relation between electric currents and charges.

Inside the volume, the electric charges may have a volume charge density, or a surface charge density, or a line charge density, or even an isolated multiple point charge distribution. Even if all these distributions are present simultaneously, they can be accounted for by summing their individual contributions. The total charge contained in the complete volume V is given by expression (2.1.4)

$$q(t) = \iiint_V \rho_V \, dv \tag{2.2.1}$$

$\rho_V(t)$: volume electric charge density, in coulombs per cubic meter.

Under the influence of an externally applied field, the electric charges can move out of the closed surface at a certain velocity. Referring to Figure 2.7, a net electric current diverges to cross the closed surface S in all directions. The total current coming out of the closed surface can be calculated if one knows the volume current density at each point on the surface S. The net electric current crossing the closed surface S from the inside region to the outside region is obtained by

$$I(t) = \oiint_S \bar{J}_V \bullet d\bar{s} \tag{2.2.2}$$

$\bar{J}_V$: volume current density, in amperes per square meter.

As the electric current flows from the inside region to the outside region in all directions, gradually the net electric charge distribution inside the closed volume decreases. If any electric current flows from the outside region to the inside region, it should be accounted for as well. Hence, the net electric current flowing out of the closed surface S bounding the volume V is identical to the rate of decrease of the net electric charges inside the volume V. This statement is referred to as the *continuity equation* or conservation of charges, and it is expressed mathematically as

$$I(t) = - \frac{\partial q(t)}{\partial t} \tag{2.2.3}$$

On substituting the expressions (2.2.1) and (2.2.2)

$$\oiint_S \bar{J}_V \bullet d\bar{s} = - \frac{\partial}{\partial t} \iiint_V \rho_V \, dv \tag{2.2.4}$$

The integral form of the continuity equation (2.2.4) can be converted into the equivalent differential or point form by utilizing the Gauss-divergence theorem, Appendix A. The lefthand-side closed surface integral can be replaced in terms of the volume integral of divergence of electric current density. Thus,

$$\iiint_V \nabla \bullet \bar{J}_V \, dv = -\frac{\partial}{\partial t} \iiint_V \rho_V \, dv \tag{2.2.5}$$

Hence, the continuity equation in the differential form states that at any point the divergence of the time-dependent volume electric current density is equal to the rate of decrease of the volume charge density at the same point. This is given by

$$\nabla \bullet \bar{J}_V(\bar{r}',t) = -\frac{\partial \rho_V(\bar{r}',t)}{\partial t} \tag{2.2.6}$$

Referring to continuity equation (2.2.6), if the volume charge density is zero at any point, then the divergence of the electric current density does not exist. The divergence of the electric current density is always taken as positive if the electric current is leaving or diverging out of the point under consideration and negative if the electric current is converging to the point. Following this procedure, expression (2.2.6) can be conveniently modified if other types, such as the surface and line electric current and charge distributions, are present.

2.3 MAXWELL'S EQUATIONS – TIME DOMAIN

Faraday's law represents one of Maxwell's equations. It is discussed in this section for the case of a general medium. Let us consider a three-dimensional region that is linear, homogeneous and isotropic medium. In this medium, the time-dependent electric current and electric charge sources are distributed. The two types of sources are related through the current-charge continuity equation, expression (2.2.6). The electric current and charge sources produce the time-varying electric and magnetic field distributions in the medium. Since the two sources are interrelated, the time-varying electromagnetic fields can be viewed as produced solely by either the time-varying electric charge sources or the time-varying electric current sources. Referring to Figure 2.8, let

$\bar{E}(\bar{r},t)$: electric field distribution, in volts per meter;
$\bar{D}(\bar{r},t)$: electric flux density distribution, coulombs per square meter;
$\bar{H}(\bar{r},t)$: magnetic field distribution, in amperes per meter;
$\bar{B}(\bar{r},t)$: magnetic flux density distribution, in webers per square meter.

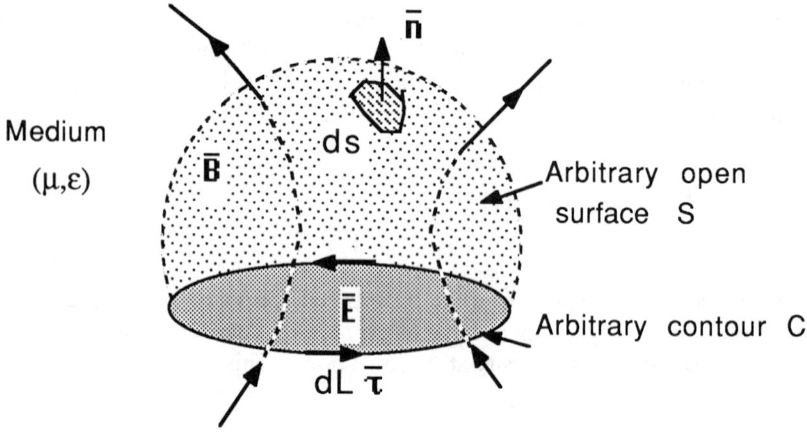

Figure 2.8 Generalized Faraday's law and Maxwell's first equation.

These four types of electromagnetic field quantities are set up in a three-dimensional medium. Faraday's law gives a relationship between the time-varying electric field and the corresponding time-varying magnetic flux density or the magnetic field itself. Referring to Figure 2.8, let us consider a closed arbitrary shaped contour *C* located along the edge of an arbitrary shaped open surface *S*. There exists an electric field distribution at every point on the contour. The total electric potential around the closed contour can be obtained by taking the line integral of the time-varying electric field. Let

$d\bar{L}(\bar{r})$: elemental length along the contour *C*

$$= dL\hat{\tau} \tag{2.3.1a}$$

The incremental electric potential along the incremental length of the contour is

$$d\Phi(t) = \bar{E}(\bar{r},t) \bullet d\bar{L}(\bar{r}) \tag{2.3.1b}$$

The total electric potential, in volts, around the closed contour *C* is given by

$$\Phi(t) = \oint_C \bar{E}(\bar{r},t) \bullet d\bar{L}(\bar{r}) \tag{2.3.1c}$$

Further, referring to Figure 2.8, the contour *C* forms the edge of an arbitrary virtual open surface *S*. The magnetic flux density distribution is everywhere in the medium, but only a part of the total magnetic flux distribution links the closed loop *C* and crosses the open surface S. Suppose the magnetic flux density distribution is known at each point on the open surface S, then the total magnetic flux linking the loop can be calculated by considering surface integral of the magnetic flux density over the complete open surface S. The incremental magnetic flux crossing the incremental surface in the perpendicular direction is given by

$$d\Omega(t) = \bar{B}(\bar{r},t) \bullet d\bar{s}(\bar{r}) \tag{2.3.2a}$$

$d\bar{s}$: incremental surface area on the open surface S.

Hence, the total magnetic flux linking contour C is given by

$$\Omega(t) = \iint_S \bar{B}(\bar{r},t) \bullet d\bar{s}(\bar{r}) \tag{2.3.2b}$$

According to Faraday's law, the total electric potential around the closed contour C is equal to the negative of the time rate of change of total magnetic flux linking the contour. Hence,

$$\Phi(t) = -\frac{\partial}{\partial t}\Omega(t) \tag{2.3.3}$$

On substituting expressions (2.3.1c) and (2.3.2b) into the preceding relationship, the integral form of Faraday's law takes the form

$$\oint_C \bar{E}(\bar{r},t) \bullet d\bar{L}(\bar{r}) = -\frac{\partial}{\partial t}\iint_S \bar{B}(\bar{r},t) \bullet d\bar{s}(\bar{r}) \tag{2.3.4}$$

Expression (2.3.4) is quite useful when electromagnetic field quantities are to be studied in a large region of space. As an example, a large region of space may contain a number of boundary surfaces separating different media, having different permittivity and permeability parameters. This situation is discussed in a later section, based on additional boundary conditions. In fact, the required boundary conditions are implicit in the preceding integral relationship and can be extracted by studying field quantities at or near a boundary surface. Using Stoke's theorem, Appendix A, the integral form of the expression (2.3.4) can be converted into the differential form. Stoke's theorem yields

$$\oint_C \bar{E}(\bar{r},t) \bullet d\bar{L}(\bar{r}) = \iint_S [\nabla \times \bar{E}(\bar{r},t)] \bullet d\bar{s}(\bar{r}) \tag{2.3.5}$$

Stoke's theorem connects a closed contour integration of the electric field quantity to the corresponding open surface integration of curl of the same electric field quantity, where C is a boundary contour along the edge of the open surface S as depicted in the Figure 2.8. Hence, the lefthand-side term of expression (2.3.4) can be replaced by the righthand-side term of expression (2.3.5) to yield the following integral relationship:

$$\iint_S [\nabla \times \bar{E}(\bar{r},t)] \bullet d\bar{s}(\bar{r}) = -\frac{\partial}{\partial t} \iint_S \bar{B}(\bar{r},t) \bullet d\bar{s}(\bar{r}) \tag{2.3.6}$$

By examining the righthand side of the integral expression (2.3.6), the partial derivative with respect to the time variable can be taken inside, since the surface integration does not depend on the time variable t. On comparing the two integrands on both sides of expression (2.3.6), the following differential form of Faraday's law is obtained:

$$\nabla \times \bar{E}(\bar{r},t) = -\frac{\partial \bar{B}(\bar{r},t)}{\partial t} \tag{2.3.7}$$

In the literature, expression (2.3.4) or (2.3.7) is generally referred to as *Maxwell's first equation.* In fact, Ampere's law represents another Maxwell's equation. It is discussed in this section for the case of a general medium.

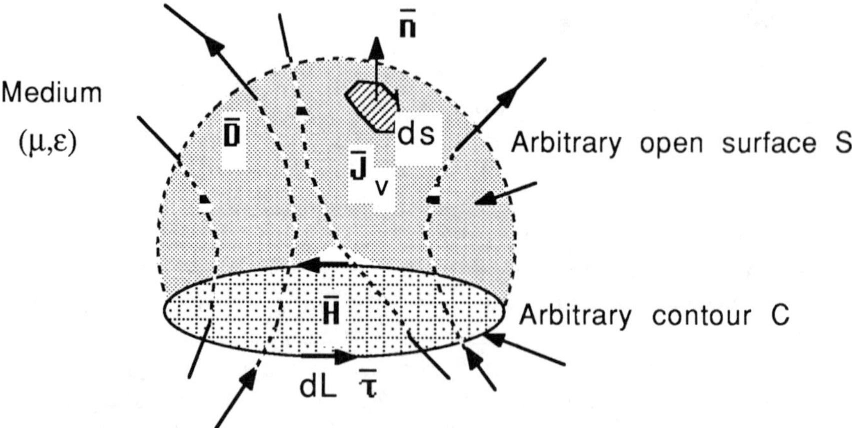

Figure 2.9 Generalization of Ampere's law and Maxwell's second equation.

Referring to Figure 2.9, let us consider a three-dimensional region that is linear, homogeneous and isotropic medium. In this medium, the time-dependent electric current and electric charge sources are distributed. As stated earlier, the two types of sources are not completely independent; in fact, they are related through the electric current–charge continuity equation, expression (2.2.6), discussed in the Section 2.2. The electric current and charge sources produce the time-varying electric and magnetic field distributions in the medium. In Figure 2.9, let

$\bar{J}_V(\bar{r},t)$: volume electric current density distribution, in amperes per square meter.

Ampere's law gives a relationship between the time-varying magnetic field and the corresponding time-varying electric current density and electric flux density distribution. Referring to Figure 2.9, let us consider a closed arbitrary shaped contour C located along the edge of an arbitrary shaped open surface S. The magnetic field exists at every point on the contour. The total magnetic potential around the closed contour can be obtained by taking line integral of the time-varying magnetic field. The incremental magnetic potential along the incremental length of contour is

$$d\Psi(t) = \bar{H}(\bar{r},t) \bullet d\bar{L}(\bar{r}) \qquad (2.3.8a)$$

The total magnetic potential, in ampere–turns, around the closed contour C is

$$\Psi(t) = \oint_C \bar{H}(\bar{r},t) \bullet d\bar{L}(\bar{r}) \qquad (2.3.8b)$$

Further, referring to Figure 2.9, contour C forms the edge of an arbitrary virtual open surface S. The electric current density and the electric flux density distributions are everywhere in the medium, but only a part of the total electric current density and a part of the electric flux density distributions link the closed loop C and cross the open surface S. Suppose the electric flux density distribution is known at each point on the open surface S, then the electric flux linking the loop can be calculated by considering surface integral of the electric flux density over the complete open surface S.

The incremental electric flux crossing the incremental surface in the perpendicular direction is

$$d\Pi(t) = \bar{D}(\bar{r},t) \bullet d\bar{s}(\bar{r}) \qquad (2.3.9a)$$

Hence, the total electric flux linking contour C is given by

$$\Pi(t) = \iint_S \bar{D}(\bar{r},t) \bullet d\bar{s}(\bar{r}) \qquad (2.3.9b)$$

Similarly, the total time-varying electric current linking the contour C is obtained as follows. The incremental electric current crossing the incremental surface ds in the perpendicular direction is given by

$$di(t) = \bar{J}_V(\bar{r},t) \bullet d\bar{s}(\bar{r}) \tag{2.3.9c}$$

Hence, the total electric current linking contour C is given by

$$i(t) = \iint_S \bar{J}_V(\bar{r},t) \bullet d\bar{s}(\bar{r}) \tag{2.3.9d}$$

According to Ampere's law, the total magnetic potential around the closed contour C is equal to sum of the electric current and the *displacement current* given by the rate of change of electric flux linking the contour C. Hence,

$$\Psi(t) = \frac{\partial}{\partial t}\Pi(t) + i(t) \tag{2.3.10}$$

On substituting expressions (2.3.8b), (2.3.9b) and (2.3.9d) into the above relationship, the integral form of Ampere's law takes the form

$$\oint_C \bar{H}(\bar{r},t) \bullet d\bar{L}(\bar{r}) = \frac{\partial}{\partial t}\iint_S \bar{D}(\bar{r},t) \bullet d\bar{s}(\bar{r}) + \iint_S \bar{J}_V(\bar{r},t) \bullet d\bar{s}(\bar{r}) \tag{2.3.11}$$

Expression (2.3.11) is quite useful when electromagnetic field quantities are to be studied in a large region of space. As an example, a large region of space may contain a number of boundary surfaces separating different media having different permittivity and permeability parameters. This situation is discussed in a later section based on boundary conditions. In fact, boundary conditions are implicit in the above integral relationship and can be extracted by studying field quantities at or near a boundary surface. Using Stoke's theorem, Appendix A, the integral form of expression (2.3.11) can be converted into the differential form. By Stoke's theorem

$$\oint_C \bar{H}(\bar{r},t) \bullet d\bar{L}(\bar{r}) = \iint_S [\nabla \times \bar{H}(\bar{r},t)] \bullet d\bar{s}(\bar{r}) \tag{2.3.12a}$$

Stoke's theorem connects a closed contour integration of magnetic field quantity to the corresponding open surface integration of curl of the same magnetic field quantity, where C is a closed boundary contour along the edge of an open surface S as depicted in the

Figure 2.9. Hence, the lefthand-side term of expression (2.3.11) can be replaced by the righthand-side term of expression (2.3.12a) to yield the following integral relationship

$$\iint_S [\nabla \times \bar{H}(\bar{r},t)] \bullet d\bar{s}(\bar{r}) = \frac{\partial}{\partial t} \iint_S \bar{D}(\bar{r},t) \bullet d\bar{s}(\bar{r}) + \iint_S \bar{J}_V(\bar{r},t) \bullet d\bar{s}(\bar{r}) \tag{2.3.12b}$$

By examining the righthand-side of the integral expression (2.3.12b), the partial derivative with respect to the time variable can be taken inside, because the surface integration does not depend on the time variable t. On comparing the integrands on both sides of expression (2.3.12b), the following differential form of Ampere's law is obtained:

$$\nabla \times \bar{H}(\bar{r},t) = \frac{\partial \bar{D}(\bar{r},t)}{\partial t} + \bar{J}_V(\bar{r},t) \tag{2.3.13}$$

In the literature, expression (2.3.11) or (2.3.13) is generally referred to as *Maxwell's second equation*. As can be seen in this expression, the curl of the time-varying magnetic field is equal to the source volume electric current density distribution plus an additional electric polarization density or displacement current density equal to the rate of change of electric flux density distribution. In general, the righthand side should include all possible types of electric current density distributions that may be present in the medium under consideration. Suppose the three-dimensional medium has conductivity, then additional conducting-type electric current density distribution exists, based on the generalized Ohm's law. Expression (2.3.13) is rewritten as

$$\nabla \times \bar{H}(\bar{r},t) = \frac{\partial \bar{D}(\bar{r},t)}{\partial t} + \bar{J}_g(\bar{r},t) + \bar{J}_c(\bar{r},t) \tag{2.3.14}$$

$\bar{J}_g(\bar{r},t)$: current density due to the source generator;
$\bar{J}_c(\bar{r},t)$: current density due to the conductivity of the medium.

2.4 SUMMARY OF ELECTROMAGNETIC FIELD EQUATIONS – TIME DOMAIN

A complete summary of *Maxwell's equations in the time-domain* is given in Tables 2.1 and 2.2. These are called time-domain equations because the sources and the corresponding electric and magnetic field quantities are functions of the time variable.

Table 2.1 lists all the relevant classical electromagnetic field equations in differential form. Similarly, Table 2.2 lists all the relevant electromagnetic field equations in integral form. Using Stoke's theorem and Gauss-divergence theorem, the differential form of expressions presented in Table 2.1 can be converted to the corresponding integral form of expressions presented in Table 2.2.

The various time-varying electromagnetic vector field quantities are given by

$\bar{E}(\bar{r},t)$: electric field distribution, in volts per meter;
$\bar{D}(\bar{r},t)$: electric flux density distribution, coulombs per square meter;
$\bar{H}(\bar{r},t)$: magnetic field distribution, in amperes per meter;
$\bar{B}(\bar{r},t)$: magnetic flux density distribution, in webers per square meter.

The electromagnetic field equations are applicable for very large three-dimensional lossy or lossless linear media. The medium parameters are assumed to be constants, and thus, they do not depend on a coordinate variable (homogeneous medium) nor on the direction of the electric and magnetic field quantities (isotropic medium):

ε : electric permittivity of the medium, in farads per meter;
μ : magnetic permeability of the medium, in henrys per meter;
σ : electric conductivity of the medium, in mhos per meter.

The electric current density and electric charge density sources are given by

$\bar{J}_V(\bar{r},t)$: volume electric current density, in amperes per square meter;
$\rho_V(\bar{r},t)$: volume electric charge density, in coulombs per cubic meter.

In Tables 2.1 and 2.2, the basic electromagnetic field equations are given by Faraday's law and Ampere's law. Expressions (2.4.1) and (2.4.2) are a set of coupled partial differential equations. These two curl equations, in fact, form an independent set of coupled relationships between the time-varying electric field and magnetic field quantities. The additional field relationships stated by the Gauss's law are discussed in the next section. Gauss law expressions (2.4.4) and (2.4.5) do not form an independent set of equations. They can be deduced directly from the two Maxwell's curl equations.

Tables 2.1 and 2.2 can be extended even for inhomogeneous medium by defining the medium parameters as a function of space coordinate variable; and similarly, by defining as tensors, the equations in tables can be applied for anisotropic medium.

Table 2.3 lists relevant expressions for the special case of statics. For electrostatics and magnetostatics, the time variable does not exist and the time derivative of all field quantities is taken as zero. In fact, various electromagnetic field expressions decouple for the two special static cases.

Table 2.1
Time-domain Maxwell's equations in differential form –
Electric current sources

Faraday's law:

$$\nabla \times \bar{E}(\bar{r},t) = -\frac{\partial \bar{B}(\bar{r},t)}{\partial t} \qquad (2.4.1)$$

Ampere's law:

$$\nabla \times \bar{H}(\bar{r},t) = \bar{J}_v(\bar{r},t) + \frac{\partial \bar{D}(\bar{r},t)}{\partial t} \qquad \text{lossless case} \qquad (2.4.2)$$

$$\nabla \times \bar{H}(\bar{r},t) = \bar{J}_g(\bar{r},t) + \frac{\partial \bar{D}(\bar{r},t)}{\partial t} + \bar{J}_c(\bar{r},t) \qquad \text{lossy case} \qquad (2.4.3)$$

Gauss's law:

$$\nabla \bullet \bar{B}(\bar{r},t) = 0 \qquad (2.4.4)$$

$$\nabla \bullet \bar{D}(\bar{r},t) = \rho_v(\bar{r},t) \qquad (2.4.5)$$

Constitutive relations:

$$\bar{D}(\bar{r},t) = \varepsilon \bar{E}(\bar{r},t) \qquad (2.4.6)$$

$$\bar{B}(\bar{r},t) = \mu \bar{H}(\bar{r},t) \qquad (2.4.7)$$

Generalized Ohm's law:

$$\bar{J}_c(\bar{r},t) = \sigma \bar{E}(\bar{r},t) \qquad (2.4.8)$$

Continuity equation:

$$\nabla \bullet \bar{J}_v(\bar{r},t) = -\frac{\partial \rho_v(\bar{r},t)}{\partial t} \qquad (2.4.9)$$

Table 2.2
Time-domain Maxwell's equations in integral form –
Electric current sources

Faraday's law:

$$\oint_C \bar{E}(\bar{r},t) \bullet d\bar{L}(\bar{r}) = -\frac{\partial}{\partial t}\iint_S \bar{B}(\bar{r},t) \bullet d\bar{s}(\bar{r}) \qquad (2.4.10)$$

Ampere's law:

$$\oint_C \bar{H}(\bar{r},t) \bullet d\bar{L}(\bar{r}) = \iint_S \bar{J}_V(\bar{r},t) \bullet d\bar{s}(\bar{r}) + \frac{\partial}{\partial t}\iint_S \bar{D}(\bar{r},t) \bullet d\bar{s}(\bar{r}) \qquad \text{lossless case} \qquad (2.4.11)$$

$$\oint_C \bar{H}(\bar{r},t) \bullet d\bar{L}(\bar{r}) = \iint_S \bar{J}_g(\bar{r},t) \bullet d\bar{s}(\bar{r}) + \frac{\partial}{\partial t}\iint_S \bar{D}(\bar{r},t) \bullet d\bar{s}(\bar{r}) + \iint_S \bar{J}_C(\bar{r},t) \bullet d\bar{s}(\bar{r}) \qquad \text{lossy case} \qquad (2.4.12)$$

Gauss's law:

$$\oiint_S \bar{B}(\bar{r},t) \bullet d\bar{s} = 0 \qquad (2.4.13)$$

$$\oiint_S \bar{D}(\bar{r},t) \bullet d\bar{s} = \iiint_V \rho_V \, dv \qquad (2.4.14)$$

Constitutive relations:

$$\bar{D}(\bar{r},t) = \varepsilon \bar{E}(\bar{r},t) \qquad (2.4.15)$$

$$\bar{B}(\bar{r},t) = \mu \bar{H}(\bar{r},t) \qquad (2.4.16)$$

Generalized Ohm's law:

$$\bar{J}_C(\bar{r},t) = \sigma \bar{E}(\bar{r},t) \qquad (2.4.17)$$

Continuity equation:

$$\oiint_S \bar{J}_V \bullet d\bar{s} = -\frac{\partial}{\partial t}\iiint_V \rho_V \, dv \qquad (2.4.18)$$

Table 2.3
Electrostatic and magnetostatic Maxwell's equations

ELECTROSTATIC EQUATIONS

Differential form:

$$\nabla \times \bar{E}(\bar{r}) = 0 \qquad (2.4.19)$$

$$\nabla \bullet \bar{D}(\bar{r}) = \rho_V(\bar{r}) \qquad (2.4.20)$$

$$\bar{D}(\bar{r}) = \varepsilon \bar{E}(\bar{r},t) \qquad (2.4.21)$$

Integral form:

$$\oint_C \bar{E}(\bar{r}) \bullet d\bar{L}(\bar{r}) = 0 \qquad (2.4.22)$$

$$\oiint_S \bar{D}(\bar{r}) \bullet d\bar{s} = \iiint_V \rho_V \, dv \qquad (2.4.23)$$

MAGNETOSTATIC EQUATIONS

Differential form:

$$\nabla \times \bar{H}(\bar{r}) = \bar{J}_V(\bar{r}) \qquad (2.4.24)$$

$$\nabla \bullet \bar{B}(\bar{r}) = 0 \qquad (2.4.25)$$

$$\bar{B}(\bar{r}) = \mu \bar{H}(\bar{r}) \qquad (2.4.26)$$

Integral form:

$$\oint_C \bar{H}(\bar{r}) \bullet d\bar{L}(\bar{r}) = \iint_S \bar{J}_V(\bar{r}) \bullet d\bar{s}(\bar{r}) \qquad (2.4.27)$$

$$\oiint_S \bar{B}(\bar{r}) \bullet d\bar{s} = 0 \qquad (2.4.28)$$

2.5 GAUSS'S LAW FOR ELECTROMAGNETIC FIELDS

The divergence property of the magnetic flux density distribution can be derived directly from Maxwell's first curl equation. Taking divergence on both sides of expression (2.4.1),

$$\nabla \bullet \nabla \times \bar{E}(\bar{r},t) = - \nabla \bullet \frac{\partial \bar{B}(\bar{r},t)}{\partial t} \tag{2.5.1a}$$

Referring to the vector identity, Appendix A, the divergence of curl of electric field quantity is always zero. Hence, the righthand-side expression becomes

$$\nabla \bullet \frac{\partial \bar{B}(\bar{r},t)}{\partial t} = 0 \tag{2.5.1b}$$

The differential operator depends on only the spatial coordinate variables and not on the partial derivative with respect to time. Expression (2.5.1b) can be rewritten as

$$\frac{\partial}{\partial t}[\nabla \bullet \bar{B}(\bar{r},t)] = 0 \tag{2.5.1c}$$

Hence, with zero initial conditions, we have the divergence relationship for the magnetic flux density in the differential form:

$$\nabla \bullet \bar{B}(\bar{r},t) = 0 \tag{2.5.2a}$$

Using Gauss-divergence theorem, Appendix A, the differential form of Gauss's law can be converted into the integral form. Hence, Gauss's law for the magnetic flux density in the integral form is obtained as

$$\oiint_S \bar{B}(\bar{r},t) \bullet d\bar{s} = 0 \tag{2.5.2b}$$

Similarly, the divergence property of the electric flux density distribution can be derived directly from Maxwell's second curl equation. Taking divergence on both sides of expression (2.4.2),

$$\nabla \bullet \nabla \times \bar{H}(\bar{r},t) = \nabla \bullet \frac{\partial \bar{D}(\bar{r},t)}{\partial t} + \nabla \bullet \bar{J}_V(\bar{r},t) \tag{2.5.3a}$$

Referring to the vector identity, Appendix A, the divergence of curl of magnetic field quantity is always zero. Hence, the righthand-side expression becomes

$$\nabla \bullet \frac{\partial \bar{D}(\bar{r},t)}{\partial t} + \nabla \bullet \bar{J}_V(\bar{r},t) = 0 \tag{2.5.3b}$$

But the electric current density and the electric charge density are related through the continuity equation (2.4.9). On substituting the continuity equation, expression (2.5.3b) takes the form

$$\nabla \bullet \frac{\partial \bar{D}(\bar{r},t)}{\partial t} - \frac{\partial \rho_V(\bar{r},t)}{\partial t} = 0 \tag{2.5.3c}$$

The differential operator depends on only the spatial coordinate variables and not on the partial derivative with respect to time. Expression (2.5.3c) can be rewritten as

$$\frac{\partial}{\partial t}[\nabla \bullet \bar{D}(\bar{r},t) - \rho_V(\bar{r},t)] = 0 \tag{2.5.3d}$$

Assuming zero initial conditions, the bracketed term is equated to zero. Hence, we have the divergence relationship for the electric flux density in terms of electric charge density in the differential form

$$\nabla \bullet \bar{D}(\bar{r},t) = \rho_V(\bar{r},t) \tag{2.5.4a}$$

Using Gauss-divergence theorem, Appendix A, the differential form of Gauss's law can be converted into the integral form. Hence, Gauss's law for the electric flux density in the integral form is obtained as

$$\oint\!\!\oint_S \bar{D}(\bar{r},t) \bullet d\bar{s} = \iiint_V \rho_V(\bar{r},t)\, dv \tag{2.5.4b}$$

2.6 ELECTROMAGNETIC BOUNDARY CONDITIONS

The various electromagnetic field equations derived in the previous sections are applicable for a very large three-dimensional region that is linear, homogeneous and isotropic medium. The medium parameters, such as permeability, permittivity, and conductivity, are assumed to be constant. In many practical situations, this may not be the case. The three-dimensional region may consist of two or more media, having different media parameters, separated by boundary layers. The permeability, permittivity, and conductivity for each medium may be constant, but the parameters vary across the boundary layer or interface. Maxwell's equations stated earlier can be applied separately for each medium to obtain a solution for the electric and magnetic field quantities. For the field solutions to be unique, they should satisfy additional conditions, called the *electromagnetic boundary conditions* at the boundary surface separating the two different media. In fact, field quantities should satisfy boundary conditions at all boundary surfaces separating different media with different media parameters.

To obtain the electromagnetic boundary conditions, Maxwell's equations in integral form are utilized. It should be noted again that the electromagnetic boundary conditions are implicit conditions in Maxwell's equations. The boundary conditions are extracted based on a detailed boundary analysis to determine right behavior of the electric and magnetic fields at the boundary layer.

To study the behavior of the electric and magnetic fields at a boundary layer, the time-varying electric and magnetic fields at any point on one side (medium 1) of the boundary layer separating two media, 1 and 2, are resolved into two sets of orthogonal components – tangential components of the electric field and magnetic field that are parallel to the boundary surface, and normal components of the electric field and magnetic field that are perpendicular to the boundary surface. Similarly, the time-varying electric and magnetic fields at any point on the other side (medium 2) of the boundary layer are also resolved into two other sets of orthogonal components, consisting of tangential and normal components of the electric and magnetic fields that are parallel and perpendicular to the boundary surface. Depending on the permeability, permittivity and conductivity characteristics of media 1 and 2, it can be shown that the two tangential components of the time-varying electric field are interrelated. Similarly, the two normal components of the time-varying electric field are also interrelated. For the time-varying magnetic field components, similar relationships also can be obtained. Once the tangential and the normal components of the electric field and magnetic field are known at the boundary surface, the complete electric field and magnetic field vectors at the boundary surface can be easily constructed.

In the following sections, detailed boundary conditions are deduced for the tangential and normal components of electric and magnetic field distributions. They are, in fact, applicable also for multi boundary layers.

2.6.1 BOUNDARY CONDITIONS FOR TANGENTIAL COMPONENTS

Let us consider behavior of the tangential components of a time-varying electric field near a boundary surface separating the two media, 1 and 2. The integral form of Faraday's law, expression (2.4.10), is utilized in the boundary condition analysis.

Expression (2.4.10) is repeated in following:

$$\oint_C \bar{E}(\bar{r},t) \bullet d\bar{L}(\bar{r}) = -\frac{\partial}{\partial t}\iint_S \bar{B}(\bar{r},t) \bullet d\bar{s}(\bar{r}) \quad (2.6.1)$$

This integral relationship is applicable to any given region, where C represents an arbitrary closed contour and S represents an open surface with contour C confined along its edge. The two media 1 and 2 can be lossy as well.

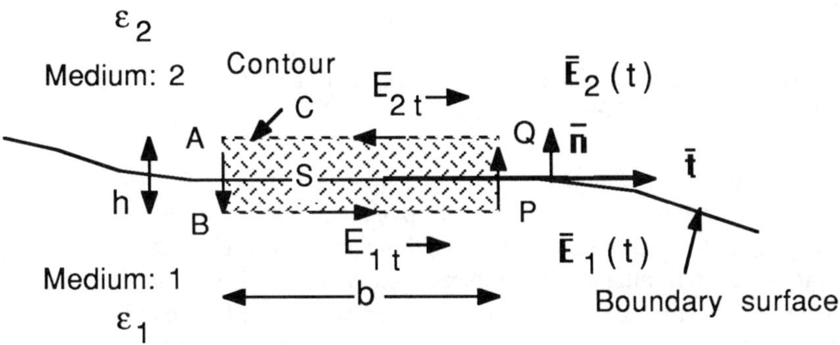

Figure 2.10 Boundary relationship for tangential tomponents electric field.

Referring to Figure 2.10,

$\bar{E}_1(\bar{r},t)$: electric field distribution in medium 1;
$\bar{E}_2(\bar{r},t)$: electric field distribution in medium 2;
$\bar{B}_1(\bar{r},t)$: magnetic flux density distribution in medium 1;
$\bar{B}_2(\bar{r},t)$: magnetic flux density distribution in medium 2;
$\hat{t}$: tangential unit vector along the boundary surface;
$\hat{n}$: normal unit vector to the boundary surface;
ε_1 : permittivity of the medium 1;
ε_2 : permittivity of the medium 2.

To determine a boundary relationship for the electric field distribution, the arbitrary contour path C is selected to be the rectangular path *BPQAB* drawn to cover both media 1 and 2. The electric field in media 1 and 2 are resolved into tangential and normal components. Along the contour C, for the two media

E_{1t}: tangential component of electric field along the line *BP*

$$= \bar{E}_1 \bullet \hat{t}$$

E_{2t}: tangential component of electric field along the line *AQ*

$$= \bar{E}_2 \bullet \hat{t}$$

The close contour integration on the lefthand side of expression (2.6.1) can be simplified as

$$\oint_{BPQAB} \bar{E} \bullet d\bar{L} = \int_{BP} \bar{E} \bullet d\bar{L} + \int_{PQ} \bar{E} \bullet d\bar{L} + \int_{QA} \bar{E} \bullet d\bar{L} + \int_{AB} \bar{E} \bullet d\bar{L} \qquad (2.6.2)$$

It is assumed that the lengths of vertical segments *PQ* and *AB* are very small, each equal to *h*; and similarly, the lengths of horizontal segments *BP* and *QA* are also very small, each equal to *b*. In a limit as the length *h* approaches zero, the second and fourth line integrals on the righthand side of expression (2.6.2) vanish. Since the segments *BP* and *QA* are very small, the distributions of tangential electric field can be assumed to be constant. Expression (2.6.2) reduces to the following expression in a limit:

$$\lim_{h \to 0} \oint_{BPQAB} \bar{E} \bullet d\bar{L} = \int_{BP} \bar{E} \bullet d\bar{L} + \int_{QA} \bar{E} \bullet d\bar{L} \qquad (2.6.3a)$$

$$= E_{1t} b - E_{2t} b \qquad (2.6.3b)$$

Further, the righthand side of expression (2.6.1) given by the surface integral over magnetic flux density, can be easily simplified. In a limit as h tends to zero, the area of surface *S* tends to zero. Hence,

$$\lim_{h \to 0} \iint_{S} \bar{B}(\bar{r},t) \bullet d\bar{s}(\bar{r}) = 0 \qquad (2.6.3c)$$

On equating expressions (2.6.3b) and (2.6.3c), we have a relationship that the two tangential components of time-varying electric field should be continuous across the boundary surface separating media 1 and 2 having different permittivity characteristics,

$$E_{1t} = E_{2t} \qquad (2.6.4a)$$

$$\hat{n} \times (\bar{E}_2 - \bar{E}_1) = 0 \qquad (2.6.4b)$$

If medium 1 is perfectly conducting, then the net electric field in the medium 1 is zero; and thus the tangential component of electric field at the boundary surface in the medium 2 is also zero. Hence,

$$E_{2t} = 0 \tag{2.6.4c}$$

Similarly, let us consider behavior of the tangential components of a time-varying magnetic field near the boundary surface separating media 1 and 2. The integral form of Ampere's law, expression (2.4.11), is utilized in the boundary condition analysis. Expression (2.4.11) is repeated in following:

$$\oint_C \bar{H}(\bar{r},t) \bullet d\bar{L}(\bar{r}) = \iint_S \bar{J}_V(\bar{r},t) \bullet d\bar{s}(\bar{r}) + \frac{\partial}{\partial t}\iint_S \bar{D}(\bar{r},t) \bullet d\bar{s}(\bar{r}) \tag{2.6.5}$$

This integral relationship is applicable for any given region, where C represents an arbitrary closed contour and S represents an open surface with contour C confined along its edge. The two media 1 and 2 can be lossy as well.

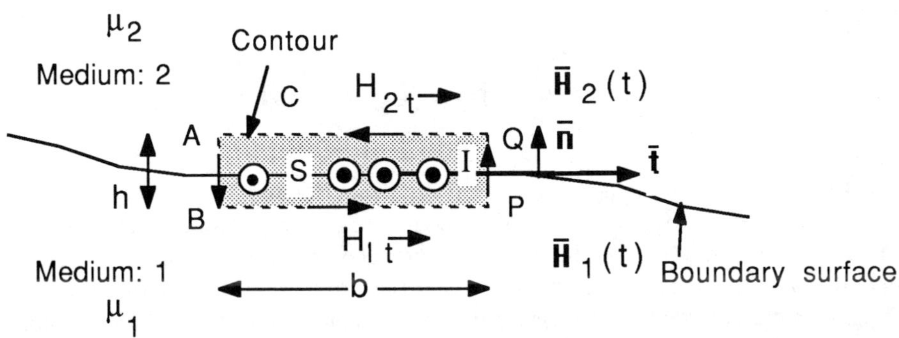

Figure 2.11 Boundary relationship for tangential components of magnetic field.

Referring to Figure 2.11,

$\bar{H}_1(\bar{r},t)$: magnetic field distribution in medium 1;
$\bar{H}_2(\bar{r},t)$: magnetic field distribution in medium 2;
$\bar{D}_1(\bar{r},t)$: electric flux density distribution in medium 1;

$\bar{D}_2(\bar{r},t)$: electric flux density distribution in medium 2;

μ_1 : permeability of the medium 1;

μ_2 : permeability of the medium 2;

$\bar{J}_V(\bar{r},t)$: volume electric current density crossing the surface S, in amperes per square meter.

To determine a boundary relationship for the magnetic field distribution, the arbitrary contour path C is selected to be the rectangular path $BPQAB$ drawn to cover both media 1 and 2. The magnetic field distributions in media 1 and 2 are resolved into tangential and normal components. Along the contour C, for the two media,

H_{1t}: tangential component of magnetic field along the line BP
$= \bar{H}_1 \bullet \hat{t}$

H_{2t}: tangential component of magnetic field along the line AQ
$= \bar{H}_2 \bullet \hat{t}$

The close contour integration on the lefthand side of expression (2.6.5) can be simplified as

$$\oint_{BPQAB} \bar{H} \bullet d\bar{L} = \int_{BP} \bar{H} \bullet d\bar{L} + \int_{PQ} \bar{H} \bullet d\bar{L} + \int_{QA} \bar{H} \bullet d\bar{L} + \int_{AB} \bar{H} \bullet d\bar{L} \qquad (2.6.6)$$

It is assumed that the lengths of vertical segments PQ and AB are very small, each equal to h; and similarly the lengths of horizontal segments BP and QA are also small, each equal to b. In a limit as the length h approaches zero, the second and fourth line integrals vanish. Since the segments BP and QA are very small, the distributions of tangential magnetic field can be assumed to be constant. Expression (2.6.6) reduces to the following expression in a limit:

$$\lim_{h \to 0} \oint_{BPQAB} \bar{H} \bullet d\bar{L} = \int_{BP} \bar{H} \bullet d\bar{L} + \int_{QA} \bar{H} \bullet d\bar{L} \qquad (2.6.7a)$$

$$= H_{1t}b - H_{2t}b \qquad (2.6.7b)$$

Further, the two surface integrals on the righthand side of expression (2.6.5) can be evaluated in a limit. Let us consider the surface integral over an electric flux density in a limit as h tends to zero. As h tends to zero, the area of surface S approaches to zero, and the integral expression simplifies to

$$\lim_{h \to 0} \iint_S \bar{D}(\bar{r},t) \bullet d\bar{s}(\bar{r}) = 0 \tag{2.6.7c}$$

The surface S becomes a contour of length b in a limit as h approaches to zero. Hence, the surface integral over the electric current density simplifies to

$$\lim_{h \to 0} \iint_S \bar{J}_V(\bar{r},t) \bullet d\bar{s}(\bar{r}) = \int_{length\ b} \bar{J}_S(\bar{r},t) \bullet d\bar{L}(\bar{r}) \tag{2.6.7d}$$

$$= I \tag{2.6.7e}$$

where

$\bar{J}_S$: surface electric current density at the boundary surface, in amperes per meter;

I : total current, in amperes, crossing the contour of length b in a perpendicular direction in a limit as h tends to zero.

On equating expressions (2.6.7b) and (2.6.7e), we have a relationship that the difference between the two tangential components of a time-varying magnetic field should be equal to the surface electric current density at the boundary surface separating the two media, 1 and 2, having different permeability characteristics, and is given by

$$H_{1t} - H_{2t} = \frac{I}{b} \tag{2.6.8a}$$

$$= J_S \tag{2.6.8b}$$

$$\hat{n} \times (\bar{H}_2 - \bar{H}_1) = \bar{J}_S \tag{2.6.8c}$$

If medium 1 is perfectly conducting, then the net magnetic field in medium 1 is zero; and thus the tangential component of magnetic field at the boundary surface is equal to the surface electric current density at the interface itself. Hence,

$$\hat{n} \times \bar{H}_2 = \bar{J}_S \tag{2.6.8d}$$

Referring to expression (2.6.8a), if the surface electric is zero at the boundary surface, then the tangential components of the magnetic field are continuous.

2.6.2 BOUNDARY CONDITIONS FOR NORMAL COMPONENTS

Let us consider behavior of the normal components of a time-varying electric field and magnetic field near the boundary surface separating the two media, 1 and 2. To establish a relationship between the two normal components of the electric field, the integral form of Gauss's law, expression (2.4.14), is utilized. In a given region, Gauss's law states that the total number of electric field lines coming out of any arbitrary closed surface is equivalent to the total charge enclosed by the closed surface. If the total number of electric field lines sum to zero on any closed surface, then the closed surface encloses no net electric charge distribution. Hence, for the time-varying electric field Gauss's law gives the relationship

$$\Pi(t) = Q(t) \tag{2.6.9a}$$

$$\oiint_S \bar{D}(\bar{r},t) \bullet d\bar{s} = \iiint_V \rho_V(\bar{r},t)\, dv \tag{2.6.9b}$$

where

Π: total number of electric field lines;

Q: total electric charges enclosed, in coulombs.

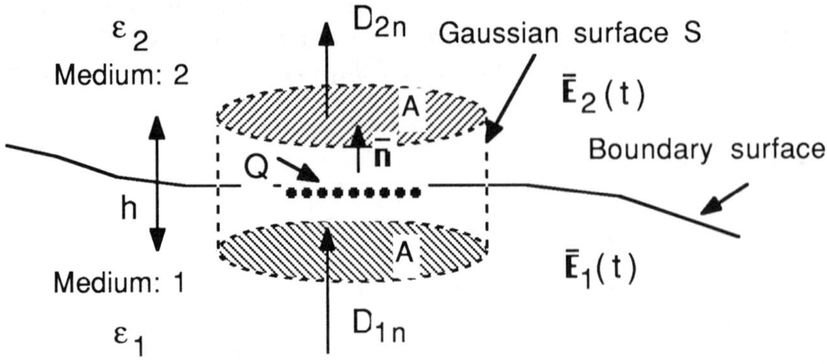

Figure 2.12 Boundary relationship for normal components of electric field.

Referring to Figure 2.12, let

$\bar{D}_1(\bar{r},t)$: electric flux density distribution in medium 1

$= D_{1n}\hat{n}$

$$D_{1n} = \varepsilon_1 E_{1n} \tag{2.6.10a}$$

where

D_{1n} : normal component of electric flux density on bottom surface;

E_{1n} : normal component of the corresponding electric field;

$\hat{n}$: normal unit vector to the boundary surface;

ε_1 : permittivity parameter of the medium 1

$= \varepsilon_0 \varepsilon_{r1}$;

ε_{r1} : relative permittivity of the medium 1.

And similarly

$\bar{D}_2(\bar{r},t)$: electric flux density distribution in medium 2

$= D_{2n}\hat{n}$

$$D_{2n} = \varepsilon_2 E_{2n} \tag{2.6.10b}$$

where

D_{2n} : normal component of electric flux density on top surface;

E_{2n} : normal component of the corresponding electric field;

ε_2 : permittivity parameter of the medium 2

$= \varepsilon_0 \varepsilon_{r2}$;

ε_{r2} : relative permittivity of the medium 2.

To determine the relationship between the two normal components of a time-varying electric flux density distribution, an arbitrary Gaussian surface S is selected to be a finite circular cylinder, as shown in the Figure 2.12, drawn to cover both media 1 and 2. The total electric field lines crossing the closed Gaussian surface S is obtained by performing closed surface integration. Assuming the electric flux density exists at each point on the closed surface S,

$$\Pi = \oiint_S \bar{D}(\bar{r},t) \bullet d\bar{s} \tag{2.6.11a}$$

$$= \iint_{\text{TOP SURFACE}} \bar{D} \bullet d\bar{s} + \iint_{\text{BOTTOM SURFACE}} \bar{D} \bullet d\bar{s} + \iint_{\text{CYLINDRICAL SURFACE}} \bar{D} \bullet d\bar{s} \tag{2.6.11b}$$

It is assumed that height h of the finite circular cylinder is very small. In a limit, as height h approaches zero, the third surface integral over the cylindrical surface on righthand side of expression (2.6.11b) vanishes. Selecting the areas of the top and bottom circular surfaces to be very small, each equal to A, the distribution of normal components of the electric flux density and the corresponding electric field can be assumed to be constant. Expression (2.6.11a) reduces to the following in a limit:

$$\lim_{h \to 0} \oiint_S \bar{D}(\bar{r},t) \bullet d\bar{s} = \iint_{TOP} \bar{D} \bullet d\bar{s} + \iint_{BOT} \bar{D} \bullet d\bar{s} \tag{2.6.12a}$$

$$= D_{2n}A - D_{1n}A \tag{2.6.12b}$$

$$= Q \quad \text{(total charge enclosed by } S\text{)} \tag{2.6.12c}$$

where the total electric charge Q is enclosed by the finite circular cylindrical Gaussian surface S. Hence, the boundary relationship states that the difference between the two normal components of time-varying electric flux density should be equal to the electric surface charge density at the boundary surface:

$$D_{2n} - D_{1n} = \frac{Q}{A} \tag{2.6.13a}$$

$$\hat{n} \bullet (\bar{D}_2 - \bar{D}_1) = \rho_s \tag{2.6.13b}$$

ρ_s : surface electric charge density, in coulombs per square meter

Expressing this in terms of the media permittivity characteristics,

$$\varepsilon_2 E_{2n} - \varepsilon_1 E_{1n} = \rho_s \tag{2.6.13c}$$

Suppose medium 1 is a perfectly conducting, then the electric field in medium 1 is zero, and the expression for the boundary condition (2.6.13a) becomes

$$D_{2n} = \rho_s \tag{2.6.13d}$$

It is interesting to note based on expression (2.6.13d), if the surface charge density is zero at the boundary surface, the normal component of the electric flux density and the corresponding electric field in medium 2 at the boundary surface are also zero. For the general two-media case, if the surface charge density is zero at the boundary surface, then the boundary relationship states that the two normal components of the time-varying electric flux density should be continuous across the boundary surface separating the two media having different permittivity characteristics

$$D_{2n} = D_{1n} \tag{2.6.14a}$$

$$\varepsilon_2 E_{2n} = \varepsilon_1 E_{1n} \tag{2.6.14b}$$

Similarly, let us consider behavior of the normal components of the time-varying magnetic field near the boundary surface separating media 1 and 2. The preceding analysis procedure can be repeated. To establish a relationship between the two normal components of magnetic field, the integral form of Gauss's law, expression (2.4.13), is utilized.

In a given region, Gauss's law states that the total number of magnetic field lines coming out of any arbitrary closed surface is equal to zero. Hence,

$$\Omega(t) = 0 \tag{2.6.15a}$$

Ω: total number of magnetic field lines

$$\oiint_S \bar{B}(\bar{r},t) \bullet d\bar{s} = 0 \tag{2.6.15b}$$

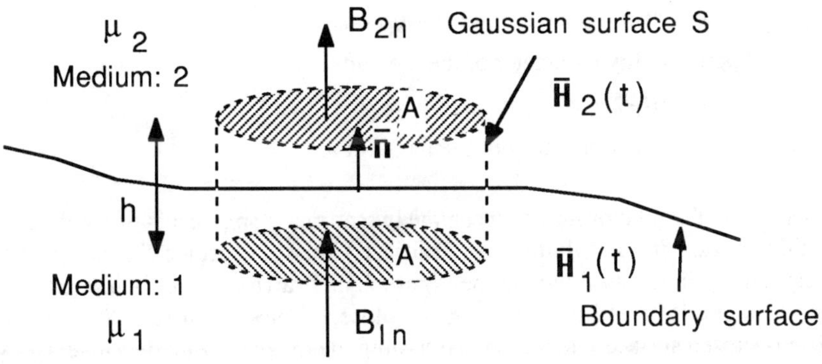

Figure 2.13 Boundary relationship for normal components of magnetic field.

Referring to Figure 2.13, let

$\bar{B}_1(\bar{r},t)$: magnetic flux density distribution in medium 1

$= B_{1n}\hat{n}$

$$B_{1n} = \mu_1 H_{1n} \tag{2.6.16a}$$

where

B_{1n} : normal component of magnetic flux density on the bottom surface;

H_{1n} : normal component of the corresponding magnetic field;

μ_1 : permeability parameter of the medium 1

$= \mu_0 \mu_{r1}$;

μ_{r1} : relative permeability of the medium 1.

And similarly

$\bar{B}_2(\bar{r},t)$: magnetic flux density distribution in medium 2

$= B_{2n}\hat{n}$

$$B_{2n} = \mu_2 H_{2n} \tag{2.6.16b}$$

where

B_{2n} : normal component of magnetic flux density on the top surface;

H_{2n} : normal component of the corresponding magnetic field;

μ_2 : permeability parameter of the medium 2

$= \mu_0 \mu_{r2}$;

μ_{r2} : relative permeability of the medium 2.

To determine the relationship between the two normal components of the time-varying magnetic flux density distribution, an arbitrary Gaussian surface S is selected to be a finite circular cylinder, as shown in the Figure 2.13, drawn to cover both media 1 and 2. The total magnetic field lines crossing the closed Gaussian surface S is obtained by performing closed surface integration. Assuming the magnetic flux density exists at each point on the closed surface S

$$\Omega = \oint\!\!\oint_S \bar{B}(\bar{r},t) \bullet d\bar{s} \tag{2.6.17a}$$

$$= \iint_{TOP\ SURFACE} \bar{B} \bullet d\bar{s} + \iint_{BOTTOM\ SURFACE} \bar{B} \bullet d\bar{s} + \iint_{CYLINDRICAL\ SURFACE} \bar{B} \bullet d\bar{s} \qquad (2.6.17b)$$

It is assumed that the height h of the finite circular cylinder is very small. In a limit, as the height h approaches zero, the third surface integral over the cylindrical surface vanishes. Selecting the areas of the top and bottom circular surfaces to be very small, each equal to A, the distribution of normal components of the magnetic flux density and the corresponding magnetic field can be assumed to be constant. Expression (2.6.17a) reduces in a limit:

$$\lim_{h \to 0} \oiint_S \bar{B}(\bar{r},t) \bullet d\bar{s} = \iint_{TOP} \bar{B} \bullet d\bar{s} + \iint_{BOT} \bar{B} \bullet d\bar{s} \qquad (2.6.18a)$$

$$= B_{2n}A - B_{1n}A \qquad (2.6.18b)$$

$$= 0 \qquad (2.6.18c)$$

Hence, the boundary relationship states that the two normal components of the time-varying magnetic flux density should be continuous across the boundary surface:

$$B_{2n} = B_{1n} \qquad (2.6.19a)$$

$$\hat{n} \bullet (\bar{B}_2 - \bar{B}_1) = 0 \qquad (2.6.19b)$$

Expressing this in terms of the media permeability characteristics,

$$\mu_2 H_{2n} = \mu_1 H_{1n} \qquad (2.6.19c)$$

Suppose medium 1 is a perfectly conducting, then the magnetic field in medium 1 is zero, and the expression for boundary condition (2.6.19a) becomes

$$B_{2n} = 0 \qquad (2.6.19d)$$

2.7 POWER AND ENERGY STORED

The previous sections presented a complete description of the time-varying electric and magnetic fields generated in a very large three-dimensional medium. The time-varying electric current and charge sources generate time-varying electric and magnetic fields, which satisfy the classical Maxwell's equations, Tables 2.1 and 2.2. Further, the time-varying sources generate electromagnetic energy and radiate in all directions away from the source. In fact, the radiated energy is carried away and stored in the surrounding medium by the electric and magnetic fields. Depending on how the sources and the fields are simulated, part of the total energy is stored in the electric field, part of the total energy is stored in the magnetic field, and the remaining energy is dissipated in the medium. In the following, mathematical expressions are developed for the energy stored in the electromagnetic fields and the energy dissipated in the medium.

Let us consider a large three-dimensional lossy region that is linear, homogeneous and isotropic medium. The parameters (ε, μ, σ) are the constant permittivity, permeability, and conductivity of the medium. The electric current density and electric charge density sources are given by $\bar{J}_v(\bar{r}, t)$ and $\rho_v(\bar{r}, t)$. The sources generate various time-varying electromagnetic field quantities which are functions of the dependent spatial and time coordinate variables, and given by

$\bar{E}(\bar{r},t)$: electric field distribution, in volts per meter;
$\bar{H}(\bar{r},t)$: magnetic field distribution, in amperes per meter.

Referring to the Table 2.1 and using the constitutive relationships (2.4.6) and (2.4.7), the electric and magnetic field quantities satisfy the following two Maxwell's equations in the lossy source-free region:

$$\nabla \times \bar{E}(\bar{r},t) = -\mu \frac{\partial \bar{H}(\bar{r},t)}{\partial t} \tag{2.7.1}$$

$$\nabla \times \bar{H}(\bar{r},t) = \varepsilon \frac{\partial \bar{E}(\bar{r},t)}{\partial t} + \sigma \bar{E}(\bar{r},t) \tag{2.7.2}$$

The electric field is in terms of volts per meter and the magnetic field is in terms of amperes per meter. The product of electric and magnetic fields gives a quantity having a unit of watts per square meter representing the power density. If the power density on a given surface is known, then by performing surface integration over the power density, the net power crossing a given surface can be calculated. By definition, the power quantity is the rate of work done or energy expended, and these aspects are discussed in the following. Referring to Appendix A, let us consider the following vector identity:

$$\nabla \bullet [\bar{E}(\bar{r},t) \times \bar{H}(\bar{r},t)] = \bar{H}(\bar{r},t) \bullet [\nabla \times \bar{E}(\bar{r},t)] - \bar{E}(\bar{r},t) \bullet [\nabla \times \bar{H}(\bar{r},t)] \tag{2.7.3}$$

The lefthand-side term in this expression represents the divergence of a certain vector quantity (power density). Expression (2.7.3) is in differential form and valid at any point $\bar{r} = (x, y, z)$. Referring to Figure 2.14, electric and magnetic fields exist throughout the volume V bounded by the closed surface S. Let

dv : incremental volume surrounding the point (x, y, z);
$d\bar{s}$: incremental surface on the closed surface S.

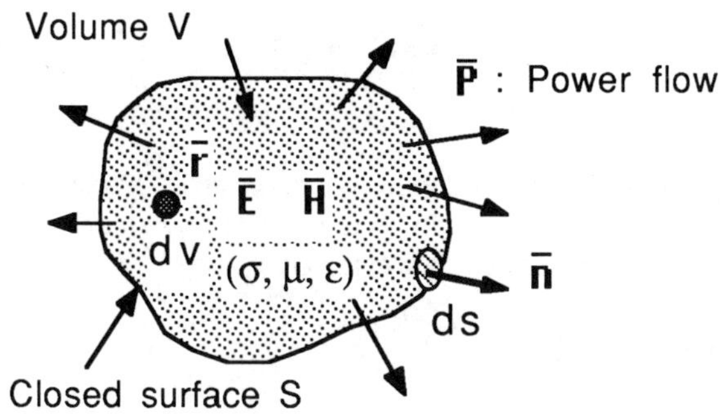

Figure 2.14 Power and Poynting vector theorem.

The righthand-side terms involving the curl of an electric field and the curl of a magnetic field in expression (2.7.3) can be replaced, using Maxwell's equations (2.7.1) and (2.7.2), to yield following relationship:

$$\nabla \bullet [\bar{E} \times \bar{H}] = \bar{H} \bullet [-\mu \frac{\partial \bar{H}(\bar{r},t)}{\partial t}] - \bar{E} \bullet [\varepsilon \frac{\partial \bar{E}(\bar{r},t)}{\partial t} + \sigma \bar{E}(\bar{r},t)] \tag{2.7.4a}$$

$$\nabla \bullet [\bar{E} \times \bar{H}] = -\mu[\bar{H} \bullet \frac{\partial \bar{H}(\bar{r},t)}{\partial t}] - \varepsilon[\bar{E} \bullet \frac{\partial \bar{E}(\bar{r},t)}{\partial t}] - [\bar{E} \bullet \sigma \bar{E}(\bar{r},t)] \tag{2.7.4b}$$

The first bracketed term on the righthand side can be simplified as

$$\frac{\partial}{\partial t}[\bar{H} \bullet \bar{H}] = \bar{H} \bullet \frac{\partial \bar{H}}{\partial t} + \frac{\partial \bar{H}}{\partial t} \bullet \bar{H} \tag{2.7.4c}$$

Hence,

$$\nabla \bullet [\bar{E} \times \bar{H}] = -\frac{\mu}{2}\frac{\partial}{\partial t}(\bar{H} \bullet \bar{H}) - \frac{\varepsilon}{2}\frac{\partial}{\partial t}(\bar{E} \bullet \bar{E}) - \sigma(\bar{E} \bullet \bar{E}) \tag{2.7.4d}$$

The expression (2.7.4d) is now multiplied throughout by an incremental volume dv and integrated throughout the complete volume V to yield the following relationship:

$$\iiint_V \nabla \bullet [\bar{E} \times \bar{H}]\, dv = \iiint_V -\frac{\mu}{2}\frac{\partial}{\partial t}(\bar{H} \bullet \bar{H})\, dv + \iiint_V -\frac{\varepsilon}{2}\frac{\partial}{\partial t}(\bar{E} \bullet \bar{E})\, dv - \iiint_V \sigma(\bar{E} \bullet \bar{E})\, dv \tag{2.7.4e}$$

On the lefthand side of this integral expression, the cross product between the electric field and the magnetic field vectors can be represented by an arbitrary vector quantity; namely, the instantaneous power density vector given by

$$\bar{P}(\bar{r},t) = \bar{E}(\bar{r},t) \times \bar{H}(\bar{r},t) \tag{2.7.5a}$$

Hence, expression (2.7.4e) can be rewritten as

$$\iiint_V \nabla \bullet \bar{P}\, dv = \frac{\partial}{\partial t}\iiint_V -\frac{\mu}{2}(\bar{H} \bullet \bar{H})\, dv + \frac{\partial}{\partial t}\iiint_V -\frac{\varepsilon}{2}(\bar{E} \bullet \bar{E})\, dv - \iiint_V \sigma(\bar{E} \bullet \bar{E})\, dv \tag{2.7.5b}$$

Based on Gauss-divergence theorem, Appendix A, the lefthand-side term of expression (2.7.5b) takes the following form:

$$\oiint_S \bar{P} \bullet d\bar{s} = \iiint_V \nabla \bullet \bar{P}\, dv \tag{2.7.5c}$$

Therefore, expression (2.7.5b) can be rewritten as

$$\oiint_S -\bar{P} \bullet d\bar{s} = \frac{\partial}{\partial t}\iiint_V \frac{\mu}{2}(\bar{H} \bullet \bar{H})\, dv + \frac{\partial}{\partial t}\iiint_V \frac{\varepsilon}{2}(\bar{E} \bullet \bar{E})\, dv + \iiint_V \sigma(\bar{E} \bullet \bar{E})\, dv \tag{2.7.5d}$$

The lefthand-side integral over the power density vector quantity P, expression (2.7.5c), may be interpreted as the power flux diverging out of the volume V bounded by the closed surface S. The negative sign on the lefthand side of integral expression (2.7.5d) represents the total power flux converging into the volume V bounded by the closed surface S. Hence, the righthand-side integral terms also represent the total power flux or power flow into the volume V. Since the volume V is bounded by a closed surface S, all the power that has converged into the closed region is just stored in the medium. Expression (2.7.5d) is generally referred to as the *Poynting vector theorem* related to the concept of power flow in a time-varying field. The power density vector defined in expression (2.7.5a) is called as the *Poynting vector*.

By definition, power is the rate of work done or energy expended. Hence, the first integral represents the instantaneous energy change in the magnetic field, and the second integral represents the instantaneous energy change in the electric field. The third integral represents the instantaneous power dissipation in the volume due to the conductivity of the medium. Therefore, referring to expression (2.7.5d), the instantaneous power flow into the volume V is given by

$$P(t) = \frac{\partial}{\partial t} w_m + \frac{\partial}{\partial t} w_e + p_d \tag{2.7.6a}$$

where

w_m: instantaneous energy density stored in magnetic field inside volume V

$$= \frac{\mu}{2} (\bar{H} \bullet \bar{H}) \tag{2.7.6b}$$

$$= \frac{\mu}{2} H^2(\bar{r},t) \qquad \text{joules per cubic meter} \tag{2.7.6c}$$

w_e: instantaneous energy density stored in electric field inside volume V

$$= \frac{\varepsilon}{2} (\bar{E} \bullet \bar{E}) \tag{2.7.6d}$$

$$= \frac{\varepsilon}{2} E^2(\bar{r},t) \qquad \text{joules per cubic meter} \tag{2.7.6e}$$

p_d: rate of energy density dissipated in volume V

$$= \sigma(\bar{E} \bullet \bar{E}) \tag{2.7.6f}$$

$$= \sigma E^2(\bar{r},t) \quad \text{joules per second per cubic meter} \tag{2.7.6g}$$

2.8 PLANE WAVE SOLUTION

A direct time domain solution for the electric and magnetic fields in a very large three-dimensional medium is discussed in this section. The medium has linear, homogeneous, and isotropic properties with known electric current source distributions. The sources are assumed to be confined to a small localized region, and they produce electric and magnetic fields both inside and outside the source region. The time domain Maxwell's equations can be systematically analyzed for determining the fields in the two regions. It should be noted at this point, the Maxwell's equations are a set of coupled partial differential equations. The electric and magnetic fields that satisfy the coupled equations are vectors and functions of the three spatial coordinate variables and the time coordinate variable. The solution for the fields, in fact, completely depends on the specific type of electric current source distribution, and the specific characteristics of the medium.

To get insight into the solution technique for electromagnetic field problems, simple analytical techniques are introduced here for the case of a three-dimensional problem, which involves solution to a second-order partial differential equation. It is shown that the electric and magnetic fields satisfy similar types of vector wave equations. The solution of the vector wave equation in the rectangular coordinate system is a *plane wave function*, representing the electric and magnetic fields. Some elegant properties of the uniform plane wave solution including its propagation characteristics in a three-dimensional medium, are discussed in detail. The relationship between the plane wave electric and magnetic fields, their orientation properties, and impedance characteristics are also derived. Further, the three-dimensional problem is also specialized to the case of a one-dimensional problem where the electric and magnetic fields vary only with respect to a single spatial coordinate variable.

In the analysis of electromagnetic time-dependent field problem, the concept of time causality and time reference is briefly discussed. Generally, the time reference is completely arbitrary, and any instant of time can be taken as reference in the solution technique. Further, the sources and the corresponding electric and magnetic fields are to be treated as causal. As long as the sources are not turned on, the electric and magnetic fields at each point in the medium are zero, assuming no previous residual field effects in the medium. Suppose, the sources are turned on at a reference time, for example at $t = 0$, the effect of the sources is first observed only in the immediate vicinity of the source region. There is a certain time interval during which the electromagnetic energy from the source region propagates and gradually establishes the field distributions at every point in the medium. This time interval completely depends on the medium permeability and permittivity characteristics and also on the physical distance between the source and field observation points. The property of field causality states that the sources can never establish the electric and magnetic fields instantaneously at all points. These concepts are exposed in a case study involving propagation properties of the plane wave electromagnetic fields.

2.8.1 WAVE EQUATION

A complete summary of the *Maxwell's equations in time domain* is given in Tables 2.1 and 2.2. Table 2.1 gives all the relevant electromagnetic field equations in differential form, and similarly; Table 2.2 gives all the relevant electromagnetic field equations in integral form.

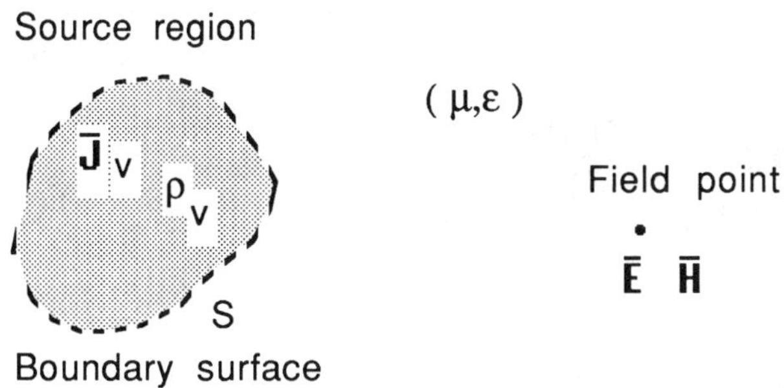

Figure 2.15 Three-dimensional unbounded medium.

In Figure 2.15, let us consider a very large three-dimensional unbounded region that is a linear, homogeneous, and isotropic lossless medium. The time-varying electric current and charge sources produce the time-varying electric and magnetic field distributions both inside and outside the source region.

At any point in the medium, the sources and the electromagnetic field quantities are given by

$\rho_V(\bar{r},t)$: volume electric charge density, in coulombs per cubic meter;

$\bar{J}_V(\bar{r},t)$: volume current density, in amperes per square meter;

$\bar{E}(\bar{r},t)$: electric field distribution, in volts per meter;

$\bar{D}(\bar{r},t)$: electric flux density distribution, coulombs per square meter;

$\bar{H}(\bar{r},t)$: magnetic field distribution, in amperes per meter;

$\bar{B}(\bar{r},t)$: magnetic flux density distribution, in webers per square meter.

Referring to Table 2.1, in the source free region, the electric and magnetic fields satisfy Maxwell's equations in differential form

$$\nabla \times \bar{E}(\bar{r},t) = -\frac{\partial \bar{B}(\bar{r},t)}{\partial t} \qquad (2.8.1a)$$

$$\nabla \times \bar{H}(\bar{r},t) = \frac{\partial \bar{D}(\bar{r},t)}{\partial t} \tag{2.8.1b}$$

$$\nabla \bullet \bar{D}(\bar{r},t) = 0 \tag{2.8.1c}$$

$$\nabla \bullet \bar{B}(\bar{r},t) = 0 \qquad \bar{r} \text{ outside source region} \tag{2.8.1d}$$

For all observation points within the source region itself, on referring to Table 2.1, the electric and magnetic fields satisfy the following Maxwell's equations in differential form

$$\nabla \times \bar{E}(\bar{r},t) = -\frac{\partial \bar{B}(\bar{r},t)}{\partial t} \tag{2.8.2a}$$

$$\nabla \times \bar{H}(\bar{r},t) = \bar{J}_V(\bar{r},t) + \frac{\partial \bar{D}(\bar{r},t)}{\partial t} \tag{2.8.2b}$$

$$\nabla \bullet \bar{D}(\bar{r},t) = \rho_V(\bar{r},t) \tag{2.8.2c}$$

$$\nabla \bullet \bar{B}(\bar{r},t) = 0 \qquad \bar{r} \text{ inside source region} \tag{2.8.2d}$$

The set of equations defined for the source-free region, expressions (2.8.1a – d) and for the source region, expressions (2.8.2a – d) can be solved separately. The electric and magnetic fields so obtained for the two regions should satisfy appropriate boundary conditions for all points on the boundary surface separating the two regions. Particularly at the boundary surface separating the source region and the source-free region, the integral form of Maxwell's equations are convenient to apply. In the following, Maxwell's equations for the source-free region are analyzed in detail. In equations (2.8.1a – d), there are four unknown field quantities; namely, electric field, electric flux density, magnetic field and magnetic flux density. Theoretically, it should be possible to solve for all the field quantities, but Gauss's law, expressions (2.8.1c) and (2.8.1d), are not an independent set of equations. As discussed in Section 2.5, they can be derived directly from the two independent curl equations by taking divergence on both sides of expressions (2.8.1a) and (2.8.1b). Hence, we have only two independent coupled curl equations with four unknown field quantities. Two other additional relationships are required before an attempt can be made to solve the equations. In fact, the additional relationships are obtained based on the medium constitutive relationships, expressions (2.4.6) and (2.4.7), connecting the electric and magnetic fields with respect to their corresponding density distributions. Further, the constitutive relationships completely depend on the properties of the medium under consideration. For a three-dimensional unbounded region that is assumed to be linear, homogeneous, and isotropic, the two constitutive relationships are given by

$$\bar{D}(\bar{r},t) = \varepsilon\bar{E}(\bar{r},t) \tag{2.8.3}$$

$$\bar{B}(\bar{r},t) = \mu\bar{H}(\bar{r},t) \tag{2.8.4}$$

ε : constant permittivity of the medium, in farads per meter

μ : constant permeability of the medium, in henrys per meter

Constitutive relationships (2.8.3) and (2.8.4) are now substituted into Maxwell's equations (2.8.1a – d) for the source-free region. As mentioned earlier, the permeability and permittivity parameters of the three-dimensional medium are taken as constants, and thus they depend on neither the time parameter variable nor on the spatial coordinate variables. Hence, the electric and magnetic fields in the source-free region satisfy the following Maxwell's equations in the differential form:

$$\nabla \times \bar{E}(\bar{r},t) = -\mu \frac{\partial \bar{H}(\bar{r},t)}{\partial t} \tag{2.8.5a}$$

$$\nabla \times \bar{H}(\bar{r},t) = \varepsilon \frac{\partial \bar{E}(\bar{r},t)}{\partial t} \tag{2.8.5b}$$

$$\nabla \bullet \bar{E}(\bar{r},t) = 0 \tag{2.8.5c}$$

$$\nabla \bullet \bar{H}(\bar{r},t) = 0 \qquad \bar{r} \text{ outside source region} \tag{2.8.5d}$$

The two independent curl equations, (2.8.5a) and (2.8.5b), are coupled partial differential equations with two unknowns. They are to be solved for the electric and magnetic field quantities subject to the known boundary conditions. In fact, there are number of analytical and numerical approaches one can adopt to solve for the fields. In the following, a straightforward procedure is presented, based on a differential equation technique. The magnetic field term in the first curl equation (2.8.5a) is first eliminated by using the second curl equation (2.8.5b) to obtain a partial differential equation in terms of only electric field distribution.

On taking curl operation on both sides of expression (2.8.5a),

$$\nabla \times [\nabla \times \bar{E}(\bar{r},t)] = -\mu \nabla \times \frac{\partial \bar{H}(\bar{r},t)}{\partial t} \tag{2.8.6}$$

In the righthand side of this expression, the partial time derivative and the spatial differential operator can be interchanged. Referring to Appendix A, we have the vector identity

$$\nabla \times [\nabla \times \bar{E}(\bar{r},t)] = \nabla[\nabla \bullet \bar{E}(\bar{r},t)] - \nabla^2 \bar{E}(\bar{r},t) \tag{2.8.7}$$

On replacing the lefthand-side term of (2.8.6) based on this vector identity and the righthand-side term, the curl of the magnetic field, by expression (2.8.5b),

$$\nabla[\nabla \bullet \bar{E}(\bar{r},t)] - \nabla^2 \bar{E}(\bar{r},t) = -\mu \frac{\partial}{\partial t}\left[\varepsilon \frac{\partial \bar{E}(\bar{r},t)}{\partial t}\right] \tag{2.8.8a}$$

Based on Gauss's law, relationship (2.8.5c), the divergence of the electric field is always zero in the source-free region, and hence expression (2.8.8a) takes the form

$$\nabla^2 \bar{E}(\bar{r},t) = \mu\varepsilon \frac{\partial^2 \bar{E}(\bar{r},t)}{\partial t^2} \tag{2.8.8b}$$

Expression (2.8.8b) is generally referred to as the *vector wave equation* for the electric field distribution. This vector wave equation is a second-order partial differential equation with dependent space and time coordinate variables. The electric field can be obtained by first solving this partial differential equation subject to the known boundary conditions. The magnetic field can be obtained by substituting back the known electric field solution into either expression (2.8.5a) or (2.8.5b).

It is also possible to obtain a separate and similar type of vector wave equation for the magnetic field distribution. The electric field term in the second curl equation (2.8.5b) can be eliminated by using the first curl equation (2.8.5a) to obtain a partial differential equation in terms of only magnetic field distribution:

$$\nabla^2 \bar{H}(\bar{r},t) = \mu\varepsilon \frac{\partial^2 \bar{H}(\bar{r},t)}{\partial t^2} \tag{2.8.8c}$$

Expression (2.8.8c) is also referred to as the *vector wave equation* for the magnetic field distribution. Referring to the expressions (2.8.8b) and (2.8.8c), the electric field and magnetic field distributions satisfy similar types of vector wave equations. The solution of the vector wave equation is quite complicated. Hence, the two vector wave equations are correspondingly reduced to six scalar wave equations by expressing the vector electric and magnetic fields in terms of their component forms.

For example, the vector wave equations can be simplified using the rectangular coordinate system:

$$\bar{E}(\bar{r},t) = E_x(\bar{r},t)\hat{x} + E_y(\bar{r},t)\hat{y} + E_z(\bar{r},t)\hat{z} \tag{2.8.9}$$

$$\bar{H}(\bar{r},t) = H_x(\bar{r},t)\hat{x} + H_y(\bar{r},t)\hat{y} + H_z(\bar{r},t)\hat{z} \tag{2.8.10}$$

These two fields are now substituted into the vector wave equations, (2.8.8b) and (2.8.8c), to reduce them to six simple scalar wave equations, given by

$$\nabla^2 E_i(\bar{\mathbf{r}},t) = \mu\varepsilon \frac{\partial^2 E_i(\bar{\mathbf{r}},t)}{\partial t^2} \tag{2.8.11a}$$

$$\nabla^2 H_i(\bar{\mathbf{r}},t) = \mu\varepsilon \frac{\partial^2 H_i(\bar{\mathbf{r}},t)}{\partial t^2} \qquad i = x,y,z \tag{2.8.11b}$$

In these wave equations, the permeability and permittivity can be rewritten as

$$\mu = \mu_0 \mu_r \tag{2.8.12a}$$

$$\varepsilon = \varepsilon_0 \varepsilon_r \tag{2.8.12b}$$

where

μ_r : relative permeability of the magnetic medium;

ε_r : relative permittivity of the dielectric medium;

μ_0 : permeability of the free-space medium

$= 4\pi \times 10^{-7}$ henry per meter; (2.8.13a)

ε_0 : permittivity of the free-space medium

$= 8.854 \times 10^{-12}$ farad per meter. (2.8.13b)

Hence,

$$\begin{aligned}\mu\varepsilon &= \mu_r \varepsilon_r \mu_0 \varepsilon_0 \\ &= \frac{1}{c^2} \\ &= \frac{\mu_r \varepsilon_r}{c_0^2}\end{aligned} \tag{2.8.14}$$

where

c : velocity of electromagnetic wave in the material medium,

$= (\mu\varepsilon)^{-1/2}$ (2.8.15a)

c_0 : velocity of electromagnetic wave in the free-space medium,

$$= (\mu_0 \varepsilon_0)^{-1/2} \tag{2.8.15b}$$

$$= 2.998 \times 10^8 \quad \text{meters per second.}$$

2.9 SOLUTION TO THE WAVE EQUATION

A detailed solution technique for the scalar wave equations, expressions (2.8.11a) and (2.8.11b), is discussed in the following. In total, six similar scalar wave equations satisfy the three electric field components and three magnetic field components. It is just required to solve one scalar wave equation, and the solution for the remaining five equations can be written down by inspection. Referring to expression (2.8.11a), for the source-free region, the scalar wave equation in rectangular coordinate system can be written as

$$\frac{\partial^2 U(x,y,z,t)}{\partial x^2} + \frac{\partial^2 U(x,y,z,t)}{\partial y^2} + \frac{\partial^2 U(x,y,z,t)}{\partial z^2} = \frac{1}{c^2}\frac{\partial^2 U(x,y,z,t)}{\partial t^2} \tag{2.9.1}$$

where the scalar field component *U(x, y, z, t)* can represent any rectangular component of either the electric field or the magnetic field. Expression (2.9.1) is a second-order, partial differential equation with three spatial coordinate variables *(x, y, z)* and one time coordinate variable, *t*. The discussion concerning the boundary conditions is taken up later. To solve partial differential equation (2.9.1), the method of *separation of variables* can be adopted. In this approach, the total solution is written as a product of four partial factored solutions, which are individually functions of a single coordinate variable.

The total solution for the scalar field component *U* is written as

$$U(x,y,z,t) = \Xi(x)\Psi(y)Z(z)T(t) \tag{2.9.2}$$

where

$\Xi(x)$: partial solution of *U* in terms of only the of *x* coordinate;
$\Psi(y)$: partial solution of *U* in terms of only the *y* coordinate;
$Z(z)$: partial solution of *U* in terms of only the *z* coordinate;
$T(t)$: partial solution of *U* in terms of only the *t* coordinate.

On differentiating this total solution twice with respect to each dependent variable *(x, y, z, t)*, the various partial derivative terms in expression (2.9.1) can be rewritten as

$$\frac{\partial^2 U(x,y,z,t)}{\partial x^2} = \Psi Z T \frac{d^2 \Xi}{dx^2} \tag{2.9.3a}$$

$$\frac{\partial^2 U(x,y,z,t)}{\partial y^2} = \Xi Z T \frac{d^2 \Psi}{dy^2} \tag{2.9.3b}$$

$$\frac{\partial^2 U(x,y,z,t)}{\partial z^2} = \Xi \Psi T \frac{d^2 Z}{dz^2} \tag{2.9.3c}$$

$$\frac{\partial^2 U(x,y,z,t)}{\partial t^2} = \Xi \Psi Z \frac{d^2 T}{dt^2} \tag{2.9.3d}$$

On substituting expressions (2.9.3a – d) into scalar wave equation (2.9.1) and dividing throughout by the total solution, expression (2.9.2), the following differential equation is obtained:

$$\frac{c^2}{\Xi}\frac{d^2 \Xi}{dx^2} + \frac{c^2}{\Psi}\frac{d^2 \Psi}{dy^2} + \frac{c^2}{Z}\frac{d^2 Z}{dz^2} = \frac{1}{T}\frac{d^2 T}{dt^2} \tag{2.9.4a}$$

It is interesting to note that the first term in this expression (2.9.4a) is a pure function of x, the second term is a pure function of y, the third term is a pure function of z, and the sum of these three terms is equal to another term that is a pure function of t. This relationship can be satisfied only when the righthand side of expression (2.9.4a) is equal to a constant and, similarly, the lefthand side is equal to the same constant. Hence,

$$\frac{c^2}{\Xi}\frac{d^2 \Xi}{dx^2} + \frac{c^2}{\Psi}\frac{d^2 \Psi}{dy^2} + \frac{c^2}{Z}\frac{d^2 Z}{dz^2} = -\omega^2 \tag{2.9.4b}$$

$$\frac{1}{T}\frac{d^2 T}{dt^2} = -\omega^2 \tag{2.9.4c}$$

where $-\omega^2$ is the separation constant.

Expressions (2.9.4b) and (2.9.4c) are two ordinary differential equations, and their solutions are attempted in the following. Considering differential equation (2.9.4c), the roots are given by $j\omega$ and $-j\omega$. The harmonic functions $\exp(j\omega t)$ and $\exp(-j\omega t)$ satisfy separately differential equation (2.9.4c), and thus, they represent independent solutions.

Further, a linear combination of these two harmonic functions is also a general solution to differential equation (2.9.4c). Therefore, the solution is given by

$$T(t) = \mathcal{P}\, e^{j\omega t} + \mathcal{Q}\, e^{-j\omega t} \tag{2.9.5a}$$

where $\mathcal{P}$ and $\mathcal{Q}$ are the constants to be determined. These two constants completely depend upon the initial condition and also the final condition of the solution as a function of the time coordinate variable t. It should be noted that the composite total solution represented in expression (2.9.2) is a *real* function with respect to the spatial and time coordinate variables, which eventually depends upon the type of source excitation. A number of considerations are now made concerning causality of the field solution at a given field point (x, y, z) in the medium, which clearly depends on the reference time selected. Using the causality, suppose the time instant $t = 0$ is selected as a reference time in the source region, then there exist two specific time intervals given by $t > 0$ and $t < 0$. For all time intervals $t < 0$, no solution exists at the field point in the medium, and for all practical considerations the solution at the field point exists only for time interval $t > 0$. In fact, the first term in the solution given by expression (2.9.5a) accounts for the real positive time, $t > 0$, while the second term accounts for the negative time, $t < 0$.

Hence, the second term in the partial solution of expression (2.9.5a) is completely dropped, and thus the solution takes the form

$$T(t) = \mathcal{P}\, e^{j\omega t} \tag{2.9.5b}$$

The constant P is such that the final field solution is eventually a real function in the time domain similar to the source excitation which is also a real function. Now, the solution to the other differential equation (2.9.4b) is attempted. On rearranging the terms, expression (2.9.4b) can be written as

$$\frac{1}{\Xi}\frac{d^2\Xi}{dx^2} + \frac{1}{\Psi}\frac{d^2\Psi}{dy^2} + \frac{1}{Z}\frac{d^2 Z}{dz^2} = -\frac{\omega^2}{c^2} \tag{2.9.6a}$$

Referring to the lefthand side of expression (2.9.6a), first term is a pure function of x, second term is a pure function of y, third term is a pure function of z, and the sum of these three terms is equal to a constant on the righthand side. This relationship can be satisfied only when each term on the lefthand side of expression (2.9.6a) is equal to a constant. Hence,

$$\frac{1}{\Xi}\frac{d^2\Xi}{dx^2} = -\beta_x^2 \tag{2.9.6b}$$

$$\frac{1}{\Psi}\frac{d^2\Psi}{dy^2} = -\beta_y^2 \tag{2.9.6c}$$

$$\frac{1}{Z}\frac{d^2Z}{dz^2} = -\beta_z^2 \tag{2.9.6d}$$

where β_x^2, β_y^2, β_z^2 are separation constants to be determined.

On comparing expressions (2.9.6a – d), the separation constants are interrelated through the relationship

$$\beta_x^2 + \beta_y^2 + \beta_z^2 = \beta^2 \tag{2.9.6e}$$

where β^2 is the composite separation constant

$$\beta^2 = \omega^2\mu\varepsilon \tag{2.9.6f}$$

The expressions (2.9.6b – d) are three ordinary differential equations, and their solutions are attempted in the following. Consider differential equation (2.9.6b); it has two roots given by $j\beta_x$ and $-j\beta_x$. Hence, the harmonic functions given by $\exp(j\beta_x x)$ and $\exp(-j\beta_x x)$ separately satisfy differential equation (2.9.6b), and thus they represent independent solutions. Further, a linear combination of these two harmonic functions is also a general solution to differential equation (2.9.6b). Similar solutions in terms of harmonic functions can be obtained for the other two differential equations, (2.9.6c) and (2.9.6d). Hence, the partial solutions to these three ordinary differential equations can be written as

$$\Xi(x) = \mathcal{P}_1 e^{j\beta_x x} + \mathcal{Q}_1 e^{-j\beta_x x} \tag{2.9.7a}$$

$$\Psi(y) = \mathcal{P}_2 e^{j\beta_y y} + \mathcal{Q}_2 e^{-j\beta_y y} \tag{2.9.7b}$$

$$Z(z) = \mathcal{P}_3 e^{j\beta_z z} + \mathcal{Q}_3 e^{-j\beta_z z} \tag{2.9.7c}$$

where $\mathcal{P}_1$, $\mathcal{P}_2$, $\mathcal{P}_3$ and $\mathcal{Q}_1$, $\mathcal{Q}_2$, $\mathcal{Q}_3$ are just constants to be determined. Again, these constants are such that the eventual solution for the scalar field quantity represented in expression (2.9.2) is a real function with respect to the spatial and time coordinate variables. The complete solution for the scalar field $U(x,y,z,t)$ is obtained by substituting all the partial factored solutions. On substituting expressions (2.9.5b), (2.9.7a), (2.9.7b), and (2.9.7c) into expression (2.9.2),

$$U(x,y,z,t) = \Xi(x)\Psi(y)Z(z)T(t) \tag{2.9.8a}$$

$$\begin{aligned} &= \mathcal{P}e^{j\omega t}[\mathcal{P}_1 e^{j\beta_x x} + Q_1 e^{-j\beta_x x}] \\ &\quad [\mathcal{P}_2 e^{j\beta_y y} + Q_2 e^{-j\beta_y y}] \\ &\quad [\mathcal{P}_3 e^{j\beta_z z} + Q_3 e^{-j\beta_z z}] \end{aligned} \tag{2.9.8b}$$

Certain physical meanings of the various terms appearing in this solution are discussed in the following. For example, let us consider the partial product solution given by the first and the third terms in the expression:

$$\Psi(y)T(t) = \mathcal{P}\mathcal{P}_2\, e^{j(\omega t+\beta_y y)} + \mathcal{P}Q_2\, e^{j(\omega t-\beta_y y)} \tag{2.9.9a}$$

The exponential terms in expression (2.9.9a) represent phase factors. As stated earlier, the final time domain solution to the wave equation should be a real function, and hence, only either the real part or the imaginary part of expression (2.9.8b) or (2.9.9a) represents the complete solution, depending upon the type of excitation of the sources. Let us analyze the second exponential term in expression (2.9.9a), which can be rewritten as

$$e^{j(\omega t-\beta_y y)} = \cos(\omega t-\beta_y y) + j\sin(\omega t-\beta_y y) \tag{2.9.9b}$$

The sine and cosine terms in expression (2.9.9b) are harmonic functions that are interrelated and differ by a phase angle of 90°. Since either the real part or the imaginary part is required to represent the final real functional solution to the wave equation, it is sufficient to study either the sine function or the cosine function to understand the behavior of the exponential phase term. The variation of the phase term is studied as a function of the time parameter t and the rectangular coordinate variable y. In table 2.4, the imaginary part of expression (2.9.9b) is calculated for various values of time t.

These cases are plotted in Figure 2.16 to show how the argument of the function $F(y,t)$ varies as a function of coordinate variable y for fixed values of time parameter t. As can be seen from the plot shown in Figure 2.16, the phase of the function $F(y,t)$ shifts in the $+y$ coordinate direction as the time t increases. This is evident by inspecting the location of peak points a, b, and c on the curves. In fact, any arbitrary point on these curves can be tracked to see how it shifts in spatial location as a function of time t. By noticing the peak of the curves, as t increases, the peak point shifts its location in the $+y$ direction. This phenomena is recognized as the traveling wave property of the electromagnetic fields.

Table 2.4
Function $F(y,t)$ for fixed values of time variable t

	t	$F(y,t) = \sin(\omega t - \beta_y y)$
CASE: A	$t = 0$	$\sin(-\beta_y y) = -\sin(\beta_y y)$
CASE: B	$t = \frac{\pi}{2\omega}$	$\sin(\frac{\pi}{2} - \beta_y y) = \cos(\beta_y y)$
CASE: C	$t = \frac{\pi}{\omega}$	$\sin(\pi - \beta_y y) = \sin(\beta_y y)$

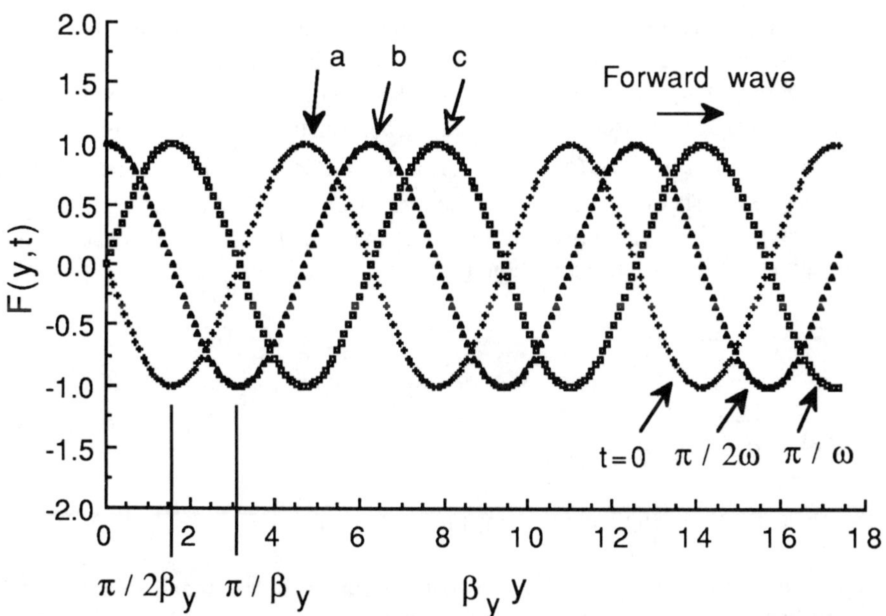

Figure 2.16 Plot of $\sin(\omega t - \beta_y y)$ as a function of y for fixed values of t.

The traveling wave, in fact, travels forward in the $+y$ direction at a constant velocity that completely depends on the permeability and permittivity characteristics of the medium. At any fixed distance from the origin, there is a perpendicular plane in which the variation of the phase function is always constant. Hence,

$$\omega t - \beta_y y = \text{constant} \tag{2.9.10a}$$

On taking derivative of relationship (2.9.10a) with respect to time t,

$$\omega - \beta_y \frac{dy}{dt} = 0 \tag{2.9.10b}$$

Hence, the phase velocity, in meters per second, in the $+y$ direction is given by

$$v_{py} = \frac{dy}{dt} \tag{2.9.10c}$$

$$= \frac{\omega}{\beta_y} \tag{2.9.10d}$$

This phase velocity result can also be obtained directly from Figure 2.16. A point located at the maxima moves to the next location in $+y$ direction in a definite time interval. The ratio of the change in distance to the corresponding change in time gives the phase velocity obtained in expression (2.9.10d). The constant β_y can be determined by referring to Figure 2.16. The function $F(y,t)$ is a periodic function and repeats itself for every 2π radians. The distance between the two successive peak points or the length of one full wave cycle is generally referred to as the *wavelength*. In this case, it is the wavelength in the y coordinate direction, given by

$$\beta_y \lambda_y = 2\pi \tag{2.9.11a}$$

$$\lambda_y = \frac{2\pi}{\beta_y} \tag{2.9.11b}$$

where λ_y is the wave length along the y direction, in meters.

Referring to expression (2.9.9b), the cosine term on the righthand side can also be analyzed based on the same procedure. Hence, the lefthand-side term of expression (2.9.9b) represents *a positive or forward traveling plane wave* that moves in the $+y$ direction at a phase velocity given by result (2.9.10d).

Similarly, let us analyze the first exponential term in expression (2.9.9a), which can be rewritten as

$$e^{j(\omega t+\beta_y y)} = \cos(\omega t+\beta_y y) + j\,\sin(\omega t+\beta_y y) \tag{2.9.12a}$$

The only difference is in the argument of phase term which has a sign change. Again, the sine and cosine terms in expression (2.9.12a) are harmonic functions that are interrelated and differ by a phase angle of 90°. The variation of sine or cosine of the phase term $(\omega t+\beta_y y)$ can be similarly studied as a function of parameters (y,t). This results in the same phase velocity discussed earlier, except the phase function moves in the $-y$ coordinate direction. Hence, the lefthand-side term of expression (2.9.12a) represents *a negative or backward traveling plane wave* that moves in the $-y$ coordinate direction at a phase velocity given by result (2.9.10d).

Therefore, it can be concluded that the partial solution given by expression (2.9.9a) consists of the superposition of two types of traveling wave functions, one traveling in the forward direction and the other traveling in the backward direction, both with the same phase velocity. Hence,

$\Psi(y)T(t)$ = (backward traveling wave) + (forward traveling wave)

backward traveling wave => $\mathcal{P}\mathcal{P}_2\, e^{j(\omega t+\beta_y y)}$ (2.9.12b)

forward traveling wave => $\mathcal{P}\mathcal{Q}_2\, e^{j(\omega t-\beta_y y)}$ (2.9.12c)

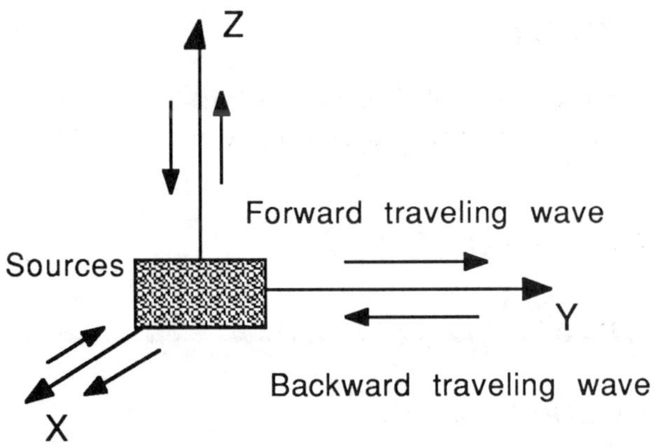

Figure 2.17 Traveling waves in the three-dimensional unbounded medium.

Referring to Figure 2.17, it is assumed that the origin of the rectangular coordinates is shifted to the source region containing the electric current and charge sources. This shifted origin may be recognized as a localized reference point. The electric current and charge sources produce electromagnetic energy in all directions. Considering the rectangular y coordinate direction, the electromagnetic field sources radiate energy in the y coordinate direction by the forward traveling wave propagating away from the sources. The three-dimensional region under study is an unbounded linear, homogeneous, and isotropic medium with no other electromagnetic field sources present. Hence, the backward traveling waves are not in the medium, and if present, propagates toward the sources. For an unbounded medium, expression (2.9.12b) is completely ignored.

In fact, similar discussions can be carried out for the x and z coordinate directions. With respect to the sources depicted in Figure 2.17, only the forward traveling waves propagate in the $+x$ and $+z$ coordinate directions.

Retaining only the forward traveling wave terms, the total solution for the scalar field obtained in expression (2.9.8a) reduces to the following form:

$$U_f(x,y,z,t) = \Xi(x)\Psi(y)Z(z)T(t) \tag{2.9.13a}$$

$$= \mathcal{P} e^{j\omega t}[Q_1 e^{-j\beta_x x}][Q_2 e^{-j\beta_y y}][Q_3 e^{-j\beta_z z}] \tag{2.9.13b}$$

$$= U_1 e^{j\omega t}[e^{-j(\beta_x x+\beta_y y+\beta_z z)}] \tag{2.9.13c}$$

where the composite constant is given by

U_1 : amplitude of forward traveling plane wave

$$= \mathcal{P}Q_1Q_2Q_3 \tag{2.9.13d}$$

Similarly by proper source simulation, if only backward traveling waves are present, the total solution for the scalar field obtained in expression (2.9.8a) reduces to the following form:

$$U_b(x,y,z,t) = \Xi(x)\Psi(y)Z(z)T(t) \tag{2.9.14a}$$

$$= \mathcal{P} e^{j\omega t}[\mathcal{P}_1 e^{j\beta_x x}][\mathcal{P}_2 e^{j\beta_y y}][\mathcal{P}_3 e^{j\beta_z z}] \tag{2.9.14b}$$

$$= U_2 e^{j\omega t}[e^{j(\beta_x x+\beta_y y+\beta_z z)}] \tag{2.9.14c}$$

where the composite constant is given by

U_2 : amplitude of backward traveling plane wave

$$= \mathcal{P}\mathcal{P}_1\mathcal{P}_2\mathcal{P}_3 \tag{2.9.14d}$$

It should be noted again that only either the real part or the imaginary part represents the proper solution to the scalar wave equation. In these two scalar field solutions, expressions (2.9.13c) and (2.9.14c), U_1 and U_2 are unknown amplitude constants that represent level of excitation of the field components. Interestingly, these amplitudes are independent of the spatial coordinate variables. Eventually, they are determined based on the known source distribution by enforcing boundary conditions at the interface separating the source region and the source-free region. The two scalar field solutions can be further written in a compact form by expressing the exponent term in terms of a vector notation.

With respect to a global origin, let

$\bar{r}$: spatial vector corresponding to the field point (x, y, z)

$$= x\hat{x} + y\hat{y} + z\hat{z} \tag{2.9.15a}$$

$\bar{\beta}$: propagation vector of the traveling waves

$$= \beta_x\hat{x} + \beta_y\hat{y} + \beta_z\hat{z} \tag{2.9.15b}$$

$$\bar{\beta} \bullet \bar{r} = \beta_x x + \beta_y y + \beta_z z \tag{2.9.15c}$$

$$= 2\pi \left[\frac{x}{\lambda_x} + \frac{y}{\lambda_y} + \frac{z}{\lambda_z}\right] \tag{2.9.15d}$$

$\lambda_x, \lambda_y, \lambda_z$: wavelength of traveling plane wave in the x, y, z directions.

With either the real or imaginary part assumed, the complete solution for the scalar plane wave field quantity as obtained in expressions (2.9.13c) and (2.9.14c) reduces to the following compact representation:

Positive or forward traveling wave is

$$U_f(\bar{r},t) = U_1\, e^{j(\omega t - \bar{\beta} \bullet \bar{r})} \tag{2.9.16a}$$

Negative or backward traveling wave is

$$U_b(\bar{r},t) = U_2\, e^{j(\omega t + \bar{\beta} \bullet \bar{r})} \tag{2.9.16b}$$

2.10 VECTOR PLANE WAVE FIELDS

Expressions (2.8.8b) and (2.8.8c), the electric and magnetic field distributions in a source-free three-dimensional region in a linear, homogeneous, and isotropic medium, satisfy similar vector wave equations. As mentioned earlier, the solution of the direct vector wave equations is quite complicated, and thus the two vector wave equations are reduced to the six independent scalar wave equations by expressing the vector electric and magnetic fields in terms of their rectangular component forms. The electric and magnetic fields in terms of rectangular components are given by

$$\bar{E}(\bar{r},t) = E_x(\bar{r},t)\hat{x} + E_y(\bar{r},t)\hat{y} + E_z(\bar{r},t)\hat{z} \tag{2.10.1}$$

$$\bar{H}(\bar{r},t) = H_x(\bar{r},t)\hat{x} + H_y(\bar{r},t)\hat{y} + H_z(\bar{r},t)\hat{z} \tag{2.10.2}$$

The three components of the electric field and three components of the magnetic field satisfy similar types of scalar wave equations, given by

$$\nabla^2 E_i(\bar{r},t) = \mu\varepsilon \frac{\partial^2 E_i(\bar{r},t)}{\partial t^2} \tag{2.10.3a}$$

$$\nabla^2 H_i(\bar{r},t) = \mu\varepsilon \frac{\partial^2 H_i(\bar{r},t)}{\partial t^2} \qquad i = x,\ y,\ z. \tag{2.10.3b}$$

In Section 2.9, a complete solution was obtained for the scalar field component by rigorously solving the scalar wave equation, which is a function of the three spatial coordinate variables *(x,y,z)* and time parameter *t*. Referring to expression (2.9.16a), the solution for the electric and magnetic field components are given by

$$E_i(\bar{r},t) = E_{0i}\, e^{j(\omega t - \bar{\beta} \bullet \bar{r})} \tag{2.10.4a}$$

$$H_i(\bar{r},t) = H_{0i}\, e^{j(\omega t - \bar{\beta} \bullet \bar{r})} \qquad i = x,\ y,\ z. \tag{2.10.4b}$$

Defining the following constant vectors for the electric and magnetic fields,

$$\bar{E}_0 = E_{0x}\hat{x} + E_{0y}\hat{y} + E_{0z}\hat{z} \tag{2.10.5a}$$

$$\bar{H}_0 = H_{0x}\hat{x} + H_{0y}\hat{y} + H_{0z}\hat{z} \tag{2.10.5b}$$

the complete solution for the vector electric and magnetic fields based on only the forward traveling plane waves is given by

$$\bar{E}(\bar{r},t) = \bar{E}_0\, e^{j(\omega t - \bar{\beta} \bullet \bar{r})} \tag{2.10.6a}$$

$$\bar{H}(\bar{r},t) = \bar{H}_0\, e^{j(\omega t - \bar{\beta} \bullet \bar{r})} \tag{2.10.6b}$$

The parameter $\bar{\beta}$ is called as the *propagation vector*, which gives the actual direction of the forward traveling wave and, referring to expression (2.9.6f), is given by

$$\bar{\beta} = \beta\hat{\beta} \tag{2.10.6c}$$

where $\hat{\beta}$ is the actual direction of the plane wave propagation, and

β : propagation constant for homogeneous, isotropic medium

$$= \omega(\mu\varepsilon)^{1/2}. \tag{2.10.6d}$$

Further, referring to expressions (2.10.6a) and (2.10.6b), the term in the exponent represents the phase of the forward traveling plane wave vector. This phase term completely depends on the coordinate variables and the time parameter variable. At any point along the propagation vector, if a plane is drawn perpendicular to the direction of propagation vector, the phase of the plane wave vector always remains constant.

Hence,

$$\omega t - \bar{\beta} \bullet \bar{r} = \text{constant} \tag{2.10.7a}$$

$$\omega t - \beta r\hat{\beta} \bullet \hat{r} = \text{constant} \tag{2.10.7b}$$

$$\omega t - \beta r \cos\psi = \text{constant} \tag{2.10.7c}$$

where ψ is the angle between the direction of the propagation vector and the spatial coordinate vector, as shown in Figure 2.18. On taking derivative of the relationship (2.10.7c) with respect to time t

$$\omega - \beta \frac{dr}{dt} \cos\psi = 0 \tag{2.10.7d}$$

Referring to Figure 2.18 and expression (2.10.7d), the phase velocity of the vector plane wave in meters per second is given by

$$v_p = \frac{dr}{dt} \tag{2.10.7e}$$

$$= \frac{\omega}{\beta \cos\psi} \tag{2.10.7f}$$

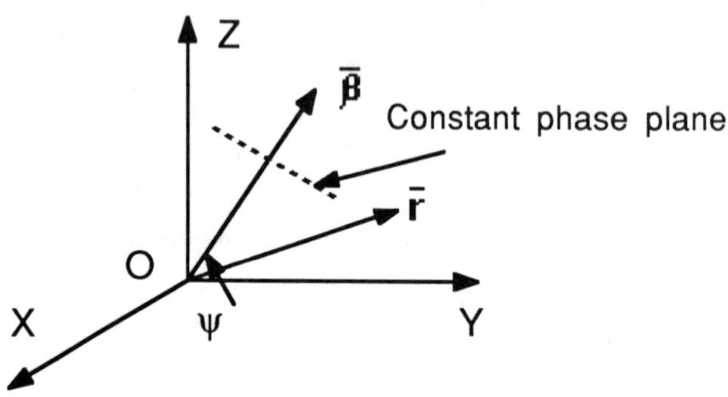

Figure 2.18 Phase velocity of vector plane wave.

As can be seen from expression (2.10.7f), the phase velocity is a function of the relative angle ψ, which basically dictates the actual velocity of the plane wave in a specific direction. The phase velocity of the plane wave along the direction of propagation is obtained by choosing angle ψ to be zero, and the expression (2.10.7f) reduces to

$$v_p = \frac{\omega}{\beta} \tag{2.10.7g}$$

$$= (\mu\varepsilon)^{-1/2} \qquad \text{meters per second} \tag{2.10.7h}$$

The plane wave propagates in the $\bar{\boldsymbol{\beta}}$ direction and is a periodic function that repeats itself every 2π radians. The distance between two successive peak points on the periodic wave or the length of one full wave cycle is the wavelength of the plane wave. In this case, it is the wavelength in the propagation direction, given by

$$\beta\lambda = 2\pi \tag{2.10.7i}$$

$$\lambda = \frac{2\pi}{\beta} \tag{2.10.7j}$$

λ : wave length of the plane wave, in meters

$$= [\lambda_x^2 + \lambda_y^2 + \lambda_z^2]^{1/2} \tag{2.10.7k}$$

2.11 PROPERTIES OF PLANE WAVE FIELDS

Interesting plane wave properties of the vector electric and magnetic fields in the source-free region are derived in this section. It can be easily shown that the electric and magnetic fields obtained in expressions (2.10.6a) and (2.10.6b) are not independent, but are closely related field quantities. Further, the electric field vector and the magnetic field vector are mutually perpendicular to the direction of propagation. Referring to expression (2.10.6a), the plane wave electric field is given by

$$\bar{E}(\bar{r},t) = \bar{E}_0 \, e^{j(\omega t - \bar{\beta} \bullet \bar{r})} \tag{2.11.1a}$$

Gauss's law for electric fields, expression (2.8.5c), in the source-free region is

$$\nabla \bullet \bar{E}(\bar{r},t) = 0 \tag{2.11.1b}$$

On substituting the plane wave electric field into the above relationship

$$\nabla \bullet \left[\bar{E}_0 \, e^{j(\omega t - \bar{\beta} \bullet \bar{r})}\right] = 0 \tag{2.11.1c}$$

According to the vector identity, Appendix A, the divergence of the product of a vector function and a scalar function is given by

$$\nabla \bullet [\bar{E}_0 \Phi] = [\nabla \bullet \bar{E}_0]\Phi + \nabla\Phi \bullet \bar{E}_0 \tag{2.11.1d}$$

where it is assumed that the scalar function

$$\Phi = e^{j(\omega t - \bar{\beta} \bullet \bar{r})} \tag{2.11.1e}$$

The quantity $\bar{E}_0$ is a constant vector independent of the coordinate variables, and thus its divergence is zero. Expression (2.11.1c) simplifies to

$$\nabla\left[e^{j(\omega t - \bar{\beta} \bullet \bar{r})}\right] \bullet \bar{E}_0 = 0 \tag{2.11.1f}$$

$$\nabla[e^{-j\bar{\beta}\bullet\bar{r}}]\bullet\bar{E}_0 = 0 \tag{2.11.1g}$$

On referring to the relationship (2.9.15c), the gradient term can be rewritten as

$$[\frac{\partial}{\partial x}\hat{x} + \frac{\partial}{\partial y}\hat{y} + \frac{\partial}{\partial z}\hat{z}][e^{-j(\beta_x x+\beta_y y+\beta_z z)}]\bullet\bar{E}_0 = 0 \tag{2.11.1h}$$

$$[-j\beta_x\hat{x} - j\beta_y\hat{y} - j\beta_z\hat{z}][e^{-j(\beta_x x+\beta_y y+\beta_z z)}]\bullet\bar{E}_0 = 0 \tag{2.11.1i}$$

$$[-j\bar{\beta}\bullet\bar{E}_0][e^{-j\bar{\beta}\bullet\bar{r}}] = 0 \tag{2.11.1j}$$

Therefore, from expression (2.11.1j), the scalar product between the propagation vector and the vector plane wave electric field is zero:

$$\bar{\beta}\bullet\bar{E}_0 = 0 \tag{2.11.1k}$$

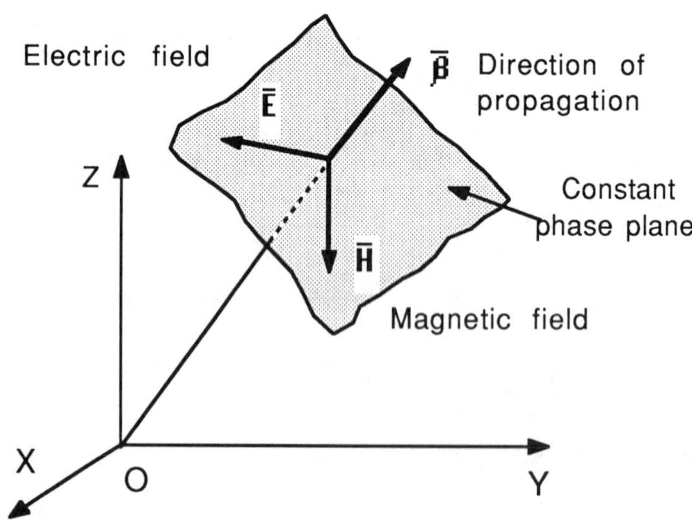

Figure 2.19 Orthogonal properties of vector plane waves.

The result obtained from expression (2.11.1k) is shown in Figure 2.19. It basically states that the plane wave electric field vector is always perpendicular to the direction of propagation of the plane wave. As stated earlier, any plane perpendicular to the direction of propagation is a constant phase plane. Hence, the plane wave electric field vector lies in the constant phase plane.

This property can be shown to be true even for the plane wave magnetic field. Referring to expression (2.10.6b), the plane wave magnetic field is given by

$$\bar{H}(\bar{r},t) = \bar{H}_0 \, e^{j(\omega t - \bar{\beta} \bullet \bar{r})} \tag{2.11.1ℓ}$$

Gauss's law for the magnetic field, expression (2.8.5d), in a source-free region is given by

$$\nabla \bullet \bar{H}(\bar{r},t) = 0 \tag{2.11.1m}$$

On substituting the plane wave magnetic field into this relationship and following the earlier simplification, it is can be again shown that the scalar product between the propagation vector and the vector plane wave magnetic field is zero:

$$\bar{\beta} \bullet \bar{H}_0 = 0 \tag{2.11.1n}$$

This is another interesting property shown in the Figure 2.19, similar in nature to the electric field case.

2.11.1 INTRINSIC IMPEDANCE OF THE MEDIUM

The plane wave electric field and magnetic field obtained in the expressions (2.10.6a) and (2.10.6b) are closely related through the intrinsic or characteristic impedance of the medium. Also, it is shown in the following analysis that the electric and magnetic field vectors are mutually perpendicular to each other, as depicted in the Figure 2.19. In fact, this can be proved by using either of Maxwell's curl equations for the source-free region. Ampere's law, expression (2.8.5b), in the source-free region is given by

$$\nabla \times \bar{H}(\bar{r},t) = \varepsilon \frac{\partial \bar{E}(\bar{r},t)}{\partial t} \tag{2.11.2a}$$

On substituting for the plane wave electric field and the magnetic field vectors, expressions (2.11.1a) and (2.11.1ℓ), into the previous curl equation

$$\nabla \times \left[\bar{H}_0 \, e^{j(\omega t - \bar{\beta} \bullet \bar{r})}\right] = \varepsilon \frac{\partial}{\partial t}\left[\bar{E}_0 \, e^{j(\omega t - \bar{\beta} \bullet \bar{r})}\right] \tag{2.11.2b}$$

According to the vector identity, Appendix A, the curl of the product of a vector function and a scalar function is given by

$$\nabla \times [\bar{H}_0 \Phi] = [\nabla \times \bar{H}_0]\Phi + \nabla\Phi \times \bar{H}_0 \tag{2.11.2c}$$

where it is assumed again that the scalar function

$$\Phi = e^{j(\omega t - \bar{\beta} \bullet \bar{r})} \tag{2.11.2d}$$

The quantity $\bar{\mathbf{H}}_0$ is a constant vector independent of the coordinate variables, and thus its curl is zero. Expression (2.11.2b) simplifies to

$$\nabla\left[e^{j(\omega t - \bar{\beta} \bullet \bar{r})}\right] \times \bar{H}_0 = j\omega\varepsilon\left[\bar{E}_0\, e^{j(\omega t - \bar{\beta} \bullet \bar{r})}\right] \tag{2.11.2e}$$

$$\nabla\left[e^{-j\bar{\beta} \bullet \bar{r}}\right] \times \bar{H}_0 = j\omega\varepsilon\left[\bar{E}_0\, e^{-j\bar{\beta} \bullet \bar{r}}\right] \tag{2.11.2f}$$

Using relationship (2.9.15c), this gradient term can be rewritten as

$$\left[\frac{\partial}{\partial x}\hat{x} + \frac{\partial}{\partial y}\hat{y} + \frac{\partial}{\partial z}\hat{z}\right]\left[e^{-j(\beta_x x+\beta_y y+\beta_z z)}\right] \times \bar{H}_0 = j\omega\varepsilon\left[\bar{E}_0\, e^{-j\bar{\beta} \bullet \bar{r}}\right] \tag{2.11.2g}$$

$$\left[-j\beta_x\hat{x} - j\beta_y\hat{y} - j\beta_z\hat{z}\right]\left[e^{-j(\beta_x x+\beta_y y+\beta_z z)}\right] \times \bar{H}_0 = j\omega\varepsilon\left[\bar{E}_0\, e^{-j\bar{\beta} \bullet \bar{r}}\right] \tag{2.11.2h}$$

$$\left[-j\bar{\beta} \times \bar{H}_0\right]\left[e^{-j\bar{\beta} \bullet \bar{r}}\right] = j\omega\varepsilon\bar{E}_0\left[e^{-j\bar{\beta} \bullet \bar{r}}\right] \tag{2.11.2i}$$

Therefore, from expression (2.11.2i), the vector product between the plane wave magnetic field and the propagation vector is equal to the vector plane wave electric field

$$\bar{H}_0 \times \bar{\beta} = \omega\varepsilon\bar{E}_0 \tag{2.11.2j}$$

According to expressions (2.10.6c) and (2.10.6d), the propagation vector can be written as

$$\bar{\beta} = \beta\hat{\beta} \tag{2.11.2k}$$

$\hat{\beta}$: direction of the plane wave propagation;

β : propagation constant for a homogeneous, isotropic medium
$= \omega(\mu\varepsilon)^{1/2}$.

On substituting representation (2.11.2k) into the vector relationship (2.11.2j),

$$\bar{H}_0 \times \hat{\beta} = \frac{\omega\varepsilon}{\beta}\bar{E}_0 \tag{2.11.3a}$$

Assuming the following representation for the electric and magnetic field vectors

$\bar{E}_0$: constant electric field vector

$$= E_0\hat{e} \tag{2.11.3b}$$

$\bar{H}_0$: constant magnetic field vector

$$= H_0\hat{h} \tag{2.11.3c}$$

Expression (2.11.3a) takes the following form:

$$H_0\hat{h} \times \hat{\beta} = \frac{\omega\varepsilon}{\beta}E_0\hat{e} \tag{2.11.3d}$$

After separating the magnitude and direction, this expression can be rewritten as

$$\hat{e} \times \hat{h} = \hat{\beta} \tag{2.11.3e}$$

$$\frac{E_0}{H_0} = \eta \tag{2.11.3f}$$

Expressions (2.11.1k), (2.11.1n) and (2.11.3e) depict the property that the plane wave electric field and magnetic field unit vectors are mutually orthogonal, and the cross product between their unit vectors gives the direction of propagation. In expression (2.11.3f), the magnitude of electric field is in terms of volts per meter, the magnitude of magnetic field is in terms of amperes per meter, and thus the ratio of the two has the unit of ohms. This ratio is generally referred to as the plane wave *intrinsic impedance* or *characteristic impedance* of the three-dimensional medium. The medium impedance depends primarily on the permittivity and permeability characteristics of the medium and also on the *plane wave nature* of the forward or backward traveling wave. Hence, in expression (2.11.3f),

η : intrinsic impedance of the medium, in ohms

$$= \frac{\beta}{\omega\epsilon} \tag{2.11.4a}$$

$$= \left(\frac{\mu}{\varepsilon}\right)^{1/2} \tag{2.11.4b}$$

Thus, for a free-space medium

η_0 : intrinsic impedance of the free-space medium, in ohms

$$= \left(\frac{\mu_0}{\varepsilon_0}\right)^{1/2} \tag{2.11.5a}$$

$$= 120\pi \tag{2.11.5b}$$

2.12 ONE-DIMENSIONAL FIELD PROBLEM

In the one-dimensional time-dependent boundary value problem, the electromagnetic sources and the corresponding fields vary only with respect to a single spatial coordinate variable and the time parameter variable. All the sources including the electric and magnetic fields are assumed to be independent of the other two spatial coordinate variables. As will be noted, the study of the one-dimensional field problem is quite useful in understanding various practical aspects of the electric and magnetic fields generated by the sources.

A complete summary of *Maxwell's equations in time domain* is given in Tables 2.1 and 2.2. Table 2.1 gives all the relevant electromagnetic field equations in differential form, and Table 2.2 gives all the relevant electromagnetic field equations in integral form. Referring to Figure 2.20, let us consider a three-dimensional unbounded region that is linear, homogeneous and isotropic lossless medium. A constant electric current source is confined to a thin sheet at $y = 0$ and polarized in the $-z$ coordinate direction, given by

$\bar{J}_S(\bar{r},t)$: surface current density, in amperes per meter

$$= -\,J_0(t)\hat{z} \tag{2.12.1}$$

The current source is completely confined to the xz plane and varies as a function of time parameter t, but it is independent of the spatial coordinate variables (x, y, z). This electric current distribution is an idealistic source distribution, which is quite useful for studying one-dimensional electromagnetic field problem. The electric current source produces the electric and magnetic field distributions both in the source region $y = 0$ and also outside the source region $|y| > 0$.

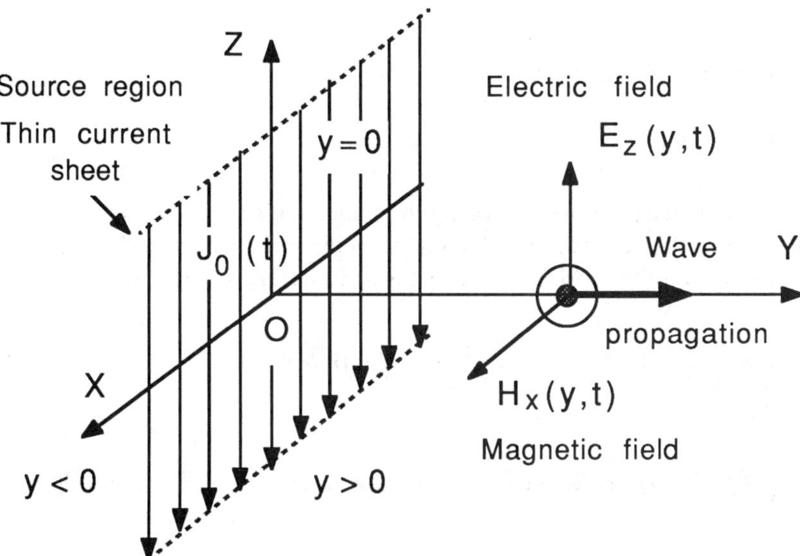

Figure 2.20 One-dimensional electromagnetic fields.

Using the rectangular coordinate system, at any field point in the medium, the electromagnetic field quantities are given by

$\bar{E}(\bar{r},t)$: electric field distribution, in volts per meter

$$= E_x(\bar{r},t)\hat{x} + E_y(\bar{r},t)\hat{y} + E_z(\bar{r},t)\hat{z} \tag{2.12.2a}$$

$\bar{H}(\bar{r},t)$: magnetic field distribution, in amperes per meter

$$= H_x(\bar{r},t)\hat{x} + H_y(\bar{r},t)\hat{y} + H_z(\bar{r},t)\hat{z} \tag{2.12.2b}$$

The complete three-dimensional medium, shown in Figure 2.20, is divided into three regions, consisting of the source region $(x, y = 0, z)$, and two source-free regions defined by $(x, y > 0, z)$ and $(x, y < 0, z)$.

Referring to Table 2.1, the electric and magnetic fields satisfy Maxwell's curl equations:

$$\nabla \times \bar{E}(\bar{r},t) = -\mu \frac{\partial \bar{H}(\bar{r},t)}{\partial t} \tag{2.12.3a}$$

$$\nabla \times \bar{H}(\bar{r},t) = \varepsilon \frac{\partial \bar{E}(\bar{r},t)}{\partial t} + \bar{J}_V(\bar{r},t) \tag{2.12.3b}$$

and the source electric current density

$$\bar{J}_V(\bar{r},t) = 0 \qquad \text{for region } |y| > 0 \tag{2.12.3c}$$

Since the boundary surface separating the source region and the source-free region is just a thin sheet of electric current, for analyzing the source region fields it is convenient to use the integral form of Maxwell's equations. Referring to Table 2.2,

$$\oint_C \bar{E}(\bar{r},t) \bullet d\bar{L}(\bar{r}) = -\mu \frac{\partial}{\partial t} \iint_S \bar{H}(\bar{r},t) \bullet d\bar{s}(\bar{r}) \tag{2.12.4a}$$

$$\oint_C \bar{H}(\bar{r},t) \bullet d\bar{L}(\bar{r}) = \iint_S \bar{J}_V(\bar{r},t) \bullet d\bar{s}(\bar{r}) + \varepsilon \frac{\partial}{\partial t} \iint_S \bar{E}(\bar{r},t) \bullet d\bar{s}(\bar{r}) \tag{2.12.4b}$$

for the region $y = 0$.

In these equations only the z component of the electric current density is present. The two curl equations, (2.12.3a) and (2.12.3b), are the coupled partial differential equations, and they are to be solved for the electric and magnetic field quantities subject to the boundary conditions. In the following, the two curl equations are written in a scalar form using the rectangular coordinate system. After substituting (2.12.2a) and (2.12.2b) into the curl equations (2.12.3a) and (2.12.3b),

$$\frac{\partial E_z}{\partial y} - \frac{\partial E_y}{\partial z} = -\mu \frac{\partial H_x}{\partial t} \tag{2.12.5a}$$

$$\frac{\partial E_x}{\partial z} - \frac{\partial E_z}{\partial x} = -\mu \frac{\partial H_y}{\partial t} \tag{2.12.5b}$$

$$\frac{\partial E_y}{\partial x} - \frac{\partial E_x}{\partial y} = -\mu \frac{\partial H_z}{\partial t} \tag{2.12.5c}$$

$$\frac{\partial H_z}{\partial y} - \frac{\partial H_y}{\partial z} = \varepsilon \frac{\partial E_x}{\partial t} + J_x \tag{2.12.6a}$$

$$\frac{\partial H_x}{\partial z} - \frac{\partial H_z}{\partial x} = \varepsilon \frac{\partial E_y}{\partial t} + J_y \tag{2.12.6b}$$

$$\frac{\partial H_y}{\partial x} - \frac{\partial H_x}{\partial y} = \varepsilon \frac{\partial E_z}{\partial t} + J_z \tag{2.12.6c}$$

In these scalar equations, the x and y components of the electric current density do not exist. The z component of the electric current density is in the xz plane polarized in the negative z coordinate direction. Further, the electric current density does not vary with respect to x and z coordinate variables and has constant distribution. Hence, at any point in the three-dimensional medium, the electric field components and the magnetic field components do not vary with respect to the x and z coordinate variables. In the preceding six scalar equations,

$$\frac{\partial}{\partial x} \rightarrow 0 \qquad \text{and} \qquad \frac{\partial}{\partial z} \rightarrow 0 \tag{2.12.7}$$

and the field components vary only with respect to the y coordinate variable and time parameter t. The six scalar equations are rewritten in the following by eliminating terms that are not present. Expression (2.12.5a) reduces to

$$\frac{\partial E_z}{\partial y} = -\mu \frac{\partial H_x}{\partial t} \tag{2.12.8a}$$

Referring to expression (2.12.5b), the derivatives on the lefthand side do not exist and the lefthand-side terms are zero, which yields on the righthand side the magnetic field component $H_y = 0$. Further, expression (2.12.5c) reduces to

$$\frac{\partial E_x}{\partial y} = \mu \frac{\partial H_z}{\partial t} \tag{2.12.8b}$$

Similarly, expression (2.12.6a) reduces to

$$\frac{\partial H_z}{\partial y} = \varepsilon \frac{\partial E_x}{\partial t} \tag{2.12.8c}$$

Referring to expression (2.12.6b), the derivatives on the lefthand side do not exist and the lefthand-side terms are zero, which yields on the righthand side the electric field component $E_y = 0$. Further, expression (2.12.6c) reduces to

$$- \frac{\partial H_x}{\partial y} = \varepsilon \frac{\partial E_z}{\partial t} + J_z \tag{2.12.8d}$$

In the preceding differential equations, expressions (2.12.8b) and (2.12.8c), form a coupled set of partial differential equations between the field components E_x and H_z. These two field components, in fact, are not coupled to the other two differential equations, (2.12.8a) and (2.12.8d). Further, there exists only the z component of electric current distribution. Based on Ampere's law, an electric current polarized in the z direction cannot produce a component of the magnetic field in the z coordinate direction. Hence, the H_z component of magnetic field does not exist, which by inspection states that there can be no E_x component of electric field distribution. It should be noted that the remaining two independent equations given by (2.12.8a) and (2.12.8d) are the valid scalar expressions for one-dimensional electromagnetic problem. Hence, the specific component of electric current source correspondingly generate the components of the electric and magnetic fields.

Referring to Figure 2.20, with the electric current density polarized in the z coordinate direction, basically the E_z component of electric field and the H_x component of magnetic field are generated. Hence, the two nonzero components of the electromagnetic fields satisfy the two scalar coupled partial differential equations:

$$\frac{\partial E_z(y,t)}{\partial y} = - \mu \frac{\partial H_x(y,t)}{\partial t} \tag{2.12.9a}$$

$$- \frac{\partial H_x(y,t)}{\partial y} = \varepsilon \frac{\partial E_z(y,t)}{\partial t} + J_z(y = 0,t) \tag{2.12.9b}$$

The same two coupled equations, (2.12.9a) and (2.12.9b), are applicable for the source-free region $|y| > 0$ with the electric current density term equated to zero. The electromagnetic fields near boundary surface $y = 0$ is first analyzed in the following. For this, the integral form of Maxwell's equations, expressions (2.12.4a) and (2.12.4b), are convenient to use.

On retaining only the nonzero electric and magnetic field components, the following two coupled integral form of equations are obtained for the source region:

$$\oint_C E_z(y,t)d\ell = - \mu \frac{\partial}{\partial t} \iint_S H_x(y,t)ds \tag{2.12.10a}$$

$$\oint_C H_x(y,t)d\ell = \iint_S J_z(y,t)ds + \varepsilon \frac{\partial}{\partial t}\iint_S E_z(y,t)ds \qquad (2.12.10b)$$

for region $y = 0$.

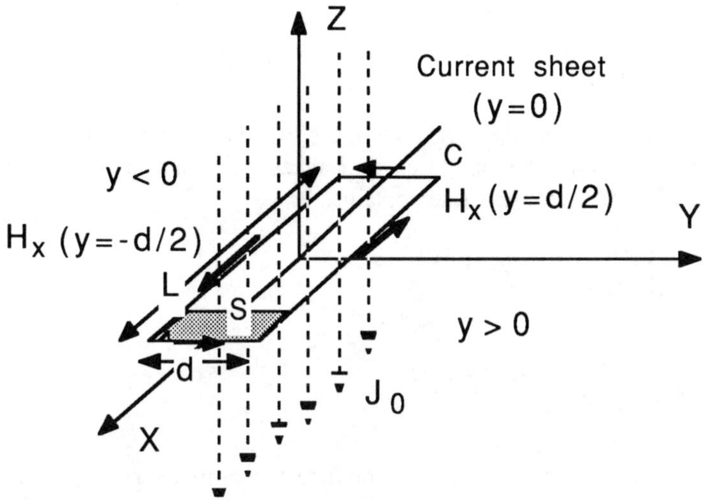

Figure 2.21 Magnetic field component near a current sheet.

In these two integral relationships, C is an arbitrary contour along the edge of surface S. As discussed earlier, the electric and magnetic field components do not vary with respect to x and z coordinate variables. Thus, in any horizontal plane parallel to the xy coordinate plane, the electric and magnetic field distributions are the same. Referring to Figure 2.21, contour C is selected as a rectangle with its surface area S lying in the xy plane so as to enclose certain portion of the thin sheet electric current distribution located at $y = 0$. Using Ampere's law obtained in expression (2.12.10b), the relationship between the magnetic field component and the electric current density distribution can be obtained. The contour integral and the two surface integrals are evaluated in a limit as width d approaches zero.

In the limit as d approaches zero, the surface area S formed by the rectangle is zero, hence the integrals on the righthand side of expression (2.12.10b) simplify to

$$\lim_{d \to 0} \iint_S J_z(y = 0,t)ds = - J_0 L \qquad (2.12.11a)$$

$$\lim_{d \to 0} \iint_S E_z(y = 0,t)ds = 0 \tag{2.12.11b}$$

Along the contour, the magnetic field distribution is uniform, and integral expression (2.12.10b) yields

$$- H_x(y = \frac{d}{2},t)L + H_x(y = -\frac{d}{2},t)L = - J_0(y = 0,t)L \tag{2.12.11c}$$

In a limit as the width d approaches zero, the two magnetic field components on the lefthand and righthand sides of the current sheet are equal, and the components of magnetic field for the source region are obtained as

$$H_x(y=0+,t) = \frac{J_0(y=0,t)}{2} \tag{2.12.12a}$$

$$H_x(y=0-,t) = -\frac{J_0(y=0,t)}{2} \tag{2.12.12b}$$

These two relationships are utilized as boundary conditions to determine the electric and magnetic field components in the source-free regions. As can be seen, the tangential components of the magnetic field are discontinuous across the source electric current sheet. For the region outside the source electric current sheet, referring to expressions (2.12.9a) and (2.12.9b), the two nonzero components of the electromagnetic fields satisfy the following two scalar wave equations:

$$\frac{\partial^2 E_z(y,t)}{\partial y^2} = \mu\varepsilon \frac{\partial^2 E_z(y,t)}{\partial t^2} \tag{2.12.13}$$

$$\frac{\partial^2 H_x(y,t)}{\partial y^2} = \mu\varepsilon \frac{\partial^2 H_x(y,t)}{\partial t^2} \tag{2.12.14}$$

A detailed solution technique for solving a second-order, partial differential scalar wave equation is discussed in Section 2.9 based on the separation of variables. In Section 2.9, expressions (2.9.12b–c), since the electric field component is a function of only y and t variables, the solutions can be written as

$$E_z(y>0,t) = E_{p+}\, e^{j(\omega t-\beta y)} \tag{2.12.15a}$$

$$E_z(y<0,t) = E_{p-}\, e^{j(\omega t+\beta y)} \tag{2.12.15b}$$

where E_{p+} and E_{p-} are unknown amplitude constants. The constants can be determined based on the boundary conditions at the source region. The magnetic field can be determined by substituting the electric field obtained in expression (2.12.15a) into the one-dimensional Maxwell's equations defined in (2.12.9a) or (2.12.9b). Hence, in expression (2.12.9b), the one-dimensional magnetic field and electric field components are related through the equation

$$\frac{\partial H_x(y,t)}{\partial y} = -\,\varepsilon\,\frac{\partial E_z(y,t)}{\partial t} \tag{2.12.16a}$$

$$\frac{\partial H_x(y,t)}{\partial y} = -\,\varepsilon\,\frac{\partial}{\partial t} E_{p+}\, e^{j(\omega t-\beta y)} \tag{2.12.16b}$$

$$\frac{\partial H_x(y,t)}{\partial y} = -\,j\omega\varepsilon E_{p+}\, e^{j(\omega t-\beta y)} \tag{2.12.16c}$$

On integrating expression (2.12.16c) with respect y, the forward traveling magnetic field plane wave in the $y > 0$ region is given by

$$H_x(y,t) = \frac{\omega\varepsilon}{\beta} E_{p+}\, e^{j(\omega t-\beta y)} \tag{2.12.16d}$$

$$H_x(y,t) = \frac{E_{p+}}{\eta}\, e^{j(\omega t-\beta y)} \tag{2.12.16e}$$

where η is the intrinsic impedance of the medium, in ohms. Similarly, the forward traveling magnetic field plane wave in the $y < 0$ region is

$$H_x(y<0,t) = -\,\frac{E_{p-}}{\eta}\, e^{j(\omega t+\beta y)} \tag{2.12.17}$$

Expressions (2.12.15 a,b), (2.12.16e), and (2.12.17) together form a solution for the electric and magnetic field components in the source-free region $|y| > 0$. The constants representing the amplitude of the fields can be determined by using the boundary conditions obtained in expressions (2.12.12a) and (2.12.12b). Let us consider the electric current source located at the $y = 0$ plane. Let

$$J_0(y=0,t) = J_0\, e^{j\omega t} \tag{2.12.18a}$$

Using boundary condition (2.12.12a) and magnetic field solution (2.12.16e)

$$H_x(y=0+,t) = \frac{J_0(y=0,t)}{2} \tag{2.12.18b}$$

$$= \frac{J_0}{2}\, e^{j\omega t} \tag{2.12.18c}$$

$$= \frac{E_{p+}}{\eta}\, e^{j(\omega t-\beta y)} \tag{2.12.18d}$$

Hence, the constant coefficients associated with the outgoing traveling waves in the two regions are given by

$$E_{p+} = \frac{\eta J_0}{2} \tag{2.12.18e}$$

$$E_{p-} = \frac{\eta J_0}{2} \tag{2.12.18f}$$

Therefore, for the one-dimensional problem with sources represented by a thin current sheet located at the plane $y = 0$, the electric and magnetic fields are

$$E_z(y>0,t) = \frac{\eta J_0}{2}\, e^{j(\omega t-\beta y)} \tag{2.12.19a}$$

$$E_z(y<0,t) = \frac{\eta J_0}{2}\, e^{j(\omega t+\beta y)} \tag{2.12.19b}$$

and

$$H_x(y>0,t) = \frac{J_0}{2}\, e^{j(\omega t-\beta y)} \tag{2.12.20a}$$

$$H_x(y<0,t) = -\,\frac{J_0}{2}\, e^{j(\omega t+\beta y)} \tag{2.12.20b}$$

It can be concluded based on this study that an infinite sheet of electric current distribution generates plane waves propagating normal to the plane containing the source electric current distribution. Only either the real part or the imaginary part of these solutions are valid, depending upon the type source excitation.

2.12.1 ELECTROMAGNETIC FIELDS IN A LOSSY MEDIUM

The study of a one-dimensional, time-dependent, boundary value problem was discussed in the previous section. The medium is assumed to be an ideal lossless case with medium parameters satisfying the linear, homogeneous and isotropic properties. Suppose the medium has some conductivity, which represents the lossy situation in the medium, then the analysis is not quite straightforward in the time domain. Even for the case of a simple one-dimensional time-domain problem, the time varying electric and magnetic fields satisfy a differential equation quite elaborate to analyze.

In the time domain, the general treatment of the current sources in a conducting medium is quite complicated. Some explanation concerning the source excitation term in Ampere's law equation (2.12.3b) is clarified in the following. The source term should not be mistaken as purely a current generator that produces the electric and magnetic fields in the surrounding region. In addition, there dependent or secondary sources can be distributed throughout the medium. The driving source term in Maxwell's equations, in fact, should include all existing electric current distributions, both of the primary and secondary types. In the case of a lossy medium, for example, the medium has some conductivity. According to the generalized Ohm's law, expression (2.1.11b), conduction currents are flowing (or leaking everywhere) in the medium, which are given by the product of the conductivity and the electric field distribution. These currents are referred to as the *dependent* or *secondary electric currents*. Even though the medium is divided into two separate regions consisting of the source and the source-free regions, for the case of a source-free region the secondary sources, the conducting electric currents, should be included in Maxwell's second equation based on the generalized Ohm's law. Hence, for the lossy medium, referring to Table 2.1,

$$\nabla \times \bar{E}(\bar{r},t) = -\frac{\partial \bar{B}(\bar{r},t)}{\partial t} \tag{2.12.21a}$$

$$\nabla \times \bar{H}(\bar{r},t) = \frac{\partial \bar{D}(\bar{r},t)}{\partial t} + \bar{J}_c(\bar{r},t) + \bar{J}_g(\bar{r},t) \tag{2.12.21b}$$

where for the source-free region the primary source electric current density is

$$\bar{J}_g(\bar{r},t) = 0 \tag{2.12.21c}$$

and the secondary source due to the conduction electric current density is given by

$$\bar{J}_C(\bar{r},t) = \sigma(\bar{r})\bar{E}(\bar{r},t) \tag{2.12.21d}$$

Hence, the following differential form of the Maxwell's equations are to be solved for a linear, inhomogeneous, and isotropic conducting medium:

$$\nabla \times \bar{E}(\bar{r},t) = -\frac{\partial \bar{B}(\bar{r},t)}{\partial t} \tag{2.12.22a}$$

$$\nabla \times \bar{H}(\bar{r},t) = \frac{\partial \bar{D}(\bar{r},t)}{\partial t} + \sigma(\bar{r})\bar{E}(\bar{r},t) \tag{2.12.22b}$$

where for the lossy medium under consideration

$\varepsilon(\bar{r})$: permittivity of the medium, in farads per meter;
$\mu(\bar{r})$: permeability of the medium, in henrys per meter;
$\sigma(\bar{r})$: conductivity of the medium, in mhos per meter.

Following the procedure discussed in the previous section for the one-dimensional problem, the field components and their densities vary only with respect to the y coordinate variable and the time parameter t. The six scalar equations based on expressions (2.12.22a,b) can be reduced to two scalar equations by eliminating the electric and magnetic field components, the electric and magnetic flux density components not present in the one-dimensional field problem. As discussed earlier, only two field components are nonzero. They are given by the electric field and its density in the z coordinate direction and the magnetic field and its density in the x coordinate direction. Hence, the following two coupled scalar differential equations are obtained:

$$\frac{\partial E_z(y,t)}{\partial y} = -\frac{\partial B_x(y,t)}{\partial t} \tag{2.12.23a}$$

$$-\frac{\partial H_x(y,t)}{\partial y} = \frac{\partial D_z(y,t)}{\partial t} + \sigma E_z(y,t) \tag{2.12.23b}$$

In these two partial differential equations, the density components can be replaced in terms of their field components by making use of constitutive relationships (2.8.3) and (2.8.4). Assuming the various medium parameters are constants and do not vary with respect any coordinate variable, expressions (2.12.22a) and (2.12.22b) simplify to the following equations for the case of a linear, homogeneous, and isotropic medium:

$$\frac{\partial E_z(y,t)}{\partial y} = -\mu \frac{\partial H_x(y,t)}{\partial t} \tag{2.12.24a}$$

$$-\frac{\partial H_x(y,t)}{\partial y} = \varepsilon \frac{\partial E_z(y,t)}{\partial t} + \sigma E_z(y,t) \tag{2.12.24b}$$

For the inhomogeneous medium, in the one-dimensional case, it is required to choose that the various medium parameters, such as the permeability, permittivity, and conductivity, vary only with respect to the y coordinate variable. The field components are completely independent of the x and z coordinates. Hence, it is enough to analyze $E_z(y,t)$ component of electric field and $H_x(y,t)$ component of magnetic field along the y coordinate axis. The two expressions (2.12.23a) and (2.12.23b) simplify to the following form:

$$\frac{\partial H_x(y,t)}{\partial t} = -\frac{1}{\mu(y)} \frac{\partial E_z(y,t)}{\partial y} \tag{2.12.25a}$$

$$\frac{\partial E_z(y,t)}{\partial t} = -\frac{1}{\varepsilon(y)} \frac{\partial H_x(y,t)}{\partial y} - \frac{\sigma(y)}{\varepsilon(y)} E_z(y,t) \tag{2.12.25b}$$

2.12.2 FINITE-DIFFERENCE NUMERICAL SOLUTION

Instead of attempting a direct analytical solution for the preceding two coupled partial differential equations, which is quite elaborate even for the one-dimensional case, a numerical solution technique based on time-marching and finite-difference approximation is considered in the following. Since the two equations are valid for every values of y and t, it is assumed that the two components of electric and magnetic fields are continuous or at least piecewise continuous with respect to the y and t variables. Let

Δy : discretized spatial increment, in meters;
Δt : discretized time increment, in seconds.

Referring to Figure 2.22, an attempt is made to enforce the validity of equations (2.12.25a) and (2.12.25b) and analyze them for the two field components at discrete m spatial points and for discrete n time steps. For convenience in the finite-difference numerical calculation, the electric field and the magnetic field components are not calculated at the same spatial locations, but are staggered alternately as shown.

Referring to the Figure 2.22, the medium is first discretized uniformly into M number of thin layers each having constant spatial width Δy. The layers have been designated with appropriate layer numbers. The final layer index M is such as to cover the complete region of interest. It is assumed that the thickness of each layer is very small, and for all practical purposes each layer can be assumed to be a linear, homogeneous, and isotropic region. Further, the electric field distribution within each layer is assumed to be constant. Thus, the midpoint of each layer is selected under spatial discretization and the component of electric field is defined at the midpoint of each layer. The constant permittivity and conductivity for a given layer is selected to be the permittivity and conductivity at the midpoint of layer itself where the electric field component is located. Since the electric and the magnetic field components are staggered, the calculation of magnetic field is performed at the spatial locations in between any two adjacent electric field components. The spatial discretization for the magnetic field, in fact, overlaps half a cell on either side. The constant permeability for a given layer is selected to be the permeability where the magnetic field is located. This natural arrangement of the field components is ideal for defining first-order central differences required for solving the coupled differential equations, (2.12.25a) and (2.12.25b).

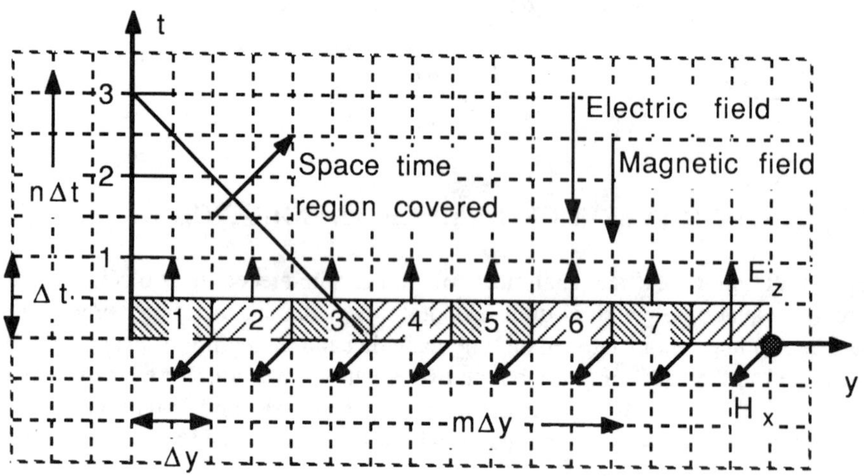

Figure 2.22 Space-time resolution for one-dimensional fields.

The differential terms in equations (2.12.25a) and (2.12.25b) are now converted into corresponding central differences based on the spatial and the time discretization. The solution approach discussed in the following is generally referred to as the *finite-difference time-domain* technique, which is quite popular whenever coupled partial differential equations are to be solved based on an approximate numerical time-stepping scheme. The limitation at this stage seems to be the discretization of complete space. For the case of a one-dimensional problem under consideration, the y coordinate variable runs from $-\infty$ to ∞. Hence, proper field terminations are required on either side where space discretization is discontinued. As noted later, proper field boundary conditions are enforced based on a zero-reflection coefficient at the terminating layers so that the electric field and the magnetic field plane waves propagating along the y coordinate direction do not return, but continue to propagate into the unbounded medium. Referring to Figure 2.23, let us define the field components as

$$E_z(y,t) = E_z^n(m\Delta y, n\Delta t) \tag{2.12.26a}$$

$$= E_z^n(m,n) \tag{2.12.26b}$$

$$H_x(y,t) = H_x^n(m\Delta y, n\Delta t) \tag{2.12.26c}$$

$$= H_x^n(m,n) \tag{2.12.26d}$$

for space discretization $\qquad m = 1, 2, 3, \ldots, M$

for time discretization $\qquad n = 1, 2, 3, \ldots, N_p$

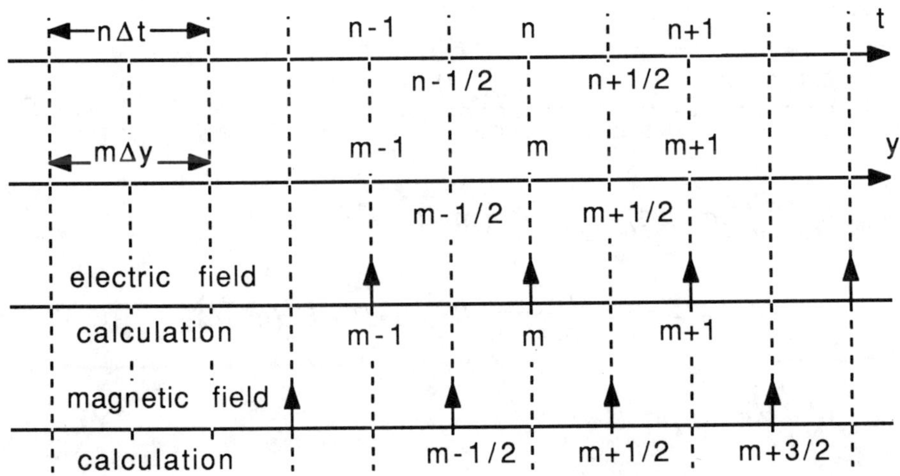

Figure 2.23 Finite-difference calculation of fields.

For the partial derivatives, a centered finite-difference approximation with second-order accuracy can be introduced with respect to either the y coordinate variable or the t time variable as shown in Figure 2.23. The following finite-difference approximations for the derivatives of electric and magnetic fields are accurate up to the second power of their corresponding discretizations. The derivatives are given by

$$\frac{\partial E_z^n(m)}{\partial t} = [E_z^{n+1/2}(m) - E_z^{n-1/2}(m)]/\Delta t \tag{2.12.27a}$$

$$\frac{\partial H_x^n(m+1/2)}{\partial t} = [H_x^{n+1/2}(m+1/2) - H_x^{n-1/2}(m+1/2)]/\Delta t \tag{2.12.27b}$$

$$\frac{\partial E_z^n(m+1/2)}{\partial y} = [E_z^n(m+1) - E_z^n(m)]/\Delta y \tag{2.12.27c}$$

$$\frac{\partial H_x^n(m)}{\partial y} = [H_x^n(m+1/2) - H_x^n(m-1/2)]/\Delta y \tag{2.12.27d}$$

Using these finite-difference approximations, at m+1/2 and n, expression (2.12.25a) can be written as

$$\frac{\partial H_x(y,t)}{\partial t} = -\frac{1}{\mu_0 \mu_r(y)} \frac{\partial E_z(y,t)}{\partial y} \tag{2.12.28a}$$

$$[H_x^{n+1/2}(m+1/2) - H_x^{n-1/2}(m+1/2)]/\Delta t = -\frac{1}{\mu_0 \mu_r(m+1/2)} [E_z^n(m+1) - E_z^n(m)]/\Delta y \tag{2.12.28b}$$

$$H_x^{n+1/2}(m+1/2) = H_x^{n-1/2}(m+1/2) + R_b [E_z^n(m) - E_z^n(m+1)] \tag{2.12.28c}$$

$$R_b = \frac{\Delta t}{\mu_0 \mu_r(m+1/2)\Delta y} \tag{2.12.28d}$$

Expression (2.12.28c) forms a convenient time-stepping algorithm for calculating the magnetic field component. Further, using the finite-difference approximations defined in (2.12.27), at *m* and *n*+1/2, expression (2.12.25b) can also be written as

$$\frac{\partial E_z(y,t)}{\partial t} = -\frac{1}{\varepsilon_0 \varepsilon_r(y)} \frac{\partial H_x(y,t)}{\partial y} - \frac{\sigma(y)}{\varepsilon_0 \varepsilon_r(y)} E_z(y,t) \tag{2.12.29a}$$

$$[E_z^{n+1}(m) - E_z^n(m)]/\Delta t$$

$$= -\frac{1}{\varepsilon_0 \varepsilon_r(m)} [H_x^{n+1/2}(m+1/2) - H_x^{n+1/2}(m-1/2)]/\Delta y$$

$$-\frac{\sigma(m)}{\varepsilon_0 \varepsilon_r(m)} [E_z^{n+1/2}(m)] \tag{2.12.29b}$$

The last term in this expression can be treated to be an average over time increments (*n*+1) and *n*, and expression (2.12.29b) can be rewritten as

$$[E_z^{n+1}(m) - E_z^n(m)]$$

$$= -\frac{\Delta t}{\varepsilon_0 \varepsilon_r(m)\Delta y} [H_x^{n+1/2}(m+1/2) - H_x^{n+1/2}(m-1/2)]$$

$$-\frac{\sigma(m)\Delta t}{\varepsilon_0 \varepsilon_r(m)} [E_z^{n+1}(m) + E_z^n(m)]/2 \tag{2.12.29c}$$

Hence,

$$E_z^{n+1}(m)\left[1 + \frac{\sigma(m)\Delta t}{2\varepsilon_0 \varepsilon_r(m)}\right] = E_z^n(m)\left[1 - \frac{\sigma(m)\Delta t}{2\varepsilon_0 \varepsilon_r(m)}\right]$$

$$-\frac{\Delta t}{\varepsilon_0 \varepsilon_r(m)\Delta y} [H_x^{n+1/2}(m+1/2) - H_x^{n+1/2}(m-1/2)] \tag{2.12.29d}$$

Expression (2.12.29d) is rewritten by dividing throughout by the bracketed term on the lefthand side to yield the following time-stepping difference equation:

$$E_z^{n+1}(m) = C_a E_z^n(m) + C_b \left[-H_x^{n+1/2}(m+1/2) + H_x^{n+1/2}(m-1/2) \right] \tag{2.12.30a}$$

$$C_{na} = 1 - \frac{\sigma(m)\Delta t}{2\varepsilon_0\varepsilon_r(m)}$$

$$C_d = 1 + \frac{\sigma(m)\Delta t}{2\varepsilon_0\varepsilon_r(m)}$$

$$C_a = \frac{C_{na}}{C_d} \tag{2.12.30b}$$

$$C_{nb} = \frac{\Delta t}{\varepsilon_0\varepsilon_r(m)\Delta y}$$

$$C_b = \frac{C_{nb}}{C_d} \tag{2.12.30c}$$

Expressions (2.12.28c) and (2.12.30a) are the *time-stepping algorithms* for the calculation of the electric and magnetic field components. With the field solutions obtained in the two space-time difference equations, the field components for a given time step can be calculated iteratively. The updated new value of a field component at any layer point depends only on its value in the previous time step and the previous values of the components of the other field at the adjacent spatial points. Hence, at any given time step, the computation of a field component will proceed one point at a time. The choice of space increment Δy and time increment Δt is dictated by the reasons of accuracy and algorithm stability, respectively. To ensure the stability of the computed fields, Δt is chosen to satisfy the inequality for the one-dimensional layer model as

$$\Delta t \leq \frac{\Delta y}{c_{max}} \tag{2.12.31}$$

c_{max} : maximum wave velocity within the model.

2.12.3 PROPAGATION OF A HALF-SINE PULSE

The finite-difference time-domain numerical approach just discussed is quite useful for studying one-dimensional, time-dependent electric and magnetic fields. The medium can have any conductivity, permittivity, and permeability characteristics. Even the case of a one-dimensional inhomogeneous medium, in the form of a layered medium, can be conveniently modeled by just specifying medium parameters at appropriate spatial points. In this approach, there is no restriction on the selection of the type of time-dependent excitation. To illustrate the basic concept of the numerical simulation of waves, the propagation of a half-sinusoidal time pulse in a lossless, free-space medium is first considered.

Let us consider a very large region of one-dimensional space that is linear, homogeneous, and isotropic. Referring to Figure 2.22, the z component of the electric field and the x component of the magnetic field vary with respect to the coordinate variable y. Along the y coordinate, the one-dimensional region is divided into number of equally spaced spatial cells of width Δy. The two finite-difference coupled expressions (2.12.28c) and (2.12.30a) are solved at each spatial cell for every time step increment. For convenience in the numerical algorithm development, the spatial cells are numbered $m = 1, 2, 3, \ldots, M$.

In this example, it is assumed that each spatial cell by itself is lossless, homogeneous, and isotropic, having the medium parameters

$$\varepsilon_r(m) = 1 \tag{2.12.32a}$$

$$\mu_r(m) = 1 \tag{2.12.32b}$$

$$\sigma(m) = 0 \tag{2.12.32c}$$

The electric field components are calculated at the midpoint of each cell and the magnetic field components are calculated at the boundary point between two adjacent cells. In this example, an electric field source generator is located at the center of $m = 5$ cell and is turned on at $t = 0$. The source generator develops a half-sinusoidal pulse given by

$$E_g(t) = E_0 \sin(\omega t)\, \delta(m-5) \tag{2.12.33a}$$

$$= E_0 \sin(2\pi f\, n\Delta t)\, \delta(m-5) \tag{2.12.33b}$$

$$n = 1, 2, 3, \ldots, N_p$$

where the delta function

$$\delta(m\text{-}5) = 1 \qquad \text{for } m = 5$$

$$= 0 \qquad \text{for } m \neq 5$$

E_0 : amplitude of half-sine wave excitation, for $N_p \leq 50$

$= 1{,}000$ volts per meter

$= 0$ for $N_p > 50$ (2.12.33c)

f : frequency of excitation

$= 3 \times 10^8$ hertz

c_0 : velocity of wave in the free-space medium

$= 3 \times 10^8$ meters per second

λ : wavelength of the excitation

$= 1$ meter

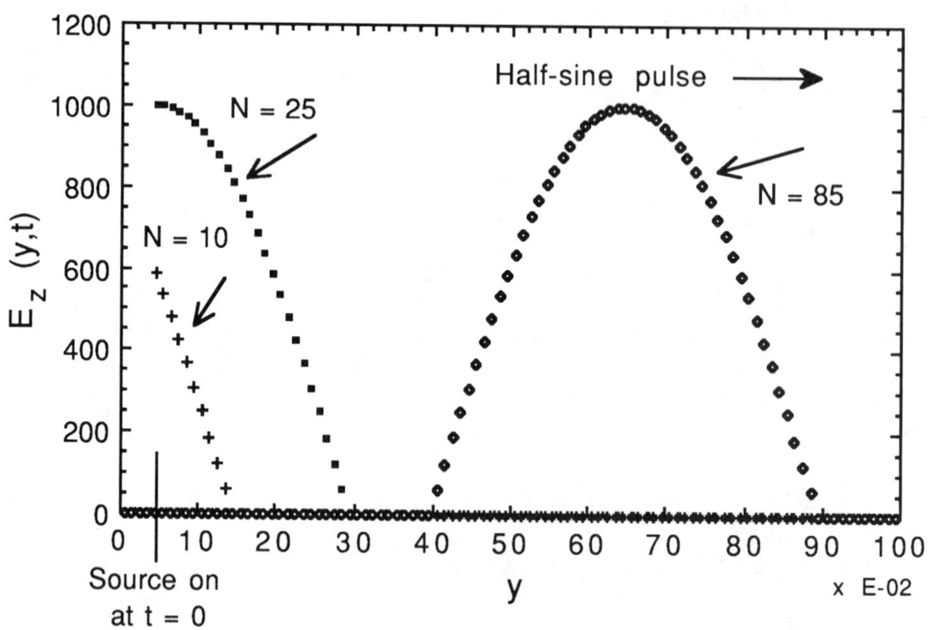

Figure 2.24 Propagation of a half-wave sine pulse in a lossless medium.

The spatial width of the half-wave sine pulse is 0.5 meter. Hence, by selecting a spatial cell resolution of $\Delta y = 0.01$, 50 cells are required to span totally the complete excitation of the half-wave sine pulse. Further, the time-step resolution is selected based on expression (2.12.31), which is given by

$$\Delta t = \frac{\Delta y}{c_0} \tag{2.12.33d}$$

The solution for spatial distribution of the half-wave sine pulse is shown in Figure 2.24. Only the first 100 spatial cells are displayed. For $t < 0$, the excitation is not turned on. According to the time causality, the electric and magnetic field distributions are zero everywhere in the medium. The excitation is turned on at $t = 0$, which is located at the spatial cell $m = 5$, and is turned off as soon as the half-wave pulse generation is completed. This occurs at the time step for the pulse $N_p = 50$. The Figure 2.24 shows three curves depicting spatial distribution of the z component of electric field in the medium at the end of specific time step $N = 10$, 25, and 85. For time step $N = 10$, only a small portion of the wavefront has appeared in the medium. For time step $N = 25$, only the first half of the half-wave pulse has appeared in the medium. For time step $N = 85$, the complete half-wave sine time pulse has appeared in the medium and, in fact, has already traveled certain distance in the positive y direction away from the source location. Since the medium is homogeneous and isotropic, there is no change in the medium parameters at any cell, and thus the half-wave time pulse travels in the forward direction with no change in amplitude or shape at a constant velocity given by the speed of light in the medium.

The propagation of the half-wave sine pulse is now considered in a conducting lossy medium. This case is similar except certain conductivity is introduced into the medium. It is assumed that each spatial cell by itself is lossy, homogeneous, and isotropic, having the medium parameters

$$\varepsilon_r(m) = 1 \tag{2.12.34a}$$

$$\mu_r(m) = 1 \tag{2.12.34b}$$

$$\sigma(m) = 0.005 \tag{2.12.34c}$$

The modeling of the media parameters is similar to the earlier numerical example. The conductivity parameter is specified at the spatial locations where the electric field components are located. Figure 2.25 shows the z component of the electric field distribution at specific spatial points plotted as a function of time steps. The excitation is again turned on at $t = 0$ at the location $m = 5$ cell. The two finite-difference coupled expressions (2.12.28c) and (2.12.30a) are solved at each spatial cell for every time step.

Figure 2.25 shows the distribution of the electric field as the half-wave sine time pulse passes the spatial locations $y = 0.05, 0.20, 0.35$, and 0.50 corresponding to the location of spatial cells $m = 5, 20, 35$, and 50. Due to the conductivity of the medium, the wave amplitude gradually attenuates as the wave propagates along the $+y$ coordinate direction.

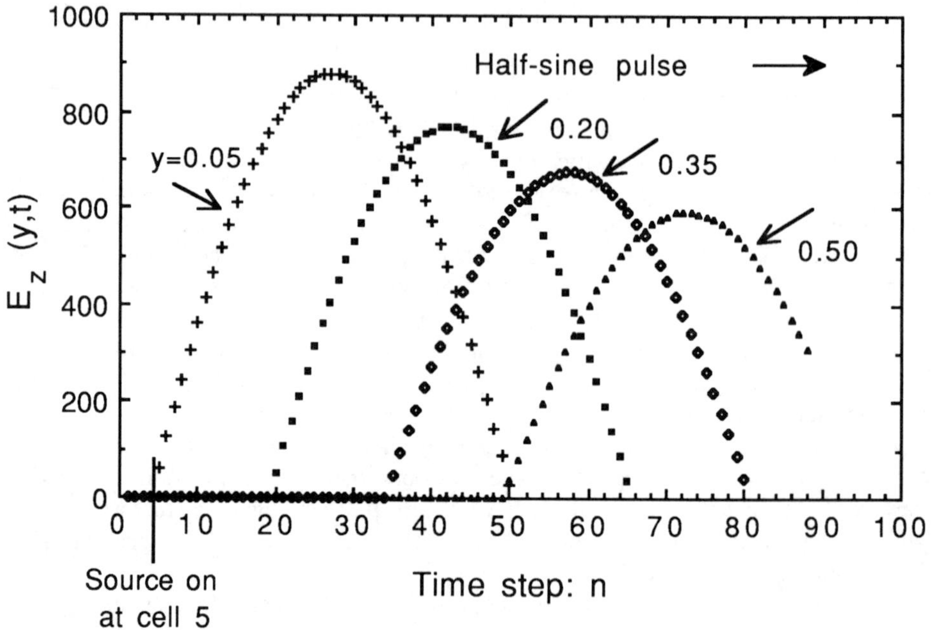

Figure 2.25 Propagation of a half-wave sine pulse in a conducting medium.

Even though the results are shown for the case of a homogeneous and isotropic medium, in fact, the time-stepping algorithm is directly useful for studying wave propagation in a lossy, inhomogeneous, and isotropic layered medium. In particular, the finite-difference time-domain numerical algorithm developed in this section is suitable for simulating arbitrary time-domain pulses for studying electromagnetic wave propagation and interaction in the one-dimensional layered medium.

2.13 FREQUENCY DOMAIN

One major difficulty in obtaining an analytical solution in the time domain for the electromagnetic fields is the dependence of fields on the time parameter variable. In this section, an alternative approach for modeling the electric and magnetic fields is discussed, based on a transform technique. The concept of the frequency domain is introduced here wherein the time parameter variable is correspondingly replaced by a frequency parameter variable using the direct *Fourier transform.* This results in a new set of Maxwell's equations that are functions of only the spatial coordinate variables. The electric and magnetic fields are no longer real quantities, but are phasor quantities possessing both magnitude and phase. If the solution for the electric and magnetic fields is desired in the time domain, then the phasor fields in the frequency domain are transformed back using the inverse Fourier transform. Basically, the solution for the electric and magnetic fields in the frequency domain is just a special case of a very specific source distributions having a specified time dependence. The electric charges and currents are assumed to vary according to a *time harmonic variation* or as a phasor quantity having both magnitude and phase distributions. Such a specific time variation correspondingly produces the time harmonic electric and magnetic field distributions. In the frequency domain, it is shown that the electric and magnetic fields satisfy the partial differential equation relationship referred to as the *Helmholtz equation.* A general solution to the vector and scalar Helmholtz equations in the rectangular coordinate system is also presented in the following sections.

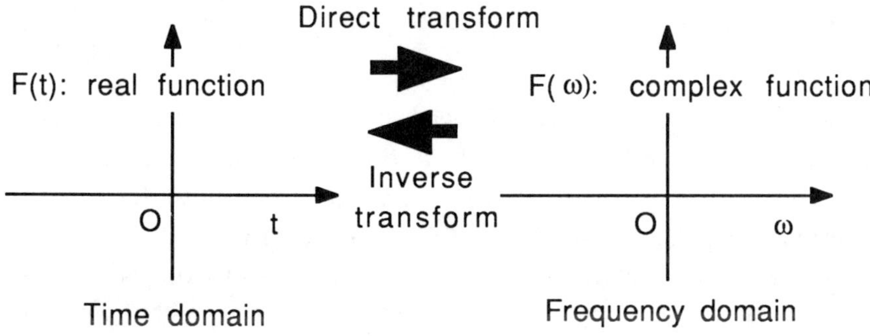

Figure 2.26 Time and frequency domains.

The concept of the Fourier transform domain is briefly discussed in the following. Referring to Figure 2.26, let $F(t)$ represent an arbitrary real function in the time domain with t as the time parameter variable, in seconds. It is assumed that the function $F(t)$ has the following properties:

1. The time domain variable t varies from $-\infty$ to ∞.
2. The function $F(t)$ and its derivatives are continuous or piecewise continuous in every finite time interval.

3. The function $F(t)$ is absolutely integrable in the range $(-\infty,\infty)$ or

$$\int_{-\infty}^{\infty} |F(t)|dt \qquad \text{converges} \tag{2.13.1}$$

Then, the real function $F(t)$ can be transformed into the frequency domain function $F(\omega)$ based on the following direct Fourier integral transform

$$\mathcal{F}[F(t)] = F(\omega) = \int_{-\infty}^{\infty} F(t)e^{-j\omega t}\, dt \tag{2.13.2}$$

where the parameter ω is the frequency domain variable, in radians per second. The function $F(\omega)$ is always a complex function referred to as a *phasor* having both magnitude and phase variation as a function of the frequency parameter ω in the range varying from $-\infty$ to ∞. The original real function $F(t)$ can be recovered back by an inverse Fourier integral transform

$$\mathcal{F}^{-1}[F(\omega)] = F(t) = \frac{1}{2\pi}\int_{-\infty}^{\infty} F(\omega)e^{j\omega t}\, d\omega \tag{2.13.3}$$

The function $F(t)$ is a real function of t, while the function $F(\omega)$ is a complex function of ω. The only way a real function can be recovered back through an inverse Fourier transform is when the function $F(\omega)$ has certain special properties. It can be verified easily that the real part is an even function with respect ω, and similarly, the imaginary part is an odd function with respect to ω. In otherwords, by looking at the frequency spectrum, its distribution is complex conjugate for negative values of the frequency.

Let

Re $F(\omega)$: real part of the complex function
Im $F(\omega)$: imaginary part of the complex function

$$F(\omega) = \text{Re } F(\omega) + j\,\text{Im } F(\omega) \tag{2.13.4a}$$

$$\text{Re } F(\omega) = \text{Re } F(-\omega) \tag{2.13.4b}$$

$$\text{Im } F(\omega) = -\,\text{Im } F(-\omega) \tag{2.13.4c}$$

Number of other properties concerning a function in the time domain and the corresponding function in the transformed frequency domain can be deduced based on the Fourier integral pair (2.13.2) and (2.13.3). Suppose $F(t)$ is a real function of variable t and its Fourier transform is known, then the Fourier transform of a derivative of the function $F(t)$ is given by the following relationship. According to the basic definition of the Fourier transform, expression (2.13.2),

$$F(\omega) = \mathcal{F}[F(t)] = \int_{-\infty}^{\infty} F(t) e^{-j\omega t}\, dt \tag{2.13.5a}$$

$$F_1(\omega) = \mathcal{F}\left[\frac{d}{dt} F(t)\right] = \int_{-\infty}^{\infty} \left[\frac{d}{dt} F(t)\right] e^{-j\omega t}\, dt \tag{2.13.5b}$$

After integrating by parts by applying the product rule,

$$F_1(\omega) = \left[F(t) e^{-j\omega t}\right]_{-\infty}^{\infty} - \int_{-\infty}^{\infty} [F(t)]\left[-j\omega e^{-j\omega t}\right] dt \tag{2.13.5c}$$

Assuming that the real function $F(t)$ tends to zero in a limit as t tends to $-\infty$ and also as t tends to ∞, then the first term in this expression is zero. Thus, the Fourier transform of the first derivative of a real function is given by

$$\mathcal{F}\left[\frac{d}{dt} F(t)\right] = j\omega F(\omega) \tag{2.13.6}$$

Similarly, the Fourier transform of the higher order derivatives can be deduced. The Fourier transform of the nth derivative of function $F(t)$ is given by

$$\mathcal{F}\left[\frac{d^n}{dt^n} F(t)\right] = [j\omega]^n F(\omega) \tag{2.13.7}$$

The Fourier transform of integral of the function $F(t)$ can be calculated based on the following procedure. According to the basic definition of the Fourier transform, expression (2.13.2),

$$F(\omega) = \mathcal{F}[F(t)] = \int_{-\infty}^{\infty} F(t) e^{-j\omega t}\, dt \tag{2.13.8a}$$

$$F_2(\omega) = \mathcal{F}[\int F(t)] = \int_{-\infty}^{\infty} [\int F(t)] e^{-j\omega t}\, dt \tag{2.13.8b}$$

On integrating by parts by applying the product rule

$$F_2(\omega) = \left[F(t)\,\frac{e^{-j\omega t}}{-j\omega}\right]_{-\infty}^{\infty} - \int_{-\infty}^{\infty} [F(t)]\left[\frac{e^{-j\omega t}}{-j\omega}\right] dt \tag{2.13.8c}$$

Assuming the real function $F(t)$ tends to zero in a limit as t tends to $-\infty$ and also as t tends to ∞, then the first term in this expression is zero. Thus, the Fourier transform of the integral of a real function is given by

$$\mathcal{F}[\int F(t)] = \frac{1}{j\omega}\, F(\omega) \tag{2.13.9}$$

Similarly, the Fourier transform of higher order integrals can be deduced. It can be concluded from the cases just discussed, the derivative with respect to time t and also the integral with respect to time t can be treated just like an operator. The Fourier transform of these operators is simply given by

$$\frac{d}{dt} \Leftrightarrow j\omega \tag{2.13.10}$$

$$\int dt \Leftrightarrow \frac{1}{j\omega} \tag{2.13.11}$$

In the following, a brief discussion on the transformation of various electric charge and current sources, electric and magnetic fields are considered. Based on the Fourier transformation, all the relevant electromagnetic laws and equations are converted into the frequency domain. It should be noted at this stage, the basic transformation from the time domain to the Fourier transform frequency domain involves defining various electromagnetic sources and their corresponding fields to vary as cosinusoidal or time-harmonic dependence. Only for the case of time-harmonic steady state, the fields can be conveniently assumed to have the following representations

$$\bar{\mathcal{E}}(\bar{r},t) = Re\, \bar{E}(\bar{r},\omega)\, e^{j\omega t} \tag{2.13.12a}$$

$$\bar{\mathcal{H}}(\bar{r},t) = Re\, \bar{H}(\bar{r},\omega)\, e^{j\omega t} \tag{2.13.12b}$$

2.14 SOURCES IN FREQUENCY DOMAIN

The sources in the frequency domain can be treated almost the same way as the sources in the time domain, except that these sources are assumed to have a time harmonic variation. To account for the charges quantitatively, the charge density distributions can be defined on the macroscopic level in an average sense and are quite convenient in electromagnetic modeling studies in the frequency domain. Let us consider the volume distribution of charges in a large three-dimensional region that is linear, homogeneous and isotropic medium, Figure 2.1. To account for the charge distribution at each point within the source region, an arbitrary source point (x', y', z') is selected, which is enclosed within a small incremental volume Δv. Let

$\Delta q(x',y',z',t)$: net charge in incremental volume, in coulombs

$$= \Delta q(x',y',z',\omega)\, e^{j\omega t} \tag{2.14.1}$$

$\Delta v(x',y',z')$: incremental volume, in cubic meter;

ω : frequency variable, in radians per second.

Then the volume charge density at the point (x', y', z') is given by

$$\rho_v(\bar{r}',\omega) = \frac{\partial q(\bar{r}',\omega)}{\partial v(\bar{r}')} \qquad \text{coulombs per cubic meter} \tag{2.14.2a}$$

Using the definition of (2.14.2a), the total charge contained in the volume V is

$$q(\omega) = \iiint_V \rho_v(\omega)\, dv \tag{2.14.2b}$$

where the triple integration is performed over the complete volume containing the source volume charge density distribution. Similarly, this can be extended to the surface charge and the line charge distributions based on the discussions presented in Section 2.1.

2.14.1 ELECTRIC CURRENT DENSITY

As discussed earlier in the time domain, the electric current is defined as the time rate of change of electric charges. The mathematical models discussed previously in Section 2.1 can be applied as well in the frequency domain. Particularly inside the conducting medium, *conduction electric current* flows in the medium and is given by the generalized Ohm's law

$$\bar{J}_C(x,y,z,\omega) = \sigma \bar{E}(x,y,z,\omega) \tag{2.14.3a}$$

$\bar{J}_C(t)$: conduction current density, in amperes per square meter

$$= \bar{J}_C(\omega)\, e^{j\omega t} \tag{2.14.3b}$$

σ : conductivity of the medium, in mhos per meter

$\bar{E}(\omega)$: applied electric field, in volts per meter

Generally, the electric currents distribute throughout a given region of space depending on the medium characteristics. To account for the electric current distribution at every point within a given source volume, let us consider an arbitrary source point (x', y', z') enclosed within an incremental volume with an incremental cross-sectional area Δs. Referring to the Figure 2.4a,

$\Delta I(t)$: electric current crossing incremental surface, in amperes

$$= \Delta I(\omega)\, e^{j\omega t} \tag{2.14.4a}$$

Δs : area of incremental surface, in square meter;

$\hat{\tau}$: unit vector along current flow.

The frequency-dependent volume current density at the point (x', y', z') is given by

$$\bar{J}_V(\bar{r}',\omega) = J_V \hat{\tau} \tag{2.14.4b}$$

where the magnitude of volume current density in amperes per square meter is

$$J_V = \lim_{\Delta s \to 0} \frac{\Delta I}{\Delta s} \tag{2.14.4c}$$

Hence, if the frequency-dependent volume current density is known at each point within a given volume, Figure 2.4b, then the total electric current crossing the arbitrary open surface S is

$$I(\omega) = \iint_S \bar{J}_V \bullet d\bar{s} \tag{2.14.4d}$$

The case of frequency-dependent surface electric currents confined to an arbitrary surface and the line currents confined to an arbitrary contour can be modeled in a similar way, as discussed in Section 2.1.2.

2.14.2 CONTINUITY EQUATION

The continuity equation, discussed rigorously in Section 2.2, relates the electric charge and current density distributions in a volume bounded by a closed surface. It can be written in the frequency domain as well.

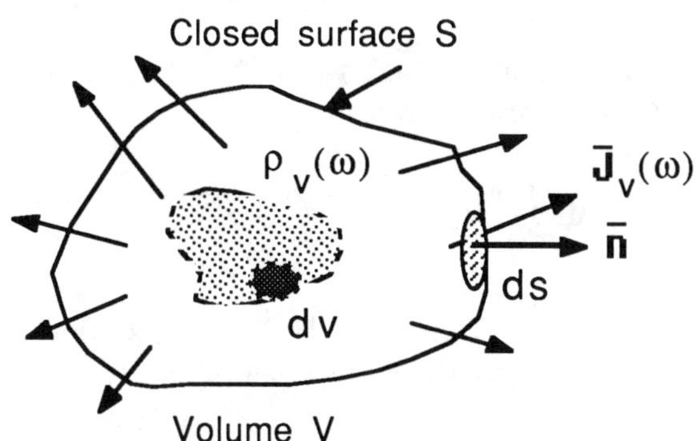

Figure 2.27 Continuity relation between electric currents and charges.

Referring to the Figure 2.27, the total charge contained in the complete volume V is given by

$$q(\omega) = \iiint_V \rho_V(\omega)\, dv \tag{2.14.5a}$$

where

$\rho_V(t)$: volume electric charge density, in coulombs per cubic meter

$$= \rho_V(\omega)\, e^{j\omega t} \tag{2.14.5b}$$

The net electric current crossing the complete closed surface S is obtained by

$$I(\omega) = \oiint_S \bar{J}_V(\omega) \bullet d\bar{s} \tag{2.14.6a}$$

where

$\bar{J}_V(t)$: volume current density, in amperes per square meter

$$= \bar{J}_V(\omega)\, e^{j\omega t} \tag{2.14.6b}$$

As the electric current flows from the inside region to outside, gradually the net electric charge inside the closed volume decreases. Hence, the *continuity equation* is given by

$$I(t) = -\frac{\partial}{\partial t} q(t) \tag{2.14.7}$$

$$I(\omega) = -j\omega q(\omega) \tag{2.14.8}$$

On substituting expressions (2.14.5a) and (2.14.6a), the frequency-dependent continuity equation in integral form is given by

$$\oiint_S \bar{J}_V \cdot d\bar{s} = -j\omega \iiint_V \rho_V \, dv \tag{2.14.9}$$

The integral form of continuity equation (2.14.9) can be converted into the equivalent differential form by utilizing the Gauss-divergence theorem, Appendix A. Hence,

$$\nabla \cdot \bar{J}_V(\bar{r}',\omega) = -j\omega\rho_V(\bar{r}',\omega) \tag{2.14.10}$$

2.15 MAXWELL'S EQUATIONS IN THE FREQUENCY DOMAIN

The time harmonic electric current and electric charge sources set up corresponding time harmonic electric and magnetic fields. Tables 2.1 and 2.2 give a complete summary of Maxwell's equations in the time-domain. These equations can be converted into the frequency-domain Maxwell's equations by using the Fourier transform integrals, expressions (2.13.2) and (2.13.3). In fact, the time derivatives of the electric and magnetic field quantities can be treated as simply an operator and replaced by the term $j\omega$ as discussed earlier in expression (2.13.10). Let

$\bar{E}(\bar{r},t)$: electric field distribution, in volts per meter

$$= \bar{E}(\bar{r},\omega)\, e^{j\omega t} \tag{2.15.1a}$$

$\bar{D}(\bar{r},t)$: electric flux density distribution, coulombs per square meter

$$= \bar{D}(\bar{r},\omega)\, e^{j\omega t} \tag{2.15.1b}$$

$\bar{H}(\bar{r},t)$: magnetic field distribution, in amperes per meter

$$= \bar{H}(\bar{r},\omega)\, e^{j\omega t} \tag{2.15.2a}$$

$\bar{B}(\bar{r},t)$: magnetic flux density distribution, in webers per square meter

$$= \bar{B}(\bar{r},\omega)\, e^{j\omega t} \tag{2.15.2b}$$

Table 2.5
Frequency-domain Maxwell's equations in differential form -- Electric current sources

---

Faraday's Law:

$$\nabla \times \bar{E}(\bar{r},\omega) = -j\omega\bar{B}(\bar{r},\omega) \quad (2.15.3)$$

Ampere's Law:

$$\nabla \times \bar{H}(\bar{r},\omega) = \bar{J}_V(\bar{r},\omega) + j\omega\bar{D}(\bar{r},\omega) \qquad \text{lossless case} \quad (2.15.4)$$

$$\nabla \times \bar{H}(\bar{r},\omega) = \bar{J}_g(\bar{r},\omega) + j\omega\bar{D}(\bar{r},\omega) + \bar{J}_c(\bar{r},\omega) \qquad \text{lossy case} \quad (2.15.5)$$

Gauss's Law:

$$\nabla \bullet \bar{B}(\bar{r},\omega) = 0 \quad (2.15.6)$$

$$\nabla \bullet \bar{D}(\bar{r},\omega) = \rho_V(\bar{r},\omega) \quad (2.15.7)$$

Constitutive Relations:

$$\bar{D}(\bar{r},\omega) = \varepsilon\bar{E}(\bar{r},\omega) \quad (2.15.8)$$

$$\bar{B}(\bar{r},\omega) = \mu\bar{H}(\bar{r},\omega) \quad (2.15.9)$$

Generalized Ohm's Law:

$$\bar{J}_c(\bar{r},\omega) = \sigma\bar{E}(\bar{r},\omega) \quad (2.15.10)$$

Continuity Equation:

$$\nabla \bullet \bar{J}_V(\bar{r},\omega) = -j\omega\rho_V(\bar{r},\omega) \quad (2.15.11)$$

Table 2.6
Frequency-domain Maxwell's equations in integral form – Electric current sources

---

Faraday's Law:

$$\oint_C \bar{E}(\bar{r},\omega) \bullet d\bar{L}(\bar{r}) = -j\omega \iint_S \bar{B}(\bar{r},\omega) \bullet d\bar{s}(\bar{r}) \tag{2.15.12}$$

Ampere's Law:

$$\oint_C \bar{H}(\bar{r},\omega) \bullet d\bar{L}(\bar{r}) = \iint_S \bar{J}_V(\bar{r},\omega) \bullet d\bar{s}(\bar{r}) + j\omega \iint_S \bar{D}(\bar{r},\omega) \bullet d\bar{s}(\bar{r}) \quad \text{lossless case} \tag{2.15.13}$$

$$\oint_C \bar{H}(\bar{r},\omega) \bullet d\bar{L}(\bar{r}) = \iint_S \bar{J}_g(\bar{r},\omega) \bullet d\bar{s}(\bar{r}) + j\omega \iint_S \bar{D}(\bar{r},\omega) \bullet d\bar{s}(\bar{r}) + \iint_S \bar{J}_C(\bar{r},\omega) \bullet d\bar{s}(\bar{r}) \quad \text{lossy case} \tag{2.15.14}$$

Gauss's Law:

$$\oiint_S \bar{B}(\bar{r},\omega) \bullet d\bar{s} = 0 \tag{2.15.15}$$

$$\oiint_S \bar{D}(\bar{r},\omega) \bullet d\bar{s} = \iiint_V \rho_V(\bar{r},\omega)\, dv \tag{2.15.16}$$

Constitutive Relations:

$$\bar{D}(\bar{r},\omega) = \varepsilon \bar{E}(\bar{r},\omega) \tag{2.15.17}$$

$$\bar{B}(\bar{r},\omega) = \mu \bar{H}(\bar{r},\omega) \tag{2.15.18}$$

Generalized Ohm's Law:

$$\bar{J}_C(\bar{r},\omega) = \sigma \bar{E}(\bar{r},\omega) \tag{2.15.19}$$

Continuity Equation:

$$\oiint_S \bar{J}_V(\bar{r},\omega) \bullet d\bar{s} = -j\omega \iiint_V \rho_V(\bar{r},\omega)\, dv \tag{2.15.20}$$

In Tables 2.5 and 2.6, the basic electromagnetic field equations are given by Faraday's law and Ampere's law. Expressions (2.15.3) and (2.15.4) are a set of coupled partial differential equations. These two curl equations, in fact, form an independent set of coupled relationships between time-harmonic varying electric field and magnetic field quantities. The additional field relationships stated by the Gauss's law, expressions (2.15.6) and (2.15.7), do not form independent set of equations. They can be directly deduced from the two Maxwell's curl equations as discussed in Section 2.5.

The electromagnetic field quantities in the time-domain can be recovered by the Fourier inverse transform integral. To accomplish this, the complete spectrum of electromagnetic field quantities is required in the range of ω from $-\infty$ to ∞.

2.16 BOUNDARY CONDITIONS IN THE FREQUENCY DOMAIN

The various frequency domain electromagnetic field equations obtained in Section 2.15 are applicable for a large three-dimensional region which is linear, homogeneous and isotropic medium. In many practical situations, the three-dimensional region may consist of two or more media having different media parameters separated by boundary layers. The Maxwell's equations derived earlier can be separately applied for each medium to obtain a solution for the time-harmonic electric and magnetic field quantities. For the field solutions to be unique, the field solutions should satisfy additional conditions at the boundary layer. In Section 2.6, boundary conditions have been obtained at a boundary layer when both time-varying electric and magnetic field distributions are present. These boundary condition expressions can be directly transformed into the frequency domain. In fact, identical boundary conditions are obtained in the frequency domain as well.

To study behavior of the electric and magnetic fields at a boundary layer, the time-harmonic electric and magnetic fields at any point on one side (medium 1) of the boundary layer separating two media 1 and 2, are resolved into two sets of orthogonal components – tangential components of the electric and magnetic fields that are parallel to the boundary surface, and normal components of the electric and magnetic fields that are perpendicular to the boundary surface. Similarly, the time-harmonic electric and magnetic fields at any point on the other side (medium 2) of the boundary layer are also resolved into two other set of orthogonal components consisting of tangential components and normal components of the electric and magnetic fields, which are parallel and perpendicular to the boundary surface.

As discussed in Section 2.6, depending on the permeability, permittivity, and conductivity characteristics of media 1 and 2, it can be shown that the two tangential components of the time-harmonic fields are interrelated. Similarly, the two normal components of the time-varying field densities are also interrelated. These relationships are summarized in the following.

Referring to the Figure 2.10,

$\bar{E}_1(\bar{r},\omega)$: electric field distribution in medium 1;
$\bar{E}_2(\bar{r},\omega)$: electric field distribution in medium 2;
$\hat{t}$: tangential unit vector along the boundary surface;
$\hat{n}$: normal unit vector along the boundary surface.

The two tangential components of the time-harmonic electric field should be continuous across the boundary surface, separating the two media having different permittivity characteristics:

$$E_{1t} = E_{2t} \tag{2.16.1a}$$

$$\hat{n} \times (\bar{E}_2 - \bar{E}_1) = 0 \tag{2.16.1b}$$

If medium 1 is perfectly conducting, then the net electric field in medium 1 is always zero, and thus the tangential component of the electric field at the boundary surface in the medium 2 is also zero.

Similarly, referring to the Figure 2.11,

$\bar{H}_1(\bar{r},\omega)$: magnetic field distribution in medium 1;
$\bar{H}_2(\bar{r},\omega)$: magnetic field distribution in medium 2;
$\bar{J}_S(\bar{r},\omega)$: electric current density at the boundary surface S, in amperes per meter.

The difference between the two tangential components of the time-harmonic magnetic field should be equal to the surface electric current flowing at the boundary surface separating the two media having different permeability characteristics:

$$H_{1t} - H_{2t} = J_S \tag{2.16.2a}$$

$$\hat{n} \times (\bar{H}_2 - \bar{H}_1) = \bar{J}_S \tag{2.16.2b}$$

If medium 1 is perfectly conducting, then the net magnetic field in medium 1 is always zero, and thus the tangential component of magnetic field at the boundary surface is equal to the surface electric current density at the interface itself. Hence,

$$\hat{n} \times \bar{H}_2 = \bar{J}_S \tag{2.16.2c}$$

Similarly, let us consider behavior of the normal components of the time-harmonic electric and magnetic fields near the boundary surface separating the two media.

Referring to the Figure 2.12,

$\bar{D}_1(\bar{r},\omega)$: electric flux density distribution in medium 1;
$\bar{D}_2(\bar{r},\omega)$: electric flux density distribution in medium 2;
ρ_s : surface electric charge density, in coulombs per square meter.

The boundary relationship states that the difference between two normal components of time-harmonic electric flux density should be equal to the electric surface charge density:

$$\hat{n} \cdot (\bar{D}_2 - \bar{D}_1) = \rho_s \tag{2.16.3}$$

And expressing this in terms of the media permittivity characteristics,

$$\varepsilon_2 E_{2n} - \varepsilon_1 E_{1n} = \rho_s \tag{2.16.4a}$$

Suppose medium 1 is perfectly conducting, then the electric field in medium 1 is zero, and the expression for the boundary condition (2.16.4a) becomes

$$D_{2n} = \rho_s \tag{2.16.4b}$$

Further, referring to the Figure 2.13,

$\bar{B}_1(\bar{r},\omega)$: magnetic flux density distribution in medium 1;
$\bar{B}_2(\bar{r},\omega)$: magnetic flux density distribution in medium 2.

The boundary relationship states that the two normal components of time-harmonic magnetic flux density should be continuous across the boundary surface:

$$\hat{n} \cdot (\bar{B}_2 - \bar{B}_1) = 0 \tag{2.16.5}$$

And expressing this in terms of the media permeability characteristics,

$$\mu_2 H_{2n} = \mu_1 H_{1n} \tag{2.16.6a}$$

Suppose medium 1 is perfectly conducting, then the magnetic field in the medium 1 is zero, and the expression for boundary condition (2.16.6a) gives

$$B_{2n} = 0 \tag{2.16.6b}$$

2.17 POWER AND ENERGY STORED IN FREQUENCY DOMAIN

The time-harmonic sources generate electromagnetic energy and radiate in all directions away from the sources. In fact, the radiated energy is carried away and stored in the surrounding medium by the electric and magnetic fields. Depending on how the sources and the fields are simulated, part of the total energy is stored in the electric field, part of the total energy is stored in the magnetic field, and the remaining energy is dissipated in the medium. Let us consider a large three-dimensional lossy region that is linear, homogeneous, and isotropic medium. The parameters ε, μ, σ are the constant permittivity, permeability, and conductivity of the medium. The time-harmonic electric current density and electric charge density sources are given by $\bar{J}_V(\bar{r},\omega)$ and $\rho_V(\bar{r},\omega)$. The time-harmonic electromagnetic field quantities are functions of the spatial coordinate variables and the frequency parameter, given by

$\bar{E}(\bar{r},t)$: electric field distribution, in volts per meter

$$= \text{Re } \bar{E}(\bar{r},\omega)\, e^{j\omega t} \tag{2.17.1a}$$

$\bar{H}(\bar{r},t)$: magnetic field distribution, in amperes per meter

$$= \text{Re } \bar{H}(\bar{r},\omega)\, e^{j\omega t} \tag{2.17.1b}$$

A complete discussion is presented in Section 2.7 concerning instantaneous energy stored and instantaneous power dissipated in the medium. Further, the Poynting vector is given by the cross product between the electric field vector and the magnetic field vector. It can be represented by the instantaneous power density vector at any field point, given by

$$\bar{P}(\bar{r},t) = \bar{E}(\bar{r},t) \times \bar{H}(\bar{r},t) \tag{2.17.2a}$$

$$\bar{P}(\bar{r},t) = P(\bar{r},t)\hat{p} \tag{2.17.2b}$$

This expression can be extended even for the case of time-harmonic electric and magnetic field quantities. It should be noted that the electric and magnetic field quantities in expressions (2.17.1a) and (2.17.1b) are real quantities having the cosine functional variation represented by the real part of the complex quantity. In all previous discussions, the exponential term has been carried over so that either the real part or the imaginary part of the function can be selected, depending on the type of excitation. From expressions (2.17.2a) and (2.17.2b), the instantaneous power density is given by

$$\bar{P}(\bar{r},t) = \text{Re } \bar{E}(\bar{r},\omega)\, e^{j\omega t} \times \text{Re } \bar{H}(\bar{r},\omega)\, e^{j\omega t} \tag{2.17.3a}$$

Assuming a certain phase angle for the electric field and the magnetic field phasors,

$$\text{Re}\,\bar{E}(\bar{r},\omega)\,e^{j\omega t} = \text{Re}\,\bar{E}_m(\bar{r},\omega)\,e^{j\phi}e^{j\omega t} \tag{2.17.3b}$$

$$= E_m\,\cos(\omega t+\phi)\hat{e} \tag{2.17.3c}$$

$$\text{Re}\,\bar{H}(\bar{r},\omega)\,e^{j\omega t} = \text{Re}\,\bar{H}_m(\bar{r},\omega)\,e^{j\psi}e^{j\omega t} \tag{2.17.3d}$$

$$= H_m\,\cos(\omega t+\psi)\hat{h} \tag{2.17.3e}$$

E_m: peak value of the electric field;

H_m: peak value of the magnetic field;

ϕ: phase angle of the electric field with respect to an arbitrary reference;

ψ: phase angle of the magnetic field with respect to the same arbitrary reference.

From expression (2.17.3a), the instantaneous power density is given by

$$P(\bar{r},t) = E_m\cos(\omega t+\phi)\,H_m\cos(\omega t+\psi) \tag{2.17.4a}$$

$$= \frac{E_m H_m}{2}\left[\cos(\phi-\psi) + \cos(2\omega t+\phi+\psi)\right] \tag{2.17.4b}$$

The field quantities are periodic functions, and their wave cycles repeat every 2π radians. Hence, the average power density over a cycle is given by

$$P_{av}(\bar{r}) = \frac{1}{2\pi}\int_{\omega t=0}^{2\pi} P(\bar{r},t)\,d(\omega t) \tag{2.17.4c}$$

$$= \frac{1}{2\pi}\int_{\omega t=0}^{2\pi} \frac{E_m H_m}{2}\left[\cos(\phi-\psi) + \cos(2\omega t+\phi+\psi)\right] d(\omega t) \tag{2.17.4d}$$

$$= \frac{E_m H_m}{2}\cos(\phi-\psi) \tag{2.17.4e}$$

Hence, for the time-harmonic field quantities, the average power density is given by one-half of the product of the peak value of the electric field, the peak value of the magnetic field, and the cosine of the difference between the two phase angles. The Poynting vector given by (2.17.3a) is rewritten for the average power density as

$$\bar{P}_{av}(\bar{r}) = \frac{1}{2}\mathrm{Re}\ \bar{E}(\bar{r},\omega) \times \bar{H}^*(\bar{r},\omega) \tag{2.17.5}$$

where the symbol (*) represents the complex conjugate. Now the stored energy densities and power dissipation for time harmonic fields in a large three-dimensional medium are calculated as follows. Referring to Table 2.5, the time-harmonic electric and magnetic field quantities satisfy the following frequency domain Maxwell's equations in the source free lossy region:

$$\nabla \times \bar{E}(\bar{r},t) = -j\omega\mu\bar{H}(\bar{r},\omega) \tag{2.17.6}$$

$$\nabla \times \bar{H}(\bar{r},\omega) = j\omega\varepsilon\bar{E}(\bar{r},\omega) + \sigma\bar{E}(\bar{r},\omega) \tag{2.17.7}$$

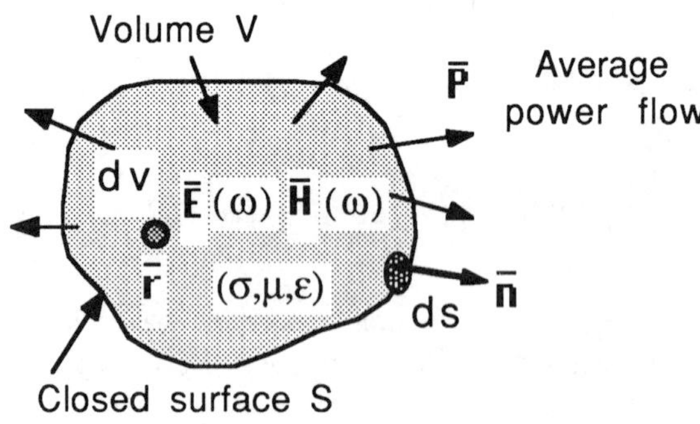

Figure 2.28 Average power flow and Poynting vector theorem.

The unit of electric field is in terms of volts per meter and the unit of magnetic field is in terms of amperes per meter, and the product of the electric and magnetic fields gives a quantity having a unit of watts per square meter representing the average power density. If the time average power density is known on a given surface, then the net time average power crossing a given surface can be calculated. Referring to Appendix A, let us consider the following vector identity:

$$\begin{aligned}\nabla \bullet [\bar{E}(\bar{r},\omega) \times \bar{H}^*(\bar{r},\omega)] &= \bar{H}^*(\bar{r},\omega) \bullet [\nabla \times \bar{E}(\bar{r},\omega)] \\ &\quad - \bar{E}(\bar{r},\omega) \bullet [\nabla \times \bar{H}^*(\bar{r},\omega)]\end{aligned} \tag{2.17.8}$$

The lefthand-side term in this expression represents the divergence of a power density. Referring to Figure 2.28, the electric and magnetic field distributions exist throughout the volume V bounded by the closed surface S. Let

dv : incremental volume surrounding the point $\bar{r}$;
$d\bar{s}$: incremental surface on the closed surface S.

The righthand-side terms involving the curl of the electric field and the curl of the conjugate magnetic field in expression (2.17.8) can be replaced using Maxwell's equations (2.17.6) and (2.17.7) to yield

$$\nabla \bullet [\bar{E} \times \bar{H}^*] = \bar{H}^* \bullet [- j\omega\mu\bar{H}(\bar{r},\omega)] - \bar{E} \bullet [- j\omega\varepsilon\bar{E}^*(\bar{r},\omega) + \sigma\bar{E}^*(\bar{r},\omega)] \qquad (2.17.9a)$$

$$\nabla \bullet [\bar{E} \times \bar{H}^*] = - j\omega\mu[\bar{H} \bullet \bar{H}^*] + j\omega\varepsilon[\bar{E} \bullet \bar{E}^*] - \sigma[\bar{E} \bullet \bar{E}^*] \qquad (2.17.9b)$$

Expression (2.17.9b) is now multiplied on both sides by an incremental volume dv and then is integrated throughout the complete volume V to yield the following relationship:

$$\iiint_V \nabla \bullet [\bar{E} \times \bar{H}^*]\, dv = \iiint_V - j\omega\mu(\bar{H} \bullet \bar{H}^*)\, dv + \iiint_V j\omega\varepsilon(\bar{E} \bullet \bar{E}^*)\, dv - \iiint_V \sigma(\bar{E} \bullet \bar{E}^*)\, dv \qquad (2.17.9c)$$

On the lefthand side of this integral expression, the cross product between the electric field vector and the magnetic field vector is written in the following in terms of the average power density vector given by expression (2.17.5). Hence, expression (2.17.9c) can be rewritten as

$$\iiint_V \nabla \bullet \bar{P}_{av}\, dv = \iiint_V \nabla \bullet \frac{1}{2}\mathrm{Re}\, \bar{E}(\bar{r},\omega) \times \bar{H}^*(\bar{r},\omega)\, dv$$

$$= \iiint_V - \frac{j\omega\mu}{2}\mathrm{Re}(\bar{H} \bullet \bar{H}^*)\, dv + \iiint_V \frac{j\omega\varepsilon}{2}\mathrm{Re}(\bar{E} \bullet \bar{E}^*)\, dv - \iiint_V \frac{\sigma}{2}\mathrm{Re}(\bar{E} \bullet \bar{E}^*)\, dv \qquad (2.17.9d)$$

Based on Gauss-divergence theorem, Appendix A, the lefthand-side term takes the form given by

$$\iint_S -\bar{P}_{av} \bullet d\bar{s} = j2\omega \iiint_V \left[\frac{\mu}{4}\mathrm{Re}(\bar{H} \bullet \bar{H}^*) - \frac{\varepsilon}{4}\mathrm{Re}(\bar{E} \bullet \bar{E}^*)\right] dv$$

$$+ \iiint_V \frac{\sigma}{2}\mathrm{Re}(\bar{E} \bullet \bar{E}^*)\, dv \qquad (2.17.9e)$$

The vector quantity $\bar{P}_{av}$ may be interpreted as the average power flux density diverging out of volume *V* which is bounded by the closed surface *S*. Further, introducing negative sign on the lefthand side of integral expression (2.17.9e), the closed surface integral represents the total average power flux converging into the volume *V* bounded by the closed surface *S*. Since the volume *V* is bounded by a closed surface, the time average power, which has converged into the inside region, remains in the medium. The first integral represents net average reactive energy stored in the magnetic and electric fields. The second integral represents net average power dissipated in the volume due to conduction currents in the medium. Hence, referring back to expression (2.17.9e), the total average power flow into the volume *V* is given by

$$P(\omega) = j2\omega[w_m - w_e] + p_d \qquad (2.17.10a)$$

where

w_m : average reactive energy density stored in the magnetic field inside volume *V*

$$= \frac{\mu}{4}\mathrm{Re}(\bar{H} \bullet \bar{H}^*) \qquad (2.17.10b)$$

$$= \frac{\mu}{4} H^2(\bar{r},\omega) \quad \text{joules per cubic meter} \qquad (2.17.10c)$$

w_e : average reactive energy density stored in the electric field inside volume *V*

$$= \frac{\varepsilon}{4}\mathrm{Re}(\bar{E} \bullet \bar{E}^*) \qquad (2.17.10d)$$

$$= \frac{\varepsilon}{4} E^2(\bar{r},\omega) \quad \text{joules per cubic meter} \qquad (2.17.10e)$$

and

p_d: average real power density dissipated inside volume V

$$= \frac{\sigma}{2}\mathrm{Re}(\bar{E} \bullet \bar{E}^*) \tag{2.17.10f}$$

$$= \frac{\sigma}{2} E^2(\bar{r},\omega) \quad \text{joules per second per cubic meter.} \tag{2.17.10g}$$

2.18 HELMHOLTZ EQUATION

A direct analytical solution for the time-harmonic electric and magnetic fields in a large three-dimensional medium is discussed in this section. The medium is assumed to have linear, homogeneous, and isotropic properties with known time-harmonic electric current and charge source distributions. The sources are assumed to be confined to a small region and produce time-harmonic electric and magnetic fields both inside and outside the source region. The frequency domain Maxwell's equations, Tables 2.5 and 2.6, are utilized for determining the electric and magnetic fields. Again, it is noted that Maxwell's equations are a set of coupled partial differential equations whose solutions are the frequency-dependent electric and magnetic fields. To get clear insight into the frequency-domain solution technique for the electromagnetic field problems, simple analytical techniques are developed based on a vector Helmholtz equation.

In the analysis of the electromagnetic frequency-dependent problem, the concept of frequency and the frequency spectrum of electric and magnetic field distribution play an important role. Generally, the solutions are to be determined for the complete frequency spectrum ranging from zero to a very large value. Based on the frequency parameter variable, the complete frequency spectrum can be divided into three ranges: the low frequency range, the middle or resonant frequency range, and the high frequency range. Different analytical approaches are generally required depending on the frequency range of the analysis. In such cases, one has to consider the electrical size of the electromagnetic boundary value problem rather than the physical dimension itself. The analysis discussed in the following is similar in procedure to the derivation of the vector wave equation studied in Section 2.8. The vector Helmholtz equation is developed including its solution in terms of plane waves. Dual cases are also discussed later for incorporating general media parameters.

Referring to Figure 2.29, let us consider a large three-dimensional unbounded region which is a linear, homogeneous, and isotropic *lossy* medium. The various sources and the corresponding fields have a time-harmonic dependence according to $exp(j\omega t)$. The time-harmonic sources are confined to a small source region, given by

$\bar{J}_g(\bar{r},\omega)$: primary source current density, in amperes per square meter;

$\bar{J}_c(\bar{r},\omega)$: conduction current density, in amperes per square meter.

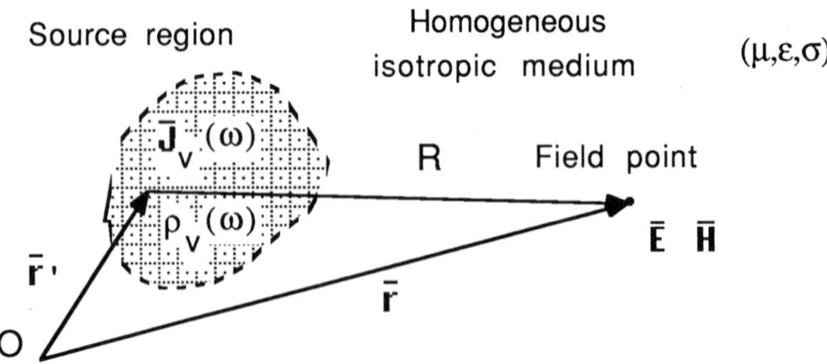

Figure 2.29 Three-dimensional medium with frequency dependent sources.

The sources produce the electric and magnetic field distributions both inside and outside the source region. At any field point in the medium, the time harmonic electromagnetic field quantities are given by

$\bar{E}(\bar{r},\omega)$: electric field distribution, in volts per meter;
$\bar{D}(\bar{r},\omega)$: electric flux density distribution, in coulombs per square meter;
$\bar{H}(\bar{r},\omega)$: magnetic field distribution, in amperes per meter;
$\bar{B}(\bar{r},\omega)$: magnetic flux density distribution, in webers per square meter.

Referring to Table 2.5, in the source-free region, the electric and magnetic fields satisfy the following frequency-dependent Maxwell's equations in differential form:

$$\nabla \times \bar{E}(\bar{r},\omega) = -j\omega\bar{B}(\bar{r},\omega) \tag{2.18.1a}$$

$$\nabla \times \bar{H}(\bar{r},\omega) = j\omega\bar{D}(\bar{r},\omega) + \bar{J}_c(\bar{r},\omega) \tag{2.18.1b}$$

$\bar{r}$ outside source region

As can be seen in expression (2.18.1b), the effect of conductivity of the lossy medium is accounted for in terms of the conduction electric current density. For all observation points within the source region itself, referring to Table 2.5,

$$\nabla \times \bar{E}(\bar{r},\omega) = -j\omega\bar{B}(\bar{r},\omega) \tag{2.18.2}$$

$$\nabla \times \bar{H}(\bar{r},\omega) = \bar{J}_g(\bar{r},\omega) + \bar{J}_c(\bar{r},\omega) + j\omega\bar{D}(\bar{r},\omega) \tag{2.18.3}$$

$\bar{r}$ inside source region

The set of time-harmonic equations defined for the source-free region by the expressions (2.18.1a, b) and also for the source region by expressions (2.18.2) and (2.18.3) can be separately solved. The electric and magnetic field distributions so obtained for the two regions should satisfy appropriate boundary conditions for all points at the boundary surface separating the two regions. Particularly, at the boundary surface separating the source region and the source free region, it is always convenient to use the time-harmonic integral form of Maxwell's equations discussed in Table 2.6.

The frequency-dependent Maxwell's equations for the source-free region are analyzed in the following. On referring to equations (2.18.1a) and (2.18.1b), there are four unknown field quantities, given by the electric field and the electric flux density, the magnetic field and the magnetic flux density. But, there are only two independent coupled curl equations, (2.18.1a) and (2.18.1b), for the four unknown field quantities. Two other additional relationships are obtained based on the constitutive relationships. The constitutive relationships completely depend on the properties of the medium under consideration. The conduction current density, which is also an unknown quantity, can be replaced in terms of the unknown electric field distribution based on the generalized Ohm's law stated in expression (2.15.10). Hence, for the case of a three-dimensional unbounded lossy region that is assumed to be linear, homogeneous and isotropic, the various constitutive relationships are given by

$$\bar{D}(\bar{r},\omega) = \varepsilon\bar{E}(\bar{r},\omega) \tag{2.18.4a}$$

$$\bar{B}(\bar{r},\omega) = \mu\bar{H}(\bar{r},\omega) \tag{2.18.4b}$$

$$\bar{J}_C(\bar{r},\omega) = \sigma\bar{E}(\bar{r},\omega) \tag{2.18.4c}$$

where

ε : permittivity of the medium, in farads per meter;

μ : permeability of the medium, in henrys per meter;

σ : conductivity of the medium, in mhos per meter.

Constitutive relationships (2.18.4a) and (2.18.4b) and generalized Ohm's law (2.18.4c) are now substituted into the frequency-dependent Maxwell's equations, (2.18.1a) and (2.18.1b), for the source-free region. Hence, the electric and magnetic fields in the source-free region satisfy the following frequency-dependent Maxwell's equations in the differential form:

$$\nabla \times \bar{E}(\bar{r},\omega) = -j\omega\mu\bar{H}(\bar{r},\omega) \tag{2.18.5a}$$

$$\nabla \times \bar{H}(\bar{r},\omega) = j\omega\varepsilon\bar{E}(\bar{r},\omega) + \sigma\bar{E}(\bar{r},\omega) \tag{2.18.5b}$$

Further, expression (2.18.5b) can rewritten as

$$\nabla \times \bar{H}(\bar{r},\omega) = j\omega\varepsilon' \bar{E}(\bar{r},\omega) \tag{2.18.5c}$$

where the complex effective permittivity of the medium is defined as

$$\varepsilon' = \varepsilon\left[1 - j\frac{\sigma}{\omega\varepsilon}\right] \tag{2.18.5d}$$

The electric field and the magnetic field divergence relationships can be easily derived by utilizing Maxwell's equations (2.18.5a) and (2.18.5c). After taking the divergence on both sides of expressions (2.18.5a) and (2.18.5c) and noting that the divergence of the curl of any vector quantity, Appendix A, is always equal to zero, the following Gauss's law relationships are obtained:

$$\nabla \bullet \bar{E}(\bar{r},\omega) = 0 \tag{2.18.5e}$$

$$\nabla \bullet \bar{H}(\bar{r},\omega) = 0 \tag{2.18.5f}$$

The two independent curl equations, (2.18.5a) and (2.18.5c), are the coupled partial differential equations with two unknowns. They are to be solved for the electric and magnetic field quantities subject to the known boundary conditions. The magnetic field term in the first curl equation (2.18.5a) is first eliminated by using the second curl equation (2.18.5c) to obtain a partial differential equation only in terms of electric field distribution. Taking curl operation on both sides of expression (2.18.5a),

$$\nabla \times [\nabla \times \bar{E}(\bar{r},\omega)] = -j\omega\mu \nabla \times \bar{H}(\bar{r},\omega) \tag{2.18.6}$$

The lefthand-side term in above expression (2.18.6) can be simplified easily in the rectangular coordinate system. Referring to the vector identity, Appendix A,

$$\nabla \times [\nabla \times \bar{E}(\bar{r},\omega)] = \nabla[\nabla \bullet \bar{E}(\bar{r},\omega)] - \nabla^2\bar{E}(\bar{r},\omega) \tag{2.18.7}$$

After replacing the lefthand-side term of (2.18.6) based on the preceding vector identity and the righthand-side curl term by expression (2.18.5c),

$$\nabla[\nabla \bullet \bar{E}(\bar{r},\omega)] - \nabla^2\bar{E}(\bar{r},\omega) = -j\omega\mu[j\omega\varepsilon' \bar{E}(\bar{r},\omega)] \tag{2.18.8a}$$

Based on Gauss's law relationship (2.18.5e), the divergence of the electric field is zero in the source-free region, and hence, expression (2.18.8a) simplifies to the form

$$\nabla^2 \bar{E}(\bar{r},\omega) = \gamma^2 \bar{E}(\bar{r},\omega) \tag{2.18.8b}$$

where the square of the complex propagation constant

$$\gamma^2 = (j\omega\mu)(j\omega\varepsilon') \tag{2.18.8c}$$

Expression (2.18.8b) is generally referred to as the *vector Helmholtz equation* for the electric field distribution. This Helmholtz equation is a second-order, partial differential equation with dependent spatial coordinate variables. The electric field can be obtained by first solving this partial differential equation subject to known boundary conditions. The magnetic field can be obtained by substituting back the known electric field solution into expression (2.18.5a). It is also possible to obtain a separate and similar vector Helmholtz equation for the magnetic field distribution. The electric field term in the second curl equation (2.18.5c) can be eliminated by using the first curl equation (2.18.5a) to obtain a partial differential equation only in terms of the magnetic field distribution:

$$\nabla^2 \bar{H}(\bar{r},\omega) = \gamma^2 \bar{H}(\bar{r},\omega) \tag{2.18.9}$$

Expression (2.18.9) is also referred to as the *vector Helmholtz equation* for the magnetic field distribution. Referring to expressions (2.18.8b) and (2.18.9), the electric field and magnetic field distributions satisfy similar types of vector equations. The general solution of these vector Helmholtz equations is quite complicated. Hence, the two vector equations are correspondingly reduced to six scalar Helmholtz equations by expressing the vector electric and magnetic fields in terms of their component form. For simplicity the vector equations are simplified using the rectangular coordinate system. Suppose the frequency-dependent electric and magnetic fields are expressed in terms of their rectangular components given by

$$\bar{E}(\bar{r},\omega) = E_x(\bar{r},\omega)\hat{x} + E_y(\bar{r},\omega)\hat{y} + E_z(\bar{r},\omega)\hat{z} \tag{2.18.10a}$$

$$\bar{H}(\bar{r},\omega) = H_x(\bar{r},\omega)\hat{x} + H_y(\bar{r},\omega)\hat{y} + H_z(\bar{r},\omega)\hat{z} \tag{2.18.10b}$$

These two field representations are now substituted into the vector equations, (2.18.8b) and (2.18.9), to reduce them to six scalar Helmholtz equations:

$$\nabla^2 E_i(\bar{r},\omega) = \gamma^2 E_i(\bar{r},\omega) \tag{2.18.11a}$$

$$i = x,y,z$$

$$\nabla^2 H_i(\bar{r},\omega) = \gamma^2 H_i(\bar{r},\omega) \tag{2.18.11b}$$

and

$$\gamma^2 = -\omega^2\mu\varepsilon' \tag{2.18.11c}$$

2.19 SOLUTION TO HELMHOLTZ EQUATION

A brief solution procedure for the scalar Helmholtz equations, expressions (2.18.11a) and (2.18.11b), is discussed in the following. In total, six similar scalar Helmholtz equations satisfy the frequency-dependent three electric field components and three magnetic field components. It is sufficient to solve only one scalar equation, and the solution for the remaining five scalar equations can be written down by inspection. Referring to expression (2.18.11a), for the case of source-free region, the scalar Helmholtz equation in rectangular coordinate system can be written as

$$\frac{\partial^2 U(x,y,z,\omega)}{\partial x^2} + \frac{\partial^2 U(x,y,z,\omega)}{\partial y^2} + \frac{\partial^2 U(x,y,z,\omega)}{\partial z^2} = \gamma^2 U(x,y,z,\omega) \tag{2.19.1}$$

where the scalar field component $U(x, y, z, \omega)$ can represent any frequency-dependent component of either the electric field or the magnetic field. To solve the scalar partial differential equation (2.19.1), the method of *separation of variables* is adopted as discussed in Section 2.9. The total solution for the scalar field U can be written as

$$U(x,y,z,\omega) = \Xi(x)\Psi(y)Z(z) \tag{2.19.2}$$

where

$\Xi(x)$: partial solution of U in terms of only the x coordinate;

$\Psi(y)$: partial solution of U in terms of only the y coordinate;

$Z(z)$: partial solution of U in terms of only the z coordinate.

On differentiating the preceding total solution (2.19.2) twice with respect to (x, y, z) and substituting them into the scalar Helmholtz equation (2.19.1), the following three separate ordinary differential equations are obtained

$$\frac{1}{\Xi}\frac{d^2\Xi}{dx^2} = \gamma_x^2 \tag{2.19.3a}$$

$$\frac{1}{\Psi}\frac{d^2\Psi}{dy^2} = \gamma_y^2 \tag{2.19.3b}$$

$$\frac{1}{Z}\frac{d^2Z}{dz^2} = \gamma_z^2 \tag{2.19.3c}$$

where γ_x^2, γ_y^2, γ_z^2 are separation constants to be determined:

$$\gamma_x^2 + \gamma_y^2 + \gamma_z^2 = \gamma^2 \tag{2.19.4}$$

Referring to Section 2.9, retaining only the forward traveling plane wave terms, the total solution for the scalar field reduces to the following form:

$$U_f(x,y,z,\omega) = \Xi(x)\Psi(y)Z(z) \tag{2.19.5a}$$

$$= [Q_1\, e^{-\gamma_x x}][Q_2\, e^{-\gamma_y y}][Q_3\, e^{-\gamma_z z}] \tag{2.19.5b}$$

$$= U_1[e^{-(\gamma_x x+\gamma_y y+\gamma_z z)}] \tag{2.19.5c}$$

U_1 : amplitude of the forward traveling wave

$$= Q_1 Q_2 Q_3 \tag{2.19.5d}$$

Similarly, if only backward traveling plane waves are present, the total solution for the scalar field reduces to the following form:

$$U_b(x,y,z,\omega) = \Xi(x)\Psi(y)Z(z) \tag{2.19.6a}$$

$$= [\mathcal{P}_1\, e^{\gamma_x x}][\mathcal{P}_2\, e^{\gamma_y y}][\mathcal{P}_3\, e^{\gamma_z z}] \tag{2.19.6b}$$

$$= U_2[e^{(\gamma_x x+\gamma_y y+\gamma_z z)}] \tag{2.19.6c}$$

U_2 : amplitude of the backward traveling wave

$$= \mathcal{P}_1\mathcal{P}_2\mathcal{P}_3 \tag{2.19.6d}$$

With respect to a global origin, let

$\bar{r}$: spatial vector corresponding to the field point *(x, y, z)*

$$= x\hat{x} + y\hat{y} + z\hat{z} \tag{2.19.7a}$$

$\bar{\gamma}$: complex propagation vector of the traveling plane waves

$$= \gamma_x \hat{x} + \gamma_y \hat{y} + \gamma_z \hat{z} \tag{2.19.7b}$$

Then the scalar product between these two vectors is given by

$$\bar{\gamma} \bullet \bar{r} = \gamma_x x + \gamma_y y + \gamma_z z \tag{2.19.7c}$$

Hence, the total solution for the scalar plane wave field reduces to the following representation:

Positive or forward traveling wave

$$U_f(\bar{r},\omega) = U_1\, e^{-\bar{\gamma} \bullet \bar{r}} \tag{2.19.8a}$$

Negative or backward traveling wave

$$U_b(\bar{r},\omega) = U_2\, e^{\bar{\gamma} \bullet \bar{r}} \tag{2.19.8b}$$

It should be noted again that expressions (2.19.8a) and (2.19.8b) represent positive and negative traveling plane waves in the frequency domain. The direction of propagation of the forward plane waves is given by the direction of the complex propagation constant traveling away from the sources. If plane wave solutions are required in the time domain, the time-harmonic variation can be included in the righthand side of the exponential terms and either the real or imaginary part is selected, depending on the excitation. Hence, in the time domain, the time-harmonic plane wave expressions take the following form:

Positive or forward traveling wave

$$U_f(\bar{r},t) = U_1\, e^{-\bar{\gamma} \bullet \bar{r}} e^{j\omega t} \tag{2.19.9a}$$

Negative or backward traveling wave

$$U_b(\bar{r},t) = U_2\, e^{\bar{\gamma} \bullet \bar{r}} e^{j\omega t} \tag{2.19.9b}$$

As noted earlier, suppose the electric and magnetic fields are expressed in terms of rectangular components, expressions (2.18.10a, b), the three components of the electric field and the three components of the magnetic field satisfy similar frequency-dependent scalar Helmholtz equations (2.18.11a, b).

Referring to expressions (2.19.8a, b), the solutions for the electric and magnetic field components are given by

$$E_i(\bar{r},\omega) = E_{0i}\, e^{-\bar{\gamma} \bullet \bar{r}} \tag{2.19.10a}$$

$$H_i(\bar{r},\omega) = H_{0i}\, e^{-\bar{\gamma} \bullet \bar{r}} \tag{2.19.10b}$$

$$i = x,\ y,\ z$$

Defining the following *constant vectors* for the electric and magnetic fields as

$$\bar{E}_0 = E_{0x}\hat{x} + E_{0y}\hat{y} + E_{0z}\hat{z} \tag{2.19.11a}$$

$$\bar{H}_0 = H_{0x}\hat{x} + H_{0y}\hat{y} + H_{0z}\hat{z} \tag{2.19.11b}$$

then the complete frequency domain solutions for the vector electric and magnetic fields based on the forward traveling plane waves are given by

$$\bar{E}(\bar{r},\omega) = \bar{E}_0\, e^{-\bar{\gamma} \bullet \bar{r}} \tag{2.19.12}$$

$$\bar{H}(\bar{r},\omega) = \bar{H}_0\, e^{-\bar{\gamma} \bullet \bar{r}} \tag{2.19.13}$$

where

$$\bar{\gamma} = \gamma\hat{\gamma} \tag{2.19.14}$$

γ: complex propagation constant for a homogeneous and isotropic medium;

$\hat{\gamma}$: actual direction of the plane wave propagation.

Similar to the discussion in Section 2.11, interesting plane wave field properties of the frequency-dependent electric and magnetic fields in a source-free region can be derived. It can be easily shown that the electric field and the magnetic field plane wave solutions obtained in expressions (2.19.12) and (2.19.13) are not independent, but are closely related field quantities. Further, the electric field vector and the magnetic field vector are mutually perpendicular to the direction of propagation. On substituting the preceding plane wave solutions into Gauss's law for the electric field and the magnetic field, expressions (2.18.5e) and (2.18.5f), and following relationships are obtained:

$$\bar{\gamma} \bullet \bar{E}_0 = 0 \tag{2.19.15}$$

$$\bar{\gamma} \bullet \bar{H}_0 = 0 \tag{2.19.16}$$

The frequency-dependent plane wave electric field vector is always perpendicular to the direction of propagation of the plane wave, and similarly, the frequency-dependent plane wave magnetic field vector is always perpendicular to the direction of propagation. Further, referring to the discussion in Section 2.11.1, the electric field vector and the magnetic field vector are mutually perpendicular to each other. This can be proved by using either of Maxwell's curl equations for the source-free region. The frequency-dependent Ampere's law, expression (2.18.5c), in the source-free region is given by

$$\nabla \times \bar{H}(\bar{r},\omega) = j\omega\varepsilon' \bar{E}(\bar{r},\omega) \tag{2.19.17a}$$

After substituting the plane wave electric and magnetic field vectors, expressions (2.19.12) and (2.19.13), into this relationship,

$$\nabla \times \left[\bar{H}_0 \, e^{-\bar{\gamma} \bullet \bar{r}}\right] = j\omega\varepsilon' \left[\bar{E}_0 \, e^{-\bar{\gamma} \bullet \bar{r}}\right] \tag{2.19.17b}$$

which simplifies to the following form after substituting expression (2.19.14)

$$\bar{H}_0 \times \hat{\gamma} = \frac{j\omega\varepsilon'}{\gamma} \bar{E}_0 \tag{2.19.18a}$$

Further, assuming the following representation for the frequency dependent electric and magnetic field constant vectors:

$\bar{E}_0$: constant electric field vector

$$= E_0 \hat{e} \tag{2.19.18b}$$

$\bar{H}_0$: constant magnetic field vector

$$= H_0 \hat{h} \tag{2.19.18c}$$

and expression (2.19.18a) can be rewritten as

$$\hat{e} \times \hat{h} = \hat{\gamma} \tag{2.19.18d}$$

$$\frac{E_0}{H_0} = \eta \tag{2.19.18e}$$

Again, expressions (2.19.15), (2.19.16), and (2.19.18d) depict an important property, that the plane wave electric field unit vector and the plane wave magnetic field unit vector are both mutually orthogonal, and the cross product between the two unit vectors gives the direction of propagation. In expression (2.19.18e), the magnitude of the electric field is in terms of volts per meter, the magnitude of the magnetic field is in terms of amperes per meter, and thus the ratio of the two is in ohms. As discussed in Section 2.11.1, this ratio is generally referred to as the *complex intrinsic impedance* of the three-dimensional lossy medium. In addition to the frequency of excitation, the medium impedance primarily depends on the conductivity, permittivity, and permeability characteristics of the medium. Hence, in expression (2.19.18e),

η : intrinsic impedance of the medium, in ohms

$$= \frac{\gamma}{j\omega\varepsilon'} \tag{2.19.19a}$$

$$= \left(\frac{\mu}{\varepsilon'}\right)^{1/2} \tag{2.19.19b}$$

The intrinsic impedance based on expression (2.19.19b) is a complex quantity having both magnitude and phase. Thus, the electric and magnetic field vectors will not be in phase.

2.20 PROPAGATION CONSTANT

The complex propagation constant depends primarily on the frequency of excitation and the three medium parameters. Let us consider a large region that is a linear, homogeneous and isotropic medium. In this medium,

μ : constant permeability, in henrys per meter

$$= \mu_0\mu_r$$

ε : constant permittivity, in farads in meter

$$= \varepsilon_0\varepsilon_r$$

σ : constant conductivity, in mhos per meter.

The derivation of the Helmholtz equation and the complex propagation constant is treated in Section 2.18. Referring to expression (2.18.8c),

$$\gamma^2 = -k^2$$

$$= -\omega^2\mu\varepsilon' \quad (2.20.1a)$$

$$= -\omega^2\mu\varepsilon\left[1 - j\frac{\sigma}{\omega\varepsilon}\right] \quad (2.20.1b)$$

$$= j\omega\mu(\sigma + j\omega\varepsilon) \quad (2.20.1c)$$

The real and imaginary parts of the complex propagation constant can be obtained by taking the square root of this expression, which has two roots. For the case of forward traveling plane wave, the proper root is selected based on

$$\gamma = \alpha + j\beta \quad (2.20.2)$$

where

α : attenuation constant, in nepers per meter;

β : phase constant, in radians per meter.

To separate out the real and imaginary parts, either the rectangular or polar representation can be used to simplify expression (2.20.1c). Hence,

$$\gamma^2 = -\omega^2\mu\varepsilon + j\omega\mu\sigma \quad (2.20.3a)$$

$$\gamma = [(\omega^2\mu\varepsilon)^2 + (\omega\mu\sigma)^2]^{1/4} e^{j(\pi - \delta)/2} \quad (2.20.3b)$$

$$\gamma = \omega(\mu\varepsilon)^{1/2}[1 + (\tan\delta)^2]^{1/4} e^{j(\pi - \delta)/2} \quad (2.20.3c)$$

$$\delta = \tan^{-1}\left(\frac{\sigma}{\omega\varepsilon}\right) \quad (2.20.3d)$$

2.21 ELECTROMAGNETIC POTENTIALS

The time-harmonic electric current and charge sources generate correspondingly time-harmonic electromagnetic fields. In the previous sections, a direct analytical solution for the time-harmonic electric and magnetic fields in a large three-dimensional medium is discussed based on the vector Helmholtz equation. Assuming the electric current and charge sources are confined to a small source region, the plane wave electric and magnetic field solutions are obtained for a source-free medium having linear, homogeneous, and isotropic properties. In a given boundary value problem, plane waves are not always generated. In general, the electric and magnetic fields generated in a given medium are quite complicated and depend completely upon the distribution and orientation of the electromagnetic sources.

For insight into the electromagnetic scattering and radiation boundary value problems, simple point-type elementary current sources can be considered. In Sections 2.1 and 2.14, detailed mathematical source models were considered for the electric current and electric charge sources based on their density distributions. By using these source models, any complicated distribution of the electric currents and charges can be treated by invoking the linearity and superposition principle. As can be noted, a simple point-type electric current or charge forms an elementary electromagnetic radiating source. If the electric and magnetic fields generated by the point source are known in a given medium, by applying linearity and superposition the complete distribution of the electric and magnetic fields can be calculated for the general source distribution.

Instead of the direct *field's* approach as discussed in the previous sections, an alternative formulation is discussed in the following based on a concept of the electromagnetic potentials. The electric and magnetic fields are expressed in terms of an arbitrary magnetic vector potential and an arbitrary electric scalar potential. Maxwell's equations are simplified using these auxiliary potentials to obtain corresponding Helmholtz type differential equations for the two arbitrary potential functions. The solutions for the two potential functions are obtained for a given point source by solving an appropriate Helmholtz differential equation. Once the solution is known for the point-type source, a superposition principle is invoked to obtain general potential integrals in terms of the known source distribution. Further, the electric and magnetic fields are calculated using the known magnetic vector potential and the electric scalar potential.

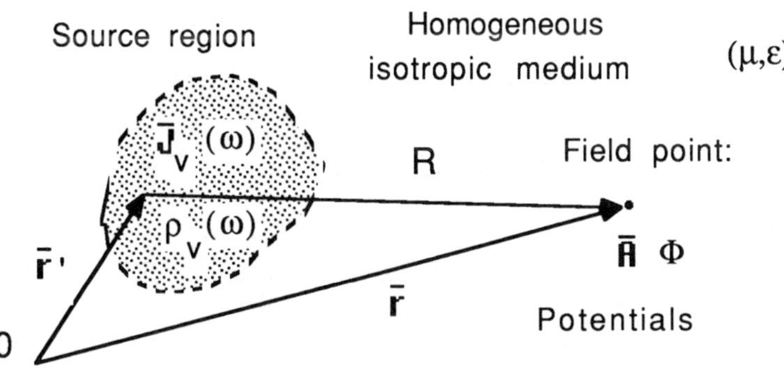

Figure 2.30 Three-dimensional unbounded medium.

A complete summary of *Maxwell's equations in frequency domain* is given in Tables 2.5 and 2.6. Table 2.5 gives all the relevant electromagnetic field equations in differential form. Referring to Figure 2.30, let us consider a large three-dimensional unbounded region that is linear, homogeneous and isotropic lossless medium. It is assumed that the various sources and the corresponding fields have a time-harmonic dependence according to $exp(j\omega t)$. The time harmonic sources are confined to a small source region, given by

$\rho_V(\bar{\mathbf{r}},\omega)$: source electric charge density, in coulombs per cubic meter;

$\bar{\mathbf{J}}_V(\bar{\mathbf{r}},\omega)$: source electric current density, in amperes per square meter.

The sources produce the electric and magnetic field distributions both inside and outside the source region. At any field point in the medium, the time-harmonic electromagnetic field quantities are given by

$\bar{\mathbf{E}}(\bar{\mathbf{r}},\omega)$: electric field distribution, in volts per meter;
$\bar{\mathbf{D}}(\bar{\mathbf{r}},\omega)$: electric flux density distribution, in coulombs per square meter;
$\bar{\mathbf{H}}(\bar{\mathbf{r}},\omega)$: magnetic field distribution, in amperes per meter;
$\bar{\mathbf{B}}(\bar{\mathbf{r}},\omega)$: magnetic flux density distribution, in webers per square meter.

Referring to Table 2.5, in the source-free region, the electric and magnetic fields satisfy the following frequency-dependent Maxwell's equations in differential form:

$$\nabla \times \bar{\mathbf{E}}(\bar{\mathbf{r}},\omega) = - j\omega\bar{\mathbf{B}}(\bar{\mathbf{r}},\omega) \tag{2.21.1a}$$

$$\nabla \times \bar{\mathbf{H}}(\bar{\mathbf{r}},\omega) = j\omega\bar{\mathbf{D}}(\bar{\mathbf{r}},\omega) \tag{2.21.1b}$$

$$\nabla \bullet \bar{\mathbf{D}}(\bar{\mathbf{r}},\omega) = 0 \tag{2.21.1c}$$

$$\nabla \bullet \bar{\mathbf{B}}(\bar{\mathbf{r}},\omega) = 0 \tag{2.21.1d}$$

$\bar{\mathbf{r}}$ outside source region

For all observation points within the source region, referring to Table 2.5, the electric and magnetic fields satisfy the following Maxwell's equations in differential form:

$$\nabla \times \bar{\mathbf{E}}(\bar{\mathbf{r}},\omega) = - j\omega\bar{\mathbf{B}}(\bar{\mathbf{r}},\omega) \tag{2.21.2a}$$

$$\nabla \times \bar{\mathbf{H}}(\bar{\mathbf{r}},\omega) = \bar{\mathbf{J}}_V(\bar{\mathbf{r}},\omega) + j\omega\bar{\mathbf{D}}(\bar{\mathbf{r}},\omega) \tag{2.21.2b}$$

$$\nabla \cdot \bar{D}(\bar{r},\omega) = \rho_V(\bar{r},\omega) \tag{2.21.2c}$$

$$\nabla \cdot \bar{B}(\bar{r},\omega) = 0 \tag{2.21.2d}$$

$\bar{r}$ inside source region

The set of time-harmonic field equations defined for the source-free region by the expressions (2.21.1a–d) and also for the source region by the expressions (2.21.2a–d) can be solved separately. The electric and magnetic field distributions so obtained for the two regions should satisfy boundary conditions at the boundary surface separating the two regions.

In field equations (2.21.2a–d), four unknown field quantities are given by the electric field and the electric flux density, the magnetic field and the magnetic flux density. But there are only two independent coupled curl equations, (2.21.2a) and (2.21.2b). Two additional relationships are obtained based on the constitutive relationships. For a three-dimensional unbounded lossless region that is assumed to be linear, homogeneous and isotropic, the constitutive relations are given by

$$\bar{D}(\bar{r},\omega) = \varepsilon\bar{E}(\bar{r},\omega) \tag{2.21.3a}$$

$$\bar{B}(\bar{r},\omega) = \mu\bar{H}(\bar{r},\omega) \tag{2.21.3b}$$

ε : constant permittivity of the medium, in farads per meter;

μ : constant permeability of the medium, in henrys per meter.

Constitutive relationships (2.21.3a) and (2.21.3b) are now substituted into the frequency-dependent Maxwell's equations (2.21.2a–d), which are valid for the source region. Referring to the Figure 2.30, the electric and magnetic fields in the source region satisfy the following frequency-dependent Maxwell's equations:

$$\nabla \times \bar{E}(\bar{r},\omega) = -j\omega\mu\bar{H}(\bar{r},\omega) \tag{2.21.4a}$$

$$\nabla \times \bar{H}(\bar{r},\omega) = j\omega\varepsilon\bar{E}(\bar{r},\omega) + \bar{J}_V(\bar{r},\omega) \tag{2.21.4b}$$

$$\nabla \cdot \varepsilon\bar{E}(\bar{r},\omega) = \rho_V(\bar{r},\omega) \tag{2.21.4c}$$

$$\nabla \cdot \mu\bar{H}(\bar{r},\omega) = 0 \tag{2.21.4d}$$

To solve coupled equations (2.21.4a) and (2.21.4b), the electric field and the magnetic field are first expressed in terms of two potential functions given by an auxiliary magnetic vector potential and an auxiliary electric scalar potential. In fact, these two potential functions are not independent, but are related through an electromagnetic *gauge condition*. Only for static fields, it can be shown that the two potentials are decoupled and so are the electric and magnetic field distributions. Gauss's law states that the divergence of the time-harmonic magnetic flux density is always equal to zero. According to expression (2.21.4d), let us choose an arbitrary representation for the magnetic flux density in the following form:

$$\bar{B}(\bar{r},\omega) = \nabla \times \bar{A}(\bar{r},\omega) \tag{2.21.5a}$$

$\bar{A}(\bar{r},\omega)$: magnetic vector potential function at the field point

$$= A_x(\bar{r},\omega)\hat{x} + A_y(\bar{r},\omega)\hat{y} + A_z(\bar{r},\omega)\hat{z} \tag{2.21.5b}$$

and constitutive relationship (2.21.3b) yields an expression for the magnetic field:

$$\bar{H}(\bar{r},\omega) = \frac{1}{\mu} \nabla \times \bar{A}(\bar{r},\omega) \tag{2.21.5c}$$

The representation of (2.21.5a) satisfies rigorously the Gauss's law relationship that the divergence of the curl of magnetic vector potential quantity is equal to zero, Appendix A. This expression is now substituted into Faraday's law expression (2.21.4a) to yield the following vector relationship:

$$\nabla \times \bar{E}(\bar{r},\omega) = - j\omega \nabla \times \bar{A}(\bar{r},\omega) \tag{2.21.6}$$

$$\nabla \times [\bar{E}(\bar{r},\omega) + j\omega\bar{A}(\bar{r},\omega)] = 0 \tag{2.21.7}$$

Referring to Appendix A, the curl of the gradient of an electric scalar potential function is always zero. Hence,

$$\nabla \times [\nabla\Phi(\bar{r},\omega)] = 0 \tag{2.21.8}$$

$\Phi(\bar{r},\omega)$: electric scalar potential function at the field point.

This identity is valid for any arbitrary scalar point function. Now vector expression (2.21.7) is equated to vector expression (2.21.8) with an arbitrary negative sign to yield the following relationship:

$$\nabla \times [\bar{E}(\bar{r},\omega) + j\omega\bar{A}(\bar{r},\omega)] = \nabla \times [- \nabla\Phi(\bar{r},\omega)] \tag{2.21.9a}$$

Based on this vector relationship, if the arbitrary magnetic vector potential and the arbitrary electric scalar potential are known, then the electric field can be obtained as

$$\bar{E}(\bar{r},\omega) = -j\omega\bar{A}(\bar{r},\omega) - \nabla\Phi(\bar{r},\omega) \tag{2.21.9b}$$

This completes the representation for the electric and magnetic fields in terms of the two arbitrary potential functions. The magnetic vector potential and the electric scalar potential are still unknown. To find these two potentials, expressions (2.21.5c) and (2.21.9b) are substituted into Ampere's law relationship (2.21.4b) to obtain

$$\nabla \times \frac{1}{\mu} \nabla \times \bar{A}(\bar{r},\omega) = j\omega\varepsilon[-j\omega\bar{A}(\bar{r},\omega) - \nabla\Phi(\bar{r},\omega)] + \bar{J}_V(\bar{r},\omega) \tag{2.21.10a}$$

$$\nabla \times \nabla \times \bar{A}(\bar{r},\omega) = \omega^2\mu\varepsilon\bar{A}(\bar{r},\omega) - j\omega\mu\varepsilon\nabla\Phi(\bar{r},\omega) + \mu\bar{J}_V(\bar{r},\omega) \tag{2.21.10b}$$

The lefthand-side term of expression (2.21.10b) can be simplified in the rectangular coordinate system. According to the vector identity, Appendix A, we have the relationship

$$\nabla \times [\nabla \times \bar{A}] = \nabla(\nabla \bullet \bar{A}) - \nabla^2\bar{A} \tag{2.21.10c}$$

Using this identity, expression (2.21.10b) is rewritten as

$$\nabla[\nabla \bullet \bar{A}(\bar{r},\omega)] - \nabla^2\bar{A}(\bar{r},\omega)$$

$$= \omega^2\mu\varepsilon\bar{A}(\bar{r},\omega) - j\omega\mu\varepsilon\nabla\Phi(\bar{r},\omega) + \mu\bar{J}_V(\bar{r},\omega) \tag{2.21.10d}$$

But the magnetic vector potential and the electric scalar potential are just arbitrary potential point functions. It is possible to seek *a solution* for the magnetic vector potential so that

$$\nabla[\nabla \bullet \bar{A}(\bar{r},\omega)] = -j\omega\mu\varepsilon\nabla\Phi(\bar{r},\omega) \tag{2.21.11}$$

Relationship (2.21.11) is generally referred to as the *Lorentz gauge condition.* By enforcing such a gauge condition on relationship (2.21.10d), the following vector *Helmholtz* partial differential equation is obtained for the magnetic vector potential:

$$\nabla^2\bar{A}(\bar{r},\omega) + \beta^2\bar{A}(\bar{r},\omega) = -\mu\bar{J}_V(\bar{r},\omega) \tag{2.21.12}$$

$$\beta^2 = \omega^2 \mu\varepsilon \tag{2.21.13a}$$

β : propagation constant of the medium, in radians per meter

$$= \omega(\mu\varepsilon)^{1/2} \tag{2.21.13b}$$

Hence, we have a systematic procedure for analyzing electromagnetic field problems. The first step consists of solving the preceding Helmholtz's vector differential equation for the solution to the vector magnetic potential. In the second step, the magnetic field distribution is calculated by using expression (2.21.5c). In the third step, the electric field distribution can be calculated by using Ampere's law expression (2.21.4b) or using expression (2.21.9b) which requires the calculation of the electric scalar potential. Referring back to the Lorentz gauge expression (2.21.11), the electric scalar potential is given by

$$\Phi(\bar{r},\omega) = \frac{j\omega}{\beta^2} \nabla \bullet \bar{A}(\bar{r},\omega) \tag{2.21.14}$$

As can be seen in the vector Helmholtz equation, expression (2.21.12), the magnetic vector potential depends completely on a single driving source term that is the electric current density distribution. In fact, a similar relationship can also be obtained for the electric scalar potential with a single driving source in terms of the electric charge density distribution.

To obtain an Helmholtz equation for the electric scalar potential, first the divergence of the expression (2.21.9b) is taken

$$\nabla \bullet \bar{E}(\bar{r},\omega) = - j\omega \nabla \bullet \bar{A}(\bar{r},\omega) - \nabla \bullet \nabla\Phi(\bar{r},\omega) \tag{2.21.15a}$$

The lefthand-side term given by the divergence of the electric field is replaced in terms of the source electric charge density by using Gauss's law expression (2.21.4c). The righthand-side term given by the divergence of the magnetic vector potential is replaced using the gauge condition (2.21.14). Hence, expression (2.21.15a) simplifies to the following Helmholtz scalar equation

$$\nabla^2\Phi(\bar{r},\omega) + \beta^2\Phi(\bar{r},\omega) = - \frac{\rho_v(\bar{r},\omega)}{\varepsilon} \tag{2.21.15b}$$

The electric scalar potential can be obtained by solving this partial differential equation. It is interesting to note that the magnetic vector potential depends only on the electric current density source distribution, and similarly, the electric scalar potential depends only on the electric charge density source distribution. The two sources are not independent, but are related through the current-charge continuity equation. Similarly, the two auxiliary potentials are not independent, but are related through the Lorentz gauge condition. As discussed in Section 2.18, the vector Helmholtz equation is difficult to solve. It can be reduced to the corresponding three scalar Helmholtz equations. Using the rectangular components defined in expression (2.21.5b), the vector Helmholtz equation (2.21.12) can be rewritten in the following scalar form:

$$\nabla^2 A_x(\bar{r},\omega) + \beta^2 A_x(\bar{r},\omega) = -\mu J_x(\bar{r},\omega) \tag{2.21.16a}$$

$$\nabla^2 A_y(\bar{r},\omega) + \beta^2 A_y(\bar{r},\omega) = -\mu J_y(\bar{r},\omega) \tag{2.21.16b}$$

$$\nabla^2 A_z(\bar{r},\omega) + \beta^2 A_z(\bar{r},\omega) = -\mu J_z(\bar{r},\omega) \tag{2.21.16c}$$

It is interesting to note from these equations that the x-directed electric current density produces only the x component of the magnetic vector potential. Similarly, y- and z-directed electric current densities correspondingly produce y and z components of the magnetic vector potentials, respectively. Instead of the volume electric current density distributions, suppose either surface or line electric current distributions are present, then they are appropriately replaced in the righthand side of the scalar differential equations. Hence, to obtain the complete solution for the magnetic vector potential, the three scalar equations (2.21.16a–c) are to be solved. There are numerous different approaches one can take to solve these scalar equations. For example, the spectral domain and Green's function approach is introduced in later sections along with linearity and superposition to construct scalar type potential integrals for an arbitrary electric current source distribution.

2.21.1 STATIC EQUATIONS

The two auxiliary potential functions, the magnetic vector potential and the electric scalar potential, for the analysis of electromagnetic field problems is also consistent with the static field equations. For either electrostatics or magnetostatics, the frequency of excitation is zero, and hence, the propagation constant is also zero. Expression (2.21.12) reduces to the vector Poisson's equation for the static magnetic vector potential, and expression (2.21.15b) reduces to the scalar Poisson's equation for the static electric scalar potential.

Hence, in the static magnetic field case,

$$\nabla^2 \bar{A}(\bar{r}) = -\mu \bar{J}_v(\bar{r}) \tag{2.21.17a}$$

$$\bar{H}(\bar{r}) = \frac{1}{\mu} \nabla \times \bar{A}(\bar{r}) \tag{2.21.17b}$$

Similarly, in the static electric field case,

$$\nabla^2 \Phi(\bar{r}) = -\frac{\rho_V(\bar{r},\omega)}{\varepsilon} \tag{2.21.18a}$$

$$\bar{E}(\bar{r}) = -\nabla \Phi(\bar{r},\omega) \tag{2.21.18b}$$

In the source-free cases, the two Poisson's differential equations, expressions (2.21.17a) and (2.21.18a), reduce to the corresponding Laplace's differential equations. The relevant solutions for the differential equations are discussed in the next section.

2.22 SOLUTION FOR POTENTIALS

The Helmholtz differential equations derived in the preceding section are rigorously solved for various types of source electric current and charge distributions. Detailed source modeling cases, such as the linear distribution of electric current sources, the surface distribution of electric current sources, and the volume distribution of electric current sources have been considered in Sections 2.1 and 2.14. To obtain the complete solution for the magnetic vector potential with arbitrary source distribution, first the solution for the Helmholtz scalar differential equation is obtained for an isolated point or elemental current source. Since the preceding scalar equations are similar in mathematical form, various components of the potentials can be conveniently written down by inspection. Once the solution is known for the case of a point-type elemental source, by invoking the concept of linearity and the superposition principle, the complete solution for the two potentials can be constructed as the superposition integrals in terms of the corresponding source distributions. By knowing the distribution of the two auxiliary potentials, the electric and magnetic field distributions in the given medium can be calculated based on the analysis carried out in the previous section.

An isolated point-type electric current element is located in an isotropic and homogeneous free-space medium. The source current element has a time-harmonic variation of $exp(j\omega t)$ and is located at the origin of a three-dimensional coordinate system. Referring to Figure 2.31, let

I_Z : electric current of the point source, in amperes, located at the source point O;

Δz : incremental length of the electric current element, in meters;

$\hat{z}$: unit vector in the direction of electric current element;

$\bar{r}$: position vector corresponding to a general field point P(x, y, z)

$$= r\hat{r}$$

$$= x\hat{x} + y\hat{y} + z\hat{z} \qquad (2.22.1a)$$

r: the radial distance between the source and field points, in meters

$$= [x^2+y^2+z^2]^{1/2} \qquad (2.22.1b)$$

ω : angular frequency of excitation, in radians per second.

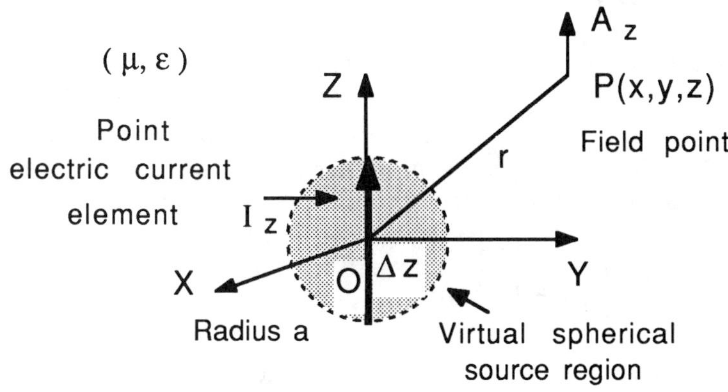

Figure 2.31 Isolated electric current point source.

The point-type ideal elemental current source is shown in the Figure 2.31. It can be simulated, for example, using a piece of conducting wire with two spherical end caps. At the center of conducting wire an ideal electric current source generator is connected. The current generator, depending upon the frequency of oscillation, charges and discharges the two conducting shells and thus maintains a constant source current along the elemental length of conducting wire. From a practical modeling consideration, the length of the current element is very small compared to wavelength of the source current excitation. This arrangement is generally viewed as an ideal *electric current dipole* and constitutes an elemental point source for simulating arbitrary distribution of electric current densities.

Referring to the Figure 2.31, the point-type current source distribution is directed along the z coordinate axis. There exists only a z component of the magnetic vector potential, and it satisfies the Helmholtz's differential equation (2.21.16c) discussed in the previous section. The z-directed electric current element located at the origin is now enclosed by a small virtual sphere of radius a. In a limit as the radius a of the sphere approaches to zero, the electric current element shrinks to a point. In such a limiting process, it is possible to assume that the total current associated with the point-type source current element to be uniform over the small spherical surface. Now the complete three-dimensional free-space medium is divided into two specific regions, the source region for $r \leq a$ and the source-free region for $r > a$. For these two regions, the z component of the magnetic vector potential satisfies the following partial differential equations:

$$\nabla^2 A_z(r,\omega) + \beta^2 A_z(r,\omega) = -\mu I_z(r,\omega) \qquad r \leq a \qquad (2.22.2a)$$

$$\nabla^2 A_z(r,\omega) + \beta^2 A_z(r,\omega) = 0 \qquad r \geq a_+ \qquad (2.22.2b)$$

The solution for the z component of magnetic vector potential at any point for $r > a$ has spherical symmetry. This is due mainly to the nature of the point-type of electric current source located at the origin. The solution for A_z is independent of the θ and ϕ coordinate variables. Referring to the Appendix A, the Laplacian operator in expression (2.22.2b) can be written down in terms of spherical coordinate variable r to obtain

$$\frac{1}{r^2}\frac{d}{dr}\left[r^2\frac{dA_z}{dr}\right] + \beta^2 A_z = 0 \qquad (2.22.3)$$

Since A_z depends only on the radial r coordinate variable, in expression (2.22.3) regular derivatives are introduced instead of the partial derivatives. Let

$$A_z(r) = \frac{G(r)}{r} \qquad (2.22.4a)$$

where $G(r)$ is the auxiliary function to be determined. Further,

$$\frac{d}{dr}\left[\frac{G(r)}{r}\right] = \frac{1}{r}\frac{d}{dr}G(r) - \frac{1}{r^2}G(r) \qquad (2.22.4b)$$

$$\frac{d}{dr}\left[r^2\frac{d}{dr}\left(\frac{G(r)}{r}\right)\right] = r\frac{d^2G(r)}{dr^2} \qquad (2.22.4c)$$

Hence, expression (2.22.3) simplifies to

$$\frac{d^2G(r)}{dr^2} + \beta^2 G(r) = 0 \qquad (2.22.4d)$$

Expression (2.22.4d) is a second-order ordinary differential equation. The two roots of this homogeneous equation are given by $j\beta$ and $-j\beta$. Using these two roots, the solution for $G(r)$ is obtained as

$$G(r) = C\, e^{j\beta r} + D\, e^{-j\beta r} \tag{2.22.4e}$$

where C and D are constant coefficients to be determined based on the boundary conditions. Therefore, using relationship (2.22.4a), the solution for the magnetic vector potential can be written as

$$A_z(r,\omega) = \frac{C}{r} e^{j\beta r} + \frac{D}{r} e^{-j\beta r} \tag{2.22.4f}$$

The solution for the magnetic vector potential obtained in expression (2.22.4f) consists of two terms. The second term represents a forward traveling spherical wave propagating in the radial direction away from the origin, where the point source is located. Similarly, the first term represents a backward traveling spherical wave propagating in the radial direction toward the origin, where the point source is located. Since the source propagates energy in all directions away from its location, the proper solution for the magnetic vector potential consists of only the second term. Hence,

$$A_z(r,\omega) = \frac{D}{r} e^{-j\beta r} \qquad r > a_+ \tag{2.22.5}$$

This result is also valid (in a limit) just outside the surface of sphere. Hence, the unknown coefficient D is determined by substituting the result (2.22.5) into differential equation (2.22.2a):

$$\nabla^2\left[\frac{D}{r} e^{-j\beta r}\right] + \beta^2 \frac{D}{r} e^{-j\beta r} = -\mu I_z(r,\omega) \tag{2.22.6a}$$

After performing volume integration over the complete source sphere of radius a, expression (2.22.6a) reduces to the following relationship

$$\iiint_{sphere} \nabla \cdot \nabla\left[\frac{D}{r} e^{-j\beta r}\right] dv + \iiint_{sphere} \beta^2 \frac{D}{r} e^{-j\beta r}\, dv = \iiint_{sphere} -\mu I_z\, dv \tag{2.22.6b}$$

The volume integrals are now evaluated in a limit as radius a of the sphere approaches zero. On applying the Gauss-divergence theorem, Appendix A, the first integral on the lefthand side of relationship (2.22.6b) simplifies to a surface integral over the sphere

$$\iiint_{\substack{\text{sphere}\\\text{volume}}} \nabla \bullet \nabla\left[\frac{D}{r}\, e^{-j\beta r}\right] dv \;=\; \iint_{\substack{\text{sphere}\\\text{surface}}} \nabla\left[\frac{D}{r}\, e^{-j\beta r}\right] \bullet d\bar{s} \qquad (2.22.6c)$$

The integrand on the righthand side of this integral can be simplified as

$$\nabla\left[\frac{D}{r}\, e^{-j\beta r}\right] = \frac{d}{dr}\left[\frac{D}{r}\, e^{-j\beta r}\right] \hat{r}$$

$$= D e^{-j\beta r}\left[-\frac{1}{r^2} - \frac{j\beta}{r}\right] \hat{r} \qquad (2.22.6d)$$

$$d\bar{s} = r^2 \sin\theta \; d\theta \; d\phi \; \hat{r} \qquad (2.22.6e)$$

On substituting relations (2.22.6d) and (2.22.6e), expression (2.22.6c) simplifies in a limit as the radius of the sphere a tends to zero:

$$\iint_{\substack{\text{sphere}\\\text{surface}}} \nabla\left[\frac{D}{r}\, e^{-j\beta r}\right] \bullet d\bar{s}$$

$$= \lim_{a \to 0} \iint_{\text{sphere}} D e^{-j\beta a}\left[-\frac{1}{a^2} - \frac{j\beta}{a}\right] a^2 \sin\theta \; d\theta \; d\phi \qquad (2.22.6f)$$

$$= -\,4\pi D \qquad (2.22.6g)$$

The second volume integral on the lefthand side of relationship (2.22.6b) simplifies to zero in a limit as the radius of sphere a tends zero. Further, referring to the righthand-side term of expression (2.22.6b), the volume of virtual source sphere contains only the line current element. The volume integral becomes just a line integral. The distribution of the source electric current is assumed to be completely uniform along the elemental length of the source. The righthand side of integral expression (2.22.6b) simplifies to:

$$\iiint_{\substack{\text{sphere}\\\text{volume}}} -\,\mu I_z \, dv \;=\; \int_{\substack{\text{element}\\\text{length}}} -\,\mu I_z \, dz \qquad (2.22.6h)$$

$$= -\,\mu I_z \Delta z \qquad (2.22.6i)$$

After substituting the results, expressions (2.22.6g) and (2.22.6i), the unknown coefficient D is obtained as

$$D = \frac{\mu I_z \Delta z}{4\pi} \tag{2.22.7}$$

Hence, the solution for the z component of magnetic vector potential, expression (2.22.5), due to an elemental electric current located at the origin is

$$A_z(r,\omega) = \frac{\mu I_z \Delta z}{4\pi r} e^{-j\beta r} \tag{2.22.8}$$

Suppose the electric current element is located at an arbitrary source point (x', y', z') instead at the origin of the coordinate system. Then the result obtained, expression (2.22.8), for the z component of magnetic vector potential is to be suitably modified to take into account the actual distance between the source point and the field observation point. In such a case,

$\bar{r}'$: position vector corresponding to the general source point

$$= r'\hat{r}'$$
$$= x'\hat{x} + y'\hat{y} + z'\hat{z} \tag{2.22.9a}$$

R : actual distance between source and field points, in meters

$$= |\bar{r} - \bar{r}'| \tag{2.22.9b}$$

The expression for the z component of magnetic vector potential is identical to the expression derived in (2.22.8), except the distance term r is replaced by the relative distance between the source point and the field observation point. Hence,

$$A_z(\bar{r}) = \frac{\mu I_z \Delta z}{4\pi R} e^{-j\beta R} \tag{2.22.10}$$

Suppose all the three components of source electric current are present, then the partial differential equations (2.21.16a–c) are solved separately to obtain the corresponding components of the magnetic vector potential. Let us consider an electric current source with an arbitrary polarization given by

$$\bar{I} = I\hat{L} \tag{2.22.11a}$$

I : electric current of the elemental source, in amperes,
at the source point (x', y', z');
ΔL : incremental length of the elemental source, in meters;
$\hat{L}$: unit vector along the direction of the electric current element;
R : actual distance between source and field points, in meters
$= |\bar{r} - \bar{r}'|$

Then, the expression for magnetic vector potential at the field point (x, y, z) in the direction of source electric current element can be written as

$$\bar{A}(\bar{r}) = A_L(\bar{r})\hat{L} \tag{2.22.11b}$$

$$A_L(\bar{r}) = \frac{\mu I \Delta L}{4\pi R} e^{-j\beta R} \tag{2.22.11c}$$

If the magnetic vector potential is required to be calculated in an arbitrary direction, a unit vector is first specified in the direction, and then a scalar product of expression (2.22.11b) is taken with respect to the unit vector in the specified direction.

2.22.1 POTENTIAL SUPERPOSITION INTEGRALS

The discussion given previously for the calculation of the magnetic vector potential distribution due to an isolated point-type electric current element can be easily extended to multiple isolated point-type electric current elements located in a very large three-dimensional medium. This is accomplished based on the linearity and superposition of the various individual magnetic vector potential contribution. Let

$$\dot{I}_n = I_n \hat{L}_n \tag{2.22.12a}$$

I_n : electric current located at n^{th} source point (x_n, y_n, z_n),
in amperes, where $n = 1, 2, 3, \ldots, N$;
ΔL_n : incremental length of the nth electric current element, in meters;
$\hat{L}_n$: unit vector along the nth electric current element;

$\bar{r}$: position vector corresponding to the field point (x, y, z)
$= x\hat{x} + y\hat{y} + z\hat{z}$ (2.22.12b)

$\bar{r}_n$: position vector corresponding to the source current element I_n

$$= x_n\hat{x} + y_n\hat{y} + z_n\hat{z} \tag{2.22.12c}$$

R_n : distance between the field point and the nth source location

$$= |\bar{r} - \bar{r}_n| \tag{2.22.12d}$$

The principle of superposition can be applied to calculate the net magnetic vector potential distribution at the field point P. Let

$\bar{A}_n(\bar{r})$: n^{th} partial magnetic vector potential due to the electric current element $I_n \Delta L_n$;

$\bar{A}(\bar{r})$: net magnetic vector potential at the field point

$$= \sum_{n=1}^{N} \bar{A}_n \tag{2.22.13a}$$

where the partial magnetic vector potential at the field point (x, y, z) due to the presence of nth electric current element directed along the nth unit vector can be written as

$$\bar{A}_n(\bar{r}) = A_n(\bar{r})\hat{L}_n \tag{2.22.13b}$$

$$A_n(\bar{r}) = \frac{\mu I_n \Delta L_n}{4\pi R_n} e^{-j\beta R_n} \tag{2.22.13c}$$

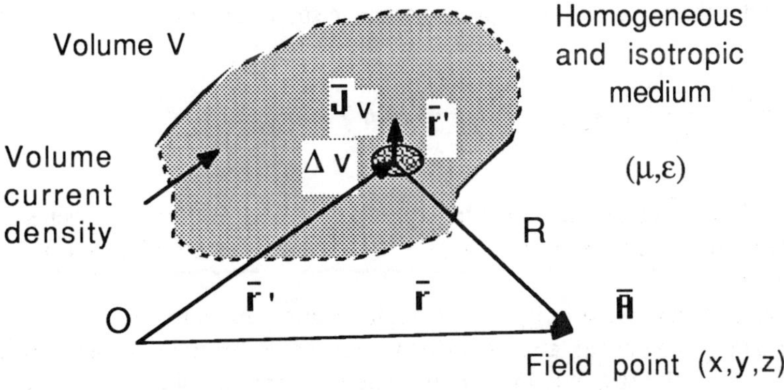

Figure 2.32 Magnetic vector potential due to volume current distribution.

The discussion now turns to calculating the magnetic vector potential due to a continuous time-harmonic volume electric current density distribution. Referring to Figure 2.32, a volume electric current distribution is confined to an arbitrary volume *V* located in a three-dimensional isotropic and homogeneous medium. Let

Δv : incremental volume enclosing the source point (x', y', z'), in cubic meter;

$\bar{J}_V(\bar{r}')$: volume current density distribution in the volume *V*, in amperes per square meters;

$\Delta\bar{J}_V$: incremental electric current element at the source point

$$= \bar{J}_V(\bar{r}')\Delta v(\bar{r}') \tag{2.22.14a}$$

R : distance between the line joining the source point to the field point

$$= |\bar{r} - \bar{r}'| \tag{2.22.14b}$$

Following the procedure discussed earlier, first the incremental magnetic vector potential at the field point is calculated due to the incremental volume electric current at the source point. In the second step, based on the concept of linearity, the principle of superposition is applied for calculating the total magnetic vector potential at the field point. The incremental magnetic vector potential at the field point is given by

$$d\bar{A}(\bar{r}) = \frac{\mu}{4\pi}\left[\frac{\bar{J}_V dv(\bar{r}')}{R}\right] e^{-j\beta R} \tag{2.22.15a}$$

The total magnetic vector potential at the field point *P(x, y, z)* is given by the triple integral of the above expression over the complete source distribution in the volume *V*:

$$\bar{A}(\bar{r}) = \frac{\mu}{4\pi}\iiint_V \bar{J}_V(\bar{r}')\,\frac{e^{-j\beta R}}{R}\,dv(\bar{r}') \tag{2.22.15b}$$

The superposition integral obtained in expression (2.22.15b) forms the general solution to the Helmholtz vector equation (2.21.12) with the arbitrary distribution of volume current density.

Similarly, a general solution to the scalar Helmholtz equation (2.21.15b) for the electric scalar potential can also be written. After comparing the Helmholtz equations (2.21.12) and (2.21.15b), only the forcing function terms are different. Hence, the total electric scalar potential at the field point *P(x, y, z)* is given by the volume integral over the complete charge density distribution in the volume *V*:

$$\Phi(\bar{r}) = \frac{1}{4\pi\varepsilon} \iiint_V \rho_v(\bar{r}') \frac{e^{-j\beta R}}{R} dv(\bar{r}') \tag{2.22.15c}$$

If other types of source distributions, such as the surface currents and charges and the line currents and charges are present, the potential integrals (2.22.15b) and (2.22.15c) are to be correspondingly modified with respect to surface and line integrals.

2.23 ELECTRIC DIPOLE FIELDS

An isolated point-type electric current element is located at the origin and polarized in the *z* coordinate direction in a three-dimensional medium, as shown in Figure 2.33. Since the electric dipole source is polarized along the *z* coordinate direction, there exists only the *z* component of the magnetic vector potential. In the following, a detailed analysis is presented to obtain relevant components of the electric and magnetic field distributions.

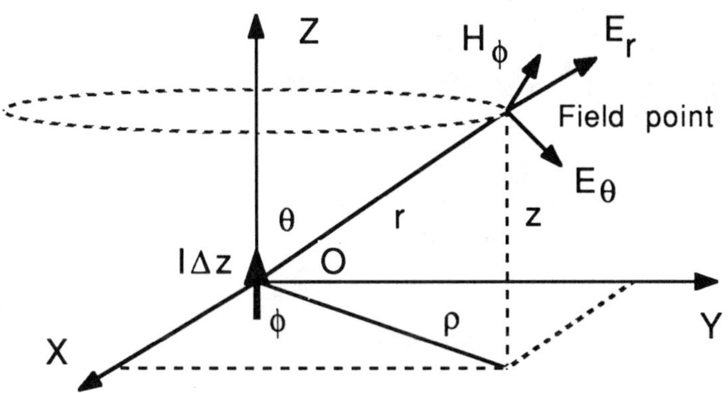

Figure 2.33 Electric and magnetic fields due to a dipole.

According to expression (2.22.8)

$$A_z(r,\omega) = \frac{\mu I \Delta z}{4\pi r} e^{-j\beta r} \tag{2.23.1a}$$

r: distance between the source point and the field point

$$= [x^2+y^2+z^2]^{1/2} \tag{2.23.1b}$$

The magnetic field distribution at any field point in terms of the magnetic vector potential is given by expression (2.21.5c):

$$\bar{H}(r,\theta,\phi) = \frac{1}{\mu} \nabla \times \bar{A}(r,\theta,\phi) \quad (2.23.2a)$$

$$= \frac{1}{\mu} \nabla \times A_z(r,\theta,\phi)\hat{z} \quad (2.23.2b)$$

$$= \frac{1}{\mu} \nabla A_z(r,\theta,\phi) \times \hat{z} \quad (2.23.2c)$$

Since the electric dipole is located at the origin, the magnetic vector potential is independent of the θ and ϕ coordinate variables. Using the spherical coordinate system,

$$\bar{H}(r,\theta,\phi) = \frac{d}{dr}\left[\frac{I\Delta z}{4\pi}\frac{e^{-j\beta r}}{r}\right] \hat{r} \times \hat{z} \quad (2.23.3a)$$

$$\hat{r} \times \hat{z} = \hat{r} \times [\hat{r} \cos\theta - \hat{\theta} \sin\theta] \quad (2.23.3b)$$

Based on expressions (2.23.3a) and (2.23.3b), there exists only the ϕ component of magnetic field which encircles the z component of electric current element. The magnetic field distribution at any field point is given by

$$\bar{H}(r,\theta,\phi) = H_\phi(r,\theta,\phi)\hat{\phi} \quad (2.23.4a)$$

$$H_\phi(r,\theta,\phi) = \frac{I\Delta z}{4\pi}\left[\frac{j\beta}{r} + \frac{1}{r^2}\right] \sin\theta \; e^{-j\beta r} \quad (2.23.4b)$$

The electric field distribution can be calculated using expression (2.21.9b), which requires both the magnetic vector potential and the electric scalar potential. But the electric scalar potential can be replaced in terms of the magnetic vector potential, which is obtained from expression (2.21.14):

$$\Phi(\bar{r},\omega) = \frac{j\omega}{\beta^2}\frac{\partial A_z(\bar{r},\omega)}{\partial z} \quad (2.23.5a)$$

After substituting expression (2.23.5a) into expression (2.21.9b), the electric field distribution is given by

$$\bar{E}(\bar{r},\omega) = -j\omega A_z(\bar{r},\omega)\hat{z} - \frac{j\omega}{\beta^2}\nabla\left[\frac{\partial A_z(\bar{r},\omega)}{\partial z}\right] \tag{2.23.5b}$$

First the derivative of A_z is calculated, and then the gradient operation is performed. Expressing the electric field distribution in terms of the spherical components,

$$\bar{E}(\bar{r},\omega) = E_r(\bar{r},\omega)\hat{r} + E_\theta(\bar{r},\omega)\hat{\theta} + E_\phi(\bar{r},\omega)\hat{\phi} \tag{2.23.5c}$$

Referring to Appendix A, the gradient of a scalar potential function in the spherical coordinates is given by the following expression:

$$\nabla\Phi(r,\theta,\phi) = \frac{\partial\Phi}{\partial r}\hat{r} + \frac{1}{r}\frac{\partial\Phi}{\partial\theta}\hat{\theta} + \frac{1}{r\,\sin\theta}\frac{\partial\Phi}{\partial\phi}\hat{\phi} \tag{2.23.5d}$$

It may be noted in expression (2.23.1a), the *z* component of the magnetic vector potential varies only with respect to radial coordinate variable *r* and is independent of the two angular spherical coordinate variables. Hence, expression (2.23.5b) for the electric field distribution at any field point simplifies to

$$\begin{aligned}\bar{E}(\bar{r},\omega) = -j\omega A_z\hat{z} &- \frac{j\omega}{\beta^2}\frac{\partial}{\partial r}\left[\frac{\partial A_z}{\partial z}\right]\hat{r} \\ &- \frac{j\omega}{\beta^2}\frac{1}{r}\frac{\partial}{\partial\theta}\left[\frac{\partial A_z}{\partial z}\right]\hat{\theta}\end{aligned} \tag{2.23.6a}$$

The components of the electric field are obtained by taking the scalar product of expression (2.23.6a) with the corresponding unit vectors. For the *z*-directed electric current element, there exist only the *r* and *θ* components of the electric field distribution. Hence, the electric field distributions are given by

$$\begin{aligned}E_r(r,\theta,\phi) = &-j\omega\left[\frac{\mu I\Delta z}{4\pi r}e^{-j\beta r}\right]\hat{r}\bullet\hat{z} \\ &- \frac{\omega}{\beta^2}\frac{d}{dr}\left[\frac{d}{dr}\frac{\mu I\Delta z}{4\pi r}e^{-j\beta r}\right]\left[\frac{dr}{dz}\right]\end{aligned} \tag{2.23.6b}$$

$$E_r(r,\theta,\phi) = \frac{I\Delta z}{2\pi}\left[\frac{\eta}{r^2} + \frac{\eta}{j\beta r^3}\right]\cos\theta \; e^{-j\beta r} \tag{2.23.6c}$$

and

$$E_\theta(r,\theta,\phi) = -\,j\omega\left[\frac{\mu I\Delta z}{4\pi r}\,e^{-j\beta r}\right]\hat{\theta}\bullet\hat{z}$$

$$-\,\frac{j\omega}{\beta^2}\,\frac{1}{r}\,\frac{d}{d\theta}\left[\frac{d}{dr}\frac{\mu I\Delta z}{4\pi r}\,e^{-j\beta r}\right]\left[\frac{dr}{dz}\right] \tag{2.23.6d}$$

$$E_\theta(r,\theta,\phi) = \frac{I\Delta z}{4\pi}\left[\frac{j\beta\eta}{r} + \frac{\eta}{r^2} + \frac{\eta}{j\beta r^3}\right]\sin\theta \; e^{-j\beta r} \tag{2.23.6e}$$

where

β : propagation constant of the medium

$= \omega(\mu\varepsilon)^{1/2}$

η : intrinsic impedance of the medium

$= (\mu/\varepsilon)^{1/2}$

Expressions (2.23.4b), (2.23.6c) and (2.23.6e) give the complete time-harmonic field distributions for an ideal electric dipole source. The electric and magnetic fields are independent of the ϕ coordinate variable. The terms with $1/r$ are called the *distant* or *far-radiation fields*. The terms with higher inverse powers of r are generally referred to as *near-field distributions*. The terms with $1/r^2$ are called the *induction field distributions* and the terms with $1/r^3$ are called as the *quasi-static field distributions*. For very large values of r, the terms containing higher inverse powers of r can be neglected.

Hence, the far fields or the radiation fields for the ideal electric dipole are given by

$$H_\phi(r,\theta,\phi) = \frac{I\Delta z}{4\pi}\,\frac{j\beta}{r}\,\sin\theta \; e^{-j\beta r} \tag{2.23.7a}$$

$$E_\theta(r,\theta,\phi) = \frac{I\Delta z}{4\pi}\,\frac{j\beta\eta}{r}\,\sin\theta \; e^{-j\beta r} \tag{2.23.7b}$$

As can be noted, only the ϕ component of magnetic field and θ component of the electric field exist in the far field region. On taking the magnitude ratio,

$$\frac{E_\theta}{H_\phi} = \eta \tag{2.23.7c}$$

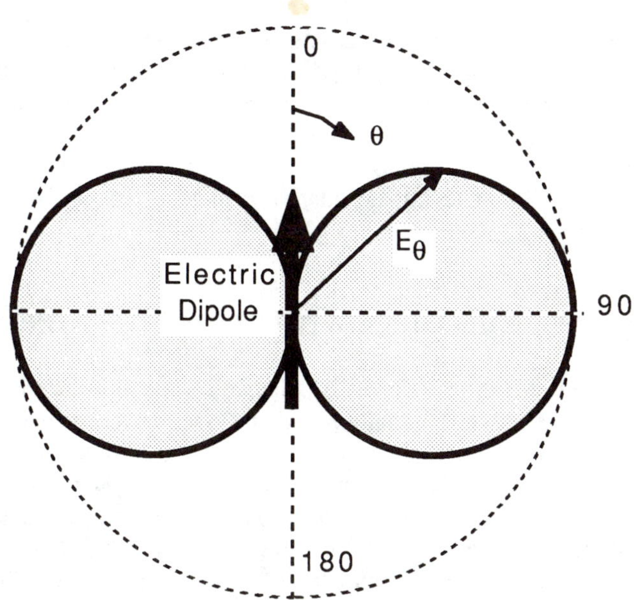

Figure 2.34 Radiation pattern of a electric dipole source.

The near-field distributions of the electric and magnetic fields are quite complicated. The electric and magnetic fields in the far radiation zone behave like spherical waves propagating along the radial coordinate direction. In the radiation zone, only the θ component of the electric field and the ϕ component of the magnetic field distributions exist. On a sphere of constant radius with the dipole source located at the center of sphere, the phase distributions of electric and magnetic fields are constants at any point on the surface of the sphere. Further, the electric and magnetic fields are polarized such that they are tangential to the surface of large sphere and propagate away in the radial direction. Similar results can be obtained even if the electric dipole is polarized along other coordinate directions by properly accounting for its polarization and coordinate shift.

Figure 2.34 shows the electric field radiation pattern of the electric dipole source in a far-field region plotted as a function of spherical angle θ. The angle θ varies from 0 to π, and the pattern is independent of the angle ϕ. Only the normalized magnitude pattern is shown in this figure. In fact, the electric and magnetic fields decay as $1/r$ and this factor has been removed by normalizing the field pattern. It should be noted again that the results obtained in the far-field region, expressions (2.23.7a) and (2.23.7b), are for the case of an ideal point-type electric dipole located at the origin of the coordinate system. Thus, the origin represents a reference point at which the phase of the field distribution is zero. This reference point is generally referred to as the *phase center*. Suppose the ideal electric dipole is shifted to certain a distance from the origin; the analysis for the electric and magnetic fields is complicated, but can be repeated in a way similar to that discussed above. For far-field distribution, the analysis is quite straightforward, and the same far field expressions (2.23.7a) and (2.23.7b) are applicable with a corresponding phase shift to take into account the change in distance between the source point and the field point.

2.24 MAXWELL'S EQUATIONS WITH MAGNETIC SOURCES

In the previous sections, various forms of Maxwell's equations were discussed. In Section 2.3, classical Maxwell's equations are discussed in terms of the time-domain variable. Similarly, in Section 2.15, the same classical Maxwell's equations are discussed in terms of the frequency-domain variable. It should be noted that the sources for the coupled electromagnetic fields are the electric currents and charges. Referring to Maxwell's equations in Tables 2.5 and 2.6, the electric current source term appears in Ampere's law, expression (2.15.4), and the electric charge source term appears in Gauss's law, expression (2.15.7). Thus, with respect to the driving source terms, Maxwell's equations are the asymmetrical partial differential equations. This is also evident from the electromagnetic boundary relationships discussed in Section 2.16. The difference between the tangential components of the total magnetic field at an interface separating two media should be equivalent to surface current density. On the other hand, the tangential components of the total electric field at an interface separating two media should be continuous.

In many electromagnetic boundary value problems involving dielectric and magnetic materials, it is possible to analyze the electromagnetic field interactions by formulating in terms of the dependent types of sources. For example, in a region where there is discontinuity in the tangential magnetic fields, they can be conveniently modeled by introducing the dependent-type electric current sources. Similarly, it is also possible to introduce *hypothetical dependent magnetic current sources* in a given region where there is discontinuity in the tangential electric fields. The virtual magnetic current and their corresponding magnetic charge sources are, in fact, just mathematical entities introduced to accomplish a straightforward formulation of the electromagnetic boundary value problem. Further, it should be noted that the classical Maxwell's equations exhibit an ideal symmetry with respect to the sources. From a modeling point of view, the magnetic current sources can be viewed as dual sources producing dual electric and magnetic field distributions.

In the following section, classical Maxwell's equations are presented with the virtual magnetic current and the corresponding virtual magnetic charge sources. Based on the concept of linearity and superposition, it can be easily shown that the electric and magnetic fields at a given point in the region consist of the sum of two types of partial field contributions. Suppose the electric and magnetic fields are known for the electric current sources. Then, based upon the duality concept, the corresponding expressions for the electric and magnetic fields can be written down for the magnetic current sources.

2.24.1 MAGNETIC SOURCE DISTRIBUTIONS

The magnetic current and charge sources in the frequency domain can be treated the same way as modeling of the electric source distributions discussed in the earlier sections. Referring to Sections 2.1 and 2.14, the magnetic currents and charges generally distribute arbitrarily throughout a given region of space depending on the medium characteristics. To account for these quantitatively, the corresponding magnetic current and magnetic charge density distributions can be defined on the macroscopic level in an average sense.

Let us consider the volume distribution of magnetic charges in a large three-dimensional region that is linear, homogeneous, and isotropic medium. Referring to Figure 2.1, to account for the magnetic charge distribution at each point within the source region, an arbitrary source point (x', y', z') is selected which is enclosed within an incremental volume Δv. If the total magnetic charge inside the incremental volume is known, then the volume magnetic charge density at the source point is given by

$$\rho_{vm}(\bar{r}',\omega) = \lim_{\Delta v \to 0} \frac{\Delta q_m(\bar{r}',\omega)}{\Delta v(\bar{r}')}$$

$$= \frac{\partial q_m(\bar{r}',\omega)}{\partial v(\bar{r}')} \tag{2.24.1a}$$

Then, the total magnetic charge contained in the volume V is

$$q_m(\omega) = \iiint_V \rho_{vm}(\omega)\, dv \tag{2.24.1b}$$

where the triple integration is performed over the complete volume containing the magnetic charge distribution. Similarly, this can be extended to the magnetic surface and line charge distributions based on the discussions presented in Section 2.1.

Further, the magnetic current can be defined as the time rate of change of magnetic charges. The mathematical models discussed previously in Sections 2.1 and 2.14 can be applied as well to define the corresponding magnetic current density distributions. Particularly inside the magnetic conducting medium *conduction type of magnetic current* flows in the medium, given by the generalized Ohm's law

$$\bar{M}_c(x,y,z,\omega) = \sigma_m \bar{H}(x,y,z,\omega) \qquad (2.24.2a)$$

$\bar{M}_c(t)$: magnetic conduction current density, in volts per square meter

$$= \bar{M}_c(\omega)\, e^{j\omega t} \qquad (2.24.2b)$$

σ_m : magnetic conductivity of the medium, in ohms per meter;

$\bar{H}(\omega)$: external applied magnetic field, in amperes per meter.

Generally, the magnetic currents distribute throughout a given region of space depending on the medium characteristics. To account for the magnetic current distribution at every point within a given source volume, let us consider an arbitrary source point (x', y', z') which is enclosed within an incremental volume with an incremental cross-sectional area Δs. Referring to Figure 2.4a,

$\Delta I_m(t)$: magnetic current crossing incremental surface, in volts

$$= \Delta I_m(\omega)\, e^{j\omega t} \qquad (2.24.3a)$$

Δs : area of incremental surface, in square meters;

$\hat{\tau}$: unit vector along the magnetic current flow.

Then, the frequency-dependent volume current density M_v at the point (x', y', z') is given by

$$\bar{M}_v(\bar{r}',\omega) = M_v \hat{\tau} \qquad (2.24.3b)$$

where the magnitude of the volume current density in volts per square meter is

$$M_v = \lim_{\Delta s \to 0} \frac{\Delta I_m}{\Delta s} \qquad (2.24.3c)$$

Hence, if the frequency-dependent volume current density is known at every point within a given volume, Figure 2.4b, then the total magnetic current crossing the arbitrary open surface S is given by

$$I_m(\omega) = \iint_S \bar{M}_v \cdot d\bar{s} \qquad (2.24.3d)$$

The frequency-dependent surface magnetic currents confined to an arbitrary surface *S* and the line magnetic currents confined to an arbitrary contour *C* can also be modeled in a similar way.

2.24.2 CONTINUITY EQUATION

The continuity equation, discussed in Section 2.2, relates the electric charge and current density distributions in a volume bounded by a closed surface. A similar relationship also holds for magnetic currents and charges. The frequency-dependent magnetic currents oscillate at a given time-harmonic rate depending on the harmonic rate of the magnetic charges.

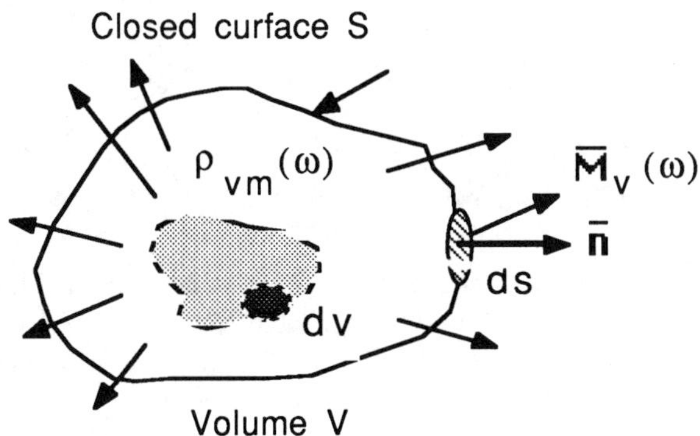

Figure 2.35 Continuity relation between magnetic currents and charges.

Referring to Figure 2.35, the total magnetic charge contained in the complete volume *V* is given by

$$q_m(\omega) = \iiint_V \rho_{vm}(\omega)\, dv \tag{2.24.4a}$$

$\rho_{vm}(t)$: volume magnetic charge density

$$= \rho_{vm}(\omega)\, e^{j\omega t} \tag{2.24.4b}$$

When the magnetic charges move out of the closed surface at a certain velocity, referring to the Figure 2.35, the resulting magnetic current crossing the closed surface *S* from the inside region to the outside region is obtained by

$$I_m(\omega) = \oiint_S \bar{M}_v(\omega) \bullet d\bar{s} \tag{2.24.5a}$$

$\bar{M}_v(t)$: volume current density, in volts per square meter

$$= \bar{M}_v(\omega)\, e^{j\omega t} \tag{2.24.5b}$$

As the magnetic current flows from the inside region to the outside, gradually the net magnetic charge inside the closed volume decreases. Hence, the continuity equation for the magnetic currents and charges has the form

$$I_m(t) = -\frac{\partial}{\partial t} q_m(t) \tag{2.24.6}$$

After substituting expressions (2.24.4a) and (2.24.5a) into this relationship,

$$\oiint_S \bar{M}_v(\omega) \bullet d\bar{s} = -j\omega \iiint_V \rho_{vm}\, dv \tag{2.24.7}$$

The integral form of continuity equation (2.24.7) can be converted into the equivalent differential form by utilizing the Gauss-divergence theorem, Appendix A. The lefthand-side closed surface integral can be replaced in terms of the volume integral of divergence of magnetic current density, thus

$$\iiint_V \nabla \bullet \bar{M}_v\, dv = -j\omega \iiint_V \rho_{vm}\, dv \tag{2.24.8}$$

Hence,

$$\nabla \bullet \bar{M}_v(\bar{r}',\omega) = -j\omega\rho_{vm}(\bar{r}',\omega) \tag{2.24.9}$$

Referring to continuity equation (2.24.9), if the volume magnetic charge density is zero at any point, then the divergence of magnetic current density is zero.

2.25 MAXWELL'S EQUATIONS IN FREQUENCY DOMAIN

The time-harmonic magnetic currents and charges set up correspondingly time-harmonic electric and magnetic fields. Tables 2.5 and 2.6 give the complete summary of the Maxwell's equations in the frequency domain with the electric current and electric charge sources. Similarly, another set of Maxwell's equations can be written in the frequency domain with the magnetic current and magnetic charge sources.

Let us consider a large three-dimensional region which is linear, homogeneous, and isotropic medium. The various electromagnetic field quantities and their corresponding sources can be written in terms of the following time-harmonic representation given by

$\bar{E}(\bar{r},t)$: electric field distribution, in volts per meter

$$= \bar{E}(\bar{r},\omega)\, e^{j\omega t} \tag{2.25.1a}$$

$\bar{D}(\bar{r},t)$: electric flux density distribution, coulombs per square meter

$$= \bar{D}(\bar{r},\omega)\, e^{j\omega t} \tag{2.25.1b}$$

$\bar{H}(\bar{r},t)$: magnetic field distribution, in amperes per meter

$$= \bar{H}(\bar{r},\omega)\, e^{j\omega t} \tag{2.25.2a}$$

$\bar{B}(\bar{r},t)$: magnetic flux density distribution, in webers per square meter

$$= \bar{B}(\bar{r},\omega)\, e^{j\omega t} \tag{2.25.2b}$$

In Tables 2.7 and 2.8, the basic electromagnetic field equations are given by Faraday's law and Ampere's law. These two curl equations, in fact, form an independent set of coupled relationships between time-harmonic varying electric field and magnetic field quantities. The additional field relationships stated by Gauss's law, expressions (2.25.6) and (2.25.7), do not form an independent set of equations. They can be directly deduced from the two Maxwell's curl equations as discussed in Section 2.5. In fact, the expressions in Tables 2.7 and 2.8 with the magnetic current and magnetic charge sources form a dual set of Maxwell's equations compared to the expressions in Tables 2.5 and 2.6 with the electric current and electric charge sources.

2.25.1 ELECTROMAGNETIC BOUNDARY CONDITIONS

The various frequency-domain electromagnetic field equations obtained in Section 2.25 are applicable for the case of a large three-dimensional region that is linear, homogeneous, and isotropic medium. In many practical situations, the three-dimensional region may consist of two or more media having different media parameters separated by boundary layers. Maxwell's equations stated in Tables 2.7 and 2.8, can be separately applied for each medium to obtain a solution for the time-harmonic electric and magnetic field quantities. For the field solutions to be unique, the field solutions should satisfy additional conditions at the boundary layer.

Table 2.7
Frequency-domain Maxwell's equations in differential form – Magnetic current sources

---

Faraday's Law:

$$\nabla \times \bar{E}(\bar{r},\omega) = -\bar{M}_v(\bar{r},\omega) - j\omega\bar{B}(\bar{r},\omega) \quad \text{lossless case} \quad (2.25.3)$$

$$\nabla \times \bar{E}(\bar{r},\omega) = -\bar{M}_g(\bar{r},\omega) - j\omega\bar{B}(\bar{r},\omega) - \bar{M}_c(\bar{r},\omega) \quad \text{lossy case} \quad (2.25.4)$$

Ampere's Law:

$$\nabla \times \bar{H}(\bar{r},\omega) = j\omega\bar{D}(\bar{r},\omega) \quad (2.25.5)$$

Gauss's Law:

$$\nabla \bullet \bar{B}(\bar{r},\omega) = \rho_{vm}(\bar{r},\omega) \quad (2.25.6)$$

$$\nabla \bullet \bar{D}(\bar{r},\omega) = 0 \quad (2.25.7)$$

Constitutive Relations:

$$\bar{D}(\bar{r},\omega) = \varepsilon\bar{E}(\bar{r},\omega) \quad (2.25.8)$$

$$\bar{B}(\bar{r},\omega) = \mu\bar{H}(\bar{r},\omega) \quad (2.25.9)$$

Generalized Ohm's Law:

$$\bar{M}_c(\bar{r},\omega) = \sigma_m\bar{H}(\bar{r},\omega) \quad (2.25.10)$$

Continuity Equation:

$$\nabla \bullet \bar{M}_v(\bar{r},\omega) = -j\omega\rho_{vm}(\bar{r},\omega) \quad (2.25.11)$$

Table 2.8
Frequency-domain Maxwell's equations in integral form –
Magnetic current sources

Faraday's Law:

$$\oint_C \bar{E}(\bar{r},\omega) \bullet d\bar{L}(\bar{r}) = - \iint_S \bar{M}_V(\bar{r},\omega) \bullet d\bar{s}(\bar{r}) - j\omega \iint_S \bar{B}(\bar{r},\omega) \bullet d\bar{s}(\bar{r}) \qquad \text{lossless case} \qquad (2.25.12)$$

$$\oint_C \bar{E}(\bar{r},\omega) \bullet d\bar{L}(\bar{r}) = - \iint_S \bar{M}_g(\bar{r},\omega) \bullet d\bar{s}(\bar{r}) - j\omega \iint_S \bar{B}(\bar{r},\omega) \bullet d\bar{s}(\bar{r}) - \iint_S \bar{M}_c(\bar{r},\omega) \bullet d\bar{s}(\bar{r}) \qquad \text{lossy case} \qquad (2.25.13)$$

Ampere's Law:

$$\oint_C \bar{H}(\bar{r},\omega) \bullet d\bar{L}(\bar{r}) = j\omega \iint_S \bar{D}(\bar{r},\omega) \bullet d\bar{s}(\bar{r}) \qquad (2.25.14)$$

Gauss's Law:

$$\oiint_S \bar{B}(\bar{r},\omega) \bullet d\bar{s} = \iiint_V \rho_{vm}(\bar{r},\omega)\, dv \qquad (2.25.15)$$

$$\oiint_S \bar{D}(\bar{r},\omega) \bullet d\bar{s} = 0 \qquad (2.25.16)$$

Constitutive Relations:

$$\bar{D}(\bar{r},\omega) = \varepsilon \bar{E}(\bar{r},\omega) \qquad (2.25.17)$$

$$\bar{B}(\bar{r},\omega) = \mu \bar{H}(\bar{r},\omega) \qquad (2.25.18)$$

Generalized Ohm's Law:

$$\bar{M}_c(\bar{r},\omega) = \sigma_m \bar{H}(\bar{r},\omega) \qquad (2.25.19)$$

Continuity Equation:

$$\oiint_S \bar{M}_V(\bar{r},\omega) \bullet d\bar{s} = - j\omega \iiint_V \rho_{vm}(\bar{r},\omega)\, dv \qquad (2.25.20)$$

In Sections 2.6 and 2.16, considering the electric current and charge sources, the boundary conditions have been obtained at a boundary layer when both time-varying electric and magnetic field distributions are present. In fact, similar types of electromagnetic boundary conditions at the boundary layer can also be deduced for the electric and magnetic field distributions when the magnetic current and magnetic charge types of sources are present.

As discussed earlier in Sections 2.6 and 2.16, to study behavior of the electric and magnetic fields at a boundary layer, the time-harmonic electric and magnetic fields at any point on one side (medium 1) of the boundary layer separating two media, 1 and 2, are resolved into two sets of orthogonal components – tangential components of the electric and magnetic fields parallel to the boundary surface, and normal components of the electric and magnetic fields perpendicular to the boundary surface. Similarly, the time-harmonic electric and magnetic fields at any point on the other side (medium 2) of the boundary layer are also resolved into two other sets of orthogonal components consisting of tangential components and normal components of the electric and magnetic fields parallel and perpendicular to the boundary surface.

As discussed in the earlier sections, depending upon the permeability, permittivity, and conductivity characteristics of the two media, it can be shown that the two tangential components of the time-harmonic fields are interrelated. Similarly, the two normal components of the time-harmonic fields are also interrelated. These relationships are presented in the following.

Referring to the Figures 2.10 and 2.11,

$\bar{E}_1(\bar{r},\omega)$: electric field distribution in medium 1;
$\bar{E}_2(\bar{r},\omega)$: electric field distribution in medium 2;
$\bar{D}_1(\bar{r},\omega)$: electric flux density distribution in medium 1;
$\bar{D}_2(\bar{r},\omega)$: electric flux density distribution in medium 2;

$\bar{H}_1(\bar{r},\omega)$: magnetic field distribution in medium 1;
$\bar{H}_2(\bar{r},\omega)$: magnetic field distribution in medium 2;
$\bar{B}_1(\bar{r},\omega)$: magnetic flux density distribution in medium 1;
$\bar{B}_2(\bar{r},\omega)$: magnetic flux density distribution in medium 2;

$\hat{t}$: tangential unit vector along the boundary surface;
$\hat{n}$: unit vector normal to the boundary surface;

and

$\bar{M}_S(\bar{r},\omega)$: magnetic current density at the boundary surface S;
ρ_{sm} : magnetic charge density at the boundary surface S.

Let us consider behavior of the tangential components of the time-harmonic electric and magnetic fields near the boundary surface separating the two media. Using expression (2.25.12), the boundary condition states that the difference between two tangential components of the time-harmonic electric fields should be identically equal to the surface magnetic current density distribution at the boundary surface separating the two media having different medium characteristics:

$$\hat{n} \times (\bar{E}_2 - \bar{E}_1) = - \bar{M}_s \tag{2.25.21}$$

If the medium 1 is perfectly conducting, the net electric field in the medium 1 is always zero, and thus the tangential component of electric field at the boundary surface in the medium 2 is equal to the surface magnetic current density distribution. Further, using expression (2.25.14), the two tangential components of time-harmonic magnetic fields should be continuous at the boundary surface separating the two media having different medium characteristics:

$$\hat{n} \times (\bar{H}_2 - \bar{H}_1) = 0 \tag{2.25.22}$$

Similarly, let us consider the behavior of the normal components of time-harmonic electric and magnetic fields near the boundary surface separating media 1 and 2. Referring to Figures 2.12 and 2.13 and using expression (2.25.15), the boundary relationship states that the difference between two normal components of time-harmonic magnetic flux density should be equal to the magnetic surface charge density:

$$\hat{n} \bullet (\bar{B}_2 - \bar{B}_1) = \rho_{sm} \tag{2.25.23}$$

ρ_{sm} : surface magnetic charge density

Further, using expression (2.25.16), the boundary relationship states that the two normal components of time-harmonic electric flux density should be continuous across the boundary surface:

$$\hat{n} \bullet (\bar{D}_2 - \bar{D}_1) = 0 \tag{2.25.24}$$

2.26 HELMHOLTZ EQUATIONS WITH MAGNETIC SOURCES

Referring to the procedure in Section 2.18, a direct analytical solution for the time-harmonic electric and magnetic fields with magnetic current and charge sources in a large three-dimensional medium is discussed in the following. The medium is assumed to have linear, homogeneous, and isotropic properties with known time-harmonic magnetic current and charge source distributions. The sources are assumed to be confined to a small region and produce time-harmonic electric and magnetic fields both inside and outside the source region. The frequency domain Maxwell's equations, Tables 2.7 and 2.8, are now utilized for determining the electric and magnetic fields. Again, it is noted that the Maxwell's equations are a set of coupled partial differential equations whose solutions are the frequency-dependent electric and magnetic fields.

The analysis is similar in procedure to the derivation of Helmholtz equations studied in Section 2.18. Referring to Figure 2.36, let us consider a large three-dimensional unbounded region that is a linear, homogeneous, and isotropic magnetic *lossy* medium. The various sources and the corresponding fields have a time-harmonic dependence according to *exp(jωt)*. The time harmonic sources are confined to a small source region, given by

$\bar{M}_g(\bar{r},\omega)$: primary source current density, in volts per square meter;
$\bar{M}_c(\bar{r},\omega)$: magnetic conduction current density, in volts per square meter.

The sources produce the electric and magnetic field distributions both inside and outside the source region. At any field point in the medium, the time-harmonic electromagnetic field quantities are given by

$\bar{E}(\bar{r},\omega)$: electric field distribution, in volts per meter;
$\bar{D}(\bar{r},\omega)$: electric flux density distribution, in coulombs per square meter;
$\bar{H}(\bar{r},\omega)$: magnetic field distribution, in amperes per meter;
$\bar{B}(\bar{r},\omega)$: magnetic flux density distribution, in webers per square meter.

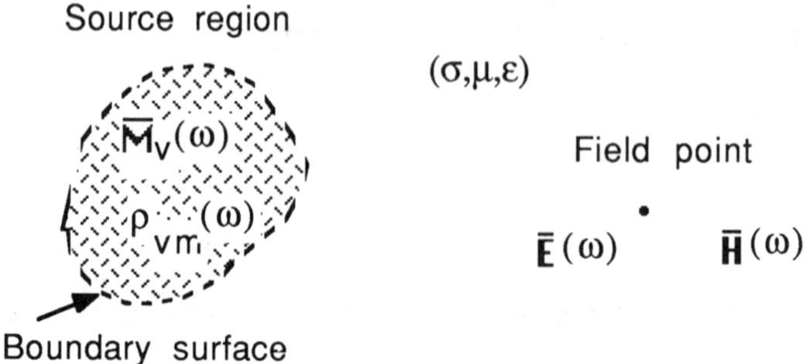

Figure 2.36 Three-dimensional unbounded medium with magnetic sources.

Referring to Table 2.7, in the source-free region, the electric and magnetic fields satisfy the following frequency-dependent Maxwell's equations in differential form:

$$\nabla \times \bar{E}(\bar{r},\omega) = -j\omega\bar{B}(\bar{r},\omega) - \bar{M}_c(\bar{r},\omega) \tag{2.26.1a}$$

$$\nabla \times \bar{H}(\bar{r},\omega) = j\omega\bar{D}(\bar{r},\omega) \qquad \bar{r} \text{ outside source region} \tag{2.26.1b}$$

In expression (2.26.1a), the effect of magnetic conductivity of the lossy medium is accounted for in terms of the magnetic conduction current density. For all observation points within the source region itself, referring to Table 2.7, the electric and magnetic fields satisfy the following Maxwell's equations in differential form:

$$\nabla \times \bar{E}(\bar{r},\omega) = -j\omega\bar{B}(\bar{r},\omega) - \bar{M}_g(\bar{r},\omega) - \bar{M}_c(\bar{r},\omega) \tag{2.26.2a}$$

$$\nabla \times \bar{H}(\bar{r},\omega) = j\omega\bar{D}(\bar{r},\omega) \qquad \bar{r} \text{ inside source region} \tag{2.26.2b}$$

The set of time-harmonic equations defined for the source-free region by expressions (2.26.1a,b) and for the source region by expressions (2.26.2a,b) can be solved separately. The electric and magnetic field distributions so obtained for the two regions should satisfy appropriate boundary conditions for all points on the boundary surface separating the two regions. Particularly, at the boundary surface separating the source region and the source-free region, it is always convenient to use the time-harmonic integral form of Maxwell's equations.

The frequency dependent Maxwell's equations for the source-free region are analyzed in the following. In equations (2.26.1a,b), there are four unknown field quantities, given by the electric field and electric flux density, the magnetic field and magnetic flux density. But there are only two independent coupled curl equations (2.26.1a,b) for the four unknown field quantities. Two other additional relationships are obtained based on the constitutive relationships. The conduction magnetic current density, which is also an unknown quantity, can be replaced in terms of the unknown magnetic field distribution based on the generalized Ohm's law stated in expression (2.25.10).

Hence, a three-dimensional unbounded lossy region assumed to be linear, homogeneous, and isotropic, the various relationships are given by

$$\bar{D}(\bar{r},\omega) = \varepsilon\bar{E}(\bar{r},\omega) \tag{2.26.3a}$$

$$\bar{B}(\bar{r},\omega) = \mu\bar{H}(\bar{r},\omega) \tag{2.26.3b}$$

$$\bar{M}_c(\bar{r},\omega) = \sigma_m\bar{H}(\bar{r},\omega) \tag{2.26.4}$$

ε : permittivity of the medium, in farads per meter

μ : permeability of the medium, in henrys per meter

σ_m : magnetic conductivity of the medium, in ohms per meter

Constitutive relationships (2.26.3a) and (2.26.3b) and generalized Ohm's law (2.26.4) are now substituted into the frequency-dependent Maxwell's equations (2.26.1a,b) for the source-free region. Hence, the electric and magnetic fields in the source-free region satisfy the following frequency dependent Maxwell's equations:

$$\nabla \times \bar{E}(\bar{r},\omega) = - j\omega\mu\bar{H}(\bar{r},\omega) - \sigma_m\bar{H}(\bar{r},\omega) \qquad (2.26.5a)$$

$$\nabla \times \bar{H}(\bar{r},\omega) = j\omega\varepsilon\bar{E}(\bar{r},\omega) \qquad (2.26.5b)$$

Expression (2.26.5a) can be rewritten as

$$\nabla \times \bar{E}(\bar{r},\omega) = - j\omega\mu'\bar{H}(\bar{r},\omega) \qquad (2.26.5c)$$

where the complex effective permeability of the medium is defined as

$$\mu' = \mu\left[1 - j\,\frac{\sigma_m}{\omega\mu}\right] \qquad (2.26.5d)$$

The electric and the magnetic field divergence relationships can be easily derived by utilizing Maxwell's equations (2.26.5b) and (2.26.5c). On taking divergence on both sides of expressions (2.26.5b) and (2.26.5c) and noting that the divergence of curl of any vector quantity, Appendix A, is always equal to zero, the following Gauss's law relationships are obtained

$$\nabla \bullet \bar{E}(\bar{r},\omega) = 0 \qquad (2.26.5e)$$

$$\nabla \bullet \bar{H}(\bar{r},\omega) = 0 \qquad (2.26.5f)$$

The two curl equations (2.26.5b) and (2.26.5c), are the coupled partial differential equations with two unknowns. They are to be solved for the electric and magnetic field quantities subject to the known boundary conditions. For the electric and magnetic fields, the following Helmholtz equations are obtained:

$$\nabla^2\bar{E}(\bar{r},\omega) = \gamma^2\bar{E}(\bar{r},\omega) \qquad (2.26.6a)$$

$$\nabla^2 \bar{H}(\bar{r},\omega) = \gamma^2 \bar{H}(\bar{r},\omega) \quad (2.26.6b)$$

$$\gamma^2 = (j\omega\mu')(j\omega\varepsilon) \quad (2.26.6c)$$

Suppose the frequency-dependent electric and magnetic fields are expressed in terms of their rectangular components as

$$\bar{E}(\bar{r},\omega) = E_x(\bar{r},\omega)\hat{x} + E_y(\bar{r},\omega)\hat{y} + E_z(\bar{r},\omega)\hat{z} \quad (2.26.7a)$$

$$\bar{H}(\bar{r},\omega) = H_x(\bar{r},\omega)\hat{x} + H_y(\bar{r},\omega)\hat{y} + H_z(\bar{r},\omega)\hat{z} \quad (2.26.7b)$$

These two field representations can now be substituted into vector equations (2.26.6a) and (2.26.6b) to reduce them to six scalar Helmholtz equations given by

$$\nabla^2 E_i(\bar{r},\omega) = \gamma^2 E_i(\bar{r},\omega) \quad (2.26.8a)$$

$$i = x,\ y,\ z$$

$$\nabla^2 H_i(\bar{r},\omega) = \gamma^2 H_i(\bar{r},\omega) \quad (2.26.8b)$$

A brief solution procedure for the scalar Helmholtz equations, expressions (2.26.8a) and (2.26.8b), is discussed in Section 2.19 based on the method of separation of variables. The total solution in terms of the scalar plane wave field reduces to the following representations for the electric and magnetic fields.

Electric field positive traveling wave

$$E_i(\bar{r},\omega) = E_0\, e^{-\bar{\gamma} \bullet \bar{r}} \quad (2.26.9a)$$

Magnetic field positive traveling wave

$$H_i(\bar{r},\omega) = H_0\, e^{-\bar{\gamma} \bullet \bar{r}} \quad (2.26.9b)$$

It should be noted again that expressions (2.26.9a) and (2.26.9b) represent positive and negative traveling plane waves in the frequency domain. The direction of the propagation of the forward plane waves is given by the direction of complex propagation constant traveling away from the sources. Based on representations (2.26.7a) and (2.26.7b), the vector electric and magnetic fields can be constructed.

2.27 ELECTROMAGNETIC POTENTIALS WITH MAGNETIC SOURCES

The time-harmonic magnetic current and charge sources generate correspondingly time-harmonic electromagnetic fields. In the previous section, a direct analytical solution for the time-harmonic electric and magnetic fields in a large three-dimensional medium is discussed based on the vector Helmholtz equation. Assuming the magnetic current and charge sources are confined to a small source region, the electric and magnetic plane wave solutions are obtained for a source-free medium having homogeneous and isotropic properties. In a given boundary value problem, plane waves are not always generated. In general, the electric and magnetic fields generated in a given medium are quite complicated and depend upon the distribution and orientation of electromagnetic sources. Similar to the discussion in Section 2.21, to get insight into the electromagnetic scattering and radiation type boundary value problems, simple point-type elementary magnetic current sources can be considered. In Section 2.24, detailed mathematical models were considered for the magnetic current and magnetic charge sources based on their density distributions. By using the source models, any complicated distribution of the magnetic currents and charges can be simulated by invoking the linearity and superposition principle. If the electric and magnetic fields generated by the point source are known, by applying superposition, the complete electric and magnetic fields can be constructed for the general source distribution.

Instead of the direct approach, using dual representation, an *alternative formulation* is discussed in the following based on the potential concept introduced rigorously in Section 2.21. The electric and magnetic fields can be expressed in terms of an arbitrary dual electric vector potential and an arbitrary magnetic scalar potential. Maxwell's equations are simplified using the auxiliary potentials to obtain Helmholtz-type differential equations for the two arbitrary potential functions. The solutions for the potential functions are analyzed for a given point source by solving the Helmholtz differential equations. Once the solution is known for the point-type source, the superposition principle can be invoked to obtain general potential integrals in terms of the known source distribution. Further, the electric and magnetic fields can be calculated using the known electric vector potential and the magnetic scalar potential. As will be noted later on, the potential representations of the partial electromagnetic fields with the magnetic-type sources exhibit a dual property as compared to the corresponding potential representations of the partial electromagnetic fields with the electric-type sources.

A complete summary of *Maxwell's equations in the frequency domain* is given in Tables 2.7 and 2.8. The Table 2.7 gives all the relevant electromagnetic field equations in differential form with magnetic current sources. Referring to Figure 2.37, let us consider a large three-dimensional unbounded region that is linear, homogeneous, and isotropic lossless medium. It is assumed that the various sources and the corresponding fields have a time-harmonic dependence according to $exp(j\omega t)$. The time-harmonic sources are confined to a small source region, given by

$\rho_{vm}(\bar{r},\omega)$: source magnetic charge density
$\bar{M}_v(\bar{r},\omega)$: source magnetic current density

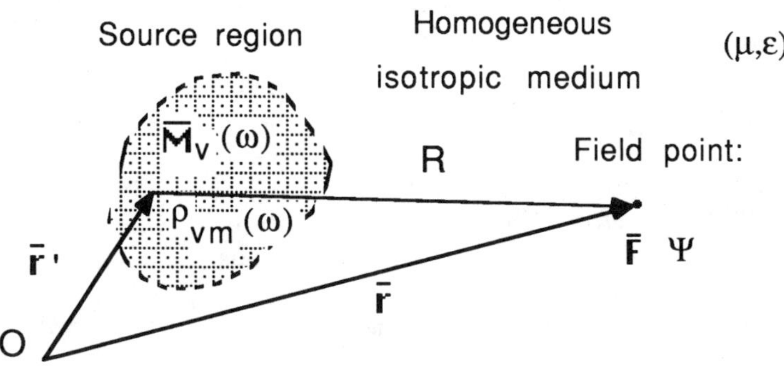

Figure 2.37 Potentials with magnetic sources.

The sources produce the electric and magnetic field distributions both inside and outside the source regions. At any field point in the medium, the time-harmonic electromagnetic field quantities are given by

$\bar{E}(\bar{r},\omega)$: electric field distribution, in volts per meter;
$\bar{D}(\bar{r},\omega)$: electric flux density distribution, in coulombs per square meter;
$\bar{H}(\bar{r},\omega)$: magnetic field distribution, in amperes per meter;
$\bar{B}(\bar{r},\omega)$: magnetic flux density distribution, in webers per square meter.

Referring to Table 2.7, in the source-free region, the electric and magnetic fields satisfy the following frequency-dependent Maxwell's equations in differential form:

$$\nabla \times \bar{E}(\bar{r},\omega) = -j\omega\bar{B}(\bar{r},\omega) \quad (2.27.1a)$$

$$\nabla \times \bar{H}(\bar{r},\omega) = j\omega\bar{D}(\bar{r},\omega) \quad (2.27.1b)$$

$$\nabla \cdot \bar{D}(\bar{r},\omega) = 0 \quad (2.27.1c)$$

$$\nabla \cdot \bar{B}(\bar{r},\omega) = 0 \quad (2.27.1d)$$

$\bar{r}$ outside source region

For all observation points within the source region itself, referring to Table 2.7, the electric and magnetic fields also satisfy the following Maxwell's equations:

$$\nabla \times \bar{E}(\bar{r},\omega) = - j\omega\bar{B}(\bar{r},\omega) - \bar{M}_V(\bar{r},\omega) \tag{2.27.2a}$$

$$\nabla \times \bar{H}(\bar{r},\omega) = j\omega\bar{D}(\bar{r},\omega) \tag{2.27.2b}$$

$$\nabla \bullet \bar{D}(\bar{r},\omega) = 0 \tag{2.27.2c}$$

$$\nabla \bullet \bar{B}(\bar{r},\omega) = \rho_{vm}(\bar{r},\omega) \tag{2.27.2d}$$

$\bar{r}$ inside source region

For the case of a three-dimensional unbounded lossless region assumed to be linear, homogeneous, and isotropic, the constitutive relations are given by

$$\bar{D}(\bar{r},\omega) = \varepsilon\bar{E}(\bar{r},\omega) \tag{2.27.3a}$$

$$\bar{B}(\bar{r},\omega) = \mu\bar{H}(\bar{r},\omega) \tag{2.27.3b}$$

ε : constant permittivity of the medium, in farads per meter;
μ : constant permeability of the medium, in henrys per meter.

Constitutive relationships (2.27.3a) and (2.27.3b) are now substituted into the frequency dependent Maxwell's equations, (2.27.2a) to (2.27.2d), which are valid for the source region

$$\nabla \times \bar{E}(\bar{r},\omega) = - j\omega\mu\bar{H}(\bar{r},\omega) - \bar{M}_V(\bar{r},\omega) \tag{2.27.4a}$$

$$\nabla \times \bar{H}(\bar{r},\omega) = j\omega\varepsilon\bar{E}(\bar{r},\omega) \tag{2.27.4b}$$

$$\nabla \bullet \varepsilon\bar{E}(\bar{r},\omega) = 0 \tag{2.27.4c}$$

$$\nabla \bullet \mu\bar{H}(\bar{r},\omega) = \rho_{vm}(\bar{r},\omega) \tag{2.27.4d}$$

$\bar{r}$ inside source region

To solve this set of coupled equations, the electric and the magnetic fields are first expressed in terms of two dual-potential functions given by an auxiliary electric vector potential and an auxiliary magnetic scalar potential. In fact, these two potential functions are not independent, but are related through an electromagnetic *gauge condition*. The analysis procedure discussed later is identical to the steps presented in Section 2.21. Gauss's law states that the divergence of a time-harmonic electric flux density is always equal to zero. According to expression (2.27.4c), let us choose an arbitrary representation for the electric flux density in the following form:

$$\bar{D}(\bar{r},\omega) = - \nabla \times \bar{F}(\bar{r},\omega) \tag{2.27.5a}$$

$\bar{F}(\bar{r},\omega)$: electric vector potential function at the field point

$$= F_x(\bar{r},\omega)\hat{x} + F_y(\bar{r},\omega)\hat{y} + F_z(\bar{r},\omega)\hat{z} \tag{2.27.5b}$$

and constitutive relationship (2.27.3a) yields

$$\bar{E}(\bar{r},\omega) = - \frac{1}{\varepsilon} \nabla \times \bar{F}(\bar{r},\omega) \tag{2.27.5c}$$

The representation of (2.27.5a) satisfies rigorously the Gauss's law relationship that the divergence of the curl of electric vector potential is equal to zero, Appendix A. Expression (2.27.5a) is now substituted into Ampere's law, expression (2.27.4b), to yield the following vector relationship:

$$\nabla \times \bar{H}(\bar{r},\omega) = - j\omega \nabla \times \bar{F}(\bar{r},\omega) \tag{2.27.6}$$

$$\nabla \times \left[\bar{H}(\bar{r},\omega) + j\omega\bar{F}(\bar{r},\omega)\right] = 0 \tag{2.27.7}$$

Referring to the Appendix A, the curl of the gradient of any scalar point function is always zero. Hence

$$\nabla \times \left[\nabla \Psi(\bar{r},\omega)\right] = 0 \tag{2.27.8}$$

$\Psi(\bar{r},\omega)$: magnetic scalar potential function at the field point

This identity is valid for any arbitrary scalar potential function. Hence, vector expression (2.27.7) is equated to another vector expression (2.27.8) with an arbitrary negative sign to yield the following relationship:

$$\nabla \times \left[\bar{H}(\bar{r},\omega) + j\omega\bar{F}(\bar{r},\omega)\right] = \nabla \times \left[- \nabla \Psi(\bar{r},\omega)\right] \tag{2.27.9a}$$

Based on this relationship, if the arbitrary electric vector potential and the arbitrary magnetic scalar potential are known, the magnetic field distribution can be obtained as

$$\bar{H}(\bar{r},\omega) = - j\omega\bar{F}(\bar{r},\omega) - \nabla\Psi(\bar{r},\omega) \tag{2.27.9b}$$

This completes the representation of electric and magnetic fields in terms of the arbitrary dual-potential functions. The electric vector potential and the magnetic scalar potential are still unknown. To find these two potentials, expressions (2.27.5c) and (2.27.9b) are substituted into Faraday's law (2.27.4a) to obtain

$$\nabla \times \frac{1}{\varepsilon} \nabla \times \bar{F}(\bar{r},\omega) = j\omega\mu\left[- j\omega\bar{F}(\bar{r},\omega) - \nabla\Psi(\bar{r},\omega)\right] + \bar{M}_V(\bar{r},\omega) \tag{2.27.10a}$$

$$\nabla \times \nabla \times \bar{F}(\bar{r},\omega) = \omega^2\mu\varepsilon\bar{F}(\bar{r},\omega) - j\omega\mu\varepsilon\nabla\Psi(\bar{r},\omega) + \varepsilon\bar{M}_V(\bar{r},\omega) \tag{2.27.10b}$$

Using the rectangular coordinate system, the lefthand-side term of expression (2.27.10b) can be simplified. According to the vector identity, Appendix A,

$$\nabla \times [\nabla \times \bar{F}] = \nabla(\nabla \bullet \bar{F}) - \nabla^2\bar{F} \tag{2.27.10c}$$

Now, expression (2.27.10b) can be rewritten as

$$\nabla[\nabla \bullet \bar{F}(\bar{r},\omega)] - \nabla^2\bar{F}(\bar{r},\omega) =$$

$$\omega^2\mu\varepsilon\bar{F}(\bar{r},\omega) - j\omega\mu\varepsilon\nabla\Psi(\bar{r},\omega) + \varepsilon\bar{M}_V(\bar{r},\omega) \tag{2.27.10d}$$

The electric vector potential and the magnetic scalar potential are arbitrary potential functions. Hence, it is possible to seek *a solution* for the electric vector potential so that

$$\nabla[\nabla \bullet \bar{F}(\bar{r},\omega)] = - j\omega\mu\varepsilon\nabla\Psi(\bar{r},\omega) \tag{2.27.11}$$

Relationship (2.27.11) is generally referred to as the dual *Lorentz gauge condition.* By enforcing such a gauge condition the relationship (2.27.10d), the following vector *Helmholtz* partial differential equation is obtained for the electric vector potential:

$$\nabla^2\bar{F}(\bar{r},\omega) + \beta^2\bar{F}(\bar{r},\omega) = - \varepsilon\bar{M}_V(\bar{r},\omega) \tag{2.27.12}$$

$$\beta^2 = \omega^2\mu\varepsilon \tag{2.27.13a}$$

β : propagation constant of the medium, in radians per meter

$$= \omega(\mu\varepsilon)^{1/2} \tag{2.27.13b}$$

Again, we have a systematic procedure for analyzing the electromagnetic dual-field problem. The first step consists of solving the preceding Helmholtz's vector differential equation for the solution of the vector magnetic potential. In the second step, the electric field distribution is calculated by using expression (2.27.5c). In the third step, the magnetic field distribution can be calculated by using Faraday's law expression (2.27.4a) or expression (2.27.9b) which requires the calculation of the magnetic scalar potential. Referring back to the Lorentz gauge expression (2.27.11), the magnetic scalar potential is given by

$$\Psi(\bar{r},\omega) = \frac{j\omega}{\beta^2} \nabla \bullet \bar{F}(\bar{r},\omega) \tag{2.27.14}$$

As can seen in the vector Helmholtz equation, expression (2.27.12), the electric vector potential depends completely on a single driving-source term that is the magnetic current density distribution. In fact, a similar relationship can also be obtained for the magnetic scalar potential with a single driving-source term corresponding to the magnetic charge density distribution. To obtain an Helmholtz equation for the magnetic scalar potential, the divergence of the expression (2.27.9b) is taken

$$\nabla \bullet \bar{H}(\bar{r},\omega) = - j\omega \nabla \bullet \bar{F}(\bar{r},\omega) - \nabla \bullet \nabla \Psi(\bar{r},\omega) \tag{2.27.15a}$$

The lefthand-side term given by the divergence of the magnetic field is replaced in terms of the source magnetic charge density by using Gauss's law expression (2.27.4d). The righthand-side term given by the divergence of the electric vector potential is replaced using gauge condition (2.27.14). Hence, expression (2.27.15a) simplifies to the following Helmholtz scalar equation:

$$\nabla^2 \Psi(\bar{r},\omega) + \beta^2 \Psi(\bar{r},\omega) = - \frac{\rho_{vm}(\bar{r},\omega)}{\mu} \tag{2.27.15b}$$

The magnetic scalar potential can be obtained by solving this differential equation. It is interesting to note that the electric vector potential depends on the magnetic current density source distribution, and the magnetic scalar potential depends on the magnetic charge density source distribution. The two sources are not independent, but are related through the magnetic current-charge continuity equation. Similarly, the two auxiliary potentials are not independent, but are related through the dual Lorentz gauge condition. As discussed in earlier sections, the vector Helmholtz equation is difficult to solve. It can be reduced to the corresponding three scalar Helmholtz equations. Using the rectangular components defined in expression (2.27.5b), the vector Helmholtz equation (2.27.12) can be rewritten in terms of the following scalar form:

$$\nabla^2 F_x(\bar{r},\omega) + \beta^2 F_x(\bar{r},\omega) = -\varepsilon M_x(\bar{r},\omega) \quad (2.27.16a)$$

$$\nabla^2 F_y(\bar{r},\omega) + \beta^2 F_y(\bar{r},\omega) = -\varepsilon M_y(\bar{r},\omega) \quad (2.27.16b)$$

$$\nabla^2 F_z(\bar{r},\omega) + \beta^2 F_z(\bar{r},\omega) = -\varepsilon M_z(\bar{r},\omega) \quad (2.27.16c)$$

It is interesting to note from these equations that the x-directed magnetic current density produces only the x component of the electric vector potential. Similarly, y- and z-directed magnetic current densities respectively produce y and z components of the electric vector potentials. Instead of the volume magnetic current density distributions, suppose either surface or line magnetic current distributions are present, then they are appropriately replaced in the righthand side of the preceding scalar differential equations. Hence, to obtain the complete solution for the electric vector potential, the above three scalar equations (2.27.16a–c) are to be solved.

2.28 SOLUTION FOR POTENTIALS WITH MAGNETIC SOURCES

Similar to the discussion in Section 2.22, the Helmholtz differential equations derived in that section can be rigorously solved for the various sources of magnetic current and charge distributions. To obtain the complete solution for the electric vector potential with an arbitrary source distribution, first the solution for the Helmholtz scalar differential equation is obtained for an isolated point or elemental magnetic current source. Since the scalar equations are similar in mathematical form, various components of the potentials can be conveniently specified by inspection. Once the solution is known for a point-type elemental source, by invoking the concept of linearity and the superposition principle, the complete solution for the two dual potentials are written as the superposition integrals in terms of the corresponding source distributions. By knowing the distribution of the two dual auxiliary potentials, the electric and magnetic field distributions in the given medium can be calculated based on the analysis carried out in the previous section.

Let us consider an isolated point-type magnetic current element located in an isotropic and homogeneous free-space medium. The source current element has a time-harmonic variation of $exp(j\omega t)$ and is located at the origin of a three-dimensional coordinate system. Referring to Figure 2.31, let

I_{zm} : magnetic current elemental source, in volts located at the source point O(0,0,0);

Δz : incremental length of the current element, in meters;

$\hat{z}$: unit vector in the direction of the current element;

$\bar{r}$: position vector corresponding to a general field point $P(x, y, z)$

$$= r\hat{r}$$

r: radial distance between the source and field points, in meters

$$= [x^2+y^2+z^2]^{1/2} \quad (2.28.1)$$

ω: angular frequency of excitation, in radians per second.

Referring to the mathematical procedure discussed in Section 2.22, the solution for the z component of the magnetic vector potential due to an elemental magnetic current located at the origin has the following form:

$$F_z(r,\omega) = \frac{\varepsilon I_{zm}\Delta z}{4\pi r} e^{-j\beta r} \quad (2.28.2)$$

Suppose the magnetic current element is located at an arbitrary source point (x', y', z') instead at the origin of the coordinate system. Then the result obtained in expression (2.28.2) for the z component of magnetic vector potential is to be suitably modified to take into account the actual distance between the source point and the field observation point. For such a case,

$\bar{r}'$: position vector corresponding to the general source point

$$= r'\hat{r}'$$

$$= x'\hat{x} + y'\hat{y} + z'\hat{z} \quad (2.28.3a)$$

R: actual distance between source and field points, in meters

$$= |\bar{r} - \bar{r}'| \quad (2.28.3b)$$

The z component of the magnetic vector potential is given by

$$F_z(\bar{r}) = \frac{\varepsilon I_{zm}\Delta z}{4\pi R} e^{-j\beta R} \quad (2.28.4)$$

In the following, a discussion is presented for calculating the electric vector potential due to the presence of a continuous time-harmonic volume magnetic current density distribution. Referring to Figure 2.38, a volume magnetic current source distribution is confined to an arbitrary volume V located in a three-dimensional isotropic and homogeneous medium. Let

Δv : incremental volume enclosing the source point (x', y', z')

$\bar{r}'$: position vector corresponding to the source point (x', y', z')

$$= x'\hat{x} + y'\hat{y} + z'\hat{z} \tag{2.28.5a}$$

$\bar{M}_V(\bar{r}')$: volume magnetic current density distribution in volume V, in volts per square meter

$\bar{r}$: position vector corresponding to the field point (x, y, z)

$$= x\hat{x} + y\hat{y} + z\hat{z} \tag{2.28.5b}$$

R : distance of the line joining the source point to the field point

$$= |\bar{r} - \bar{r}'| \tag{2.28.5c}$$

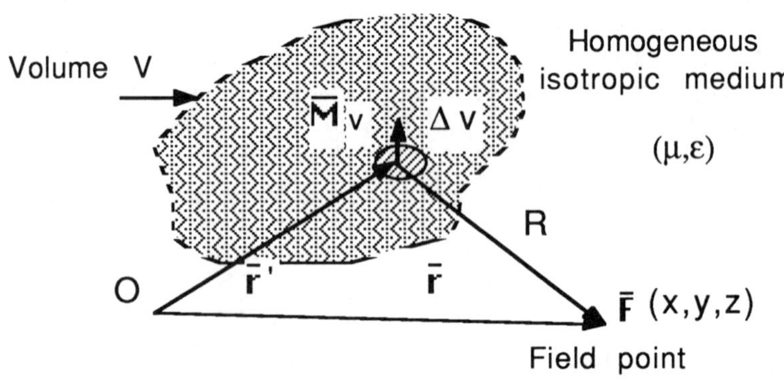

Figure 2.38 Electric vector potential due to volume magnetic current.

Following the procedure discussed earlier, first the incremental electric vector potential at the field point is calculated due to the incremental volume magnetic current at the source point. In the second step, based on the concept of linearity, the principle of superposition is applied for calculating the total electric vector potential at the field point. The differential electric vector potential at the field point is given by

$$d\bar{F}(\bar{r}) = \frac{\varepsilon}{4\pi}\left[\frac{\bar{M}_V dv(\bar{r}')}{R}\right] e^{-j\beta R} \tag{2.28.6a}$$

The total electric vector potential at the field point is given by the triple integral of the above expression over the complete source magnetic current distribution in the volume *V*:

$$\bar{F}(\bar{r}) = \frac{\varepsilon}{4\pi} \iiint_V \bar{M}_v(\bar{r}') \frac{e^{-j\beta R}}{R} dv(\bar{r}') \tag{2.28.6b}$$

Similarly, a general solution to the scalar Helmholtz equation (2.27.15b) for the electric scalar potential can also be written. The total magnetic scalar potential at the field point *P(x, y, z)* is given by the volume integral over the complete source magnetic charge density distribution in the volume *V*:

$$\Psi(\bar{r}) = \frac{1}{4\pi\mu} \iiint_V \rho_{vm}(\bar{r}') \frac{e^{-j\beta R}}{R} dv(\bar{r}') \tag{2.28.6c}$$

If other types of magnetic source distributions, such as the surface magnetic currents and charges and the line magnetic currents and charges, are present in the medium, then the preceding dual-potential integrals, (2.28.6b) and (2.28.6c), are to be correspondingly modified with respect to the surface and line integrals.

In the electromagnetic interaction studies dealing with material environment, both the electric current and the magnetic current sources are to be considered simultaneously. In such cases, superposition is invoked to combine the electromagnetic fields generated by the individual sources.

In the next chapters, two-dimensional electromagnetic scattering and interaction studies are introduced in the frequency domain based on the integral equation and the method of moments numerical technique. Even though the integral equations can be formulated based on number of different approaches using field equations, only the step-by-step formulation utilizing the electromagnetic potentials and boundary conditions, is discussed extensively.

Chapter 3

Two-Dimensional Perfectly Conducting Objects: TM Polarization

The study of two-dimensional conducting objects plays an important role in understanding various physical phenomena associated with electromagnetic scattering, penetration, and interaction. As discussed in the Chapter 2, general electromagnetic fields are vector quantities. In a given medium, they are functions of three spatial coordinate variables and the frequency of excitation of the electric current and charge sources. Reducing the dimensionality of a boundary value problem from three to two results in the dependence of various electromagnetic field quantities on only two coordinate variables. This specific situation generally simplifies analysis of electric and magnetic field quantities considerably. In fact, for certain specialized cases, vector nature of electromagnetic field problem can be resolved into two separate, uncoupled scalar boundary value problems by choosing appropriate polarization of external incident excitation. With reduction in dimensionality, the interaction physics of two-dimensional electric and magnetic fields are easy to understand and visualize and in many instances, lead to clear insight into an obvious extension to the three-dimensional boundary value problem.

In this chapter, analysis of electromagnetic scattering and interaction by an arbitrary shaped, two-dimensional conducting object is studied in a systematic manner starting from the classical Maxwell's equations. Figure 3.1 shows a canonical shape and configuration of a two-dimensional object. The geometry of a perfectly conducting object is assumed to be uniform and infinite in extent along a specific coordinate variable, such as, the z coordinate axis. Further, as shown in the Figure 3.1, a cross-sectional area of the scattering geometry is uniform along its axis, which is assumed to coincide with the z axis of coordinate system. In general, such a two-dimensional object that is uniform and infinite in length along its axis is difficult to simulate. But, this special object forms quite a valuable boundary value geometry for understanding various aspects of the electromagnetic analytical formulation and its subsequent solution based upon numerical techniques. Various physics of electromagnetic interaction due to material of the geometry, shape, and configuration of the scatterer, such as straight edges and curved arbitrary edges, sharp wedge-type corners and smooth rounded corners, narrow slots and apertures, can be extracted conveniently from the two-dimensional study.

The analysis of a two-dimensional boundary value problem can be carried out either in terms of a rectangular or cylindrical coordinate system. It is assumed that the perfectly

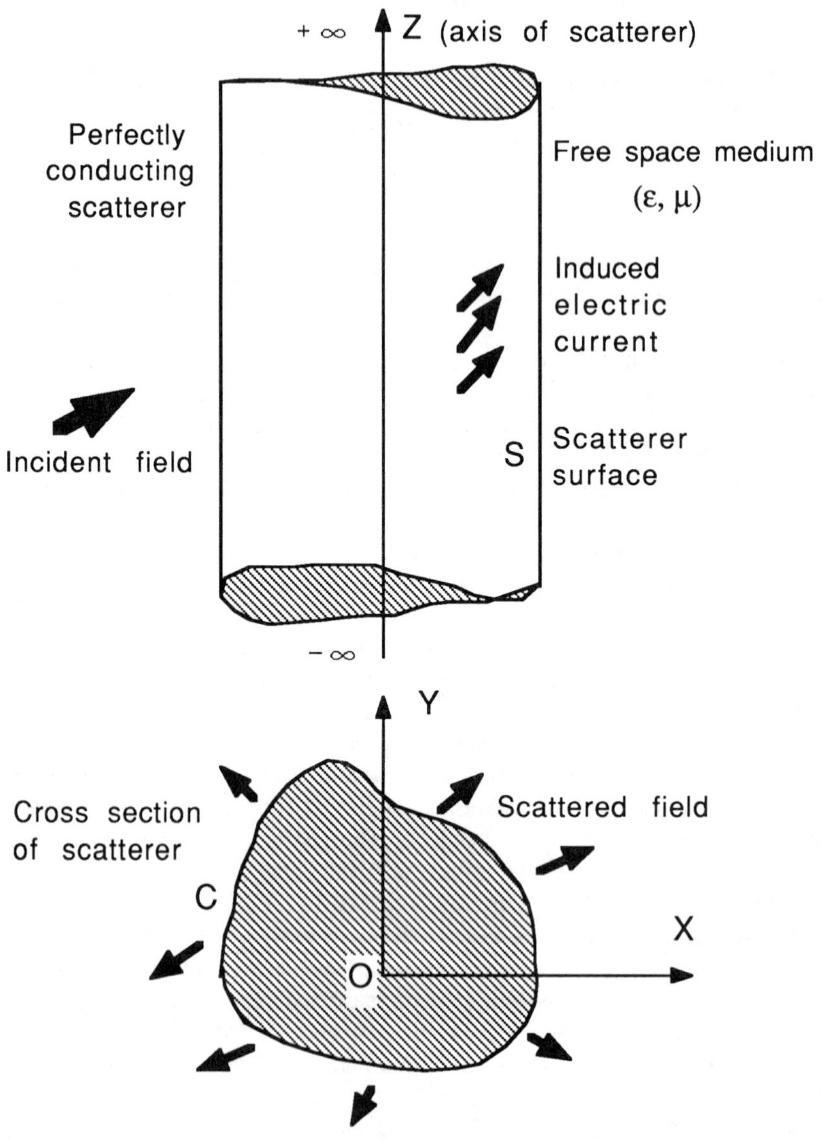

Figure 3.1 Two-dimensional perfectly conducting scatterer.

conducting object is located in a large free-space medium that is linear, homogeneous, and isotropic, having constant permittivity and permeability. The time-harmonic incident electric and magnetic excitation fields are simulated and propagated in the free-space medium by their corresponding time-harmonic primary electric current and charge sources, which are located at a large distance from the interacting object under study. For all practical modeling considerations, the incident excitation fields are taken to be those field distributions existing in the free-space medium in the absence of an interacting object. When the two-dimensional conducting object is introduced into the free-space medium, the incident fields interact to induce secondary electric currents and charges on the conducting object. These induced electric currents and charges produce and propagate scattered electric and magnetic fields in all directions. In the free-space medium, the total fields now consist of the sum of the incident field and the scattered field. Further, complete electric and magnetic field distributions in the free-space medium are such that they satisfy familiar electromagnetic boundary conditions. According to the electromagnetic boundary conditions, the total tangential electric field is zero on the surface of perfectly conducting object. Similarly, the total tangential magnetic field on the surface of perfectly conducting object gives the induced electric current distribution. Since the object under study is perfectly conducting, skin depth can be taken as zero, and all induced electric currents and charges can be assumed to reside just on the surface of conducting object.

In most of the electromagnetic scattering and interaction studies, the induced electric currents on the surface of conducting object is an unknown quantity. Hence, for the unknown induced electric current distribution, a boundary value problem is set up either in terms of a differential equation or an integral equation or an integro-differential equation subject to the appropriate boundary conditions. Generally, the boundary value equations can be solved directly using a straightforward analytical method based on an appropriate eigenfunction expansion for canonical type of scatterer, such as a circular conducting cylinder. But, for an arbitrary shaped two-dimensional conducting object, the boundary value equations can be conveniently solved using an appropriate numerical technique suitable for the differential equations or the integral equations. Once the induced electric currents are known, then the Maxwell's equations can be directly utilized to obtain the scattered electric and magnetic fields in the region of interest. The relevant electromagnetic scattering properties, such as near total field distribution, far scattered field distribution, scattered power distribution, and radar cross section, can be obtained easily.

3.1 GENERAL FIELD EQUATIONS

A complete summary of *Maxwell's equations in the frequency domain* is given in Tables 2.5 and 2.6 with electric current and charge sources. Table 2.5 gives all relevant electromagnetic field equations in differential form, and similarly, Table 2.6 gives all relevant electromagnetic field equations in integral form. Referring to Figure 3.2, let us consider a large unbounded region which is a linear, homogeneous, and isotropic lossless medium. Let the electric current and charge sources be confined to a small source region, given by

$\rho_V(\bar{r},\omega)$: volume electric charge density, in coulombs per cubic meter;
$\bar{J}_V(\bar{r},\omega)$: volume electric current density, in amperes per square meter.

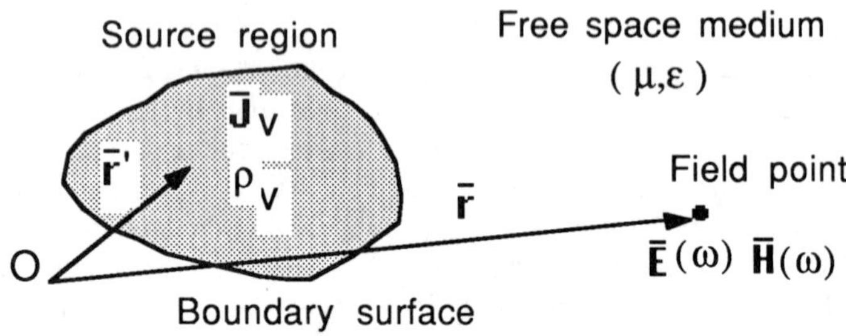

Figure 3.2 Sources in an unbounded free-space medium.

The sources produce electric and magnetic field distributions both inside and outside the source region. At any field point in the medium, the electromagnetic field quantities are given by

$\bar{E}(\bar{r},\omega)$: electric field distribution, in volts per meter;
$\bar{D}(\bar{r},\omega)$: electric flux density distribution, coulombs per square meter;
$\bar{H}(\bar{r},\omega)$: magnetic field distribution, in amperes per meter;
$\bar{B}(\bar{r},\omega)$: magnetic flux density distribution, in webers per square meter.

Referring to Table 2.5, in the source-free region, the electric and magnetic fields and their corresponding densities satisfy the following frequency-dependent Maxwell's equations in differential form:

$$\nabla \times \bar{E}(\bar{r},\omega) = - j\omega\bar{B}(\bar{r},\omega) \qquad (3.1.1a)$$

$$\nabla \times \bar{H}(\bar{r},\omega) = j\omega\bar{D}(\bar{r},\omega) \qquad (3.1.1b)$$

$\bar{r}$ outside source region

For all observation points within the source region itself, after referring to Table 2.5, the electric and magnetic fields also satisfy the following Maxwell's equations in differential form:

$$\nabla \times \bar{E}(\bar{r},\omega) = -j\omega\bar{B}(\bar{r},\omega) \tag{3.1.2a}$$

$$\nabla \times \bar{H}(\bar{r},\omega) = \bar{J}_V(\bar{r},\omega) + j\omega\bar{D}(\bar{r},\omega) \tag{3.1.2b}$$

$\bar{r}$ inside source region

The set of time-harmonic equations defined for the source-free region by expressions (3.1.1a) and (3.1.1b) and also for the source region by expressions (3.1.2a) and (3.1.2b) can be separately solved. The electric and magnetic field distributions so obtained for the two regions should satisfy appropriate boundary conditions for all points at the boundary surface separating the two regions. Particularly, at the boundary surface separating the source region and the source-free region, it is always convenient and useful to solve the time-harmonic integral form of Maxwell's equations. Referring to Table 2.6,

$$\oint_C \bar{E}(\bar{r},\omega) \bullet d\bar{L}(\bar{r}) = -j\omega \iint_S \bar{B}(\bar{r},\omega) \bullet d\bar{s}(\bar{r}) \tag{3.1.3a}$$

$$\oint_C \bar{H}(\bar{r},\omega) \bullet d\bar{L}(\bar{r}) = \iint_S \bar{J}_V(\bar{r},\omega) \bullet d\bar{s}(\bar{r}) + j\omega \iint_S \bar{D}(\bar{r},\omega) \bullet d\bar{s}(\bar{r}) \tag{3.1.3b}$$

The frequency-dependent Maxwell's equations for the source-free region are analyzed in the following. In equations (3.1.1a) and (3.1.1b), there are four unknown field quantities, given by the electric field and electric flux density, the magnetic field and magnetic flux density. But there are only two independent coupled curl equations (3.1.1a) and (3.1.1b) for the four unknown field quantities. Two additional relationships are obtained based on the constitutive relationships.

Hence, for an unbounded lossless free-space medium that is linear, homogeneous, and isotropic, various relationships are given by

$$\bar{D}(\bar{r},\omega) = \varepsilon\bar{E}(\bar{r},\omega) \tag{3.1.4a}$$

$$\bar{B}(\bar{r},\omega) = \mu\bar{H}(\bar{r},\omega) \tag{3.1.4b}$$

ε : permittivity of the free-space medium, in farads per meter;

μ : permeability of the free-space medium, in henrys per meter.

The constitutive relationships (3.1.4a) and (3.1.4b) are now substituted into the frequency-dependent Maxwell's equations (3.1.2a) and (3.1.2b) for the source region. It is noted again that the permeability and permittivity parameters of the medium are constants and do not depend on the frequency parameter variable nor on the spatial coordinate variables. Hence, the electric and magnetic fields in the source region satisfy the following frequency dependent Maxwell's equations in differential form:

$$\nabla \times \bar{E}(\bar{r},\omega) = -j\omega\mu\bar{H}(\bar{r},\omega) \tag{3.1.5a}$$

$$\nabla \times \bar{H}(\bar{r},\omega) = j\omega\varepsilon\bar{E}(\bar{r},\omega) + \bar{J}_V(\bar{r},\omega) \tag{3.1.5b}$$

For the electric and magnetic fields, their corresponding divergence relationships can be easily derived by utilizing Maxwell's equations (3.1.2a) and (3.1.2b). The divergence on both sides of expressions (3.1.2a) and (3.1.2b) is first taken; and invoking continuity equation (2.15.11) and noting that the divergence of curl of any vector quantity, Appendix A, is always zero, the following Gauss's law relationships are obtained

$$\nabla \bullet \varepsilon\bar{E}(\bar{r},\omega) = \rho_V(\bar{r},\omega) \tag{3.1.5c}$$

$$\nabla \bullet \mu\bar{H}(\bar{r},\omega) = 0 \tag{3.1.5d}$$

The set of equations defined for the source region, expressions (3.1.5a) and (3.1.5b), are the coupled vector equations. They can be reduced to their scalar form of coupled equations by assuming rectangular coordinate system. Using the rectangular coordinate system (x, y, z), at any field point in the medium, the relationship between various scalar components of the electromagnetic field quantities are given by Table 3.1. The following rectangular components are substituted into the curl equations (3.1.5a) and (3.1.5b):

$$\bar{E}(\bar{r},\omega)= E_x(\bar{r},\omega)\hat{x} + E_y(\bar{r},\omega)\hat{y} + E_z(\bar{r},\omega)\hat{z} \tag{3.1.6a}$$

$$\bar{H}(\bar{r},\omega) = H_x(\bar{r},\omega)\hat{x} + H_y(\bar{r},\omega)\hat{y} + H_z(\bar{r},\omega)\hat{z} \tag{3.1.6b}$$

$$\bar{J}_V(\bar{r},\omega) = J_x(\bar{r},\omega)\hat{x} + J_y(\bar{r},\omega)\hat{y} + J_z(\bar{r},\omega)\hat{z} \tag{3.1.6c}$$

Similarly, the two divergence relationships (3.1.5c) and (3.1.5d) can also be written in terms of scalar form. In fact, the six scalar coupled equations, Table 3.1, are also applicable for the source-free region with the source electric current components completely removed.

Table 3.1
Electric and magnetic fields in a rectangular coordinate system

$$\frac{\partial E_z}{\partial y} - \frac{\partial E_y}{\partial z} = -j\omega\mu H_x \tag{3.1.7a}$$

$$\frac{\partial E_x}{\partial z} - \frac{\partial E_z}{\partial x} = -j\omega\mu H_y \tag{3.1.7b}$$

$$\frac{\partial E_y}{\partial x} - \frac{\partial E_x}{\partial y} = -j\omega\mu H_z \tag{3.1.7c}$$

$$\frac{\partial H_z}{\partial y} - \frac{\partial H_y}{\partial z} = j\omega\varepsilon E_x + J_x \tag{3.1.8a}$$

$$\frac{\partial H_x}{\partial z} - \frac{\partial H_z}{\partial x} = j\omega\varepsilon E_y + J_y \tag{3.1.8b}$$

$$\frac{\partial H_y}{\partial x} - \frac{\partial H_x}{\partial y} = j\omega\varepsilon E_z + J_z \tag{3.1.8c}$$

Specifically for the source-free region, the three components of the electric field distribution (E_x, E_y, E_z) and the three components of the magnetic field distribution (H_x, H_y, H_z) satisfy the scalar Helmholtz equations, expressions (2.18.11a) and (2.18.11b), as derived in Section 2.18. Hence, for a linear, homogeneous, and isotropic lossless source-free region, the rectangular components of the electric and magnetic fields satisfy the following scalar Helmholtz equations:

$$\nabla^2 E_i(\bar{r},\omega) + k^2 E_i(\bar{r},\omega) = 0 \tag{3.1.9a}$$

$$\nabla^2 H_i(\bar{r},\omega) + k^2 H_i(\bar{r},\omega) = 0 \tag{3.1.9b}$$

for $i = x, y, z$

and

$$\gamma = jk \tag{3.1.9c}$$

k : propagation constant of the lossless free-space medium
$= \omega(\mu\varepsilon)^{1/2}$ (3.1.9d)

The field expressions given in the Table 3.1 can also be expressed in terms of other orthogonal coordinate systems, such as cylindrical and spherical coordinate systems. The set of equations, (3.1.5a) and (3.1.5b), are reduced to the scalar form of coupled equations by assuming cylindrical coordinate components, given by

$$\bar{E}(\bar{r},\omega) = E_\rho(\bar{r},\omega)\hat{\rho} + E_\phi(\bar{r},\omega)\hat{\phi} + E_z(\bar{r},\omega)\hat{z} \tag{3.1.10a}$$

$$\bar{H}(\bar{r},\omega) = H_\rho(\bar{r},\omega)\hat{\rho} + H_\phi(\bar{r},\omega)\hat{\phi} + H_z(\bar{r},\omega)\hat{z} \tag{3.1.10b}$$

$$\bar{J}_V(\bar{r},\omega) = J_\rho(\bar{r},\omega)\hat{\rho} + J_\phi(\bar{r},\omega)\hat{\phi} + J_z(\bar{r},\omega)\hat{z} \tag{3.1.10c}$$

Table 3.2
Electric and magnetic fields in a cylindrical coordinate system

---

$$\frac{1}{\rho}\frac{\partial E_z}{\partial \phi} - \frac{\partial E_\phi}{\partial z} = -j\omega\mu H_\rho \tag{3.1.11a}$$

$$\frac{\partial E_\rho}{\partial z} - \frac{\partial E_z}{\partial \rho} = -j\omega\mu H_\phi \tag{3.1.11b}$$

$$\frac{1}{\rho}\frac{\partial(\rho E_\phi)}{\partial \rho} - \frac{1}{\rho}\frac{\partial E_\rho}{\partial \phi} = -j\omega\mu H_z \tag{3.1.11c}$$

$$\frac{1}{\rho}\frac{\partial H_z}{\partial \phi} - \frac{\partial H_\phi}{\partial z} = j\omega\varepsilon E_\rho + J_\rho \tag{3.1.12a}$$

$$\frac{\partial H_\rho}{\partial z} - \frac{\partial H_z}{\partial \rho} = j\omega\varepsilon E_\phi + J_\phi \tag{3.1.12b}$$

$$\frac{1}{\rho}\frac{\partial(\rho H_\phi)}{\partial \rho} - \frac{1}{\rho}\frac{\partial H_\rho}{\partial \phi} = j\omega\varepsilon E_z + J_z \tag{3.1.12c}$$

Using the cylindrical coordinate system (ρ, ϕ, z), at any field point in the medium, the relationship between various components of the electromagnetic field quantities are given by Table 3.2. The six scalar equations obtained in Table 3.2 in terms of the cylindrical coordinate system are generally difficult to analyze. Except for the z coordinate direction, the two unit vectors depict different directions from point to point. For the three cylindrical components of electric field distribution and the corresponding three cylindrical components of magnetic field distribution, there is *no simple form* of Helmholtz-type scalar equations similar to the expressions defined in (3.1.9a) and (3.1.9b) for the rectangular coordinate system.

Alternatively, the electric and magnetic fields can be represented in terms of an arbitrary vector and scalar potential distributions. Further, in later sections, specific source orientations are also selected to obtain insight into the study of two-dimensional field distributions.

3.1.1 FIELDS IN TERMS OF POTENTIALS

Referring to the discussion in Section 2.21, a complete solution for the electric and magnetic field quantities can also be expressed in terms of two arbitrary potential functions, given by a vector magnetic potential and a scalar electric potential. Gauss's law states that the divergence of the time-harmonic magnetic flux density is always equal to zero. According to expression (3.1.5d), the magnetic flux density is expressed as

$$\bar{B}(\bar{r},\omega) = \nabla \times \bar{A}(\bar{r},\omega) \tag{3.1.13a}$$

$\bar{A}(\bar{r},\omega)$: magnetic vector potential function at the field point

$$= A_x(\bar{r},\omega)\hat{x} + A_y(\bar{r},\omega)\hat{y} + A_z(\bar{r},\omega)\hat{z} \tag{3.1.13b}$$

and constitutive relationship (3.1.4b) yields an expression for the magnetic field distribution:

$$\bar{H}(\bar{r},\omega) = \frac{1}{\mu} \nabla \times \bar{A}(\bar{r},\omega) \tag{3.1.13c}$$

Expression (3.1.13a) is now substituted into Faraday's law, expression (3.1.5a), to obtain an expression for the electric field distribution:

$$\bar{E}(\bar{r},\omega) = -j\omega\bar{A}(\bar{r},\omega) - \nabla\Phi(\bar{r},\omega) \tag{3.1.13d}$$

$\Phi(\bar{r},\omega)$: electric scalar potential function at the field point.

This completes the representation for the electric and magnetic fields in terms of the arbitrary potential functions. The magnetic vector and the electric scalar potentials are still unknown. To find these two potentials, expressions (3.1.13c) and (3.1.13d) are substituted into Ampere's law relationship (3.1.5b) to choose

$$\nabla[\nabla \bullet \bar{\mathbf{A}}(\bar{\mathbf{r}},\omega)] = - j\omega\mu\varepsilon \nabla\Phi(\bar{\mathbf{r}},\omega) \tag{3.1.14}$$

which is the *Lorentz gauge condition.* By enforcing such a gauge condition, the following vector *Helmholtz* partial differential equation is obtained for the magnetic vector potential:

$$\nabla^2\bar{\mathbf{A}}(\bar{\mathbf{r}},\omega) + k^2\bar{\mathbf{A}}(\bar{\mathbf{r}},\omega) = - \mu\bar{\mathbf{J}}_V(\bar{\mathbf{r}},\omega) \tag{3.1.15}$$

$$k^2 = \omega^2\mu\varepsilon \tag{3.1.16}$$

Hence, we have a systematic procedure for analyzing electromagnetic field problems. The first step consists of solving the preceding Helmholtz vector differential equation for the vector magnetic potential. In the second step, the magnetic field distribution is calculated using expression (3.1.13c). In the third step, the electric field distribution can be calculated using Ampere's law expression (3.1.5b), or expression (3.1.13d), which requires the calculation of electric scalar potential. Referring back to the Lorentz gauge expression (3.1.14), the electric scalar potential is given by

$$\Phi(\bar{\mathbf{r}},\omega) = \frac{j\omega}{k^2} \nabla \bullet \bar{\mathbf{A}}(\bar{\mathbf{r}},\omega) \tag{3.1.17}$$

In the vector Helmholtz equation, expression (3.1.15), the magnetic vector potential depends only on a single driving-source term – the electric current density distribution. In fact, a similar relationship can also be obtained for the electric scalar potential with a single driving-source term based on the electric charge density distribution. Taking divergence of expression (3.1.13d) and substituting relationships (3.1.5c) and (3.1.17), an Helmholtz equation for the electric scalar potential is obtained as

$$\nabla^2\Phi(\bar{\mathbf{r}},\omega) + k^2\Phi(\bar{\mathbf{r}},\omega) = - \frac{\rho_V(\bar{\mathbf{r}},\omega)}{\varepsilon} \tag{3.1.18}$$

As discussed in Section 2.18, the vector Helmholtz equation is difficult to solve. It can be reduced to the corresponding three scalar Helmholtz equations. Using the rectangular components defined in expression (3.1.13b), the vector Helmholtz equation (3.1.15) can be rewritten in terms of the following scalar form:

$$\nabla^2 A_x(\bar{r},\omega) + k^2 A_x(\bar{r},\omega) = -\mu J_x(\bar{r},\omega) \tag{3.1.19a}$$

$$\nabla^2 A_y(\bar{r},\omega) + k^2 A_y(\bar{r},\omega) = -\mu J_y(\bar{r},\omega) \tag{3.1.19b}$$

$$\nabla^2 A_z(\bar{r},\omega) + k^2 A_z(\bar{r},\omega) = -\mu J_z(\bar{r},\omega) \tag{3.1.19c}$$

It is interesting to note from these equations that the x-directed electric current density produces only x component of the magnetic vector potential. Similarly, the y- and z-directed electric current densities produce only the y and z components of the magnetic vector potentials, respectively. Instead of the volume electric current density distributions, suppose either surface or line electric current distributions are present, then they are appropriately replaced in the righthand side of the scalar differential equations.

Section 2.22 gives a method to solve for the scalar Helmholtz differential equation. Referring to the Figure 3.2, consider the volume electric current and charge source distributions confined to an arbitrary volume V and located in a three-dimensional isotropic and homogeneous medium. Let

$\bar{J}_V(\bar{r}')$: volume current density distribution in the volume V

$\bar{r}'$: position vector corresponding to the source point (x', y', z')

$$= x'\hat{x} + y'\hat{y} + z'\hat{z} \tag{3.1.20a}$$

$\bar{r}$: position vector corresponding to the field point (x, y, z)

$$= x\hat{x} + y\hat{y} + z\hat{z} \tag{3.1.20b}$$

R : distance between the source point and the field point

$$= |\bar{r} - \bar{r}'| \tag{3.1.20c}$$

The total magnetic vector potential at the field point is given by the volume integral over the complete source current density in volume V:

$$A_i(\bar{r}) = \frac{\mu}{4\pi} \iiint_V J_i(\bar{r}') \, \frac{e^{-jkR}}{R} \, dv(\bar{r}') \qquad i = x, y, z \tag{3.1.21}$$

and the total electric scalar potential at the field point is given by the volume integral over the complete source charge density in volume V:

$$\Phi(\bar{r}) = \frac{1}{4\pi\varepsilon} \iiint_V \rho_V(\bar{r}') \, \frac{e^{-jkR}}{R} \, dv(\bar{r}') \tag{3.1.22}$$

If other types of source distributions, such as the surface currents and charges and the line currents and charges, are present the two potential integrals (3.1.21) and (3.1.22) are to be correspondingly modified with respect to the surface and line integrals. The components of the electric and magnetic field distributions based on the two arbitrary vector and scalar potentials are given in Table 3.3 using the rectangular coordinate components:

$$\bar{E}(\bar{r},\omega) = E_x(\bar{r},\omega)\hat{x} + E_y(\bar{r},\omega)\hat{y} + E_z(\bar{r},\omega)\hat{z} \tag{3.1.23a}$$

$$\bar{H}(\bar{r},\omega) = H_x(\bar{r},\omega)\hat{x} + H_y(\bar{r},\omega)\hat{y} + H_z(\bar{r},\omega)\hat{z} \tag{3.1.23b}$$

$$\bar{A}(\bar{r},\omega) = A_x(\bar{r},\omega)\hat{x} + A_y(\bar{r},\omega)\hat{y} + A_z(\bar{r},\omega)\hat{z} \tag{3.1.23c}$$

Table 3.3

Electric and magnetic fields in terms of potentials in the rectangular coordinate system

---

$$H_x = \frac{1}{\mu}\left[\frac{\partial A_z}{\partial y} - \frac{\partial A_y}{\partial z}\right] \tag{3.1.24a}$$

$$H_y = \frac{1}{\mu}\left[\frac{\partial A_x}{\partial z} - \frac{\partial A_z}{\partial x}\right] \tag{3.1.24b}$$

$$H_z = \frac{1}{\mu}\left[\frac{\partial A_y}{\partial x} - \frac{\partial A_x}{\partial y}\right] \tag{3.1.24c}$$

$$E_x = \frac{-j\omega}{k^2}\left[k^2 A_x + \frac{\partial^2 A_x}{\partial x^2} + \frac{\partial^2 A_y}{\partial x \partial y} + \frac{\partial^2 A_z}{\partial x \partial z}\right] \tag{3.1.25a}$$

$$E_y = \frac{-j\omega}{k^2}\left[k^2 A_y + \frac{\partial^2 A_y}{\partial y^2} + \frac{\partial^2 A_x}{\partial y \partial x} + \frac{\partial^2 A_z}{\partial y \partial z}\right] \tag{3.1.25b}$$

$$E_z = \frac{-j\omega}{k^2}\left[k^2 A_z + \frac{\partial^2 A_z}{\partial z^2} + \frac{\partial^2 A_x}{\partial z \partial x} + \frac{\partial^2 A_y}{\partial z \partial y}\right] \tag{3.1.25c}$$

3.2 TRANSVERSE MAGNETIC POLARIZATION

In Section 3.1, general expressions for the electric and magnetic field distributions were obtained corresponding to the components of source electric current density distribution. For insight into the electromagnetic field problem, a specific component along with specific orientation of the source electric current distribution, such as the z component of the source electric current density, is now considered. Referring to Figure 3.3, let us consider an ideal time-harmonic source electric current distribution, consisting of only the z-directed line current distribution located along the z coordinate axis from $-\infty$ to ∞. The medium is assumed to be linear, homogeneous, and isotropic. Then

$\bar{J}_V(\bar{r}',\omega)$: source electric current distribution

$$= J_z(\bar{r}',\omega)\hat{z} \tag{3.2.1a}$$

$$= I_z(z',\omega)\delta(x')\delta(y')\hat{z} \tag{3.2.1b}$$

At any field point, there is only the A_z component of the magnetic vector potential. The other two components A_x and A_y do not exist.

$\bar{A}(\bar{r},\omega)$: magnetic vector potential function at the field point

$$= A_z(\bar{r},\omega)\hat{z} \tag{3.2.2}$$

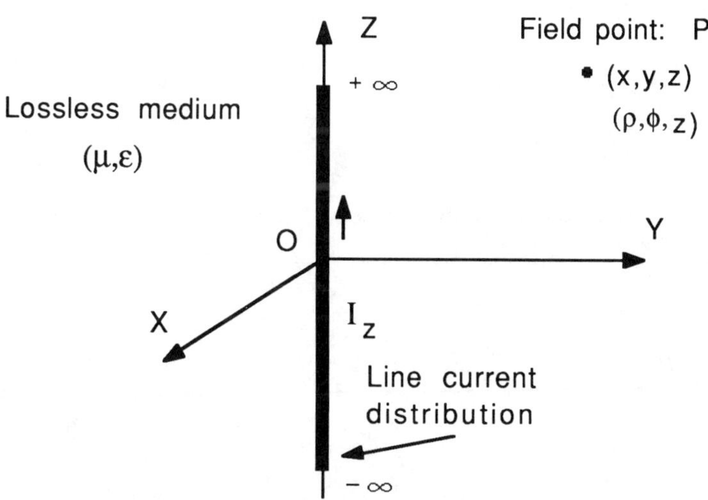

Figure 3.3 Ideal z-directed line source electric current distribution.

Further, referring to expression (3.1.19c), the z component of the magnetic vector potential satisfies the scalar Helmholtz equation:

$$\nabla^2 A_z(\bar{r},\omega) + k^2 A_z(\bar{r},\omega) = -\mu J_z(\bar{r},\omega) \tag{3.2.3}$$

For the line electric current source distribution which is located along the z axis

$\bar{r}'$: position vector corresponding to the source point $(x'=0, y'=0, z')$

$$= z'\hat{z} \tag{3.2.4a}$$

$\bar{r}$: position vector corresponding to the field point (x, y, z)

$$= x\hat{x} + y\hat{y} + z\hat{z} \tag{3.2.4b}$$

R: distance of the line joining the source point to the field point

$$= [(x^2 + y^2 + (z-z')^2]^{1/2} \tag{3.2.4c}$$

The corresponding electric field and magnetic field distributions at any general field point in the free-space medium are listed in Table 3.4a using the rectangular coordinate (x, y, z) variables:

$$\bar{E}(\bar{r},\omega) = E_x(\bar{r},\omega)\hat{x} + E_y(\bar{r},\omega)\hat{y} + E_z(\bar{r},\omega)\hat{z} \tag{3.2.5a}$$

$$\bar{H}(\bar{r},\omega) = H_x(\bar{r},\omega)\hat{x} + H_y(\bar{r},\omega)\hat{y} + H_z(\bar{r},\omega)\hat{z} \tag{3.2.5b}$$

$$\bar{A}(\bar{r},\omega) = A_z(x,y,z,\omega)\hat{z} \tag{3.2.5c}$$

Table 3.4a

Transverse magnetic fields in terms of potential in the rectangular coordinate system

---

$$H_x = \frac{1}{\mu}\frac{\partial A_z}{\partial y} \tag{3.2.6a}$$

$$H_y = -\frac{1}{\mu}\frac{\partial A_z}{\partial x} \tag{3.2.6b}$$

$$H_z = 0 \tag{3.2.6c}$$

$$E_x = \frac{-j\omega}{k^2} \frac{\partial^2 A_z}{\partial x \partial z} \tag{3.2.7a}$$

$$E_y = \frac{-j\omega}{k^2} \frac{\partial^2 A_z}{\partial y \partial z} \tag{3.2.7b}$$

$$E_z = \frac{-j\omega}{k^2}\left[k^2 A_z + \frac{\partial^2 A_z}{\partial z^2}\right] \tag{3.2.7c}$$

Referring to integral expression (3.1.21), the total magnetic vector potential at the field point $P(x,y,z)$ is given by the line integral over the complete line source electric current distribution along z axis:

$$A_z(x,y,z) = \frac{\mu}{4\pi} \int_{z'=-\infty}^{\infty} I_z(z') \frac{e^{-jkR}}{R} dz' \tag{3.2.8a}$$

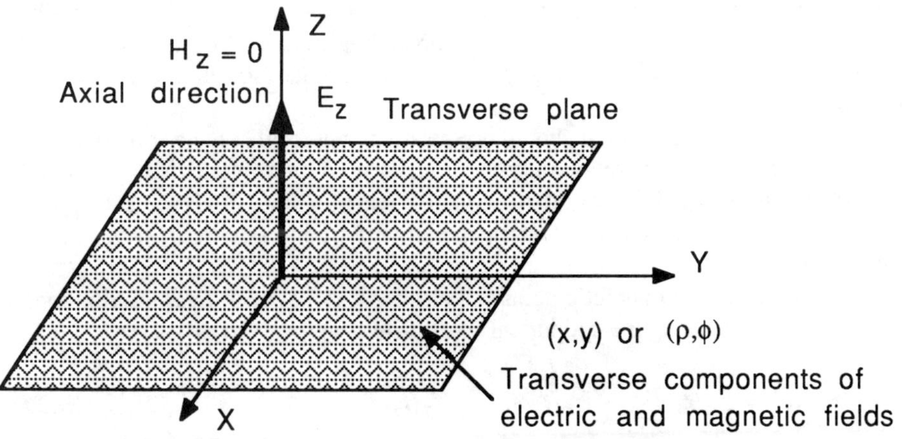

Figure 3.4 Transverse magnetic (TM) fields.

Integral expression (3.2.8a) can be simplified in number of ways if the distribution of the line source electric current is known. For a special case, where the line source electric current is constant and has no variation with the z' coordinate variable, the magnetic vector potential distribution will be independent of the z coordinate variation. In such a case, at $z = 0$, the preceding integral expression can be simplified as

$$A_z(x,y,z=0) = \frac{\mu I_z}{4\pi}\left[\int_{z'=-\infty}^{\infty} \frac{e^{-jkR}}{R}\, dz'\right] \tag{3.2.8b}$$

$$R = [x^2 + y^2 + (z')^2]^{1/2} \tag{3.2.8c}$$

$$A_z(x,y) = \frac{\mu I_z}{4j} H_0^{(2)}\left[k(x^2+y^2)^{1/2}\right] \tag{3.2.8d}$$

$H_0^{(2)}$: Hankel function of second kind and zero order.

Referring to Appendix B, the bracketed integral term in expression (3.2.8b), can be recognized as the well-known outgoing cylindrical wave function given by the Hankel function of second kind and zero order. Further details of the solution based on cylindrical wave functions are presented in the next section using Green's function technique. In fact, the ideal z-directed electric current line source, figure 3.3, can be extended to simulate two-dimensional z-directed arbitrary electric current sources based on the superposition principle. Using the preceding relationship, the corresponding electric and magnetic field distributions in the rectangular coordinate system can be calculated.

Referring to the Table 3.4a, it should be specifically noted that the z component of the magnetic field is zero. There are only the *transverse to z* components of the magnetic field, which are oriented perpendicular to the z axis as shown in Figure 3.4. The field distributions obtained in Table 3.4a are generally referred to as the *transverse magnetic (TM) fields*.

The same transverse magnetic polarized electric and magnetic field distributions are listed in Table 3.4b, using the cylindrical coordinate system, given by

$$\bar{E}(\bar{r},\omega) = E_\rho(\bar{r},\omega)\hat{\rho} + E_\phi(\bar{r},\omega)\hat{\phi} + E_z(\bar{r},\omega)\hat{z} \tag{3.2.9a}$$

$$\bar{H}(\bar{r},\omega) = H_\rho(\bar{r},\omega)\hat{\rho} + H_\phi(\bar{r},\omega)\hat{\phi} + H_z(\bar{r},\omega)\hat{z} \tag{3.2.9b}$$

$$\bar{A}(\bar{r},\omega) = A_z(\rho,\phi,z,\omega)\hat{z} \tag{3.2.9c}$$

Table 3.4b

Transverse magnetic fields in terms of potential in the cylindrical coordinate system

---

$$H_\rho = \frac{1}{\mu}\frac{1}{\rho}\frac{\partial A_z}{\partial \phi} \tag{3.2.10a}$$

$$H_\phi = -\frac{1}{\mu}\frac{\partial A_z}{\partial \rho} \tag{3.2.10b}$$

$$H_z = 0 \tag{3.2.10c}$$

$$E_\rho = \frac{-j\omega}{k^2}\frac{\partial^2 A_z}{\partial \rho \partial z} \tag{3.2.11a}$$

$$E_\phi = \frac{-j\omega}{k^2}\frac{1}{\rho}\frac{\partial^2 A_z}{\partial \phi \partial z} \tag{3.2.11b}$$

$$E_z = \frac{-j\omega}{k^2}\left[k^2 A_z + \frac{\partial^2 A_z}{\partial z^2}\right] \tag{3.2.11c}$$

3.3 FIELD EQUATIONS IN SPECTRAL DOMAIN

The transverse magnetic field distributions are obtained in Tables 3.4a and 3.4b, for an ideal line current distribution located along the z coordinate axis. In the analysis discussed in Section 3.2, it is assumed that the line source electric current has only the z coordinate dependence and is polarized along the z axis. The magnetic vector potential and the electric and magnetic field quantities are functions of the spatial coordinates (x, y, z) or (ρ, ϕ, z). It is possible to introduce further simplification into the analysis by transforming all the sources and their field quantities from the z spatial coordinate domain to a corresponding spectral transform domain. This process basically converts the three-dimensional sources and their field quantities to the corresponding two dimensional sources and their field quantities. This is similar to defining an equivalent Fourier integral transform to convert all the sources, potentials, and fields with (x, y, z) dependence to the corresponding (x, y, k_z) dependence, where k_z is the spectral transform parameter. To get back to the original field quantities with (x, y, z)

dependence, the inverse integral transform is performed for the spectral domain field quantities.

The concept of the spectral transform domain is briefly discussed in the following. Let $F(z)$ represent an arbitrary function in the z coordinate domain. It is assumed that the function $F(z)$ has the following properties:

1. The coordinate variable z varies from $-\infty$ to ∞.
2. The function $F(z)$ and its derivatives are continuous or piecewise continuous in every finite coordinate interval.
3. The function $F(z)$ is absolutely integrable in the range $(-\infty,\infty)$ or

$$\int_{-\infty}^{\infty} |F(z)|\, dz \quad \text{converges.} \tag{3.3.1}$$

Then, the function $F(z)$ can be transformed into the spectral domain function $F(k_z)$ based on the following direct spectral integral transform:

$$\mathcal{S}[F(z)] = F(k_z) = \frac{1}{\sqrt{2\pi}} \int_{-\infty}^{\infty} F(z)\, e^{jk_z z}\, dz \tag{3.3.2}$$

where k_z is the spectral domain variable or a parameter similar to the phase propagation constant along the z coordinate axis, in radians per meter. The function $F(k_z)$ is always a complex function, referred to as a *phasor* quantity. It is a function of the spectral parameter variable k_z in the range varying from $-\infty$ to ∞. The original function $F(z)$ can be recovered back by performing inverse integral transform:

$$\mathcal{S}^{-1}[F(k_z)] = F(z) = \frac{1}{\sqrt{2\pi}} \int_{-\infty}^{\infty} F(k_z)\, e^{-jk_z z}\, dk_z \tag{3.3.3}$$

Based on these integral transforms, all the relevant electromagnetic equations can be converted into the spectral domain. It should be noted at this stage that the basic transformation from the z coordinate domain to the spectral transform domain involves defining various electromagnetic sources and their corresponding fields to vary as

$$\bar{J}(x,y,z) = \bar{J}(x,y,k_z)\, e^{-jk_z z} \tag{3.3.4a}$$

$$\bar{E}(x,y,z) = \bar{E}(x,y,k_z)\, e^{-jk_z z} \tag{3.3.4b}$$

$$\bar{H}(x,y,z) = \bar{H}(x,y,k_z)\, e^{-jk_z z} \tag{3.3.4c}$$

Hence, the various electromagnetic equations can be transformed to the spectral domain using the following operator substitutions given by

$$\frac{\partial}{\partial z} \Leftrightarrow -jk_z \tag{3.3.5a}$$

$$\frac{\partial^2}{\partial z^2} \Leftrightarrow -k_z^2 \tag{3.3.5b}$$

$$\int dz \Leftrightarrow \frac{1}{-jk_z} \tag{3.3.5c}$$

Referring to Figure 3.3, in terms of the spectral domain, let

$\bar{J}_V(\bar{r}',\omega)$: source electric current distribution

$$= J_z(x',y',k_z,\omega)\, e^{-jk_z z'}\hat{z} \tag{3.3.6a}$$

$$= I_z(k_z,\omega)\delta(x')\delta(y')\, e^{-jk_z z'}\hat{z} \tag{3.3.6b}$$

$\bar{A}(\bar{r},\omega)$: magnetic vector potential at the field point

$$= A_z(x,y,k_z,\omega)\, e^{-jk_z z}\hat{z} \tag{3.3.7}$$

Further transforming expression (3.2.3) into the spectral domain, the z component of the magnetic vector potential satisfies the scalar Helmholtz equation:

$$\nabla_\tau^2 A_z(x,y,k_z,\omega) + k_\tau^2 A_z(x,y,k_z,\omega) = -\mu I_z(k_z,\omega)\delta(x)\delta(y) \tag{3.3.8a}$$

where

$$k_\tau^2 = k^2 - k_z^2 \tag{3.3.8b}$$

k_τ : transverse propagation constant.

The scalar Helmholtz equation (3.3.8a) is to be solved first for the z component of the magnetic vector potential. Then, Table 3.5a gives all the components of electric and magnetic fields in the spectral transform domain in terms of the z component of magnetic vector potential written in transverse rectangular coordinates (x, y):

$$\bar{E}(x,y,k_z) = E_x(x,y,k_z)\hat{x} + E_y(x,y,k_z)\hat{y} + E_z(x,y,k_z)\hat{z} \tag{3.3.9a}$$

$$\bar{H}(x,y,k_z) = H_x(x,y,k_z)\hat{x} + H_y(x,y,k_z)\hat{y} + H_z(x,y,k_z)\hat{z} \tag{3.3.9b}$$

$$\bar{A}(x,y,k_z) = A_z(x,y,k_z)\hat{z} \tag{3.3.9c}$$

Table 3.5a

Transverse magnetic fields in the spectral domain in rectangular coordinates

---

$$H_x = \frac{1}{\mu}\frac{\partial A_z}{\partial y} \tag{3.3.10a}$$

$$H_y = -\frac{1}{\mu}\frac{\partial A_z}{\partial x} \tag{3.3.10b}$$

$$H_z = 0 \tag{3.3.10c}$$

$$E_x = \frac{-\omega k_z}{k^2}\frac{\partial A_z}{\partial x} \tag{3.3.11a}$$

$$E_y = \frac{-\omega k_z}{k^2}\frac{\partial A_z}{\partial y} \tag{3.3.11b}$$

$$E_z = \frac{-j\omega}{k^2}[k^2 - k_z^2]A_z \tag{3.3.11c}$$

The transverse magnetic (TM) field expressions in the Table 3.5a can be written conveniently in terms of transverse cylindrical coordinates (ρ, ϕ) which are given in Table 3.5b:

$$\bar{E}(\rho,\phi,k_z) = E_\rho(\rho,\phi,k_z)\hat{\rho} + E_\phi(\rho,\phi,k_z)\hat{\phi} + E_z(\rho,\phi,k_z)\hat{z} \quad (3.3.12a)$$

$$\bar{H}(\rho,\phi,k_z) = H_\rho(\rho,\phi,k_z)\hat{\rho} + H_\phi(\rho,\phi,k_z)\hat{\phi} + H_z(\rho,\phi,k_z)\hat{z} \quad (3.3.12b)$$

$$\bar{A}(\rho,\phi,k_z) = A_z(\rho,\phi,k_z)\hat{z} \quad (3.3.12c)$$

Table 3.5b

Transverse magnetic fields in the spectral domain in cylindrical coordinates

---

$$H_\rho = \frac{1}{\mu}\frac{1}{\rho}\frac{\partial A_z}{\partial \phi} \quad (3.3.13a)$$

$$H_\phi = -\frac{1}{\mu}\frac{\partial A_z}{\partial \rho} \quad (3.3.13b)$$

$$H_z = 0 \quad (3.3.13c)$$

$$E_\rho = \frac{-\omega k_z}{k^2}\frac{\partial A_z}{\partial \rho} \quad (3.3.14a)$$

$$E_\phi = \frac{-\omega k_z}{k^2}\frac{1}{\rho}\frac{\partial A_z}{\partial \phi} \quad (3.3.14b)$$

$$E_z = \frac{-j\omega}{k^2}[k^2 - k_z^2]A_z \quad (3.3.14c)$$

It should be noted that the field quantities obtained in Tables 3.5a and 3.5b are only functions of the two-dimensional spatial coordinate variables either (x, y) or (ρ, ϕ). In the study of electromagnetic scattering and interaction problems, irrespective of which transverse spatial coordinates are adopted, the field quantities should be determined in the complete spectrum of the transform parameter k_z varying from $-\infty$ to ∞. In the following section, a solution for the z component of the magnetic vector potential is discussed in terms of the two-dimensional Green's function.

3.4 SOLUTION FOR THE MAGNETIC VECTOR POTENTIAL

Referring to the Figure 3.3, the line source electric current distribution is located along the z coordinate axis. Hence,

$$\bar{I}(\rho'=0,k_z,\omega) = I_z(k_z,\omega)\delta(\rho')\, e^{-jk_z z'}\hat{z} \tag{3.4.1a}$$

$\bar{r}'$: position vector corresponding to the line current source

$$= z'\hat{z} \tag{3.4.1b}$$

$\bar{r}$: position vector corresponding to the field point

$$= \rho\hat{\rho} + z\hat{z} \tag{3.4.1c}$$

Further, from expression (3.3.8a), in the spectral transform domain the z component of magnetic vector potential satisfies the following scalar Helmholtz equation:

$$\nabla_\tau^2 A_z(\rho,\phi,k_z) + k_\tau^2 A_z(\rho,\phi,k_z) = -\mu I_z(k_z)\delta(\rho) \tag{3.4.2a}$$

$$k_\tau^2 = k^2 - k_z^2 \tag{3.4.2b}$$

where k_τ is the transverse propagation constant. Referring to the Figure 3.3 and the righthand-side term in expression (3.4.2a), the spectral distribution of the line source electric current is located along the z coordinate axis and has both magnitude and harmonic phase variation. Since the Helmholtz equation is independent of z variable, the observation point can selected in the xy transverse plane.

In the following, a solution for the z component of the magnetic vector potential is presented in the cylindrical coordinate system. Expressing the transverse Laplacian operator in the cylindrical coordinate system, Appendix A, for all observation points outside the line source electric current, the scalar Helmholtz equation (3.4.2a) reduces to

$$\frac{1}{\rho}\frac{\partial}{\partial\rho}\left[\rho\,\frac{\partial A_z(\rho,\phi,k_z)}{\partial\rho}\right] + \frac{1}{\rho^2}\,\frac{\partial^2 A_z(\rho,\phi,k_z)}{\partial\phi^2} + k_\tau^2\, A_z(\rho,\phi,k_z) = 0 \tag{3.4.3a}$$

To solve the scalar partial differential equation (3.4.3a), the method of *separation of variables* is adopted, which is discussed in the Section 2.9. The total solution for the magnetic vector potential A_z can be written as the product of two terms, since A_z depends on separable two coordinate variables:

$$A_z(\rho,\phi,k_z) = P(\rho)\Phi(\phi) \tag{3.4.3b}$$

$P(\rho)$: partial solution in terms of the ρ coordinate only;
$\Phi(\phi)$: partial solution in terms of the ϕ coordinate only.

On substituting for A_z and then dividing by A_z, the scalar Helmholtz equation (3.4.3a) simplifies to the following form:

$$\frac{\rho}{P}\frac{\partial}{\partial\rho}\left[\rho\frac{\partial P}{\partial\rho}\right] + k_\tau^2\rho^2 = -\frac{1}{\Phi}\frac{\partial^2\Phi}{\partial\phi^2} \tag{3.4.3c}$$

In relationship (3.4.3c), a pure function of ρ is equal to a pure function of ϕ. This is possible when the lefthand-side and the righthand-side terms are both equal to a constant. Hence, expression (3.4.3c) reduces to

$$\rho\frac{d}{d\rho}\left[\rho\frac{dP}{d\rho}\right] + [k_\tau^2\rho^2 - n^2]\,P = 0 \tag{3.4.3d}$$

$$\frac{1}{\Phi}\frac{d^2\Phi}{d\phi^2} = -n^2 \tag{3.4.3e}$$

where n is a separation constant. The expression (3.4.3e) is an ordinary second-order differential equation whose solutions are harmonic functions. The roots of this equation are given by jn and $-jn$. Hence, the harmonic functions of the type $sin(n\phi)$ and $cos(n\phi)$ or $exp(jn\phi)$ and $exp(-jn\phi)$ satisfy the differential equation (3.4.3e). Even the linear combination of these two harmonic functions also satisfies differential equation (3.4.3e).

Further, the solution for the second-order differential equation given by expression (3.4.3d) is quite complicated. But, its solution can still be obtained in the form of a convergent power series, Appendix B. In fact, expression (3.4.3d) can be recognized as the familiar *Bessel differential equation* of order *n*, whose solutions are the well-known *Bessel functions* of the first and second kind.

Referring to the Appendix B, possible solutions to the Bessel differential equation are given by the following form of Bessel functions:

$$B_n(k_\tau\rho) \Rightarrow J_n(k_\tau\rho),\ Y_n(k_\tau\rho)$$
$$\text{or} \qquad H_n^{(1)}(k_\tau\rho),\ H_n^{(2)}(k_\tau\rho) \tag{3.4.4a}$$

where

$k_\tau\rho$: argument of the Bessel function;

$J_n(k_\tau\rho)$: Bessel function of first kind and order n;

$Y_n(k_\tau\rho)$: Bessel function of second kind or Neumann function and order n

and

$H_n^{(1)}(k_\tau\rho)$: Hankel function of first kind and order n

$$= J_n(k_\tau\rho) + jY_n(k_\tau\rho) \tag{3.4.4b}$$

$H_n^{(2)}(k_\tau\rho)$: Hankel function of second kind and of order n

$$= J_n(k_\tau\rho) - jY_n(k_\tau\rho) \tag{3.4.4c}$$

The solution in terms of the Bessel functions (3.4.4a) can be verified by substituting them into differential equation (3.4.3d). In general, any two linear combination of the Bessel functions also satisfy the differential equation (3.4.3d). Hence, the total solution (3.4.3b) for the magnetic vector potential can be constructed in terms of these partial terms. It is advantageous to know the behavior of these special functions in order to choose the right set of functions for specific representation of field quantities in a given boundary value problem.

It may be noted, the Bessel functions of the first kind J_n are finite and well behaved in a limit as argument tends to zero, and hence are suitable for representing the field behavior in the neighborhood $\rho \to 0$. The Bessel functions of the second kind Y_n are not finite and possess logarithmic singularity in a limit as argument tends to zero, and thus are suitable for representing fields with singular behavior. By studying asymptotic behavior with large arguments, in fact, cylindrical wave properties can be extracted from the Bessel functions. In particular, the Hankel functions of the second kind represent radially outgoing cylindrical wave functions, and similarly, the Hankel functions of the first kind represent radially incoming cylindrical wave functions. Thus, the Hankel functions are suitable for representing propagating field distributions possessing cylindrical symmetry with respect to the z-directed line current sources.

Since the z-directed line current source located along the z axis radiate fields away in all radial directions, a typical solution for the magnetic vector potential A_z is represented only in terms of the outgoing cylindrical waves, which can be written as

$$A_z(\rho,\phi,k_z,\omega) = P(\rho)\Phi(\phi) \qquad \text{for all integer values of } n \tag{3.4.5a}$$

$$= M_n H_n^{(2)}(k_\tau\rho)\, e^{jn\phi} \qquad -\infty < \text{integer } n < \infty \tag{3.4.5b}$$

M_n : amplitude coefficient of the nth cylindrical wave function.

The linear combination of the various solutions obtained in expression (3.4.5b) also forms the complete solution for the z component of the magnetic vector potential. Referring to Figure 3.3 and the Helmholtz scalar equation (3.4.2a), the ideal line source electric current distribution is restricted to the z coordinate axis and has no ϕ angular dependence. Hence, $n = 0$ is just sufficient to represent the proper solution. Further, the total solution for A_z takes the form

$$A_z(\rho,\phi,k_z,\omega) = M_0 H_0^{(2)}(k_\tau\rho) \tag{3.4.6}$$

The amplitude constant M_0 in expression (3.4.6) can be evaluated by applying the integral form of Ampere's law, which gives the line integral of the encircling magnetic field equal to the amplitude of line current source. Choosing a circular path in Figure 3.5, only an angular component of the magnetic field distribution is encircling the line source electric current.

Let us consider a field point at radial distance r from the line current axis at which the ϕ component of magnetic field is to be determined.

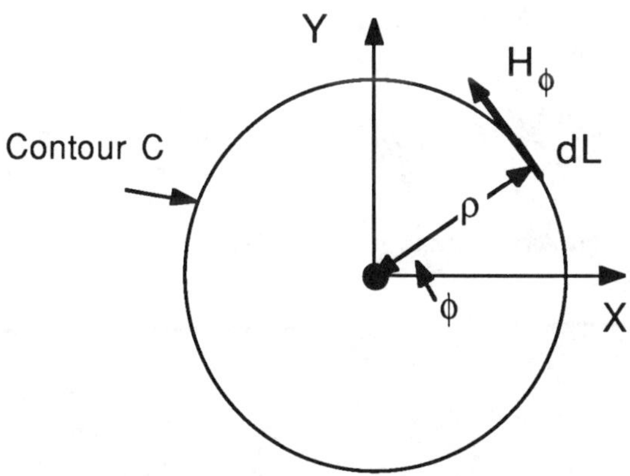

Figure 3.5 Magnetic field due to line current source.

Referring to the Figure 3.5, a closed circular contour path is drawn to pass through the field point with its center located at the line current axis. According to Ampere's law, expression (2.15.14),

$$\lim_{\rho \to 0} \oint_C \bar{H} \bullet d\bar{L} = I_z \tag{3.4.7a}$$

At the field point and along the circular contour path C, expression (3.4.7a) takes the form

$$\bar{H} = H_\rho \hat{\rho} + H_\phi \hat{\phi} \tag{3.4.7b}$$

$$d\bar{L} = \rho \, d\phi \hat{\phi} \tag{3.4.7c}$$

$$\lim_{\rho \to 0} \int_{\phi=0}^{2\pi} H_\phi \rho \, d\phi = I_z \tag{3.4.7d}$$

Along the circular path, the angular component of magnetic field is constant, and it can be pulled out of the integral. Referring to expression (3.3.13b) in the Table 3.5b,

$$H_\phi = - \frac{1}{\mu} \frac{\partial A_z}{\partial \rho} \tag{3.4.7e}$$

$$= - \frac{1}{\mu} \frac{\partial}{\partial \rho} [M_0 H_0^{(2)}(k_\tau \rho)] \tag{3.4.7f}$$

$$= - \frac{M_0}{\mu} k_\tau [- J_1(k_\tau \rho) + jY_1(k_\tau \rho)] \tag{3.4.7g}$$

On choosing a low-argument approximation for the Bessel functions in a limit as the argument $k_\tau \rho$ tends to zero,

$$J_1(k_\tau \rho) \approx 0 \tag{3.4.7h}$$

$$Y_1(k_\tau \rho) \approx \frac{-2}{\pi k_\tau \rho} \tag{3.4.7i}$$

After substituting expressions (3.4.7g), (3.4.7h) and (3.4.7i), integral expression (3.4.7d) yields the constant coefficient:

$$M_0 = \frac{\mu I_z}{4j} \tag{3.4.7j}$$

Hence, referring to the Figure 3.3, for the case of a z-directed line electric current distribution, expression (3.4.6) for the A_z component of the magnetic vector potential takes the following form:

$$A_z(\rho,k_z) = \frac{\mu I_z}{4j} H_0^{(2)}(k_\tau \rho) \tag{3.4.8}$$

This solution for the Helmholtz differential equation (3.4.2a) with a *unit* delta function-type line current source is generally referred to as the two-dimensional *Green's function.* This expression for the magnetic vector potential can be conveniently modified if the z-directed line source electric current is not located along the z coordinate axis, but is shifted to an arbitrary source location (x', y') and is still oriented parallel to the z coordinate axis. Based on this procedure, z-directed general source current distributions can be considered.

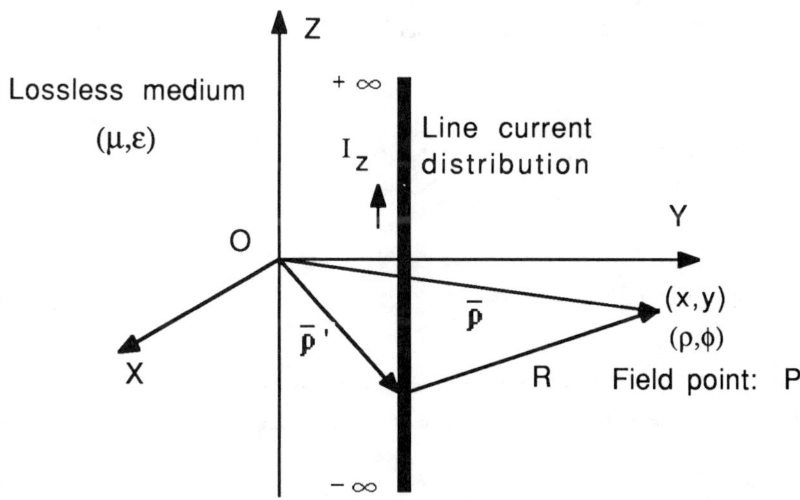

Figure 3.6 Shifted line source electric current distribution.

Referring to Figure 3.6, let us consider an ideal time-harmonic source electric current distribution consisting of only the z-directed line current located parallel to the z coordinate axis from $-\infty$ to ∞. The medium is assumed to be linear, homogeneous, and isotropic. Then

$\bar{J}_V(x',y',k_z)$: line source electric current distribution

$$= I_z(k_z,\omega)\delta(x-x')\delta(y-y')\; e^{-jk_z z'}\hat{z} \tag{3.4.9a}$$

At any field point, there is only the z component of the magnetic vector potential. In the spectral transform domain, expression (3.4.8) for the A_z component of the magnetic vector potential in terms of the two-dimensional Green's function takes the following form:

$$A_z(\rho,\phi,k_z,\omega) = \frac{\mu I_z}{4j} H_0^{(2)}(k_\tau R) \tag{3.4.9b}$$

R : distance between the source point and the field observation point

$$= |\bar{\rho} - \bar{\rho}'| \tag{3.4.9c}$$

Using the Table 3.5b, various components of electric and magnetic field distributions can now be calculated. It should be noted that these fields are produced entirely by the z polarized line source electric current. Thus, the axial component of scattered electric field and the angular component of scattered magnetic field are given by

$$E_z(\bar{\rho}) = \frac{-\omega\mu I_z}{4} \frac{k_\tau^2}{k^2} H_0^{(2)}(k_\tau R) \tag{3.4.10a}$$

$$H_\phi(\bar{\rho}) = - \frac{\partial}{\partial\rho} \frac{I_z}{4j} H_0^{(2)}(k_\tau R) \tag{3.4.10b}$$

The electric and magnetic field expressions obtained here are quite useful in the study of electromagnetic scattering and interaction by the two-dimensional conducting object.

3.5 ELECTRIC FIELD INTEGRAL EQUATION – TM CASE

Now a detailed analysis procedure is presented for the case of electromagnetic scattering by a two-dimensional conducting object based on an electric field integral equation formulation. Referring to Figure 3.7a, a perfectly conducting (material conductivity of scatterer, $\sigma \to \infty$) two-dimensional object is oriented with its axis coinciding with the z axis of cylindrical coordinate system. Along the axis, the scattering object is infinitely long and has a uniform arbitrary cross section. It is placed in a linear, homogeneous, and isotropic lossless medium and is excited externally by a time-harmonic transverse magnetic (TM to z) polarized plane wave.

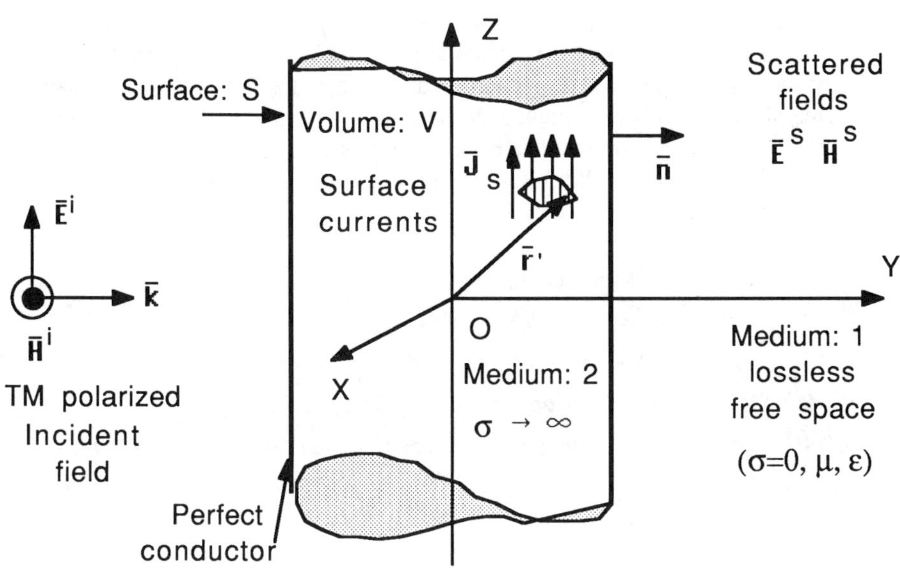

Figure 3.7a Geometry of perfectly conducting scatterer – TM excitation.

As shown in the Figure 3.7a, the TM polarized plane wave is incident on the conducting object at an arbitrary angle of incidence with its incident electric field polarized parallel and the corresponding incident magnetic field polarized perpendicular to the z coordinate axis. It should be clearly noted that there is no propagation of the incident plane wave parallel to the axis of two-dimensional object. For this special case of time-harmonic transverse magnetic *normal* excitation, there is only an axial component of the induced electric currents on the surface of conducting object. The induced electric currents produce only the axial scattered electric field and the transverse magnetic field distributions in the surrounding medium. Hence, in the free-space medium outside the conducting object, the total field distribution consists of the sum of the incident field and the scattered field. These total electric and magnetic field distributions near the surface of the conducting object are such that they satisfy appropriate electromagnetic boundary conditions. The induced electric current distribution on the conducting object is still an unknown quantity and can be determined by enforcing the familiar boundary condition that the total tangential component of the electric field on the surface of conducting object is zero.

The classical approach is to treat the induced axial electric currents on the surface of conducting object as unknown and set up a boundary value integral equation. Even though the integral equation can be derived in a number of different ways, the discussion concerning the boundary value integral equation presented in this section is based upon the electromagnetic equivalence principle. This integral equation formulation is suited for the case of both closed and open type of conducting objects. The Figure 3.7a shows geometry of the conducting scatterer located in an isotropic lossless free-space medium. The scatterer has a volume V contained in medium 2 and is bounded by a surface S. Outside the volume of the medium 2 is medium 1, representing the free-space region, and the externally excited TM to z-polarized incident plane wave field is contained in it.

Referring to the Figures 3.7a and 3.7b, let

ε : permittivity of free-space medium 1;

μ : permeability of free-space medium 1;

$\sigma = 0$: conductivity of free-space medium 1;

$(\bar{E}^i, \bar{H}^i)$: the electric and magnetic incident fields in medium 1;

$(\bar{E}^S, \bar{H}^S)$: the electric and magnetic scattered fields in medium 1;

$(\bar{E}, \bar{H})$: the electric and magnetic total fields in medium 1

$= (\bar{E}^i, \bar{H}^i) + (\bar{E}^S, \bar{H}^S)$

Referring to the conducting boundary condition discussed in the Section 2.16, there is no electric or magnetic field inside the perfectly conducting region. On the surface of perfectly conducting object, the following boundary conditions are satisfied

$$\bar{J}_S(\bar{r}') = \hat{n}' \times \left[\bar{H}^i(\bar{r}') + \bar{H}^S(\bar{r}')\right] \qquad \bar{r}' \text{ on } S \tag{3.5.1a}$$

$$\bar{E}(\bar{r})\Big|_{\tan} = \hat{n} \times \left[\bar{E}^i(\bar{r}) + \bar{E}^S(\bar{r})\right] = 0 \qquad \bar{r} \text{ on } S \tag{3.5.1b}$$

where

tan : tangential component of the field;

$\hat{n}$: unit vector normal to the surface S.

Using expressions (3.3.4) and (3.3.6a), in the spectral transform domain, the representation for these harmonic field quantities is given by

$\bar{J}_S(\bar{r}')$: induced electric current on the surface S, in amperes per meter

$$= J_z(\rho',\phi',k_z)\, e^{-jk_z z'}\hat{z} \tag{3.5.2a}$$

$\bar{E}(\bar{r})$: total electric field in the free-space medium 1

$$= \bar{E}(\rho,\phi,k_z,\omega)\, e^{-jk_z z} \tag{3.5.2b}$$

$\bar{H}(\bar{r})$: total magnetic field in the free-space medium 1

$$= \bar{H}(\rho,\phi,k_z,\omega)\, e^{-jk_z z} \tag{3.5.2c}$$

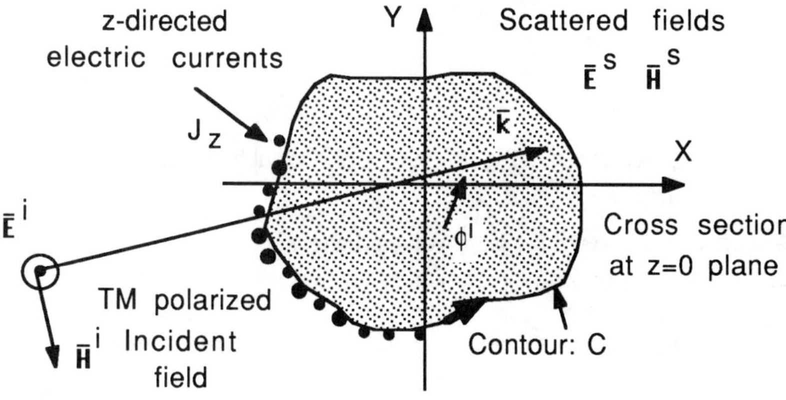

Figure 3.7b Cross section of perfectly conducting scatterer – TM excitation.

Referring to the Figure 3.7b and the discussion in Section 2.19 regarding plane wave fields in the frequency domain, the incident plane wave electric and magnetic fields can be written as follows. The plane wave incident electric field vector is polarized along the z axis, and the incident magnetic field vector is polarized in a plane perpendicular to the z axis. There is no propagation of the fields parallel to the z coordinate axis, indicating that the incident excitation fields are uniform and completely independent of the z coordinate variable. In fact, the corresponding field quantities, such as the induced electric current, the scattered electric and magnetic field distributions are also independent of the z coordinate variable. According to expressions (2.19.11a) and (2.19.11b)

$$\bar{E}_0 = E_0\hat{z} \tag{3.5.3a}$$

$$\bar{H}_0 = H_{0x}\hat{x} + H_{0y}\hat{y} \tag{3.5.3b}$$

In the frequency domain, the time-harmonic incident electric and magnetic fields based on the forward traveling plane waves are given by

$$\bar{E}^i(\bar{\rho},\omega) = E_z^i(\bar{\rho},\omega)\hat{z} \tag{3.5.4a}$$

$$E_z^i(\bar{\rho},\omega) = E_0\, e^{-j\bar{k}\cdot\bar{\rho}} \tag{3.5.4b}$$

$$\bar{H}^i(\bar{\rho},\omega) = \bar{H}_0\, e^{-j\bar{k}\cdot\bar{\rho}} \tag{3.5.4c}$$

$$\bar{k} = k\hat{k} \tag{3.5.4d}$$

where

k : propagation constant for homogeneous, isotropic medium

$$= \omega(\mu\varepsilon)^{1/2} \tag{3.5.4e}$$

$\hat{k}$: actual direction of the plane wave propagation;

ω : frequency of excitation, in radians per second.

With respect to a global origin O, which represents reference phase center, let

$\bar{\rho}$: spatial vector corresponding to the field point (x, y)

$$= x\hat{x} + y\hat{y} \tag{3.5.5a}$$

$$= \rho\ \cos\phi\hat{x} + \rho\ \sin\phi\hat{y} \tag{3.5.5b}$$

$\bar{k}$: propagation vector of the traveling plane wave

$$= k_x\hat{x} + k_y\hat{y} \tag{3.5.5c}$$

$$= k\ \cos\phi^i\hat{x} + k\ \sin\phi^i\hat{y} \tag{3.5.5d}$$

ϕ^i : arbitrary angle of incidence of the TM plane wave field;

$k_z = 0$: propagation constant along the z coordinate direction.

Referring to expression (2.19.18d), it should be noted that the incident plane wave electric and magnetic fields satisfy the righthand rule and are related by

$$\bar{H}_0 \times \hat{k} = \frac{\omega\varepsilon}{k}\bar{E}_0 \tag{3.5.5f}$$

$$\frac{E_0}{H_0} = \eta \tag{3.5.5g}$$

η : intrinsic impedance of the free-space medium

$$= \left(\frac{\mu}{\varepsilon}\right)^{1/2} \tag{3.5.5h}$$

As discussed earlier, the normally excited plane wave is uniform in the z coordinate direction, and hence the axial electric currents induced and the scattered electric and magnetic fields are also uniform with no variation with respect to the z coordinate variable. Therefore, the various field expressions obtained in Sections 3.3 and 3.4 and Tables 3.5a and 3.5b, can be completely simplified by substituting either

$$\frac{\partial}{\partial z} = 0 \qquad \text{or} \qquad k_z = 0 \tag{3.5.6}$$

Table 3.6a gives the scattered electric and magnetic fields in rectangular components expressed in terms of the magnetic vector potential A_z:

$$\bar{E}(x,y,\omega) = E_z(x,y,\omega)\hat{z} \tag{3.5.7a}$$

$$\bar{H}(x,y,\omega) = H_x(x,y,\omega)\hat{x} + H_y(x,y,\omega)\hat{y} \tag{3.5.7b}$$

$$\bar{A}(x,y,\omega) = A_z(x,y,\omega)\hat{z} \tag{3.5.8}$$

Table 3.6a

Transverse magnetic fields for normal excitation in rectangular coordinates

---

$$H_x = \frac{1}{\mu}\frac{\partial A_z}{\partial y} \tag{3.5.9a}$$

$$H_y = -\frac{1}{\mu}\frac{\partial A_z}{\partial x} \tag{3.5.9b}$$

$$E_z = -j\omega A_z \tag{3.5.10}$$

The spectrum of the variable k_z varies from $-\infty$ to ∞, and the case of normal incident excitation yielding condition (3.5.6) is just a special case to isolate the z coordinate variable. Referring to Figures 3.7a and 3.7b, various field quantities can be obtained by

$\bar{A}(\bar{\rho},\omega)$: magnetic vector potential at the field point

$$= A_z(\bar{\rho},\omega)\hat{z} \tag{3.5.11a}$$

Further, using relationship (3.5.6), the z component of the magnetic vector potential satisfies the scalar Helmholtz differential equation:

$$\nabla_\tau^2 A_z(\bar{\rho},\omega) + k^2 A_z(\bar{\rho},\omega) = -\mu J_z(\bar{\rho},\omega) \tag{3.5.11b}$$

The scalar Helmholtz equation (3.5.11b) is to be solved first for the z component of magnetic vector potential. Then, Table 3.6a is used to calculate nonzero components of the electric and magnetic fields. In fact, the TM field expressions in the Table 3.6a can also be rewritten in terms of the cylindrical coordinates (ρ, ϕ), and they are presented in Table 3.6b for the cylindrical components of scattered electric and magnetic fields:

$$\bar{E}(\rho,\phi,\omega) = E_z(\rho,\phi,\omega)\hat{z} \tag{3.5.12a}$$

$$\bar{H}(\rho,\phi,\omega) = H_\rho(\rho,\phi,\omega)\hat{\rho} + H_\phi(\rho,\phi,\omega)\hat{\phi} \tag{3.5.12b}$$

$$\bar{A}(\rho,\phi,\omega) = A_z(\rho,\phi,\omega)\hat{z} \tag{3.5.12c}$$

Table 3.6b

Transverse magnetic fields for normal Excitation in cylindrical coordinates

---

$$H_\rho = \frac{1}{\mu}\frac{1}{\rho}\frac{\partial A_z}{\partial \phi} \tag{3.5.13a}$$

$$H_\phi = -\frac{1}{\mu}\frac{\partial A_z}{\partial \rho} \tag{3.5.13b}$$

$$E_z = -j\omega A_z \tag{3.5.14}$$

3.6 EQUIVALENT ELECTRIC CURRENT SOURCES

The righthand-side source term in the Helmholtz differential equation (3.5.11c) should include the complete distribution of z-directed surface electric currents on the conducting object. In fact, Section 3.4 discusses the solution for a magnetic vector potential satisfying a similar type of scalar differential equation, but with an ideal z-directed line source electric current located in a linear, homogeneous, and isotropic medium. The solution given by expression (3.4.9b) can now be utilized along with superposition of the line current sources to obtain the complete solution for the Helmholtz scalar equation (3.5.11c).

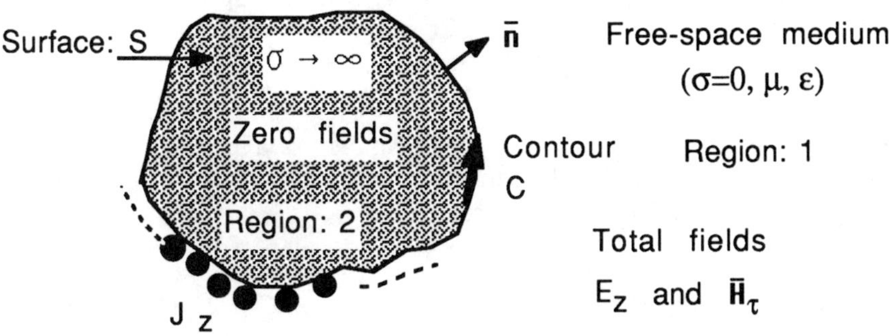

Figure 3.8a Original conducting scatterer boundary value problem.

To formulate a rigorous boundary value integral equation, an electromagnetic equivalent problem is set up using an equivalence principle, which is discussed in the following. The sources simulated in the equivalent problem are generally referred to as the equivalent or virtual electric current sources, but are located in a linear, homogeneous, and isotropic medium. The electric and magnetic fields produced by the equivalent currents are identical to the electric and magnetic fields of the original boundary value problem and also satisfy the relevant conducting surface boundary conditions. Figure 3.8a shows the original boundary value problem of a perfectly conducting scatterer. Only the cross section of the geometry at $z = 0$ plane is shown. The arbitrary shaped conducting scattering geometry has a cross-sectional area S which is bounded by a contour C. For normal excitation, the incident and scattered electric and magnetic fields are independent of the z coordinate variable.

Original Conducting Scatterer

Referring to the Figure 3.8a, let

ε : permittivity of the free-space medium, region 1;

μ : permeability of the free-space medium, region 1;

$\sigma = 0$: conductivity of the free-space medium, region 1;

$\sigma => \infty$: conductivity of the scatterer medium, region 2;

$(E_z^i, \bar{H}_\tau^i)$: electric and magnetic incident fields in the free-space medium;

$(E_z^s, \bar{H}_\tau^s)$: electric and magnetic scattered fields in the free-space medium;

$(E_z, \bar{H}_\tau)$: electric and magnetic total fields in the free space medium

$$= (E_z^i, \bar{H}_\tau^i) + (E_z^s, \bar{H}_\tau^s)$$

Referring to the conducting boundary conditions discussed in Section 2.16, no electric field or magnetic field is inside the perfectly conducting region 2. On contour C, the following boundary conditions are satisfied:

$$\bar{J}_z(\bar{\rho}') = \hat{n}' \times \left[\bar{H}_\tau^i(\bar{\rho}') + \bar{H}_\tau^s(\bar{\rho}')\right] \qquad \bar{\rho}' \text{ on } C \tag{3.6.1a}$$

$$E_z(\bar{\rho})\Big|_{\tan} = \left[E_z^i(\bar{\rho}) + E_z^s(\bar{\rho})\right] = 0 \qquad \bar{\rho} \text{ on } C \tag{3.6.1b}$$

where

tan : tangential component of the field;

$\hat{n}'$: unit vector normal to the surface C.

Equivalent Scatterer Geometry (valid for region 1 only)

In the following, a step-by-step procedure is adopted to set up an equivalent boundary value problem. Referring to Figure 3.8b, a conducting scatterer geometry is shown that is identical in shape to the original conducting scatterer, except a virtual contour C_+ is drawn just outside the scatterer boundary.

On the virtual contour C_+, a certain distribution of z-directed electric currents are simulated. In a limit, as the virtual contour C_+ tends to the original contour C, the simulated electric currents are such that they produce the same scattered electric and magnetic fields in the free-space medium that are in the original conducting scatterer problem, as shown in the Figure 3.8a. The simulated electric currents on the virtual boundary are generally referred to as the *equivalent* or *virtual electric currents*. Inside the virtual contour C+, the total electric and magnetic fields are zero. Hence, the region 2 conducting material medium can be completely removed from the inside region of the virtual contour and is totally replaced by a medium having properties identical to the region 1 free-space medium.

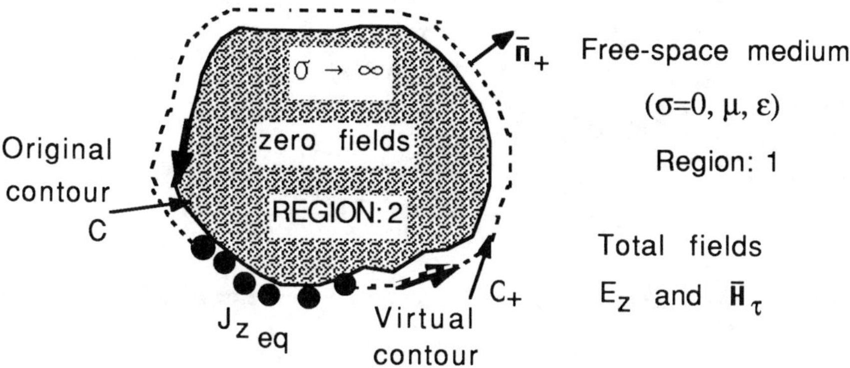

Figure 3.8b Equivalent scatterer of the original problem.

The equivalent boundary value scattering problem is shown in Figure 3.8c. Corresponding to the equivalent problem,

ε : permittivity of the free-space medium, for regions 1 and 2;

μ : permeability of the free-space medium, for regions 1 and 2;

$\sigma = 0$: conductivity of the free-space medium, for regions 1 and 2;

$(E_z^i, \bar{H}_\tau^i)$: the same electric and magnetic incident fields as in region 1;

$(E_z^s, \bar{H}_\tau^s)$: the same electric and magnetic scattered fields as in the original.

Boundary Value Problem Valid Only for the Region 1

Referring to Figure 3.8c, the equivalent electric currents on the virtual contour C_+ also satisfy the following boundary conditions:

$\bar{J}_{z_{eq}}(\bar{\rho}')$: equivalent electric currents valid only for the region 1 fields

$$= \hat{n}' \times \bar{H}_\tau(\bar{\rho}') \qquad \bar{\rho}' \text{ on } C_+ \rightarrow C \tag{3.6.2a}$$

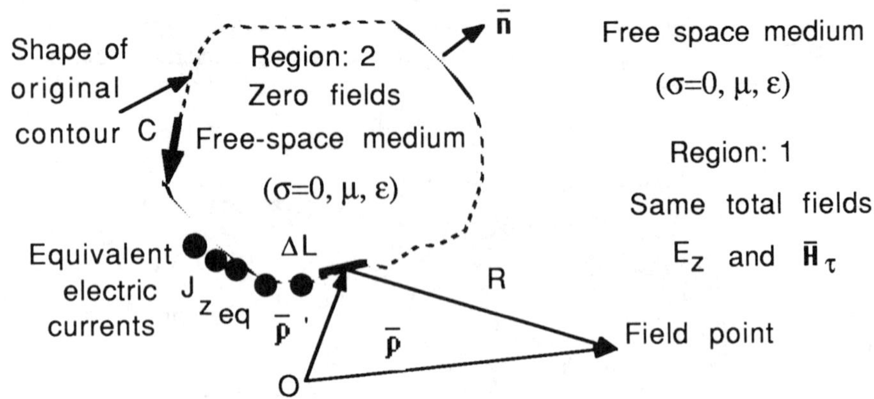

Figure 3.8c Electromagnetic equivalence for conducting scatterer valid for region 1.

and

$$E_z(\bar{\rho})\Big|_{\tan} = 0 \qquad \bar{\rho} \text{ on } C_+ \rightarrow C \tag{3.6.2b}$$

Based on the *z*-directed equivalent electric current density distribution, various scattered electric and magnetic fields valid only for region 1 can now be calculated using the expressions given in the Table 3.6b. Referring to the Figure 3.8c, the *z*-directed electric current within an incremental length ΔL on virtual contour C can be treated like an ideal line source electric current in a free-space medium. Expression (3.4.9b) for the two-dimensional Green's function is now utilized to construct a superposition integral for the total *z* component of the magnetic vector potential. Let

$dA_z(\bar{\rho},\omega)$: incremental component of the magnetic vector potential;

$J_{z_{eq}}(\bar{\rho}',\omega)dL(\bar{\rho}')$: elemental line source electric current, in amperes;

$\bar{\rho}$: field observation point in region 1 outside contour C;

$\bar{\rho}'$: source point of line electric current on contour C.

Based on expression (3.4.9b), the incremental *z* component of the magnetic vector potential is given by

$$dA_z(\bar{\rho},\omega) = \frac{\mu}{4j} J_{z_{eq}}(\bar{\rho}',\omega)\, H_0^{(2)}(k|\bar{\rho} - \bar{\rho}'|)\, dL(\bar{\rho}') \tag{3.6.3a}$$

Hence, the total magnetic vector potential at any field observation point in region 1 outside the contour C is

$$A_z(\bar{\rho},\omega) = \frac{\mu}{4j}\int_C J_{z_{eq}}(\bar{\rho}',\omega)\, H_0^{(2)}(k|\bar{\rho} - \bar{\rho}'|)\, dL(\bar{\rho}') \tag{3.6.3b}$$

Referring to the Table 3.6b, various components of the electric and magnetic field distributions can be calculated by substituting result (3.6.3b) for the magnetic vector potential. For example, the z component of the scattered electric field and the angular component of the scattered magnetic field are given by

$$E_z^s(\bar{\rho},\omega) = \frac{-\omega\mu}{4}\int_C J_{z_{eq}}(\bar{\rho}',\omega)\, H_0^{(2)}(k|\bar{\rho} - \bar{\rho}'|)\, dL(\bar{\rho}') \tag{3.6.3c}$$

$$H_\phi^s(\bar{\rho},\omega) = \frac{-1}{4j}\int_C J_{z_{eq}}(\bar{\rho}',\omega)\Big[\frac{\partial}{\partial\rho} H_0^{(2)}(k|\bar{\rho} - \bar{\rho}'|)\Big]\, dL(\bar{\rho}') \tag{3.6.3d}$$

The scattered electric and magnetic field expressions obtained here are quite useful in the study of electromagnetic scattering and interaction by the arbitrary shaped, two-dimensional conducting object. It should be noted that the results are expressed in terms of an integral expression. The integrand of this integral representation contains the distribution of unknown electric current surface density distribution. Detailed discussions are presented in the following sections to determine the unknown distribution of the electric current for certain class of canonical and arbitrary shaped cross-sectional geometries.

3.6.1 EFIE – PERFECT CONDUCTOR

An *electric field integral equation* (EFIE) in terms of the unknown equivalent electric current distribution can be obtained by enforcing the z component of the *total electric field to be zero* on contour C. At any point in region 1 outside the scatterer contour C, the tangential total electric field or the z component of the total electric field is given by the sum of the z components of incident electric field and scattered electric field:

$$E_z(\bar{\rho}) = E_z^i(\bar{\rho}) + E_z^s(\bar{\rho}) \qquad \bar{\rho} \text{ in region 1} \tag{3.6.4a}$$

After substituting for the z component of the scattered electric field derived in expression (3.6.3c),

$$E_z(\bar{\rho}) = E_z^i(\bar{\rho}) + \frac{-\omega\mu}{4}\int_C J_{z_{eq}}(\bar{\rho}',\omega)\, H_0^{(2)}(k|\bar{\rho} - \bar{\rho}'|)\, dL(\bar{\rho}') \qquad (3.6.4b)$$

$\bar{\rho}$ in region 1

The electric field integral equation for the arbitrary shaped, perfectly conducting scatterer with transverse magnetic plane wave excitation is obtained by moving the field observation point directly onto the contour C to yield

$$E_z^i(\bar{\rho},\omega) = \frac{\omega\mu}{4}\int_C J_{z_{eq}}(\bar{\rho}',\omega)\, H_0^{(2)}(k|\bar{\rho} - \bar{\rho}'|)\, dL(\bar{\rho}') \qquad (3.6.4c)$$

$\bar{\rho}$ on C

In most of the further development, the subscript *eq* on the electric current distribution is dropped, and it is still understood that the currents are the electromagnetic equivalent currents. The EFIE expression for the perfectly conducting scatterer can be written in a compact notation given by

$$E_z^i(\bar{\rho}) = \mathcal{L}[J_{z_{eq}}(\bar{\rho}')] \qquad (3.6.5a)$$

$\bar{\rho}$ on C

$$\mathcal{L} = \frac{\omega\mu}{4}\int_C H_0^{(2)}(k|\bar{\rho} - \bar{\rho}'|)\, dL(\bar{\rho}') \qquad (3.6.5b)$$

where

$\mathcal{L}$: EFIE integral operator for TM polarization.

3.7 METHOD OF MOMENTS

Invariably the electromagnetic scattering and interaction problem can be formulated in terms of an integral equation or a differential equation to yield a generalized operator type equation. In this section, a general approach to analyze an operator-type equation is discussed. Depending upon the complexity of the operator equation, the approach discussed in the following leads eventually to a numerical formalism. In the electric field integral equation, expression (3.6.5a), the current distribution $\bar{J}(\bar{r}')$ is unknown. It can be determined by defining first a scalar product given by

$$<\bar{F}, \bar{W}_m> = \int_C \bar{F}(\bar{r}) \bullet \bar{W}_m(\bar{r})\, dL(\bar{r}) \tag{3.7.1}$$

where

$\bar{W}_m(\bar{r})$: a set of independent weighting functions in the domain of C for $m = 1, 2, 3, \ldots, M$.

The weighting functions can be continuous or piecewise continuous or discrete in the domain of validity of the integral equation, in this case on boundary contour C. In fact, integral equation (3.6.5a) is valid at every point on the contour C. This validity can be enforced in the domain of contour C based on the scalar product defined above. Hence, integral equation (3.6.5a) can now be tested on both sides based on the scalar product defined in expression (3.7.1) to obtain an equivalent functional form of the equation:

$$< L[\bar{J}(\bar{r}')], \bar{W}_m(\bar{r}) > \Big|_{\tan} = < \bar{E}^i(\bar{r}), \bar{W}_m(\bar{r}) > \Big|_{\tan} \tag{3.7.2}$$

$$m = 1, 2, 3, \ldots, M.$$

Expression (3.7.2) represents M set of linear functional equations. There are a number of different approaches to solve for the unknown quantity in the functional equations. One popular method is discussed in the following based on approximating the distribution of the unknown quanity. Hence, the nature of accuracy of the solution completely depends on the type of approximate representation. Let the unknown electric current distribution on the contour C be represented in terms of the following current expansion function:

$$\bar{J}(\bar{r}') \approx \sum_{n=1}^{N} I_n \bar{P}_n(\bar{r}') \tag{3.7.3}$$

where

$\bar{P}_n(\bar{r})$: a set of independent current expansion functions in the domain of C, for $n = 1, 2, 3, \ldots, N$.

The current expansion functions can be continuous or piecewise continuous or discrete in the domain of validity, in this case on boundary contour C. On substituting current expansion (3.7.3) into functional expression (3.7.2) and assuming the integral operator to be linear, the following set of linearly independent equations is obtained:

$$< L[\sum_{n=1}^{N} I_n \bar{P}_n(\bar{r}')], \bar{W}_m(\bar{r}) >\Big|_{\tan} = < \bar{E}^i(\bar{r}), \bar{W}_m(\bar{r}) >\Big|_{\tan} \qquad (3.7.4)$$

$$m = 1, 2, 3, \ldots, M$$

$$\sum_{n=1}^{N} I_n < L[\bar{P}_n(\bar{r}')], \bar{W}_m(\bar{r}) >\Big|_{\tan} = < \bar{E}^i(\bar{r}), \bar{W}_m(\bar{r}) >\Big|_{\tan} \qquad (3.7.5)$$

$$m = 1, 2, 3, \ldots, M$$

Expression (3.7.5) represents a set of M linear simultaneous equations with N unknowns quantities. By choosing the indices $M = N$, the set of linear equation (3.7.5) reduces to the corresponding square matrix equation. Then, the resulting matrix equation can be solved for the various unknown independent current coefficients for a given type of excitation. As was noted earlier, once the electric current distribution $\bar{J}(\bar{r}')$ is known on contour C, all other scattering and interaction properties can be determined based on the electromagnetic field expressions given in the Table 3.6b. This procedure is generally referred to as the *method of moments* technique for solving a linear operator-type equation, and it is quite straightforward to implement for a given linear integral or differential or integro-differential equation. Generally, this procedure leads to an eventual matrix solution that can be numerically solved by developing an algorithm on a digital computer.

In the following sections, specific two-dimensional conducting geometries, such as circular cylindrical scatterer, square and rectangular scatterers, and thin strip scatterer are considered. These examples expose detailed insight into the selection and implementation of the weighting and expansion functions.

3.8 PERFECTLY CONDUCTING CIRCULAR CYLINDER

The electric field integral equation, expression (3.6.4c) derived in Section 3.6.1, is now applied to analyze electromagnetic scattering by a perfectly conducting, circular cylinder externally excited by a TM polarized plane wave. Since the structure is perfectly conducting, the analysis is applicable for any circular cross section, either hollow or solid geometry. The circular geometry of radius a, shown in Figure 3.9, has perfect symmetry with angular variable ϕ of cylindrical coordinate system. Hence, the scattering analysis can be carried out by just assuming the incident plane wave to be propagating along the x coordinate direction, which gives the angle of incidence to be zero. The electric current induced along circular contour C varies with respect to the angular variable and, in fact, repeats itself for every 2π radians. Hence, the electric current on the circular scatterer can be represented in terms of an infinite Fourier series.

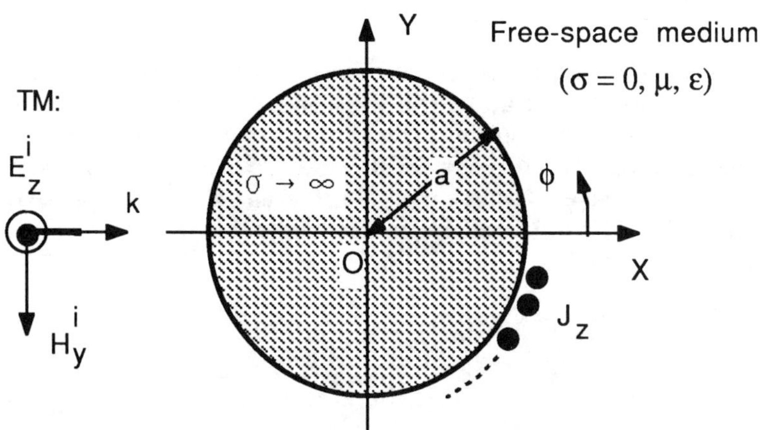

Figure 3.9 Perfectly conducting circular scatterer – TM excitation.

Referring to expression (3.6.3c), for $\bar{\rho}$ outside C the z component of scattered electric field is given by

$$E_z^S(\bar{\rho},\omega) = -\frac{\omega\mu}{4}\int_C J_z(\bar{\rho}',\omega)\, H_0^{(2)}(k|\bar{\rho}-\bar{\rho}'|)\, dL(\bar{\rho}') \tag{3.8.1a}$$

The electric current is now expanded in terms of the following Fourier series:

$$J_z(\rho'=a,\phi',\omega) = \sum_{\ell=-\infty}^{\infty} A_\ell\, e^{j\ell\phi'} \tag{3.8.1b}$$

A_ℓ : Fourier coefficient of the current expansion

Referring back to the discussion in Section 3.7, the electric current expansion just selected represents a sum over a set of independent expansion terms, and each term is an entire domain harmonic function completely covering the domain of contour C. Current expansion (3.8.1b) is now substituted into expression (3.8.1a) to obtain

$$E_z^S(\rho,\phi,\omega) = -\frac{\omega\mu}{4}\int_{\phi'=0}^{2\pi}\Big[\sum_{\ell=-\infty}^{\infty} A_\ell e^{j\ell\phi'}\Big] H_0^{(2)}(k|\bar{\rho} - \bar{\rho}'|)\, a\, d\phi' \quad (3.8.2)$$

$\bar{\rho}$ outside C

The zero-order Hankel function term in this expression represents, in fact, outgoing cylindrical waves from the electric current line source elements located on contour C. It is possible to rewrite all the outgoing cylindrical waves in terms of a set of equivalent cylindrical modes, which are all referred with respect to a single reference point (such as the origin) by applying the coordinate transformation theorem. Based on the coordinate shift, Appendix B, the Hankel function term can be written as

$$H_0^{(2)}(k|\bar{\rho} - \bar{\rho}'|) = \sum_{n=-\infty}^{\infty} J_n(k\rho')H_n^{(2)}(k\rho)\, e^{jn(\phi - \phi')} \quad (3.8.3a)$$

for $\bar{\rho} \geq \bar{\rho}'$

and

$$H_0^{(2)}(k|\bar{\rho} - \bar{\rho}'|) = \sum_{n=-\infty}^{\infty} J_n(k\rho)H_n^{(2)}(k\rho')\, e^{jn(\phi - \phi')} \quad (3.8.3b)$$

for $\bar{\rho} \leq \bar{\rho}'$

After substituting the expression (3.8.3a), expression (3.8.2) for the z component of scattered electric field takes the following form:

$$E_z^S(\rho,\phi,\omega) = -\frac{\omega\mu}{4}\int_{\phi'=0}^{2\pi}\Big[\sum_{\ell=-\infty}^{\infty} A_\ell\, e^{j\ell\phi'}\Big]$$

$$\Big\{\sum_{n=-\infty}^{\infty} J_n(ka)\, e^{-jn\phi'}\big[H_n^{(2)}(k\rho)\, e^{jn\phi}\big]\Big\}\, a\, d\phi' \quad (3.8.3c)$$

$\bar{\rho}$ outside C

In the double infinite summation, one infinite summation can be easily performed for all possible values ℓ ranging from $-\infty$ to ∞ by utilizing the orthogonality property of the harmonic functions. For all values of $\ell = n$ in the range from $-\infty$ to ∞, the integral exists and is zero for all values of $\ell \neq n$ in expression (3.8.3c). Hence, the following simplification is obtained:

$$E_z^s(\rho,\phi,\omega) = \sum_{n=-\infty}^{\infty} a_n H_n^{(2)}(k\rho)\, e^{jn\phi} \qquad \bar{\rho} \text{ outside } C \qquad (3.8.4a)$$

where the scattered electric field coefficients a_n in terms of the electric current Fourier coefficients are given by

$$a_n = -\frac{\pi\omega\mu a}{2} J_n(ka) A_n \qquad (3.8.4b)$$

To determine the scattered electric field coefficients, the total tangential electric field boundary condition, expression (3.6.2b), on the perfectly conducting cylinder is enforced:

$$E_z^i(\bar{\rho}) + E_z^s(\bar{\rho}) = 0 \qquad \bar{\rho} \text{ on } C \qquad (3.6.5a)$$

On substituting for the z component of the scattered electric field

$$E_z^i(\rho=a,\phi) = \sum_{n=-\infty}^{\infty} - a_n H_n^{(2)}(ka)\, e^{jn\phi} \qquad \bar{\rho} \text{ on } C \qquad (3.8.5b)$$

Referring to this boundary value relationship, the term on lefthand side is the excitation plane wave and the term on righthand side is the scattered electric field written in terms of a set of outgoing cylindrical wave modes. To determine the scattered field coefficients a_n, it is necessary to assess how much of the incident excitation actually couples to each outgoing scattered cylindrical mode. This can be accomplished based on a wave transformation, where the given plane wave is also expressed in terms of a set of cylindrical wave modes. Referring to Section 3.5, expressions (3.5.4) and (3.5.5), the TM polarized incident plane wave is given by

$$E_z^i(\bar{\rho},\omega) = E_0\, e^{-j\bar{k}\cdot\bar{\rho}} \qquad (3.8.6a)$$

For a plane wave propagating in the x coordinate direction, the angle of incidence of the TM plane wave field, $\phi^i = 0$. The following relationship is obtained by expressing it in terms of an infinite series of cylindrical wave modes:

$$E_z^i(\rho,\phi) = E_0\, e^{-jk\rho\cos\phi} \tag{3.8.7a}$$

$$= E_0 \sum_{m=-\infty}^{\infty} b_m J_m(k\rho)\, e^{jm\phi} \tag{3.8.7b}$$

b_m : mode amplitude to be determined.

To determine the various mode amplitudes in relationship (3.8.7b), again the orthogonality property of the harmonic functions is utilized. After multiplying both sides by $exp(-jn\phi)$ and integrating between the limits 0 to 2π,

$$\int_{\phi=0}^{2\pi} e^{-jk\rho\cos\phi} e^{-jn\phi}\, d\phi$$

$$= \int_{\phi=0}^{2\pi} \sum_{m=-\infty}^{\infty} b_m J_m(k\rho)\, e^{jm\phi}\, e^{-jn\phi}\, d\phi \tag{3.8.7c}$$

$$= 2\pi b_m J_m(k\rho) \qquad \text{for } m = n \tag{3.8.7d}$$

$$= 0 \qquad \text{for } m \neq n \tag{3.8.7e}$$

Hence,

$$\int_{\phi=0}^{2\pi} e^{-jk\rho\cos\phi} e^{-jm\phi}\, d\phi = 2\pi b_m J_m(k\rho) \tag{3.8.8a}$$

The coefficient b_m can be determined by taking the mth order derivative on both sides of expression (3.8.8a) and evaluating it in a limit as ρ tends to zero. Referring to Appendix B, this relationship can be easily recognized as the integral representation for the Bessel function of first kind and order m. Thus,

$$J_m(k\rho) = \frac{j^m}{2\pi} \int_{\phi=0}^{2\pi} e^{-j(k\rho\cos\phi + m\phi)}\, d\phi \tag{3.8.8b}$$

and

$$b_m = j^{-m} \tag{3.8.8c}$$

Hence, the plane wave excitation, expressions (3.8.7a) and (3.8.7b), propagating in the x coordinate direction can be represented in terms of an infinite series of cylindrical modes given by

$$E_z^i(\rho,\phi) = E_0 \sum_{m=-\infty}^{\infty} j^{-m} J_m(k\rho)\, e^{jm\phi} \tag{3.8.9}$$

After substituting the plane wave representation, expression (3.8.9), into boundary condition relationship (3.8.5b), the scattered electric field mode coefficients, a_n, can be determined by

$$E_0 \sum_{m=-\infty}^{\infty} j^{-m} J_m(ka)\, e^{jm\phi} = \sum_{n=-\infty}^{\infty} - a_n H_n^{(2)}(ka)\, e^{jn\phi} \tag{3.8.10a}$$

Now a set of independent weighting terms are selected to test expression (3.8.10a) on both sides, where each weighting term forms an entire domain harmonic function in the domain of scatterer contour C. To determine the scattered field mode coefficients, relationship (3.8.10a) is multiplied on both sides by the entire domain weighting function $exp(-j\ell\phi)$ and integrated between the limits 0 to 2π with respect to the angular variable ϕ. Thus, for all integer values between $\ell = -\infty$ to ∞,

$$\int_{\phi=0}^{2\pi} E_0 \sum_{m=-\infty}^{\infty} j^{-m} J_m(ka)\, e^{jm\phi} e^{-j\ell\phi}\, d\phi$$

$$= \int_{\phi=0}^{2\pi} \sum_{n=-\infty}^{\infty} - a_n H_n^{(2)}(ka)\, e^{jn\phi} e^{-j\ell\phi}\, d\phi \tag{3.8.10b}$$

Therefore, enforcing the orthogonality in tested relationship (3.8.10b), the scattered electric field coefficients are obtained as

$$a_n = - \frac{j^{-n} J_n(ka) E_0}{H_n^{(2)}(ka)} \tag{3.8.10c}$$

These scattered field coefficients are now substituted into expression (3.8.4a). The expression for the z component of the scattered electric field distribution in the free-space medium (outside the perfectly conducting circular scatterer geometry) is given by

$$E_z^s(\rho,\phi,\omega) = -E_0 \sum_{n=-\infty}^{\infty} \frac{j^{-n} J_n(ka)}{H_n^{(2)}(ka)} H_n^{(2)}(k\rho)\, e^{jn\phi} \tag{3.8.10d}$$

$\bar{\rho}$ outside C

Using the field expressions given in Table 3.6b, the radial and the angular components of the scattered magnetic field distributions in the free-space medium can be calculated. Further, the electric current Fourier expansion coefficients are obtained by expressions (3.8.4b) and (3.8.10c):

$$A_n = \frac{2j^{-n} E_0}{\pi\omega\mu a H_n^{(2)}(ka)} \tag{3.8.11a}$$

The z-directed electric current density distribution, in amperes per meter, on contour C of the perfectly conducting circular cylinder is now obtained by substituting the preceding Fourier coefficients into current expansion (3.8.1b):

$$J_z(\phi') = J_z(\rho'=a,\phi',\omega) \tag{3.8.11b}$$

$$= \frac{2E_0}{\pi\eta ka} \sum_{n=-\infty}^{\infty} \frac{j^{-n}}{H_n^{(2)}(ka)} e^{jn\phi'} \tag{3.8.11c}$$

a : radius of the circular conducting cylinder, in meters;

η : intrinsic impedance of the free-space medium, in ohms.

Expression (3.8.11c) is now calculated for various values of the electrical radius ka. The Hankel functions of order n and the real argument are well documented in the literature. Figure 3.10 shows the distribution of an induced electric current density, in amperes per meter, on a perfectly conducting circular cylinder of electrical radius $ka = 1$ plotted as a function of angular variable ϕ. The incident electric field is assumed to be 1 volt per meter and the distribution of the electric current, shown in the Figure 3.10, is normalized with respect to the amplitude of incident magnetic field. Only the distribution

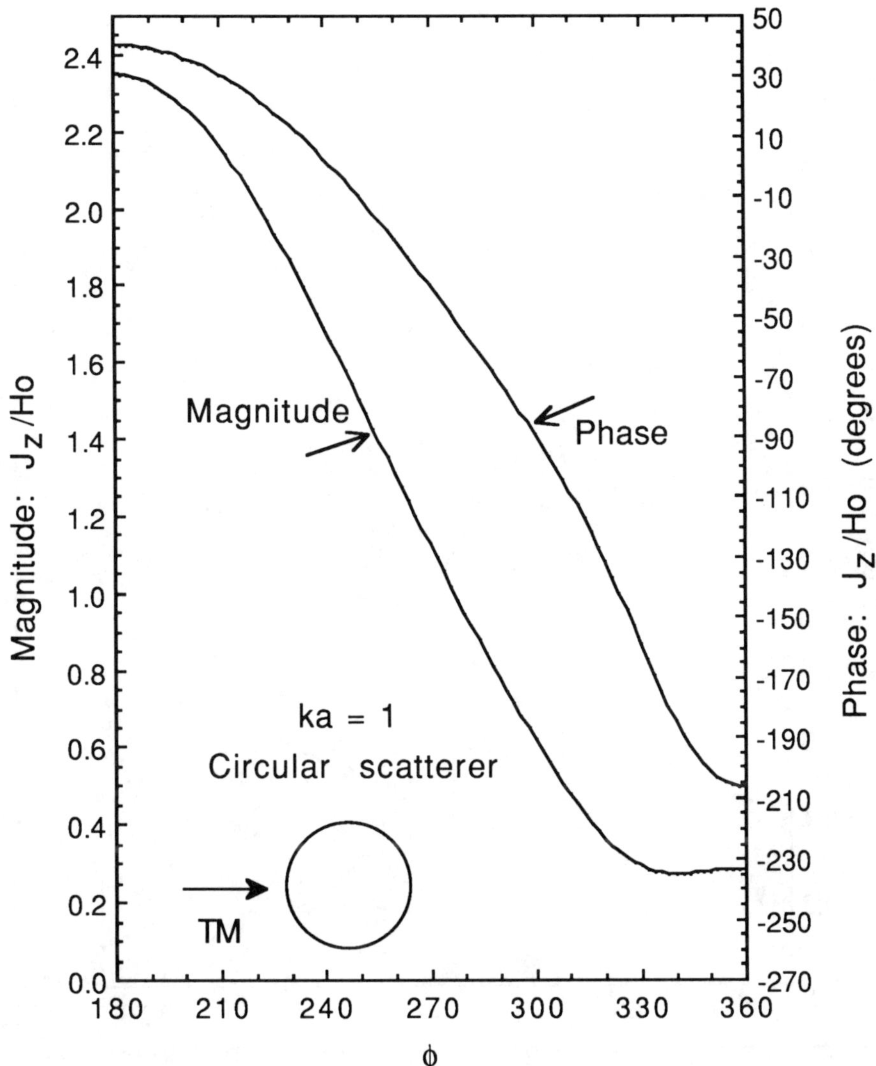

Figure 3.10 Magnitude and phase of the electric current distribution on a perfectly conducting circular cylinder – TM excitation.

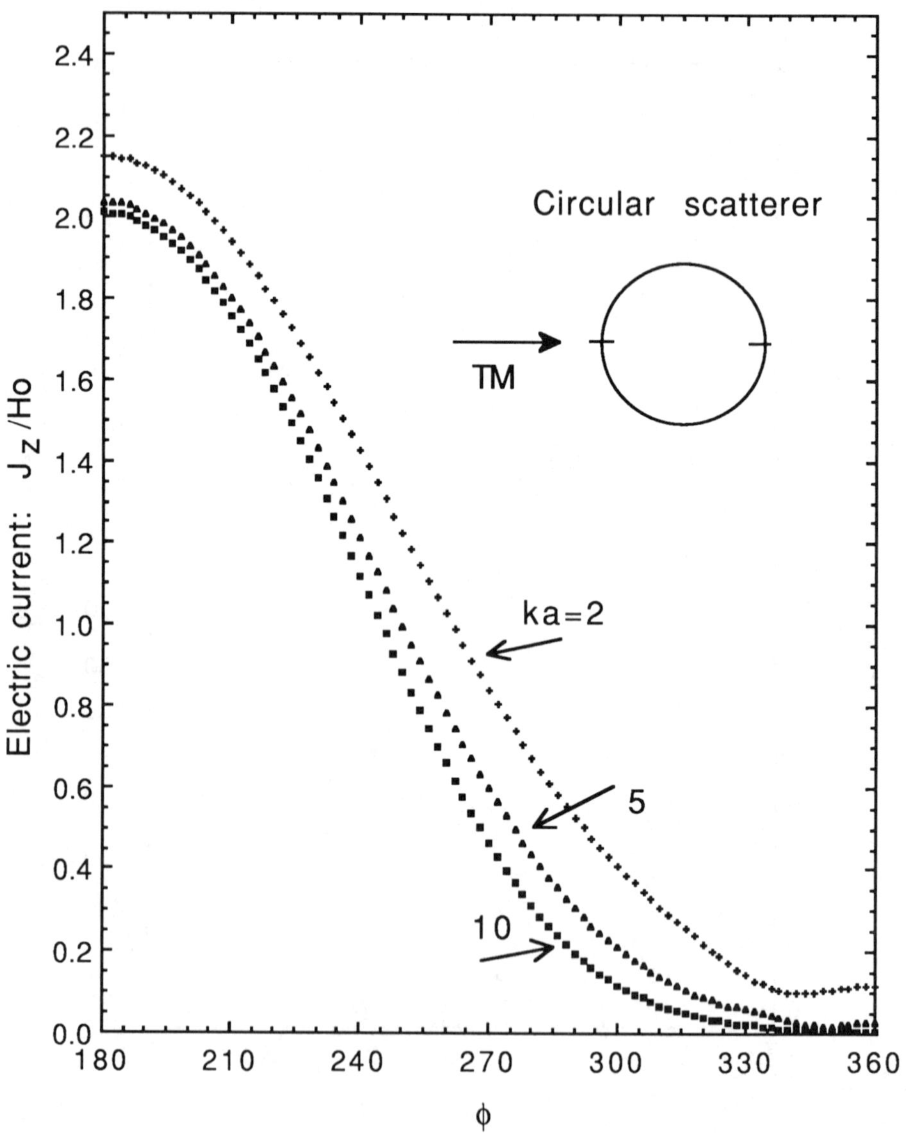

Figure 3.11 Magnitude of the electric current distribution for different radii on a perfectly conducting circular cylinder – TM excitation.

in the angular range of 180° to 360° is shown, because it is symmetrical in the remaining angular range. Specifically, the distribution of the induced electric current is symmetrical with respect to the incident angle. Referring to the Figure 3.9, the angle $\phi = 180^{o}$ corresponds to the illuminated side, and similarly, the angle $\phi = 0^{o}$ corresponds to the shadow side of the circular cylindrical scatterer. The magnitude of the electric current drops off gradually from the illuminated side toward the shadow side. Further, the phase distribution of the electric current gradually changes from the illuminated to the shadow side. For the size $ka = 1$, the phase of the electric current changes by approximately 247°.

Figure 3.11 shows, the magnitude distribution of electric current density for the electrical size $ka = 2$, 5, and 10. As can be seen, when the radius of the scatterer is gradually increased, the distribution of the electric current on the illuminated side gradually approaches to physical optics limit, and further, the induced electric current reaches very small values in the deep shadow region. In the numerical results shown, for the cases of $ka = 1$ and 2, only the first ten positive and negative cylindrical modes are included. Similarly, for the cases of $ka = 5$ and 10, only the first twenty positive and negative cylindrical modes are included in the calculation.

3.8.1 THIN CONDUCTING WIRE

When the electrical radius ka of the circular conducting cylinder is very small, referring to the Figure 3.3, the conducting cylinder behaves like a thin filament of electric current. For such a structure, there is no angular variation in the distribution of the electric current. Hence, only the $n = 0$ term is dominant in expression (3.8.11c).

Therefore, the distribution of the electric current density on an infinitely long, thin perfectly conducting wire is given by

$$J_z(a,\phi')\Big|_{ka<<1} = \frac{2E_0}{\pi\eta ka}\,\frac{1}{H_0^{(2)}(ka)} \qquad (3.8.12a)$$

where

$$H_0^{(2)}(ka) = J_0(ka) - jY_0(ka) \qquad (3.8.12b)$$

For $ka << 1$, Appendix B, the zero-order Bessel and Neumann functions are given by their small argument approximations:

$$J_0(ka) \approx 1 \qquad (3.8.13a)$$

$$Y_0(ka) \approx \frac{2}{\pi}\, \ell n\left(\frac{\gamma ka}{2}\right) \tag{3.8.13b}$$

where $\gamma = 1.781$ and $\ell n\ \gamma = 0.5772$ is Euler's constant. For very small values of *ka*, the Bessel function of second kind and zero-order can be further approximated as

$$Y_0(ka) \approx \frac{2}{\pi}\, \ell n(ka) \tag{3.8.13c}$$

Since the distribution of the current density is uniform, the *z*-directed total filamentary current on the infinitely long conducting wire is given by

$$I_z = \int_{\phi=0}^{2\pi} J_z\, a\, d\phi \tag{3.8.14a}$$

$$= \frac{2\pi j E_0}{k\eta} \frac{1}{\ell n(ka)} \tag{3.8.14b}$$

Hence, the electric current induced along the thin conducting wire is out of phase by 90° with respect to the incident plane wave excitation.

3.8.2 NEAR- AND FAR-FIELD DISTRIBUTIONS

The scattered electric and magnetic field distributions in the free-space medium outside the perfectly conducting circular scatterer are given by general expressions (3.8.10d), (3.5.13a), and (3.5.13b). It should be noted that the first and second kind of Bessel and Hankel functions of real argument and integer order are well documented in the literature. By substituting the field observation point (ρ, ϕ) and performing the summation over N cylindrical wave modes, the scattered electric field distribution given by expression (3.8.10d) can be calculated. In the numerical calculation, enough modes should be included to obtain convergence of the infinite series solution. Generally, the number of modes required for performing the summation is just more than the electrical distance of the observation point. Thus, for n positive, the upper limit N is taken to be $|k\rho|$ plus an additional five to ten modes, depending upon the numerical accuracy required. Further, the near total electric field is given by sum of the scattered field expression (3.8.10d) and the incident field expression (3.8.9).

Figure 3.12a shows the distribution of the *z* component of total electric field in the neighborhood of a perfectly conducting circular cylinder, $ka = 5$. The numerical results

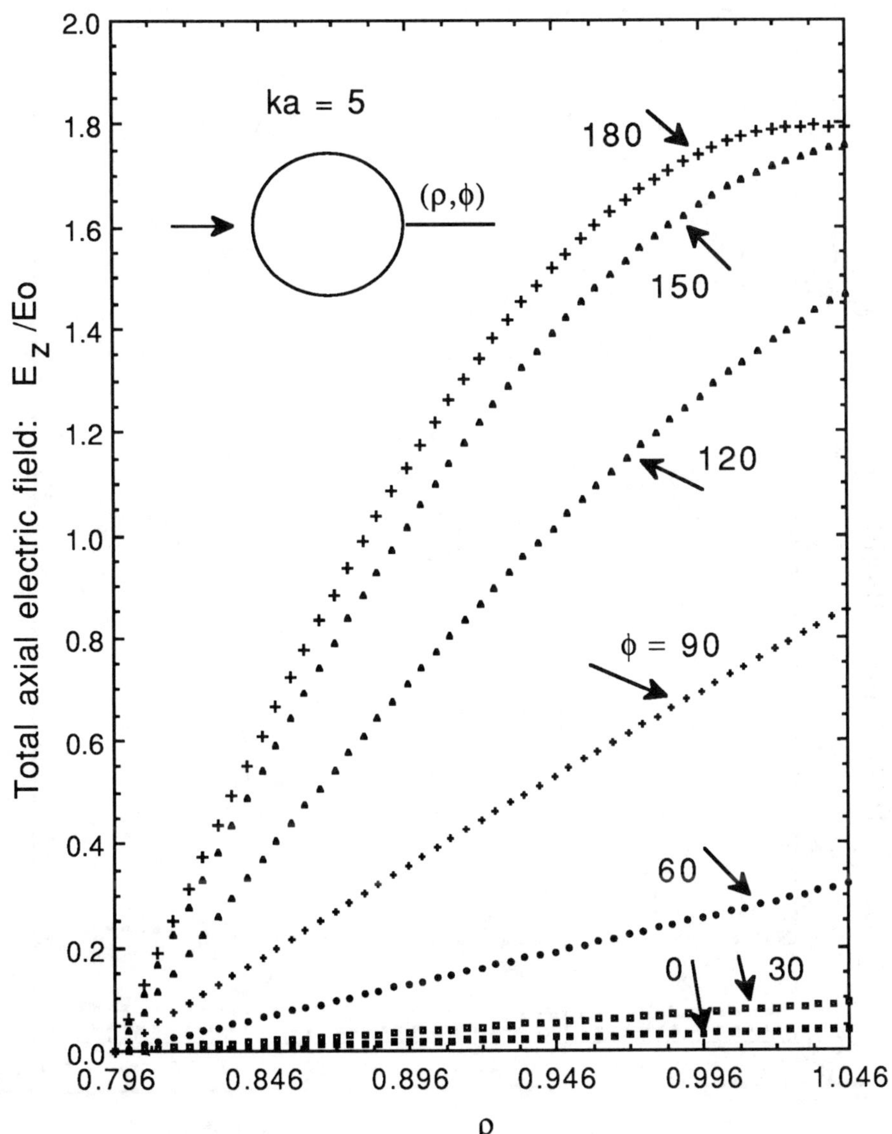

Figure 3.12a Magnitude of the total axial near electric field distribution on a perfectly conducting circular cylinder – TM excitation.

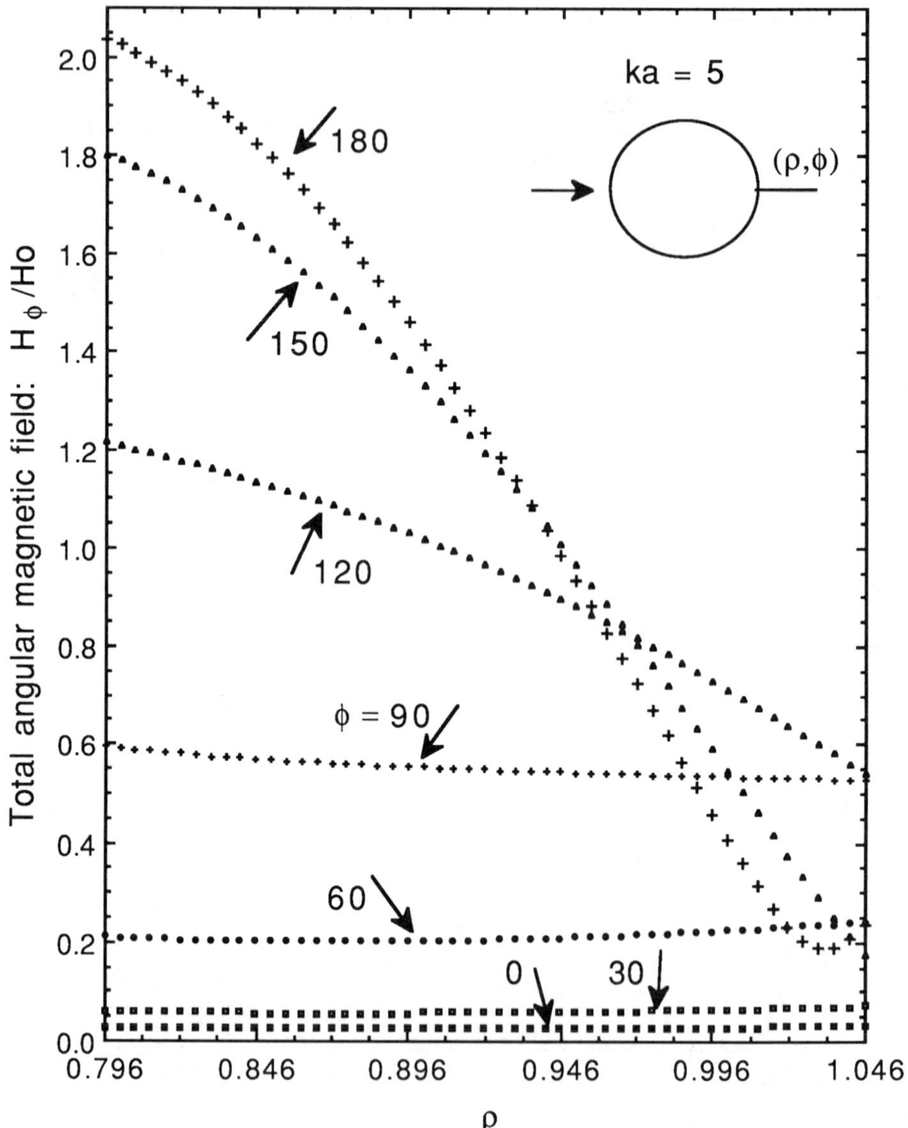

Figure 3.12b Magnitude of the total angular near magnetic field distribution on a perfectly conducting circular cylinder – TM excitation.

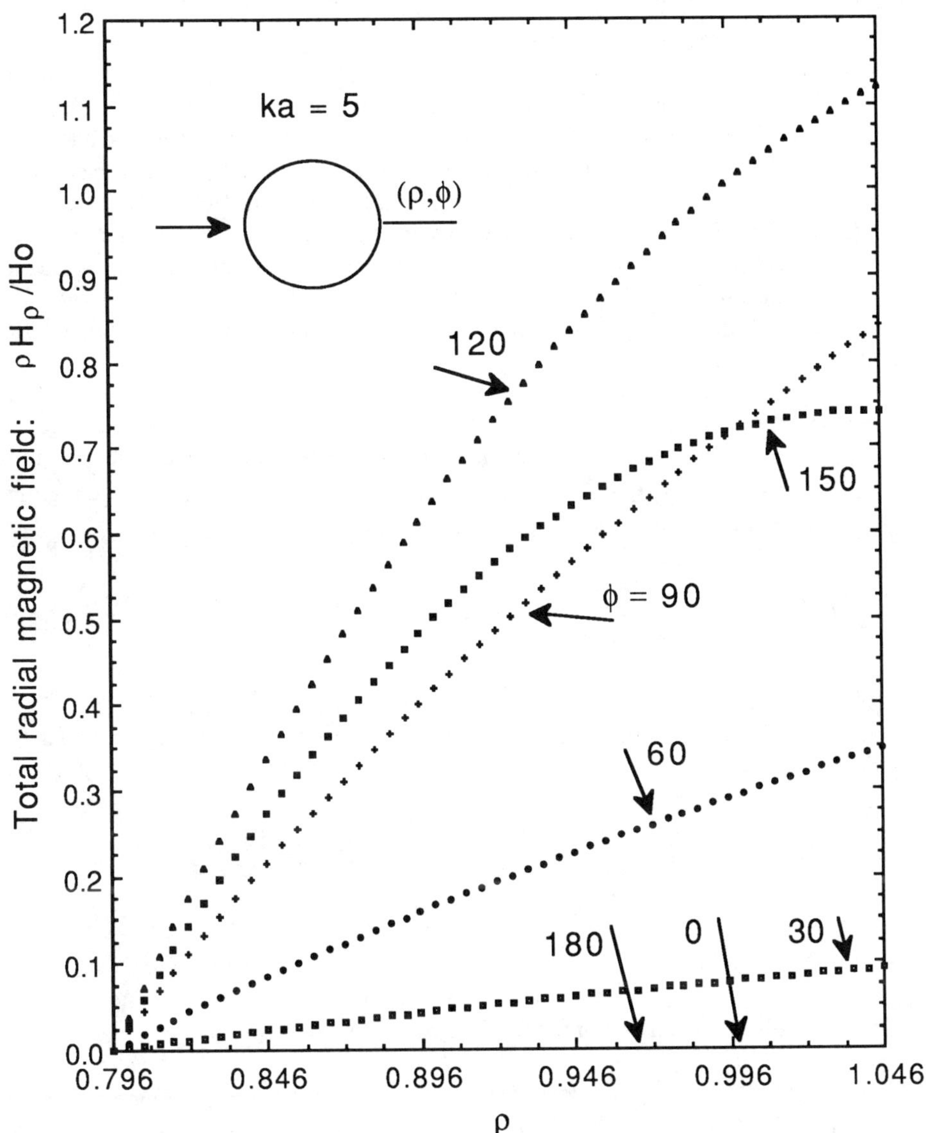

Figure 3.12c Magnitude of the total radial near magnetic field distribution on a perfectly conducting circular cylinder – TM excitation.

are normalized with respect to the amplitude of the incident electric field E_0. The numerical data shown are obtained for fixed observation angles $\phi = 0, 30, 60, 90, 120, 150$, and 180; and the radial variable ρ is varied from a point on the circular conducting scatterer in increments of 0.005 to cover a radial distance of 0.25 meter which is about a quarter wavelength corresponding to the frequency of excitation of 300 MHz. As can be noted, the total z component of the electric field is zero on the surface of scatterer and rises linearly in the vicinity of the circular conducting scatterer. Figure 3.12b shows the distribution of the angular component of the total magnetic field, calculated using expressions (3.5.13b). For points on the surface of scatterer, in fact, this distribution of the angular component of the total magnetic field directly represents the axial component of the electric current distribution, as shown in Figure 3.11.

Similarly, Figure 3.12c shows the radial component of the total magnetic field distribution.

For very large distances from the scatterer, the numerical generation of the Bessel and Hankel functions is quite slow, and it is preferable to use their large argument approximation, Appendix B. The z component of the scattered field distribution in the far-field region is obtained by substituting the argument $|k\rho| \rightarrow \infty$. In the large argument approximation, the Hankel function can be written as

$$H_n^{(2)}(k\rho) \sim \sqrt{\frac{2j}{\pi k\rho}}\; j^n\, e^{-jk\rho} \tag{3.8.15}$$

After substituting the large argument approximation given in expression (3.8.15) into the infinite series solution (3.8.10d), the z component of the scattered field in the far-field region is obtained as

$$E_z^s(\rho,\phi,\omega) \sim E_0\sqrt{\frac{2j}{\pi k\rho}}\; e^{-jk\rho} \sum_{n=-\infty}^{\infty} -\frac{J_n(ka)}{H_n^{(2)}(ka)}\, e^{jn\phi} \qquad \text{for } k\rho \rightarrow \infty \tag{3.8.16}$$

and the normalized magnitude of the far-field pattern is given by

$$\left|\frac{E_z^s(\phi,\omega)}{E_z^i(\phi,\omega)}\right| \sim \sqrt{\frac{2}{\pi k\rho}}\; \left|\sum_{n=-\infty}^{\infty} -\frac{J_n(ka)}{H_n^{(2)}(ka)}\, e^{jn\phi}\right| \qquad \text{for } k\rho \rightarrow \infty \tag{3.8.17}$$

Radar Cross Section

The radar cross section of an electromagnetic scatterer is an important scattering property of interest in many practical radar applications. Generally, the radar cross section is defined for a plane wave excitation, and it is closely related to the power radiated by an object in a given direction. It is given by the ratio of the scattered power in a given direction to the average power of the incident plane wave field assumed to be radiating isotropically in all directions.

Hence, in the two-dimensional case, the differential scattering cross section or the bistatic cross section, in meters, is defined as

$$\mathrm{RCS}(\phi) = \lim_{\rho \to \infty} 2\pi\rho \left| \frac{E_z^s(\phi,\omega)}{E_z^i(\phi,\omega)} \right|^2 \tag{3.8.18}$$

For observation angle $\phi = \phi^i$, a forward scattering cross section is obtained. Similarly, for observation angle $\phi = \pi + \phi^i$, a back or monostatic scattering cross section is obtained. Using expression (3.8.17), the bistatic radar cross section for the circular conducting cylinder for TM excitation is given by

$$\mathrm{RCS}(a,\phi) = \frac{4}{k} \left| \sum_{n=-\infty}^{\infty} - \frac{J_n(ka)}{H_n^{(2)}(ka)} e^{jn\phi} \right|^2 \tag{3.8.19}$$

This expression is numerically evaluated for various values of ϕ, and the results of the bistatic radar cross section are shown in Figure 3.13. In fact, taking the square root of the data presented in Figure 3.13 gives the magnitude distribution of the far-field pattern of conducting circular cylinder. In Figure 3.13, the radar cross section data is shown for ka = 1, 2, 5, and 10, and is normalized with respect to the electrical size ka. As the electrical size ka of the circular conducting cylinder is increased, referring to the Figure 3.13, the beamwidth of the power pattern in forward scattering direction becomes narrower.

3.8.3 AN ALTERNATIVE FORMULATION

In the preceding section, a detailed analysis was presented for determining the distribution of electric current for TM plane wave excitation by rigorously solving the electric field integral equation. In fact, an alternative procedure can be followed for the study of scattering by a circular conducting cylinder, which is well reported in the

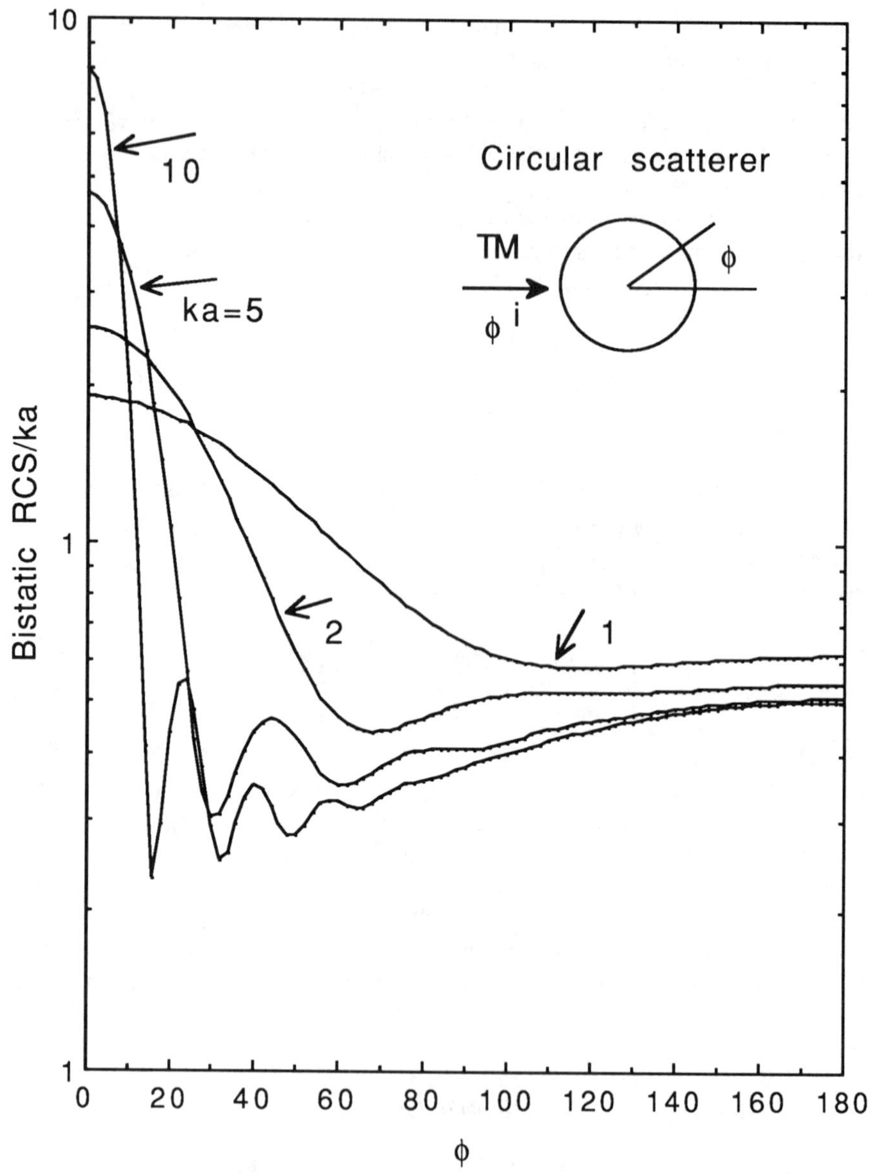

Figure 3.13 Bistatic radar cross section for various electrical sizes of a perfectly conducting circular cylinder – TM excitation.

literature. Referring to the study of electromagnetic scattering by a circular conducting cylinder for TM excitation, Section 3.8, with normal excitation there are only the axial component of the scattered electric field and the corresponding transverse components of the scattered magnetic field distributions. Starting from the EFIE, expression (3.8.4a) shows that the z component of the scattered electric field in terms of an infinite sum over a set of outgoing cylindrical wave modes is given by

$$E_z^s(\rho,\phi) = \sum_{n=-\infty}^{\infty} a_n H_n^{(2)}(k\rho)\, e^{jn\phi} \qquad |\bar{\rho}| \geq a \tag{3.8.20a}$$

where a is the radius of the circular conducting scatterer and a_n are the scattered field modal coefficients, which can be determined based on the total electric field conducting boundary condition. Referring to expression (3.8.9), the z component of TM plane wave electric field propagating along the x coordinate axis can also be written as an infinite sum over a set of Bessel function type cylindrical modes:

$$E_z^i(\rho,\phi) = E_0 \sum_{n=-\infty}^{\infty} j^{-n} J_n(k\rho)\, e^{jn\phi} \tag{3.8.20b}$$

Similar representation can be obtained even for the arbitrary angles of incidence by modifying the exponential term by $(\phi - \phi^i)$. Now, the boundary condition can be enforced that, on the surface of the perfectly conducting circular cylinder, the total electric field consisting of the sum of z components of the incident electric field and the scattered electric field is equal to zero. Hence

$$\sum_{n=-\infty}^{\infty} a_n H_n^{(2)}(ka)\, e^{jn\phi} + E_0 \sum_{n=-\infty}^{\infty} j^{-n} J_n(ka)\, e^{jn\phi} = 0 \tag{3.8.21a}$$

By utilizing orthogonality for the cylindrical modes, the preceding boundary condition yields the scattered field modal coefficients as

$$a_n = -\frac{j^{-n} J_n(ka) E_0}{H_n^{(2)}(ka)} \tag{3.8.21b}$$

which is, in fact, the same as the result obtained in expression (3.8.10c). Further, using the source-free Faraday's law stated in expression (3.1.5a),

$$- j\omega\mu \bar{H}^S(\bar{r},\omega) = \nabla \times \hat{z}E_z^S(\rho,\phi) \tag{3.8.22a}$$

and in terms of the cylindrical coordinate system,

$$- j\omega\mu[H_\rho^S\hat{\rho} + H_\phi^S\hat{\phi}] = \frac{1}{\rho}\frac{\partial}{\partial\phi}E_z^S\hat{\rho} - \frac{\partial}{\partial\rho}E_z^S\hat{\phi} \tag{3.8.22b}$$

For the circular conducting scatterer, if the angular component of the total magnetic field is known, then the induced current electric can be directly obtained. The angular component of total magnetic field is given by

$$H_\phi(\rho,\phi)\Big|_{\text{total}} = \frac{1}{j\omega\mu}\frac{\partial}{\partial\rho}\left[E_z^i(\rho,\phi) + E_z^S(\rho,\phi)\right] \tag{3.8.22c}$$

$$= \frac{1}{j\omega\mu}\frac{\partial}{\partial\rho}\sum_{n=-\infty}^{\infty}[j^{-n}J_n(k\rho)E_0 + a_nH_n^{(2)}(k\rho)]\,e^{jn\phi} \tag{3.8.22d}$$

$$= -\frac{jk}{\omega\mu}\sum_{n=-\infty}^{\infty}[j^{-n}J_n'(k\rho)E_0 + a_nH_n^{(2)'}(k\rho)]\,e^{jn\phi} \tag{3.8.22e}$$

Further, the angular component total magnetic field directly gives the induced electric current in the axial coordinate direction. Hence

$$J_z(a,\phi') = [H_\phi^i(a,\phi') + H_\phi^S(a,\phi')] \tag{3.8.23a}$$

$$= -E_0\frac{jk}{\omega\mu}\sum_{n=-\infty}^{\infty}j^{-n}\left[J_n'(ka) - \frac{J_n(ka)}{H_n^{(2)}(ka)}H_n^{(2)'}(ka)\right]e^{jn\phi'} \tag{3.8.23b}$$

The bracketed term in expression (3.8.23b) can be simplified using the Wronskian relationship for the Bessel functions, Appendix B, which is given by the following expression:

$$J_n(ka)Y_n'(ka) - J_n'(ka)Y_n(ka) = \frac{2}{\pi ka} \tag{3.8.23c}$$

After substituting relationship (3.8.23c) into expression (3.8.23b), the expression for the induced electric current takes the identical form as obtained earlier in expression (3.8.11c):

$$J_z(a,\phi') = \frac{2E_0}{\pi\eta ka} \sum_{n=-\infty}^{\infty} \frac{j^{-n}}{H_n^{(2)}(ka)} e^{jn\phi'} \tag{3.8.24}$$

3.9 ARBITRARY CROSS SECTION - TM EXCITATION

The electromagnetic scattering by a perfectly conducting circular cylinder was discussed in the previous section based on a rigorous analytical approach. The circular cylinder happens to be a special scattering geometry, wherein the induced electric current and the corresponding scattered electric and magnetic field distributions can be conveniently expanded in terms of an eigenfunction series expansion. In a two-dimensional perfectly conducting scatterer with arbitrary cross section having arbitrary edges and corners, the analysis technique is quite complicated. In the following, an approximate analysis procedure is discussed in detail to obtain a numerical solution for electromagnetic scattering by solving the electric field integral equation (3.6.4c). This is, in fact, accomplished based on the method of moments technique described in Section 3.7.

Let us consider the geometry of a two-dimensional, perfectly conducting scatterer having arbitrary cross section and placed in a linear, homogeneous, and isotropic lossless medium, shown in Figure 3.14. The scatterer is uniform along its axis, having the same arbitrary cross section. It is assumed that the z coordinate axis of the cylindrical coordinate system (ρ, ϕ, z) coincides with the axis of the scatterer. The external incident plane wave field is transverse magnetic (TM) polarized and propagating with its direction of propagation normal to the z coordinate axis. The angle ϕ^i represents the incident angle that the direction of propagation makes with the x coordinate axis. As discussed earlier in Section 3.5, for normal excitation various field quantities are independent of the z coordinate variation. Referring to Table 3.6b, there are only the z component of the scattered electric field and the x, y components of the transverse scattered magnetic fields. The TM to z polarized incident plane wave is given by

$$E_z^i(\bar{\rho},\omega) = E_0\, e^{-j\bar{k}\cdot\bar{\rho}} \tag{3.9.1}$$

Hence, referring to expressions (3.5.4) and (3.5.5), for the case of a plane wave propagating normal to the axis of scatterer with ϕ^i arbitrary angle of incidence in transverse plane, the TM plane wave electric field can be written as

$$E_z^i(\rho,\phi) = E_0\, e^{-jk\rho \cos(\phi - \phi^i)} \tag{3.9.2}$$

E_0 : amplitude of incident plane wave electric field, in volts per meter.

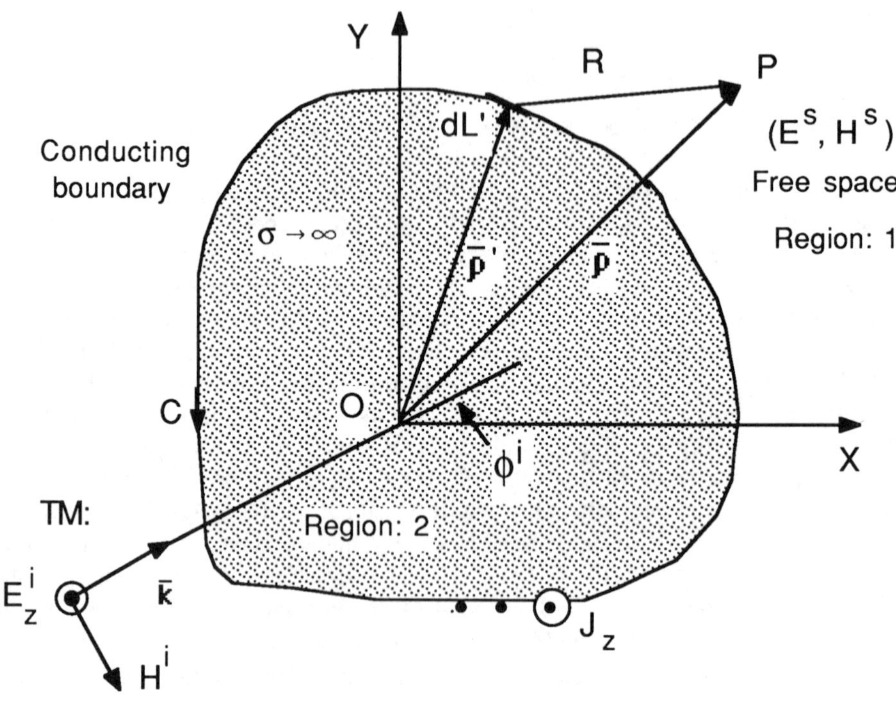

Figure 3.14 Geometry of a two-dimensional perfectly conducting scatterer with arbitrary cross section – TM excitation.

As discussed earlier in Section 3.6, there is only the *z* component of the induced electric currents on contour *C* (designated by primed coordinates) enclosing arbitrary cross section of the perfectly conducting scatterer. Now, the electromagnetic principle is invoked to replace the original induced electric currents on the perfectly conducting scatterer by the corresponding equivalent electric current distribution on a virtual contour. The geometrical shape of the virtual contour is such that it encloses an identical cross-sectional region, having the same properties as that of the external free-space medium, Figure 3.8c. Further, the equivalent electric currents produce the same scattered electric and magnetic fields of the original problem in the region outside the scatterer and also are subject to the same original electromagnetic boundary conditions along the virtual contour. Let

$J_z(\bar{\rho}')$: equivalent electric current distribution, in amperes per meter;

$E_z^s(\bar{\rho})$: scattered electric field component, in volts per meter;

$H_x^s(\bar{\rho}),\ H_y^s(\bar{\rho})$: scattered magnetic field components, in amperes per meter.

Referring to Table 3.6a,

$$E_z^s(\bar{\rho}) = -\,j\omega A_z(\bar{\rho}) \tag{3.9.3a}$$

$$H_x^s(\bar{\rho}) = \frac{1}{\mu}\frac{\partial}{\partial y}A_z(\bar{\rho}) \tag{3.9.3b}$$

$$H_y^s(\bar{\rho}) = -\,\frac{1}{\mu}\frac{\partial}{\partial x}A_z(\bar{\rho}) \tag{3.9.3c}$$

where the magnetic vector potential $A_z(\bar{\rho})$ in terms of the unknown equivalent electric current distribution on contour C is given by expression (3.6.3b):

$$A_z(\bar{\rho},\omega) = \frac{\mu}{4j}\int_C J_z(\bar{\rho}',\omega)\, H_0^{(2)}(k|\bar{\rho} - \bar{\rho}'|)\, dL(\bar{\rho}') \tag{3.9.4}$$

The unknown equivalent electric current distribution can be obtained by solving the electric field integral equation (3.6.4a):

$$E_z^i(\bar{\rho},\omega) = \frac{k\eta}{4}\int_C J_z(\bar{\rho}',\omega)\, H_0^{(2)}(k|\bar{\rho} - \bar{\rho}'|)\, dL(\bar{\rho}') \tag{3.9.5}$$

for $\bar{\rho}$ on C

In order to solve the EFIE (3.9.5), the method of moments numerical technique, Section 3.7, is applied. This is accomplished by a suitable selection of a set of independent current expansion functions and a set of independent weighting functions.

A number of modeling considerations should be understood before a suitable choice can be made to represent the unknown electric current distribution, Figure 3.15a, in terms of a set of current expansion functions, and similarly, for a suitable choice of a set of weighting function to test the EFIE, expression (3.9.5), on both sides.

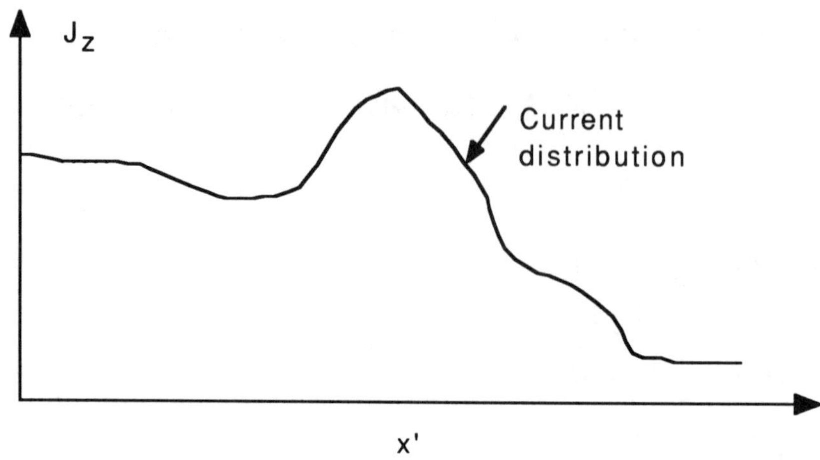

Figure 3.15a An arbitrary unknown electric current distribution.

3.9.1 EXPANSION FUNCTIONS

For convenience contour C will be stretched out and redrawn with x' variable. On contour C, the unknown electric current distribution exists and is to be determined. Initially, if one has some knowledge of the distribution of electric current, then it is easier to choose an approximate representation by a suitable sampling of the current distribution. Depending upon the electrical size of the conducting scatterer and its arbitrary cross section having arbitrary edges and corners, the electric current may have current singularities, current nulls and even a rapid phase change in its distribution. For example, let us consider the arbitrary distribution shown in Figure 3.15a. This distribution of the unknown electric current can be sampled at discrete spatial points and represented in terms of a set of independent delta functions.

Referring to Figure 3.15b, along contour C, a set of current sampling points x'_n where $n = 1, 2, 3, \ldots, N$, are selected with the corresponding unknown complex current coefficients I_n to represent approximately the magnitude as well as the phase of unknown current distribution. Then, the unknown electric current distribution can be approximated in a piecewise sense and written in terms of the delta function piecewise expansion terms as

$$J_z(x') \approx \sum_{n=1}^{N} I_n \delta_n(x' - x'_n) \tag{3.9.6}$$

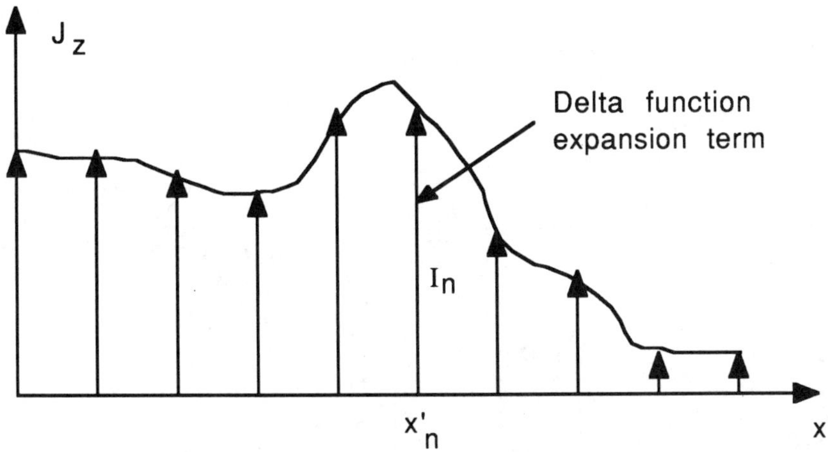

Figure 3.15b Representation in terms of delta functions.

Even though this piecewise delta function-type current expansion, expression (3.9.6), is simple and ideal for performing integration along the boundary contour to obtain the magnetic vector potential and the scattered electric and magnetic fields, the representation practically is not suitable because it looses critical information concerning the actual distribution of the electric current between adjacent sampling points. Further, the spatial derivative distribution of the current expansion is not well defined. Hence, the piecewise delta function current expansion functions are completely excluded in the context of obtaining numerical solution discussed in this section.

A more realistic approximation for the unknown electric current distribution can be obtained by expanding the distribution in terms of a set of piecewise constant or pulse expansion functions. Referring to Figure 3.15c, the boundary contour is first sampled at a number of spatial locations consisting of odd and even points. The unknown distribution of the electric current, shown in the Figure 3.15a, is now approximated by a set of staircase or piecewise pulse functions so that the area under the distribution curve is the same. In fact, various pulse functions are defined between the limits of two consecutive odd points.

For example, the piecewise pulse function P_{2n} is located so that the even point x'_{2n} represents its midlocation and is defined as between the two consecutive odd points given by

$$x'_{2n+1} = x'_{2n} + \frac{\Delta_{2n}}{2} \tag{3.9.7a}$$

$$x'_{2n-1} = x'_{2n} - \frac{\Delta_{2n}}{2} \tag{3.9.7b}$$

where the length of the constant pulse segment is

$$\Delta_{2n} = x'_{2n+1} - x'_{2n-1} \tag{3.9.7c}$$

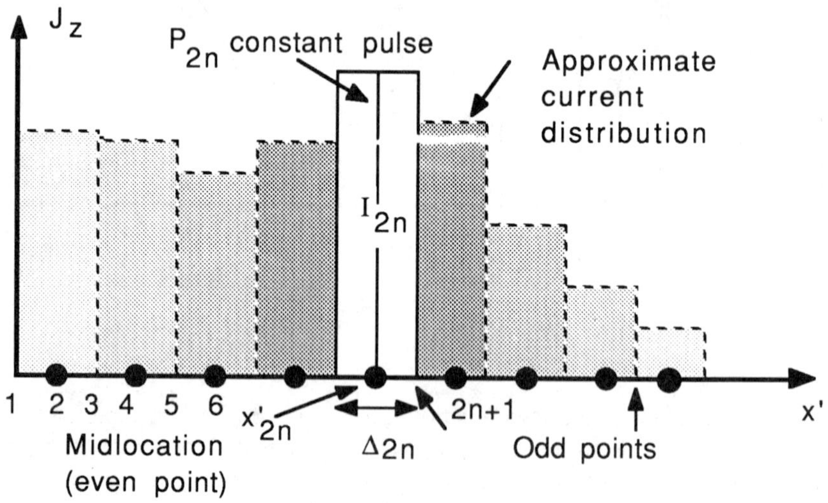

Figure 3.15c Piecewise pulse current expansion functions.

The distribution of the electric current is assumed to be constant within a pulse expansion and to correspond to its distribution at the midpoint. Referring to Figure 3.15c, the unknown complex current coefficient I_{2n} where $n = 1, 2, 3, \ldots, N$, is assumed to be constant within the piecewise pulse P_{2n} and to represent approximately the magnitude as well as the phase of unknown current distribution within the specific pulse region. Thus, the unknown electric current distribution can be approximated in a piecewise sense, expressed below

$$J_z(x') \approx \sum_{n=1}^{N} I_{2n} P_{2n}(x') \tag{3.9.8a}$$

where the piecewise pulse function

$$P_{2n}(x') = 1 \qquad \text{for } x'_{2n-1} \leq x' \leq x'_{2n+1}$$
$$= 0 \qquad \text{otherwise} \tag{3.9.8b}$$

The piecewise pulse expansion set just defined is quite useful in the numerical solution of the integral and integro-differential equations. In fact, the pulse expansion functions are quite easy to implement and ideal whenever complicated distributions having current singularities at geometrical corners of the scatterer are to be simulated. Other types of higher order expansion terms, such as piecewise linear or triangle functions and piecewise sinusoidal functions, can be utilized to improve representation of the unknown electric current distribution. In this chapter, the two-dimensional, arbitrary shaped perfectly conducting scatterer with the TM excitation is exclusively studied using the piecewise pulse expansion functions. Even though the higher order expansion terms seem to be useful, they are not discussed here and their studies are discussed in later chapters where specific electromagnetic boundary value problems are considered.

3.9.2 WEIGHTING FUNCTIONS

In the method of moments numerical technique, Section 3.7, equality of the integral equation should be enforced on both sides at each point in the domain of validity of the integral equation. This is accomplished by testing through a scalar product in the domain of validity, as defined in expression (3.7.1), that is, by weighting on both sides of the integral equation with respect to a convenient weighting function.

In general, the piecewise expansion terms just discussed, such as delta functions, pulse functions, or any other linear functions, can be adopted as weighting functions to implement the scalar product (3.7.1) to reduce the integral equation to its functional form. In a special case, when the current expansion function and the weighting function are identical, the integral equation solution technique indirectly reduces to the so-called Galerkin's technique. But, it is not always necessary to choose the same type of sampling functions for both testing the integral equation and expanding the unknown current distribution.

Referring to Figure 3.16a, along the boundary contour a set of sampling points x_m, where $m = 1, 2, 3, \ldots, M$, are selected. A simple weighting function can now be defined at these sampling points in a piecewise sense, written as a set of delta function piecewise terms:

$$W(x) = \delta_m(x - x_m) \tag{3.9.9}$$
$$m = 1, 2, 3, \ldots, M.$$

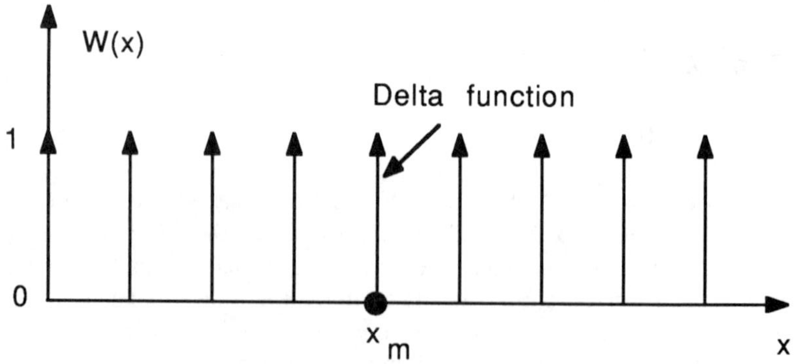

Figure 3.16a Delta function weighting functions.

The weighting function in terms of the piecewise delta functions is simple and ideal for performing testing through the scalar product, expression (3.7.1). The method of testing based on the delta functions enforces equality of the integral equation only at specific spatial points and is generally referred to as the *point matching* procedure. For complicated distributions, this type of weighting functions in practice looses critical information concerning equality in the region between adjacent sampling points, unless a large number of sampling points with close spacing are selected.

A more realistic weighting function can be defined in terms of a distribution function, such as the piecewise pulse approximation shown in Figure 3.16b. The various pulse weighting functions are defined between the limits of two consecutive odd points. For example, the piecewise pulse W_{2m} is located so that the even point x_{2m} represents its midlocation and is defined between the two consecutive odd points given by

$$x_{2m+1} = x_{2m} + \frac{\Delta_{2m}}{2} \tag{3.9.10a}$$

$$x_{2m-1} = x_{2m} - \frac{\Delta_{2m}}{2} \tag{3.9.10b}$$

where the length of the pulse segment is

$$\Delta_{2m} = x_{2m+1} - x_{2m-1} \tag{3.9.10c}$$

Then the weighting function, in a piecewise sense, can written in terms of the piecewise pulse or constant terms as

$$W(x) = W_{2m}(x) \qquad m = 1, 2, 3, \ldots, M \tag{3.9.11a}$$

and the piecewise pulse function

$$W_{2m}(x) = 1 \qquad \text{for } x_{2m-1} \leq x \leq x_{2m+1}$$
$$= 0 \qquad \text{otherwise} \tag{3.9.11b}$$

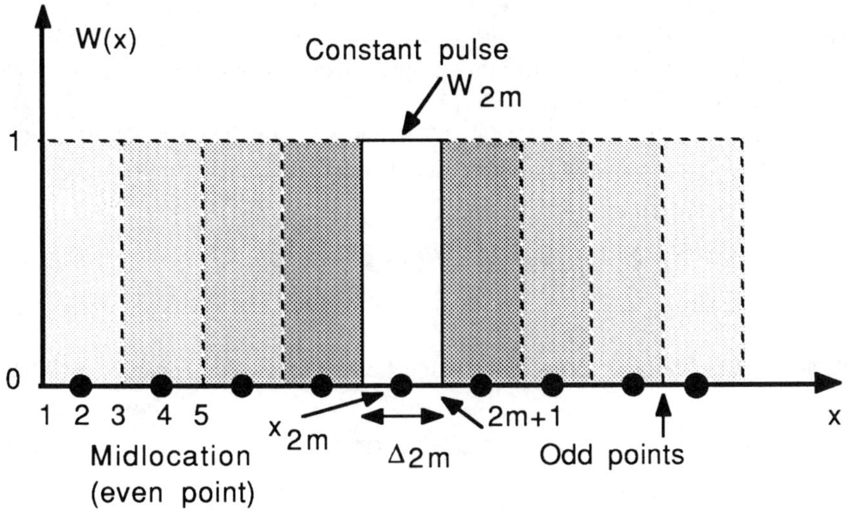

Figure 3.16b Pulse function weighting functions.

The pulse weighting function just defined is quite useful in the numerical solution of the integral and integro-differential equations. In fact, the pulse weighting functions are quite easy to implement and ideal whenever complicated field distributions are to be simulated. Other types of higher order weighting terms consisting of piecewise linear or triangle functions, or piecewise sinusoidal functions can also be utilized to improve testing procedure. These cases are taken up in later chapters.

3.9.3 REDUCTION OF EFIE TO A MATRIX EQUATION

To reduce integral equation (3.9.5) to its equivalent functional form, the integral equation is first tested on both sides using the pulse type of weighting function developed previously. If contour C of the scatterer is completely smooth, with no geometrical discontinuities, then it is possible to pick arbitrary sampling regions along contour C with their midpoints located at $\bar{\rho}_{2m}$, $m = 1, 2, 3, \ldots, M$. The pulse weighting functions developed in Figure 3.16b can now be set up on contour C with respect to various piecewise sampling regions. On the other hand, if contour C of the scatterer is not smooth, but has sharp corners as indicated in Figure 3.17a, care should be taken in properly positioning the weighting function with respect to the geometrical discontinuity. It should be noted, in fact, the distribution of electric current (for TM polarization) has singular behavior near sharp-wedge type corners. From a numerical modeling point of view, specific weighting functions close to wedge-type corners are always properly arranged so that they do *not* enclose the geometrical discontinuity points. Referring to Figure 3.17a, the midpoint of the weighting function on either side is located half a cell away from the geometrical corner.

The EFIE expression (3.9.5) is repeated here and will now be tested on both sides to reduce it to the corresponding functional form of equation. Hence, for $\bar{\rho}$ on C,

$$E_z^i(\bar{\rho},\omega) = \frac{k\eta}{4} \int_C J_z(\bar{\rho}',\omega)\, H_0^{(2)}(k|\bar{\rho} - \bar{\rho}'|)\, dL(\bar{\rho}') \tag{3.9.12a}$$

or with shorthand notation in terms of the magnetic vector potential, it can be rewritten as

$$E_z^i(\bar{\rho}) = j\omega A_z(\bar{\rho}) \qquad \bar{\rho} \text{ on } C \tag{3.9.12b}$$

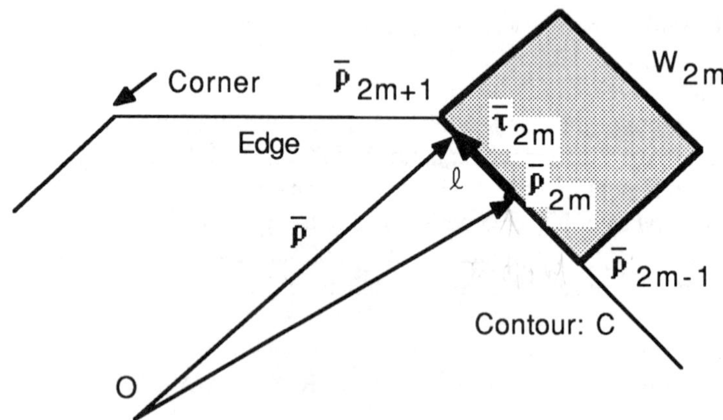

Figure 3.17a Positioning of the weighting function near discontinuities.

Referring to the development in Section 3.9.2, the pulse weighting functions are now selected between two consecutive odd points. Avoiding sharp wedge-type corners, the piecewise weighting pulse W_{2m} is located such that the even point $\bar{\rho}_{2m}$ with $m = 1, 2, 3, \ldots, M$ represents its midlocation and is defined between the limits of two consecutive odd points, given by

$$\bar{\rho}_{2m+1} = \bar{\rho}_{2m} + \frac{\Delta_{2m}}{2}\hat{\tau}_{2m} \tag{3.9.13a}$$

$$\bar{\rho}_{2m-1} = \bar{\rho}_{2m} - \frac{\Delta_{2m}}{2}\hat{\tau}_{2m} \tag{3.9.13b}$$

$\hat{\tau}_{2m}$: tangential unit vector along the segment;

Δ_{2m} : length of the weighting pulse segment

$$= \left|\bar{\rho}_{2m+1} - \bar{\rho}_{2m-1}\right| \tag{3.9.13c}$$

Then, the weighting function in terms of the piecewise pulse terms is

$$W(\bar{\rho}) = W_{2m}(\bar{\rho}) \qquad m = 1, 2, 3, \ldots, M \tag{3.9.14a}$$

and the piecewise pulse function

$$W_{2m}(\bar{\rho}) = 1 \quad \text{for } \bar{\rho}_{2m-1} \le \bar{\rho} \le \bar{\rho}_{2m+1}$$
$$= 0 \quad \text{otherwise} \tag{3.9.14b}$$

Pulse Testing. Integral expression (3.9.12b) is now multiplied on both sides by weighting function (3.9.14a) and integrated along scatterer contour C to yield the following functional relationship:

$$\int_C E_z^i(\bar{\rho}) W_{2m}(\bar{\rho})\, dL(\bar{\rho}) = \int_C j\omega A_z(\bar{\rho}) W_{2m}(\bar{\rho})\, dL(\bar{\rho}) \tag{3.9.15a}$$

$$\int_{\Delta_{2m}} E_z^i(\bar{\rho})\, dL(\bar{\rho}) = \int_{\Delta_{2m}} j\omega A_z(\bar{\rho})\, dL(\bar{\rho}) \tag{3.9.15b}$$

$\bar{\rho}$ on C, and $m = 1, 2, 3, \ldots, M$

Numerical integration can be rigorously implemented to calculate these weighted integrals. If the integrands are smooth varying functions, it can be assumed that the two integrands on the lefthand and righthand sides of expression (3.9.15b) do not vary over the limits of integration interval defined by piecewise weighting pulse function (3.9.14b). This is also the case when a sufficient number of closely spaced, piecewise weighting functions are introduced for testing the integral equation. Thus, a suitable approximation can be introduced by calculating the integrand at the midpoint of weighting function and multiplying it by the segment length of weighting function to yield the following functional form of equation:

$$E_z^i(\bar{\rho}_{2m}) = j\omega A_z(\bar{\rho}_{2m}) \qquad m = 1, 2, 3, \ldots, M \tag{3.9.15c}$$

which is, in fact, numerically equivalent to the *point matching* of integral expression (3.9.12b) at various discrete points given by the midpoints of weighting functions. Hence, on substituting for the z component of the magnetic vector potential into the tested expression (3.9.15c), the following functional relationship is obtained:

$$E_z^i(\bar{\rho}_{2m}) = \frac{k\eta}{4} \int_C J_z(\bar{\rho}')\, H_0^{(2)}(k|\bar{\rho}_{2m} - \bar{\rho}'|)\, dL(\bar{\rho}') \tag{3.9.16}$$

$$m = 1, 2, 3, \ldots, M$$

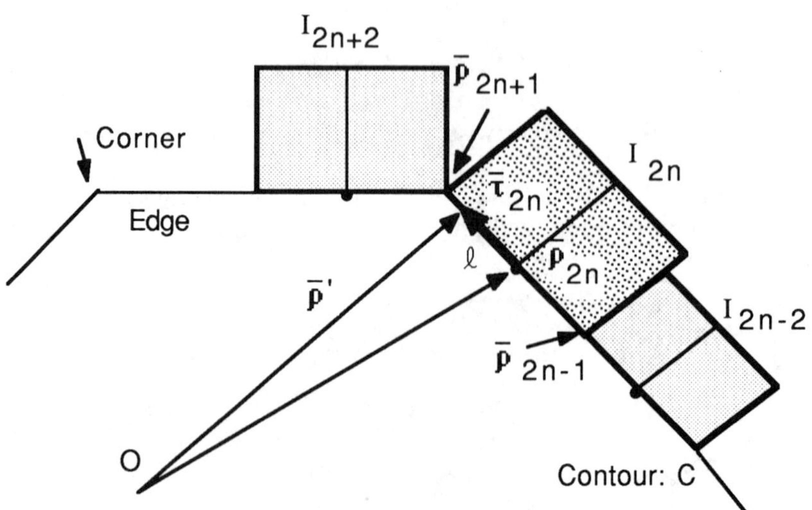

Figure 3.17b Positioning of the expansion function near discontinuities.

Current Expansion. Referring to Figure 3.17b, the unknown electric current distribution $\bar{J}_z(\bar{\rho}')$ is now expanded in terms of a set of piecewise, linearly independent pulse expansion functions given by

$$J_z(\bar{\rho}') \approx \sum_{n=1}^{N} I_{2n} P_{2n}(\bar{\rho}') \tag{3.9.17a}$$

where I_{2n} are the unknown electric current coefficients, and the piecewise pulse function

$$\begin{aligned} P_{2n}(\bar{\rho}') &= 1 \qquad \text{for } \bar{\rho}_{2n-1} \leq \bar{\rho}' \leq \bar{\rho}_{2n+1} \\ &= 0 \qquad \text{otherwise} \end{aligned} \tag{3.9.17b}$$

A number of considerations are given concerning the actual location and arrangement of the pulse expansion functions on contour C. Generally, by choosing the weighting function to be same as the expansion function, the various pulses of the expansion function can be arranged to overlap with the pulses of the weighting function. If contour *C* of the scatterer is not smooth, but has sharp wedge-type corners, as indicated in Figure 3.17b, care should to be taken in properly positioning the expansion function with respect to the geometrical discontinuities. It should be noted, in fact, the distribution of the electric current (for TM polarization) has singular behavior near the sharp wedge-type corners. Detailed numerical study is reported in a later section concerning field behavior near geometrical singularities. From the numerical modeling point of view, the specific expansion functions close to the wedge-type corners are always properly arranged so that they do *not* enclose the geometrical discontinuity points. Referring to the Figure 3.17b, the midpoint of the current expansion function on either side is located half a cell away from the geometrical corner. With respect to the *primed* coordinates, the electric current pulse expansions are located between the odd points:

$$\bar{\rho}_{2n+1} = \bar{\rho}_{2n} + \frac{\Delta_{2n}}{2} \hat{\tau}_{2n} \tag{3.9.18a}$$

$$\bar{\rho}_{2n-1} = \bar{\rho}_{2n} - \frac{\Delta_{2n}}{2} \hat{\tau}_{2n} \tag{3.9.18b}$$

$\hat{\tau}_{2n}$: tangential unit vector along the segment;

Δ_{2n} : length of the expansion pulse segment

$$= |\bar{\rho}_{2n+1} - \bar{\rho}_{2n-1}| \tag{3.9.18c}$$

After substituting electric current pulse expansion (3.9.17a) into functional equation (3.9.16), the following matrix form of equation is obtained:

$$E_z^i(\bar{\rho}_{2m}) = \frac{k\eta}{4}\int_C \Big[\sum_{n=1}^{N} I_{2n}P_{2n}(\bar{\rho}')\Big]H_0^{(2)}(k|\bar{\rho}_{2m} - \bar{\rho}'|)\, dL(\bar{\rho}') \qquad (3.9.19a)$$

$$m = 1, 2, 3, \ldots, M$$

Since the original operator equation is linear, the summation and the integration operations can be interchanged to obtain

$$E_z^i(\bar{\rho}_{2m}) = \sum_{n=1}^{N} I_{2n}\Big[\frac{k\eta}{4}\int_C P_{2n}(\bar{\rho}')H_0^{(2)}(k|\bar{\rho}_{2m} - \bar{\rho}'|)\, dL(\bar{\rho}')\Big] \qquad (3.9.19b)$$

$$m = 1, 2, 3, \ldots, M$$

Referring to expression (3.9.17b), the current expansion pulse functions are valid only over their segments and zero outside. Hence, expression (3.9.19b) simplifies to the following dense matrix representation:

$$E_z^i(\bar{\rho}_{2m}) = \sum_{n=1}^{N} I_{2n}\Big[\frac{k\eta}{4}\int_{\Delta_{2n}} H_0^{(2)}(k|\bar{\rho}_{2m} - \bar{\rho}'|)\, dL(\bar{\rho}')\Big] \qquad (3.9.19c)$$

$$m = 1, 2, 3, \ldots, M$$

The expression (3.9.19c) forms a set of M linear simultaneous algebraic equations for the N unknown electric current coefficients. A direct numerical solution can obtained by choosing $M = N$, the number of equations equal to the number of unknowns.

The expression (3.9.19c) can be rewritten in terms of a compact generalized matrix equation

$$[Z_{mn}][I_n] = [V_m] \qquad (3.9.20a)$$

where the elements of the matrix equation are giben by

Z_{mn} : generalized impedance matrix element;
I_n : element of unknown current coefficient column vector;
V_m : element of incident excitation column vector.

$$[Z_{mn}] = \begin{bmatrix} Z_{11} & Z_{12} & Z_{13} & \cdots & Z_{1n} & \cdots & Z_{1N} \\ Z_{21} & Z_{22} & Z_{23} & \cdots & Z_{2n} & \cdots & Z_{2N} \\ Z_{31} & Z_{32} & Z_{33} & \cdots & Z_{3n} & \cdots & Z_{3N} \\ \cdots & & & & & & \cdots \\ Z_{m1} & Z_{m2} & Z_{m3} & \cdots & Z_{mn} & \cdots & Z_{mN} \\ \cdots & & & & & & \cdots \\ \cdots & & & & & & \cdots \\ Z_{M1} & Z_{M2} & Z_{M3} & \cdots & Z_{Mn} & \cdots & Z_{MN} \end{bmatrix} \tag{3.9.20b}$$

where the elements of the generalized impedance matrix are given by

$$Z_{mn} = \frac{k\eta}{4} \int_{-\Delta_{2n}/2}^{\Delta_{2n}/2} H_0^{(2)}(kR)\, d\ell \qquad \text{for } m \neq n \tag{3.9.20c}$$

$$R = \left|\bar{\rho}_{2m} - (\bar{\rho}_{2n} + \ell\hat{\tau}_{2n})\right| \tag{3.9.20d}$$

For m not equal to n, the integrand of the mutual or off-diagonal matrix elements is well behaved. The matrix elements can be evaluated numerically by implementing either the Gaussian quadrature or even trapezoidal integration algorithm. When the source point and the observation point are sufficiently far apart, the off-diagonal matrix elements can be conveniently approximated as

$$Z_{mn} = \frac{k\eta}{4} H_0^{(2)}(k|\bar{\rho}_{2m} - \bar{\rho}_{2n}|)\Delta_{2n} \qquad \text{for } m \neq n \tag{3.9.20e}$$

Further, the elements of the self or diagonal terms in generalized impedance matrix (3.9.20b) are given by

$$Z_{nn} = \frac{k\eta}{2} \int_{0}^{\Delta_{2n}/2} H_0^{(2)}(k\ell)\, d\ell \qquad \text{for } m = n \tag{3.9.20f}$$

In this integral, the Hankel function integrand has a singularity at its lower limit, but this integrand has an integrable singularity. Hence, the diagonal matrix term defined in the integral can be integrated in a close form with small argument approximation substituted for the Hankel function, Appendix B. Integral expression (3.9.20f) is simplified by using the small argument result

$$H_0^{(2)}(z) \approx 1 - j\frac{2}{\pi}\ln\left(\frac{\gamma z}{2}\right) \tag{3.9.20g}$$

and

$$Z_{nn} = \frac{k\eta}{4}\Delta_{2n}\left[1 - j\frac{2}{\pi}\ln\left(\frac{\gamma k\Delta_{2n}}{4e}\right)\right] \quad \text{for } m = n \tag{3.9.20h}$$

$$\ln\gamma = 0.577215665 \tag{3.9.20i}$$

$$\gamma = 1.781072418 \tag{3.9.20j}$$

$$e = 2.718281828 \quad \text{(Euler's constant)} \tag{3.9.20k}$$

Further, in the matrix expression (3.9.20a), the excitation term for the TM polarized plane wave propagating with an angle of incidence ϕ^i is given by

$$V_m = E_z^i(\bar{\rho}_{2m}) \tag{3.9.21a}$$

$$E_z^i(\bar{\rho}_{2m}) = E_0\, e^{-jk\rho_{2m}\cos(\phi_{2m} - \phi^i)} \tag{3.9.21b}$$

The unknown electric current distribution is obtained by inverting the generalized matrix equation (3.9.20a):

$$[I_n] = [Z_{mn}]^{-1}[V_m] \tag{3.9.22}$$

It is not always necessary to the obtain the matrix inversion for numerical solution of the electric current distribution. In fact, a number of alternative procedures are available for the numerical solution of a dense matrix equation, such as the back substitution method and upper or lower triangularization method, which can be adopted for obtaining the electric current distribution.

3.9.4 SQUARE CONDUCTING CYLINDER

Based on the numerical solution discussed in the previous section, a number of two-dimensional conducting, open and closed geometries can be analyzed to determine their electromagnetic scattering properties. Figure 3.18a shows the cross section of a two-dimensional, perfectly conducting square cylinder located in a linear, homogeneous, and isotropic lossless medium. The side length of the square cylinder is selected as $ks = 2$ and excited by an external monochromatic plane wave propagating at an angle of incidence $\phi^i = 90^o$. For the TM polarization, the incident electric field is polarized parallel to the z axis and the corresponding incident magnetic field is polarized parallel to the x axis. With normal excitation on one side, Figure 3.18a, the electric current induced on the square cylinder is symmetrical with respect to the y coordinate axis.

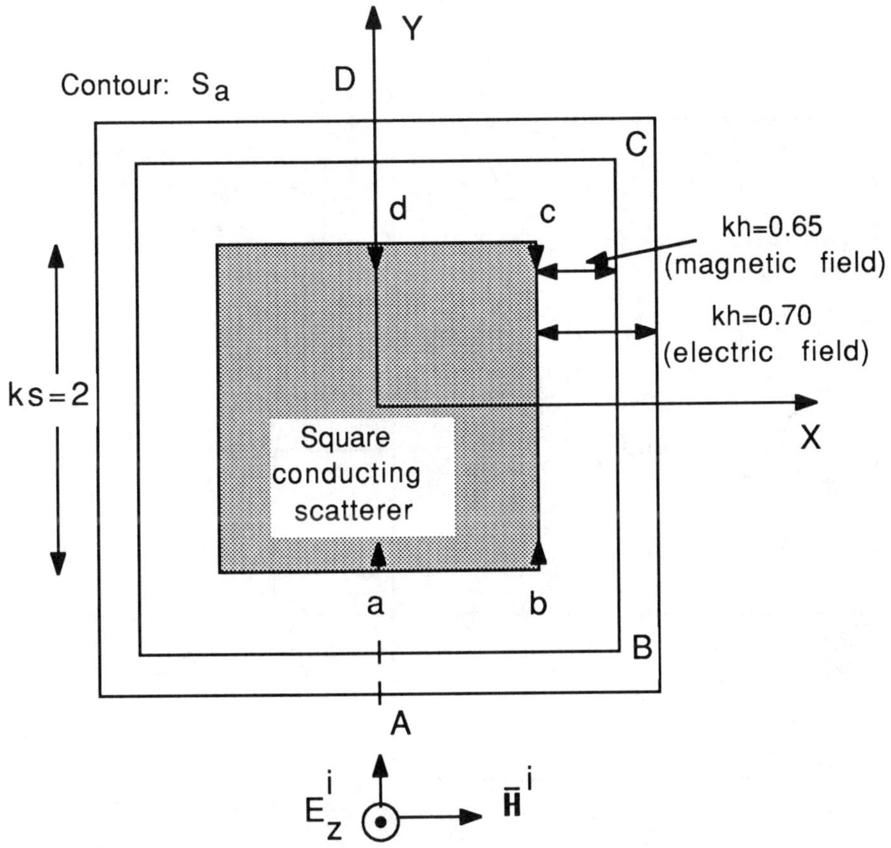

Figure 3.18a Perfectly conducting square scatterer – TM normal incidence.

Figure 3.18b shows various locations of the even and odd sampling points along with the arrangement of the electric current pulse expansion functions for a MOM analysis of the square conducting scatterer. To show the arrangement of the pulse functions, each side length of the square cylinder is divided into three equal segments so that, in the Figure 3.18b, the total number of unknown electric current pulses is selected as $N = 12$. As discussed earlier, various pulse expansion terms are arranged on the square cylinder with even points representing the midlocations of the current pulses and the consecutive odd points representing the limits of the pulse expansion terms. Further, *no current sampling pulse* encloses the right-angle corners of the square conducting cylinder, where electric current singularities exist. Correspondingly, a total number of weighting pulses, $M = 12$, is also selected to coincide with the mid (even) points of the various current expansion pulses.

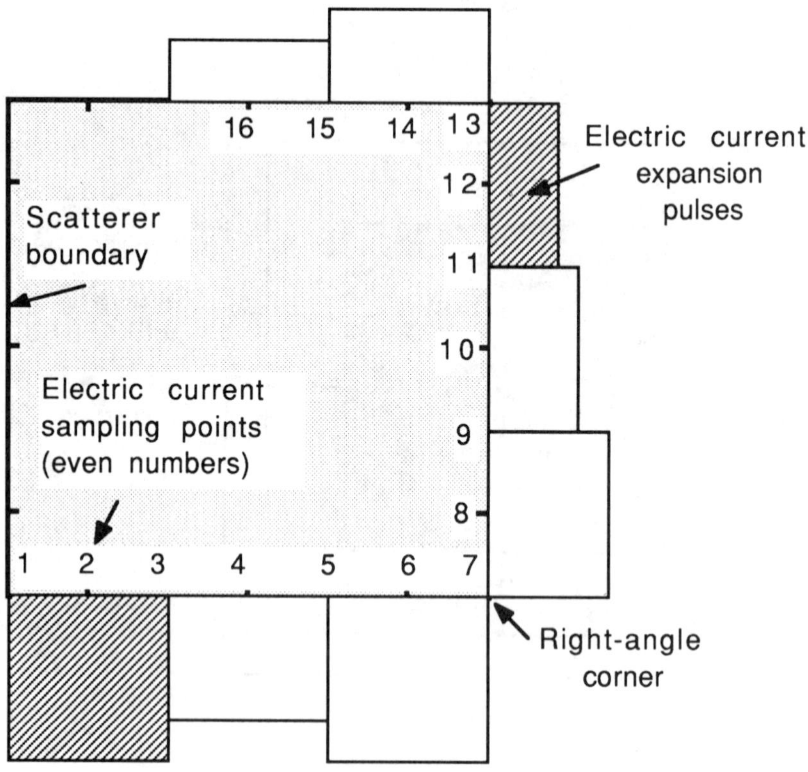

Figure 3.18b Location of the sampling points and electric current pulse expansion functions for the square scatterer – TM case.

A general computer algorithm can now be written to generate the geometry, the generalized impedance matrix, and the excitation column vector for the scatterer. According to expression (3.9.22), the impedance matrix is inverted using the Gauss-Seidel algorithm and multiplied by the excitation column vector to obtain the electric current coefficients. Figure 3.19a shows a plot of the distribution of electric current, for $N = M = 80$, along the boundary of the square conducting cylinder designated by the points *abcd*. Referring to Figure 3.18a, location *a* is at the center of the illuminated region and location *d* is at the center of the shadow region. Further, locations *b* and *c* correspond to the two right-angle corners on the square cylinder, with location *b* at the illuminated side and location *c* at the shadow side. The distribution of the electric current has pronounced current singularities at the two corners, *b* and *c*. The data shown in Figure 3.19a are normalized with respect to the amplitude of incident magnetic field. The data are also compared rigorously with respect to an alternative numerical solution based on the finite-difference time-domain technique, Chapter 1, which is discussed in detail in another book. The phase of the electric current distribution is shown in Figure 3.19b. Along the contour points *abcd* of the square conducting scatterer, following from the illuminated side toward the shadow side, the phase angle of the electric current changes by approximately 280°. Even though the magnitude of electric current has singularities at corners *b* and *c*, the phase angle of the electric current exhibits no singularity behavior and, in fact, shows a gradual phase change along the square cylinder including at the geometrical corner points.

Scattered Field Distribution

The electric current distribution in terms of the known current coefficients, as obtained by expression (3.9.22), is now substituted into the scattered field expressions (3.9.3) and (3.9.4), and the near scattered electric and magnetic field components are numerically calculated. Hence, the axial component of near scattered electric field distribution is given by

$$E_z^s(x,y) = \sum_{n=1}^{N} I_{2n}\Big[-\frac{k\eta}{4}\int_{\Delta_{2n}} H_0^{(2)}(k|\bar{\rho}-\bar{\rho}'|)\, dL(\bar{\rho}')\Big] \tag{3.9.23a}$$

$$= \sum_{n=1}^{N} I_{2n}\Big\{-\frac{k\eta}{4}\int_{-\Delta_{2n}/2}^{\Delta_{2n}/2} H_0^{(2)}\big[k|\bar{\rho}-(\bar{\rho}_{2n}+\ell\,\hat{\tau}_{2n})|\big]\, d\ell\Big\} \tag{3.9.23b}$$

The transverse x component of the near scattered magnetic field distribution is obtained by expressions (3.9.3b) and (3.9.4):

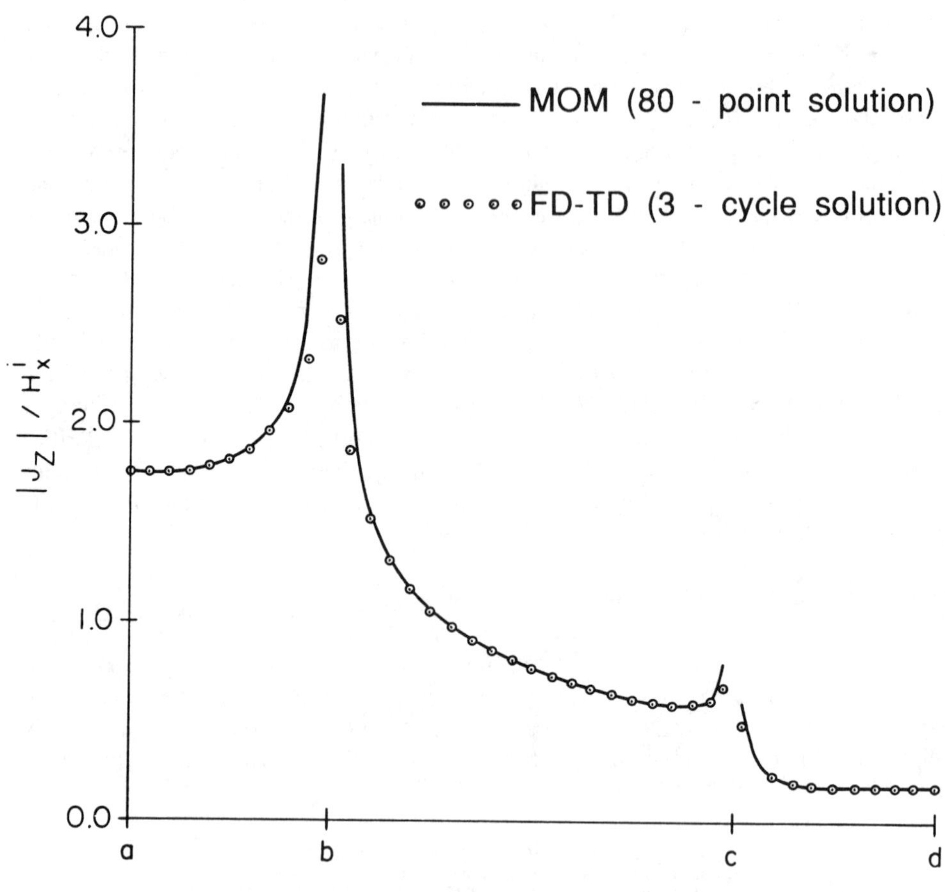

Figure 3.19a Magnitude of induced electric current distribution on a perfectly conducting square scatterer, normal incidence – TM Excitation.

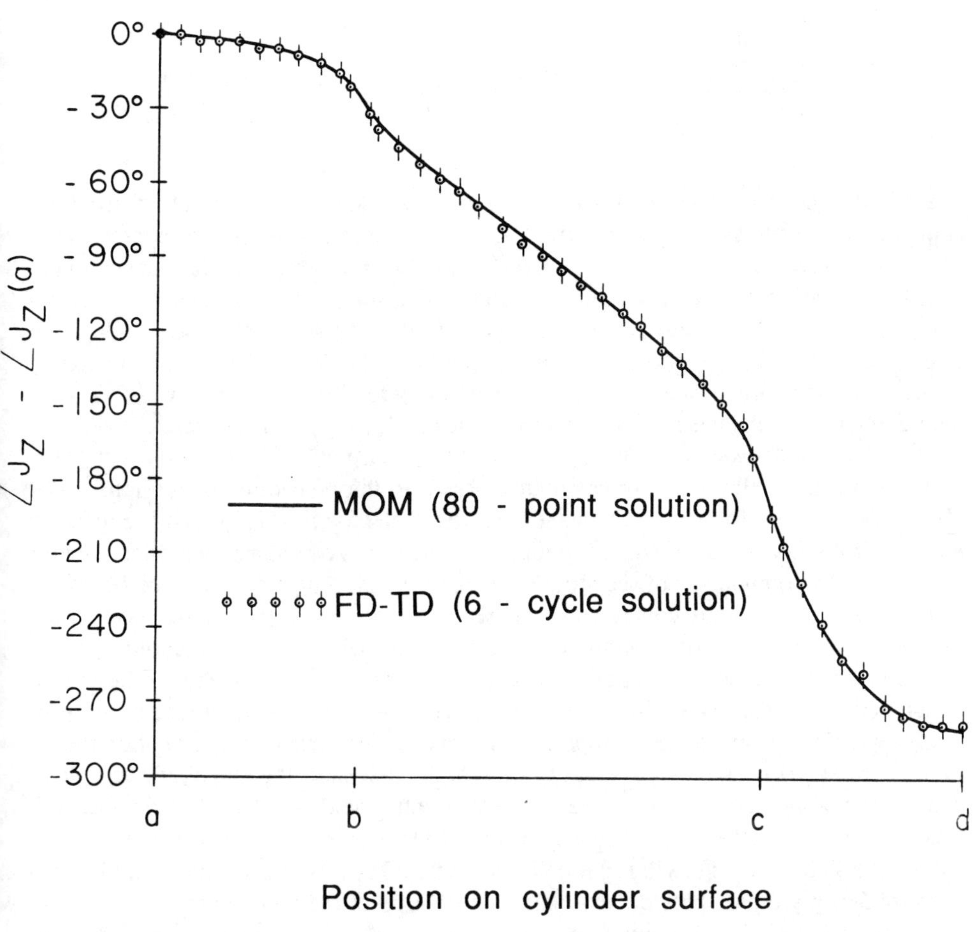

Figure 3.19b Phase of induced electric current distribution on a perfectly conducting square scatterer, normal incidence – TM Excitation.

$$H_x^S(x,y) = \sum_{n=1}^{N} I_{2n}\left\{\frac{1}{4j}\int_{\Delta_{2n}} \left[\frac{\partial}{\partial y} H_0^{(2)}(k|\bar{\rho} - \bar{\rho}'|)\right] dL(\bar{\rho}')\right\} \tag{3.9.24a}$$

$$= \sum_{n=1}^{N} I_{2n}\left\{\frac{1}{4j}\int_{-\Delta_{2n}/2}^{\Delta_{2n}/2} \left[\frac{\partial}{\partial y} H_0^{(2)}\left(k\left|\bar{\rho} - (\bar{\rho}_{2n} + \ell\hat{\tau}_{2n})\right|\right)\right] d\ell\right\} \tag{3.9.24b}$$

Similarly, the transverse y component of the near scattered magnetic field distribution is obtained by expressions (3.9.3c) and (3.9.4). The corresponding field expression is similar to expression (3.9.24b) except the integrand is differentiated with respect to the coordinate variable x with a change in sign. The distribution of near electric and magnetic fields is an important information from the standpoint of understanding the electromagnetic scattering properties. In fact, the FD-TD field distribution is obtained by subtracting the incident field from the corresponding total field, which are functions of the field point coordinates (x, y) on or outside the conducting square scatterer.

Figure 3.18a shows a contour S_a designated by the points $ABCD$ along which the scattered field distributions are calculated based on the preceding expressions. The observation contour for the z component of scattered electric field is located at a distance $kh = 0.7$ from the surface of square conducting scatterer. A similar observation contour for the x and y components of scattered magnetic field is located at a distance $kh = 0.65$ from the surface of square conducting scatterer. The special nontrivial case corresponds to $kh = 0$, for which the observation contour coincides with boundary contour C of the square scatterer. In fact, on contour C, the total z component of the electric field is zero which is the boundary condition originally enforced, and the appropriate component of total magnetic field gives the z component of induced electric current distribution shown in the Figures 3.19a and 3.19b. Figures 3.20a and 3.20b show the magnitude and phase distribution of the z component of near scattered electric field calculated based on derived field expressions (3.9.23a–b). The near electric field data is also compared with respect to the FD-TD data showing good agreement. Figures 3.20c and 3.20d show correspondingly the magnitude and phase distribution of the x or y component of near scattered magnetic field again calculated on derived field expressions (3.9.24a–b). Again, the near magnetic field data is also compared with respect to the FD-TD data showing good agreement.

It should be noted that the results of induced electric current in near scattered electric and magnetic fields completely depend upon the angle of incidence of the TM plane wave excitation. The calculations are repeated for an oblique angle of incidence as well. Figures 3.21 and 3.22 show corresponding numerical data for the electromagnetic scattering by the same perfectly conducting square cylinder with an oblique angle of incidence, $\phi^i = 45^{\circ}$. These data are also compared with respect to the numerical data

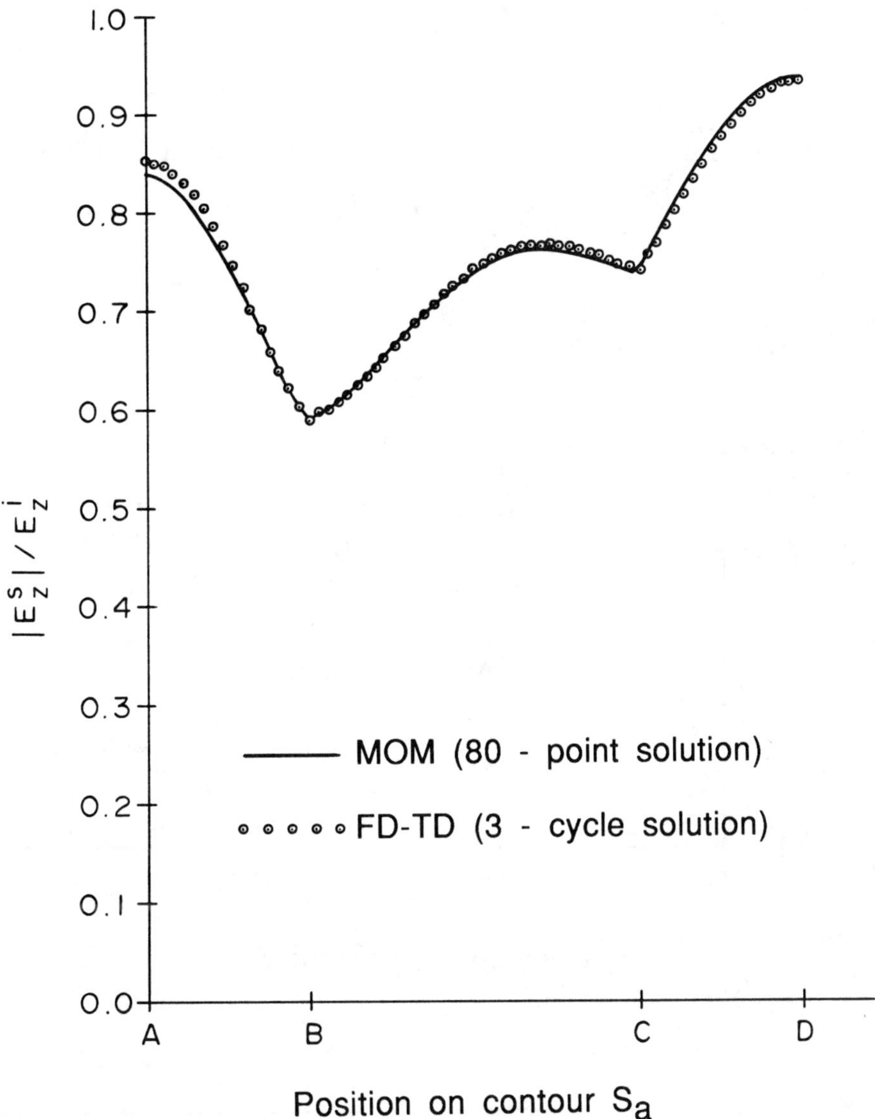

Figure 3.20a Magnitude of axial scattered electric field distribution tangential to the contour S_a, square scatterer, normal incidence – TM excitation.

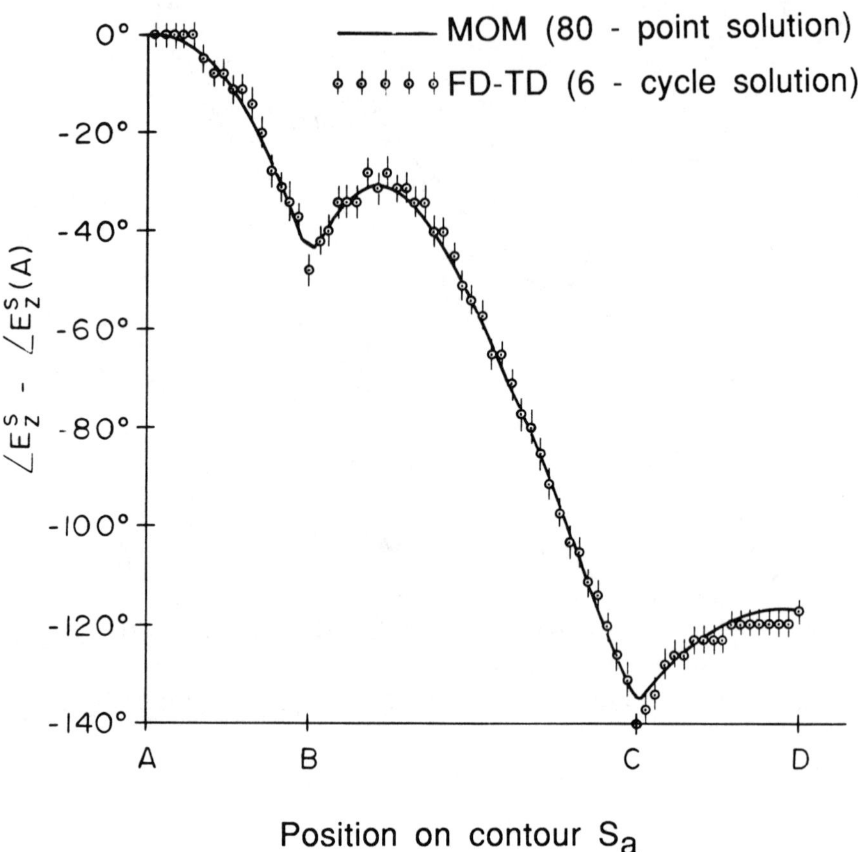

Figure 3.20b Phase of axial scattered electric field distribution tangential to the contour S_a, square scatterer, normal incidence – TM excitation.

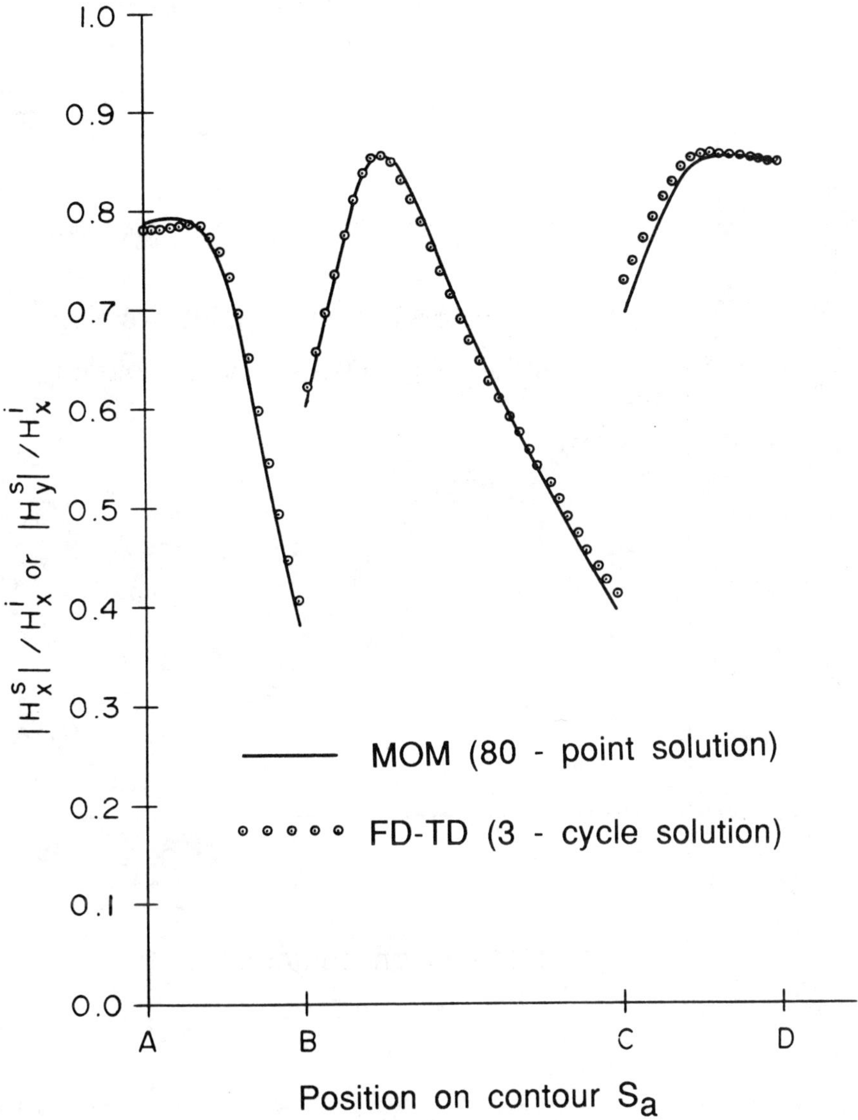

Figure 3.20c Magnitude of near scattered magnetic field distribution tangential to the contour S_a, square scatterer, normal incidence – TM excitation.

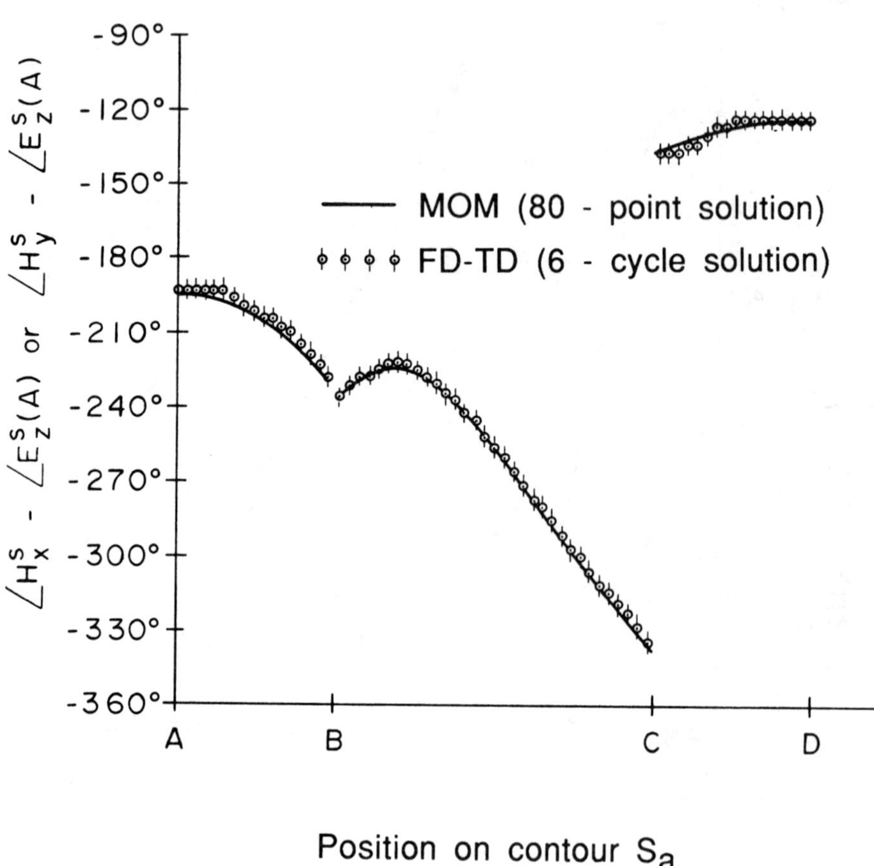

Figure 3.20d Phase of near scattered magnetic field distribution tangential to the contour S_a, square scatterer, normal incidence – TM excitation.

obtained based on the FD-TD method showing good agreement. General background of the FD-TD method is discussed in the earlier introductory chapter, and further details are covered in another book.

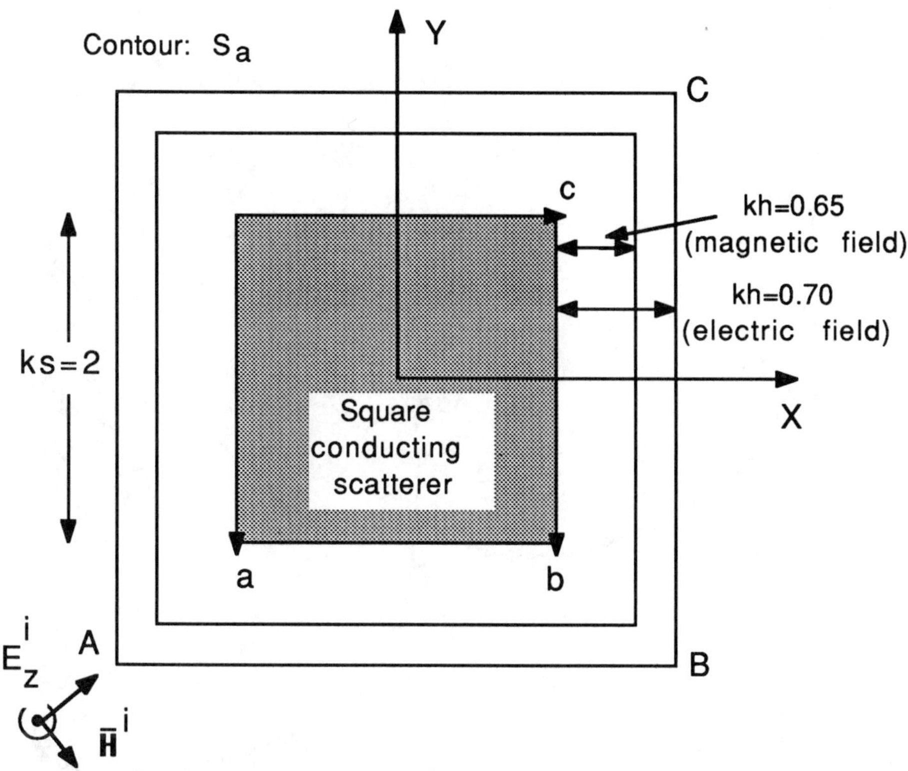

Figure 3.21a Perfectly conducting square scatterer – TM oblique incidence.

Far-Field and Radar Cross Section

The electric and magnetic far-field distributions can be obtained by using large argument approximation for the free-space Green's function. In the large argument approximation, as $|k\rho| \rightarrow \infty$, the Hankel function can be written as

$$H_0^{(2)}(k\rho) \sim \sqrt{\frac{2j}{\pi k\rho}}\; e^{-jk\rho} \tag{3.9.25a}$$

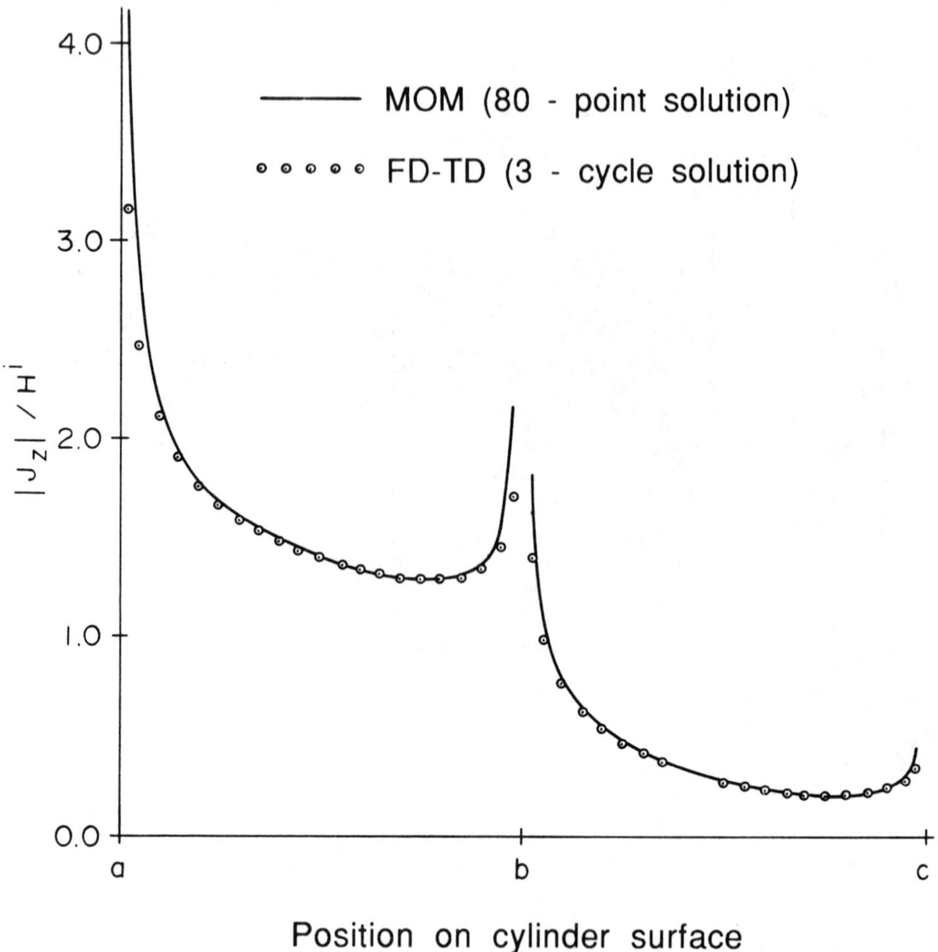

Figure 3.21b Magnitude of induced electric current distribution on a perfectly conducting square scatterer, oblique incidence – TM excitation.

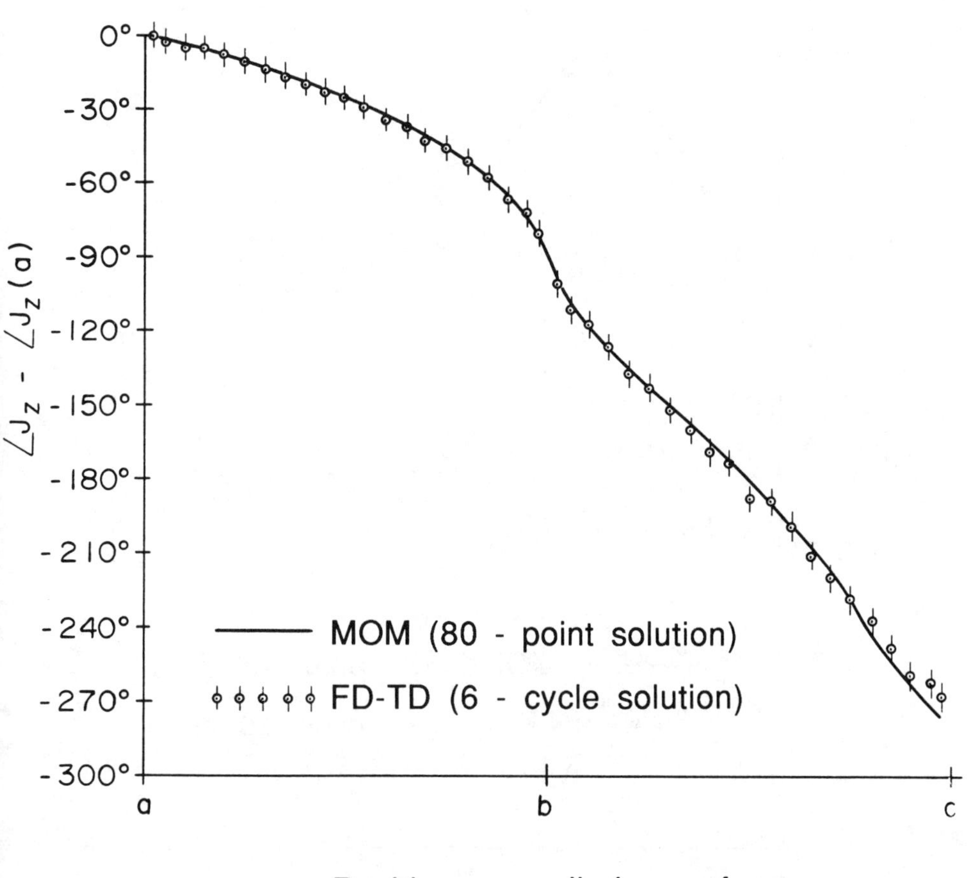

Figure 3.21c Phase of induced electric current distribution on a perfectly conducting square scatterer, oblique incidence – TM excitation.

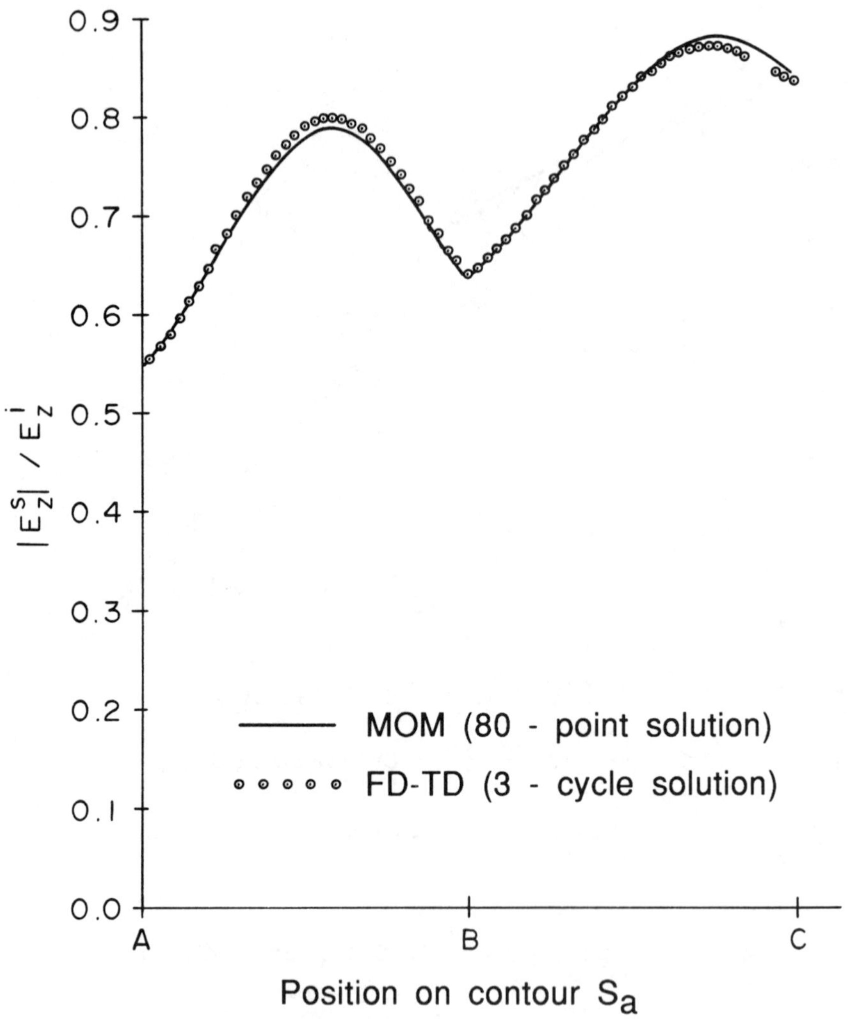

Figure 3.22a Magnitude of axial scattered electric field distribution tangential to the contour S_a, square scatterer, oblique incidence – TM excitation.

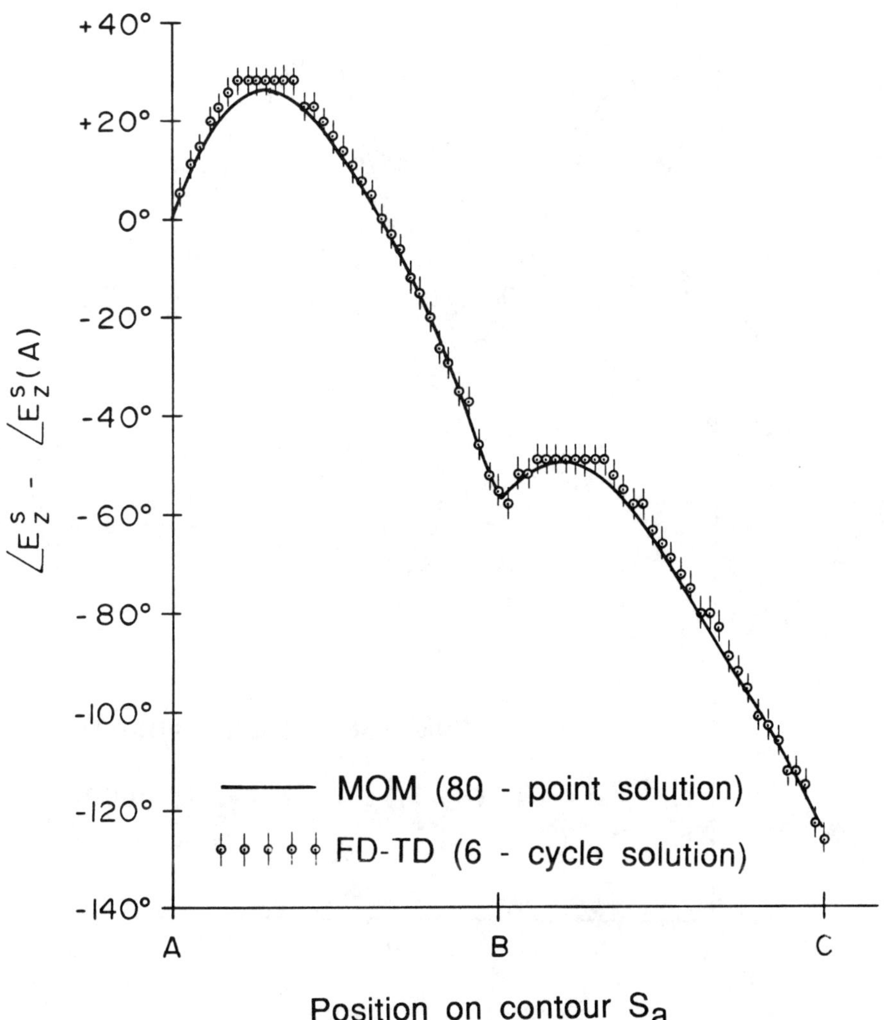

Figure 3.22b Phase of axial scattered electric field distribution tangential to the contour S_a, square scatterer, oblique incidence – TM excitation.

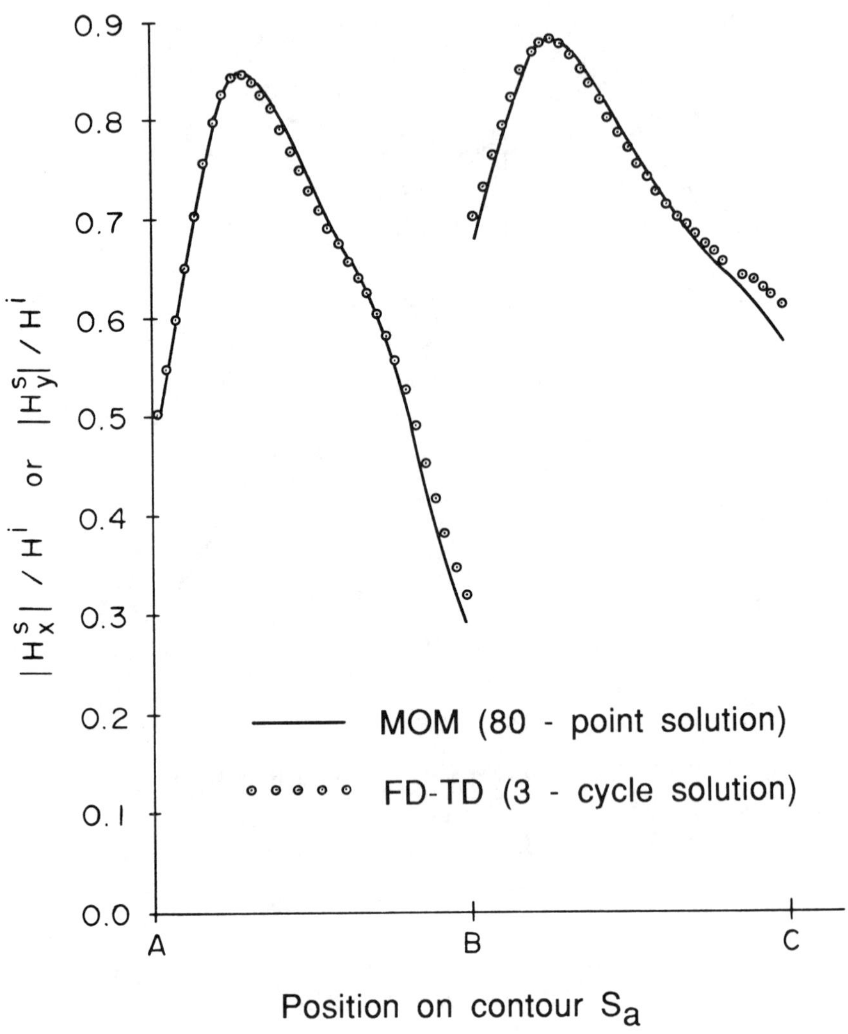

Figure 3.22c Magnitude of near scattered magnetic field distribution tangential to the contour S_a, square scatterer, oblique incidence – TM excitation.

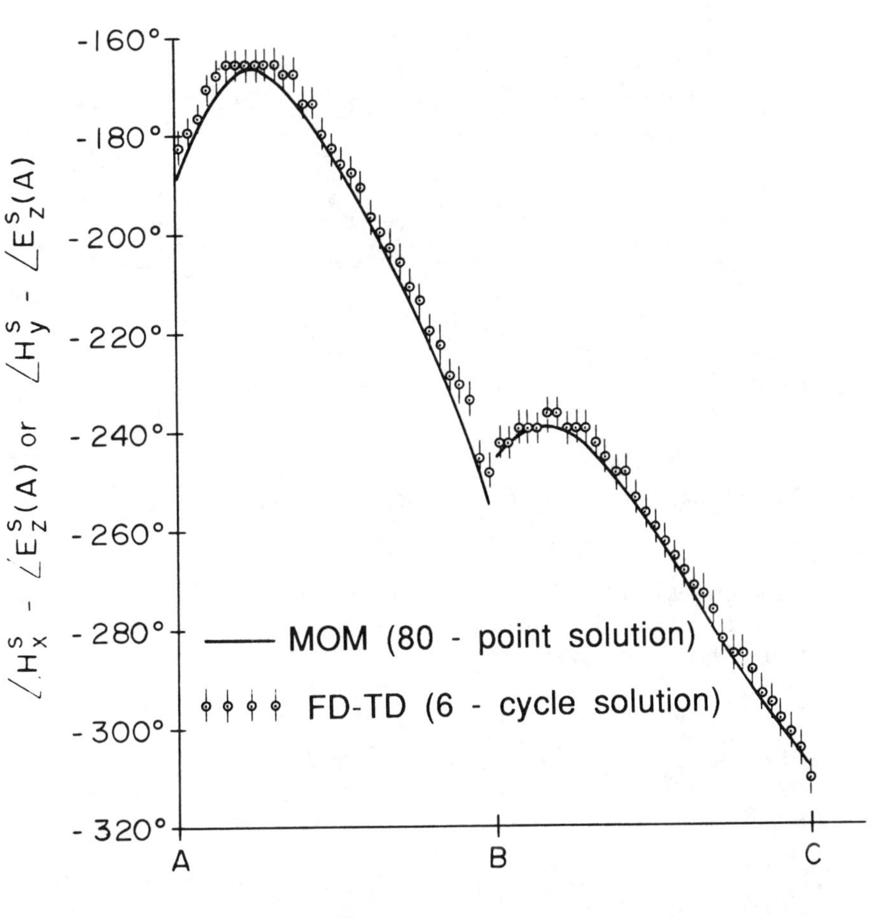

Figure 3.22d Phase of near scattered magnetic field distribution tangential to the contour S_a, square scatterer, oblique incidence – TM excitation.

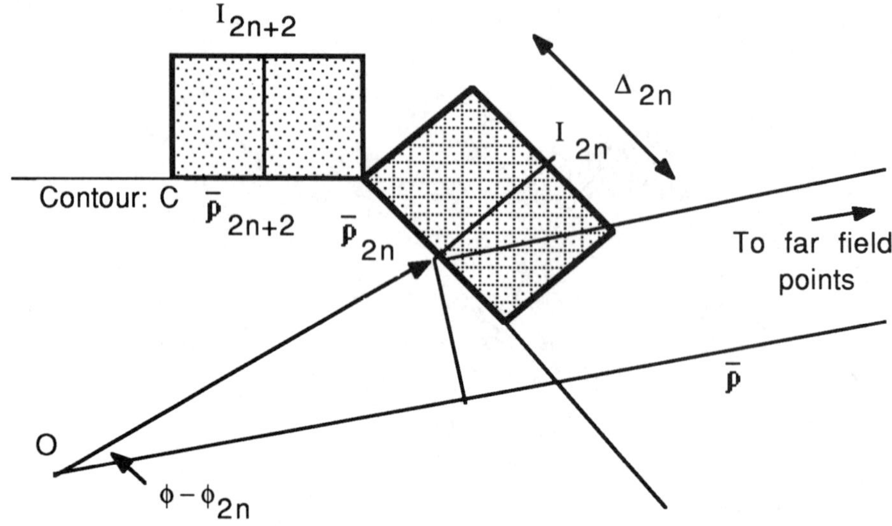

Figure 3.23 Calculation of far-field distribution.

Referring to expression (3.9.23a), the scattered electric field distribution in the far-field region can be calculated by approximating the integral and substituting expression (3.9.25a) for the Hankel function to obtain

$$E_z^s(\rho,\phi) = \sum_{n=1}^{N} I_{2n}\Big[-\frac{k\eta}{4}\, H_0^{(2)}(k|\bar{\rho} - \bar{\rho}_{2n}|)\Big]\Delta_{2n} \tag{3.9.25b}$$

$$E_z^s(\rho,\phi) \sim \sum_{n=1}^{N} I_{2n}\Big[-\frac{k\eta}{4}\sqrt{\frac{2j}{\pi(k|\bar{\rho} - \bar{\rho}_{2n}|)}}\; e^{-j(k|\bar{\rho} - \bar{\rho}_{2n}|)}\Big]\Delta_{2n} \tag{3.9.25c}$$

Referring to Figure 3.23, in the far-field region, $|k\rho| \to \infty$, the magnitude term in denominator of expression (3.9.25c) can be written as

$$|\bar{\rho} - \bar{\rho}_{2n}| \approx \rho \tag{3.9.26a}$$

and the exponential term contributing the phase distribution of the electric current can be approximated as

$$|\bar{\rho} - \bar{\rho}_{2n}| \approx \rho - \rho_{2n} \cos(\phi - \phi_{2n}) \tag{3.9.26b}$$

After substituting these far-field approximations, expressions (3.9.26a–b), the scattered electric field distribution in the far-field region reduces to the following form:

$$E_z^s(\rho,\phi) \sim \sum_{n=1}^{N} I_{2n}\, k\eta \mathcal{K} \left[e^{jk\rho_{2n}\cos(\phi - \phi_{2n})} \right] \Delta_{2n} \tag{3.9.27a}$$

$$\mathcal{K} = \frac{1}{\sqrt{8\pi k\rho}}\; e^{-jk\rho}\; e^{-j3\pi/4} \tag{3.9.27b}$$

In the far-field region, there are only the axial E_z component of the scattered electric field and the angular H_ϕ component of the scattered magnetic field distributions. According to expression (3.9.27b), the magnitude of the scattered electric field decays as a reciprocal of the square root of the radial variable, and the two far scattered field distributions behave as pure cylindrical waves propagating along the radial coordinate direction and are directly related through the medium intrinsic impedance.

Further, the bistatic radar cross section of the perfectly conducting scatterer having an arbitrary shape can be calculated using expressions (3.8.18) and (3.9.27a–b):

$$\mathrm{RCS}(\phi) = \lim_{\rho \to \infty} 2\pi\rho \left| \frac{E_z^s(\phi,\omega)}{E_z^i(\phi,\omega)} \right|^2 \tag{3.9.28}$$

Figure 3.24 shows a plot of bistatic radar cross section (RCS) of the perfectly conducting square scatterer, $ks = 2$, calculated from expressions (3.9.27a) and (3.9.28). This bistatic RCS data is numerically generated by using the induced electric current results presented in the Figures 3.19a and 3.19b for the normal incident excitation, $\phi^i = 90^o$. As can be seen, the bistatic data exhibits symmetrical distribution with respect to the direction of excitation about which the square scatterer also has geometrical symmetry. Since the bistatic RCS distribution is proportional to the scattered power, the far-field scattering field pattern is directly obtained by taking square root of the RCS distribution.

Also, Figure 3.24 shows plot of bistatic radar cross section for oblique incident excitation, $\phi^i = 45^o$, calculated using the electric current results presented in the Figures 3.21b and 3.21c. Again, the bistatic data exhibits symmetrical distribution with respect to the direction of excitation about which the scatterer also has geometrical symmetry.

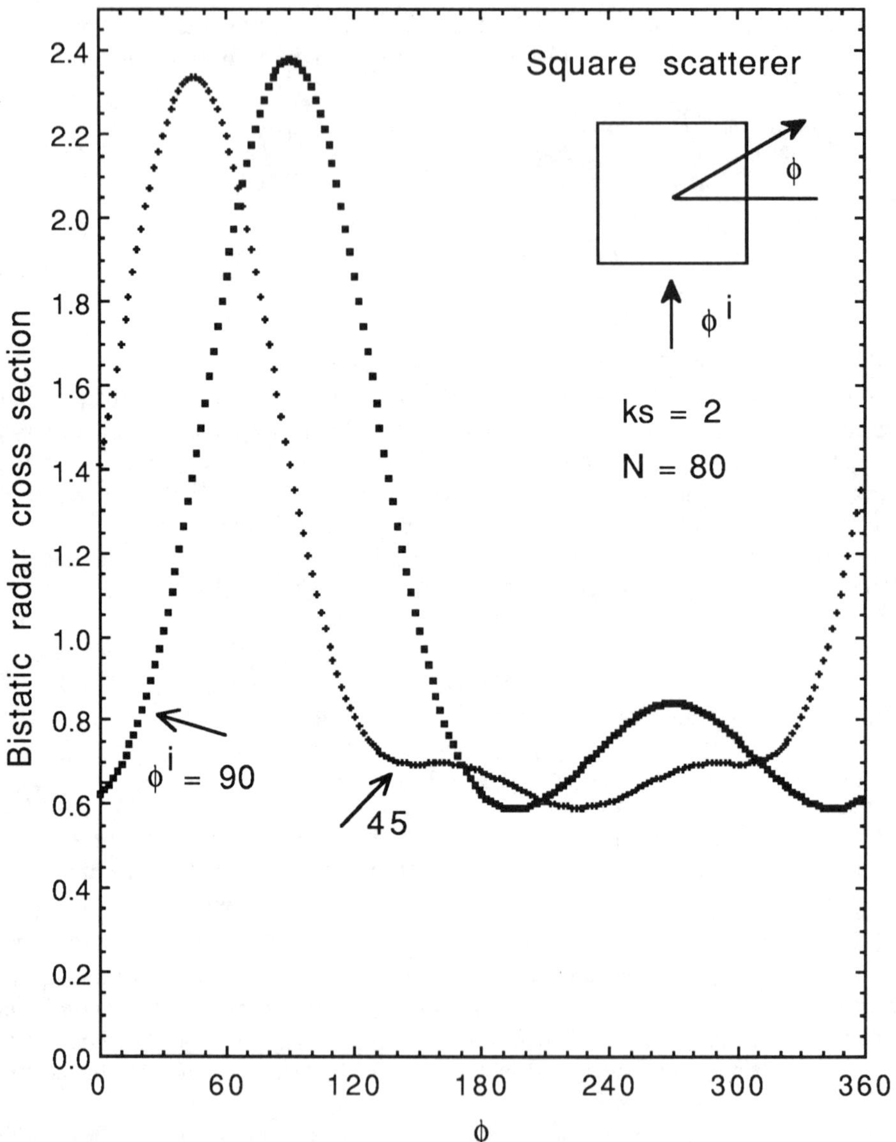

Figure 3.24 Bistatic radar cross section of a perfectly conducting square scatterer – TM excitation.

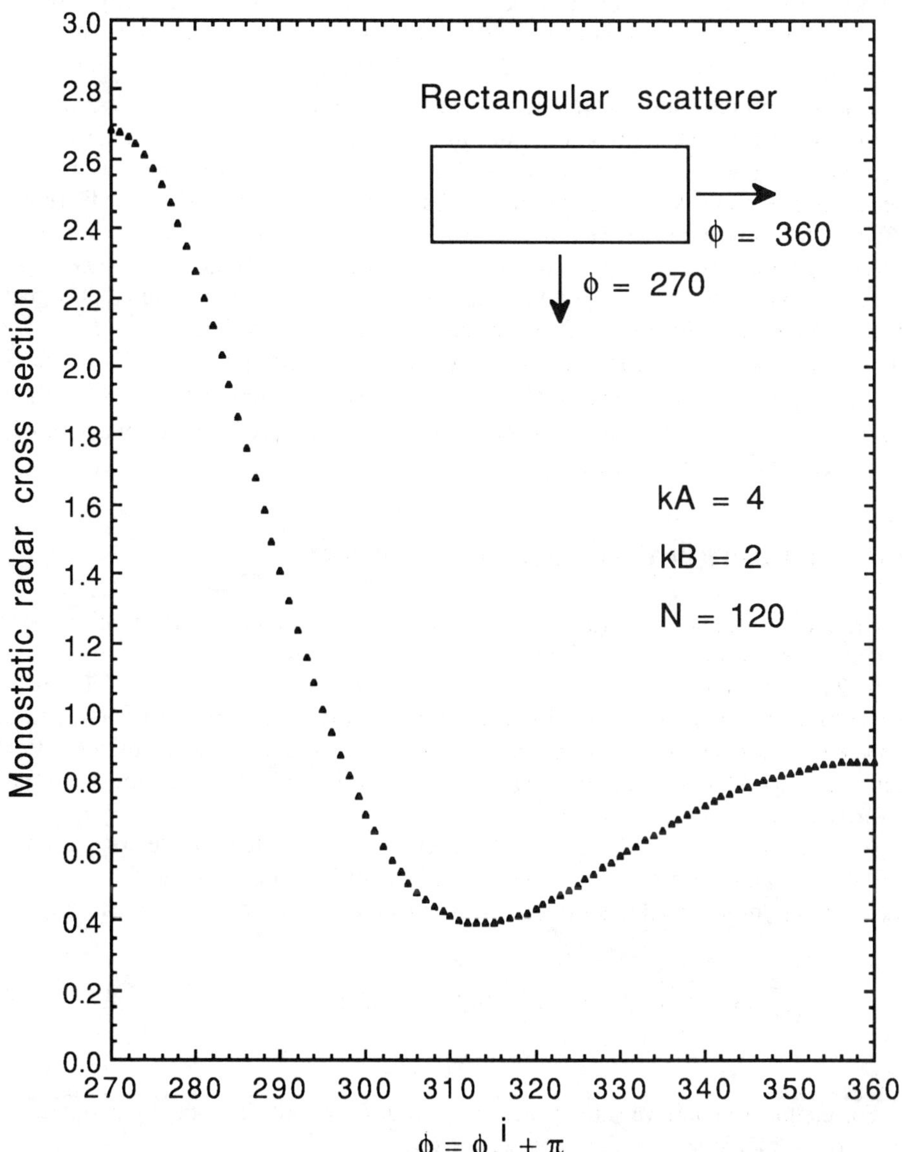

Figure 3.25 Monostatic radar cross section of a perfectly conducting rectangular scatterer – TM excitation.

In the numerical analysis just carried out, the perfectly conducting square scatterer has the electrical size $ks = 2$ for each side. This corresponds to the side length of $s/\lambda = 0.3183$. The distribution of electric current on each side has been sampled using twenty piecewise pulse functions with $N = 80$ for all four sides of the square scatterer. This gives a uniform resolution (width of each current sampling pulse) of $\Delta_{2n}/\lambda = 0.0159$.

In many practical applications, the monostatic or backscattering data are quite important, which depend upon number of parameters, such as the angle of incidence, electrical size, and shape of the geometry. To produce the monostatic RCS distribution, referring to expression (3.9.22), the inverse of the generalized impedance matrix can be stored separately and multiplied by the excitation term recalculated for every angle of incidence to obtain the corresponding induced electric current distribution. The mono-RCS data are then obtained using expression (3.9.28). Figure 3.25 shows a plot of the monostatic radar cross section of a perfectly conducting rectangular scatterer for incident angles ranging from $\phi^i = 90^o$ to 180°. The size of the scatterer is selected as length $kA = 4$ and width $kB = 2$. The monostatic data shown in the Figure 3.25 are generated with a uniform sampling with a matrix size of $N = 120$.

3.10 THIN-STRIP CONDUCTING SCATTERER

The electromagnetic scattering by a thin, perfectly conducting strip with transverse magnetic excitation can be studied using the electric field integral equation, expression (3.9.12a), followed by the method of moments technique discussed earlier. Referring to Figure 3.26, the geometry of a thin-strip scatterer is oriented along the x coordinate axis between limits O and A. For transverse magnetic excitation, the incident electric field is polarized parallel to the z coordinate axis and the corresponding incident magnetic field is polarized in the transverse xy plane.

As shown in the Figure 3.26, there exists induced axially directed electric currents are on both top and bottom sides of the thin-strip scatterer. Referring to Section 3.9, the magnetic vector potential in terms of unknown electric current distribution has the form

$$A_z(\bar{\rho}) = \frac{\mu}{4j}\int_C J_z(\bar{\rho}')\, H_0^{(2)}(k|\bar{\rho} - \bar{\rho}'|)\, dL(\bar{\rho}') \qquad (3.10.1)$$

For the top and bottom sides representing boundary contour C, the following integral equation is satisfied on the thin-strip scatterer:

$$E_z^i(\bar{\rho}) = j\omega A_z(\bar{\rho}) \qquad \bar{\rho} \text{ on C} \qquad (3.10.2)$$

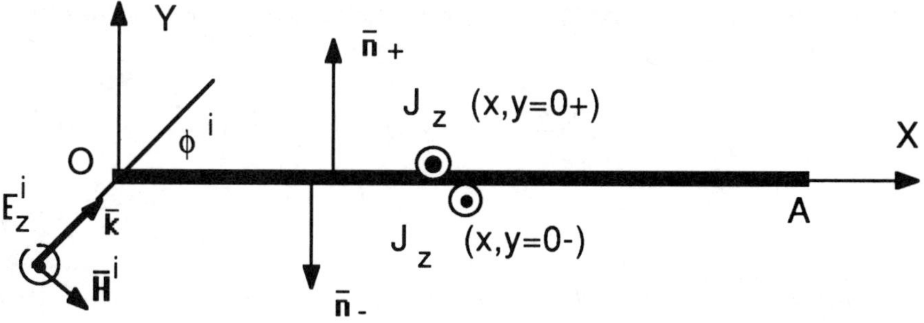

Figure 3.26 Geometry of a thin-strip scatterer.

The integral expression for the magnetic vector potential can be split into two parts, corresponding to the top and bottom surface currents on the thin-strip scatterer and further can be combined into a composite unknown electric current on the scatterer shown in Figure 3.27. For example, referring to Figure 3.26, the magnetic vector potential integral can be written as

$$A_z(\bar{\rho}) = \frac{\mu}{4j}\int_0^A J_z(x',y'=0-)\; H_0^{(2)}(k|\bar{\rho} - \bar{\rho}'|)\; dL'$$

$$+ \frac{\mu}{4j}\int_A^0 J_z(x',y'=0+)\; H_0^{(2)}(k|\bar{\rho} - \bar{\rho}'|)\; dL' \qquad (3.10.3)$$

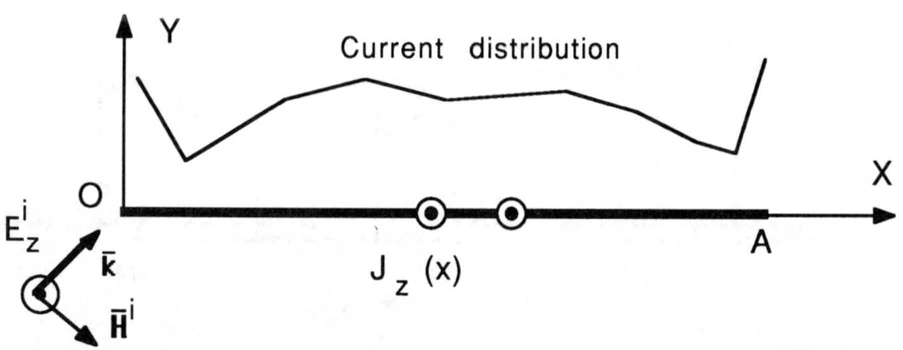

Figure 3.27 Modeling of the thin-strip scatterer.

After combining the two integrals, expression (3.10.3) reduces to the following form:

$$A_z(\bar{\rho}) = \frac{\mu}{4j}\int_0^A \left[J_z(x',y'=0-) - J_z(x',y'=0+)\right] H_0^{(2)}(k|\bar{\rho} - \bar{\rho}'|)\, dL'$$

(3.10.4a)

$$A_z(\bar{\rho}) = \frac{\mu}{4j}\int_0^A J_z(x')\, H_0^{(2)}(k|x - x'|)\, dx' \qquad (3.10.4b)$$

where $J_z(x')$ is the composite equivalent electric current on the thin-strip scatterer of length L, and integral equation (3.10.2) simplifies to

$$E_z^i(x) = \frac{\omega\mu}{4}\int_0^L J_z(x')\, H_0^{(2)}(k|x - x'|)\, dx' \qquad x \text{ on strip} \qquad (3.10.5)$$

As can be seen from this EFIE formulation for the thin-strip scatterer, the distribution of the unknown electric current, which is to be solved, is the net value of equivalent electric current on the thin-strip represented by the difference of the bottom and top currents. The numerical technique discussed in Section 3.9 can be implemented to solve for the unknown electric current. Figure 3.28 shows an arrangement of piecewise current expansion pulses for the case of thin-strip scatterer. At the two open ends, there exists current singularities and the current expansion pulses are located half a cell away. Substituting the current expansion functions and weighting integral equation (3.10.5) using similar weighting pulses, the unknown distribution of the electric current is obtained for a given angle of excitation of the TM plane wave.

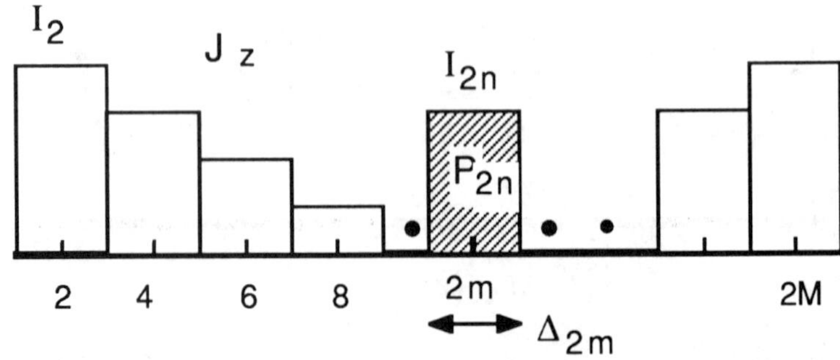

Figure 3.28 Current expansion pulses for the thin-strip scatterer – TM excitation.

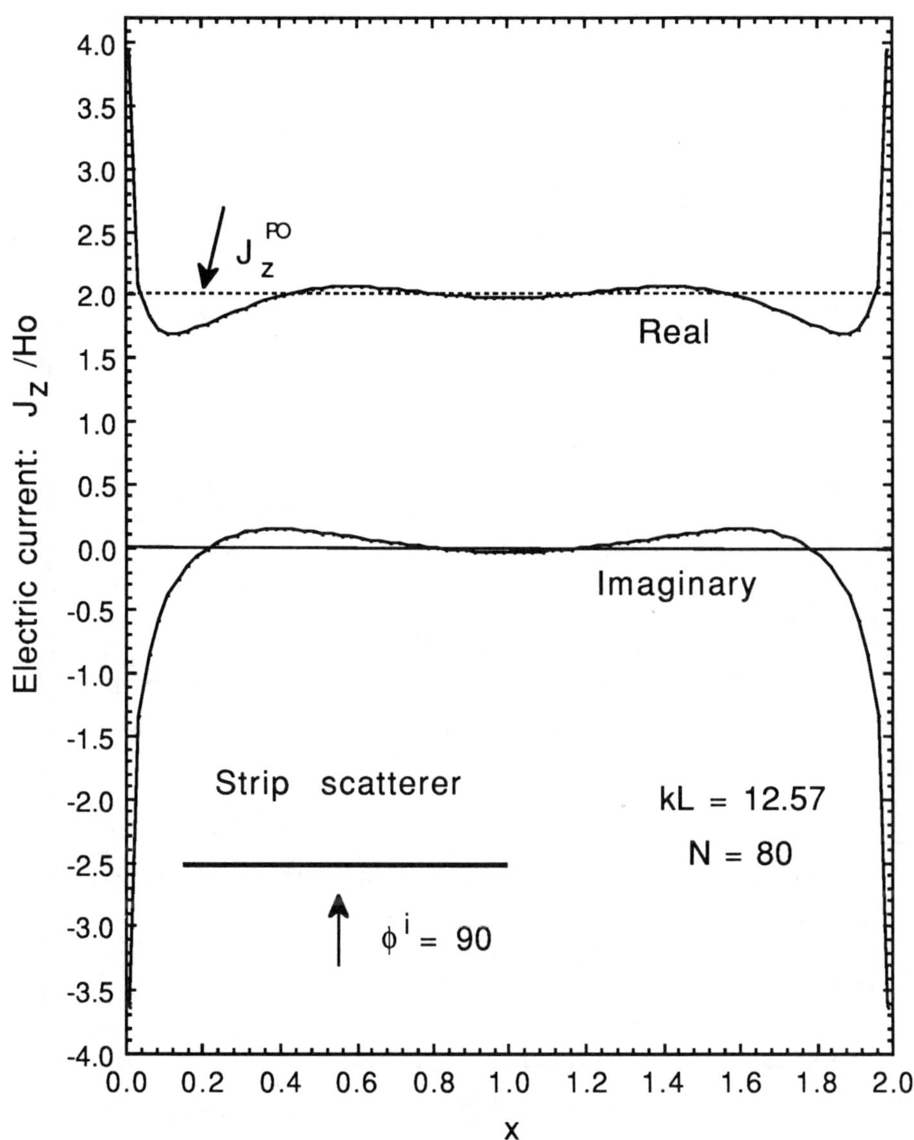

Figure 3.29 Electric current distribution on a perfectly conducting thin-strip scatterer – TM excitation.

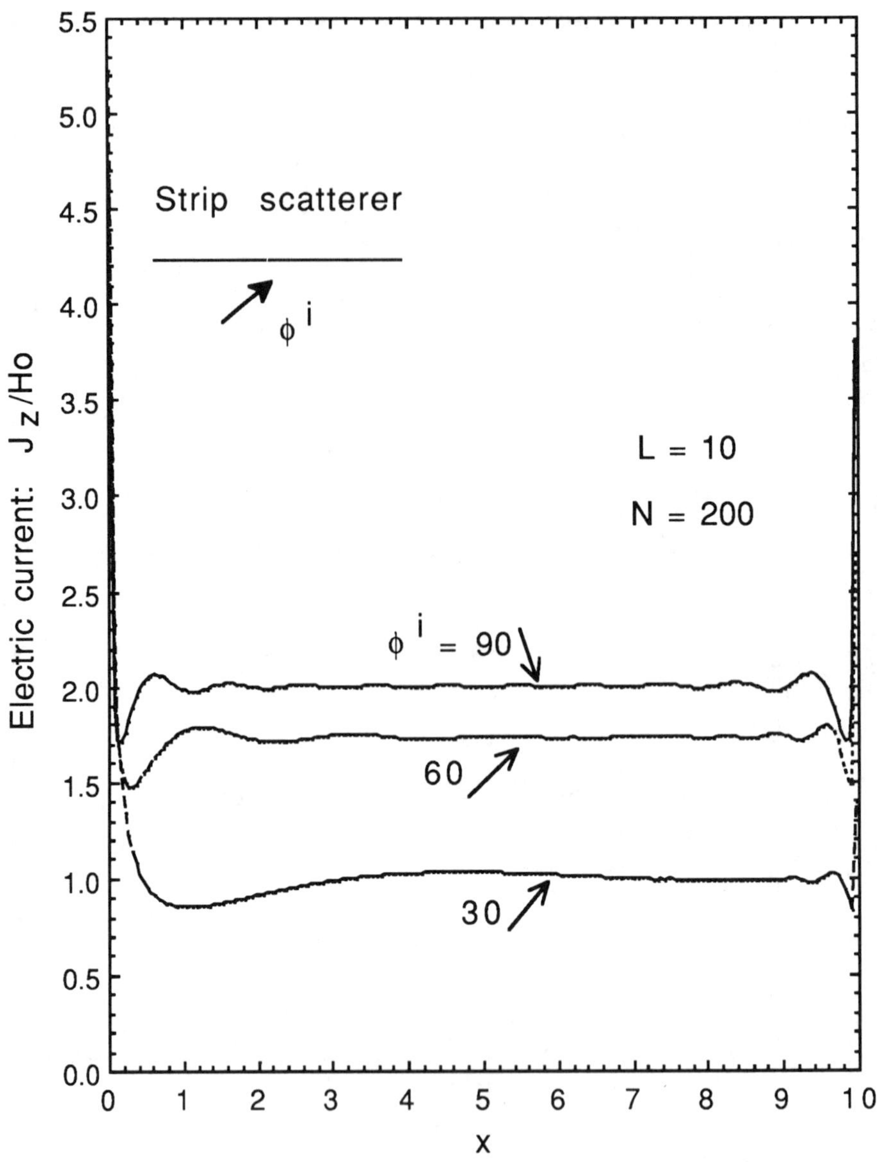

Figure 3.30 Magnitude of electric current distribution on a perfectly conducting thin-strip scatterer – TM excitation.

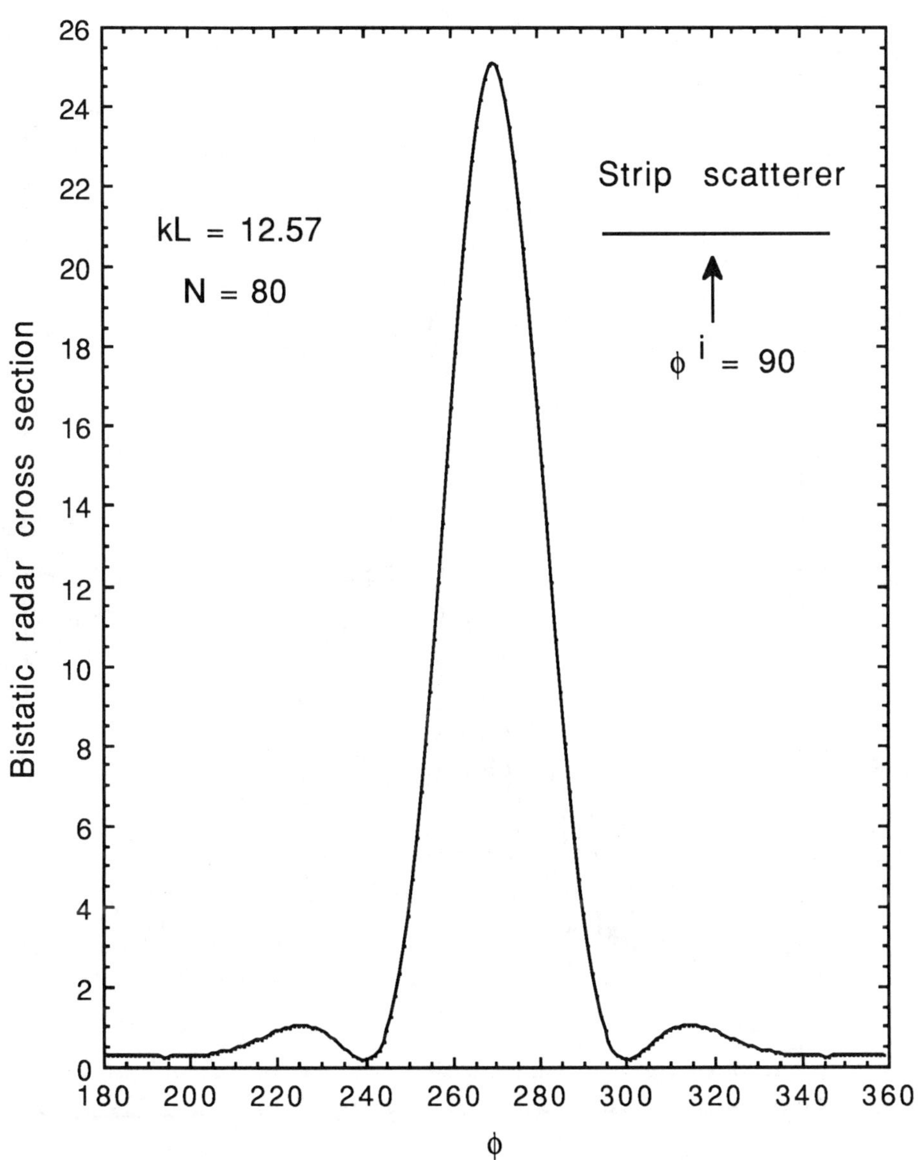

Figure 3.31 Bistatic radar cross section of a perfectly conducting thin-strip scatterer – TM excitation.

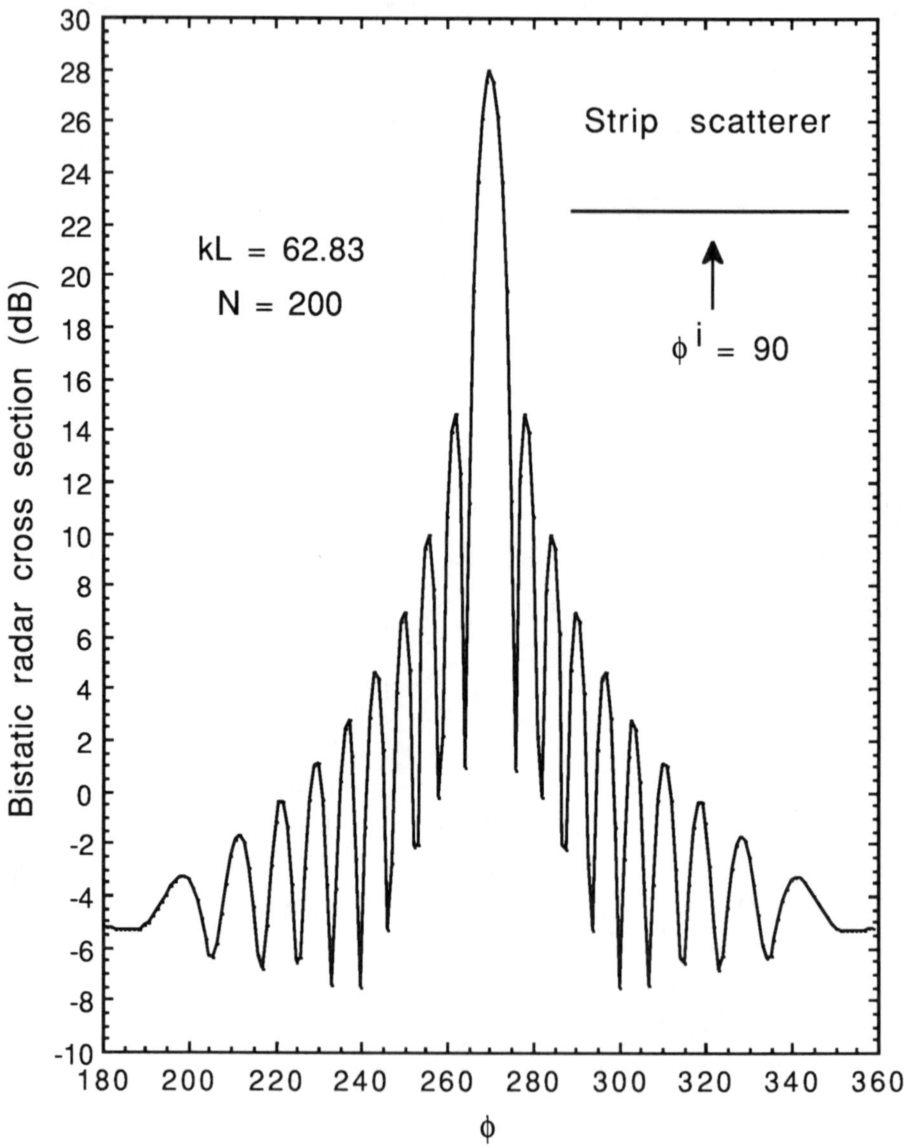

Figure 3.32 Bistatic radar cross section of a perfectly conducting thin-strip large scatterer – TM excitation.

Figure 3.29 shows the distribution of the induced electric current on a thin-strip scatterer with broadside excitation. For the results shown, the frequency of excitation is assumed to be 300 MHz, so that the wavelength of incident plane wave is 1 meter and the propagation constant in the free-space medium is $k = 2\pi$. The total length of the conducting strip is selected as $L = 2$ meters and excited normally with the plane wave magnetic field polarized along the x coordinate direction. Both the real and imaginary parts of the electric current distribution are shown in the Figure 3.29, with normal to the strip assumed in the $-y$ coordinate direction. The results are generated with the method of moments matrix size of $N = 80$ and are normalized with respect to the incident magnetic field.

The distribution of the electric current has singularity behavior at the two free ends of the thin-strip scatterer. In fact, the distribution exhibits predominant singular behavior very close to the scatterer tips and attains the well-known physical optics limit away from the two free ends. This should be obvious if one examines the distribution of electric current for a semi-infinite, perfectly conducting strip (half-plane problem) excited normally by a TM polarized plane wave. The analysis approach for the half-plane problem is not discussed here, as it is beyond the scope of this book. For the half-plane case, both the real and imaginary parts show singular behavior near the tip and exist for a range approximately beyond $kx = \pi$. Away from the tip of scatterer, the real part of the induced electric current reaches the physical optics limit and the corresponding imaginary part of the current is zero. Figure 3.30 shows the magnitude of the induced electric current for TM incident excitation on a thin-strip scatterer of length $L = 10$ meters. The result shown is generated with a matrix size of $N = 200$.

Once the distribution of the electric current is known, both the near and far scattered field distributions can be easily calculated based on expressions (3.9.23b) and (3.9.27a), and similarly, the radar cross section can be calculated based on expression (3.9.28). Figure 3.31 shows distribution of the bistatic radar cross section for the thin-strip scatterer of length $L = 2$ meters. This bistatic result is obtained using the electric current distribution shown in the Figure 3.29. For normal incidence, the backscattering return is maximum and exhibits symmetry with respect to the direction of normal excitation. Similarly, for the large thin-strip scatterer with length $L = 10$ meters, the bistatic radar cross section in decibels is shown in Figure 3.32 in the angular range of 180° to 360°. Again, the bistatic data is maximum for the broadside angle and gradually drops off to a minimum for the end on observation angles. The results are shown with a resolution of one degree angle and to identify various side lobes finer angular resolution is required.

3.10.1 CONVERGENCE DATA

A number of numerical case studies are presented in the previous section concerning the electromagnetic scattering and interaction by perfectly conducting two-dimensional objects with TM excitation. Specifically, a square conducting scatterer and a thin-strip scatterer are studied. It should be noted that the size of matrix required to obtain a converged solution completely depends not only on the electrical dimension but also on the geometrical shape of the scatterer. In the following, numerical data are presented showing convergence of the induced electrical current and radar cross section for the TM excited conducting square scatterer.

Table 3.7

Electric current convergence data for the case of a square conducting scatterer – Normal incidence on one side

Matrix size	*Induced Electric Current, J_z/H_0*			
N	*Illuminated side (center)*		*Shadow side (center)*	
	Magnitude	*Phase*	*Magnitude*	*Phase*
12	1.70875	-4.54448	0.19442	75.67725
20	1.74364	-4.96552	0.18215	76.07770
28	1.74343	-4.93848	0.17837	75.35386
36	1.74269	-4.93823	0.17644	75.01311
44	1.74189	-4.94251	0.17528	74.80623
52	1.74123	-4.94828	0.17452	74.67036
60	1.74068	-4.95419	0.17397	74.57180
68	1.74025	-4.96023	0.17357	74.50133
76	1.73990	-4.96567	0.17326	74.44418
84	1.73961	-4.97095	0.17301	74.40430

The geometry of a square conducting scatterer is shown in Figure 3.18a. The electric length of each side is $ks = 2$. Figure 3.18b shows various locations and detailed arrangements of the electric current pulse expansion functions with uniform resolution. The angle of incidence of TM plane wave excitation is selected as 90°. To observe the convergence of the numerical data, referring to Figure 3.18a, two test points are selected on the geometry – one point at center of the illuminated side, and other point at center of the shadow side. The electric current expansion pulses are uniformly distributed along the square scatterer. In Table 3.7, magnitude and phase of the electric current are listed for various values of matrix size, *N*.

As can be seen from these numerical data, the illuminated-side electric current has converged even with a matrix size $N = 36$, but the shadow-side electric current is still slowly converging. The magnitude of the shadow-side electric current is relatively very small compared to the illuminated-side electric current and, in fact, requires a slightly larger matrix size for good convergence. This convergence data are for the case of a small square conducting scatterer, $ks = 2$, in which there is strong coupling and interaction between various corners of the scatterer geometry. It may be noted, as the scatterer electrical size becomes larger, there is relatively loose coupling between sides of the scatterer, and thus, it is possible to adopt a lower resolution for the electric current sampling.

Table 3.8
Radar cross section convergence data for the case of a square conducting scatterer – Normal incidence on one side

Matrix Size	*Radar cross section*	
N	*Forward scattering*	*Back scattering*
12	2.32764	0.78974
20	2.35965	0.81908
28	2.36922	0.82893
36	2.37336	0.83359
44	2.37551	0.83621
52	2.37676	0.83786
60	2.37753	0.83897
68	2.37804	0.83975
76	2.37838	0.84034
84	2.37863	0.84078

As discussed in Section 3.9, the far-field quantities are obtained by integrating the induced electric current distribution over the free-space Green's function. Hence, due to the numerical averaging effect, in most of electromagnetic and interaction studies, the convergence rate of the far-field quantities is generally faster than the near-field quantities. For the same square geometry, Table 3.8 shows variations of forward and backscattering radar cross section listed as a function of the matrix size, N. Based on the numerical data presented, it is obvious that the convergence of the electric current distribution is slower than the convergence of far-field or radar cross section. As stated earlier, for an electrically large scatterer, it is possible to obtain far-field numerical data even with a sparse resolution of about 10 to 20 current samples per wavelength.

The direct numerical solution for the analysis of electrically large scatterers requires a large matrix size. This, in fact, poses a serious question regarding the computational burden, such as the computer storage and the computer time requirements. Further, the computational time required to generate and invert the large impedance matrix is also very large. These aspects are addressed in the next section.

3.11 ELECTRICALLY LARGE OBJECTS

In earlier sections, the boundary value EFIE and the MOM numerical technique have been utilized for the study of electromagnetic scattering by arbitrary shaped conducting objects with TM plane wave excitation. Even though the direct approach is elegant, as far as its application to analyze electrically large object is concerned, it inherently suffers from a wide range of computational difficulties. The method of moments system matrix is full and dense, requiring impractical demand on computer resources. In addition to operational numerical errors and ill-conditioning involved in the solution of a large-scale matrix equation, the direct numerical technique bears progressive degradation in accuracy of near-field solution as the size of system matrix increases. Key limitations render the direct integral equation and method moments technique unattractive beyond low and resonant frequencies, unless vast computer resources are easily available.

Dimensionally very large resources

The direct integral equation and method of moments technique generate a system of linear equations having dense, complex valued, full coefficient matrices. For a conventional matrix approach, the required computer storage is on the order of $O(P^2) + C_pP$, where P is the number of unknowns to be determined and C_p is a constant that depends upon scatterer input geometry and the desired display output of the numerical solution. Similarly, computer execution time is on the order of $O(P^2) + C_qP$ to $O(P^3) + C_rP$, where C_q and C_r are constants that depend upon the numerical model adapted to generate the system matrix and, further, solve for the unknown current coefficients. A spatial resolution requirement on the order of $\lambda/10$ to $\lambda/20$ to avoid aliasing of vital near-field magnitude and phase information implies dimensionally large computer resource requirements. This, in fact, places a cap on the direct formulation of scattering problem and their applications.

Ill-conditioning

The large system matrix (of electrically large object) generated by the direct integral equation and method of moments numerical technique tends to become highly ill-conditioned. This potentially degrades the accuracy of the computed results. Accuracy and condition number of the numerical matrix solution may also be degraded by the floating point wordlength used in a computer and specific numerical procedure adopted for computing individual elements and the eventual solution of the system matrix. Accumulating errors of these types can be troublesome, especially when hundreds of millions of matrix elements or floating point operations are involved in one modeling problem.

The apparent computational difficulties with the direct integral equation and method of moments have prompted an alternative procedure for a numerical solution based on spatial decomposition technique. Using rigorous electromagnetic equivalence, this technique virtually divides an electrically large object into a multiplicity of subzones. It permits the maximum size of the method of moments system matrix that should be inverted to be strictly limited (so that error bounds and condition numbers can be controlled) regardless of the electrical size of the large scattering object being modeled. The requirement on the computer resources is of order (N), where N is the number of spatial subzones and each subzone is electrically small, spanning in the order of few wavelengths. Few numerical examples are presented to consider possible applicability of the spatial decomposition technique to analyze two-dimensional electrically large conducting objects.

3.11.1 SPATIAL DECOMPOSITION TECHNIQUE

This section presents a preliminary formulation of the *spatial decomposition technique* (SDT) for the integral equation and method of moments. Using rigorous electromagnetic equivalences, the SDT allows one to divide an electrically large arbitrary shaped conducting object into a multiplicity of subzones. The individual subzones are separated by virtual boundaries across which cancelling tangential electric-type virtual currents are postulated. The subzones are defined as distinct scattering targets, having fully enclosing boundaries with additional unknown virtual electric currents introduced as needed to define the interfaces. The electromagnetic field boundary conditions are well preserved across the interface separating two subzones by requiring that the tangential virtual currents on one side of the interface be equal and opposite to the tangential virtual currents on the other side. By sequentially implementing the integral equation and method of moments solution for each subzone, effectively scanning the original target subzone by subzone, a rapidly convergent iterative process can be established. Figure 3.33a illustrates a two-dimensional conducting object. It is excited externally by a transverse magnetic (TM) polarized plane wave propagating in a direction normal to the z coordinate axis.

As shown in the Figure 3.33a, axially directed equivalent electric currents $\bar{J}_z(\bar{\rho}')$ reside on a virtual boundary, conforming to the physical boundary of the scatterer. Referring to Section 3.6, the EFIE for the two-dimensional scattering case is given by, for $\bar{\rho}$ on C

$$E_z^i(\bar{\rho}) = \mathcal{L}[J_z(\bar{\rho}')] \tag{3.11.1a}$$

$$\mathcal{L} = \frac{\omega\mu}{4} \int_C H_0^{(2)}(k|\bar{\rho} - \bar{\rho}'|)\, dL(\bar{\rho}') \tag{3.11.1b}$$

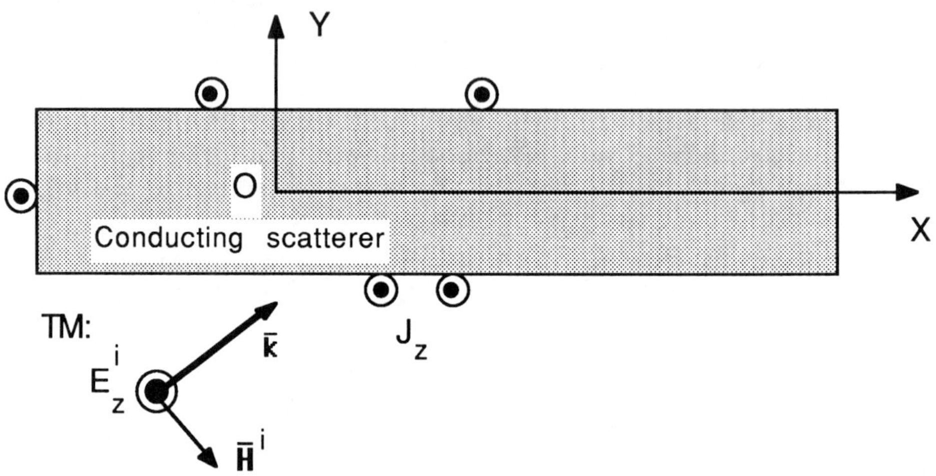

Figure 3.33a Electrically large two-dimensional perfectly conducting scatterer – TM excitation.

In the case of a direct method of moments solution, the axial electric current J_Z is expanded in terms of a suitable expansion function, such as pulse expansion functions and is tested on both sides by the weighting functions to reduce integral equation (3.11.1a) to its equivalent matrix equation. As the electrical size of the scatterer increases, the direct solution technique puts impractical demands on the computer resources, in addition to various unsettled numerical accuracy and ill-conditioning problems of the large-system matrix associated with electrically very large objects.

To circumvent this high demand on computer resources and also reduce the numerical difficulties, the formulation of boundary value problem is modified, still retaining all the physics of electromagnetic scattering and interaction as depicted schematically in the Figures 3.33a and 3.33b. Figure 3.33a represents a rectangular scatterer geometry with unknown electric current distribution on a virtual boundary. Figure 3.33b has an identical geometry, but the virtual boundary contour is modified to define (in this case) two distinct spatial subzones.

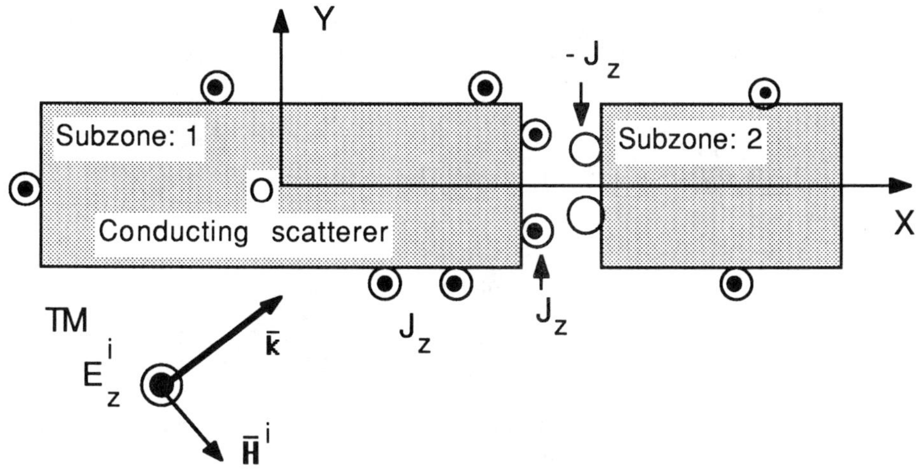

Figure 3.33b Spatial decomposition technique for an electrically large conducting scatterer – TM excitation.

A key point to note is that, at the virtual interface separating the two subzones, the tangential virtual currents on one side of the interface must be equal and opposite to the tangential virtual currents on the other side of the interface. In fact, the scatterer can be divided into an arbitrary number of N distinct spatial subzones in this manner (in Figure 3.33b, $N = 2$). Now, the usual electric field integral equation and method of moments technique is used to compute the electric currents along the enclosing boundary of one subzone. In effect, the subzone is treated as a distinct scatterer. It should be noted that a part of the subzone's boundary is the virtual interface separating it from the adjacent subzone. The excitation for this subzone consists of the original incident plane wave and additional excitation due to the electric currents residing on the boundaries of the remaining spatial subzones and radiating into the free space. Initially, the additional excitation due to the currents on the remaining spatial subzones is not known. But, these can be conveniently approximated to a zeroth order by using, for example, the physical optics solution.

Hence, for subzone 1, the modified EFIE takes the form

$$F_z^i(\bar{\rho}) = \mathcal{L}[J_{z1}(\bar{\rho}')] \qquad \text{for subzone 1, } \bar{\rho} \text{ on } C_1 \tag{3.11.2a}$$

and

$J_{z1}(\bar{\rho}')$: unknown electric current distribution for the subzone 1 where the boundary contour C_1 encloses completely spatial subzone 1, and the total excitation for spatial subzone 1 is given by plane wave excitation and additional excitation due to the remaining spatial subzones, $n = 2, 3, 4, \ldots, N$:

$$F_z^i(\bar{\rho}) = E_z^i(\bar{\rho}) - \sum_{n=2}^{N} L[J_{zn}(\bar{\rho}'_n)] \tag{3.11.2b}$$

for subzone 1, and
$\bar{\rho}'_n$ on the nth subzone boundary C_n

The analysis is now shifted to the next adjacent subzone. The excitation for this subzone consists of the original plane wave and additional excitation due to the currents on the boundaries of the remaining subzones, including the updated currents on the first subzone. In this manner, the step-by-step analysis approach can be sequentially implemented for rest of the subzones, effectively scanning the original scatterer subzone by subzone, always using the incident plane wave and the latest surface currents as the excitation for the subzone of interest. As can be seen in the SDT iterative process, the subzones are used rather than the complete structure, and the subzones are not completely related to individual blocks of the original full matrix problem, rather the subzones are defined as distinct targets having fully enclosing boundaries with additional virtual electric current unknowns (still maintaining the boundary conditions) introduced as needed to define the virtual interface consistently between the subzones.

Thus, the SDT provides a means to implement the existing integral equation and method of moments technique with a computer memory and execution time requirement of $O(N)$, where N is the number of spatial subzones. Since each subzone is electrically small, spanning few wavelengths, it is expected that conditioning of the method of moments system matrix resulting for each subzone is acceptable for numerical processing with limited demands on computer resources.

To illustrate the results of the spatial decomposition technique, the electromagnetic scattering by a perfectly conducting thin-strip and a rectangular scatterer are considered. The distribution of electric current on the thin strip scatterer can be directly obtained by applying the method of moments numerical technique, using a pulse expansion set to represent the unknown electric current distribution, Figure 3.28, and also testing the EFIE expression (3.11.1a) by the same pulses. Figure 3.34 shows a plot of the magnitude of electric current on a thin-strip scatterer of total length, $L = 25\lambda$, excited with an angle of incidence $\phi^i = 90^o$. This result is obtained using a full matrix of size (250 x 250) with a current resolution of 10 pulse samples per wavelength. To apply the spatial decomposition technique, the thin-strip scatterer is divided, for example, into five subzones, $N = 5$. Maintaining the same current resolution of 10 samples per wavelength, the matrix size is now only (50 x 50) for each subzone model. On the lefthand-side of SDT expression (3.11.2b), the total excitation for spatial subzone 1 is

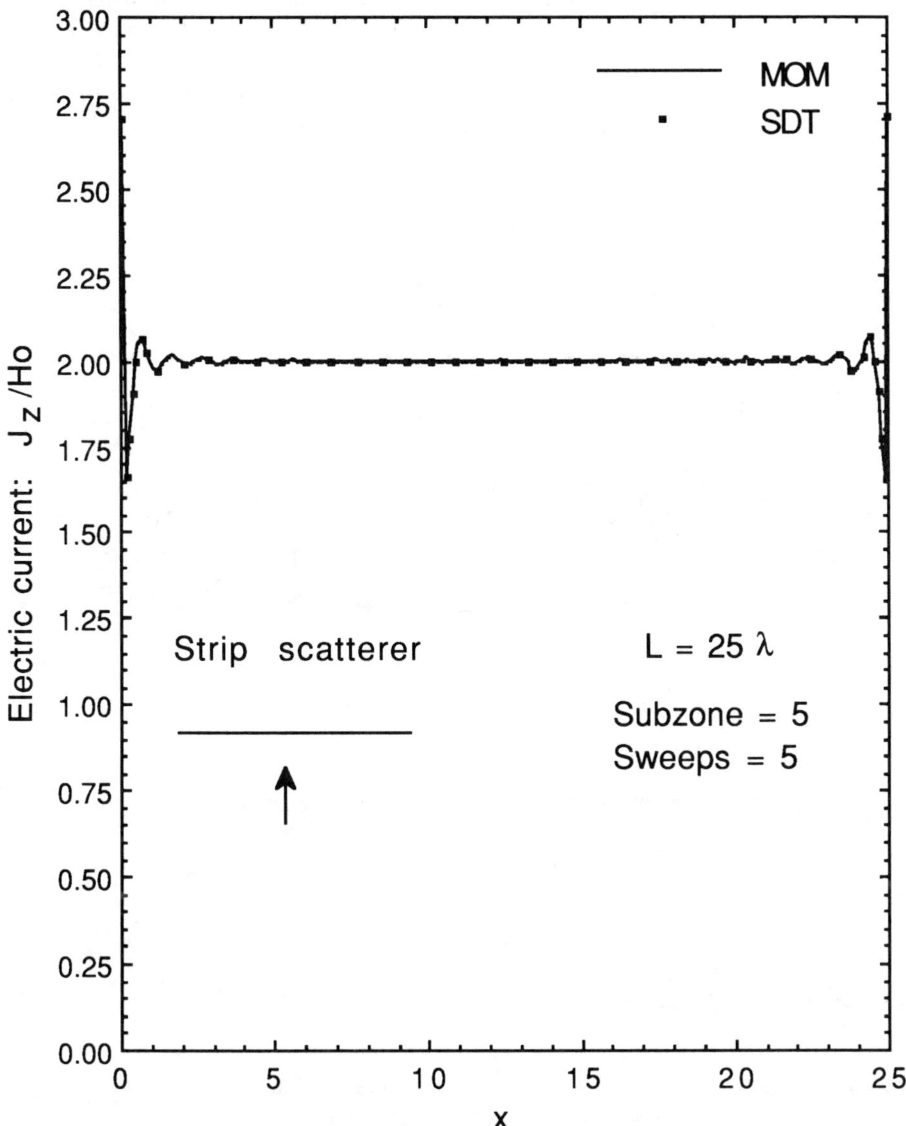

Figure 3.34 Magnitude of electric current distribution on an electrically large, thin-strip scatterer, normal incidence – TM excitation.

given by the plane wave excitation and additional excitation due to the electric currents on remaining spatial subzones, n = 2, 3, 4, and 5. But, the distribution of the electric current on the remaining subzones, n = 2, 3, 4, and 5, are not known initially. However, they can be obtained approximately using the physical optics (PO) by just considering incident and reflected fields on an infinitely large planar conducting sheet. This is accomplished by considering the boundary value problem in Figure 3.35.

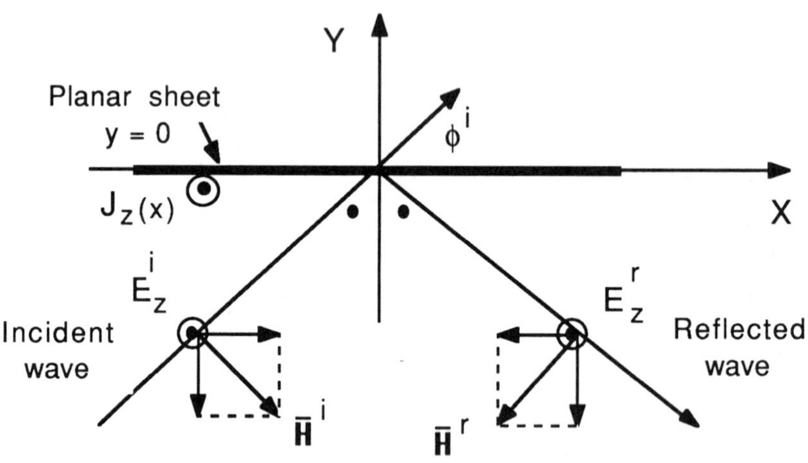

Figure 3.35 Calculation of PO currents on an infinitely large sheet.

The TM polarized incident electric and magnetic fields are given by

$$E_z^i(x,y) = E_0\, e^{-j(k_x x + k_y y)} \tag{3.11.3a}$$

$$\vec{H}^i(x,y) = \frac{E_0}{\eta}\left[\hat{x}\sin\phi^i - \hat{y}\cos\phi^i\right] e^{-j(k_x x + k_y y)} \tag{3.11.3b}$$

Using the boundary condition that the z component of total electric field is equal to zero, referring to Figure 3.35, the TM polarized reflected electric and magnetic fields are given by

$$E_z^r(x,y) = -\, E_0\, e^{-j(k_x x - k_y y)} \tag{3.11.4a}$$

$$\vec{H}^r(x,y) = -\frac{E_0}{\eta}\left[-\hat{x}\sin\phi^i - \hat{y}\cos\phi^i\right] e^{-j(k_x x - k_y y)} \tag{3.11.4b}$$

The physical optics induced electric current on the infinite planar sheet is obtained by

$$\bar{J}_z(x',y=0) = -\hat{y}' \times \left[\vec{H}^i(x',y=0) + \vec{H}^r(x',y=0)\right] \tag{3.11.5a}$$

$$J_z(x') = \frac{2E_0}{\eta}\sin\phi^i\, e^{-jkx\cos\phi^i} \tag{3.11.5b}$$

With the normal broadside excitation, $\phi^i = 90^o$, this PO expression yields an initial current with no current singularities at the thin-strip scatterer ends. In fact, the goal of the SDT is to sequentially update this initial current for each subzone. After determining subzone $n = 1$ current, the analysis is now shifted to the adjacent subzone $n = 2$. The excitation for this subzone 2 consists of the original plane wave plus additional excitation due to the approximate currents on the boundaries of remaining subzones, n = 3, 4, and 5, including the updated current on the first subzone. This step-by-step analysis approach is sequentially implemented in an iterative sweep for each subzone from one end of the scatterer to the other end. Once the first sweep is completed, a first-order approximate distribution of the electric current on the thin-strip scatterer is numerically obtained. It is noted here that no additional boundary conditions are enforced in the application of SDT. Better approximation of the distribution of magnitude and phase of the electric current on the thin-strip scatterer can be obtained by more iterative sweeps. Figure 3.34 shows the distribution of magnitude of electric current calculated based on the SDT (with five spatial subzones, $N = 5$) on the thin-strip scatterer excited with an angle of incidence $\phi^i = 90^o$. The results shown are obtained for five sweeps with less than one percent error in the region separating two adjacent subzones. It should be noted that the electric currents are valid only for the specified angle of incidence, and the SDT iterative process should be repeated again for other angles of incidence.

The distribution of the electric current can now be utilized for calculating either the near electric and magnetic field distributions or the bistatic radar cross section data. In fact, the number of successive iterative sweeps is determined based on the degree of convergence required of the electric current or the radar cross section data. Figure 3.36 shows a plot of the bistatic radar cross section obtained using SDT compared with the direct MOM solution. In the angular range of $\phi = 30^o$ to 150°, the bistatic radar cross section converges in two sweeps with less than one percent error, but for the grazing angles more sweeps are required; and for the results shown, five sweeps are utilized. Figure 3.37 shows a plot of the monostatic radar cross section obtained using SDT compared with respect to the direct MOM solution.

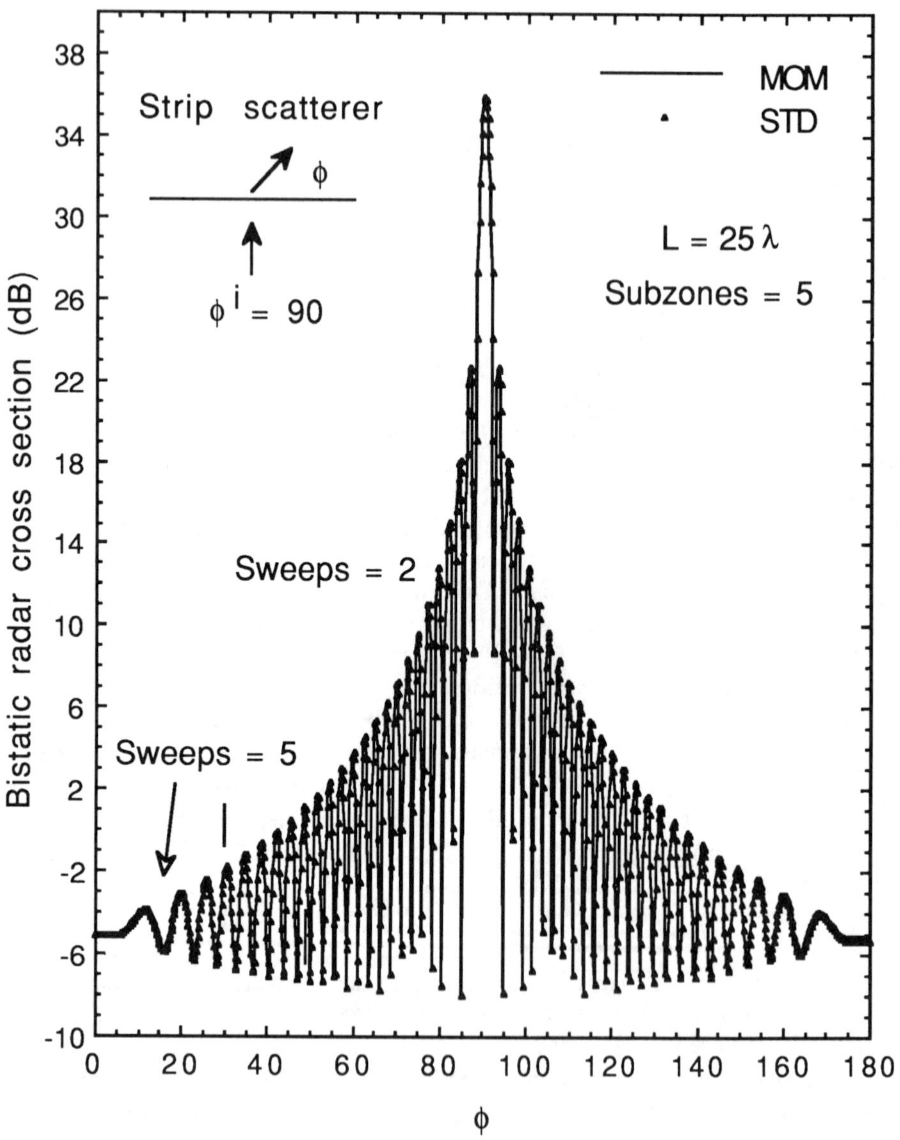

Figure 3.36 Bistatic radar cross section of the electrically large, thin-strip scatterer – TM excitation.

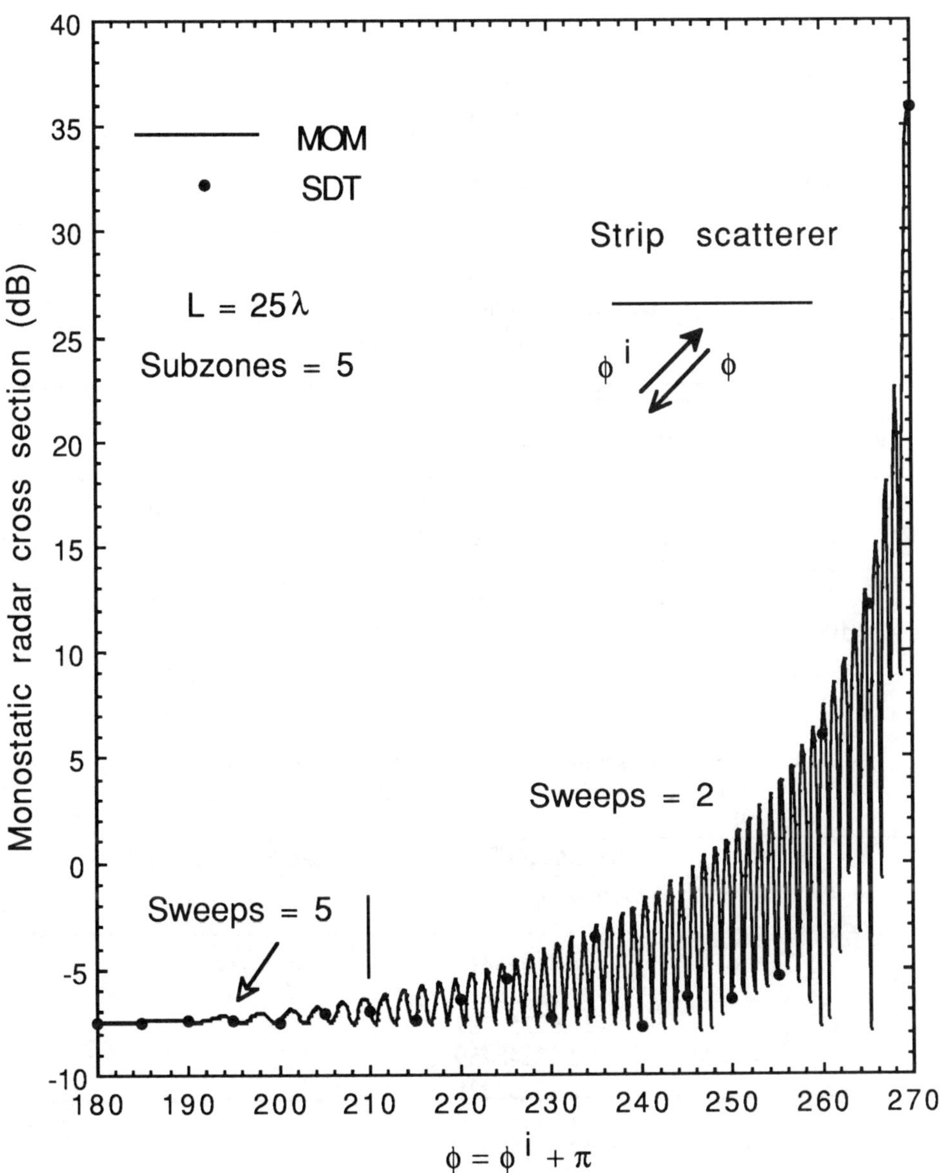

Figure 3.37 Monostatic radar cross section of the electrically large, thin-strip scatterer – TM excitation.

Table 3.9
Direct MOM solution

Full Matrix size Total	*Bistatic RCS in dB Observation angle, $\phi = 90^o$*	*Run time in seconds*
200	27.971437	205.2

Table 3.10
SDT solution, variation of number of subzones
(number of sweeps = 1)

Subzone Matrix size	*Number of subzones*	*Bistatic RCS in dB Observation angle, $\phi = 90^o$*	*Run time in seconds*
20	10	27.95345306	10.0
25	8	27.97183609	12.6
40	5	27.96068573	23.9
50	4	27.97832870	34.1
100	2	27.96839523	121.0

Table 3.11
SDT solution, variation of number of sweeps
(subzone Matrix size = 50 and number of subzones = 4)

Number of sweeps	*Bistatic RCS in dB Observation angle, $\phi = 90^o$*	*Run time in seconds*
1	27.97832870	34.1 per sweep
2	27.97270012	
3	27.97251701	
4	27.97248459	
5	27.97203225	
6	27.97153854	
7	27.97137260	
8	27.97138977	
9	27.97143173	
10	27.97144699	

In fact, numerical studies indicate smaller subzone sizes can be adapted with even lower computer resources, but require more iterative sweeps for the same degree of convergence. The Tables 3.9, 3.10, and 3.11 show the computer resources required on a Sun 4.0 workstation as a function of subzone size and number of iterative sweeps. The representative case of a thin-strip scatterer of total length $L = 10\lambda$, excited at an angle of incidence, $\phi^i = 90^\circ$, is considered with a finer current resolution of 20 pulse samples per wavelength.

These studies using the spatial decomposition technique can be easily extended for the case of electromagnetic scattering by a perfectly conducting rectangular scatterer. Figure 3.38a shows a plot of a bistatic radar cross section of a rectangular scatterer of total length $L = 24.5\lambda$ and width $W = 0.5\lambda$ excited at an angle of incidence $\phi^i = 90^\circ$. This result is obtained using a full matrix of size (500 x 500) for the direct MOM solution. Similar to the thin-strip case discussed earlier, to apply the spatial decomposition technique, the scatterer is divided into five subzones, $N = 5$, with same current resolution for each subzone model. The Figure 3.38a also shows a plot of the bistatic radar cross section obtained using SDT compared with the direct MOM solution. Both forward and backscattering data in the angular range $\phi = 90^\circ$ to 120° converge with only two iterative sweeps, and for the angular range $\phi = 120^\circ$ to 150° convergence is obtained with five sweeps with less than one percent error. However, for the grazing angles more sweeps are required, and for the bistatic results shown ten sweeps are utilized. The monostatic data, on the other hand, converges rapidly with only two iterative sweeps. Figure 3.38b shows a plot of the monostatic radar cross section obtained using SDT with two sweeps compared with the direct MOM solution.

Table 3.12

Computer resources: direct MOM vs. SDT
(Bistatic radar cross section of thin-strip)

strip size	direct MOM matrix size	MOM run time (sec)	SDT matrix size	SDT(*) run time (sec)
10λ	200	205	50	34
25λ	500	3175	50	83

(*) on Sun 4.0 Workstation run time for one SDT sweep which gives a first-order approximate result.

Details of the computational burden to obtain bistatic radar cross section of two thin, perfectly conducting strips of lengths 10λ and 25λ using the direct MOM technique and the SDT are reported in Table 3.12 for TM excitation. It can be inferred from the Table

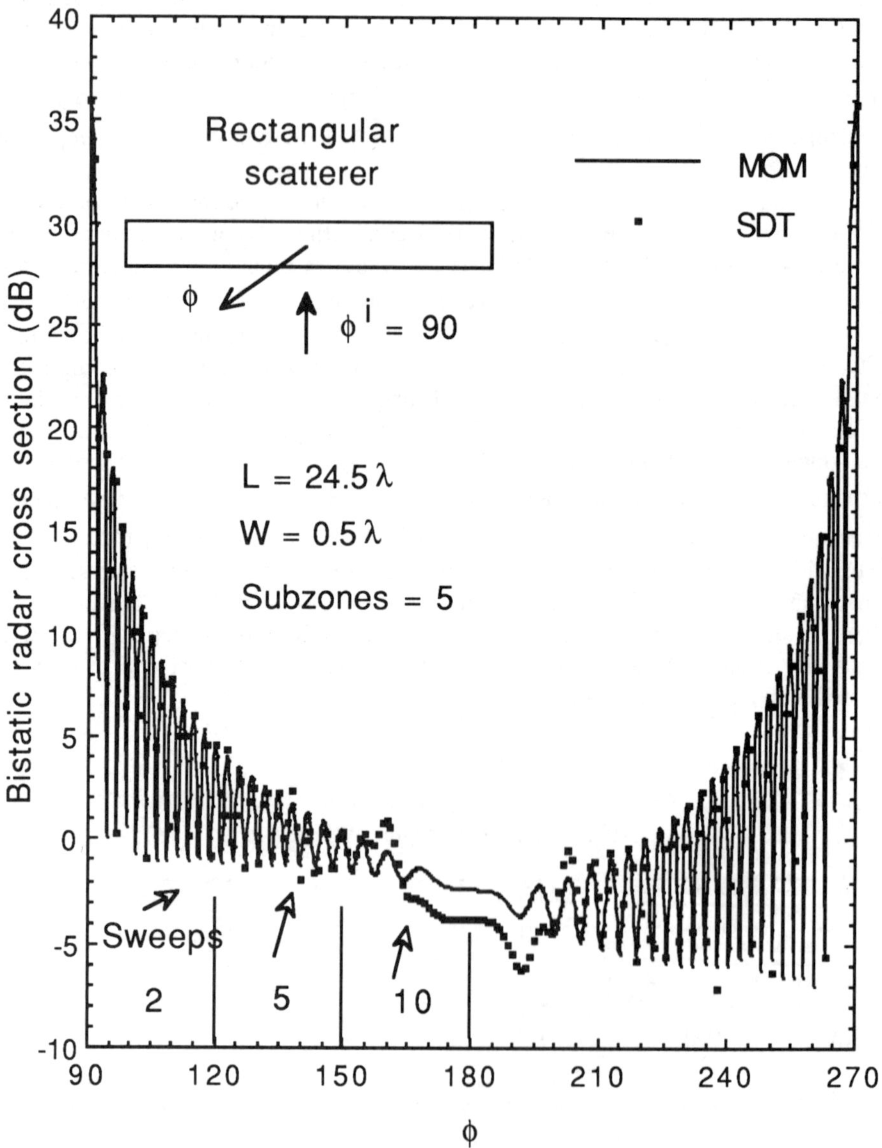

Figure 3.38a Bistatic radar cross section of the electrically large, thick rectangular scatterer – TM excitation.

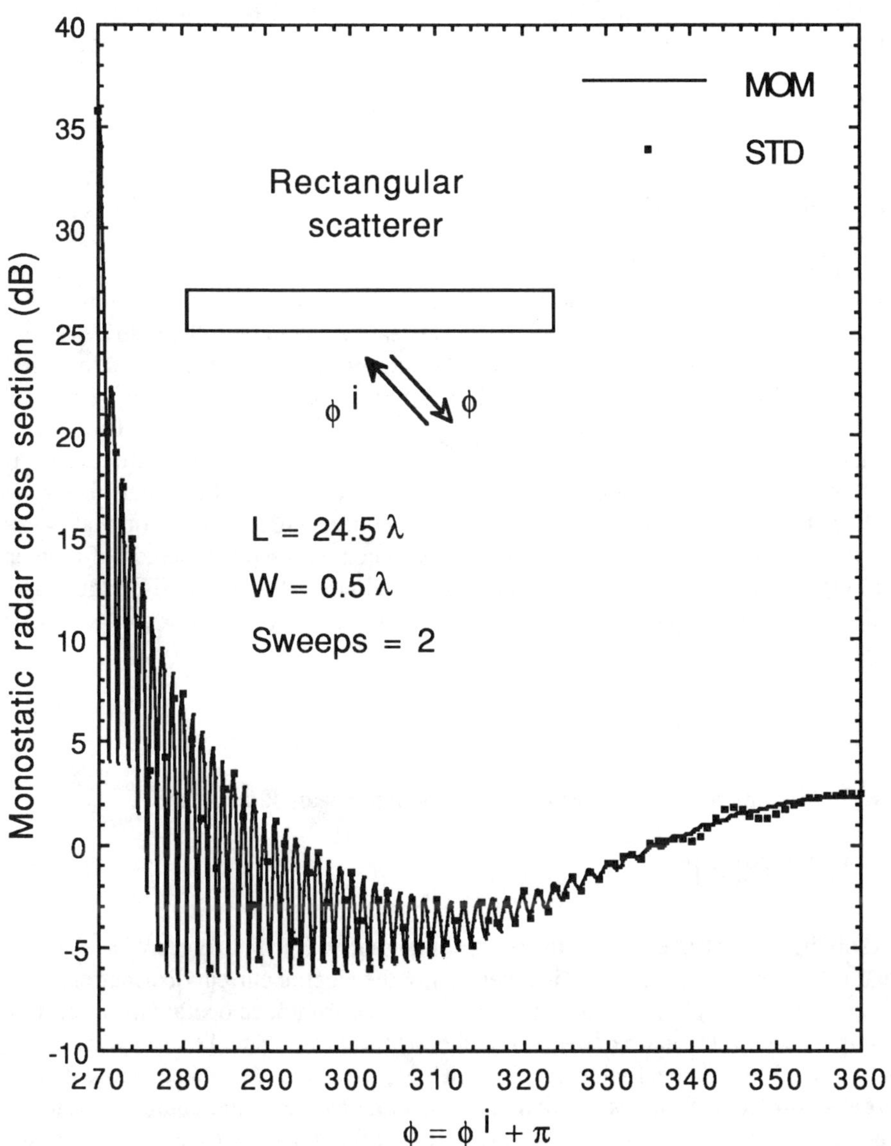

Figure 3.38b Monostatic radar cross section of the electrically large, thick rectangular scatterer – TM excitation.

3.12 that the spatial decomposition technique provides a 100:1 savings in computer storage and about 40:1 savings in run time over the direct MOM solution for larger strip geometry. This comparative savings appears to increase with the size of the scatterer. It may be noted that, as the MOM matrix size increases by a factor of R, then the direct MOM run time increases by R^3 whereas the SDT run time increases by R.

3.11.2 MATRIX CONDITION NUMBER

It is evident that the SDT offers substantial savings in computer storage and time compared to the direct integral equation and method of moments solution. From an error analysis and estimation point of view, the STD deals with only smaller matrices, arising from the division of the large scatterer into a multiplicity of subzones. Even though the Gauss-Seidel inversion of the matrices inevitably introduces rounding-off errors, the extent of which essentially depends on the sophistication of the computing device and its wordlength. In fact, the rounding-off errors are fairly random in nature and do not cancel out in a given computation, but rather tend to accumulate if later calculations are based on the earlier ones. A detailed discussion concerning relative error analysis and estimation of the degree of ill-conditioning associated with typical method of moments scattering problems can be addressed using a qualitative figure based on the "matrix condition number". A convenient measure of the condition of a matrix can be assessed by defining

$$\text{Cond}\,([Z]) = ||Z|| \cdot ||Z^{-1}|| \tag{3.11.6}$$

where considering the infinite norm, the maximum row sum of the matrix

$$||Z|| = ||Z||_{\infty} \tag{3.11.7}$$

In solving the matrix equation, the condition number of the matrix represents an upper bound on the relative uncertainty in determining the electric current distribution. In the numerical inversion alone, the direct method requires arithmetic operations proportional to a third power of the matrix size. But in the SDT only very small matrices are handled along with additional excitation terms, which are proportional to the electrical size of subzones. Figure 3.39 shows a variation of the worst-case condition number depicted as a function of number of subzones for the strip and the rectangle case studies discussed earlier. Thus, the submatrices corresponding to the subzones are better conditioned than the large full matrix of the direct method. The condition of the subzone matrix improves as it becomes smaller; however, this improvement cannot be extracted indefinitely in view of stronger coupling and interaction associated with very small subzones.

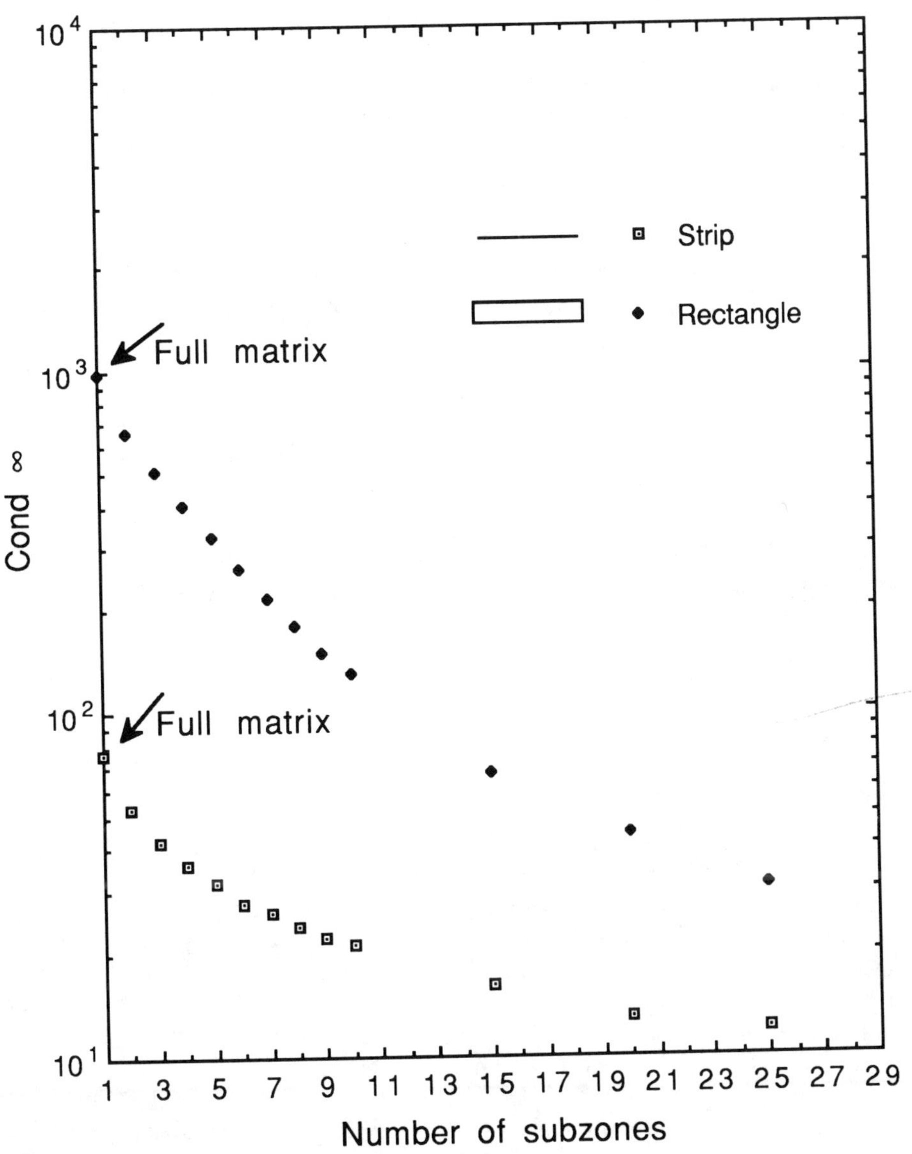

Figure 3.39 Variation of the matrix condition number as a function of subzones for thin-strip and rectangular scatterers – TM excitation.

3.12 ANALYSIS OF APERTURES

The analysis of electromagnetic scattering and interaction by an arbitrary shaped two-dimensional conducting object, either close or open type, has been studied in a systematic manner starting from the classical Maxwell's equations. Figure 3.40 shows two typical cases of canonical shape and configuration of the perfectly conducting objects. By applying the electromagnetic equivalence, the formulation of boundary value problem can be obtained in terms of the well-known electric field integral equation.

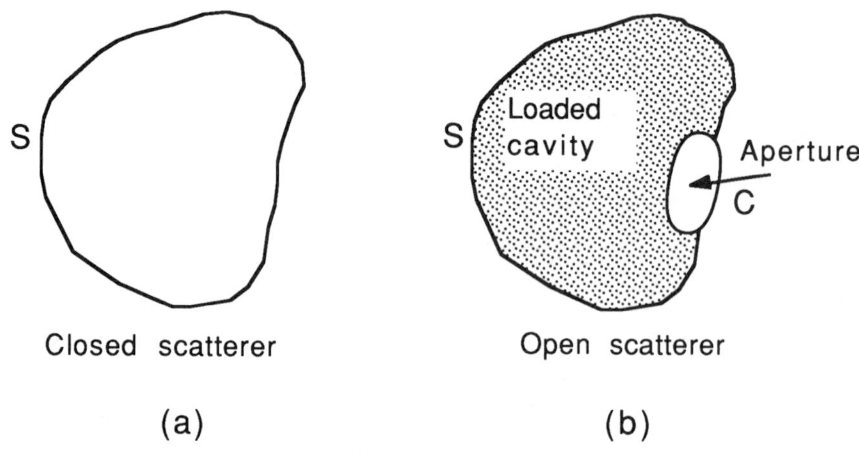

Figure 3.40 Closed and open scattering geometries.

To set up an exterior scattering boundary value problem, generally the perfectly conducting object is assumed to be located in a large free-space medium that is linear, homogeneous, and isotropic, having constant permittivity and permeability. The time-harmonic incident electric and magnetic excitation fields are simulated in the free-space medium by their corresponding time-harmonic primary electric and charge sources, which are located at a large distance from the interacting object under study. For all practical modeling considerations, the incident excitation fields are taken to be those fields existing in the free-space medium in the absence of any interacting object. When the conducting object is introduced into the free-space medium, the incident fields interact to induce secondary electric currents and charges on the conducting object. These induced electric currents and charges produce and propagate scattered electric and magnetic fields in all directions. In the free-space medium, total fields now consist of sum of the incident field and the scattered field. Further, the complete electric and magnetic field distributions in the free-space medium are such that they satisfy familiar electromagnetic boundary conditions. According to the electromagnetic boundary conditions, the total tangential electric field is zero on the surface of a perfectly conducting object. Similarly, the total tangential magnetic field on the surface of perfectly conducting object gives the induced electric current distribution.

Referring to the Figure 3.40, either for the closed or open type of scatterer, the induced electric currents on the surface of conducting object is an unknown quantity to be determined first. Once the induced electric currents are known, then Maxwell's equations can be directly utilized to obtain the scattered electric and magnetic fields in the region of interest. Specifically, in the case of open type scatterer as shown in the Figure 3.40b, an aperture region separates an interior cavity region from the exterior unbounded region. The induced electric currents on the surface of the conducting object, in fact, produce electric and magnetic fields in both interior and exterior regions, including appropriate fields in the aperture region. Even though the formulation of the boundary value problem based on the induced electric current is rigorous, from the numerical analysis point of view the shape and size of the aperture, and the loaded cavity behind it, play a critical role in the process of determining the field distributions.

At this point, it is possible to introduce the concept of electromagnetic equivalence and the duality principle to come up with an *alternative formulation* for the boundary value problem based on the distribution of fields in the aperture region. This is accomplished using the boundary conditions that the total tangential electric and magnetic field distributions in the aperture region are continuous. Initially, the distribution of electric and magnetic fields in the aperture region are unknown quantities. A boundary value problem is set up in terms of either a differential equation or an integral equation for the unknown aperture electric field distribution or the corresponding equivalent magnetic current distribution. This alternative approach basically leads to the well-known aperture theory. Even though the study of aperture and the loaded cavity behind the aperture itself is a separate field, beyond the scope of this book, the purpose of this section is to expose and apply familiar integral equations to a certain class of boundary value problems involving infinitely large perfectly conducting planar sheet with perforated aperture or hole in it. Briefly, the aperture theory is introduced in the following to derive a dual type of integral equation in terms of the unknown magnetic current distribution in the aperture. Further, relevance of the integral equation for the study of conducting object with aperture is also discussed.

3.12.1 APERTURE EQUIVALENCES

Referring to Figure 3.41, let us consider an infinitely large, perfectly conducting screen perforated with an arbitrary shaped aperture or hole, which is located in the *yz* plane separating two media, 1 and 2, having different permittivity, permeability, and conductivity properties. Medium 1 is a linear, homogeneous, and isotropic lossless region and contains incident excitation fields. In the absence of an aperture, energy contained in the incident electric and magnetic fields is completely reflected back into medium 1, and thus, the perfectly conducting screen acts like a shield separating medium 2 from medium 1. With the aperture present, as shown in the Figure 3.41, depending upon the size and shape of the aperture, only part of the incident energy is reflected back into medium 1 and the remaining energy is transmitted into medium 2. Thus, there are electric and magnetic field distributions in both media 1 and 2, including at the aperture interface separating the two media. The distribution of fields in the aperture region is important information and quite complicated to assess. Further, the electromagnetic boundary conditions are such that the tangential total electric and magnetic fields in the aperture region are continuous.

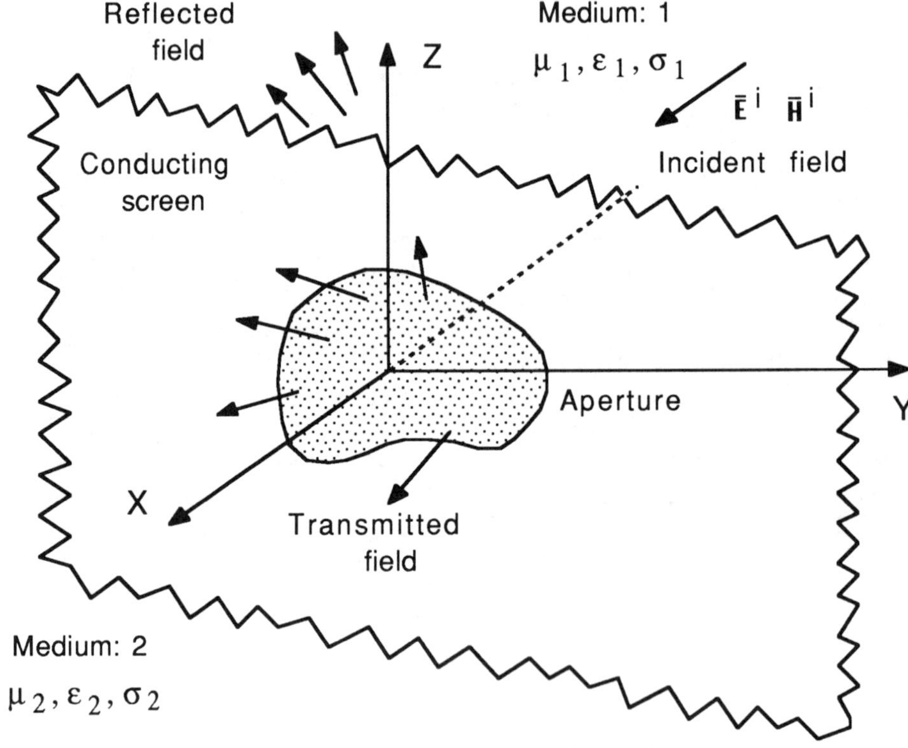

Figure 3.41 Geometry of a conducting screen perforated with an aperture.

One straightforward method to determine the electric and magnetic fields in the two media, 1 and 2, is by knowing the induced electric currents on the perfectly conducting screen excluding the aperture region. Referring to Section 3.6.1, the electromagnetic equivalence and the electric field integral equation can be applied to determine the electric currents on the perfectly conducting screen. For the aperture–conducting screen boundary value problem under consideration, this methodology is quite cumbersome. An alternative formulation is presented that expresses the field distributions in the two media in terms of the unknown tangential field distributions just in the aperture region.

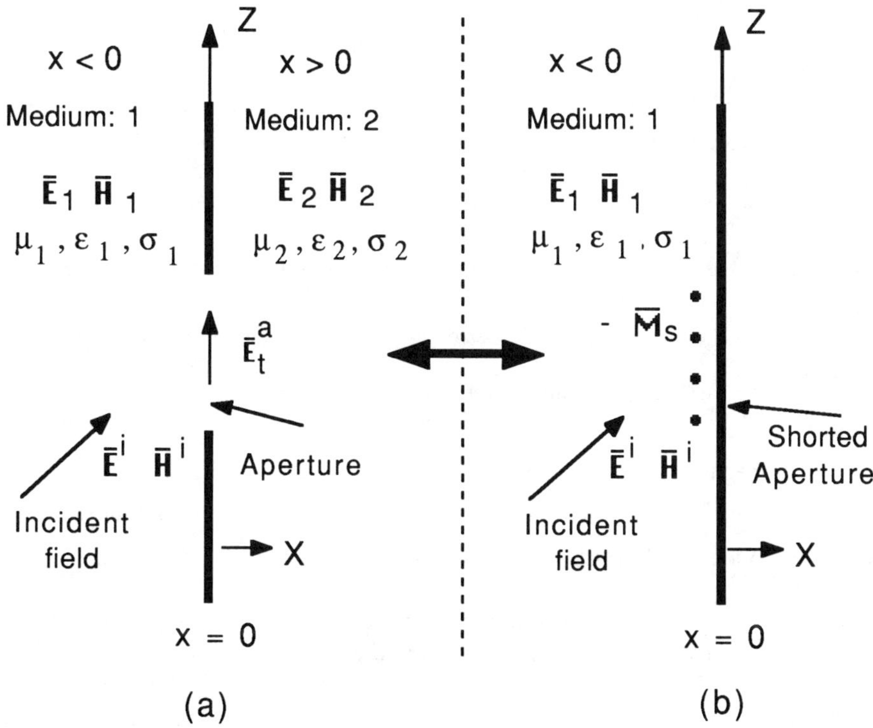

Figure 3.42 Aperture equivalence for medium 1 fields.

Original Boundary Value Problem

The original aperture–conducting screen boundary value problem is repeated in Figure 3.42a to develop separately aperture equivalences for the left half-space and the right half- space regions. For the $x < 0$ region covering medium 1, let

ε_1 : permittivity of the left half-space medium 1;

μ_1 : permeability of the left half-space medium 1;

$\sigma_1 = 0$: conductivity of the left half-space medium 1;

$(\bar{E}^i, \bar{H}^i)$: electric and magnetic incident fields in the left half-space medium 1;

$(\bar{E}_1, \bar{H}_1)$: electric and magnetic total fields in the left half-space medium 1.

For the $x > 0$ region covering medium 2, let

ε_2 : permittivity of the right half-space medium 2;

μ_2 : permeability of the right half-space medium 2;

σ_2 : conductivity of the right half-space medium 2;

$(\bar{E}_2, \bar{H}_2)$: electric and magnetic total fields in right half space medium 2;

and for the $x = 0$ region, where the conducting screen and the aperture are located, let

$\sigma \Rightarrow \infty$: conductivity of the perfectly conducting screen located at $x = 0$;

$\bar{E}_t^a$: electric field distribution in the aperture region;

$\bar{H}_t^a$: magnetic field distribution in the aperture region.

Left Half-Space – Medium 1 Equivalence

To obtain the electric and magnetic field distributions in the left half-space medium 1, an equivalent boundary value problem is now set up that is valid only for the $x < 0$ region. A virtual boundary surface is introduced at a distance $x < 0$- in front of the screen-aperture geometry. The total tangential components of electric and magnetic fields on the virtual boundary surface are converted into the following equivalent magnetic and electric currents based on the relationships

$$\bar{M}_1(\bar{r}') = -\bar{E}_1(\bar{r}')\big|_{\tan} \times \hat{n}' \qquad \bar{r}' \text{ on virtual surface at } x < 0- \qquad (3.12.1a)$$

$$\bar{J}_1(\bar{r}') = -\hat{n}' \times \bar{H}_1(\bar{r}')\big|_{\tan} \qquad \bar{r}' \text{ on virtual surface at } x < 0- \qquad (3.12.1b)$$

The normal to the virtual surface is directed in the negative x direction. In a limit, except in the region in front of the aperture, the equivalent magnetic current distribution reduces to zero by invoking the boundary condition that the total tangential component of the electric field is zero on the perfectly conducting screen.

The aperture region is now short-circuited or physically closed using a perfectly conducting planar sheet having an identical shape and size of the aperture, so that medium 2 is completely isolated from medium 1. Referring to Figure 3.42b, this process changes the basic structure of the aperture–conducting screen boundary value problem and, thus alters the original electric and magnetic field distributions in medium 1. To maintain the distribution of the original electric and magnetic fields in medium 1, the effect of aperture radiation is taken into account in terms of equivalent magnetic currents given by

$$\bar{M}_S(\bar{r}') = -\bar{E}_t^a(\bar{r}=0-) \times \hat{n}' \tag{3.12.2}$$

Hence, these equivalent magnetic current sources are introduced just in front of the short-circuited aperture, and correspondingly, the equivalent electric currents are everywhere on the shorted screen. These two equivalent currents are such that they produce the same original field distributions. The total fields in medium 1 can be obtained as the sum of radiated field from the equivalent magnetic currents just in front of the shorted-circuited aperture region and the so-called short-circuited field consisting of the incident field plus the reflected field from the uniform perfectly conducting screen. Hence

$$\bar{E}_1(\bar{r}) = \bar{E}_1^{sc}(\bar{r}) + \bar{E}_1^{s}(\bar{r}) \tag{3.12.3a}$$

$$\bar{H}_1(\bar{r}) = \bar{H}_1^{sc}(\bar{r}) + \bar{H}_1^{s}(\bar{r}) \tag{3.12.3b}$$

where the superscript *sc* stands for the short-circuited fields and the superscript *s* corresponds to the radiated fields. The equivalent boundary value problem shown in the Figure 3.42b can be further modified to determine the fields radiated from the equivalent magnetic currents located very close to the shorted aperture region. In fact, the equivalent magnetic currents are tangential to the infinitely large planar conducting screen. The conducting screen is eliminated by introducing an image magnetic current distribution in the plane of aperture. Figure 3.43 shows the equivalent boundary value problem for calculating the radiated fields in medium 1. The floating magnetic currents, in fact, are located in a linear, homogeneous, and isotropic medium. The aperture radiated fields can now be calculated in medium 1 using the equivalent magnetic currents.

A complete summary of *Maxwell's equations in the frequency domain* is given in Tables 2.7 and 2.8. Table 2.7 gives all the relevant electromagnetic field equations in differential form with magnetic current sources. Referring to the discussion in Sections

2.27 and 2.28, the electric and magnetic radiated field quantities can be expressed in terms of two arbitrary potential functions given by a vector electric potential and a scalar magnetic potential. Thus, in terms of the dual-potential functions,

$\bar{F}_1(\bar{r},\omega)$: electric vector potential function at the field point

$$= F_{1x}(\bar{r},\omega)\hat{x} + F_{1y}(\bar{r},\omega)\hat{y} + F_{1z}(\bar{r},\omega)\hat{z} \tag{3.12.4}$$

$\Psi_1(\bar{r},\omega)$: magnetic scalar potential function at the field point;

and the electric and magnetic radiated fields are given by

$$\bar{E}_1^S(\bar{r},\omega) = -\frac{1}{\varepsilon'_1}\nabla \times \bar{F}_1(\bar{r},\omega) \tag{3.12.5a}$$

$$\bar{H}_1^S(\bar{r},\omega) = -j\omega\bar{F}_1(\bar{r},\omega) - \nabla\Psi_1(\bar{r},\omega) \tag{3.12.5b}$$

The electric vector potential and the magnetic scalar potential are interrelated through

$$\nabla[\nabla \bullet \bar{F}_1(\bar{r},\omega)] = -j\omega\mu_1\varepsilon'_1\nabla\Psi_1(\bar{r},\omega) \tag{3.12.6}$$

which is the Lorentz gauge condition. Further, the following vector *Helmholtz* partial differential equation is satisfied by the electric vector potential:

$$\nabla^2\bar{F}_1(\bar{r},\omega) + k_1^2\bar{F}_1(\bar{r},\omega) = -\varepsilon'_1[-2\bar{M}_S(\bar{r},\omega)] \tag{3.12.7a}$$

where

ε'_1 : effective permittivity of medium 1

$$= \varepsilon_1\left(1 - j\frac{\sigma_1}{\omega\varepsilon_1}\right) \tag{3.12.7b}$$

k_1 : propagation constant of medium 1

$$= \omega(\mu_1\varepsilon'_1)^{1/2} \tag{3.12.7c}$$

Section 2.28 gives a method to solve for Helmholtz differential equation (3.12.7a) in terms of the three-dimensional Green's function. Referring to the discussion in Section 2.28 and Figure 3.43, consider the surface magnetic current source distributions confined to the aperture region S and located in a three-dimensional isotropic and homogeneous medium.

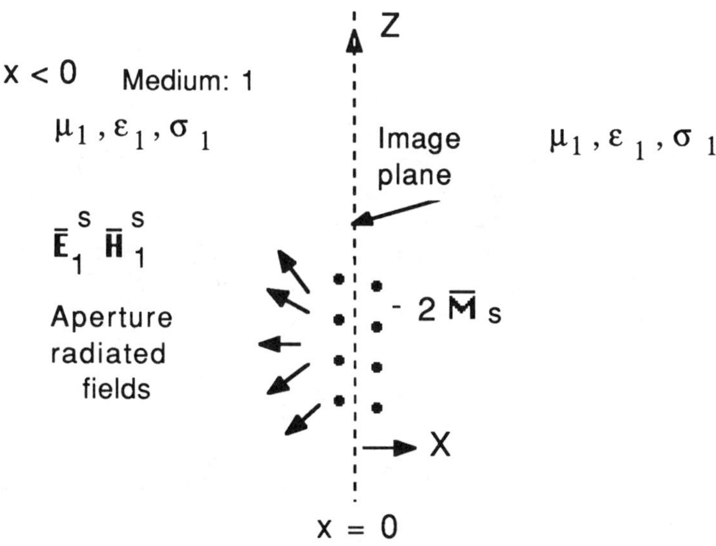

Figure 3.43 Radiated fields into medium 1 from the aperture region.

Referring to Figure 3.43, let

$-2\bar{M}_S(\bar{r}')$: surface current density distribution, in volts per meter

$\bar{r}'$: position vector corresponding to the source point (x', y', z')

$$= y'\hat{y} + z'\hat{z} \tag{3.12.8a}$$

$\bar{r}$: position vector corresponding to the field point (x, y, z) in $x < 0$ region, medium 1

$$= x\hat{x} + y\hat{y} + z\hat{z} \tag{3.12.8b}$$

R : distance between the source point and the field point

$$= |\bar{r} - \bar{r}'| \tag{3.12.8c}$$

The total electric vector potential at the field point is given by the surface integral over the complete magnetic current density in the source region S:

$$\bar{F}_1(\bar{r}) = \frac{\varepsilon'_1}{4\pi} \iint_S - 2\bar{M}_S(\bar{r}') \frac{e^{-jk_1R}}{R} ds(\bar{r}') \tag{3.12.9}$$

for $\bar{r}$ in $x < 0$ region

Right Half-Space – Medium 2 Equivalence

To obtain the electric and magnetic field distributions in the right half-space, medium 2, an equivalent boundary value problem is now set up that is valid only for the $x > 0$ region. A virtual boundary surface is introduced at a distance $x > 0+$ in front of the screen-aperture geometry. The total tangential components of electric and magnetic fields on the virtual boundary surface are converted into the following equivalent magnetic and electric currents based on the relationships

$$\bar{M}_2(\bar{r}') = \bar{E}_2(\bar{r}')\Big|_{\tan} \times \hat{n}' \qquad \bar{r}' \text{ on virtual surface at } x > 0+ \tag{3.12.10a}$$

$$\bar{J}_2(\bar{r}') = \hat{n}' \times \bar{H}_2(\bar{r}')\Big|_{\tan} \qquad \bar{r}' \text{ on virtual surface at } x > 0+ \tag{3.12.10b}$$

The normal to the virtual surface is directed in the positive x direction. In a limit, except in the region in front of the aperture, the equivalent magnetic current distribution reduces to zero by invoking the boundary condition that the total tangential component of the electric field is zero on the perfectly conducting screen.

The aperture region is now short-circuited or physically closed using a perfectly conducting planar sheet having an identical shape and size as the aperture, so that medium 1 is completely isolated from medium 2. Referring to the Figure 3.44b, this process again changes the basic structure of the aperture-conducting screen boundary value problem and, thus alters the original electric and magnetic field distributions in medium 2. To maintain the distribution of the original electric and magnetic fields in medium 2, the effect of the aperture radiation is taken into account in terms of equivalent magnetic currents given by

$$\bar{M}_S(\bar{r}') = \bar{E}_t^a(\bar{r}=0+) \times \hat{n}' \tag{3.12.11}$$

Hence, these equivalent magnetic current sources are introduced just in front of the short-circuited aperture. The equivalent magnetic currents are such that they produce the original field distributions. No external excitation fields are present in medium 2, and if present, they could be accounted in terms of short-circuited fields. The total fields in medium 2 are obtained directly as the radiated fields from the equivalent magnetic currents just in front of the short-circuited aperture region. Hence

$$\bar{E}_2(\bar{r}) = \bar{E}_2^S(\bar{r}) \tag{3.12.12a}$$

$$\bar{H}_2(\bar{r}) = \bar{H}_2^S(\bar{r}) \tag{3.12.12b}$$

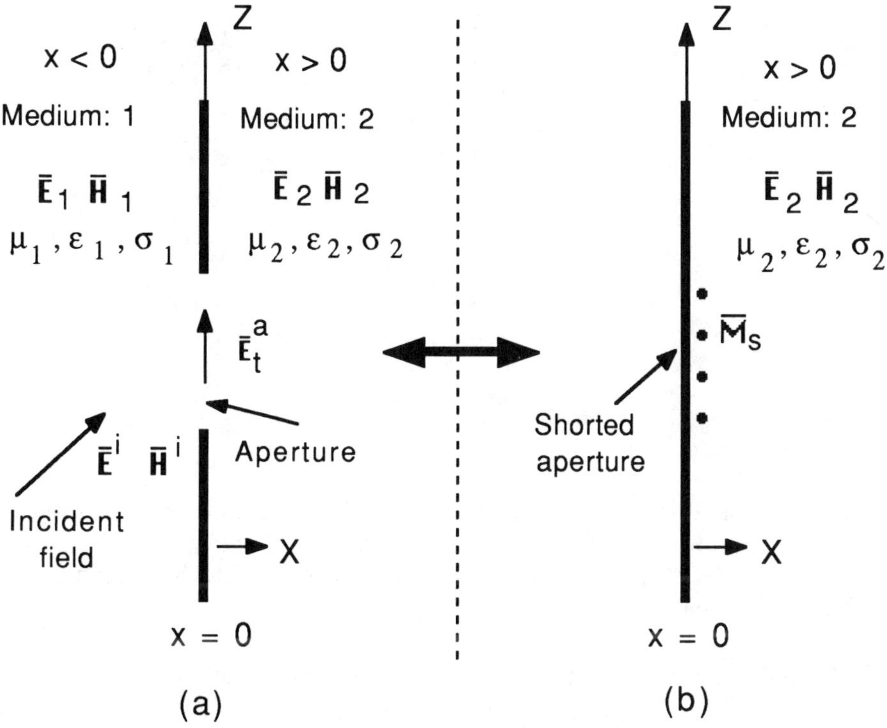

Figure 3.44 Aperture equivalence for medium 2 fields.

The equivalent boundary value problem, shown in the Figure 3.44b, can be further modified to determine the fields radiated from the equivalent magnetic currents located close to the shorted aperture region. In fact, the equivalent magnetic currents are tangential to the infinitely large planar conducting screen. The conducting screen is eliminated by introducing an image magnetic current distribution in the plane of aperture. Figure 3.45 shows the equivalent boundary value problem for calculating the radiated fields in medium 2. The floating magnetic currents, in fact, are located in a linear, homogeneous, and isotropic medium.

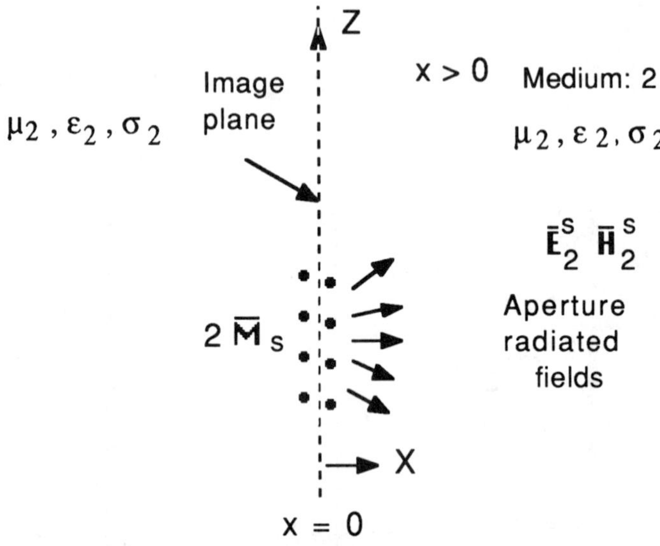

Figure 3.45 Radiated fields into medium 2 from the aperture region.

Referring to the discussion in Sections 2.27 and 2.28, the electric and magnetic radiated field quantities can be expressed in terms of two arbitrary potential functions, given by a vector electric potential and a scalar magnetic potential. Thus, in terms of the dual-potential functions,

$\bar{F}_2(\bar{r},\omega)$: electric vector potential function at the field point

$$= F_{2x}(\bar{r},\omega)\hat{x} + F_{2y}(\bar{r},\omega)\hat{y} + F_{2z}(\bar{r},\omega)\hat{z} \qquad (3.12.13)$$

$\Psi_2(\bar{r},\omega)$: magnetic scalar potential function at the field point;

and the electric and magnetic radiated fields are given by

$$\bar{E}_2^S(\bar{r},\omega) = -\frac{1}{\varepsilon'_2}\nabla \times \bar{F}_2(\bar{r},\omega) \qquad (3.12.14a)$$

$$\bar{H}_2^S(\bar{r},\omega) = -j\omega\bar{F}_2(\bar{r},\omega) - \nabla\Psi_2(\bar{r},\omega) \tag{3.12.14b}$$

The electric vector potential and the magnetic scalar potential are interrelated through

$$\nabla[\nabla \bullet \bar{F}_2(\bar{r},\omega)] = -j\omega\mu_2\varepsilon'_2\nabla\Psi_2(\bar{r},\omega) \tag{3.12.15}$$

which is the Lorentz gauge condition. Further, the following vector *Helmholtz* partial differential equation is satisfied by the electric vector potential:

$$\nabla^2\bar{F}_2(\bar{r},\omega) + k_2^2\bar{F}_2(\bar{r},\omega) = -\varepsilon'_2[2\bar{M}_S(\bar{r},\omega)] \tag{3.12.16a}$$

ε'_2 : effective permittivity of the medium 2

$$= \varepsilon_2\left(1 - j\frac{\sigma_2}{\omega\varepsilon_2}\right) \tag{3.12.16b}$$

k_2 : propagation constant of the medium 2

$$= \omega(\mu_2\varepsilon'_2)^{1/2} \tag{3.12.16c}$$

Section 2.28 gives a method to solve for Helmholtz differential equation (3.12.16a) in terms of the three-dimensional Green's function. Referring to the discussion in Section 2.28 and Figure 3.44, consider the surface magnetic current source distributions confined to the aperture region S and located in a three-dimensional isotropic and homogeneous medium. Let

$2\bar{M}_S(\bar{r}')$: surface current density distribution, in volts per meter;

$\bar{r}'$: position vector corresponding to the source point (x', y', z')

$$= y'\hat{y} + z'\hat{z} \tag{3.12.17a}$$

$\bar{r}$: position vector corresponding to the field point (x, y, z) in the $x > 0$ region, medium 2

$$= x\hat{x} + y\hat{y} + z\hat{z} \tag{3.12.17b}$$

R : distance between the source point and the field point

$$= |\bar{r} - \bar{r}'| \tag{3.12.17c}$$

The total electric vector potential at the field point is given by the surface integral over the complete magnetic current density in the source region S:

$$\bar{F}_2(\bar{r}) = \frac{\varepsilon'_2}{4\pi} \iint_S 2\bar{M}_s(\bar{r}') \frac{e^{-jk_2R}}{R} ds(\bar{r}') \tag{3.12.18}$$

for $\bar{r}$ in the $x > 0$ region.

3.13 TWO–DIMENSIONAL APERTURE

The preceding field expressions are applicable for the three-dimensional problem wherein a perfectly conducting screen with a perforated aperture separates two media having different medium properties. In the following, these field expressions are specialized to the case of a two-dimensional problem. This can be accomplished by introducing spectral domain transformation as discussed previously in Section 3.3.

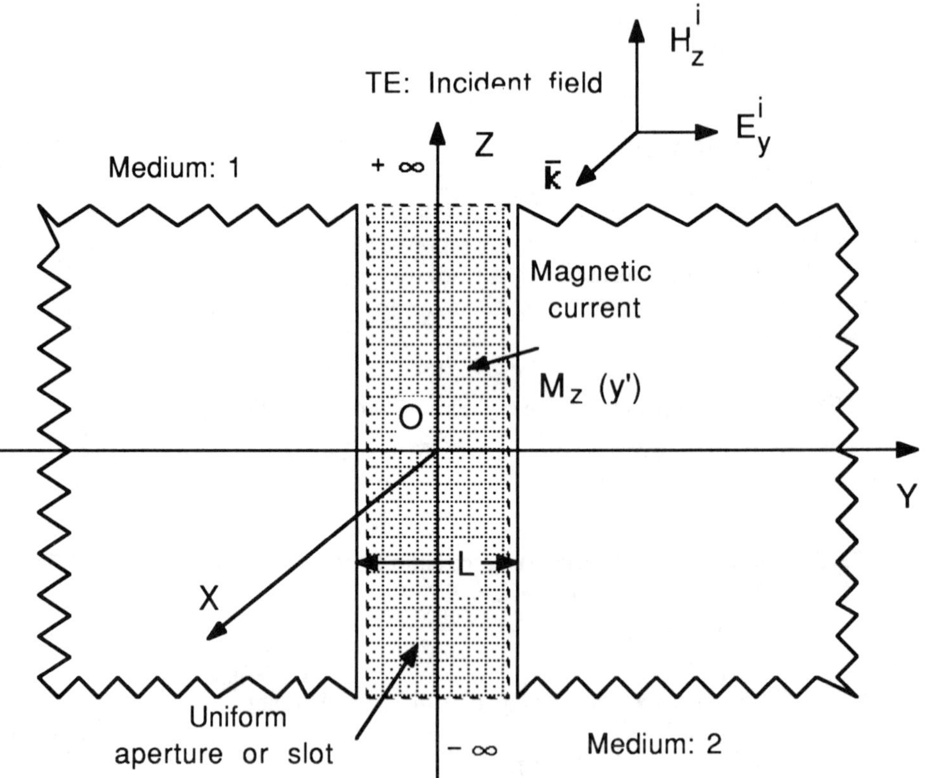

Figure 3.46 Geometry of the two-dimensional aperture-screen.

In the following, an analysis procedure is presented for the case of electromagnetic scattering and penetration by a two-dimensional aperture-conducting screen based on an integral equation formulation. Referring to Figure 3.46, an infinitely large perfectly conducting screen with a uniform aperture or slot is located in the yz plane. The aperture is oriented with its axis coinciding with the z axis of the coordinate system. Along the axis, the aperture is infinitely long and has a uniform cross section. It is placed in the $x = 0$ plane and it separates media 1 ($x < 0$) and 2 ($x > 0$). In medium 1, the aperture-screen is excited externally by a time-harmonic *transverse electric* (TE to z) polarized plane wave. As shown in the Figure 3.46, the TE polarized plane wave is incident at an arbitrary angle of incidence with its incident magnetic field polarized parallel and the corresponding incident electric field polarized perpendicular to the z coordinate axis. It should be clearly noted that there is no propagation of the incident plane wave parallel to the axis of two-dimensional aperture. For this special case of the time-harmonic transverse electric *normal* excitation, there is only an axial component of the induced magnetic currents on the surface $x = 0$ of aperture region. The induced magnetic currents produce only the axial scattered or radiated magnetic field and the corresponding transverse electric field distributions in the two media. The total electric and magnetic field distributions near the surface of the aperture are such that they satisfy appropriate electromagnetic boundary conditions. The distribution of induced magnetic currents on the aperture is still an unknown quantity and can be determined by enforcing the familiar boundary condition that the total tangential component of the magnetic field should be continuous in the aperture region $x = 0$.

Similar to the classical approach discussed in Section 3.12 based on the aperture equivalences, the distribution of induced axial magnetic currents in the aperture region is treated as unknown and a boundary value integral equation is formulated. It is shown that the integral equation for the aperture distribution with TE excitation has similarity with respect to the integral equation of the perfectly conducting thin-strip scatterer with TM excitation. The two cases form dual boundary value problems so that if the results for one case is known, then results for the other case can be obtained by duality.

Referring to Figure 3.46 and the discussion in Section 2.19 regarding plane wave fields in the frequency domain, the incident plane wave electric and magnetic fields can be written as follows. The plane wave incident magnetic field vector is polarized along the z axis and the incident electric field vector is polarized in a plane perpendicular to the z axis. There is no propagation of the fields parallel to the z coordinate axis, indicating that the incident excitation fields are uniform and completely independent of the z coordinate variable. In fact, the corresponding field quantities, such as the induced magnetic currents, the scattered or radiated electric and magnetic field distributions in the two media, are also independent of the z coordinate variable.

According to expressions (2.19.11a–b), the TE incident fields are given by

$$\bar{H}_0 = H_0\hat{z} \tag{3.13.1a}$$

$$\bar{E}_0 = E_{0x}\hat{x} + E_{0y}\hat{y} \tag{3.13.1b}$$

$$\bar{H}^i(\bar{\rho},\omega) = H_z^i(\bar{\rho},\omega)\hat{z} \tag{3.13.2a}$$

$$H_z^i(\bar{\rho},\omega) = H_0\, e^{-j\bar{k}\bullet\bar{\rho}} \tag{3.13.2b}$$

$$\bar{E}^i(\bar{\rho},\omega) = \bar{E}_0\, e^{-j\bar{k}\bullet\bar{\rho}} \tag{3.13.3}$$

$$\bar{k} = k_1\hat{k} \tag{3.13.4}$$

k_1 : propagation constant for homogeneous, isotropic medium

$$= \omega(\mu_1\varepsilon_1)^{1/2} \tag{3.13.5a}$$

$\hat{k}$: actual direction of the plane wave propagation;
ω : frequency of excitation, in radians per second.

With respect to a global origin O, which represents reference phase center, let

$\bar{\rho}$: spatial vector corresponding to the field point (x, y)

$$= \rho\ \cos\phi\hat{x} + \rho\ \sin\phi\hat{y} \tag{3.13.5b}$$

$\bar{k}$: propagation vector of the traveling plane wave

$$= k_1\ \cos\phi^i\hat{x} + k_1\ \sin\phi^i\hat{y} \tag{3.13.5c}$$

ϕ^i : arbitrary angle of incidence of the TE plane wave field;
$k_z = 0$: propagation constant along the z coordinate direction

As discussed earlier, the normally excited plane wave is uniform in the z coordinate direction, and hence, the axial induced magnetic currents and the radiated electric and magnetic fields in the two media,1 and 2, are also uniform with no variation with respect to the z coordinate variable. Hence, the various field expressions obtained in Section 3.12 can be completely simplified by substituting either

$$\frac{\partial}{\partial z} \Rightarrow 0 \tag{3.13.6a}$$

or

$$k_z = 0 \tag{3.13.6b}$$

It is interesting to note that the spectrum of the spectral variable k_z varies from $-\infty$ to ∞, and the case of normal incident excitation yielding condition (3.13.6a–b) is just a special case to isolate the dependence of various field quantities with respect to the z coordinate variable.

Table 3.13

Transverse electric fields for normal excitation, rectangular coordinates

---

$$\bar{H}^s_m(x,y,\omega) = H^s_{mz}(x,y,\omega)\hat{z} \tag{3.13.7a}$$

$$\bar{E}^s_m(x,y,\omega) = E^s_{mx}(x,y,\omega)\hat{x} + E^s_{my}(x,y,\omega)\hat{y} \tag{3.13.7b}$$

$$\bar{F}(x,y,\omega) = F_{mz}(x,y,\omega)\hat{z} \tag{3.13.8}$$

$$E^s_{mx} = -\frac{1}{\varepsilon'_m}\frac{\partial F_{mz}}{\partial y} \tag{3.13.9a}$$

$$E^s_{my} = \frac{1}{\varepsilon'_m}\frac{\partial F_{mz}}{\partial x} \tag{3.13.9b}$$

$$H^s_{mz} = -j\omega F_{mz} \tag{3.13.10}$$

for medium $m = 1, x < 0-$

for medium $m = 2, x > 0+$

Table 3.13 gives various scattered or radiated field components corresponding to the TE excitation. Referring to Figures 3.46 and either expressions (3.12.5a–b) or (3.12.14a–b), various field components in the two media, 1 and 2, can be obtained by substituting

$\bar{F}(x,y,\omega)$: electric vector potential at the field point

$$= F_{1z}(x,y,\omega)\hat{z} \qquad \text{for } x < 0- \qquad (3.13.11a)$$

$$= F_{2z}(x,y,\omega)\hat{z} \qquad \text{for } x > 0+ \qquad (3.13.11b)$$

where the z component of electric vector potential satisfies the scalar Helmholtz differential equation based on relationship (3.12.7a) or (3.12.16a). The solutions for the electric vector potentials are given by expressions (3.12.9) and (3.12.18). Based on condition (3.13.6a–b), the incident fields, short-circuited fields, aperture radiated fields, and the aperture unknown magnetic current distribution are independent of the z coordinate variable. Hence, various field quantities are determined just in the xy transverse plane or $z = 0$ plane.

3.13.1 APERTURE INTEGRAL EQUATION – TE CASE

Following the aperture equivalences discussed in Section 3.12, relevant two-dimensional expressions for the total electric and magnetic fields can be written in terms of the unknown aperture magnetic current distribution for the two media, 1 and 2. By enforcing the boundary condition that the tangential component of the total magnetic field should be continuous in the aperture region, an integral equation can be deduced for the unknown aperture magnetic current distribution.

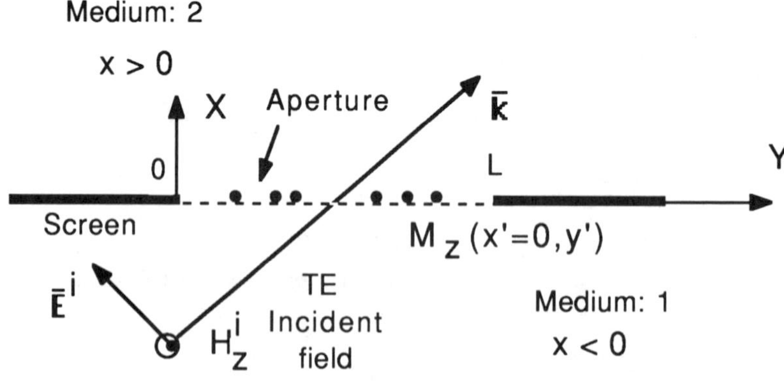

Figure 3.47 Geometry of TE excited aperture.

Figure 3.47 shows geometry of a TE excited aperture or slot. In fact, the geometry is derived directly from Figure 3.46 with a transverse cut at the $z = 0$ plane. The total length of the aperture is L and it separates two media, 1 and 2. For medium 1 ($x < 0$), the effect of aperture field distribution is taken into account in terms of the z component of equivalent magnetic current given by

$$\bar{M}_z = - \bar{E}^a_y(x=0-,y,z) \times \hat{x}' \tag{3.13.12}$$

The total magnetic field in medium 1 can be obtained as the sum of radiated field from the equivalent magnetic current and the short-circuited field. Hence

$$\bar{H}_1(x,y) = \bar{H}_1^{SC}(x,y) - j\omega F_{1z}(x,y)\hat{z} \tag{3.13.13}$$

Referring to the Figure 3.47,

$-2M_z(x'=0-,y')$: magnetic current density distribution, in volts per meter;

$\bar{r}'$: position vector corresponding to the source point $(x'=0, y', z')$

$$= y'\hat{y} + z'\hat{z} \tag{3.13.14a}$$

$\bar{r}$: position vector corresponding to the field point $(x, y, z=0)$ in $x < 0$ region, medium 1

$$= x\hat{x} + y\hat{y} \tag{3.13.14b}$$

The z component of electric vector potential at the field point in the xy plane is given by the surface integral over the complete magnetic current density in the aperture region:

$$F_{1z}(\bar{r}) = \frac{\varepsilon'_1}{4\pi} \iint_S -2M_z(x'=0,y') \frac{e^{-jk_1|\bar{r}-\bar{r}'|}}{|\bar{r}-\bar{r}'|} dy'dz' \tag{3.13.15a}$$

$$F_{1z}(\bar{r}) = \frac{\varepsilon'_1}{4\pi} \int_0^L -2M_z(y') \left[\int_{z'=-\infty}^{z'=\infty} \frac{e^{-jk_1R_1}}{R_1} dz' \right] dy' \tag{3.13.15b}$$

$\bar{r}$ in $x < 0$, $z = 0$ plane

$$R_1 = \left[(x)^2+(y-y')^2+z'^2\right]^{1/2} \tag{3.13.15c}$$

The bracketed term in expression (3.13.15b) is the well-known integral representation for the Hankel function of the zero order and second kind, Appendix B,

$$H_0^{(2)}(k_1R) = \frac{j}{\pi} \int_{z'=-\infty}^{\infty} \frac{e^{-jk_1R_1}}{R_1} dz' \tag{3.13.16}$$

After substituting this relationship for the Hankel function into potential integral representation (3.13.15b), the z component of vector electric potential for the two dimensional case takes the form

$$F_{1z}(x,y) = \frac{\varepsilon'_1}{4j} \int_0^L - 2M_z(y')\, H_0^{(2)}(k_1R)\, dy' \tag{3.13.17a}$$

$$R = \left[(x)^2+(y-y')^2\right]^{1/2} \tag{3.13.17b}$$

for (x, y) in medium 1, $z = 0$ plane

Similarly, for medium 2 ($x > 0$), the effect of aperture field distribution is taken into account in terms of the z component of equivalent magnetic current given by

$$\bar{M}_z = \bar{E}_y^a(x=0+,y,z) \times \hat{x}' \tag{3.13.18}$$

The total magnetic field in medium 2 can be obtained as the sum of radiated field from the equivalent magnetic current. Hence

$$\bar{H}_2(x,y) = - j\omega F_{2z}(x,y)\hat{z} \tag{3.13.19}$$

Referring to the Figure 3.47,

$2M_z(x'=0+,y')$: magnetic current density distribution, in volts per meter;

$\bar{r}'$: position vector corresponding to the source point $(x'=0, y', z')$

$$= y'\hat{y} + z'\hat{z} \tag{3.13.20a}$$

$\bar{r}$: position vector corresponding to the field point $(x, y, z=0)$ in $x > 0$ region, medium 2

$$= x\hat{x} + y\hat{y} \tag{3.13.20b}$$

The z component of electric vector potential at the field point in the xy plane is given by the surface integral over the complete magnetic current density in the aperture region

$$F_{2z}(\bar{r}) = \frac{\varepsilon'_2}{4\pi} \iint_S 2M_z(x'=0,y') \frac{e^{-jk_2|\bar{r}-\bar{r}'|}}{|\bar{r}-\bar{r}'|} dy'dz' \tag{3.13.21a}$$

$$F_{2z}(\bar{r}) = \frac{\varepsilon'_2}{4\pi} \int_0^L 2M_z(y') \left[\int_{z'=-\infty}^{z'=\infty} \frac{e^{-jk_2R_2}}{R_2} dz' \right] dy' \tag{3.13.21b}$$

$\bar{r}$ in $x > 0$, $z = 0$ plane

$$R_2 = \left[(x)^2 + (y-y')^2 + z'^2\right]^{1/2} \tag{3.13.21c}$$

The bracketed term in expression (3.13.21b) is the well-known integral representation for the Hankel function of the zero order and second kind, Appendix B, and is given by expression (3.13.16). On substituting relationship (3.13.16) for the Hankel function into potential integral representation (3.13.21b), the z component of vector electric potential for the two-dimensional case takes the form

$$F_{2z}(x,y) = \frac{\varepsilon'_2}{4j} \int_0^L 2M_z(y') H_0^{(2)}(k_2R)\, dy' \tag{3.13.22a}$$

$$R = \left[(x)^2 + (y-y')^2\right]^{1/2} \tag{3.13.22b}$$

for (x, y) in medium 2, $z = 0$ plane

To obtain the integral equation for the aperture region, the z components of the total magnetic field expressions (3.13.13) and (3.13.19) are equated at the $x = 0$ plane where the two-dimensional aperture is located. Thus

$$\bar{H}_1(x=0,y) \bullet \hat{z} = \bar{H}_2(x=0,y) \bullet \hat{z} \tag{3.13.23a}$$

$$-j\omega F_{1z}(x=0,y) + j\omega F_{2z}(x=0,y) = -\bar{H}_1^{SC}(x=0,y) \bullet \hat{z} \tag{3.13.23b}$$

The short-circuited field on the perfected conducting screen located on the $x = 0$ plane is given by negative of twice the incident magnetic field based on the boundary conditions. Hence, boundary condition (3.13.23b) yields the following integral equation for the TE-excited aperture in terms of the unknown z component of the magnetic current distribution:

$$\frac{\omega \varepsilon'_1}{4} \int_0^L M_z(y') H_0^{(2)}(k_1|y - y'|)\, dy'$$

$$+ \frac{\omega \varepsilon'_2}{4} \int_0^L M_z(y') H_0^{(2)}(k_2|y - y'|)\, dy' = H_z^i(x=0,y) \qquad (3.13.24a)$$

For a special case, medium 2 is same as medium 1, having identical permittivity and permeability properties:

$$\varepsilon = \varepsilon_1 = \varepsilon_2$$

$$\mu = \mu_1 = \mu_2$$

$$k = k_1 = k_2$$

Then, the integral equation for the TE-excited aperture reduces to the following form:

$$\frac{\omega \varepsilon}{4} \int_0^L 2M_z(y')\, H_0^{(2)}(k|y - y'|)\, dy' = H_z^i(x=0,y) \qquad (3.13.24b)$$

This integral equation is in a dual form compared to the integral equation of the thin-strip perfectly conducting scatterer located in a free-space medium as discussed in Section 3.10. Referring to expression (3.10.5), the EFIE for the thin-strip scatterer is given by the following:

$$E_z^i(x) = \frac{\omega \mu}{4} \int_0^L J_z(x')\, H_0^{(2)}(k|x - x'|)\, dx' \qquad (3.13.24c)$$

Based on the concept of duality, it may noted that by replacing the permittivity by permeability, and incident magnetic field by incident electric field, integral equation (3.13.24b) for the TE-excited aperture is identical to the integral equation (3.13.24c) for the TM-excited thin strip scatterer. As far as the unknown quantities are concerned, twice the aperture magnetic current distribution is to be treated similar to the net electric current distribution on the thin-strip scatterer.

The numerical computer algorithm developed for the thin-strip scatterer can be directly utilized to determine the aperture magnetic current distribution. The numerical results reported in Section 3.10 for the thin-strip scatterer can be correspondingly interpreted for the TE-excited two-dimensional aperture. It should be noted, to use and interpret the same numerical data obtained in Section 3.10, the electric current distribution in the thin-strip scatterer is always normalized with respect to the incident magnetic field, while the correspondingly magnetic current distribution in the aperture is normalized with respect to the incident electric field. Similarly, based on the duality principle, the scattered electric field and the scattered magnetic field distributions for the TM-excited thin-strip scatterer are to be treated like the radiated magnetic field and the radiated electric field distributions for the TE-excited aperture with appropriate normalizations.

Far-Field and Radar Cross Section

Following the discussions in Section 3.9, the electric and magnetic far-field aperture distributions can be obtained by using a large argument approximation for the Green's function. In the large argument approximation, as $|k\rho| \to \infty$, the Hankel function can be written as

$$H_0^{(2)}(k\rho) \sim \sqrt{\frac{2j}{\pi k\rho}}\; e^{-jk\rho} \tag{3.13.25a}$$

Referring to expression (3.13.19), the scattered magnetic field distribution in the far field (region 2) can be calculated by approximating the integral and substituting expression (3.9.25a) for the Hankel function to obtain

$$H_z^s(\rho,\phi) = \sum_{n=1}^{N} I_{2n}\Big[-\frac{k}{4\eta}H_0^{(2)}(k|\bar{\rho}-\bar{\rho}_{2n}|)\Big]\Delta_{2n} \tag{3.13.25b}$$

$$H_z^s(\rho,\phi) \sim \sum_{n=1}^{N} I_{2n}\Big[-\frac{k}{4\eta}\sqrt{\frac{2j}{\pi(k|\bar{\rho}-\bar{\rho}_{2n}|)}}\; e^{-j(k|\bar{\rho}-\bar{\rho}_{2n}|)}\Big]\Delta_{2n} \tag{3.13.25c}$$

Referring to Figure 3.23, in the far-field region, $|k\rho| \to \infty$, the magnitude term in denominator of the expression (3.13.25c) can be written as

$$|\bar{\rho}-\bar{\rho}_{2n}| \approx \rho \tag{3.13.26a}$$

and the exponential term contributing to the phase distribution of the magnetic current can be approximated as

$$|\bar{\rho} - \bar{\rho}_{2n}| \approx \rho - \rho_{2n} \cos(\phi - \phi_{2n}) \tag{3.13.26b}$$

After substituting the far-field approximations, expressions (3.13.26a–b), the scattered magnetic field distribution in the far-field region reduces to the following form:

$$H_z^s(\rho,\phi) \sim \sum_{n=1}^{N} I_{2n} \frac{k}{\eta} \mathcal{K} \left[e^{jk\rho_{2n} \cos(\phi - \phi_{2n})} \right] \Delta_{2n} \tag{3.13.27a}$$

$$\mathcal{K} = \frac{1}{\sqrt{8\pi k\rho}} \, e^{-jk\rho} \, e^{-j3\pi/4} \tag{3.13.27b}$$

In the far-field region, there are only the axial H_z component of the scattered magnetic field and the angular E_ϕ component of the scattered electric field distributions. According to expression (3.13.27b), the magnitude of the scattered magnetic field decays as reciprocal of the square root of the radial variable, and the two far scattered field distributions behave as pure cylindrical waves propagating along the radial coordinate direction, and they are directly related through the medium (region 2) intrinsic impedance.

Further, the bistatic radar cross section of the TE-excited aperture in a conducting screen can be calculated using expressions (3.8.18) and (3.13.27a–b):

$$\mathrm{RCS}(\phi) = \lim_{\rho \to \infty} 2\pi\rho \left| \frac{H_z^s(\phi,\omega)}{H_z^i(\phi,\omega)} \right|^2 \tag{3.13.28}$$

Chapter 4

Two-Dimensional Perfectly Conducting Object: TE Polarization

In the previous chapter, an approach for analyzing the electromagnetic scattering and interaction by an arbitrary shaped, two-dimensional perfectly conducting object is studied in a systematic manner starting from classical Maxwell's equations for the transverse magnetic excitation. In the following sections, a similar analysis is presented corresponding to the transverse electric excitation. Again, the geometry of a perfectly conducting object is assumed to be uniform and infinite in length along the z coordinate axis. Further, as shown in the Figure 3.1, the cross-sectional area of the scattering geometry is uniform and the same along its axis, which is assumed to coincide with the z axis of the coordinate system. The scattering analysis for transverse electric excitation can be carried out either in terms of the rectangular or cylindrical coordinate system. It is also assumed that the perfectly conducting object is located in a large free-space medium that is linear, homogeneous, and isotropic, having constant permittivity and permeability. The time-harmonic incident electric and magnetic fields with transverse electric polarization are simulated and propagated in the free-space medium by their corresponding time-harmonic primary electric current and charge sources, which are located at a far distance from the interacting object. For all practical modeling considerations, the incident excitation fields are taken to be those field distributions existing in the free-space medium in the absence of any interacting object. When the two-dimensional conducting object is introduced into the medium, the incident fields interact to induce secondary electric currents and charges on the conducting object. These induced electric currents and charges produce and propagate scattered electric and magnetic fields in all directions. In the free-space medium, the total fields now consist of the sum of incident and scattered fields. As discussed earlier, the complete electric and magnetic field distributions in the free-space medium are such that they satisfy the familiar electromagnetic boundary conditions. According to the electromagnetic boundary conditions, the total tangential electric field is zero on the surface of perfectly conducting object. Similarly, the total tangential magnetic field on the surface of perfectly conducting object gives the induced electric current distribution.

Similar to the study of transverse magnetic excitation, the induced electric current on the surface of a conducting object is treated as an unknown quantity. Hence, for the unknown induced electric current distribution, a boundary value problem is set up either in terms of a differential equation or an integral equation or an integro-differential equation subject to the appropriate boundary conditions. Similar to the analysis procedure discussed in Chapter 3, the boundary value equation can be directly solved using the analytical method based on an appropriate eigenfunction expansion for the canonical type of scatterer, such as a circular conducting cylinder. But, for an arbitrary shaped, two-dimensional conducting object, the boundary value equation can be conveniently solved using the method of moments numerical technique, which is suitable for either the differential equation or the integral equation. Once the induced electric currents are known, then Maxwell's equations can be directly utilized to obtain the scattered electric and magnetic fields in the region of interest. The relevant electromagnetic scattering properties, such as the near-field and far-field distributions and the radar cross section for the transverse electric excitation, can also be obtained.

4.1 GENERAL FIELD EQUATIONS

A complete summary of *Maxwell's equations in the frequency domain* is given in Tables 2.5 and 2.6. Referring to Figure 3.2, let us consider a large unbounded region that is a linear, homogeneous, and isotropic lossless medium. The sources are confined to a small source region, given by

$\rho_V(\bar{r},\omega)$: volume electric charge density, in coulombs per cubic meter;

$\bar{J}_V(\bar{r},\omega)$: volume current density, in amperes per square meter.

These sources produce the electric and magnetic field distributions both inside and outside the source region. At any field point in the medium, the electric and magnetic fields satisfy the following frequency-dependent Maxwell's equations in differential form:

$$\nabla \times \bar{E}(\bar{r},\omega) = -j\omega\mu\bar{H}(\bar{r},\omega) \tag{4.1.1a}$$

$$\nabla \times \bar{H}(\bar{r},\omega) = j\omega\varepsilon\bar{E}(\bar{r},\omega) + \bar{J}_V(\bar{r},\omega) \tag{4.1.1b}$$

For the electric and magnetic fields, their corresponding divergence relationships are

$$\nabla \cdot \varepsilon\bar{E}(\bar{r},\omega) = \rho_V(\bar{r},\omega) \tag{4.1.1c}$$

$$\nabla \cdot \mu\bar{H}(\bar{r},\omega) = 0 \tag{4.1.1d}$$

The set of equations just defined are the coupled vector equations. They can be reduced to the scalar form of coupled equations by assuming a rectangular coordinate system. Using the rectangular coordinate system (x, y, z), at any field point in the medium, the scalar electromagnetic field equations are given in Table 3.1. The expressions in Table 3.1 can also be expressed in terms of cylindrical coordinate system (Table 3.2) or any other generalized cylindrical coordinate system.

4.1.1 FIELD EQUATIONS IN THE SPECTRAL DOMAIN

The field equations obtained earlier in Tables 3.1 and 3.2 are functions of the spatial coordinates (x, y, z) or (ρ, ϕ, z). It is possible to introduce further simplification into the analysis by transforming all the sources and the field quantities from the z spatial coordinate domain to the corresponding spectral transform domain as discussed in Section 3.3. This process basically converts the three-dimensional sources and their field quantities to the corresponding two-dimensional sources and their field quantities. As noted earlier, this process is similar to defining an equivalent Fourier integral transform to convert all the sources and their fields with (x, y, z) dependence to the corresponding (x, y, k_z) dependence, where k_z is the spectral transform parameter. To get back the original field quantities with (x, y, z) dependence, the inverse integral transform is performed over the spectral domain field quantities. Referring to Section 3.3, the function $F(z)$ can be transformed into the spectral domain function $F(k_z)$ based on the following direct spectral integral transform:

$$\mathcal{S}[F(z)] = F(k_z) = \frac{1}{\sqrt{2\pi}} \int_{-\infty}^{\infty} F(z)\, e^{jk_z z}\, dz \tag{4.1.2}$$

where k_z is the spectral domain variable and similar to the propagation phase constant along the z coordinate axis, in radians per meter. The function $F(k_z)$ is always a complex function of the spectral parameter variable k_z in the range varying from $-\infty$ to ∞. The original function $F(z)$ can be recovered by performing inverse integral transform:

$$\mathcal{S}^{-1}[F(k_z)] = F(z) = \frac{1}{\sqrt{2\pi}} \int_{-\infty}^{\infty} F(k_z)\, e^{-jk_z z}\, dk_z \tag{4.1.3}$$

Based on these integral transforms, all the relevant electromagnetic equations can be converted into the spectral domain by using operator definition

$$\frac{\partial}{\partial z} \Leftrightarrow -jk_z \tag{4.1.4}$$

It should be noted at this stage that the basic transformation from the z coordinate domain to the spectral transform domain involves defining various electromagnetic sources and corresponding fields to vary as

$$\bar{J}(x,y,z) = \bar{J}(x,y,k_z)\, e^{-jk_z z} \tag{4.1.5a}$$

$$\bar{E}(x,y,z) = \bar{E}(x,y,k_z)\, e^{-jk_z z} \tag{4.1.5b}$$

$$\bar{H}(x,y,z) = \bar{H}(x,y,k_z)\, e^{-jk_z z} \tag{4.1.5c}$$

and

$$\bar{E}(x,y,k_z) = E_x(x,y,k_z)\hat{x} + E_y(x,y,k_z)\hat{y} + E_z(x,y,k_z)\hat{z} \tag{4.1.6a}$$

$$\bar{H}(x,y,k_z) = H_x(x,y,k_z)\hat{x} + H_y(x,y,k_z)\hat{y} + H_z(x,y,k_z)\hat{z} \tag{4.1.6b}$$

$$\bar{J}(x,y,k_z) = J_x(x,y,k_z)\hat{x} + J_y(x,y,k_z)\hat{y} + J_z(x,y,k_z)\hat{z} \tag{4.1.6c}$$

Table 4.1
Electric and magnetic fields in the rectangular coordinate system – Spectral transform domain

---

$$\frac{\partial E_z}{\partial y} - (-jk_z)E_y = -j\omega\mu H_x \tag{4.1.7a}$$

$$(-jk_z)E_x - \frac{\partial E_z}{\partial x} = -j\omega\mu H_y \tag{4.1.7b}$$

$$\frac{\partial E_y}{\partial x} - \frac{\partial E_x}{\partial y} = -j\omega\mu H_z \tag{4.1.7c}$$

$$\frac{\partial H_z}{\partial y} - (-jk_z)H_y = j\omega\varepsilon E_x + J_x \tag{4.1.8a}$$

$$(-jk_z)H_x - \frac{\partial H_z}{\partial x} = j\omega\varepsilon E_y + J_y \tag{4.1.8b}$$

$$\frac{\partial H_y}{\partial x} - \frac{\partial H_x}{\partial y} = j\omega\varepsilon E_z + J_z \tag{4.1.8c}$$

Using the rectangular coordinate system, at any field point *(x, y, z)* in the medium, the electromagnetic field equations are given in Tables 3.1 and 3.2. These equations can be written in the spectral transform domain, as given by Table 4.1. In fact, the various electromagnetic equations are transformed using operator substitution (4.1.4). Further,

$$\frac{\partial^2}{\partial z^2} \Leftrightarrow -k_z^2 \tag{4.1.9a}$$

$$\int \partial z \Leftrightarrow \frac{1}{-jk_z} \tag{4.1.9b}$$

Referring to Table 4.1, the scalar coupled equations are also applicable for the source-free region with the source electric current components completely removed. Specifically for the source-free region, the three components of electric field distribution (E_x, E_y, E_z) and the three components of magnetic field distribution (H_x, H_y, H_z) satisfy the following Helmholtz equations, expressions (2.18.11a) and (2.18.11b), derived in Section 2.18. Hence, the rectangular components of electric and magnetic fields satisfy

$$\nabla^2 E_i(\bar{r},\omega) + k^2 E_i(\bar{r},\omega) = 0 \tag{4.1.10a}$$

$$i = x,\ y,\ z$$

$$\nabla^2 H_i(\bar{r},\omega) + k^2 H_i(\bar{r},\omega) = 0 \tag{4.1.10b}$$

k : propagation constant of the lossless medium

$$= \omega(\mu\varepsilon)^{1/2} \tag{4.1.10c}$$

Using the expressions (4.1.4) and (4.1.9a), in the spectral transform domain,

$$\nabla = \nabla_\tau + \frac{\partial}{\partial z}\hat{z} \tag{4.1.10d}$$

$$= \nabla_\tau - jk_z\hat{z} \tag{4.1.10e}$$

$$\nabla^2 = \nabla_\tau^2 + \frac{\partial^2}{\partial z^2} \tag{4.1.10f}$$

$$= \nabla_\tau^2 - k_z^2 \tag{4.1.10g}$$

Table 4.2
Electric and magnetic fields in the cylindrical coordinate system
Spectral transform domain

$$\frac{1}{\rho}\frac{\partial E_z}{\partial \phi} - (-jk_z)E_\phi = -j\omega\mu H_\rho \tag{4.1.13a}$$

$$(-jk_z)E_\rho - \frac{\partial E_z}{\partial \rho} = -j\omega\mu H_\phi \tag{4.1.13b}$$

$$\frac{1}{\rho}\frac{\partial(\rho E_\phi)}{\partial \rho} - \frac{1}{\rho}\frac{\partial E_\rho}{\partial \phi} = -j\omega\mu H_z \tag{4.1.13c}$$

$$\frac{1}{\rho}\frac{\partial H_z}{\partial \phi} - (-jk_z)H_\phi = j\omega\varepsilon E_\rho + J_\rho \tag{4.1.14a}$$

$$(-jk_z)H_\rho - \frac{\partial H_z}{\partial \rho} = j\omega\varepsilon E_\phi + J_\phi \tag{4.1.14b}$$

$$\frac{1}{\rho}\frac{\partial(\rho H_\phi)}{\partial \rho} - \frac{1}{\rho}\frac{\partial H_\rho}{\partial \phi} = j\omega\varepsilon E_z + J_z \tag{4.1.14c}$$

$$\nabla_\tau^2 E_i(x,y,k_z) + k_\tau^2 E_i(x,y,k_z) = 0 \tag{4.1.11a}$$

$$\nabla_\tau^2 H_i(x,y,k_z) + k_\tau^2 H_i(x,y,k_z) = 0 \tag{4.1.11b}$$

$$k_\tau^2 = k^2 - k_z^2 \tag{4.1.11c}$$

k_τ : transverse propagation constant

Using the cylindrical coordinate system, the corresponding electromagnetic field equations in the spectral transform domain takes the form given in Table 4.2. The electric and magnetic fields are expressed in terms of cylindrical components:

$$\bar{E}(\rho,\phi,k_z) = E_\rho(\rho,\phi,k_z)\hat{\rho} + E_\phi(\rho,\phi,k_z)\hat{\phi} + E_z(\rho,\phi,k_z)\hat{z} \tag{4.1.12a}$$

$$\bar{H}(\rho,\phi,k_z) = H_\rho(\rho,\phi,k_z)\hat{\rho} + H_\phi(\rho,\phi,k_z)\hat{\phi} + H_z(\rho,\phi,k_z)\hat{z} \tag{4.1.12b}$$

$$\bar{J}(\rho,\phi,k_z) = J_\rho(\rho,\phi,k_z)\hat{\rho} + J_\phi(\rho,\phi,k_z)\hat{\phi} + J_z(\rho,\phi,k_z)\hat{z} \tag{4.1.12c}$$

4.1.2 FIELDS AND POTENTIALS

Referring to the discussions in Sections 2.21 and 3.1.1, the complete solution for the electric and magnetic field quantities can be conveniently expressed in terms of two arbitrary potential functions given by a vector magnetic potential and a scalar electric potential as

$$\bar{H}(\bar{r},\omega) = \frac{1}{\mu} \nabla \times \bar{A}(\bar{r},\omega) \tag{4.1.15a}$$

$$\bar{E}(\bar{r},\omega) = - j\omega\bar{A}(\bar{r},\omega) - \nabla\Phi(\bar{r},\omega) \tag{4.1.15b}$$

where the following vector *Helmholtz* partial differential equation satisfies the magnetic vector potential:

$$\nabla^2\bar{A}(\bar{r},\omega) + k^2\bar{A}(\bar{r},\omega) = - \mu\bar{J}_V(\bar{r},\omega) \tag{4.1.16a}$$

$$k^2 = \omega^2\mu\varepsilon \tag{4.1.16b}$$

and, similarly, the electric scalar potential satisfies the scalar Helmholtz equation:

$$\nabla^2\Phi(\bar{r},\omega) + k^2\Phi(\bar{r},\omega) = - \frac{\rho_V(\bar{r},\omega)}{\varepsilon} \tag{4.1.16c}$$

As discussed in Section 2.18, the vector Helmholtz equation is difficult to solve. It can be reduced to the corresponding three scalar Helmholtz equations. Using the rectangular components defined in expression (3.1.13b), vector Helmholtz equation (4.1.16a) can be rewritten in terms of the following scalar form:

$$\nabla^2 A_x(\bar{r},\omega) + k^2 A_x(\bar{r},\omega) = - \mu J_x(\bar{r},\omega) \tag{4.1.17a}$$

$$\nabla^2 A_y(\bar{r},\omega) + k^2 A_y(\bar{r},\omega) = -\mu J_y(\bar{r},\omega) \tag{4.1.17b}$$

$$\nabla^2 A_z(\bar{r},\omega) + k^2 A_z(\bar{r},\omega) = -\mu J_z(\bar{r},\omega) \tag{4.1.17c}$$

It is interesting to note from these equations that the *x*-directed electric current density produces only the *x* component of the magnetic vector potential. Similarly, *y*- and *z*-directed electric current densities produce *y* and *z* components of the magnetic vector potentials respectively.

Section 2.22 gives a method to obtain a solution for the scalar Helmholtz differential equations based on the three-dimensional Green's function and superposition integral representation. Considering the volume electric current and electric charge source distributions confined to an arbitrary volume *V* and located in a three-dimensional isotropic and homogeneous medium, let

$\bar{r}'$: position vector corresponding to the source point (x', y', z')

$$= x'\hat{x} + y'\hat{y} + z'\hat{z} \tag{4.1.18a}$$

$\bar{r}$: position vector corresponding to the field point (x, y, z)

$$= x\hat{x} + y\hat{y} + z\hat{z} \tag{4.1.18b}$$

R : distance of the line joining the source point to the field point

$$= |\bar{r} - \bar{r}'| \tag{4.1.18c}$$

The total magnetic vector potential at the field point is given by the volume integral over the complete source current density distribution in the volume *V*:

$$A_i(\bar{r}) = \frac{\mu}{4\pi} \iiint_V J_i(\bar{r}') \frac{e^{-jkR}}{R} \, dv(\bar{r}') \qquad i = x, y, z \tag{4.1.19}$$

and the total electric scalar potential at the field point is given by the volume integral over the complete source charge density distribution in the volume *V*:

$$\Phi(\bar{r}) = \frac{1}{4\pi\varepsilon} \iiint_V \rho_v(\bar{r}') \frac{e^{-jkR}}{R} \, dv(\bar{r}') \tag{4.1.20}$$

The components of electric and magnetic field distributions based on the two arbitrary vector and scalar potentials are given in Table 3.3 in terms of the rectangular coordinate system.

4.1.3 TRANSFORMED FIELDS IN TRANSVERSE COORDINATES

The expressions given previously can be conveniently simplified for the case of two-dimensional electromagnetic scattering and interaction studies. As discussed earlier, the geometry of a two-dimensional structure is uniform and infinite in length in the z coordinate direction. Hence, the spectral transformation can be applied to the various components of the magnetic vector potential and the electric scalar potential. Let

$\bar{A}(x,y,z)$: magnetic vector potential

$$= \bar{A}(x,y,k_z)\, e^{-jk_z z} \tag{4.1.21a}$$

$\Phi(x,y,z)$: electric scalar potential

$$= \Phi(x,y,k_z)\, e^{-jk_z z} \tag{4.1.21b}$$

Referring to Figure 4.1, the magnetic vector potential can be resolved into the axial and transverse components, and written as

$$\bar{A}(x,y,z,\omega) = \left[\bar{A}_\tau(x,y,k_z) + A_z(x,y,k_z)\hat{z}\right] e^{-jk_z z} \tag{4.1.22}$$

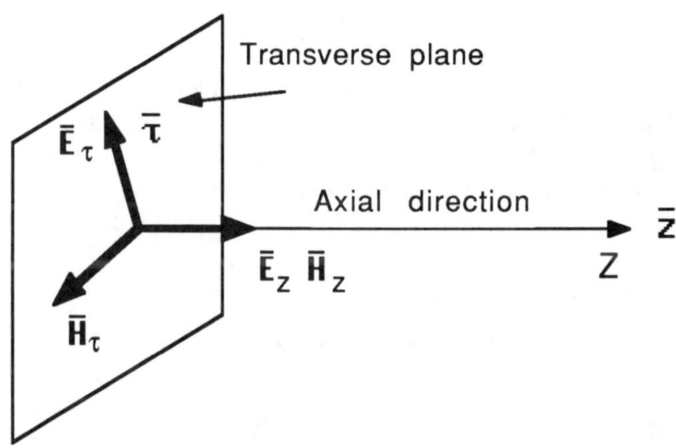

Figure 4.1 Axial and transverse fields.

The transverse components of the electric and magnetic fields in a plane perpendicular to the z coordinate axis can be expressed in the following form:

$$\bar{E}_\tau(\bar{r},\omega) = \bar{E}_\tau(x,y,k_z)\, e^{-jk_z z} \tag{4.1.23a}$$

$$\bar{H}_\tau(\bar{r},\omega) = \bar{H}_\tau(x,y,k_z)\, e^{-jk_z z} \tag{4.1.23b}$$

and similarly, the axial or z components of the electric and magnetic fields can be expressed as

$$\bar{E}_z(\bar{r},\omega) = \bar{E}_z(x,y,k_z)\, e^{-jk_z z} \tag{4.1.24a}$$

$$= \hat{z} E_z(x,y,k_z)\, e^{-jk_z z} \tag{4.1.24b}$$

$$\bar{H}_z(\bar{r},\omega) = \bar{H}_z(x,y,k_z)\, e^{-jk_z z} \tag{4.1.25a}$$

$$= \hat{z} H_z(x,y,k_z)\, e^{-jk_z z} \tag{4.1.25b}$$

$\hat{\tau}$: unit vector in the transverse plane;

$\hat{z}$: unit vector in the axial z coordinate direction.

As discussed in Sections 3.1.1 and 4.1.2, the magnetic and electric field distributions can be generated in terms of this magnetic vector potential and electric scalar potential. On substituting expression (4.1.22) into the field representations (4.1.15a) and (4.1.15b), the following relationships are obtained:

$$\left[\nabla_\tau + \frac{\partial}{\partial z}\hat{z}\right] \times \left[\bar{A}_\tau(x,y,k_z) + \bar{A}_z(x,y,k_z)\right] e^{-jk_z z}$$
$$= \mu\left[\bar{H}_\tau(x,y,k_z) + \bar{H}_z(x,y,k_z)\right] e^{-jk_z z} \tag{4.1.26a}$$

$$\nabla_\tau \times \bar{A}_\tau(x,y,k_z) + (-jk_z)\hat{z} \times \bar{A}_\tau(x,y,k_z) - \hat{z} \times \nabla_\tau A_z(x,y,k_z)$$
$$= \mu\bar{H}_\tau(x,y,k_z) + \mu\bar{H}_z(x,y,k_z) \tag{4.1.26b}$$

and

$$\bar{E}(x,y,k_z)\, e^{-jk_z z} = -\, j\omega\left[\bar{A}_\tau(x,y,k_z) + \bar{A}_z(x,y,k_z)\right] e^{-jk_z z}$$
$$-\left[\nabla_\tau + \frac{\partial}{\partial z}\hat{z}\right]\Phi(x,y,k_z)\, e^{-jk_z z} \tag{4.1.26c}$$

Table 4.3
Electric and magnetic fields in terms of potentials –
Spectral transform domain

$$\mu\bar{H}_\tau(x,y,k_z) = (-jk_z)\hat{z} \times \bar{A}_\tau(x,y,k_z) - \hat{z} \times \nabla_\tau A_z(x,y,k_z) \qquad (4.1.27a)$$

$$\mu\bar{H}_z(x,y,k_z) = \nabla_\tau \times \bar{A}_\tau(x,y,k_z) \qquad (4.1.27b)$$

$$\bar{E}_\tau(x,y,k_z) = -j\omega\bar{A}_\tau(x,y,k_z) - \nabla_\tau\Phi(x,y,k_z) \qquad (4.1.28a)$$

$$E_z(x,y,k_z) = -j\omega A_z(x,y,k_z) - (-jk_z)\Phi(x,y,k_z) \qquad (4.1.28b)$$

Table 4.3 gives the axial and transverse components of the electric and magnetic fields in terms of the components of potentials in the spectral transform domain. From expressions (4.1.16a–c), the following *Helmholtz* differential equations are obtained for the components of magnetic vector potential and electric scalar potential:

$$\nabla_\tau^2\bar{A}_\tau(x,y,k_z) + k_\tau^2\bar{A}_\tau(x,y,k_z) = -\mu\bar{J}_\tau(x,y,k_z) \qquad (4.1.29a)$$

$$\nabla_\tau^2 A_z(x,y,k_z) + k_\tau^2 A_z(x,y,k_z) = -\mu J_z(x,y,k_z) \qquad (4.1.29b)$$

$$\nabla_\tau^2\Phi(x,y,k_z) + k_\tau^2\Phi(x,y,k_z) = -\frac{\rho_v(x,y,k_z)}{\varepsilon} \qquad (4.1.29c)$$

These Helmholtz partial differential equations are to be solved first for the potentials, and then the field components can be calculated based on the expressions in Table 4.3.

4.2 TRANSVERSE ELECTRIC POLARIZATION

In the previous section, general expressions for the electric and magnetic field distributions in the spectral domain are obtained corresponding to the components of the source electric current density distribution in the axial and transverse directions. Let us consider a special case wherein there is *only* the transverse component of the electric current density distribution and the corresponding charge density distribution.

Referring to the two-dimensional geometry shown in Figure 3.1, let us assume that a volume electric current density distribution is polarized completely in the transverse plane. In fact, this volume current density is associated with its corresponding charge density distribution through the current-charge continuity equation discussed in Section 2.2, expression (2.2.6). The complete volume current density distribution can be simulated as the superposition of line sources. Figure 4.2 shows an ideal time-harmonic source electric current and charge distributions consisting of only the transverse directed (polarized parallel to the *xy* coordinate plane) line current and the corresponding charge distribution located along the *z* coordinate axis from $-\infty$ to $+\infty$. The medium is assumed to be linear, homogeneous, and isotropic. Let

$\bar{J}_V(\bar{r}',\omega)$: source electric current distribution

$$= \bar{J}_\tau(x',y',z',\omega) \tag{4.2.1a}$$

$$= I_\tau(k_z,\omega)\delta(x')\delta(y')\, e^{-jk_z z'}\hat{\tau}, \tag{4.2.1b}$$

$\rho_V(\bar{r}',\omega)$: source electric charge distribution

$$= \rho_\ell(k_z,\omega)\delta(x')\delta(y')\, e^{-jk_z z'} \tag{4.2.2}$$

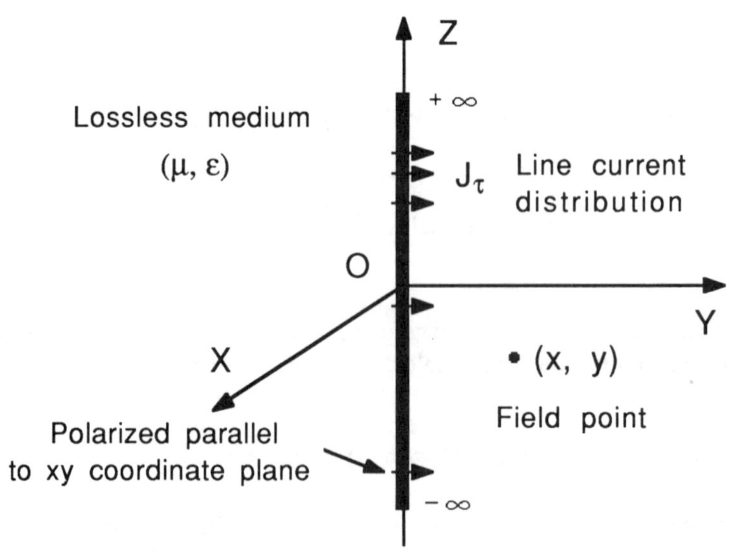

Figure 4.2 Transverse-directed line source electric current distribution.

At any field point, there is only the transverse component of magnetic vector potential, which can be defined as

$\bar{A}(\bar{r},\omega)$: magnetic vector potential function at the field point

$$= \bar{A}_\tau(x,y,z,\omega) \quad (4.2.3a)$$

$$= \bar{A}_\tau(x,y,k_z,\omega)\, e^{-jk_z z} \quad (4.2.3b)$$

Table 4.4
Electromagnetic fields in terms of potentials in the spectral domain – Transverse current excitation

$$\mu\bar{H}_\tau(x,y,k_z) = (-jk_z)\hat{z} \times \bar{A}_\tau(x,y,k_z) \quad (4.2.4a)$$

$$\mu\bar{H}_z(x,y,k_z) = \nabla_\tau \times \bar{A}_\tau(x,y,k_z) \quad (4.2.4b)$$

$$\bar{E}_\tau(x,y,k_z) = -\, j\omega\bar{A}_\tau(x,y,k_z) - \nabla_\tau\Phi(x,y,k_z) \quad (4.2.4c)$$

$$E_z(x,y,k_z) = -\, (-jk_z)\Phi(x,y,k_z) \quad (4.2.4d)$$

Based on these representations, the expressions in the Table 4.3 can be simplified. The corresponding electric and magnetic field distributions in the spectral domain are given in Table 4.4. In Table 4.4, the transverse component of magnetic vector potential and the electric scalar potential satisfy the following scalar Helmholtz equations:

$$\nabla_\tau^2\bar{A}_\tau(x,y,k_z,\omega) + k_\tau^2\bar{A}_\tau(x,y,k_z,\omega) = -\,\mu I_\tau(k_z,\omega)\delta(x)\delta(y)\hat{\tau} \quad (4.2.5a)$$

$$\nabla_\tau^2\Phi(x,y,k_z,\omega) + k_\tau^2\Phi(x,y,k_z,\omega) = -\,\frac{1}{\varepsilon}\,\rho_\ell(k_z,\omega)\delta(x)\delta(y) \quad (4.2.5b)$$

4.2.1 TRANSVERSE ELECTRIC FIELDS

Further, let us consider a special case of an excitation where the distribution of the transverse electric current is completely *independent of the z coordinate variable*. This, in fact, produces the magnetic vector potential and the electric scalar potential independent of the *z* coordinate variable. The various field expressions derived in Table 4.4 can now be further simplified for this situation. Hence

$$\frac{\partial}{\partial z} \Rightarrow 0 \tag{4.2.6a}$$

or

$$k_z = 0 \tag{4.2.6b}$$

In general, any coordinate system can be utilized to express the electric and magnetic fields, such as the rectangular system, cylindrical system, or the generalized cylindrical system in the transverse coordinate plane. After substituting the condition (4.2.6b), the electric field and magnetic field expressions in Table 4.4 reduce to the following form in Table 4.5.

Table 4.5

Transverse electric fields in terms of potentials –
Spectral transform domain

$$\bar{H}_z(x,y) = \frac{1}{\mu} \nabla_\tau \times \bar{A}_\tau(x,y) \tag{4.2.7a}$$

$$\bar{E}_\tau(x,y) = -j\omega\bar{A}_\tau(x,y) - \nabla_\tau \Phi(x,y) \tag{4.2.7b}$$

In the rectangular coordinate system, the transverse variables are given by *(x, y)* and the differential operator can be written as

$$\nabla_\tau = \frac{\partial}{\partial x}\hat{x} + \frac{\partial}{\partial y}\hat{y} \tag{4.2.8a}$$

and the transverse Laplace operator expressed in the rectangular coordinate system takes the form

$$\nabla_\tau^2 = \frac{\partial^2}{\partial x^2} + \frac{\partial^2}{\partial y^2} \tag{4.2.8b}$$

Table 4.6
Transverse electric fields in terms of potentials in the spectral domain – Rectangular coordinate system

$$\hat{z}H_z(x,y) = \frac{1}{\mu} \nabla_\tau \times \bar{A}_\tau(x,y) \tag{4.2.9a}$$

$$E_x(x,y) = -j\omega A_x(x,y) - \frac{\partial}{\partial x}\Phi(x,y) \tag{4.2.9b}$$

$$E_y(x,y) = -j\omega A_y(x,y) - \frac{\partial}{\partial y}\Phi(x,y) \tag{4.2.9c}$$

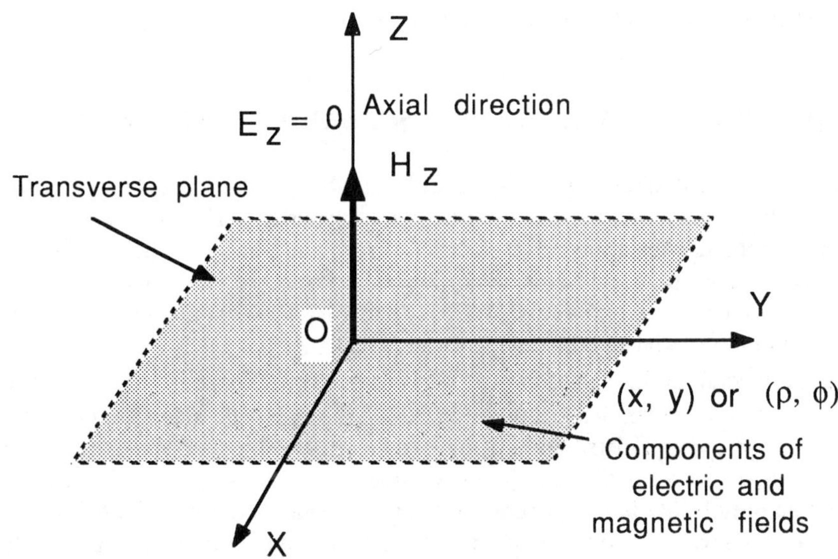

Figure 4.3 Transverse electric fields.

The field expressions given in Table 4.5 can be resolved into respective x and y components. The various components of electric and magnetic field distributions in the spectral transform domain are given in Table 4.6.

After substituting the condition stated in expression (4.2.6b), Helmholtz differential equations (4.2.5a) and (4.2.5b) also can be simplified. Hence, the transverse component of the magnetic vector potential and the electric scalar potential satisfy the following Helmholtz differential equations:

$$\nabla_{\tau}^{2}\bar{A}_{\tau}(x,y,\omega) + k^{2}\bar{A}_{\tau}(x,y,\omega) = -\mu I_{\tau}(\omega)\delta(x)\delta(y)\hat{\tau} \tag{4.2.10a}$$

$$\nabla_{\tau}^{2}\Phi(x,y,\omega) + k^{2}\Phi(x,y,\omega) = -\frac{1}{\varepsilon}\rho_{\ell}(\omega)\delta(x)\delta(y) \tag{4.2.10b}$$

As can be seen from Tables 4.5 and 4.6, and Figure 4.3, for the case of transverse electric excitation, there are only axial component of the magnetic field distribution and the transverse components of the electric field distributions. These can be calculated by first knowing the expression for the transverse component of magnetic vector potential and the corresponding electric scalar potential. In the following, a solution procedure is discussed to analyze the two Helmholtz equations (4.2.10a) and (4.2.10b).

4.3 SOLUTION FOR VECTOR AND SCALAR POTENTIALS

Referring to Figure 4.2, the transverse electric current distribution is polarized parallel to the *xy* coordinate plane and located along the *z* coordinate axis. For the special excitation case just discussed, the transverse electric current and charge distributions and the corresponding potentials and fields do not vary with respect to the *z* coordinate variable. Hence, only Helmholtz equations (4.2.10a) and (4.2.10b) should be solved for the magnetic vector potential and the electric scalar potential.

Helmholtz differential equation (4.2.10a) is solved first for the τ component of the magnetic vector potential. Referring to Figure 4.2, the righthand-side term in expression (4.2.10a) is the spectral distribution of transverse directed line current source, which is located along the z coordinate axis. The two-dimensional transverse electric current has a constant amplitude and is independent of the *z* coordinate variable. Thus, for the two-dimensional analysis, the observation point is taken in the *xy* transverse plane. In the following, a solution for the τ component of magnetic vector potential is written in the cylindrical coordinate system following the procedure discussed in Section 3.4. Expressing the transverse Laplacian operator in the cylindrical coordinate system, for all observation points outside the transverse-directed line source electric current, scalar Helmholtz equation (4.2.10a) reduces to

$$\frac{1}{\rho}\frac{\partial}{\partial\rho}\left[\rho\,\frac{\partial\bar{A}_{\tau}(\rho,\phi,\omega)}{\partial\rho}\right] + \frac{1}{\rho^{2}}\frac{\partial^{2}\bar{A}_{\tau}(\rho,\phi,\omega)}{\partial\phi^{2}} + k^{2}\bar{A}_{\tau}(\rho,\phi,\omega) = 0 \tag{4.3.1}$$

To solve the scalar partial differential equation (4.3.1), the method of *separation of variables* can be adopted, which is discussed in Sections 2.9 and 3.4. Hence, the total solution for the magnetic vector potential A_τ can be written as

$$\bar{A}_\tau(\rho,\omega) = \frac{\mu I_\tau}{4j} H_0^{(2)}(k\rho)\hat{\tau} \tag{4.3.2}$$

Again, this solution for the Helmholtz differential equation with a delta function-type source excitation is generally referred to as the two-dimensional *Green's function.* Expression (4.3.2) for the magnetic vector potential can be conveniently modified if the transverse-directed line source electric current distribution is not located along the z coordinate axis, but is shifted to an arbitrary source location (x', y') and is still polarized parallel to the xy coordinate plane.

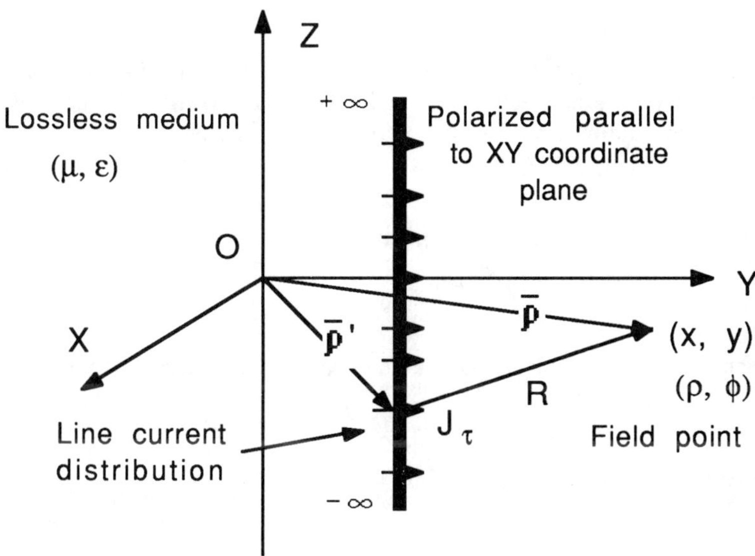

Figure 4.4 Shifted transverse line source electric current distribution.

Referring to Figure 4.4, let us consider an ideal time-harmonic line source electric current distribution consisting of only the transverse-directed electric current distribution located parallel to the z coordinate axis from $-\infty$ to $+\infty$. The medium is assumed to be linear, homogeneous, and isotropic. Then

$\bar{J}_V(x',y',\omega)$: transverse-directed source electric current distribution

$$= I_\tau(\omega)\delta(x-x')\delta(y-y')\hat{\tau}' \tag{4.3.3a}$$

At any field point in the transverse plane, there is only the τ component of the magnetic vector potential. In the spectral transform domain, expression (4.3.2) for the magnetic vector potential (or the two-dimensional Green's function with unit current excitation) takes the following form

$$\bar{A}_\tau(\rho,\phi,\omega) = \frac{\mu I_\tau(\rho',\phi')}{4j} H_0^{(2)}(kR)\hat{\tau}' \tag{4.3.3b}$$

R : distance between the source point and the field point in the transverse plane, in meters

$$= |\bar{\rho} - \bar{\rho}'| \tag{4.3.3c}$$

A similar analysis can be carried out using scalar Helmholtz equation (4.2.10b) to find an expression for the scalar electric potential. Again, referring to Figure 4.4, let us consider the corresponding time-harmonic source electric charge distribution located parallel to the z coordinate axis from $-\infty$ to $+\infty$. Then

$\rho_V(x',y',\omega)$: source electric charge distribution

$$= \rho_\ell(\omega)\delta(x-x')\delta(y-y') \tag{4.3.4a}$$

In the spectral transform domain, the electric scalar potential takes the following form:

$$\Phi(\rho,\phi,\omega) = \frac{\rho_\ell(\rho',\phi')}{4j\varepsilon} H_0^{(2)}(kR) \tag{4.3.4b}$$

The expressions for the two potentials just derived are for the filamentary two-dimensional line electric current and charge distributions. These expressions can be easily generalized for the case of arbitrary distributions of transverse directed electric currents and the corresponding electric charges by applying linearity and superposition. Further, using Table 4.5, the electric and magnetic field distributions can be calculated by substituting the results (4.3.3b) for the magnetic vector potential and (4.3.4b) for the electric scalar potential. The electric and magnetic field expressions so obtained are quite useful in the study of electromagnetic scattering and interaction from the two-dimensional conducting object with transverse electric excitation.

4.4 INTEGRAL EQUATIONS - TE CASE

Referring to Figure 4.5a, a perfectly conducting (material conductivity, $\sigma \to \infty$), two-dimensional object is oriented with its axis coinciding with the z axis of the cylindrical coordinate system. Along the axis, the scattering object is infinitely long and has a uniform arbitrary cross section. It is placed in a linear, homogeneous, and isotropic lossless medium and is excited externally by a time-harmonic transverse electric (TE to z) polarized plane wave. The TE polarized plane wave is incident on the conducting object at an arbitrary angle of incidence with its incident magnetic field polarized parallel and the corresponding incident electric field polarized perpendicular to the z coordinate axis. It should be clearly noted that there is no propagation of the incident plane wave parallel to the axis of the two-dimensional object.

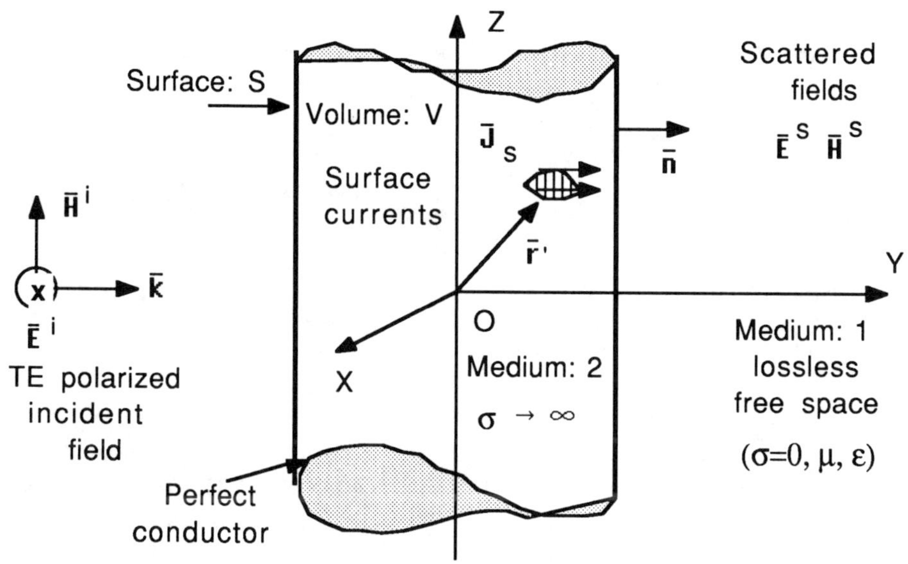

Figure 4.5a Perfectly conducting two-dimensional scatterer – TE excitation.

For this special case of time-harmonic transverse electric *normal* excitation, there is only a transverse component of the induced electric currents on the surface of conducting object. The induced electric currents produce the scattered electric and magnetic field distributions in the surrounding medium. Hence, in the free-space medium outside the conducting object, the total field distribution consists of the sum of the incident field and the scattered field. Further, the total electric and magnetic field distributions near the surface of the conducting object are such that they satisfy appropriate electromagnetic boundary conditions. The induced electric current distribution on the conducting object is still unknown, but can be determined by enforcing the appropriate boundary condition, that the total tangential component of the electric field on the surface of conducting object is zero; and similarly, the total tangential component of the magnetic field on the surface of conducting object is equal to the induced electric current distribution.

As discussed in Section 3.5, a classical approach is to treat the induced transverse electric currents on the surface of the conducting object as unknown and set up a boundary value integral equation. Based upon the electromagnetic equivalence principle, two types of integral equations – an electric field type and a magnetic field type – are presented in the following sections. The magnetic field integral equation formulation is suited for closed type of conducting object, while the electric field integral equation formulation is suited for both closed or open types of conducting objects.

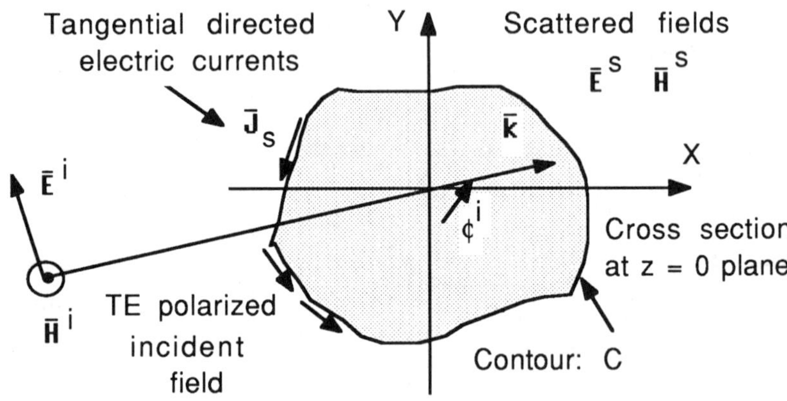

Figure 4.5b Cross-section of a perfectly conducting scatterer – TE excitation.

Figure 4.5b shows the cross-sectional geometry of the conducting scatterer located in an isotropic lossless free-space medium. The scatterer has a volume V contained in medium 2 and is bounded by a surface S. Outside the volume of medium 2 is medium 1, representing the free-space region, and the externally excited TE to z polarized incident plane wave field is contained in it.

Referring to Figures 4.5a and 4.5b, let

ε : permittivity of free-space medium 1;

μ : permeability of free-space medium 1;

$\sigma = 0$: conductivity of free-space medium 1;

$(\bar{E}^i, \bar{H}^i)$: electric and magnetic incident fields in medium 1;

$(\bar{E}^S, \bar{H}^S)$: electric and magnetic scattered fields in medium 1;

$(\bar{E}, \bar{H})$: electric and magnetic total fields in medium 1

$= (\bar{E}^i, \bar{H}^i) + (\bar{E}^S, \bar{H}^S)$

Referring to the conducting boundary condition discussed in Section 2.16, there is no electric or magnetic field inside the perfectly conducting region, and on the surface of the perfectly conducting object, the following boundary conditions are satisfied:

Magnetic field boundary condition

$$\bar{J}_S(\bar{r}') = \hat{n}' \times [\bar{H}^i(\bar{r}') + \bar{H}^S(\bar{r}')] \qquad \text{for } \bar{r}' \text{ on } S \tag{4.4.1a}$$

Electric field boundary conditions

$$\bar{E}(\bar{r})\Big|_{\tan} = \hat{n} \times [\bar{E}^i(\bar{r}) + \bar{E}^S(\bar{r})] = 0 \qquad \text{for } \bar{r} \text{ on } S \tag{4.4.1b}$$

tan : tangential component of the field.

Referring to Figures 4.5a and 4.5b and the discussion in Section 2.19 regarding plane wave fields in the frequency domain, the incident plane wave electric and magnetic fields can be written as follows. The plane wave incident magnetic field vector is polarized along the z axis, and the incident electric field vector is polarized in a plane perpendicular to the z axis. There is no propagation of the fields parallel to the z coordinate axis, indicating that the incident excitation fields are uniform and completely independent of the z coordinate variable. According to the expressions (2.19.11a–b),

$$\bar{H}_0 = H_0\hat{z} \tag{4.4.2}$$

$$\bar{E}_0 = E_{0x}\hat{x} + E_{0y}\hat{y} \tag{4.4.3}$$

In the frequency domain, the time-harmonic incident electric and magnetic fields based on the forward traveling plane waves are given by

$$\bar{H}^i(\bar{\rho},\omega) = H_z^i(\bar{\rho},\omega)\hat{z} \tag{4.4.4a}$$

$$H_z^i(\bar{\rho},\omega) = H_0\, e^{-j\bar{k}\bullet\bar{\rho}} \tag{4.4.4b}$$

$$\bar{E}^i(\bar{\rho},\omega) = \bar{E}_0\, e^{-j\bar{k}\bullet\bar{\rho}} \tag{4.4.4c}$$

$$\bar{k} = k\hat{k} \tag{4.4.4d}$$

k : propagation constant for homogeneous and isotropic medium

$$= \omega(\mu\varepsilon)^{1/2} \tag{4.4.4e}$$

$\hat{k}$: actual direction of the plane wave propagation;

ω : frequency of excitation, in radians per second.

With respect to a global origin O, that represents the reference phase center, let

$\bar{\rho}$: spatial vector corresponding to the field point (x, y)

$$= x\hat{x} + y\hat{y} \tag{4.4.5a}$$

$$= \rho\cos\phi\hat{x} + \rho\sin\phi\hat{y} \tag{4.4.5b}$$

$\bar{k}$: propagation vector of the traveling plane waves

$$= k_x\hat{x} + k_y\hat{y} \tag{4.4.5c}$$

$$= k\cos\phi^i\hat{x} + k\sin\phi^i\hat{y} \tag{4.4.5d}$$

ϕ^i : arbitrary angle of incidence of the TE plane wave field;

$k_z = 0$: propagation constant along z coordinate direction.

Referring to expression (2.19.18d), it should be noted that the incident plane wave electric and magnetic fields are related through the following form:

$$\bar{H}_0 \times \hat{k} = \frac{\omega\varepsilon}{k}\bar{E}_0 \tag{4.4.5e}$$

$$\frac{E_0}{H_0} = \eta \tag{4.4.5g}$$

η : intrinsic impedance of the free-space medium

$$= \left(\frac{\mu}{\varepsilon}\right)^{1/2} \tag{4.4.5h}$$

Since the normal plane wave excitation is uniform in the z coordinate direction, the transverse-directed electric currents induced are also uniform along the z coordinate direction. The various scattered electric and magnetic fields produced by the induced electric currents are also uniform with no variation with respect to the z coordinate variable. Hence, the various field expressions obtained in Section 4.2 and Tables 4.5 and 4.6, can be directly utilized. Further, in the following analysis,

$$\frac{\partial}{\partial z} \Rightarrow 0 \quad \text{or} \quad k_z = 0 \tag{4.4.6}$$

It is interesting to note that the spectrum of spectral variable k_z varies from $-\infty$ to ∞, and the case of normal incident excitation yielding condition (4.4.6) is just a special case to isolate dependence of the field quantities with respect to the z coordinate variable. Referring to the Figures 4.5a and 4.5b, various field quantities can be obtained from

$\bar{J}_S(\bar{\rho}',\omega)$: source electric current surface distribution, in amperes per meter

$$= J_S(\bar{\rho}',\omega)\hat{s}' \tag{4.4.7a}$$

$\bar{A}_\tau(\bar{\rho},\omega)$: magnetic vector potential function at the field point

$$= A_\tau(\bar{\rho},\omega)\hat{\tau} \tag{4.4.7b}$$

$\hat{s}'$: tangential unit vector along which the induced electric currents are directed on the surface of the scatterer.

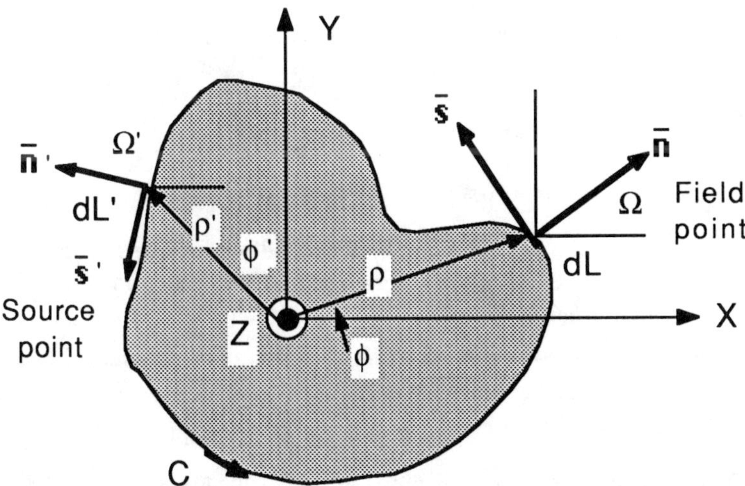

Figure 4.6 Relationships between various unit vectors.

For application to arbitrary cross-sectional geometries, it is useful to introduce and work in terms of generalized cylindrical coordinate system, as shown in Figure 4.6. On the surface of a perfectly conducting scatterer, let us consider three mutually orthogonal coordinates given by *(n, s, z)* such that the following righthand system is satisfied:

$$\hat{z} \times \hat{n} = \hat{s} \tag{4.4.8}$$

n : coordinate variable along normal to the surface;

s : coordinate variable along tangent to the surface.

Then in terms of the rectangular coordinate system

$$\hat{n} = \hat{x} \cos\Omega + \hat{y} \sin\Omega \tag{4.4.9a}$$

$$\hat{s} = - \hat{x} \sin\Omega + \hat{y} \cos\Omega \tag{4.4.9b}$$

$$\hat{n}' = \hat{x} \cos\Omega' + \hat{y} \sin\Omega' \tag{4.4.9c}$$

$$\hat{s}' = - \hat{x} \sin\Omega' + \hat{y} \cos\Omega' \tag{4.4.9d}$$

Ω, Ω' : angles the respective normals make with x coordinate axis at the field and source points.

In the Figure 4.6, unit vectors located at the field point are given by

$$\bar{n} = \hat{n} n \tag{4.4.9e}$$

$$\bar{s} = \hat{s} s \tag{4.4.9f}$$

The field expressions in Table 4.5 can be written down in terms of the generalized cylindrical coordinates. In terms of the generalized transverse coordinate variables *(n, s)*, the differential operator takes the form

$$\nabla_{\tau} = \frac{\partial}{\partial n} \hat{n} + \frac{\partial}{\partial s} \hat{s} \tag{4.4.10}$$

After substituting these relationships into Table 4.5, the various electric and magnetic field components reduce to the form given in Table 4.7.

Table 4.7
Transverse electric fields in terms of generalized cylindrical coordinates – TE excitation

$$H_z(x,y) = \frac{1}{\mu}\hat{z} \bullet \nabla_\tau \times \bar{A}_\tau(x,y) \tag{4.4.11a}$$

$$E_n(x,y) = - j\omega A_n(x,y) - \frac{\partial}{\partial n}\Phi(x,y) \tag{4.4.11b}$$

$$E_s(x,y) = - j\omega A_s(x,y) - \frac{\partial}{\partial s}\Phi(x,y) \tag{4.4.11c}$$

4.5 EQUIVALENT ELECTRIC CURRENT SOURCES

The righthand-side source term in differential equation (4.2.10a) should include the complete distribution of tangential *s*-directed surface electric currents on the conducting object. Section 4.3 discusses the solution for a magnetic vector potential satisfying the scalar differential equation but with an ideal transverse-directed line source current located in a linear, homogeneous, and isotropic medium. The solutions given by expressions (4.3.3b) and (4.3.4b) can be utilized along with linearity and superposition to obtain the complete solution in terms of the superposition potential integrals.

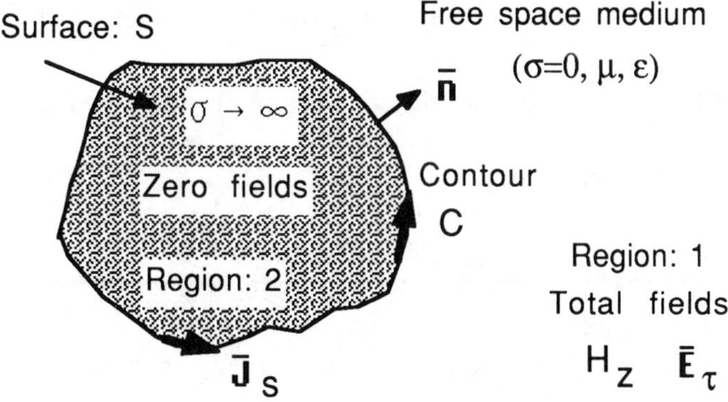

Figure 4.7a Original conducting scatterer boundary value problem.

To formulate a boundary value integral equation, an electromagnetic equivalent problem is set up using the equivalent principle as discussed in Section 3.6. The sources simulated in the equivalent problem are generally referred as the *equivalent* or *virtual electric current sources*, but these are located in a linear, homogeneous, and isotropic medium. The electric and magnetic fields produced by the equivalent currents are identical to the electric and magnetic fields of the original boundary value problem and also satisfy the relevant conducting surface boundary conditions.

Original Conducting Scatterer

Figure 4.7a shows the original boundary value problem of a perfectly conducting scatterer located in a linear, homogeneous, and isotropic medium. Only the cross section of the geometry at the $z = 0$ transverse plane is shown. The arbitrary shaped conducting scatterer geometry has a cross-sectional area S bounded by a contour C. For normal excitation, the incident field, the induced tangential-directed electric currents along contour C, and the scattered electric and magnetic fields are independent of the z coordinate variable. Referring to Figure 4.7a, let

ε : permittivity of free-space medium, region 1;

μ : permeability of free-space medium, region 1;

$\sigma = 0$: conductivity of free-space medium, region 1;

$\sigma => \infty$: conductivity of the scatterer medium, region 2;

and

$(H_z^i, \bar{E}_\tau^i)$: magnetic and electric incident fields in free-space medium;

$(H_z^s, \bar{E}_\tau^s)$: magnetic and electric scattered fields in free-space medium;

$(H_z, \bar{E}_\tau)$: magnetic and electric total fields in free-space medium

$$= (H_z^i, \bar{E}_\tau^i) + (H_z^s, \bar{H}_\tau^s)$$

Referring to the conducting boundary conditions discussed in Section 2.16, there is no electric or magnetic field inside the perfectly conducting region. On contour C, the following boundary conditions are satisfied:

$$\bar{J}_S(\bar{\rho}') = \hat{n}' \times [\bar{H}_z^i(\bar{\rho}') + \bar{H}_z^s(\bar{\rho}')] \qquad \bar{\rho}' \text{ on } C \tag{4.5.1a}$$

$$E_S(\bar{\rho}) = \hat{s} \cdot [\bar{E}^{i}_{\tau}(\bar{\rho}) + \bar{E}^{s}_{\tau}(\bar{\rho})] = 0 \qquad \bar{\rho} \text{ on } C \tag{4.5.1b}$$

Equivalent Scatterer Geometry (valid for region 1 only)

The following procedure is adopted to set up an equivalent boundary value problem. Figure 4.7b shows a conducting scatterer geometry identical in shape to the original conducting scatterer, except that a virtual contour C_+ is drawn just outside the scatterer boundary.

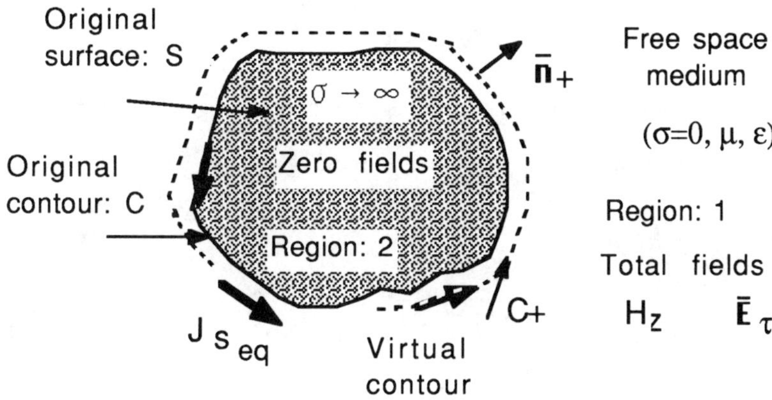

Figure 4.7b Equivalent scatterer of the original problem.

On this virtual contour C_+, certain tangential distributions of electric currents are simulated. In a limit as the virtual contour C_+ tends to the original contour C, the simulated tangential electric currents are such that they produce the same scattered electric and magnetic fields in the free-space medium as in the original conducting scatterer problem, shown in Figure 4.7a. The simulated electric currents on the virtual boundary are generally referred to as the *equivalent* or *virtual electric currents*. Inside virtual contour C_+, the total electric and magnetic fields are zero. Hence, the region 2 conducting material medium can be completely removed from the inside region of the virtual contour and replaced by a medium having properties identical to the region 1 free-space medium. The equivalent boundary value scattering problem is shown in Figure 4.7c.

Figure 4.7c shows virtual or equivalent tangential electric current distribution on contour C whose shape is identical to the original scatterer boundary contour. Let

ε : permittivity of the free-space medium, regions 1 and 2;

μ : permeability of the free-space medium, regions 1 and 2;

$\sigma = 0$: conductivity of the free-space medium, regions 1 and 2;

$(H_z^i, \bar{E}_\tau^i)$: the same magnetic and electric incident fields in region 1;

$(H_z^s, \bar{E}_\tau^s)$: the same magnetic and electric scattered fields as in region 1.

In Figure 4.7c, the equivalent electric currents on virtual contour C_+ also satisfy the following boundary conditions on the total fields:

$\bar{J}_{S_{eq}}(\bar{\rho}')$: equivalent electric currents valid only for region 1 fields

$$= \hat{n}' \times [\bar{H}_Z(\bar{\rho}')] \qquad \bar{\rho}' \text{ on } C_+ \to C \qquad (4.5.2a)$$

$$\bar{E}_\tau(\bar{\rho})\Big|_{tan} = E_S(\bar{\rho})$$

$$= \hat{s} \bullet [\bar{E}_\tau(\bar{\rho})] = 0 \qquad \bar{\rho} \text{ on } C_+ \to C \qquad (4.5.2b)$$

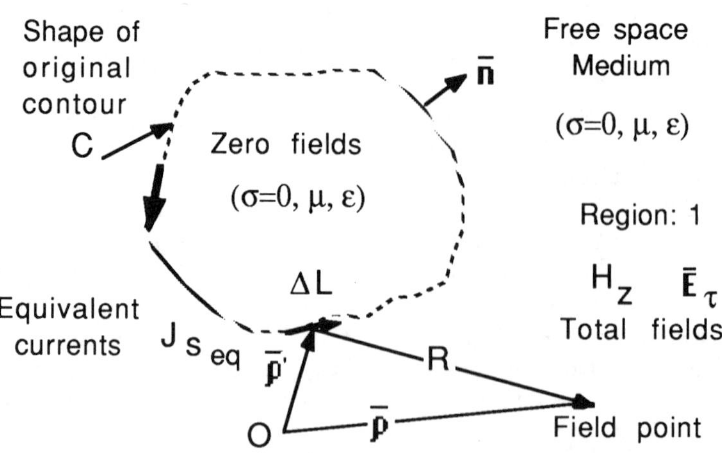

Figure 4.7c Electromagnetic equivalence for conducting scatterer valid for region 1.

Based on the s-directed equivalent electric current density distribution, various scattered electric and magnetic fields valid only for the region 1 can now be calculated using the expression obtained in Table 4.7. In Figure 4.7c, the tangential s-directed electric current in an incremental length ΔL on virtual contour C can be treated just like an ideal two-dimensional line source distribution. There exists only the transverse component of the magnetic vector potential. Expression (4.3.3b) for the two-dimensional Green's function is now utilized to construct a superposition integral for the total transverse component of the magnetic vector potential. Let

$dA_\tau(\bar{\rho},\omega)$: incremental A_τ component of the magnetic vector potential;

$J_{s_{eq}}(\bar{\rho}',\omega)\; dL(\bar{\rho}')$: elemental source electric current, in amperes;

$\bar{\rho}$: field observation point in region 1 outside contour C;

$\bar{\rho}'$: line current source point on contour C.

Based on expression (4.3.3b), the incremental transverse component of the magnetic vector potential is given by

$$d\bar{A}_\tau(\bar{\rho},\omega) = \frac{\mu}{4j}\,\hat{s}'J_{s_{eq}}(\bar{\rho}',\omega)\; H_0^{(2)}(k|\bar{\rho} - \bar{\rho}'|)\; dL(\bar{\rho}') \tag{4.5.3a}$$

Hence, the total magnetic vector potential at any field observation point in the region 1 outside the contour C takes the form

$$\bar{A}_\tau(\bar{\rho},\omega) = \frac{\mu}{4j}\int_C \hat{s}'J_{s_{eq}}(\bar{\rho}',\omega)\; H_0^{(2)}(k|\bar{\rho} - \bar{\rho}'|)\; dL(\bar{\rho}') \tag{4.5.3b}$$

Similarly, the total electric scalar potential at any field observation point in region 1 outside contour C can also be calculated as

$d\Phi(\bar{\rho},\omega)$: incremental electric scalar potential;

$\rho_{\ell_{eq}}(\bar{\rho}',\omega)\; dL(\bar{\rho}')$: elemental source electric charge, in coulombs.

Based on expression (4.3.4b), the incremental electric scalar potential is given by

$$d\Phi(\bar{\rho},\omega) = \frac{1}{4j\varepsilon}\,\rho_{\ell_{eq}}(\bar{\rho}',\omega)\; H_0^{(2)}(k|\,\bar{\rho} - \bar{\rho}'\,|)\; dL(\bar{\rho}') \tag{4.5.4a}$$

Hence, the total electric scalar potential at any field observation point in region 1 outside contour C takes the form

$$\Phi(\bar{\rho},\omega) = \frac{1}{4j\varepsilon} \int_C \rho_{\ell eq}(\bar{\rho}',\omega)\, H_0^{(2)}(k|\bar{\rho} - \bar{\rho}'|)\, dL(\bar{\rho}') \tag{4.5.4b}$$

Expression (4.5.4b) can be rewritten in terms of the transverse electric current distribution using the current-charge continuity equation as discussed in Sections 2.2 and 2.14.2. Further, referring to Table 4.7, various components of the electric and magnetic field distributions can be calculated by substituting these results for the magnetic vector potential and the electric scalar potential. These scattered field expressions are quite useful in the development of integral equations. For example, using expression (4.4.11a) for the observation points outside the conducting scatterer, the z component of scattered magnetic field is given by

$$H_z^S(\bar{\rho},\omega) = \frac{1}{\mu}\, \hat{z} \bullet \nabla_\tau \times \bar{A}_\tau(\bar{\rho}) \tag{4.5.5a}$$

$$= \frac{1}{4j}\, \hat{z} \bullet \nabla_\tau \times \int_C \hat{s}' J_{seq}(\bar{\rho}')\, H_0^{(2)}(k|\bar{\rho} - \bar{\rho}'|)\, dL' \tag{4.5.5b}$$

$$H_z^S(\bar{\rho},\omega) = \frac{1}{4j} \int_C J_{seq}(\bar{\rho}')[\hat{z} \bullet \nabla_\tau \times H_0^{(2)}(k|\bar{\rho} - \bar{\rho}'|)\, d\bar{L}'] \tag{4.5.5c}$$

Similarly, substituting expressions (4.5.3b) and (4.5.4b) into expression (4.4.11c), the tangential s component of scattered electric field is given by

$$\hat{s} \bullet \bar{E}^S(\bar{\rho},\omega) = -\, j\omega \bar{A}_s(x,y) - \frac{\partial}{\partial s} \Phi(x,y) \tag{4.5.6a}$$

$$= -\,\frac{\omega\mu}{4} \int_C (\hat{s} \bullet \hat{s}') J_{seq}(\bar{\rho}')\, H_0^{(2)}(k|\bar{\rho} - \bar{\rho}'|)\, dL'$$

$$-\, \frac{1}{4j\varepsilon}\frac{\partial}{\partial s} \int_C \rho_{\ell eq}(\bar{\rho}',\omega)\, H_0^{(2)}(k|\bar{\rho} - \bar{\rho}'|)\, dL' \tag{4.5.6b}$$

The expression for the scattered electric field can be further simplified by replacing the charge distribution in terms of the tangential directed electric current. Referring to expression (2.14.10), the electric current-charge continuity equation yields

$$\nabla \bullet \bar{J}_V(\bar{r}',\omega) = -j\omega\rho_V(\bar{r}',\omega) \tag{4.5.7a}$$

In terms of the two-dimensional transverse quantities discussed earlier, considering only the tangential s-directed electric current and the corresponding electric line charge distribution,

$$\frac{\partial}{\partial s'} J_{s_{eq}}(\bar{\rho}',\omega) = -j\omega\rho_{\ell_{eq}}(\bar{\rho}',\omega) \tag{4.5.7b}$$

After substituting relationship (4.5.7b) into expression (4.5.6b), the tangential component of scattered electric field is given by

$$\hat{s} \bullet \bar{E}^S(\bar{\rho},\omega) = -\frac{\omega\mu}{4}\int_C (\hat{s} \bullet \hat{s}')J_{s_{eq}}(\bar{\rho}')\, H_0^{(2)}(k|\bar{\rho} - \bar{\rho}'|)\, dL'$$

$$-\frac{1}{4\omega\varepsilon}\frac{\partial}{\partial s}\int_C \left[\frac{\partial}{\partial s'} J_{s_{eq}}(\bar{\rho}')\right] H_0^{(2)}(k|\bar{\rho} - \bar{\rho}'|)\, dL' \tag{4.5.7c}$$

It is not always necessary to express the scattered electric field distribution in terms of the induced tangential s-directed electric current distribution. In fact, expression (4.5.5c) gives the scattered magnetic field distribution. This result can be substituted directly into Maxwell's equations to obtain the scattered electric field distribution. Referring to Ampere's law, expression (4.1.1b), for the source-free case,

$$\nabla \times \bar{H}(\bar{r},\omega) = j\omega\varepsilon\bar{E}(\bar{r},\omega) \tag{4.5.8a}$$

For the two-dimensional case, assuming no field variations with respect to the z coordinate variable, using generalized cylindrical coordinates,

$$\nabla_\tau = \frac{\partial}{\partial n}\hat{n} + \frac{\partial}{\partial s}\hat{s} \tag{4.5.8b}$$

$$\bar{H}^S(\bar{\rho},\omega) = \hat{z}H_z^S(\bar{\rho},\omega) \tag{4.5.8c}$$

$$\bar{E}^S(\bar{\rho},\omega) = \hat{s}E_s^S(\bar{\rho},\omega) + \hat{n}E_n^S(\bar{\rho},\omega) \tag{4.5.8d}$$

After substituting these expressions into Ampere's law (4.5.8a), the components of scattered electric field distribution obtained are as given in Table 4.8.

Table 4.8
Transverse electric fields in terms of axial magnetic field

$$E_s^S(\bar{\rho},\omega) = -\frac{1}{j\omega\varepsilon}\frac{\partial}{\partial n}H_z^S(\bar{\rho},\omega) \tag{4.5.9a}$$

$$E_n^S(\bar{\rho},\omega) = \frac{1}{j\omega\varepsilon}\frac{\partial}{\partial s}H_z^S(\bar{\rho},\omega) \tag{4.5.9b}$$

The scattered electric and magnetic field expressions just obtained are quite useful in the study of electromagnetic scattering and interaction by the arbitrary shaped, two-dimensional conducting object. It should be noted that the results are expressed in terms of an integral expression. The integrand of the integral representation contains the distribution of unknown tangential-directed electric current distribution. Detailed discussions are presented in the following sections to determine the unknown distribution of electric current for a certain class of canonical and arbitrary cross-sectional geometries. Two types of integral equations – electric field type and magnetic field type – are derived by invoking the electric field boundary condition and the magnetic field boundary condition. These integral equations are further studied based on suitable numerical techniques.

4.5.1 MFIE – PERFECT CONDUCTOR

A *magnetic field integral equation* (MFIE) in terms of the unknown equivalent electric current distribution can be obtained by enforcing boundary condition related to the z component of the total magnetic field on contour C. At any point in region 1 outside scatterer contour C, the tangential total magnetic field is given by the sum of z components of the incident and scattered magnetic fields:

$$H_z(\bar{\rho}) = H_z^i(\bar{\rho}) + H_z^s(\bar{\rho}) \qquad \bar{\rho} \text{ in region 1} \tag{4.5.10a}$$

Referring to expression (4.5.5a) and substituting for the z component of scattered magnetic field,

$$H_z(\bar{\rho}) = H_z^i(\bar{\rho}) + \frac{1}{\mu}\hat{z} \bullet \nabla_\tau \times \bar{A}_\tau(\bar{\rho}) \tag{4.5.10b}$$

$$= H_z^i(\bar{\rho}) + \frac{1}{4j}\hat{z} \bullet \nabla_\tau \times \int_C \hat{s}' J_{s_{eq}}(\bar{\rho}')\, H_0^{(2)}(k|\bar{\rho} - \bar{\rho}'|)\, dL'$$

$$\bar{\rho} \text{ in region 1} \tag{4.5.10c}$$

Using expression (4.5.2a), the magnetic field integral equation for the arbitrary shaped, perfectly conducting scatterer with transverse electric excitation is obtained by moving the field observation point directly to contour C to yield

$$- J_{s_{eq}}(\bar{\rho}) = H_z^i(\bar{\rho}) + \frac{1}{\mu}\hat{z} \bullet \nabla_\tau \times \bar{A}_\tau(\bar{\rho}) \tag{4.5.11a}$$

$$= H_z^i(\bar{\rho}) + \frac{1}{4j}\hat{z} \bullet \nabla_\tau \times \int_{C_0} \hat{s}' J_{s_{eq}}(\bar{\rho}')\, H_0^{(2)}(k|\bar{\rho} - \bar{\rho}'|)\, dL'$$

$$+ \frac{1}{4j}\hat{z} \bullet \nabla_\tau \times \fint_C \hat{s}' J_{s_{eq}}(\bar{\rho}')\, H_0^{(2)}(k|\bar{\rho} - \bar{\rho}'|)\, dL'$$

$$\bar{\rho} \text{ on } C \tag{4.5.11b}$$

In integral equation (4.5.11b), the integral over contour C is separated into two separate integral terms. The first integral term over contour C_0 represents the value of the integral when the field observation point lies on the source segment covered by the contour C_0. This term gives the principal value of the integral that basically represents the scattered magnetic field very close to the perfectly conducting scatterer carrying a current sheet. Hence

$$H_z^s(\bar{\rho})\Big|_{\text{as } \bar{\rho} \to \bar{\rho}'} = - \frac{J_{s_{eq}}(\bar{\rho})}{2} \tag{4.5.12}$$

The second integral is written with a bar on the integral sign to recognize the field observation point on remaining of the contour outside source segment C_0. The kernel function in the integrand of second integral is well behaved, and thus the transverse curl operation can be taken inside the integral to yield a normal derivative term similar to expression (4.5.5c). The MFIE takes the following form:

$$- H_z^i(\bar{\rho}) = \frac{J_{s_{eq}}(\bar{\rho})}{2} + \frac{1}{4j}\, \hat{z} \bullet \nabla_\tau \times \fint_C \hat{s}' J_{s_{eq}}(\bar{\rho}')\, H_0^{(2)}(k|\bar{\rho} - \bar{\rho}'|)\, dL' \qquad \bar{\rho} \text{ on } C \tag{4.5.13a}$$

This expression can be further simplified by noting that the transverse differential operator is in terms of unprimed coordinates and the sources are located at primed coordinates

$$\nabla_\tau \times \left[\hat{s}' J_{s_{eq}}(\bar{\rho}')\, H_0^{(2)}(k|\bar{\rho} - \bar{\rho}'|)\right]$$

$$= \nabla_\tau \times \left[\bar{J}_{s_{eq}}(\bar{\rho}')\, H_0^{(2)}(k|\bar{\rho} - \bar{\rho}'|)\right] \tag{4.5.13b}$$

$$= -\left[\hat{s}' J_{s_{eq}}(\bar{\rho}') \times \nabla_\tau H_0^{(2)}(k|\bar{\rho} - \bar{\rho}'|)\right] \tag{4.5.13c}$$

After substituting expression (4.5.13c) into MFIE expression (4.5.13a), the following simplification is obtained:

$$- H_z^i(\bar{\rho}) = \frac{J_{s_{eq}}(\bar{\rho})}{2} - \frac{1}{4j}\, \hat{z} \bullet \fint_C \left[\hat{s}' J_{s_{eq}}(\bar{\rho}') \times \nabla_\tau H_0^{(2)}(k|\bar{\rho} - \bar{\rho}'|)\right] dL' \qquad \bar{\rho} \text{ on } C \tag{4.5.14}$$

In most of the further development, the subscript *eq* is dropped and it is still understood that the currents are the electromagnetic equivalent currents. The MFIE expression can be written in a compact notation given by

$$- H_z^i(\bar{\rho}) = \mathcal{L}_1[J_S(\bar{\rho}')] \qquad \bar{\rho} \text{ on } C \qquad (4.5.15a)$$

$$\mathcal{L}_1 = \frac{1}{2} - \frac{1}{4j}\,\hat{z} \bullet \oint_C \left[d\bar{L}(\bar{\rho}') \times \nabla_\tau H_0^{(2)}(k|\bar{\rho} - \bar{\rho}'|)\right] \qquad (4.5.15b)$$

where

$\mathcal{L}_1$: MFIE integral operator for TE polarization.

4.5.2 EFIE – PERFECT CONDUCTOR

Similarly, an electric field integral equation in terms of the unknown equivalent electric current distribution can be obtained by enforcing the s component of the total electric field to be zero on contour C. At any point in region 1 outside scatterer contour C, the tangential total electric field or the s component of the total electric field is given by the sum of s components of incident and scattered electric fields. Thus

$$\bar{E}_\tau(\bar{\rho})\Big|_{\tan} = \hat{s} \bullet \bar{E}_\tau(\bar{\rho}) = 0 \qquad \bar{\rho} \text{ on } C \qquad (4.5.16a)$$

$$\hat{s} \bullet \bar{E}_\tau^i(\bar{\rho}) = - \hat{s} \bullet \bar{E}_\tau^s(\bar{\rho}) \qquad \bar{\rho} \text{ on } C \qquad (4.5.16b)$$

After substituting for the s component of the scattered electric field derived in expression (4.5.7c), the electric field integral equation for the arbitrary shaped, perfectly conducting scatterer with transverse electric excitation is obtained as

$$\hat{s} \bullet \bar{E}_\tau^i(\bar{\rho}) = \frac{\omega\mu}{4}\int_C (\hat{s} \bullet \hat{s}')J_S(\bar{\rho}')\, H_0^{(2)}(k|\bar{\rho} - \bar{\rho}'|)\, dL'$$

$$+ \frac{1}{4\omega\varepsilon}\frac{\partial}{\partial s}\int_C \left[\frac{\partial}{\partial s'} J_S(\bar{\rho}')\right] H_0^{(2)}(k|\bar{\rho} - \bar{\rho}'|)\, dL' \qquad (4.5.17)$$

This EFIE expression can also be written in a compact notation given by

$$E_s^i(\bar{\rho}) = \mathcal{L}_2[J_s(\bar{\rho}')] \qquad \bar{\rho} \text{ on } C \tag{4.5.18a}$$

$$\mathcal{L}_2 = \frac{\omega\mu}{4}\int_C (\hat{s} \bullet \hat{s}')\, H_0^{(2)}(k|\bar{\rho} - \bar{\rho}'|)\, dL' + \frac{1}{4\omega\varepsilon}\frac{\partial}{\partial s}\int_C H_0^{(2)}(k|\bar{\rho} - \bar{\rho}'|)\, dL' \left[\frac{\partial}{\partial s'}\right] \tag{4.5.18b}$$

$\mathcal{L}_2$: EFIE integral operator for TE polarization

Following the procedure discussed in Section 3.7, the magnetic field integral equation or the electric field integral equation can be solved numerically by applying the method of moments technique. In the following sections, specific two-dimensional electromagnetic boundary value problems with TE excitation are considered.

4.6 PERFECTLY CONDUCTING CIRCULAR CYLINDER

The magnetic field integral equation as given by expression (4.5.15), and the electric field integral equation as given by expression (4.5.18), were derived systematically in the previous section. They can now be applied to analyze the electromagnetic scattering and interaction due to a perfectly conducting circular cylinder excited by a TE polarized plane wave incidence. Since the structure is perfectly conducting, the analysis presented is applicable for a circular cross section, which can be either hollow or solid. The circular cylindrical geometry of radius a, as shown in Figure 4.8, has perfect symmetry with respect to the angular variable ϕ of a cylindrical coordinate system. Hence, the scattering and interaction analysis can be carried out by just assuming the angle of incidence to be zero. The tangential s-directed electric current induced along circular contour C varies with respect to the angular variable and, in fact, repeats itself every 2π radians. For the circular cylinder, the tangential s direction is same as the angular ϕ direction. Hence, as in Section 3.8, the distribution of electric current can be represented in terms of an infinite Fourier series and the corresponding current modal coefficients can be determined by enforcing orthogonality condition.

Referring to expression (4.5.7c), at any field point in free-space medium 1 outside circular scatterer contour C, the tangential s component of scattered electric field in terms of the unknown induced electric current is given by

$$\hat{s} \bullet \bar{E}^S_\tau(\bar{\rho},\omega) = - \frac{\omega\mu}{4} \int_C (\hat{s} \bullet \hat{s}')J_S(\bar{\rho}')\, H_0^{(2)}(k|\bar{\rho} - \bar{\rho}'|)\, dL'$$

$$- \frac{1}{4\omega\varepsilon} \frac{\partial}{\partial s} \int_C \left[\frac{\partial}{\partial s'} J_S(\bar{\rho}')\right] H_0^{(2)}(k|\bar{\rho} - \bar{\rho}'|)\, dL' \tag{4.6.1a}$$

The distribution of tangential *s*-directed electric current is now expanded in terms of the following Fourier series:

$$J_S(\rho'=a,\phi',\omega) = \sum_{\ell=-\infty}^{\infty} B_\ell\, e^{j\ell\phi'} \tag{4.6.1b}$$

B_ℓ : Fourier coefficient of the current expansion.

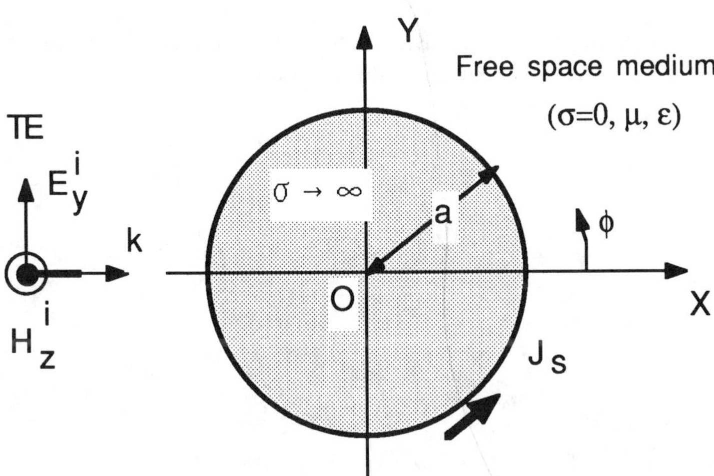

Figure 4.8 Perfectly conducting circular scatterer – TE excitation.

Referring to discussion in Section 3.7, the current expansion just selected represents a set of independent expansion terms, and each term is an entire domain harmonic expansion function. The current expansion (4.6.1b) is now substituted into expression (4.6.1a) to obtain

$$\hat{s} \bullet \bar{E}^{S}_{\tau}(\bar{\rho},\omega) = - \frac{\omega\mu}{4} \int_C \cos(\phi - \phi') \Big[\sum_{\ell=-\infty}^{\infty} B_\ell \, e^{j\ell\phi'} \Big] H_0^{(2)}(k|\bar{\rho} - \bar{\rho}'|) \, a \, d\phi'$$

$$- \frac{1}{4\omega\varepsilon\rho} \frac{\partial}{\partial\phi} \int_C \Big[\frac{\partial}{\partial\phi'} \sum_{\ell=-\infty}^{\infty} B_\ell \, e^{j\ell\phi'} \Big] H_0^{(2)}(k|\bar{\rho} - \bar{\rho}'|) \, d\phi' \qquad (4.6.2)$$

The zero-order Hankel function term in expression (4.6.2) represents outgoing cylindrical waves from the tangential s-directed electric current line source elements located on contour C. It is possible to rewrite all the outgoing cylindrical waves in terms of a set of equivalent cylindrical modes, which are all referred with respect to a single reference point (origin) by applying the coordinate shifting theorem. For all observations points outside the conducting scatterer, Appendix B,

$$H_0^{(2)}(k|\bar{\rho} - \bar{\rho}'|) = \sum_{n=-\infty}^{\infty} J_n(k\rho') H_n^{(2)}(k\rho) \, e^{jn(\phi - \phi')} \qquad (4.6.3)$$

After substituting expression (4.6.3), integral expression (4.6.2) for the tangential s-directed scattered electric field takes the following form:

$$\hat{s} \bullet \bar{E}^{S}_{\tau}(\bar{\rho},\omega) =$$

$$- \frac{\omega\mu}{4} \int_C \Big[\frac{1}{2} e^{j(\phi - \phi')} + \frac{1}{2} e^{-j(\phi - \phi')} \Big] \Big[\sum_{\ell=-\infty}^{\infty} B_\ell \, e^{j\ell\phi'} \Big]$$

$$\Big[\sum_{n=-\infty}^{\infty} J_n(ka) \, e^{-jn\phi'} \Big] \Big[H_n^{(2)}(k\rho) \, e^{jn\phi} \Big] \, a \, d\phi'$$

$$- \frac{1}{4\omega\varepsilon\rho} \frac{\partial}{\partial\phi} \int_C \Big[\frac{\partial}{\partial\phi'} \sum_{\ell=-\infty}^{\infty} B_\ell \, e^{j\ell\phi'} \Big]$$

$$\Big[\sum_{n=-\infty}^{\infty} J_n(ka) \, e^{-jn\phi'} \Big] \Big[H_n^{(2)}(k\rho) \, e^{jn\phi} \Big] \, d\phi'$$

$\bar{\rho}$ outside C (4.6.4)

Expression (4.6.4) can be further rearranged as

$$\hat{s} \bullet \bar{E}_{\tau}^{S}(\bar{\rho},\omega) = -\frac{\omega\mu}{8}\int_C \Big[\sum_{n=-\infty}^{\infty}\sum_{\ell=-\infty}^{\infty} B_\ell\, e^{j[\ell-(n+1)]\phi'}\Big] \Big[aJ_n(ka)H_n^{(2)}(k\rho)\, e^{j(n+1)\phi}\Big]\, d\phi'$$

$$-\frac{\omega\mu}{8}\int_C \Big[\sum_{n=-\infty}^{\infty}\sum_{\ell=-\infty}^{\infty} B_\ell\, e^{j[\ell-(n-1)]\phi'}\Big] \Big[aJ_n(ka)H_n^{(2)}(k\rho)\, e^{j(n-1)\phi}\Big]\, d\phi'$$

$$-\frac{1}{4\omega\varepsilon\rho}\int_C \Big[\sum_{n=-\infty}^{\infty}\sum_{\ell=-\infty}^{\infty} (jn)(j\ell)B_\ell\, e^{j[\ell-n)]\phi'}\Big] \Big[J_n(ka)H_n^{(2)}(k\rho)\, e^{jn\phi}\Big]\, d\phi'$$

$\bar{\rho}$ outside C (4.6.5)

In the double summation, expression (4.6.5), one infinite summation can be easily performed for all possible values of ℓ ranging from $-\infty$ to ∞ by utilizing the orthogonality property of the harmonic functions. For all values of $\ell = n + 1$ in the range from $-\infty$ to ∞, the first integral exists; and similarly, for all values of $\ell = n - 1$ in the range from $-\infty$ to ∞, the second integral exists. Also, the third integral exists for all values of $\ell = n$. Hence, the following simplification is obtained:

$$\hat{s} \bullet \bar{E}_{\tau}^{S}(\bar{\rho},\omega) = -\frac{\omega\mu}{8}\sum_{m=-\infty}^{\infty} B_m\Big[2\pi a J_{m-1}(ka)H_{m-1}^{(2)}(k\rho)e^{jm\phi}\Big]$$

$$-\frac{\omega\mu}{8}\sum_{m=-\infty}^{\infty} B_m\Big[2\pi a J_{m+1}(ka)H_{m+1}^{(2)}(k\rho)e^{jm\phi}\Big]$$

$$-\frac{1}{4\omega\varepsilon\rho}\sum_{m=-\infty}^{\infty} (-m^2)\, B_m\Big[2\pi J_m(ka)H_m^{(2)}(k\rho)e^{jm\phi}\Big] \qquad (4.6.6)$$

$$\hat{s} \cdot \bar{E}^{s}_{\tau}(\bar{\rho},\omega) = \sum_{m=-\infty}^{\infty} a_m \Big[J_{m-1}(ka)H^{(2)}_{m-1}(k\rho) + J_{m+1}(ka)H^{(2)}_{m+1}(k\rho) - \frac{2m^2}{k^2 a\rho} J_m(ka)H^{(2)}_m(k\rho) \Big] e^{jm\phi} \tag{4.6.7a}$$

a_m : scattered field modal coefficients

$$= -\frac{\pi\omega\mu a}{4} B_m \tag{4.6.7b}$$

To determine the scattered field modal coefficients, the total tangential electric field boundary condition, expression (4.5.16), on the perfectly conducting cylinder is enforced. Hence

$$-\hat{s} \cdot \bar{E}^{i}_{\tau}(\rho=a,\ \phi) = \sum_{m=-\infty}^{\infty} a_m \Big[J_{m-1}(ka)H^{(2)}_{m-1}(ka) + J_{m+1}(ka)H^{(2)}_{m+1}(ka) - 2\Big(\frac{m}{ka}\Big)^2 J_m(ka)H^{(2)}_m(ka) \Big] e^{jm\phi} \tag{4.6.8}$$

In this boundary relationship, the term on the lefthand side is the tangential s component of TE plane wave excitation and the term on the righthand side is the scattered electric field written in terms of a set of outgoing cylindrical modes. To determine the scattered field coefficients a_m, it is necessary to assess how much of the incident excitation actually couples to each outgoing scattered cylindrical mode. This can be accomplished based on the wave transformation, discussed in Section 3.8, where the given plane wave is expressed in terms of a set of cylindrical modes.

Figure 4.8 shows the polarization of TE plane wave excitation. The incident magnetic field is polarized in the z coordinate direction and propagating along the x coordinate axis. Correspondingly, the incident electric field is polarized in the y coordinate direction. Referring to Section 4.4, expressions (4.4.4) and (4.4.5), the TE polarized incident plane wave for zero angle of incidence is given by

$$H^{i}_{z}(\rho,\phi) = H_0 e^{-jk\rho\cos\phi} \tag{4.6.9a}$$

$$\bar{E}^{i}_{\tau}(\rho,\phi) = \hat{y} E_0 e^{-jk\rho\cos\phi} \tag{4.6.9b}$$

For the tangential s-directed plane wave propagating along the x coordinate direction, referring to representation (3.8.9), the following relationship is obtained by expressing (4.6.9b) in terms of an infinite series of cylindrical modes:

$$\hat{s} \bullet \vec{E}^{\,j}_{\tau}(\rho,\phi) = \hat{s} \bullet \hat{y} E_0\, e^{-jk\rho \cos\phi} \tag{4.6.10a}$$

$$= E_0 \cos\phi \sum_{m=-\infty}^{\infty} j^{-m} J_m(k\rho)\, e^{jm\phi} \tag{4.6.10b}$$

$$= \frac{E_0}{2} \sum_{m=-\infty}^{\infty} j^{-m} J_m(k\rho)\left[e^{j(m+1)\phi} + e^{j(m-1)\phi}\right] \tag{4.6.10c}$$

$$= \frac{E_0}{2} \sum_{m=-\infty}^{\infty} \left[j^{-(m-1)} J_{m-1}(k\rho) + j^{-(m+1)} J_{m+1}(k\rho)\right] e^{jm\phi} \tag{4.6.10d}$$

After substituting the plane wave representation, expression (4.6.10d), into the electric field boundary relationship (4.6.8), the scattered field coefficient can be conveniently determined by

$$-\frac{E_0}{2} \sum_{m=-\infty}^{\infty} \left[j^{-(m-1)} J_{m-1}(ka) + j^{-(m+1)} J_{m+1}(ka)\right] e^{jm\phi}$$

$$= \sum_{m=-\infty}^{\infty} a_m \Big[J_{m-1}(ka) H^{(2)}_{m-1}(ka) + J_{m+1}(ka) H^{(2)}_{m+1}(ka)$$

$$- 2\left(\frac{m}{ka}\right)^2 J_m(ka) H^{(2)}_m(ka)\Big]\, e^{jm\phi} \tag{4.6.11}$$

Referring to the discussion in Section 3.7, a set of independent weighting function terms is selected to test the preceding expression on both sides, where each weighting function term forms an entire domain harmonic function in the domain of scatterer contour C. Relationship (4.6.11) is now multiplied on both sides by the entire domain weighting function $\exp(-jn\phi)$ and integrated between the limits 0 to 2π with respect to ϕ. Therefore, the scattered electric field coefficients are obtained as

$$a_m = -\frac{E_0}{2}\frac{\text{NUM}(m)}{\text{DEN}(m)} \tag{4.6.12a}$$

where

$$\text{NUM}(m) = j^{-(m-1)} J_{m-1}(ka) + j^{-(m+1)} J_{m+1}(ka)$$

$$= j^{-(m-1)}\left[J_{m-1}(ka) - J_{m+1}(ka)\right]$$

$$= j2\, j^{-m} J_m'(ka) \tag{4.6.12b}$$

$$\text{DEN}(m) = J_{m-1}(ka)H_{m-1}^{(2)}(ka) + J_{m+1}(ka)H_{m+1}^{(2)}(ka)$$

$$- 2\left(\frac{m}{ka}\right)^2 J_m(ka)H_m^{(2)}(ka)$$

$$= \left[J_m'(ka) + \frac{m}{ka} J_m(ka)\right]\left[H_m^{(2)'}(ka) + \frac{m}{ka} H_m^{(2)}(ka)\right]$$

$$+ \left[- J_m'(ka) + \frac{m}{ka} J_m(ka)\right]\left[- H_m^{(2)'}(ka) + \frac{m}{ka} H_m^{(2)}(ka)\right]$$

$$- 2\left(\frac{m}{ka}\right)^2 J_m(ka)H_m^{(2)}(ka) \tag{4.6.12c}$$

$$= 2J_m'(ka)H_m^{(2)'}(ka) \tag{4.6.12d}$$

In this expression, the prime notation indicates the derivative of a corresponding Bessel or Hankel function. Further, to arrive at the simplification, various identities relating to the Bessel functions and their derivatives have been utilized, Appendix B. Thus, substituting expressions (4.6.12b) and (4.6.12d), the scattered field modal coefficients are given by

$$a_m = -j\frac{E_0}{2}\frac{j^{-m}}{H_m^{(2)'}(ka)} \tag{4.6.13}$$

Expression (4.6.7b) gives the relationship between the scattered field modal coefficients and the Fourier coefficients of the electric current expansion. Hence

$$B_m = \frac{j2H_0}{ka\pi} \frac{j^{-m}}{H_m^{(2)'}(ka)} \tag{4.6.14}$$

The tangential s-directed or angular component of the electric current density distribution, in amperes per meter, on contour C of a perfectly conducting circular cylinder, Figure 4.8, is now obtained by substituting the preceding Fourier coefficients into the current expansion (4.6.1b).

Hence, for the TE excitation, the angular component of the electric current is given by

$$J_\phi(a,\phi') = \frac{j2H_0}{\pi ka} \sum_{m=-\infty}^{\infty} \frac{j^{-m}}{H_m^{(2)'}(ka)} e^{jm\phi'} \tag{4.6.15}$$

a : radius of the circular conducting cylinder, in meters.

Expression (4.6.15) is now used to calculate the distribution of the electric current for various values of electrical radius ka. In fact, the derivative of the Hankel function of order m and the real argument are well documented and can be easily generated numerically. Figure 4.9 shows the distribution of angular-directed induced electric current density, in amperes per meter, on a perfectly conducting circular cylinder of radius $ka = 1$ plotted as a function of angular variable ϕ. The TE polarized incident magnetic field is assumed to be 1 ampere per meter and the distribution of electric current shown in the Figure 4.9 is normalized with respect to the amplitude of the incident magnetic field. Only the distribution in the angular range of 180° to 360° is shown; and in fact, it is symmetrical in the remaining angular range, specifically the distribution of induced electric current is symmetrical with respect to the incident angle. Referring to Figure 4.8, the angle $\phi = 180^o$ forms the illuminated side, and similarly, the angle $\phi = 0^o$ forms the shadow side of the circular cylindrical scatterer. The magnitude of the electric current drops off gradually from the illuminated side toward the shadow side. Further, the phase distribution of the electric current gradually changes from the illuminated to the shadow side. For the size $ka = 1$, the phase of the electric current changes by approximately 182.6^o.

Figure 4.10 shows the magnitude distribution of the electric current density for $ka =$ 2, 5, and 10. When the radius of a circular conducting scatterer is gradually increased, the electric current distribution in the illuminated side approaches the physical optics

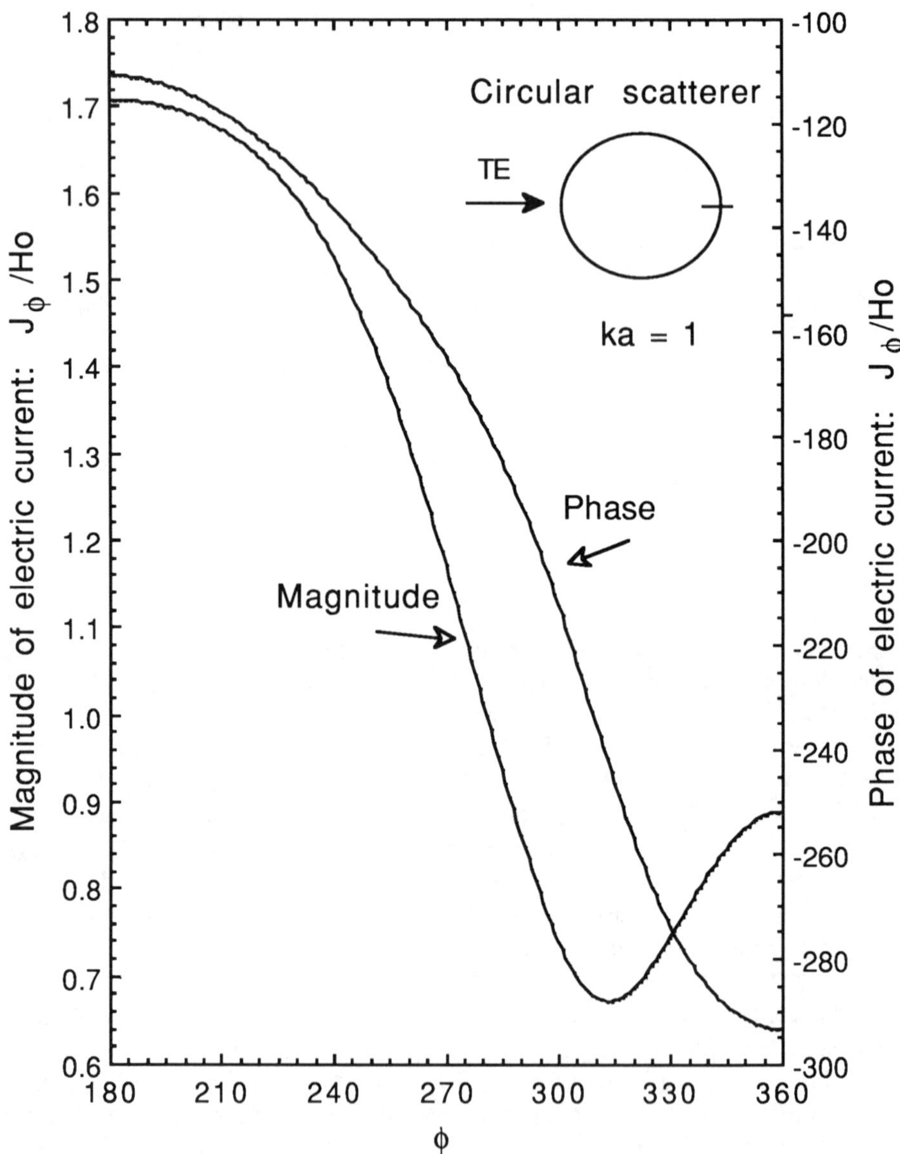

Figure 4.9 Magnitude and phase of the electric current distribution on a perfectly conducting circular scatterer – TE excitation.

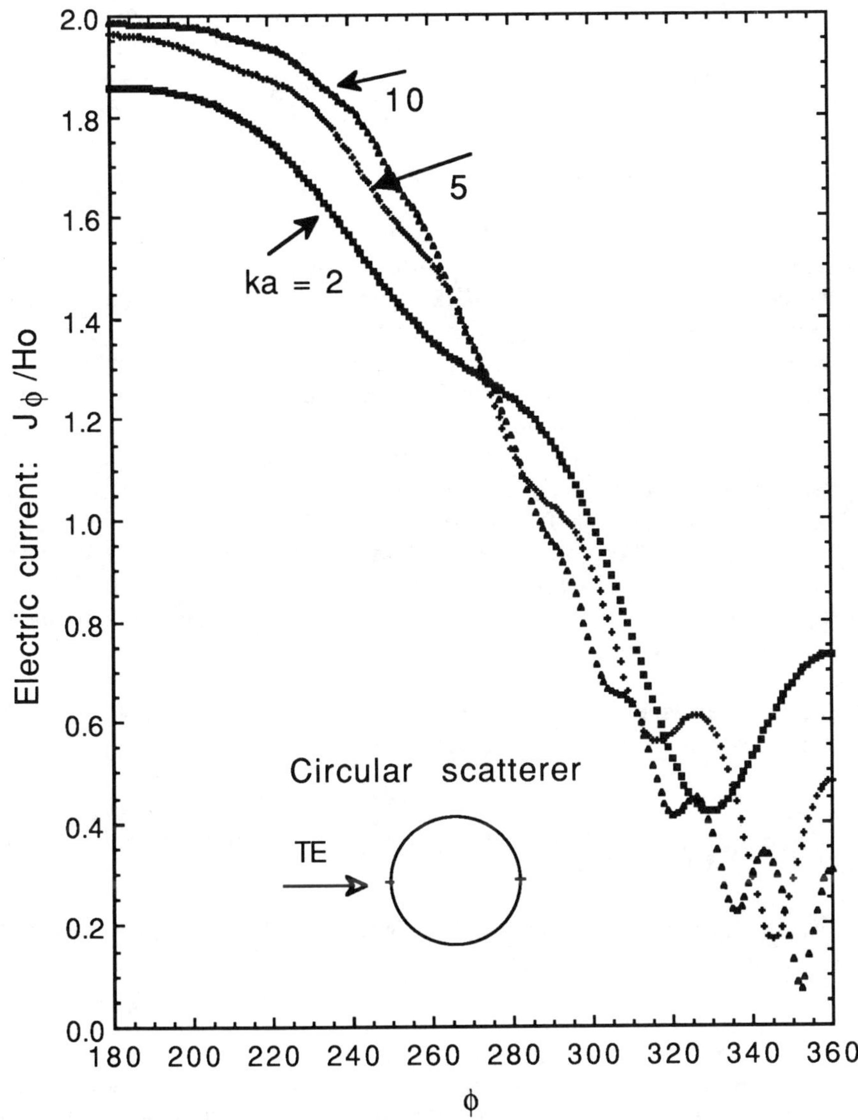

Figure 4.10 Magnitude of the electric current distribution for different radii on a perfectly conducting circular scatterer – TE excitation.

limit, and further, the induced current distribution exhibits a dominant creeping wave phenomena deep in the shadow region. In the numerical results shown, for the cases of $ka = 1$ and 2, only the first ten positive and negative cylindrical modes are included. Similarly, for the cases of $ka = 5$ and 10, only the first twenty positive and negative cylindrical modes are included in the calculation.

4.6.1 ALTERNATIVE FORMULATION

In the preceding section, a detailed analysis was presented for determining the distribution of electric current for TE excitation by solving rigorously the electric field integral equation. Instead of the electric field integral equation, it is possible to obtain the same solution using the magnetic field integral equation. In fact, an alternative and relatively simpler procedure can be followed for the study of scattering by a circular conducting cylinder. Referring to the study of electromagnetic scattering by a circular conducting cylinder for TM excitation, Section 3.8, with normal excitation, there are only the axial component of scattered electric field and the corresponding transverse components of scattered magnetic field distributions. Further, as shown in expression (3.8.4a), the axial z component of the scattered electric field can be written in terms of an infinite sum over a set of outgoing cylindrical wave modes. Hence, based on duality, it is possible to introduce a similar representation for the case of transverse electric normal excitation. In TE normal excitation, referring to Section 4.2.1, there are only the axial component of the scattered magnetic field and the corresponding transverse components of scattered electric field distributions. Hence, the axial z component of scattered magnetic field can be expressed in terms of an infinite sum over a set of outgoing cylindrical wave modes. Referring to expression (3.8.4a),

$$H_z^S(\rho,\phi,\omega) = H_0 \sum_{n=-\infty}^{\infty} T_n H_n^{(2)}(k\rho) e^{jn\phi} \tag{4.6.16}$$

where T_n are the scattered field modal coefficients. Using the source-free Ampere's law, expression (4.1.1b),

$$j\omega\varepsilon \bar{E}^S(\bar{r},\omega) = \nabla \times \hat{z} H_z^S(\rho,\phi,\omega) \tag{4.6.17a}$$

Expressing the electric and magnetic fields in terms of a cylindrical coordinate system,

$$j\omega\varepsilon[E_\rho^S \hat{\rho} + E_\phi^S \hat{\phi}] = \frac{1}{\rho}\frac{\partial}{\partial\phi} H_z^S \hat{\rho} - \frac{\partial}{\partial\rho} H_z^S \hat{\phi} \tag{4.6.17b}$$

For the circular conducting scatterer, the angular component of the total electric field is zero. Further, the angular component of the total electric field can be expressed in terms of the incident and scattered magnetic fields given by

$$E_\phi(\rho,\phi)\Big|_{\text{total}} = -\frac{1}{j\omega\varepsilon}\frac{\partial}{\partial\rho}\left[H_z^i(\rho,\phi) + H_z^s(\rho,\phi)\right] \tag{4.6.17c}$$

$$= -H_0\frac{1}{j\omega\varepsilon}\frac{\partial}{\partial\rho}\sum_{n=-\infty}^{\infty}\left[j^{-n}J_n(k\rho) + T_nH_n^{(2)}(k\rho)\right]e^{jn\phi} \tag{4.6.17d}$$

$$= H_0\frac{jk}{\omega\varepsilon}\sum_{n=-\infty}^{\infty}\left[j^{-n}J_n'(k\rho) + T_nH_n^{(2)'}(k\rho)\right]e^{jn\phi} \tag{4.6.17e}$$

On equating expression (4.6.17e) to zero and enforcing the orthogonality condition for the various cylindrical modes, the modal coefficients can be obtained. First multiplying the expression by exp$(-jm\phi)$ and integrating between the limits 0 to 2π, the various modal coefficients in the scattered field representation (4.6.16) are given by

$$T_n = -\frac{j^{-n}J_n'(ka)}{H_n^{(2)'}(ka)} \tag{4.6.18}$$

Further, the axial z component total magnetic field on the conducting circular cylinder gives directly the induced electric current in the angular direction. Hence

$$J_\phi(a,\phi',\omega) = -\left[H_z^i(a,\phi,\omega) + H_z^s(a,\phi,\omega)\right] \tag{4.6.19a}$$

$$= -H_0\sum_{n=-\infty}^{\infty}j^{-n}\left[J_n(ka) - \frac{J_n'(ka)}{H_n^{(2)'}(ka)}H_n^{(2)}(ka)\right]e^{jn\phi'} \tag{4.6.19b}$$

The Wronskian relationship, Appendix B, for the Bessel functions gives

$$J_n(ka)Y_n'(ka) - J_n'(ka)Y_n(ka) = \frac{2}{\pi ka} \tag{4.6.19c}$$

After substituting relationship (4.6.19c) into expression (4.6.19b), the expression for the angular component of induced electric current takes the form obtained in expression (4.6.15):

$$J_\phi(a,\phi',\omega) = \frac{j2H_0}{\pi ka} \sum_{n=-\infty}^{\infty} \frac{j^{-n}}{H_n^{(2)'}(ka)} e^{jn\phi'} \qquad (4.6.19d)$$

4.6.2 SCATTERED NEAR- AND FAR-FIELD DISTRIBUTIONS

The z component of scattered magnetic field distribution in the free-space medium is given by general expression (4.6.16). By substituting the field observation point (ρ, ϕ) and performing the summation over N cylindrical modes, the scattered magnetic field distribution can be easily calculated. In the numerical calculation, enough number of modes should be included to obtain convergence of the series. As pointed out earlier, generally the number of modes required for performing the summation is just more than the electrical distance of the observation point. Thus for n positive, the upper limit N is taken to be $|k\rho|$ plus an additional five to ten modes, depending on the numerical accuracy required. For very large distances from the scatterer, the numerical generation of the Bessel and Hankel functions is quite slow, and it is preferable to use their large argument approximation. After substituting the modal coefficients, the scattered magnetic field at any general point is given by expression (4.6.16):

$$H_z^S(\rho,\phi,\omega) = H_0 \sum_{n=-\infty}^{\infty} - \frac{j^{-n} J_n'(k\rho)}{H_n^{(2)'}(k\rho)} H_n^{(2)}(k\rho)\, e^{jn\phi} \qquad (4.6.20)$$

Figure 4.11a shows the distribution of the z component of total magnetic field in the neighborhood of a perfectly conducting circular cylinder, $ka = 5$. The numerical results are normalized with respect to the amplitude of the incident magnetic field H_0. The numerical data shown are obtained for fixed observation angles $\phi = 0, 30, 60, 90, 120, 150$, and 180; and the radial variable ρ is varied from a point on the circular conducting scatterer in increments of 0.005 to cover a radial distance of 0.25 meter which is about a quarter wavelength corresponding to the frequency of excitation of 300 MHz. For points on the surface of scatterer, in fact, this distribution of the axial component of the total magnetic field directly represents the angular component of the electric current distribution, as shown in Figure 4.10. Figure 4.11b shows the distribution of the angular component of the total electric field, calculated using expressions (4.6.17c). As

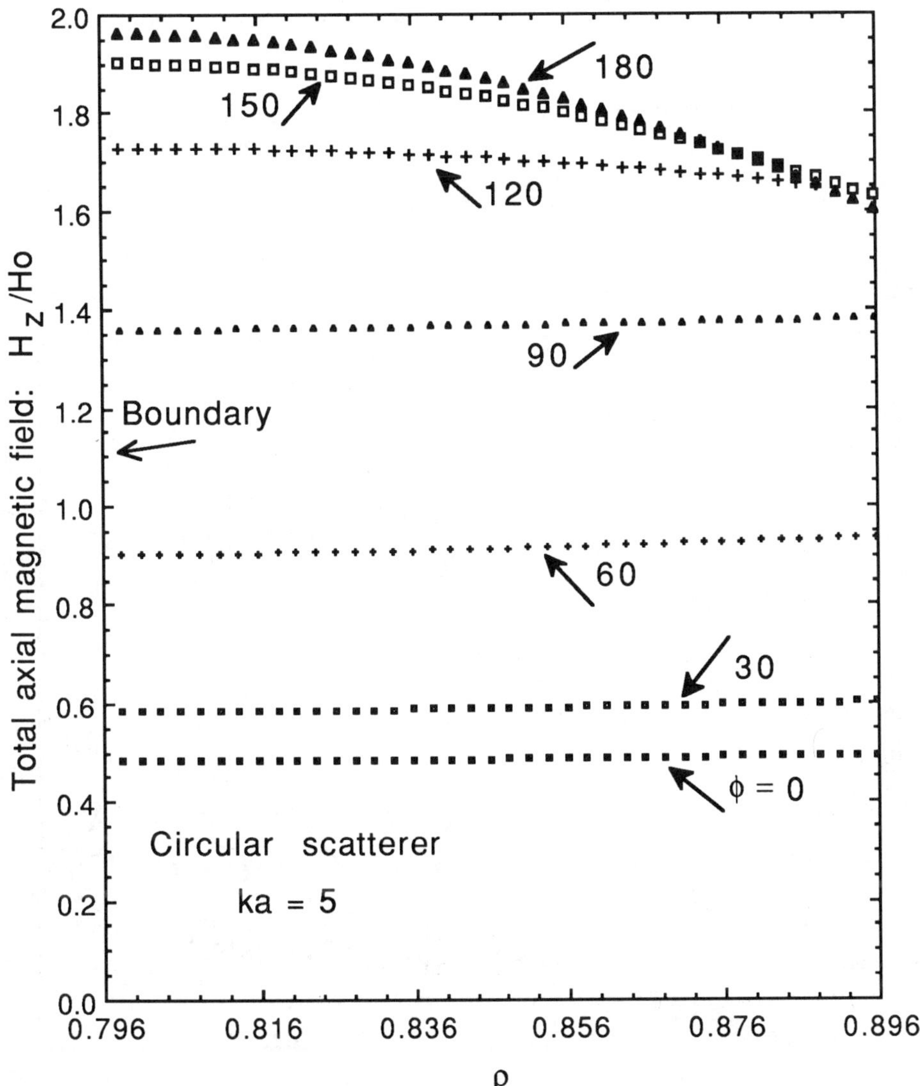

Figure 4.11a Magnitude of the total axial near magnetic field distribution on a perfectly conducting circular cylinder – TE excitation.

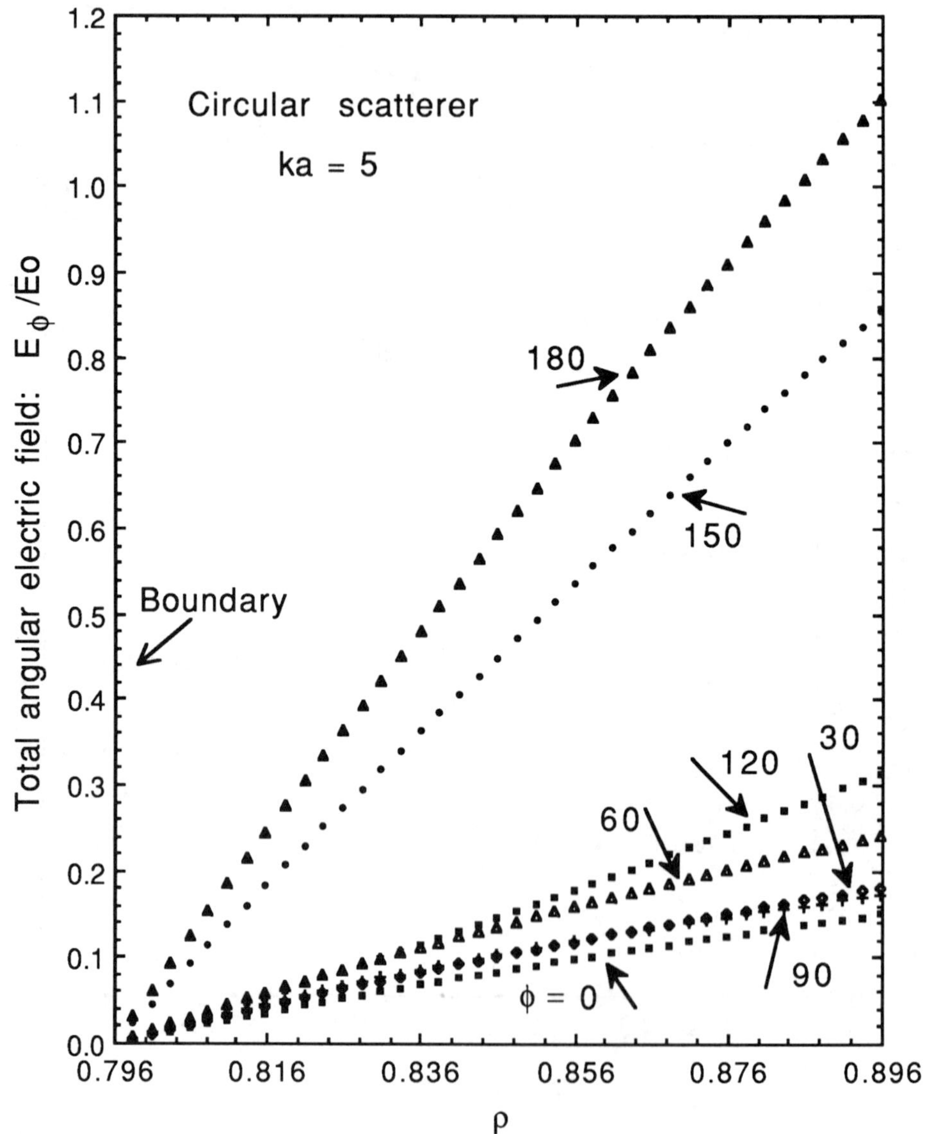

Figure 4.11b Magnitude of the total angular near electric field distribution on a perfectly conducting circular cylinder – TE excitation.

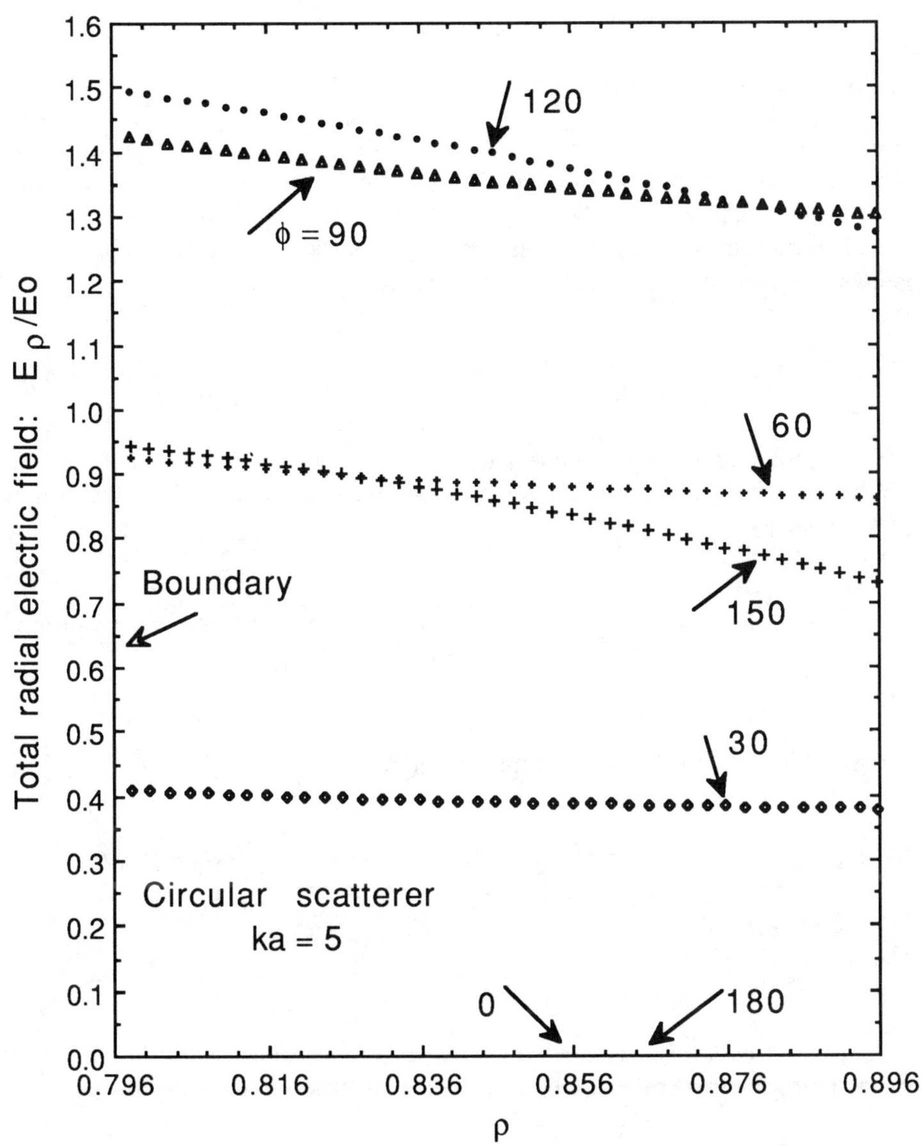

Figure 4.11c Magnitude of the total radial near electric field distribution on a perfectly conducting circular cylinder – TE excitation.

can be noted, the total angular component of the electric field is zero on the surface of scatterer and rises linearly in the vicinity of the circular conducting scatterer. Similarly, Figure 4.11c shows the radial component of the total electric field distribution.

Far-Field and Radar Cross Section

For very large distances from the scatterer, the numerical generation of the Bessel and Hankel functions is quite slow, and it is preferable to use their large argument approximation, Appendix B. The z component of the scattered field distribution in the far-field region is obtained by substituting the argument $|k\rho| \rightarrow \infty$. In the large argument approximation, the Hankel function can be written as

$$H_n^{(2)}(k\rho) \sim \sqrt{\frac{2j}{\pi k\rho}}\; j^n e^{-jk\rho} \tag{4.6.21}$$

After substituting the large argument approximation given in expression (4.6.21) into the infinite series solution (4.6.20), the z component of scattered field in the far-field region is obtained as

$$H_z^s(\rho,\phi,\omega) = H_0 \sqrt{\frac{2j}{\pi k\rho}}\; e^{-jk\rho} \sum_{n=-\infty}^{\infty} - \frac{J_n'(ka)}{H_n^{(2)'}(ka)}\; e^{jn\phi} \qquad k\rho \rightarrow \infty \tag{4.6.22}$$

and the magnitude of the far-field pattern is given by

$$\left| \frac{H_z^s(\phi,\omega)}{H_z^i(\phi,\omega)} \right| = \sqrt{\frac{2}{\pi k\rho}} \left| \sum_{n=-\infty}^{\infty} - \frac{J_n'(ka)}{H_n^{(2)'}(ka)}\; e^{jn\phi} \right| \qquad k\rho \rightarrow \infty \tag{4.6.23}$$

As defined earlier in Section 3.8 for the two-dimensional case, the differential scattering cross section or the bistatic cross section, in meters, is defined by

$$\mathrm{RCS}(\phi) = \lim_{\rho \rightarrow \infty} 2\pi\rho \left| \frac{H_z^s(\phi,\omega)}{H_z^i(\phi,\omega)} \right|^2 \tag{4.6.24}$$

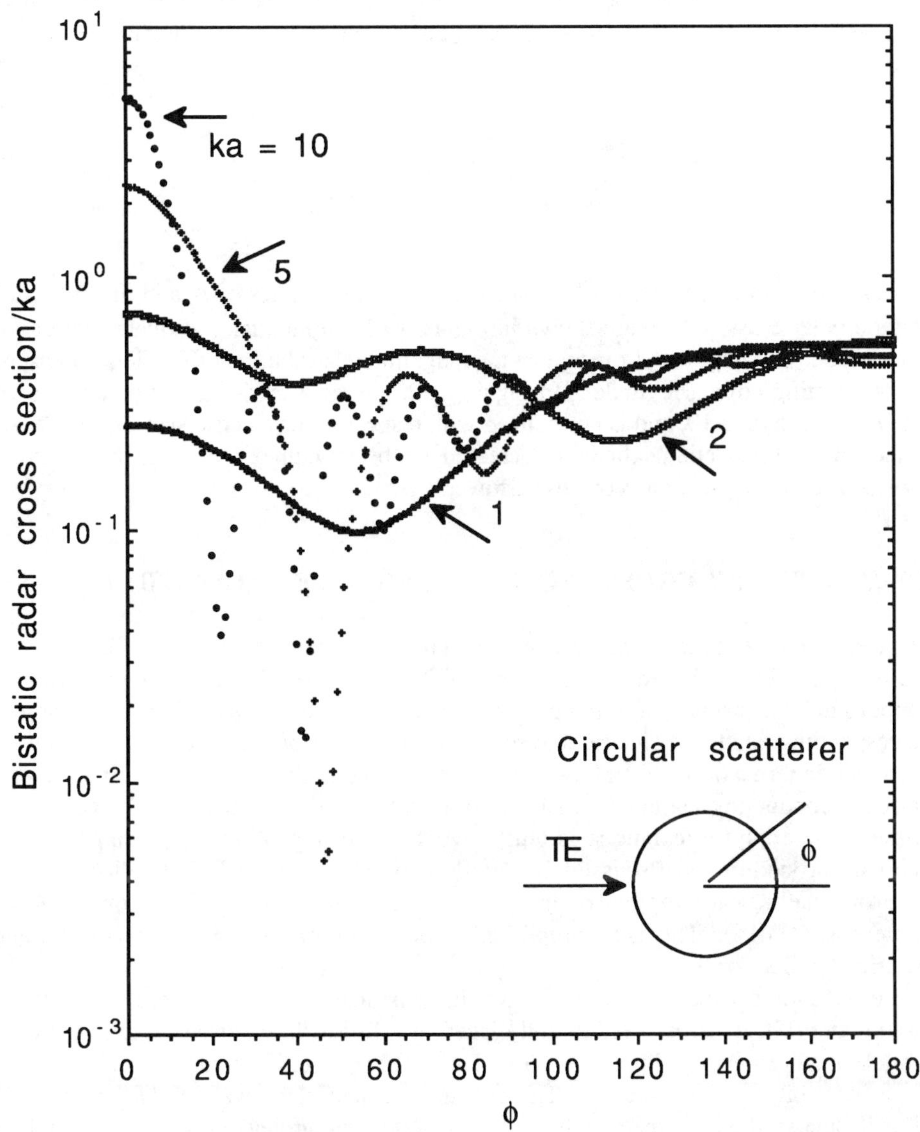

Figure 4.12 Bistatic radar cross section for various electrical sizes of a perfectly conducting circular cylinder – TE excitation.

For observation angle $\phi = \phi^i$, a forward scattering cross section is obtained. Similarly, for observation angle $\phi = \pi + \phi^i$, a back or monostatic scattering cross section is obtained. Hence, the bistatic radar cross section for the circular conducting cylinder is given by

$$\text{RCS}(\phi) = \frac{4}{k}\left| \sum_{n=-\infty}^{\infty} - \frac{J_n'(ka)}{H_n^{(2)'}(ka)} e^{jn\phi} \right|^2 \tag{4.6.25}$$

This expression is numerically evaluated for various values of ϕ, and the results of bistatic radar cross section are shown in Figure 4.12. In fact, taking square root of the data presented in Figure 4.12 results in the magnitude distribution of far-field pattern of the conducting circular cylinder. In Figure 4.12, the radar cross section data is shown for ka = 1, 2, 5, and 10, and is normalized with respect to the electric size ka. As the ka of the circular conducting cylinder is increased, the beamwidth of the power pattern in the forwardscattering direction becomes narrower.

4.7 ARBITRARY CROSS SECTION – EFIE – TE EXCITATION

The electromagnetic scattering by a perfectly conducting circular cylinder is discussed in the previous section based on a rigorous analytical approach. The circular cylinder happens to be a special scattering geometry, where the induced electric current and the corresponding scattered electric and magnetic field distributions can be conveniently expanded in terms of the eigenfunction series expansion. In case of a two-dimensional, perfectly conducting scatterer with an arbitrary cross section having arbitrary edges and corners, the analysis technique is quite involved. In the following, an approximate analysis procedure is discussed in detail to obtain a numerical solution for the electromagnetic scattering by solving either the EFIE expression (4.5.17) or the MFIE expression (4.5.14). This is accomplished based on the method of moments technique described in Section 3.7.

Let us consider the geometry of a two-dimensional, perfectly conducting scatterer having an arbitrary cross section and placed in a linear, homogeneous, and isotropic lossless medium, shown in Figure 4.13. The scatterer is uniform along its axis having the same arbitrary cross section. The external incident plane wave field is transverse electric polarized with respect to the axis of scatterer and propagates with its direction of propagation normal to the z axis. The angle ϕ^i represents the incident angle that the direction of propagation makes with the x coordinate axis. As discussed earlier, due to the normal excitation, the various field quantities are independent of the z coordinate variation. Thus, referring to Table 4.5, there are only the z component of scattered

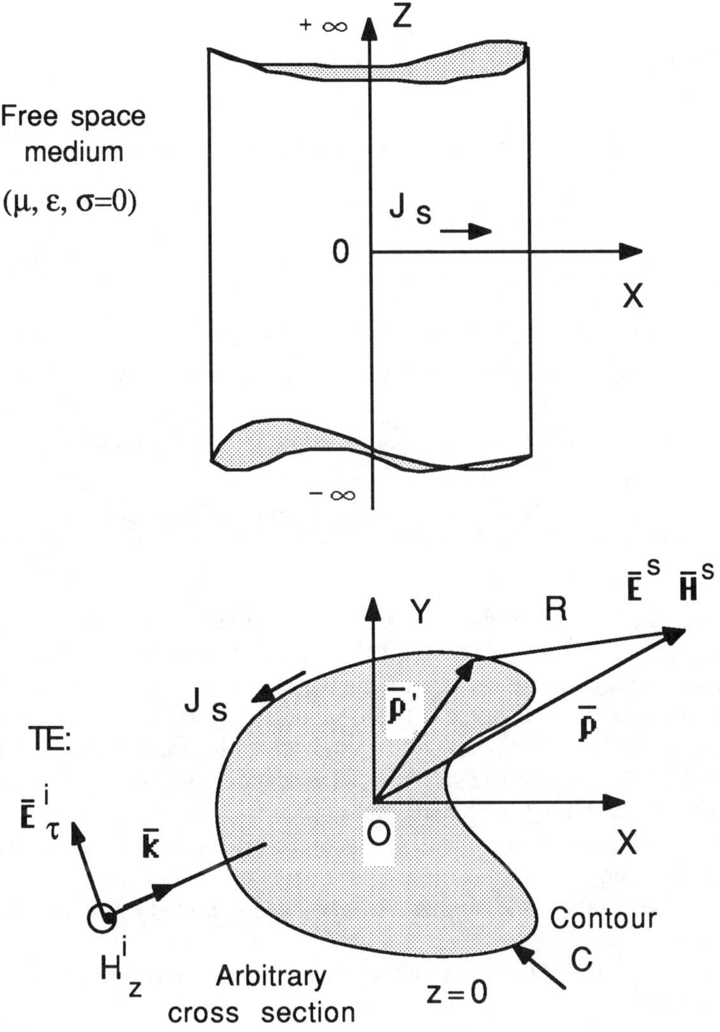

Figure 4.13 Geometry of an arbitrary shaped cross section, two-dimensional perfectly conducting scatterer – TE excitation.

magnetic field and the x, y components of transverse scattered electric field. Referring to the Figure 4.13, the TE to z polarized incident plane wave magnetic field is given by

$$H_z^i(\bar{\rho},\omega) = H_0\, e^{-j\bar{k}\bullet\bar{\rho}} \tag{4.7.1}$$

Hence, for a plane wave propagating normal to the axis of scatterer

$$H_z^i(\rho,\phi) = H_0\, e^{-jk\rho\,\cos(\phi - \phi^i)} \tag{4.7.2}$$

H_0 : amplitude of incident plane wave magnetic field, in amperes per meter;

ϕ^i : arbitrary angle of incidence of the TE plane wave field.

The corresponding transverse component of the incident electric field is given by

$$\bar{E}_\tau^i(\rho,\phi) = [-\sin\phi^i\hat{x} + \cos\phi^i\hat{y}]\, E_0\, e^{-jk\rho\,\cos(\phi - \phi^i)} \tag{4.7.3}$$

As discussed in Section 4.3, there is only the transverse component of induced electric currents on the contour C (designated by the primed coordinates) that encloses the arbitrary cross section of a perfectly conducting scatterer. Now, the electromagnetic principle is invoked to replace the original induced electric currents on the perfectly conducting scatterer by the corresponding equivalent electric current distribution on a virtual contour. The geometrical shape of the virtual contour is such that it encloses an identical cross-sectional region, having the same properties as that of the external free-space medium. Further, the equivalent electric currents produce the same scattered electric and magnetic fields of the original problem in the region outside the scatterer and also are subjected to the same original electromagnetic boundary conditions along the virtual contour.

In the following, the EFIE discussed in Section 4.5.2 is analyzed in detail based on the method of moments. Let

$\bar{J}_S(\bar{\rho}')$: unknown surface electric current distribution along contour C;

$H_z^S(\bar{\rho})$: scattered magnetic field component;

$\bar{E}_\tau^S(\bar{\rho})$: scattered electric field component.

Now referring to expression (4.5.6a), the tangential s component of scattered electric field is given by

$$\hat{s} \bullet \bar{E}^{S}_{\tau}(\bar{\rho}) = - j\omega\, \hat{s} \bullet \bar{A}_{\tau}(\bar{\rho}) - \frac{\partial}{\partial s}\Phi(\bar{\rho}) \tag{4.7.4a}$$

where the magnetic vector potential and the electric scalar potential in terms of the unknown tangential-directed surface electric current distribution have the form

$$\bar{A}_{\tau}(\bar{\rho}) = \frac{\mu}{4j}\int_{C} \hat{s}' J_S(\bar{\rho}')\, H_0^{(2)}(k|\bar{\rho} - \bar{\rho}'|)\, dL' \tag{4.7.4b}$$

$$\Phi(\bar{\rho}) = \frac{1}{4\omega\varepsilon}\int_{C} \left[\frac{\partial}{\partial s'} J_S(\bar{\rho}')\right] H_0^{(2)}(k|\bar{\rho} - \bar{\rho}'|)\, dL' \tag{4.7.4c}$$

On enforcing the electric field boundary condition on contour C of the perfectly conducting scatterer, expression (4.7.4a) yields the integro-differential equation

$$\hat{s} \bullet \bar{E}^{j}_{\tau}(\bar{\rho}) = j\omega\, \hat{s} \bullet \bar{A}_{\tau}(\bar{\rho}) + \frac{\partial}{\partial s}\Phi(\bar{\rho}) \qquad \bar{\rho} \text{ on } C \tag{4.7.5}$$

To solve electric field integral equation (4.7.5), the method of moments numerical solution, as described in Section 3.7, is applied. This is accomplished by a suitable selection of a set of independent current expansion functions and a corresponding set of independent weighting functions. A number of modeling considerations should be understood before a suitable choice can be made to represent the unknown electric current distribution in terms of a set of current expansion functions and, similarly, on the suitable choice of a set of weighting function to test the EFIE on its both sides. Various numerical aspects are discussed in the following.

4.7.1 EXPANSION FUNCTIONS

For convenience contour C is stretched out and redrawn with a x' variable. On contour C, the unknown tangential-directed electric current distribution is to be determined. Initially, if one has the knowledge of the type of distribution of electric current, then it is easier to make a choice on the approximate representation by a suitable sampling of the distribution function. Depending upon the electrical size of the conducting scatterer and its arbitrary cross section having arbitrary edges and corners, the electric current may have a well-behaved, continuous distribution with current nulls or even a rapid phase change in the distribution.

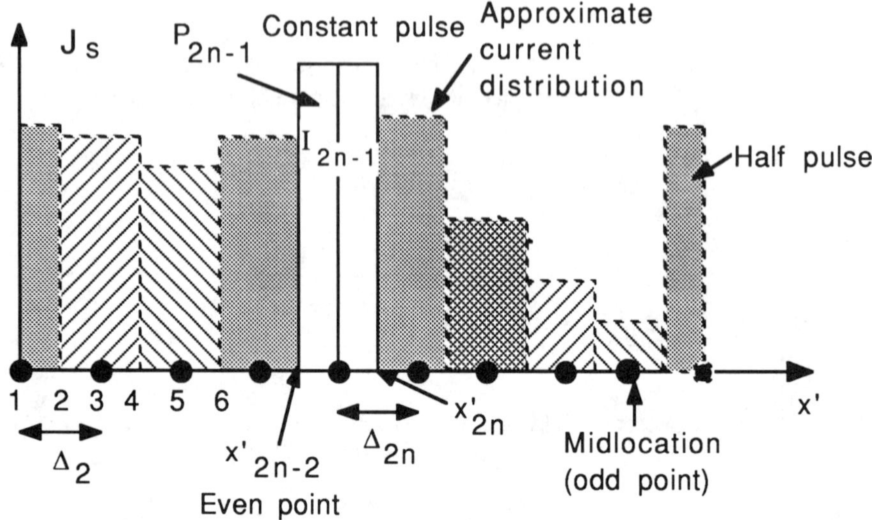

Figure 4.14a Pulse current expansion functions for the closed contour – TE case.

Similar to the TM case, a number of considerations should be given to the choice and arrangement of the current expansion functions. Further, it is necessary to draw a clear distinction between the closed contour and the open contour. The electric current modeling and the choice of expansion functions for the closed contour, as in the case of closed-type conducting scatterer, is simple and straightforward. In fact, the distribution of the tangential electric current is continuous along the closed contour, even for the scatterer with arbitrary edges and sharp corners. As shown in Figure 4.14a, for the case of a closed contour C, the distribution of the unknown electric current can be sampled at discrete spatial points and represented in terms of a set of independent pulse expansion functions. The various pulse functions are defined between two consecutive even points. As an example, the piecewise pulse function P_{2n-1} is located at the odd point x'_{2n-1} and is defined between the two consecutive even points given by

$$x'_{2n} = x'_{2n-1} + \frac{\Delta_{2n}}{2} \qquad (4.7.6a)$$

$$x'_{2n-2} = x'_{2n-1} - \frac{\Delta_{2n-2}}{2} \tag{4.7.6b}$$

where the length of straight line segments are given by

$$\Delta_{2n} = x'_{2n+1} - x'_{2n-1} \tag{4.7.6c}$$

$$\Delta_{2n-2} = x'_{2n-1} - x'_{2n-3} \tag{4.7.6d}$$

For the closed contour, the first point is same as the last sampling point. It is should be noted that a pulse current expansion term spans two halves of adjacent straight line segments. Suppose contour C of scatterer is not smooth, but has sharp corners, then the unknown current expansion pulse is arranged such that half the pulse spans along one side of the corner and the remaining half the pulse spans on the other side of corner. Referring to the Figure 4.14a, the unknown complex current coefficients I_{2n-1} is constant within the piecewise pulse P_{2n-1} and represents approximately the magnitude as well as the phase of unknown current distribution within that specific pulse region. Thus

$$J_s(x') \approx \sum_{n=1}^{N} I_{2n-1} P_{2n-1}(x') \tag{4.7.7a}$$

where

$$P_{2n-1}(x') = 1 \qquad \text{for } x'_{2n-2} \leq x' \leq x'_{2n}$$

$$= 0 \qquad \text{otherwise} \tag{4.7.7b}$$

Further, for the case of perfectly conducting open scatterer, such as a thin-strip scatterer, the tangential composite electric current goes to zero at the ends of the open geometry. For such a case, referring to Figure 4.14b, it is convenient to model the open ends by a half-pulse expansion terms having zero amplitude.

This arrangement of the current expansion pulses is quite useful to treat numerically derivative of the current distribution whether the geometry of scatterer is closed or open. In fact, the pulse expansion functions are quite easy to implement and ideal whenever complicated field distributions, such as the current distributions at geometrical edges and corners of a scatterer are to be simulated. Other types of higher order expansion terms consisting of piecewise triangle or piecewise sine or cosine functions can also be utilized to improve representation of the unknown electric current distribution. As discussed earlier in the method of moments numerical procedure, Section 3.7, equality of the integral equation should be enforced on both sides at every point in the domain of validity of the integral equation. This is accomplished by testing through the scalar product in the domain of validity, based on expression (3.7.1), on both sides of the integral equation with respect to a convenient weighting function.

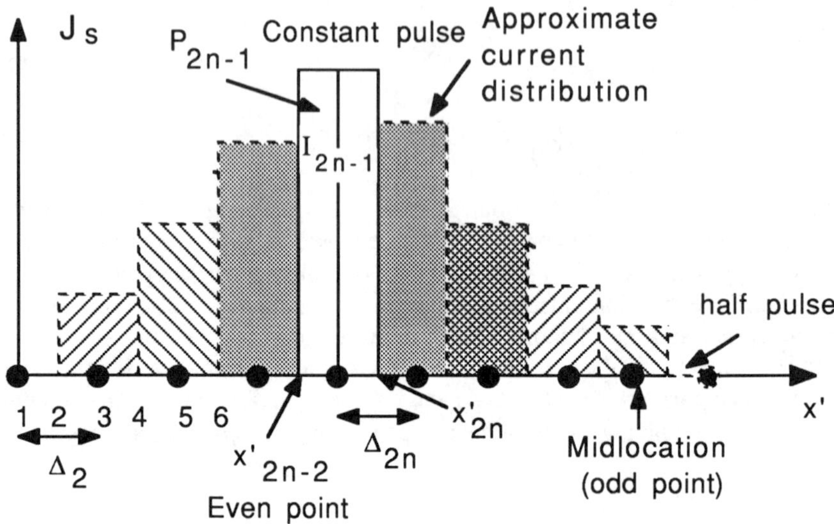

Figure 4.14b Pulse current expansion functions for the open contour – TE case.

In the numerical analysis presented in the following, the piecewise pulse expansion terms discussed earlier are directly adopted as weighting functions for the scalar product to reduce the integral equation to its functional form. In fact, the weighting functions with the corresponding variable m are arranged to coincide at the same locations as the current expansion terms.

4.8 REDUCTION TO MATRIX EQUATION

To reduce integral equation (4.7.5) to its equivalent functional form, the integral equation is first tested on both sides using the pulse type of weighting functions. If contour C of the scatterer is completely smooth with no geometrical discontinuities, then it is possible to pick arbitrary sampling regions along contour C with weighting functions located at points at $\bar{\rho}_{2m-1}$, $m = 1, 2, 3, \ldots, M$. The pulse weighting functions developed in Figure 4.14a can now be set up on contour C with respect to the various piecewise sampling regions. On the other hand, if contour C of the scatterer is not smooth, but has sharp corners, as indicated in Figure 4.15a, care should be taken in properly positioning the weighting function with respect to the geometrical discontinuity. From the numerical modeling point of view, specific weighting function close to the wedge- type corner is always properly arranged on both sides of the corner so that it *encloses* the geometrical discontinuity point.

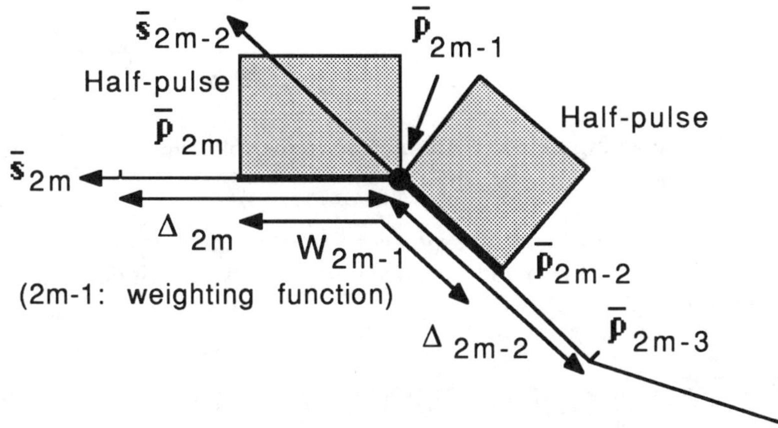

Figure 4.15a Positioning of the weighting function near discontinuities.

Referring to Figure 4.15a, the corner point of geometry is spanned by the weighting function on either side by half a cell. The piecewise weighting pulse W_{2m-1} is located such that it spans the odd point $\bar{\rho}_{2m-1}$, $m = 1, 2, 3, \ldots, M$, and is defined between the two consecutive even points given by

$$\bar{\rho}_{2m} = \bar{\rho}_{2m-1} + \frac{\Delta_{2m}}{2}\hat{s}_{2m} \tag{4.8.1a}$$

$$\bar{\rho}_{2m-2} = \bar{\rho}_{2m-1} - \frac{\Delta_{2m-2}}{2}\hat{s}_{2m-2} \tag{4.8.1b}$$

where

$\hat{s}_{2m-2}$: tangential unit vector along the length of pulse segment Δ_{2m-2};
$\hat{s}_{2m}$: tangential unit vector along the length of pulse segment Δ_{2m}.

$$\Delta_{2m} = \left|\bar{\rho}_{2m+1} - \bar{\rho}_{2m-1}\right| \tag{4.8.1c}$$

$$\Delta_{2m-2} = \left|\bar{\rho}_{2m-1} - \bar{\rho}_{2m-3}\right| \tag{4.8.1d}$$

Then the weighting function in terms of the piecewise pulse terms is given by

$$W_{2m-1}(\bar{\rho}) = 1 \qquad \text{for } \bar{\rho}_{2m-2} \le \bar{\rho} \le \bar{\rho}_{2m}$$

$$= 0 \qquad \text{otherwise} \tag{4.8.2}$$

$$m = 1, 2, 3, \ldots, M$$

These pulse weighting functions are utilized in the following to test electric field integral equation (4.7.5).

Pulse Testing EFIE integral expression (4.7.5) is now multiplied on both sides by weighting function (4.8.2) and integrated along scatterer contour C to yield the following relationship:

$$\hat{s} \bullet \bar{E}^{j}_{\tau}(\bar{\rho}) = j\omega\, \hat{s} \bullet \bar{A}_{\tau}(\bar{\rho}) + \frac{\partial}{\partial s}\Phi(\bar{\rho}) \qquad \bar{\rho} \text{ on } C \tag{4.8.3a}$$

$$\int_C \hat{s} \bullet \bar{E}^{j}_{\tau}(\bar{\rho}) W_{2m-1}(\bar{\rho})\, dL(\bar{\rho}) = \int_C j\omega\, \hat{s} \bullet \bar{A}_{\tau}(\bar{\rho}) W_{2m-1}(\bar{\rho})\, dL(\bar{\rho})$$

$$+ \int_C \left[\frac{\partial}{\partial s}\Phi(\bar{\rho})\right] W_{2m-1}(\bar{\rho})\, dL(\bar{\rho}) \tag{4.8.3b}$$

$$m = 1, 2, 3, \ldots, M$$

The expression (4.8.3b) can be simplified as

$$\int_{\bar{\rho}_{2m-2}}^{\bar{\rho}_{2m}} \hat{s} \bullet \bar{E}^{j}_{\tau}(\bar{\rho})\, dL(\bar{\rho}) = \int_{\bar{\rho}_{2m-2}}^{\bar{\rho}_{2m}} j\omega\, \hat{s} \bullet \bar{A}_{\tau}(\bar{\rho})\, dL(\bar{\rho})$$

$$+ \int_{\bar{\rho}_{2m-2}}^{\bar{\rho}_{2m}} \frac{\partial}{\partial s}\Phi(\bar{\rho})\, dL(\bar{\rho}) \tag{4.8.3c}$$

$$\bar{\rho} \text{ on } C, \text{ and } m = 1, 2, 3, \ldots, M$$

Further, these integrals can be split into two integrals corresponding to the two half straight line segments of the weighting pulse term. Numerical integration can be implemented to calculate the above weighted integrals. If the integrands are smooth varying functions, it can be assumed that the integrands in the lefthand and the righthand sides of expression (4.8.3c) do not vary over the limits of integration interval defined by the piecewise weighting pulse function. Thus, a suitable approximation can be introduced by calculating the integrand at the point $\bar{\rho}_{2m-1}$ and multiplying it by the corresponding half segment length of the weighting function. The second term on the righthand side can be replaced by its finite difference approximation to yield the following functional form of equation:

$$\hat{s}_{2m-2} \cdot \bar{E}^{i}_{\tau}(\bar{\rho}_{2m-1})\frac{\Delta_{2m-2}}{2} + \hat{s}_{2m} \cdot \bar{E}^{i}_{\tau}(\bar{\rho}_{2m-1})\frac{\Delta_{2m}}{2}$$

$$= j\omega\, \hat{s}_{2m-2} \cdot \bar{A}_{\tau}(\bar{\rho}_{2m-1})\frac{\Delta_{2m-2}}{2}$$

$$+ j\omega\, \hat{s}_{2m} \cdot \bar{A}_{\tau}(\bar{\rho}_{2m-1})\frac{\Delta_{2m}}{2}$$

$$- \Phi(\bar{\rho}_{2m-2}) + \Phi(\bar{\rho}_{2m}) \qquad (4.8.4)$$

$$m = 1, 2, 3, \ldots, M$$

Current Expansion Referring to Figure 4.15b, the unknown tangential electric current distribution $\bar{J}_S(\bar{\rho}')$ is now expanded in terms of a set of piecewise linearly independent pulse expansion functions. If contour C of scatterer is completely smooth with no geometrical discontinuities, then it is possible to pick arbitrary sampling regions along contour C with expansion functions located at points $\bar{\rho}_{2n-1}$, $n = 1, 2, 3, \ldots, N$. On the other hand, if contour C of the scatterer is not smooth, but has sharp corners, as indicated in Figure 4.15b, care should be taken in properly positioning the location of the expansion function with respect to the geometrical discontinuity. From the numerical modeling point of view, a specific expansion function close to the wedge-type corner is always properly arranged on both sides of the corner so that it *encloses* the geometrical discontinuity point. Referring to Figure 4.15b, the corner point of the geometry is spanned by the expansion function on either side by half a cell.

The piecewise expansion pulse P_{2n-1} is located such that it spans the odd point $\bar{\rho}_{2n-1}$, $n = 1, 2, 3, \ldots, N$, and is defined between the two consecutive even points given by

$$\bar{\rho}_{2n} = \bar{\rho}_{2n-1} + \frac{\Delta_{2n}}{2}\hat{s}_{2n} \qquad (4.8.5a)$$

$$\bar{\rho}_{2n-2} = \bar{\rho}_{2n-1} - \frac{\Delta_{2n-2}}{2}\hat{s}_{2n-2} \tag{4.8.5b}$$

where

$\hat{s}_{2n-2}$: the tangential unit vector along the length of pulse segment Δ_{2n-2};

$\hat{s}_{2n}$: the tangential unit vector along the length of pulse segment Δ_{2n}.

$$\Delta_{2n} = |\bar{\rho}_{2n+1} - \bar{\rho}_{2n-1}| \tag{4.8.5c}$$

$$\Delta_{2n-2} = |\bar{\rho}_{2n-1} - \bar{\rho}_{2n-3}| \tag{4.8.5d}$$

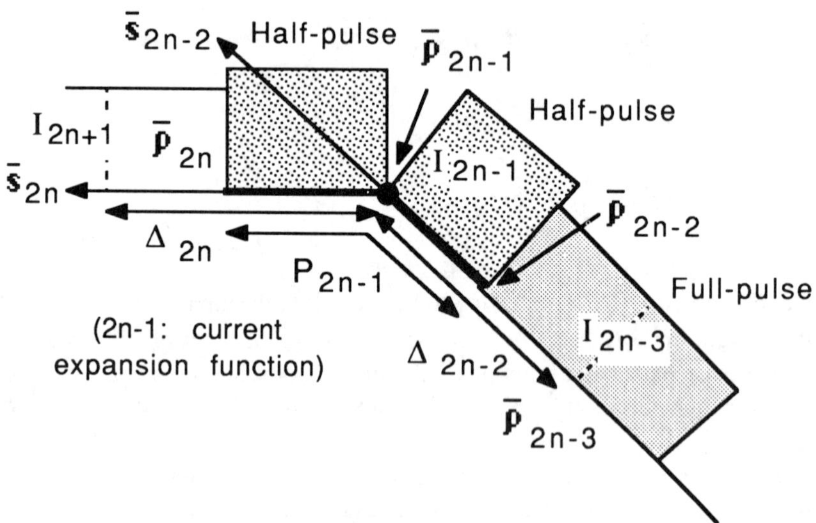

Figure 4.15b Positioning of the expansion function near discontinuities.

Then, the tangential-directed electric current distribution in terms of the piecewise pulse expansion terms is given by

$$J_S(\bar{\rho}') \approx \sum_{n=1}^{N} I_{2n-1} P_{2n-1}(\bar{\rho}') \tag{4.8.6a}$$

where

$$P_{2n-1}(\bar{\rho}') = 1 \qquad \text{for } \bar{\rho}_{2n-2} \le \bar{\rho}' \le \bar{\rho}_{2n}$$

$$= 0 \qquad \text{otherwise} \tag{4.8.6b}$$

$$n = 1, 2, 3, \ldots, N$$

and in expression (4.8.6a), I_{2n-1} are coefficients of the unknown electric current. After substituting for the magnetic vector potential and the electric scalar potential given in expressions (4.7.4b) and (4.7.4c), functional expression (4.8.4) reduces to

$$\left[\frac{\Delta_{2m-2}}{2}\hat{s}_{2m-2} + \frac{\Delta_{2m}}{2}\hat{s}_{2m}\right] \bullet \vec{E}^{i}_{\tau}(\bar{\rho}_{2m-1})$$

$$= \frac{\Delta_{2m-2}}{2}\frac{\omega\mu}{4}\int_C [\hat{s}_{2m-2} \bullet \hat{s}'] \, J_s(\bar{\rho}') \, H_0^{(2)}(k|\bar{\rho}_{2m-1} - \bar{\rho}'|) \, dL'$$

$$+ \frac{\Delta_{2m}}{2}\frac{\omega\mu}{4}\int_C [\hat{s}_{2m} \bullet \hat{s}'] \, J_s(\bar{\rho}') \, H_0^{(2)}(k|\bar{\rho}_{2m-1} - \bar{\rho}'|) \, dL'$$

$$- \frac{1}{4\omega\varepsilon}\int_C \left[\frac{\partial}{\partial s'} J_s(\bar{\rho}',\omega)\right] H_0^{(2)}(k|\bar{\rho}_{2m-2} - \bar{\rho}'|) \, dL'$$

$$+ \frac{1}{4\omega\varepsilon}\int_C \left[\frac{\partial}{\partial s'} J_s(\bar{\rho}',\omega)\right] H_0^{(2)}(k|\bar{\rho}_{2m} - \bar{\rho}'|) \, dL' \tag{4.8.7}$$

$$m = 1, 2, 3, \ldots, M$$

In functional equation (4.8.7), the distribution of electric current and its derivative are unknown which are to be determined. The functional form of the equation can be reduced to its equivalent matrix form by expanding the unknown distribution. For the unknown electric current distribution, the current expansion defined in expression (4.8.6a) can be substituted. Further, the derivative of electric current distribution, which represents the electric charge distribution, can also be expanded in terms of a set of charge expansion functions by suitably introducing the difference approximation based on the current expansion functions. This arrangement is shown in Figure 4.15c.

Hence, the derivative of the electric current is expanded as follows

$$\frac{\partial}{\partial s'} J_s(\bar{\rho}') \approx \sum_{n=1}^{N} \frac{I_{2n+1} - I_{2n-1}}{\Delta_{2n}} P'_{2n}(\bar{\rho}') \tag{4.8.8a}$$

where

$$P'_{2n}(\bar{\rho}') = 1 \qquad \text{for } \bar{\rho}_{2n-1} \le \bar{\rho}' \le \bar{\rho}_{2n+1}$$

$$= 0 \qquad \text{otherwise} \tag{4.8.8b}$$

$$n = 1, 2, 3, \ . \ . \ . \ , N$$

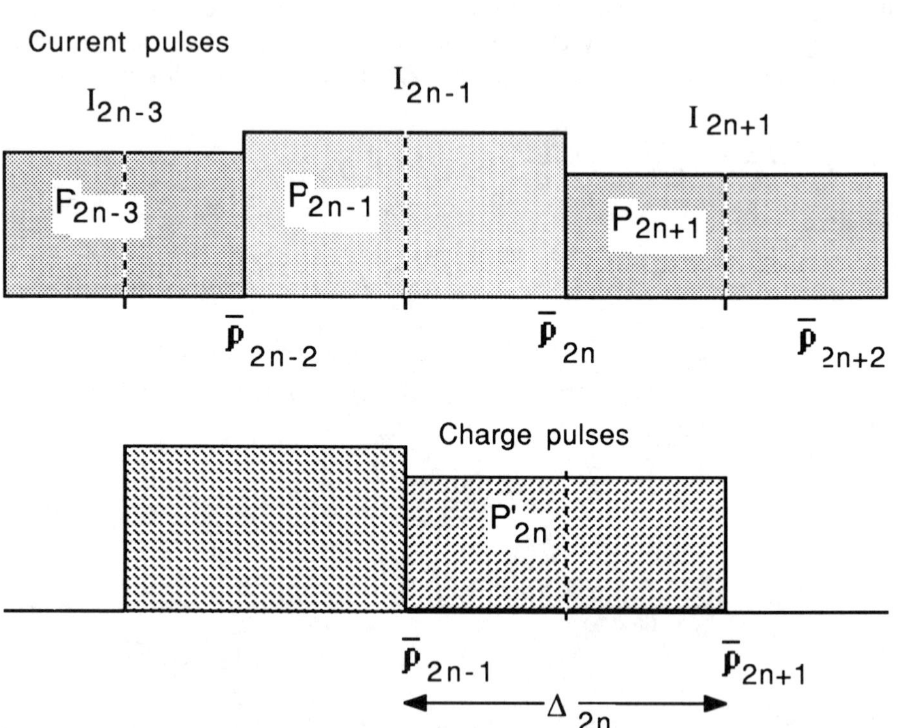

Figure 4.15c Arrangement of electric charge expansion pulses.

On substituting the pulse expansion functions (4.8.6a) and (4.8.8a) for the electric current and its derivative distributions into functional equation (4.8.7), the following matrix form of equation is obtained:

$$\left[\frac{\Delta_{2m-2}}{2}\hat{s}_{2m-2} + \frac{\Delta_{2m}}{2}\hat{s}_{2m}\right] \cdot \bar{E}^{i}_{\tau}(\bar{\rho}_{2m-1})$$

$$= \frac{\Delta_{2m-2}}{2}\,\frac{\omega\mu}{4}\int_C [\hat{s}_{2m-2}\cdot\hat{s}'][\sum_{n=1}^{N} I_{2n-1}P_{2n-1}(\bar{\rho}')]$$

$$H_0^{(2)}(k|\bar{\rho}_{2m-1}-\bar{\rho}'|)\,dL' + \frac{\Delta_{2m}}{2}\,\frac{\omega\mu}{4}\int_C [\hat{s}_{2m}\cdot\hat{s}']$$

$$[\sum_{n=1}^{N} I_{2n-1}P_{2n-1}(\bar{\rho}')]\,H_0^{(2)}(k|\bar{\rho}_{2m-1}-\bar{\rho}'|)\,dL'$$

$$-\frac{1}{4\omega\varepsilon}\int_C [\sum_{n=1}^{N}\frac{I_{2n+1}-I_{2n-1}}{\Delta_{2n}}\;P'_{2n}(\bar{\rho}')]H_0^{(2)}(k|\bar{\rho}_{2m-2}-\bar{\rho}'|)\,dL'$$

$$+\frac{1}{4\omega\varepsilon}\int_C [\sum_{n=1}^{N}\frac{I_{2n+1}-I_{2n-1}}{\Delta_{2n}}\;P'_{2n}(\bar{\rho}')]H_0^{(2)}(k|\bar{\rho}_{2m}-\bar{\rho}'|)\,dL'$$

$$m = 1, 2, 3, \ldots, M \qquad (4.8.9)$$

Since the original operator equation is linear, the summation and integration operations can be interchanged to obtain

$$[\frac{\Delta_{2m-2}}{2}\hat{s}_{2m-2} + \frac{\Delta_{2m}}{2}\hat{s}_{2m}]\cdot\bar{E}^{i}_{\tau}(\bar{\rho}_{2m-1})$$

$$= \sum_{n=1}^{N} I_{2n-1}\,\frac{\Delta_{2m-2}}{2}\,\frac{\omega\mu}{4}\int_C [\hat{s}_{2m-2}\cdot\hat{s}'][P_{2n-1}(\bar{\rho}')]$$

$$H_0^{(2)}(k|\bar{\rho}_{2m-1}-\bar{\rho}'|)\,dL' + \sum_{n=1}^{N} I_{2n-1}\,\frac{\Delta_{2m}}{2}\,\frac{\omega\mu}{4}\int_C [\hat{s}_{2m}\cdot\hat{s}']$$

$$[P_{2n-1}(\bar{\rho}')]H_0^{(2)}(k|\bar{\rho}_{2m-1}-\bar{\rho}'|)\,dL'$$

$$-\sum_{n=1}^{N}\frac{I_{2n+1}}{\Delta_{2n}}\,\frac{1}{4\omega\varepsilon}\int_C [P'_{2n}(\bar{\rho}')]H_0^{(2)}(k|\bar{\rho}_{2m-2}-\bar{\rho}'|)\,dL'$$

$$+\sum_{n=1}^{N}\frac{I_{2n-1}}{\Delta_{2n}}\,\frac{1}{4\omega\varepsilon}\int_C [P'_{2n}(\bar{\rho}')]H_0^{(2)}(k|\bar{\rho}_{2m-2}-\bar{\rho}'|)\,dL'$$

$$+ \sum_{n=1}^{N} \frac{I_{2n+1}}{\Delta_{2n}} \frac{1}{4\omega\varepsilon} \int_C [P'_{2n}(\bar{\rho}')] H_0^{(2)}(k|\bar{\rho}_{2m} - \bar{\rho}'|)\, dL'$$

$$- \sum_{n=1}^{N} \frac{I_{2n-1}}{\Delta_{2n}} \frac{1}{4\omega\varepsilon} \int_C [P'_{2n}(\bar{\rho}')] H_0^{(2)}(k|\bar{\rho}_{2m} - \bar{\rho}'|)\, dL'$$

$$m = 1, 2, 3, \ldots, M \qquad (4.8.10)$$

Referring to expression (4.8.6b) and (4.8.8b), the expansion pulse functions are valid over their segments only and zero outside. Hence

$$\left[\frac{\Delta_{2m-2}}{2} \hat{s}_{2m-2} + \frac{\Delta_{2m}}{2} \hat{s}_{2m}\right] \cdot E_\tau^j(\bar{\rho}_{2m-1})$$

$$= \sum_{n=1}^{N} I_{2n-1} \frac{\Delta_{2m-2}}{2} \frac{\omega\mu}{4} \int_{\bar{\rho}_{2n-2}}^{\bar{\rho}_{2n-1}} \cos(\Omega_{2m-2} - \Omega_{2n-2})$$

$$H_0^{(2)}(k|\bar{\rho}_{2m-1} - \bar{\rho}'|)\, dL' + \sum_{n=1}^{N} I_{2n-1} \frac{\Delta_{2m-2}}{2} \frac{\omega\mu}{4}$$

$$\int_{\bar{\rho}_{2n-1}}^{\bar{\rho}_{2n}} \cos(\Omega_{2m-2} - \Omega_{2n})\, H_0^{(2)}(k|\bar{\rho}_{2m-1} - \bar{\rho}'|)\, dL'$$

$$+ \sum_{n=1}^{N} I_{2n-1} \frac{\Delta_{2m}}{2} \frac{\omega\mu}{4} \int_{\bar{\rho}_{2n-2}}^{\bar{\rho}_{2n-1}} \cos(\Omega_{2m} - \Omega_{2n-2})$$

$$H_0^{(2)}(k|\bar{\rho}_{2m-1} - \bar{\rho}'|)\, dL' + \sum_{n=1}^{N} I_{2n-1} \frac{\Delta_{2m}}{2} \frac{\omega\mu}{4}$$

$$\int_{\bar{\rho}_{2n-1}}^{\bar{\rho}_{2n}} \cos(\Omega_{2m} - \Omega_{2n})\, H_0^{(2)}(k|\bar{\rho}_{2m-1} - \bar{\rho}'|)\, dL'$$

$$- \sum_{n=1}^{N} \frac{I_{2n-1}}{\Delta_{2n-2}} \frac{1}{4\omega\varepsilon} \int_{\bar{\rho}_{2n-3}}^{\bar{\rho}_{2n-1}} H_0^{(2)}(k|\bar{\rho}_{2m-2} - \bar{\rho}'|)\, dL'$$

$$+ \sum_{n=1}^{N} \frac{I_{2n-1}}{\Delta_{2n}} \frac{1}{4\omega\varepsilon} \int_{\bar{\rho}_{2n-1}}^{\bar{\rho}_{2n+1}} H_0^{(2)}(k|\bar{\rho}_{2m-2} - \bar{\rho}'|)\, dL'$$

$$+ \sum_{n=1}^{N} \frac{I_{2n-1}}{\Delta_{2n-2}} \frac{1}{4\omega\varepsilon} \int_{\bar{\rho}_{2n-3}}^{\bar{\rho}_{2n-1}} H_0^{(2)}(k|\bar{\rho}_{2m} - \bar{\rho}'|)\, dL'$$

$$- \sum_{n=1}^{N} \frac{I_{2n-1}}{\Delta_{2n}} \frac{1}{4\omega\varepsilon} \int_{\bar{\rho}_{2n-1}}^{\bar{\rho}_{2n+1}} H_0^{(2)}(k|\bar{\rho}_{2m} - \bar{\rho}'|)\, dL'$$

$$m = 1, 2, 3, \ldots, M \qquad (4.8.11)$$

where

Ω_{2m-2} : angle that the normal to the straight line field segment Δ_{2m-2} makes with the x coordinate axis;

Ω_{2m} : angle that the normal to the straight line field segment Δ_{2m} makes with the x coordinate axis;

Ω_{2n-2} : angle that the normal to the straight line source segment Δ_{2n-2} makes with the x coordinate axis;

Ω_{2n} : angle that the normal to the straight line source segment Δ_{2n} makes with the x coordinate axis.

Expression (4.8.11) forms a set of M linear simultaneous algebraic equations for the N unknown electric current coefficients. A direct numerical solution can be obtained by choosing $M = N$, the number of equations equal to the number of unknowns.

Expression (4.8.11) can also be rewritten in terms of a compact generalized matrix equation

$$[Z_{mn}][I_n] = [V_m] \tag{4.8.12a}$$

where

Z_{mn} : generalized impedance matrix element

I_n : element of unknown current column vector

V_m : element of excitation column vector

$$[Z_{mn}] = \begin{bmatrix} Z_{11} & Z_{12} & Z_{13} & \cdots & Z_{1n} & \cdots & Z_{1N} \\ Z_{21} & Z_{22} & Z_{23} & \cdots & Z_{2n} & \cdots & Z_{2N} \\ Z_{31} & Z_{32} & Z_{33} & \cdots & Z_{3n} & \cdots & Z_{3N} \\ \cdots & \cdots & \cdots & \cdots & \cdots & \cdots & \cdots \\ Z_{m1} & Z_{m2} & Z_{m3} & \cdots & Z_{mn} & \cdots & Z_{mN} \\ \cdots & \cdots & \cdots & \cdots & \cdots & \cdots & \cdots \\ \cdots & \cdots & \cdots & \cdots & \cdots & \cdots & \cdots \\ Z_{M1} & Z_{M2} & Z_{M3} & \cdots & Z_{Mn} & \cdots & Z_{MN} \end{bmatrix} \tag{4.8.12b}$$

where the elements of the generalized impedance matrix consist of sum of eight integral terms defined in expression (4.8.11). Each integral term should be carefully studied to see whether the observation point is outside the limits of integration or within the limits of integration. In Section 3.9.3, details concerning numerical approximation to the matrix terms is discussed. Further, in matrix expression (4.8.12a), the excitation term for the TE polarized plane wave propagating with an angle of incidence ϕ^i is given by

$$V_m = \left[\frac{\Delta_{2m-2}}{2}\hat{s}_{2m-2} + \frac{\Delta_{2m}}{2}\hat{s}_{2m}\right] \bullet \bar{E}^i_\tau(\bar{\rho}_{2m-1}) \tag{4.8.13a}$$

$$\bar{E}^i_\tau(\rho,\phi) = [-\sin\phi^i\hat{x} + \cos\phi^i\hat{y}]$$

$$E_0 e^{-jk\rho_{2m-1}\cos[\phi_{2m-1} - \phi^i]} \tag{4.8.13b}$$

On substituting for the tangential unit vectors, expression (4.8.13a) simplifies to

$$V_m = \left[\frac{\Delta_{2m-2}}{2}\cos(\Omega_{2m-2} - \phi^i) + \frac{\Delta_{2m}}{2}\cos(\Omega_{2m} - \phi^i)\right]$$

$$E_0\, e^{-jk\rho_{2m-1}\cos[\phi_{2m-1} - \phi^i]} \qquad (4.8.13c)$$

The unknown electric current distribution is obtained by inverting the generalized matrix equation (4.8.12a):

$$[I_n] = [Z_{mn}]^{-1}\,[V_m] \qquad (4.8.14)$$

4.8.1 SQUARE CONDUCTING CYLINDER

Based on the numerical solution discussed in the previous section, a number of canonical two-dimensional conducting geometries are analyzed to determine their electromagnetic scattering properties. Figure 3.18a shows the cross section of a two-dimensional perfectly conducting square cylinder located in a linear, homogeneous, and isotropic lossless medium. The side length of the square cylinder is selected as $ks = 2$ and excited by an external monochromatic plane wave propagating at an angle of incidence $\phi^i = 90^o$. For TE polarization, the incident magnetic field is polarized parallel to the z axis and the corresponding incident electric field is polarized parallel to the $-x$ axis. With normal excitation on one side, Figure 3.18a, the electric current induced on the square cylinder is symmetrical with respect to the y coordinate axis.

Figure 4.16a shows various locations of even and odd sampling points along with the arrangement of electric current pulse expansion functions for the square conducting scatterer. For the example shown, each side length of the square cylinder is divided into three equal pulse segments so that the total number of unknown electric current pulses is selected as $N = 12$. As discussed earlier, the various pulse expansion terms are arranged on the square cylinder with odd points representing the midlocations of the current pulses and the consecutive even points representing the limits of the pulse expansion terms. Further, there is *always a current sampling pulse* at the right angle corners (half pulse on either side) of the square conducting cylinder. Correspondingly, the total number of match points, $M = 12$, is also selected to coincide with the odd points of the various current expansion pulses.

A general numerical computer algorithm is now written to generate the geometry, the generalized impedance matrix, and the excitation column vector for the square conducting scatterer for TE normal excitation. According to expression (4.8.14), the impedance matrix is inverted numerically and multiplied by the excitation column vector to obtain the electric current coefficients. Figure 4.16b shows a plot of the distribution of electric current, for $N = M = 80$, along the circumference of the square conducting cylinder designated by the points *abcd*. Referring to Figure 3.18a, location *a* is at the center of illuminated region and location *d* is at the center of shadow region. Further, locations *b* and *c* correspond to the two corners on the square cylinder with location *b* on the illuminated side and location *c* on the shadow side. The distribution of the electric current is continuous at the two corners, *b* and *c*. The data shown in Figure 4.16b are normalized with respect to the amplitude of incident magnetic field. Along with the magnitude data, the phase distribution of the induced electric current is shown. Along the contour points *abcd* of the square conducting scatterer, following from the illuminated side towards the shadow side, the phase angle of the electric current has a continuous distribution and changes by approximately 219°.

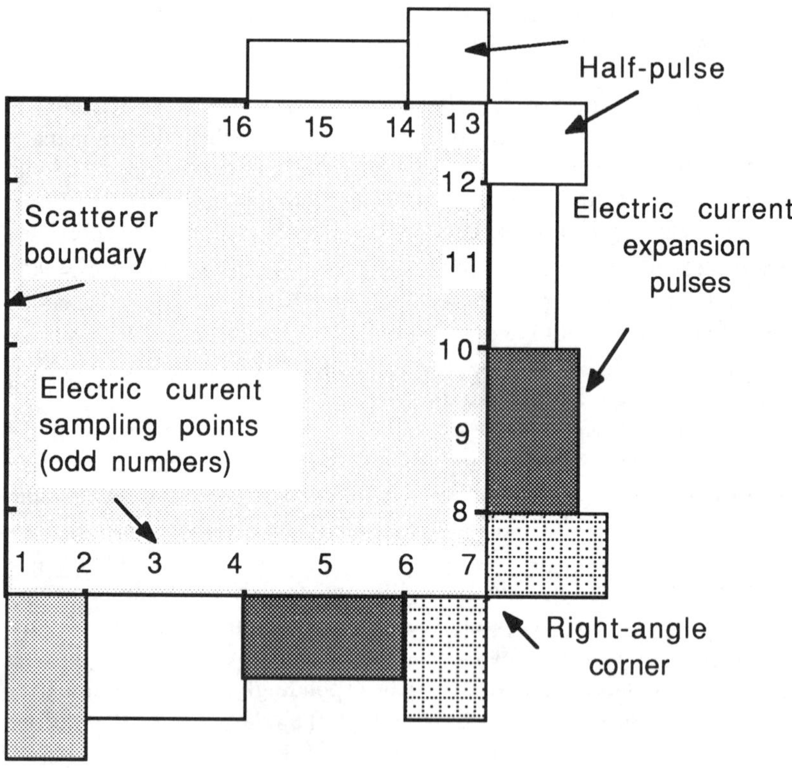

Figure 4.16a Arrangement of TE current expansion functions.

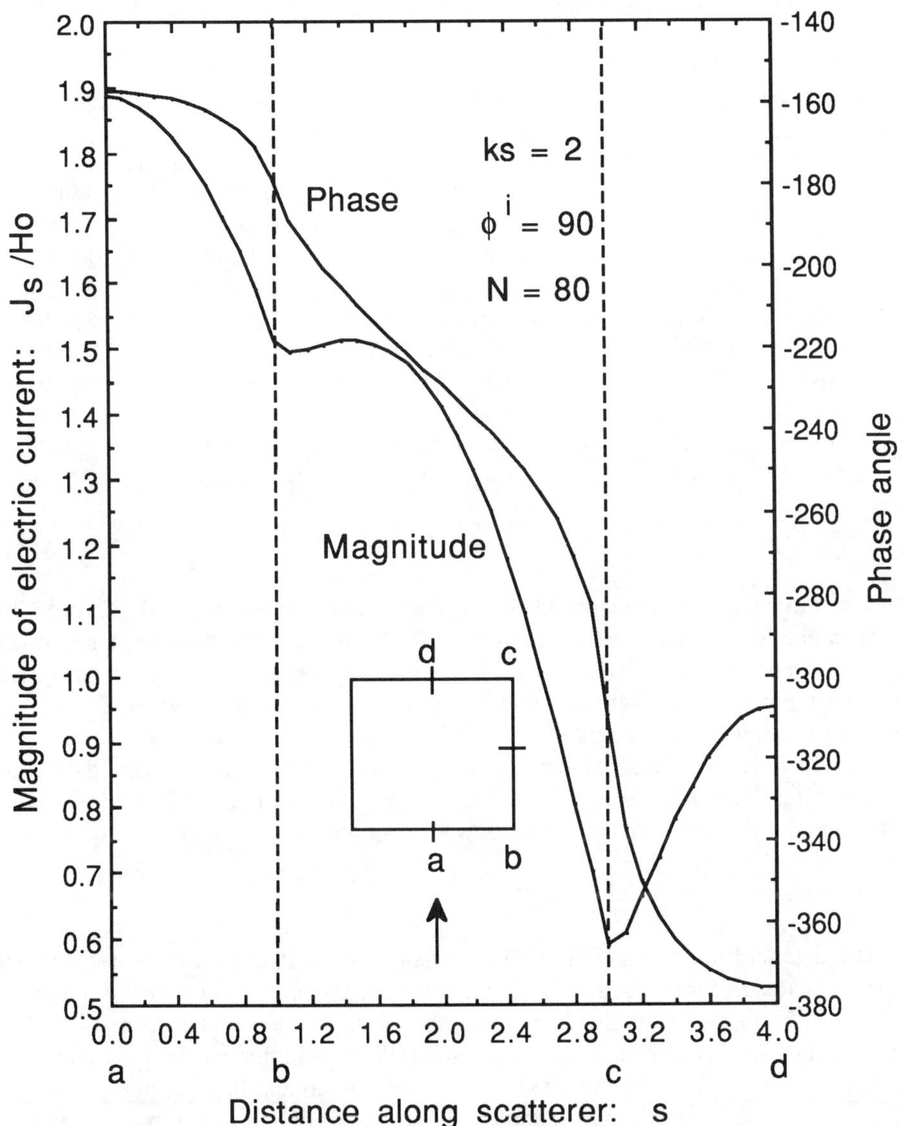

Figure 4.16b Magnitude and phase of the electric current distribution on a perfectly conducting square scatterer – TE excitation.

Table 4.9
Convergence data for a square conducting scatterer with TE excitation

N *matrix size*	*Current at illuminated point a*	*Current at shadow point d*	*forward RCS*	*Monostatic RCS*
8	2.008623	0.812016	0.38808	0.36472
16	1.895492	0.876276	0.49277	0.29487
24	1.885618	0.909602	0.52848	0.28012
32	1.883862	0.925679	0.54524	0.27426
40	1.883680	0.934873	0.55462	0.27126
48	1.883868	0.930704	0.56047	0.26948
56	1.884131	0.944678	0.56440	0.26834
64	1.884391	0.947539	0.56719	0.26755
72	1.884625	0.949681	0.56925	0.26697
80	1.884826	0.946395	0.57082	0.26654

As mentioned earlier, the induced electric current data represented in the Figure 4.16b has been obtained with a matrix size of $N = 80$. Such a large matrix size is not always required. Depending on the desired accuracy of the numerical solution, one has to look into convergence of the electric current data at various points on the scatterer. In Table 4.9, the variation of the electric current at illuminated point a and shadow point d are tabulated as function of matrix size. For just a two-decimal point accuracy, the electric current at the illuminated side point has converged even with a matrix size of $N = 32$. Relatively, the electric current at the shadow side point is still converging.

Near-Scattered Field Calculation

The calculation of near scattered magnetic and electric fields and the far-field distributions is discussed later. The electric current distribution in terms of the known current coefficients, as obtained in expression (4.8.14), is now substituted into field expression (4.7.4). The near scattered electric field distribution has only the transverse field components. To calculate the electric field component in a given direction, the following numerical procedure is adopted. First the location of a field point is selected, and then a unit vector is drawn, along which the component of the electric field is to be calculated. Along the unit vector two points are selected for efficently calculating the derivative of electric scalar potential term. Let

$\bar{\rho}$: location of field point with coordinates (ρ, ϕ);

$\hat{s}$: unit vector along the direction of electric field component;
Δ : incremental distance along unit vector enclosing the field point.

$$\bar{\rho}_+ = \bar{\rho} + \frac{\Delta}{2}\hat{s} \tag{4.8.15a}$$

$$\bar{\rho}_- = \bar{\rho} - \frac{\Delta}{2}\hat{s} \tag{4.8.15b}$$

Hence, the near scattered electric field distribution is given by

$$\begin{aligned}
-\hat{s} \bullet \bar{E}^S_\tau(\bar{\rho}) = & \sum_{n=1}^{N} I_{2n-1} \frac{\omega\mu}{4} \int_{\bar{\rho}_{2n-2}}^{\bar{\rho}_{2n-1}} [\hat{s} \bullet \hat{s}_{2n-2}]\, H_0^{(2)}(k|\bar{\rho} - \bar{\rho}'|)\, dL' \\
& + \sum_{n=1}^{N} I_{2n-1} \frac{\omega\mu}{4} \int_{\bar{\rho}_{2n-1}}^{\bar{\rho}_{2n}} [\hat{s} \bullet \hat{s}_{2n}]\, H_0^{(2)}(k|\bar{\rho} - \bar{\rho}'|)\, dL' \\
& - \sum_{n=1}^{N} \frac{I_{2n-1}}{\Delta_{2n-2}} \frac{1}{4\omega\varepsilon} \int_{\bar{\rho}_{2n-3}}^{\bar{\rho}_{2n-1}} H_0^{(2)}(k|\bar{\rho}_- - \bar{\rho}'|)\, dL' \\
& + \sum_{n=1}^{N} \frac{I_{2n-1}}{\Delta_{2n}} \frac{1}{4\omega\varepsilon} \int_{\bar{\rho}_{2n-1}}^{\bar{\rho}_{2n+1}} H_0^{(2)}(k|\bar{\rho}_- - \bar{\rho}'|)\, dL' \\
& + \sum_{n=1}^{N} \frac{I_{2n-1}}{\Delta_{2n-2}} \frac{1}{4\omega\varepsilon} \int_{\bar{\rho}_{2n-3}}^{\bar{\rho}_{2n-1}} H_0^{(2)}(k|\bar{\rho}_+ - \bar{\rho}'|)\, dL' \\
& - \sum_{n=1}^{N} \frac{I_{2n-1}}{\Delta_{2n}} \frac{1}{4\omega\varepsilon} \int_{\bar{\rho}_{2n-1}}^{\bar{\rho}_{2n+1}} H_0^{(2)}(k|\bar{\rho}_+ - \bar{\rho}'|)\, dL'
\end{aligned} \tag{4.8.16a}$$

Further, the integral terms in expression (4.8.16a) can be rewritten in the following form, which is suitable for numerical evaluation. Thus, the expression for the near scattered electric field distribution takes the form

$$- \hat{s} \bullet \bar{E}^{S}_{\tau}(\bar{\rho}) =$$

$$\sum_{n=1}^{N} I_{2n-1} \frac{\omega\mu}{4} \int_{-\Delta_{2n-2}/2}^{0} [\hat{s} \bullet \hat{s}_{2n-2}]\, H_0^{(2)}[k|\bar{\rho} - (\bar{\rho}_{2n-1} + \hat{s}_{2n-2}\ell)|]\, d\ell$$

$$+ \sum_{n=1}^{N} I_{2n-1} \frac{\omega\mu}{4} \int_{0}^{\Delta_{2n}/2} [\hat{s} \bullet \hat{s}_{2n}]\, H_0^{(2)}[k|\bar{\rho} - (\bar{\rho}_{2n-1} + \hat{s}_{2n}\ell)|]\, d\ell$$

$$- \sum_{n=1}^{N} \frac{I_{2n-1}}{\Delta_{2n-2}} \frac{1}{4\omega\varepsilon} \int_{-\Delta_{2n-2}/2}^{0} H_0^{(2)}[k|\bar{\rho}_{-} - (\bar{\rho}_{2n-1} + \hat{s}_{2n-2}\ell)|]\, d\ell$$

$$+ \sum_{n=1}^{N} \frac{I_{2n-1}}{\Delta_{2n}} \frac{1}{4\omega\varepsilon} \int_{0}^{\Delta_{2n}/2} H_0^{(2)}[k|\bar{\rho}_{-} - (\bar{\rho}_{2n-1} + \hat{s}_{2n}\ell)|]\, d\ell$$

$$+ \sum_{n=1}^{N} \frac{I_{2n-1}}{\Delta_{2n-2}} \frac{1}{4\omega\varepsilon} \int_{-\Delta_{2n-2}/2}^{0} H_0^{(2)}[k|\bar{\rho}_{+} - (\bar{\rho}_{2n-1} + \hat{s}_{2n-2}\ell)|]\, d\ell$$

$$- \sum_{n=1}^{N} \frac{I_{2n-1}}{\Delta_{2n}} \frac{1}{4\omega\varepsilon} \int_{0}^{\Delta_{2n}/2} H_0^{(2)}[k|\bar{\rho}_{+} - (\bar{\rho}_{2n-1} + \hat{s}_{2n}\ell)|]\, d\ell$$

(4.8.16b)

For field observation points sufficiently far away, these integral terms can be simplified based on single-source point calculation along a segment. Similarly, the z component of the near scattered magnetic field distribution can be calculated by using expression (4.5.5c).

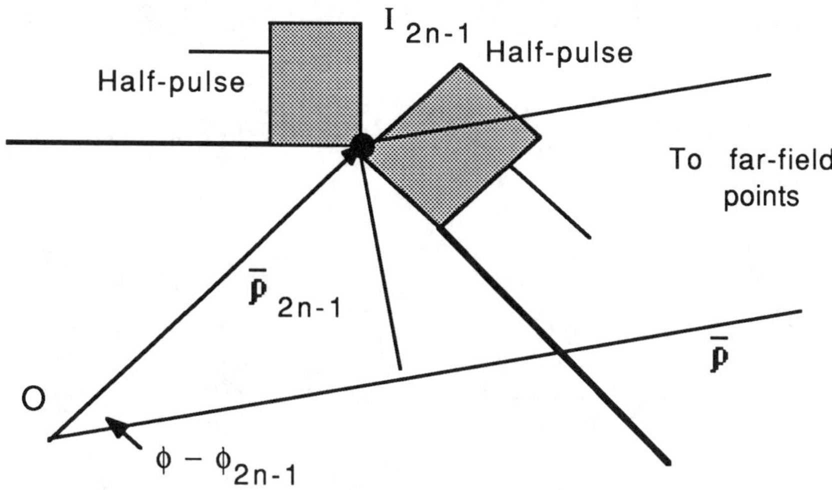

Figure 4.17 Calculation of the far- field distribution.

Far-Field and Radar Cross Section

The electric and magnetic far-field distributions can be obtained by using the large argument approximation for the free-space Green's function. In the far-field region, there are only the ϕ component of scattered electric field distribution and the corresponding z component of scattered magnetic field distribution. Further, in the far-field region, the scattered electric and magnetic fields behave as cylindrical waves with one over square root of ρ behavior, and the two components are directly related through the medium intrinsic impedance.

In expression (4.7.4a), only the magnetic vector potential term is retained, which is the dominant term for the far-field calculation. The other term, corresponding to the derivative of electric scalar potential, does not contribute to the far-field calculation. In the large argument approximation, $|k\rho| \to \infty$, the Hankel function can be written as

$$H_0^{(2)}(k\rho) \sim \sqrt{\frac{2j}{\pi k\rho}}\; e^{-jk\rho} \tag{4.8.17a}$$

Referring to expression (4.8.16a) and retaining only the terms corresponding to the magnetic vector potential, the scattered electric field distribution in the far-field region reduces to

$$E^s_\phi(\rho,\phi) = -\sum_{n=1}^{N} \frac{\Delta_{2n-2}}{2}\, I_{2n-1}\, \frac{\omega\mu}{4} \cos[\phi - \Omega_{2n-2}]\, H_0^{(2)}(k|\bar{\boldsymbol{\rho}}_{2n-1} - \bar{\boldsymbol{\rho}}'|)$$

$$-\sum_{n=1}^{N} \frac{\Delta_{2n}}{2}\, I_{2n-1}\, \frac{\omega\mu}{4} \cos[\phi - \Omega_{2n}]\, H_0^{(2)}(k|\bar{\boldsymbol{\rho}}_{2n-1} - \bar{\boldsymbol{\rho}}'|)$$

$$\sim \sum_{n=1}^{N} I_{2n-1}\Big[\frac{\Delta_{2n-2}}{2}\cos(\phi - \Omega_{2n-2}) + \frac{\Delta_{2n}}{2}\cos(\phi - \Omega_{2n})\Big]$$

$$\Big[-\frac{k\eta}{4}\sqrt{\frac{2j}{\pi(k|\bar{\boldsymbol{\rho}} - \bar{\boldsymbol{\rho}}_{2n-1}|)}}\; e^{-j(k|\bar{\boldsymbol{\rho}} - \bar{\boldsymbol{\rho}}_{2n-1}|)}\Big] \qquad (4.8.17b)$$

Referring to Figure 4.17, in the far-field region, $|k\rho| \to \infty$, the magnitude term in the denominator of expression (4.8.17b) can be written as

$$|\bar{\boldsymbol{\rho}} - \bar{\boldsymbol{\rho}}_{2n-1}| \approx \rho \qquad (4.8.18a)$$

and the exponential term contributing to the phase distribution can be approximated as in the following expression

$$|\bar{\boldsymbol{\rho}} - \bar{\boldsymbol{\rho}}_{2n-1}| \approx \rho - \rho_{2n-1}\cos(\phi - \phi_{2n-1}) \qquad (4.8.18b)$$

After substituting these far-field approximations, expressions (4.8.18a) and (4.8.18b), the scattered electric field distribution in the far-field region reduces to the following form:

$$E^s_\phi(\rho,\phi) \sim \sum_{n=1}^{N} I_{2n-1}\Big[\frac{\Delta_{2n-2}}{2}\cos(\phi - \Omega_{2n-2}) + \frac{\Delta_{2n}}{2}\cos(\phi - \Omega_{2n})\Big]$$

$$\Big[k\eta\mathcal{K}\, e^{jk\rho_{2n-1}\cos(\phi - \phi_{2n-1})}\Big] \qquad (4.8.19a)$$

where

$$\mathcal{K} = \frac{1}{\sqrt{8\pi k\rho}}\, e^{-jk\rho}\, e^{-j3\pi/4} \qquad (4.8.19b)$$

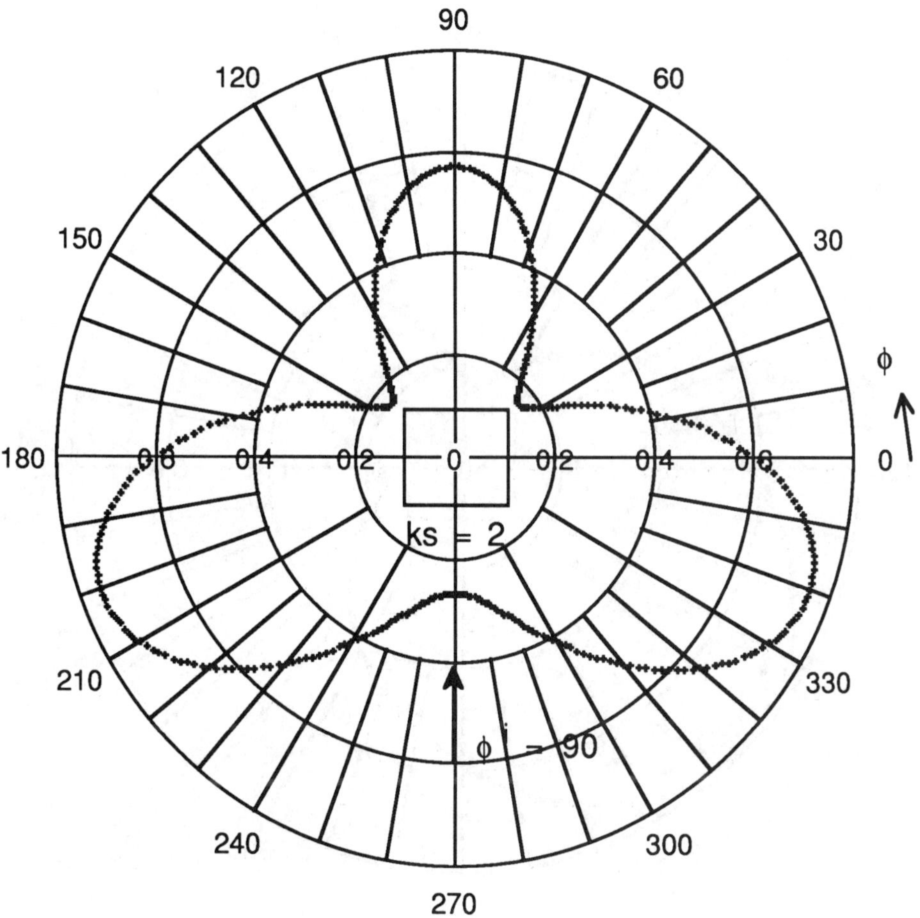

Figure 4.18 Bistatic radar cross section of a perfectly conducting square scatterer, normal incidence on a side – TE excitation.

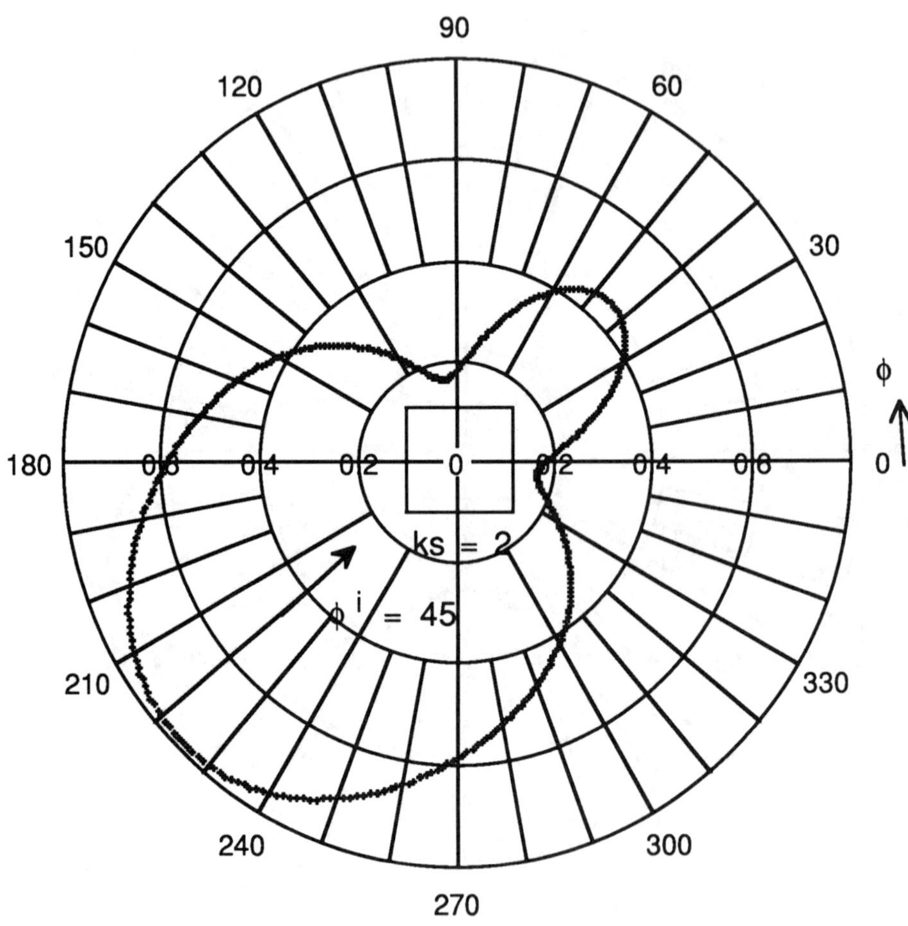

Figure 4.19 Bistatic radar cross section of a perfectly conducting square scatterer, oblique incidence on a corner – TE excitation.

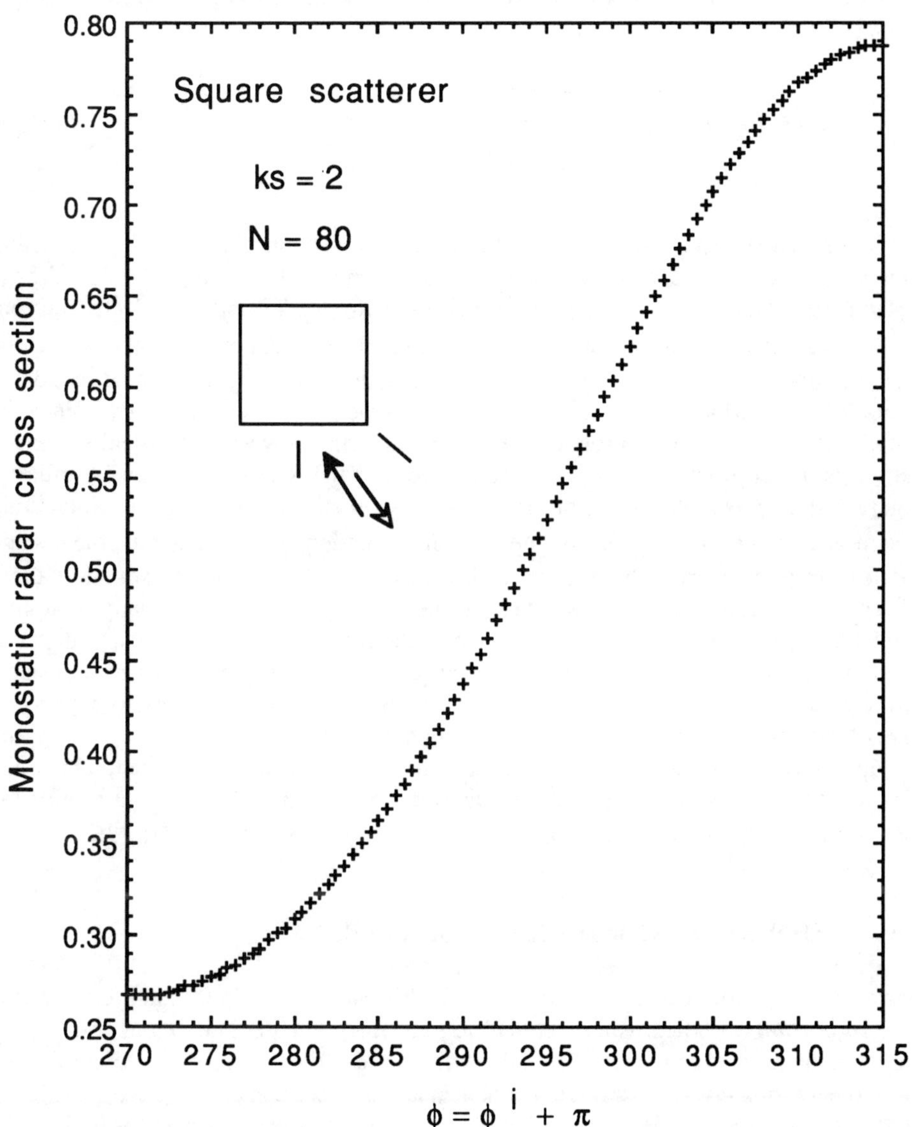

Figure 4.20 Monostatic radar cross section of a perfectly conducting square scatterer – TE excitation.

Further, for TE excitation the bistatic radar cross section of the perfectly conducting scatterer having arbitrary edges and corners can be calculated using expression (3.8.18), given by

$$\mathrm{RCS}(\phi) = \lim_{\rho \to \infty} 2\pi\rho \left| \frac{E_{\phi}^{s}(\phi,\omega)}{E_{s}^{i}(\phi,\omega)} \right|^{2} \tag{4.8.20}$$

Figure 4.18 shows a plot of the bistatic radar cross section of a perfectly conducting square scatterer, $ks = 2$, calculated based on expressions (4.8.19) and (4.8.20). These bistatic RCS data are numerically generated by using the TE-induced electric current results presented in Figure 4.16b for the normal incident excitation, $\phi^i = 90^\circ$. As can be seen, the bistatic data exhibit symmetrical distribution with respect to the direction of excitation about which the square scatterer also has the geometrical symmetry. Since the bistatic RCS distribution is proportional to the scattering power, the far-field scattering pattern is directly obtained by taking square root of the RCS distribution. Similarly, Figure 4.19 shows a plot of the bistatic radar cross section of the perfectly conducting square scatterer, $ks = 2$, for the oblique incident excitation, $\phi^i = 45^\circ$. Again, the bistatic data exhibit symmetrical distribution with respect to the direction of excitation about which the square scatterer also has geometrical symmetry. In many practical applications, the monostatic or backscattering data is quite important, which depends upon the number of parameters, such as the angle of incidence, electrical size, and shape of the geometry. To produce the monostatic RCS distribution, referring to expression (4.8.14), the inverse of the generalized impedance matrix can be stored separately and multiplied by the excitation term recalculated for every angle of incidence to obtain the corresponding induced electric current distribution. Figure 4.20 shows a plot of the monostatic radar cross section for incident angles ranging from $\phi^i = 90^\circ$ to 135°.

4.8.2 THIN-STRIP CONDUCTING SCATTERER

The electromagnetic scattering by a thin, perfectly conducting strip for transverse electric excitation can be studied using the electric field integral equation, expression (4.7.5), based on the method of moments technique discussed earlier. Referring to Figure 4.21a, the geometry of a thin-strip scatterer is oriented along x coordinate axis between the limits O and A having a total length equal to L. For the transverse electric excitation, the component of incident magnetic field is polarized parallel to the z coordinate axis and the incident electric field is polarized in the transverse xy plane. As shown, induced tangential-directed electric currents are on both the top and bottom sides of the thin-strip scatterer.

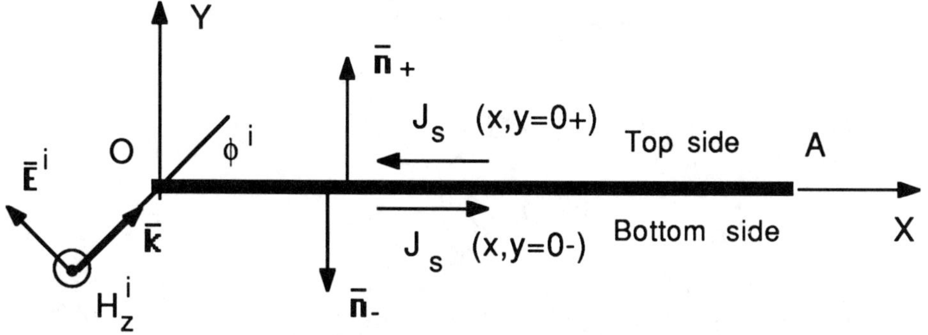

Figure 4.21a Geometry of a thin-strip scatterer.

Referring to Section 4.7, the magnetic vector and electric scalar potential integrals in terms of the unknown tangential-directed electric current distribution have the form

$$\bar{A}_\tau(\bar{\rho}) = \frac{\mu}{4j} \int_C \hat{s}' J_s(\bar{\rho}') H_0^{(2)}(k|\bar{\rho} - \bar{\rho}'|)\, dL(\bar{\rho}') \tag{4.8.21a}$$

$$\Phi(\bar{\rho}) = \frac{1}{4\omega\varepsilon} \int_C \left[\frac{\partial}{\partial s'} J_s(\bar{\rho}')\right] H_0^{(2)}(k|\bar{\rho} - \bar{\rho}'|)\, dL(\bar{\rho}') \tag{4.8.21b}$$

On the thin-strip scatterer, the following integro-differential equation is satisfied

$$\hat{s} \bullet \bar{E}_\tau^i(\bar{\rho}) = j\omega\, \hat{s} \bullet \bar{A}_\tau(\bar{\rho}) + \frac{\partial}{\partial s}\Phi(\bar{\rho}) \qquad \bar{\rho} \text{ on } C \tag{4.8.21c}$$

The boundary contour C in the potential integrals can be split into two parts corresponding to the top and bottom side currents:

$$\bar{A}_\tau(\bar{\rho}) = \frac{\mu}{4j} \int_O^A \hat{s}' J_s(x', y'=0-)\; H_0^{(2)}(k|s - s'|)\, ds'$$

$$+ \frac{\mu}{4j} \int_A^O \hat{s}' J_s(x', y'=0+)\; H_0^{(2)}(k|s - s'|)\, ds' \tag{4.8.22a}$$

After combining the two integral terms, expression (4.8.22a) reduces to

$$\bar{A}_\tau(\bar{\rho}) = \frac{\mu}{4j} \int_0^L \left[\hat{x}' J_x(x', y'=0-) + \hat{x}' J_x(x', y'=0+)\right] H_0^{(2)}(k|x - x'|)\, dx' \tag{4.8.22b}$$

$$\bar{A}_\tau(\bar{\rho}) = \frac{\mu}{4j} \int_0^L \hat{x}' J_x(x') H_0^{(2)}(k|x - x'|)\, dx' \tag{4.8.22c}$$

where $J_x(x')$ is the composite equivalent electric current on the thin-strip scatterer, and the corresponding electric scalar potential reduces to

$$\Phi(\bar{\rho}) = \frac{1}{4\omega\varepsilon} \int_0^L \left[\frac{\partial}{\partial x'} J_x(x')\right] H_0^{(2)}(k|x - x'|)\, dx' \tag{4.8.22d}$$

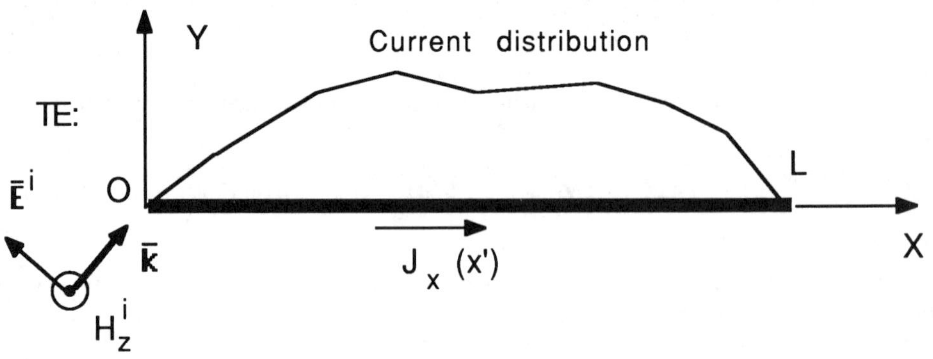

Figure 4.21b Modeling of the thin-strip scatterer.

Hence, referring to Figure 4.21b, for the thin-strip scatterer the electric field integral equation (4.8.21c) simplifies to

$$E_x^i(x) = j\omega A_x(x) + \frac{\partial}{\partial x} \Phi(x) \qquad x \text{ on strip} \tag{4.8.23a}$$

$$E_x^i(x) = \frac{\omega\mu}{4} \int_0^L J_x(x') H_0^{(2)}(k|x - x'|)\, dx'$$

$$+ \frac{1}{4\omega\varepsilon} \frac{\partial}{\partial x} \int_0^L \left[\frac{\partial}{\partial x'} J_x(x') \right] H_0^{(2)}(k|x - x'|)\, dx' \qquad (4.8.23b)$$

As can be seen from this EFIE formulation for the thin-strip scatterer, the distribution of unknown electric current, which is to be solved, is the net value of the electric current on the thin-strip represented by the sum of the top and bottom currents. Further, referring to Figure 4.21b, the net electric current distribution is zero at the two open ends of the scatterer. This boundary condition on the unknown electric current should be appropriately incorporated in the numerical solution technique. To enforce the electric current boundary condition, referring to Figure 4.14b, at the two free ends of thin-strip scatterer, half-pulse current expansion terms with zero amplitude are incorporated.

Figure 4.22a shows the distribution of induced electric current on the thin-strip scatterer for transverse electric excitation. For the results shown, the frequency of excitation is assumed to be 300 MHz, so that the wavelength of incident plane wave is 1 meter and the propagation constant in the free-space medium $k = 2\pi$. The total length of the conducting strip is selected as $L = 2$ meters and excited normally with the plane wave electric field polarized in the $-x$ coordinate direction. Both the real and imaginary parts of the electric current distribution are shown in the Figure 4.22a with normal to the strip assumed to be in the $-y$ coordinate direction. The results are generated with the method of moments matrix size of $N = 80$ and normalized with respect to the incident plane wave magnetic field. The distribution of the electric current, in fact, is zero at the free ends of thin-strip scatterer. The dominant real part of the electric current oscillates about the physical optics limit. Figure 4.22b shows the corresponding bistatic radar cross section of the perfectly conducting thin-strip scatterer.

Similarly, Figure 4.23a shows the magnitude of induced electric current for TE normal incident excitation on the thin-strip scatterer of length $L = 10$ meters. The results shown are generated with a matrix size of $N = 200$. For this structure, the monostatic radar cross section in the angular range of 180° to 270° is shown in Figure 4.23b. The monostatic data are maximum on the broadside for normal excitation and gradually drop off to zero for the end on incidence. The results shown are plotted with a resolution of 0.25^o angle. To clearly mark various sidelobes, it is necessary to generate the data with finer angular resolution.

Further case study involving electrically large thin-strip scatterer is taken up in a later section.

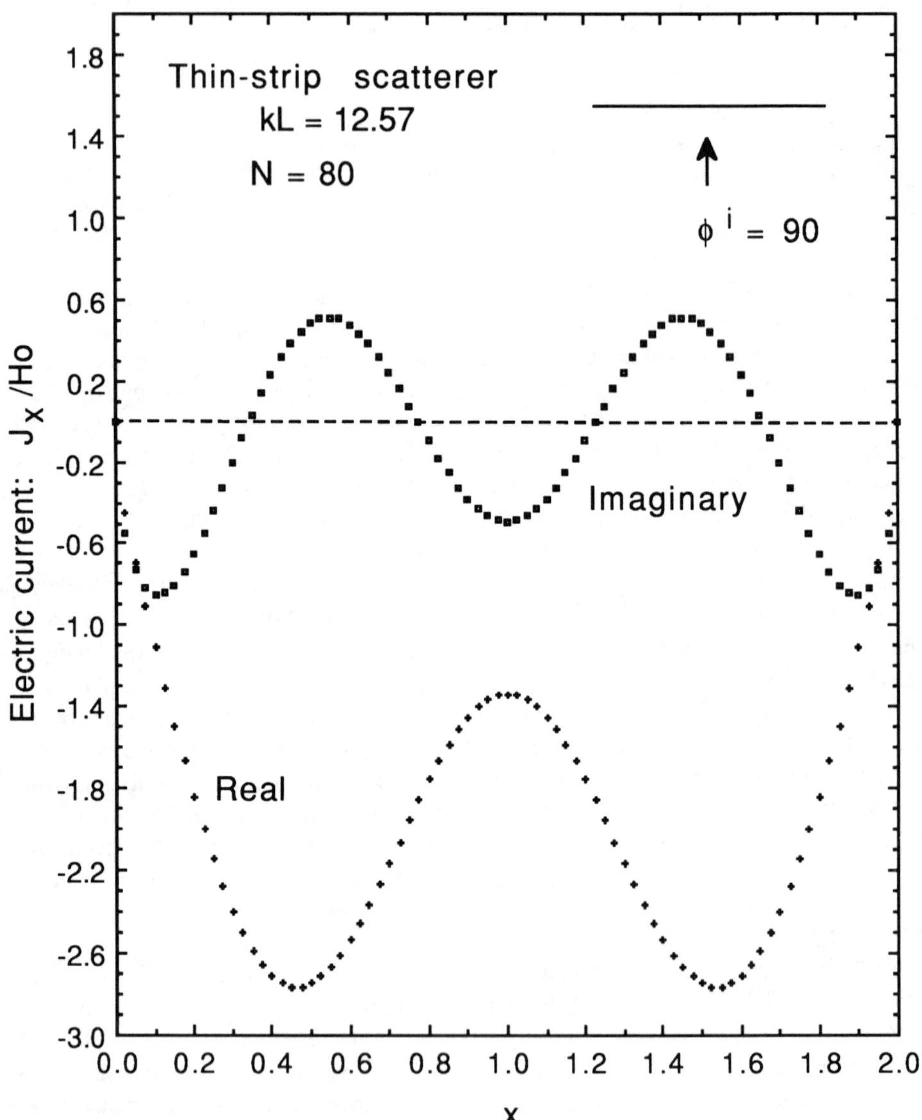

Figure 4.22a Electric current distribution on a perfectly conducting thin-strip scatterer – TE excitation.

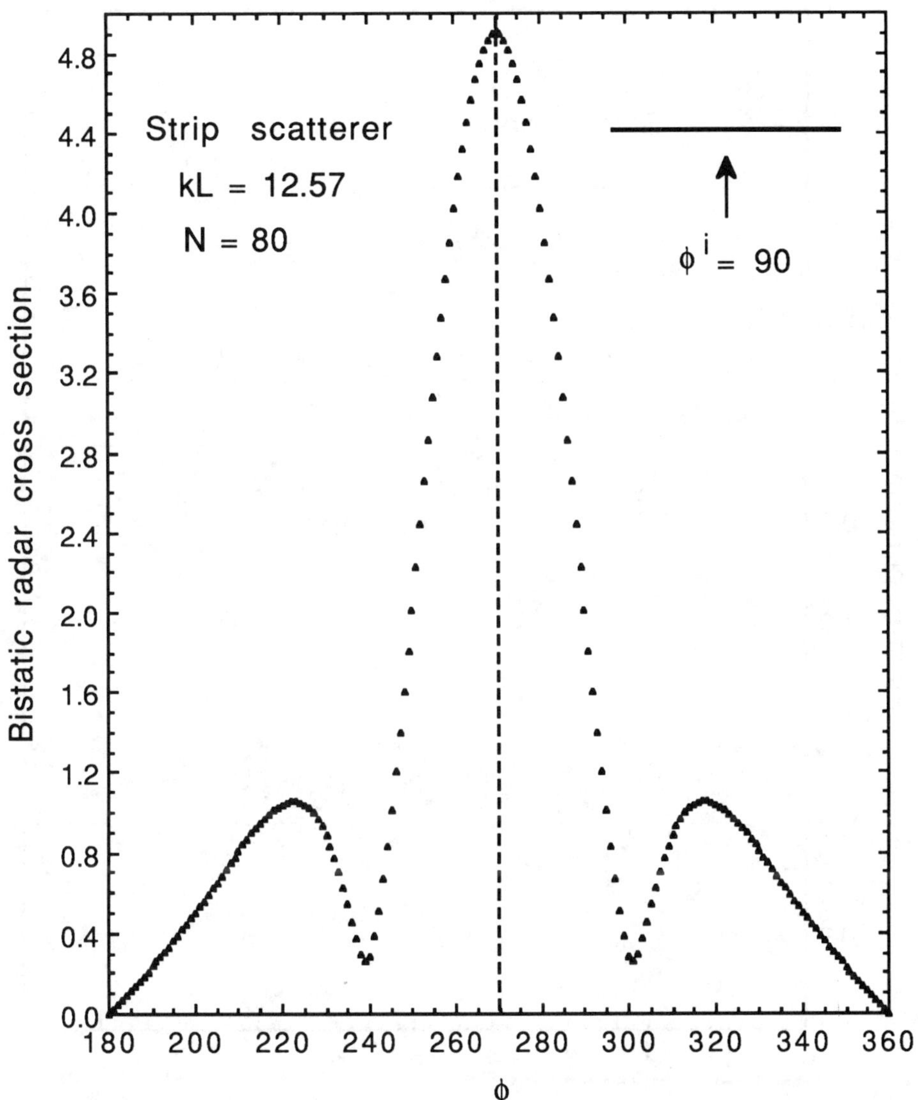

Figure 4.22b Bistatic radar cross section of a perfectly conducting thin-strip scatterer – TE excitation.

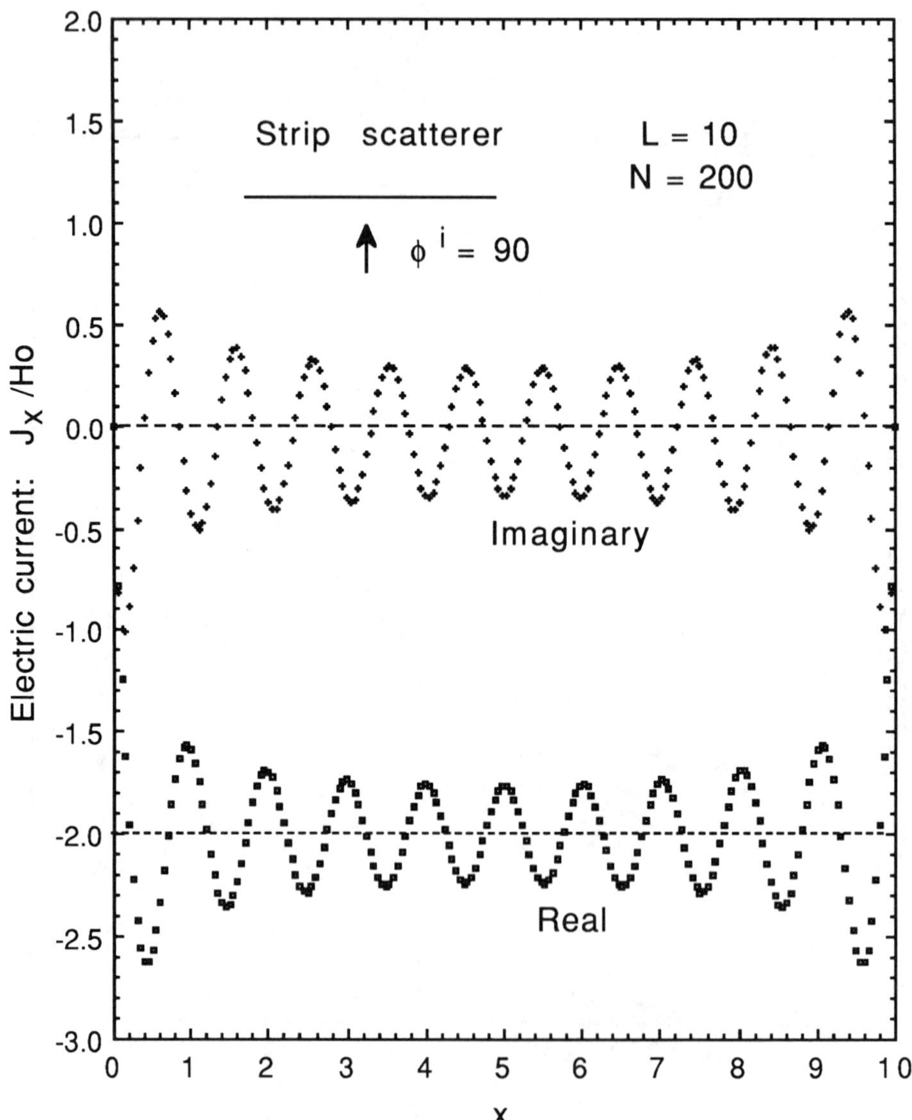

Figure 4.23a Electric current distribution on a perfectly conducting large thin-strip scatterer – TE excitation.

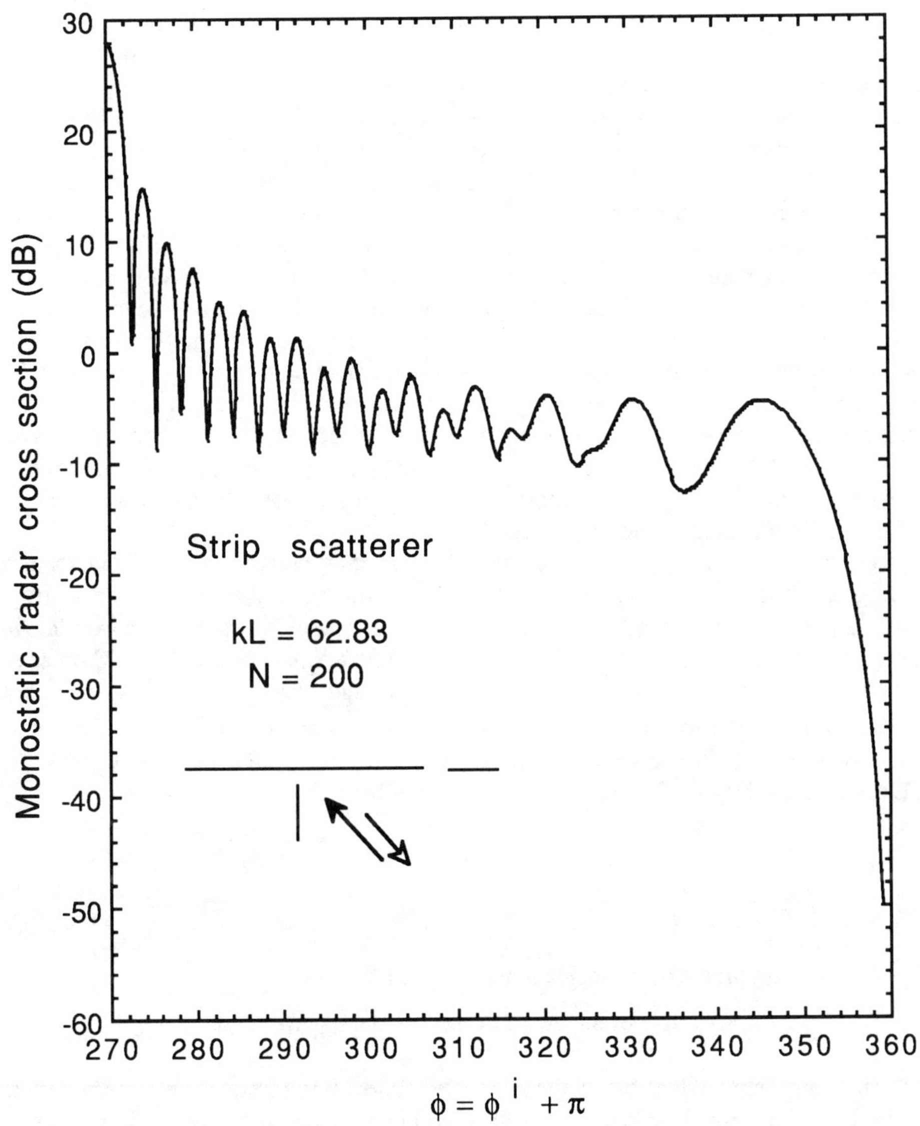

Figure 4.23b Monostatic radar cross section of a perfectly conducting large thin-strip scatterer – TE excitation.

4.9 ARBITRARY CROSS SECTION - MFIE - TE EXCITATION

The electromagnetic scattering by a two-dimensional perfectly conducting scatterer with an arbitrary cross section, having arbitrary edges and corners is discussed in the previous section based on the EFIE approach. This approach is suitable for analyzing both open and closed type of scatterers. It should be noted that the EFIE involves the calculation of both the magnetic vector potential and electric scalar potential. In the development of the numerical technique, the electric charge term is replaced in terms of the divergence of electric current using the current-charge continuity equation. This procedure basically introduces second-order derivative operation in the evaluation of EFIE. Thus, the numerical implementation of the EFIE for TE polarization is relatively more complicated than the EFIE for the TM polarization. In fact, an alternative procedure can be conveniently considered for the TE polarization using the magnetic field integral equation. Due to the specific nature of the magnetic field boundary condition, the MFIE is generally suitable for the analyzing close type of scatterers. One distinct advantage with the MFIE procedure is that the method of moments numerical technique, Section 3.7, is straightforward. In the following, an approximate analysis procedure is discussed in detail to obtain a numerical solution for the electromagnetic scattering by solving the MFIE, expression (4.5.14).

Let us consider the geometry of a two-dimensional, perfectly conducting scatterer having an arbitrary cross section and placed in a linear, homogeneous, and isotropic lossless medium, shown in Figure 4.13. The scatterer is uniform along its axis, having the same arbitrary cross section. The external incident plane wave field is TE polarized with respect to the z coordinate axis of the scatterer and propagates with its direction of propagation normal to the z axis. The angle ϕ^i represents the incident angle that the direction of propagation makes with the x coordinate axis. Referring to Figure 4.13, the TE to z polarized incident plane wave magnetic field is given by

$$H_z^i(\rho,\phi) = H_0\, e^{-jk\rho \cos(\phi - \phi^i)} \tag{4.9.1}$$

H_0 : amplitude of the incident plane wave magnetic field;

ϕ^i : arbitrary angle of incidence of the TE plane wave field.

The complete derivation of the magnetic field integral equation in terms of the unknown equivalent electric current distribution is presented in Section 4.5 by enforcing boundary condition related to the z component of the total magnetic field on contour C. At any point in region 1 outside scatterer contour C, the tangential total magnetic field is given by the sum of the z components of the incident magnetic field and the scattered magnetic field:

$$H_z(\bar{\rho}) = H_z^i(\bar{\rho}) + H_z^s(\bar{\rho}) \qquad \bar{\rho} \text{ in region 1} \qquad (4.9.2a)$$

Referring to expression (4.5.5a) and substituting for the *z* component of the scattered magnetic field,

$$H_z(\bar{\rho}) = H_z^i(\bar{\rho}) + \frac{1}{\mu}\, \hat{z} \bullet \nabla_\tau \times \bar{A}_\tau(\bar{\rho}) \qquad (4.9.2b)$$

$$= H_z^i(\bar{\rho}) + \frac{1}{4j}\, \hat{z} \bullet \nabla_\tau \times \int_C \bar{J}_S(\bar{\rho}')\, H_0^{(2)}(k|\bar{\rho} - \bar{\rho}'|)\, dL(\bar{\rho}')$$

$$\bar{\rho} \text{ in region 1} \qquad (4.9.2c)$$

Using expression (4.5.2a), the MFIE for the arbitrary shaped, perfectly conducting scatterer with transverse electric excitation is obtained by moving the field observation point directly on contour C to yield

$$- H_z^i(\bar{\rho}) = \frac{J_S(\bar{\rho})}{2} - \frac{1}{4j}\, \hat{z} \bullet \fint_C \left[\bar{J}_S(\bar{\rho}') \times \nabla_\tau H_0^{(2)}(k|\bar{\rho} - \bar{\rho}'|)\right] dL(\bar{\rho}')$$

$$\bar{\rho} \text{ on } C \qquad (4.9.3)$$

The integral is written with a bar to recognize that the field observation point on contour *C* is outside the source segment. The principle value of the integral is already separated, and thus the value of the integral is zero when the source point segment coincides with the field point segment. In this MFIE expression,

$\bar{J}_S(\bar{\rho}')$: unknown tangential electric current distribution along contour *C*, in amperes per meter

$$= \hat{s}' J_S(\bar{\rho}', \omega) \qquad (4.9.4)$$

The kernel inside the integral term of the magnetic field integral equation (4.9.3) can be further simplified to the following form:

$$- H_z^i(\bar{\rho}) = \frac{J_S(\bar{\rho})}{2} + \fint_C J_S(\bar{\rho}')\, G(k|\bar{\rho} - \bar{\rho}'|)\, dL(\bar{\rho}')$$

$$\bar{\rho} \text{ on } C \qquad (4.9.5a)$$

$$G(k|\bar{\rho} - \bar{\rho}'|) = \frac{k}{4j}\left[\frac{x-x'}{R}\cos\Omega' + \frac{y-y'}{R}\sin\Omega'\right] H_0^{(2)'}(kR) \tag{4.9.5b}$$

$$\begin{aligned} R &= |\bar{\rho} - \bar{\rho}'| \\ &= \left[(x - x')^2 + (y - y')^2\right]^{1/2} \end{aligned} \tag{4.9.5c}$$

Ω': angle that the normal at the source point makes with the x coordinate axis.

4.9.1 REDUCTION TO MATRIX EQUATION

To reduce integral equation (4.9.5a) to its equivalent functional form, the integral equation is first tested on both sides using the same pulse-type weighting functions developed in Section 4.7. Further, the unknown tangential electric current distribution is expanded in terms of a set of independent piecewise pulse expansion terms to reduce the functional form of equation to its equivalent matrix form of equation. Referring to Figure 4.15a, the geometry is properly modeled by choosing the corner point of geometry and spanning it by the weighting function on either side by half a cell. The piecewise weighting pulse W_{2m-1} is located such that it spans the odd point $\bar{\rho}_{2m-1}$ with $m = 1, 2, 3, \ldots, M$ and defined between the two consecutive even points given by

$$\bar{\rho}_{2m} = \bar{\rho}_{2m-1} + \frac{\Delta_{2m}}{2}\hat{s}_{2m} \tag{4.9.6a}$$

$$\bar{\rho}_{2m-2} = \bar{\rho}_{2m-1} - \frac{\Delta_{2m-2}}{2}\hat{s}_{2m-2} \tag{4.9.6b}$$

and the weighting function term based on the piecewise pulses is given by

$$\begin{aligned} W_{2m-1}(\bar{\rho}) &= 1 \qquad \bar{\rho}_{2m-2} \le \bar{\rho} \le \bar{\rho}_{2m} \\ &= 0 \qquad \text{otherwise} \end{aligned} \tag{4.9.7}$$

$$m = 1, 2, 3, \ldots, M$$

Pulse Testing The MFIE integral expression (4.9.5) is now multiplied on both sides by weighting function (4.9.7) and integrated along the scatterer contour C to yield the following relationship:

$$\int_C -H_z^i(\bar{\rho})W_{2m-1}(\bar{\rho})\,dL(\bar{\rho}) = \int_C \frac{J_s(\bar{\rho})}{2} W_{2m-1}(\bar{\rho})\,dL(\bar{\rho})$$

$$+\int_C \left[\fint_C J_s(\bar{\rho}')G(k|\bar{\rho}-\bar{\rho}'|)dL(\bar{\rho}')\right]W_{2m-1}(\bar{\rho})\,dL(\bar{\rho}) \tag{4.9.8a}$$

$\bar{\rho}$ on C, and $m = 1, 2, 3, \ldots, M$

Expression (4.9.8a) can be simplified as

$$\int_{\bar{\rho}_{2m-2}}^{\bar{\rho}_{2m}} -H_z^i(\bar{\rho})\,dL(\bar{\rho}) = \int_{\bar{\rho}_{2m-2}}^{\bar{\rho}_{2m}} \frac{J_s(\bar{\rho})}{2}\,dL(\bar{\rho})$$

$$+\int_{\bar{\rho}_{2m-2}}^{\bar{\rho}_{2m}} \left[\fint_C J_s(\bar{\rho}')\,G(k|\bar{\rho}-\bar{\rho}'|)dL(\bar{\rho}')\right]dL(\bar{\rho}) \tag{4.9.8b}$$

$\bar{\rho}$ on C, and $m = 1, 2, 3, \ldots, M$

Further, these integrals can be split into two integrals, corresponding to the two half straight line segments of the weighting pulse term. Numerical integration can be implemented to calculate the weighted integrals. If the integrands are smooth varying functions, the integrands on the lefthand and righthand sides of expression (4.9.8b) can be assumed not to vary over the limits of the integration interval. Thus, a suitable approximation can be introduced by calculating the integrand at the point $\bar{\rho}_{2m-1}$ and multiplying it by the corresponding half segment length of the weighting function to obtain

$$-H_z^i(\bar{\rho}_{2m-1}) = \frac{I_{2m-1}(\bar{\rho}_{2m-1})}{2}$$

$$+\fint_C J_s(\bar{\rho}')\,G(k|\bar{\rho}_{2m-1}-\bar{\rho}'|)\,dL(\bar{\rho}') \tag{4.9.9}$$

$m = 1, 2, 3, \ldots, M$

Current Expansion Referring to Figure 4.15b, the unknown tangential electric current distribution $J_S(\bar{\rho}')$ is now expanded in terms of a set of piecewise linearly independent pulse expansion functions. If contour C of the scatterer is completely smooth with no geometrical discontinuities, then it is possible to pick arbitrary sampling regions along contour C with expansion functions at points located at $\bar{\rho}_{2n-1}$, n = 1, 2, 3, . . . , N. On the other hand, if contour C of the scatterer is not smooth, but has sharp corners, as indicated in Figure 4.15b, care should to be taken in properly positioning the location of expansion function with respect to the geometrical discontinuity. From the numerical modeling point of view, the specific expansion function close to the wedge-type corner is always properly arranged on both sides of the corner so that it encloses the geometrical discontinuity point. Referring to Figure 4.15b, the corner point of geometry is spanned by the expansion function on either side by half a cell. The piecewise expansion pulse P_{2n-1} is located such that it spans the odd point $\bar{\rho}_{2n-1}$ with n = 1, 2, 3, . . . , N and defined between the two consecutive even points given by

$$\bar{\rho}_{2n} = \bar{\rho}_{2n-1} + \frac{\Delta_{2n}}{2}\hat{s}_{2n} \tag{4.9.10a}$$

$$\bar{\rho}_{2n-2} = \bar{\rho}_{2n-1} - \frac{\Delta_{2n-2}}{2}\hat{s}_{2n-2} \tag{4.9.10b}$$

$\hat{s}_{2n-2}$: the tangential unit vector along the length of pulse segment Δ_{2n-2};

$\hat{s}_{2n}$: the tangential unit vector along the length of pulse segment Δ_{2n}.

The tangential-directed electric current distribution in terms of the piecewise pulse expansion terms is given by

$$J_S(\bar{\rho}') \approx \sum_{n=1}^{N} I_{2n-1} P_{2n-1}(\bar{\rho}') \tag{4.9.10c}$$

where

$$P_{2n-1}(\bar{\rho}') = 1 \qquad \bar{\rho}_{2n-2} \le \bar{\rho}' \le \bar{\rho}_{2n}$$

$$= 0 \qquad \text{otherwise} \tag{4.9.10d}$$

$$n = 1, 2, 3, \ldots, N$$

and in expression (4.9.10c), I_{2n-1} are the coefficients of the unknown electric current. After substituting pulse expansion function (4.9.10c) into weighted functional equation (4.9.9), the following matrix form of equation is obtained:

$$- H_z^i(\bar{\rho}_{2m-1}) = \frac{I_{2m-1}(\bar{\rho}_{2m-1})}{2} + \oint_C \Big[\sum_{n=1}^{N} I_{2n-1} P_{2n-1}(\bar{\rho}')\Big] G(k|\bar{\rho}_{2m-1} - \bar{\rho}'|)\, dL(\bar{\rho}')$$

$$m = 1, 2, 3, \ldots, M \tag{4.9.11a}$$

Since the original operator equation is linear, the summation and the integration operations can be interchanged to obtain

$$- H_z^i(\bar{\rho}_{2m-1}) = \frac{I_{2m-1}(\bar{\rho}_{2m-1})}{2} + \sum_{n=1}^{N} I_{2n-1} \Big[\oint_C P_{2n-1}(\bar{\rho}') G(k|\bar{\rho}_{2m-1} - \bar{\rho}'|)\, dL(\bar{\rho}')\Big]$$

$$m = 1, 2, 3, \ldots, M \tag{4.9.11b}$$

Referring to expression (4.9.10d), the current expansion pulse functions are valid over their segments only and zero outside. Hence, expression (4.9.11b) simplifies to the following form:

$$- H_z^i(\bar{\rho}_{2m-1}) = \frac{I_{2m-1}(\bar{\rho}_{2m-1})}{2} + \sum_{n=1}^{N} I_{2n-1} \Big[\int_{\bar{\rho}_{2n-2}}^{\bar{\rho}_{2n-1}} G_1(k|\bar{\rho}_{2m-1} - \bar{\rho}'|)\, dL(\bar{\rho}') + \int_{\bar{\rho}_{2n-1}}^{\bar{\rho}_{2n}} G_2(k|\bar{\rho}_{2m-1} - \bar{\rho}'|)\, dL(\bar{\rho}')\Big]$$

$$m = 1, 2, 3, \ldots, M \tag{4.9.11c}$$

It should be noted that for all values $m = n$, the integral terms in the preceding expression are zero and the two kernels are given by

$$G_1(k|\bar{\rho}_{2m-1} - \bar{\rho}'|) = \frac{k}{4j}\Big[\frac{x_{2m-1}-x'}{R_{2m-1}}\cos\Omega'_{2n-2}$$

$$+\frac{y_{2m-1}-y'}{R_{2m-1}}\sin\Omega'_{2n-2}\Big]H_0^{(2)'}(kR_{2m-1})$$

(4.9.12a)

$$G_2(k|\bar{\rho}_{2m-1} - \bar{\rho}'|) = \frac{k}{4j}\Big[\frac{x_{2m-1}-x'}{R_{2m-1}}\cos\Omega'_{2n}$$

$$+\frac{y_{2m-1}-y'}{R_{2m-1}}\sin\Omega'_{2n}\Big]H_0^{(2)'}(kR_{2m-1})$$

(4.9.12b)

$$R_{2m-1} = |\bar{\rho}_{2m-1} - \bar{\rho}'|$$

$$= \Big[(x_{2m-1} - x')^2 + (y_{2m-1} - y')^2\Big]^{1/2} \qquad (4.9.12c)$$

Ω_{2n-2}: angle that the normal to the straight line source segment Δ_{2n-2} makes with the x coordinate axis;

Ω_{2n}: angle that the normal to the straight line source segment Δ_{2n} makes with the x coordinate axis.

Expression (4.9.12c) forms a set of M linear simultaneous algebraic equations for the N unknown electric current coefficients. A direct numerical solution can be obtained by choosing $M = N$, the number of equations equal to the number of unknowns. Expression (4.9.11c) can also be rewritten in terms of a compact generalized matrix equation

$$[Z_{mn}][I_n] = [V_m] \qquad (4.9.13a)$$

where the excitation term for the TE polarized plane wave propagating with an angle of incidence ϕ^i is given by

$$V_m = - H_0 e^{-jk\rho_{2m-1} \cos[\phi_{2m-1} - \phi^i]} \tag{4.9.13b}$$

In matrix equation (4.8.13a), the self and mutual elements are given by

$$Z_{mn} = \frac{1}{2} \qquad m = n \tag{4.9.14a}$$

$$Z_{mn} = \int_{-\Delta_{2n-2}/2}^{0} G_1\left[k\left|\bar{\rho}_{2m-1} - (\bar{\rho}_{2n-1} + \hat{s}_{2n-2}\ell)\right|\right] d\ell + \int_{0}^{\Delta_{2n}/2} G_2\left[k\left|\bar{\rho}_{2m-1} - (\bar{\rho}_{2n-1} + \hat{s}_{2n}\ell)\right|\right] d\ell \qquad m \neq n \tag{4.9.14b}$$

The unknown electric current distribution is obtained by inverting the generalized matrix equation (4.8.13a):

$$[I_n] = [Z_{mn}]^{-1}[V_m] \tag{4.9.15}$$

4.9.2 EFIE AND MFIE VALIDATIONS

In the previous sections, the electric field integral equation and the magnetic field integral equation and their numerical treatments are discussed in detail. To show validation of the two methods, in the following, canonical two-dimensional perfectly conducting scattering geometries are studied based on both the EFIE and the MFIE. Figure 4.24 shows the results of induced electric current distribution on a conducting circular cylinder located in a linear, homogeneous, and isotropic lossless medium. The radius of the cylinder is selected as $ka = 6$ and excited by an external monochromatic plane wave propagating at an angle of incidence $\phi^i = 0^o$. For the TE polarization, the incident magnetic field is polarized parallel to the z axis, and the corresponding incident electric field is polarized parallel to the y axis. In Figure 4.24, the electric current distribution based on the MFIE solution, expression (4.9.15), is illustrated. The matrix size in the MFIE solution is selected such that the number of unknowns solved is $N = 20$, 30, 40, and 50. The results are compared with respect the electric current result obtained based

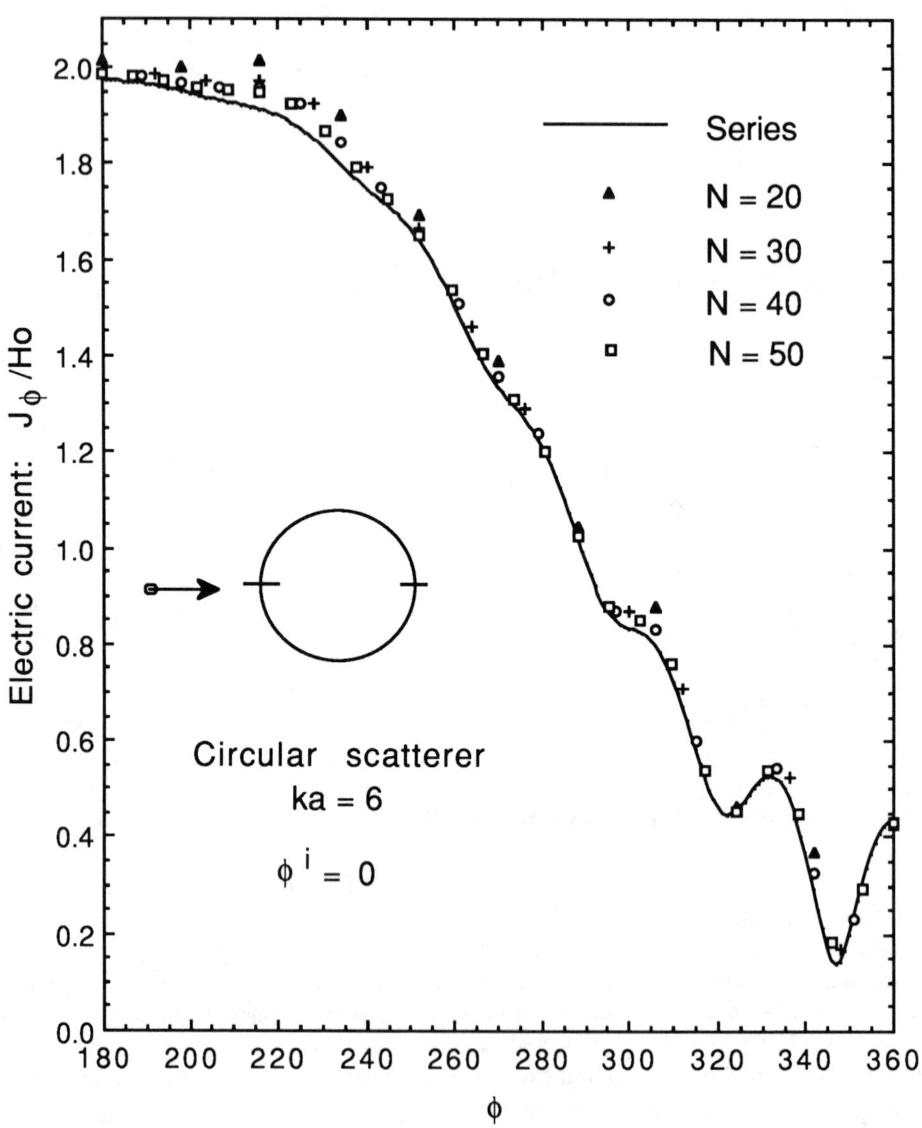

Figure 4.24 Comparison between the MFIE solution and the series solution, electric current on a circular conducting scatterer – TE excitation.

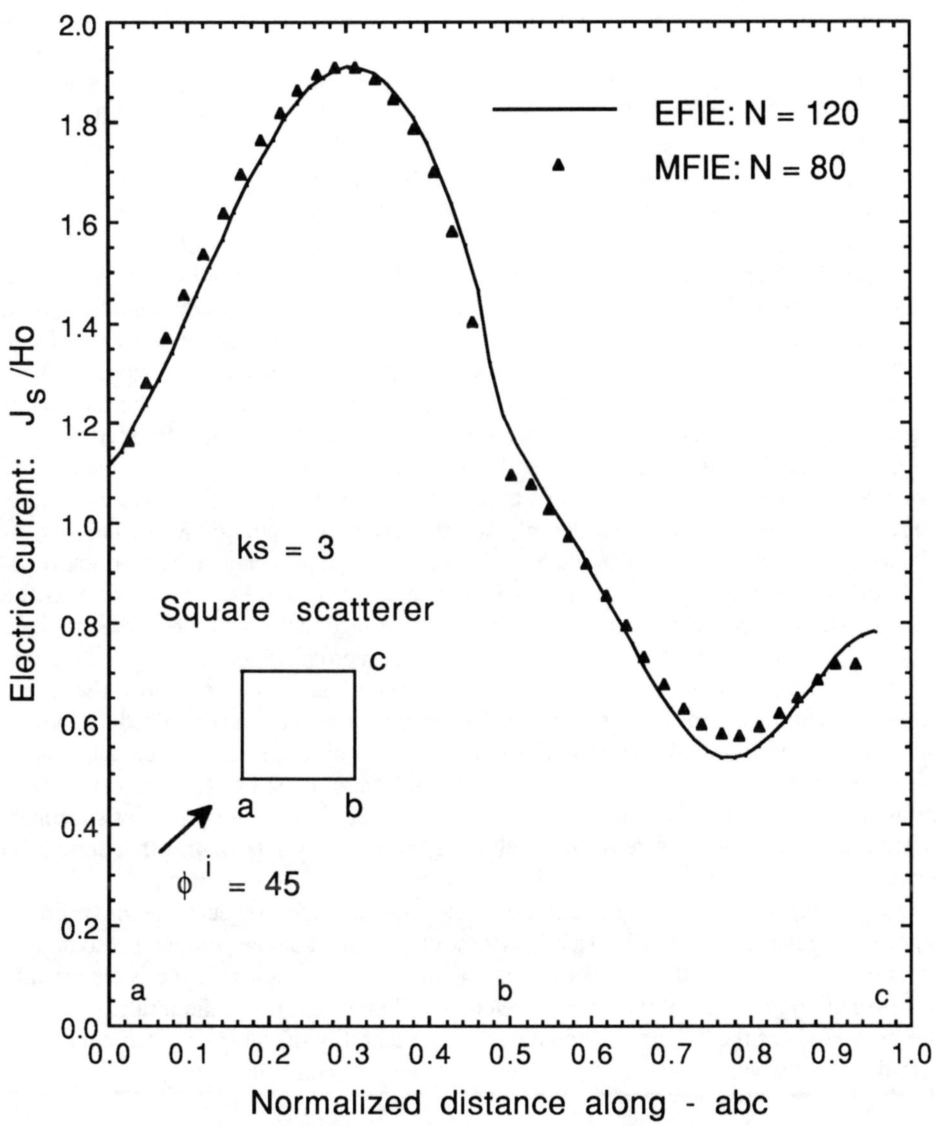

Figure 4.25 Comparison between the MFIE solution and the EFIE solution, electric current on a square conducting scatterer – TE excitation.

on the TE series solution, expression (4.6.15b). Excellent validation is observed between the two sets of results.

Figure 4.25 shows distribution of the TE electric current for the a square conducting scatterer. For the EFIE solution, each side length of the square cylinder, $ks = 3$, is divided into thirty equal pulse segments so that the total number of unknown electric current pulses is selected as $N = 120$. Similarly, for the MFIE solution, each side length is divided into 20 equal segments, and the number of unknowns is selected as $N = 80$. The square conducting scatterer is TE excited with an oblique angle of incidence equal to 45°. The distribution of the electric current along the two sides of the square scatterer, *ab* and *bc*, are shown with good validation between the two sets of results.

It should be noted again that either the EFIE or the MFIE can be utilized for analyzing the closed-type of conducting scatterers. But for the open-type conducting scatterers, only the EFIE is suitable. Based on the numerical studies, for electrically large scatterers, the MFIE solution technique converges faster than the EFIE solution.

In the earlier sections, the boundary value EFIE integral equation and the method of moments numerical technique have been utilized to study electromagnetic scattering by arbitrary shaped conducting objects with TE plane wave excitation. Even though the direct approach is elegant, as far as its application to analyze electrically large object is concerned, it inherently suffers from a wide range of computational difficulties, as discussed in Section 3.11. The method of moments complex-valued system matrix is full and dense, placing a large demand on computer resources. In addition to the operational numerical errors and ill-conditioning involved in the solution of a large-scale matrix equation, the accuracy of the near-field solution progressively degrades as the size of system matrix increases. The apparent computational difficulties with the direct integral equation and method of moments has prompted an alternative procedure to seek numerical solution based on a spatial decomposition technique. In Section 3.11, using rigorous electromagnetic equivalence, the spatial decomposition technique has been explored that virtually divides an electrically large object into a multiplicity of subzones. It permits the maximum size of the method of moments system matrix that should be inverted to be strictly limited (so that error bounds and condition numbers can be controlled) regardless of the electrical size of the large scattering object being modeled. A significant outcome of this approach is that the requirement on the computer resources is of order N, where N is the number of spatial subzones and each subzone is electrically small spanning on the order of few wavelengths. In the following, numerical examples are presented for the TE case to consider the applicability of the spatial decomposition technique to analyze two-dimensional electrically large conducting objects.

4.10 SDT APPROACH – TE CASE

This section presents TE formulation of the spatial decomposition technique for the integral equation and method of moments. Using rigorous electromagnetic equivalences, the spatial decomposition technique allows one to divide an electrically large, arbitrary shaped conducting object into a multiplicity of subzones. The individual subzones, in fact, are separated by virtual boundaries across which cancelling tangential electric-type virtual currents are postulated. The subzones are defined as distinct scattering targets, having fully enclosing boundaries, with additional unknown virtual electric currents introduced as needed to define the interfaces. The electromagnetic field boundary conditions are well preserved across the interface separating two subzones by requiring that the tangential virtual currents on one side of the interface be equal and opposite to the tangential virtual currents on the other side. By sequentially implementing the integral equation and method of moments solution for each subzone, effectively scanning the original target subzone by subzone, a rapidly convergent iterative process can be established.

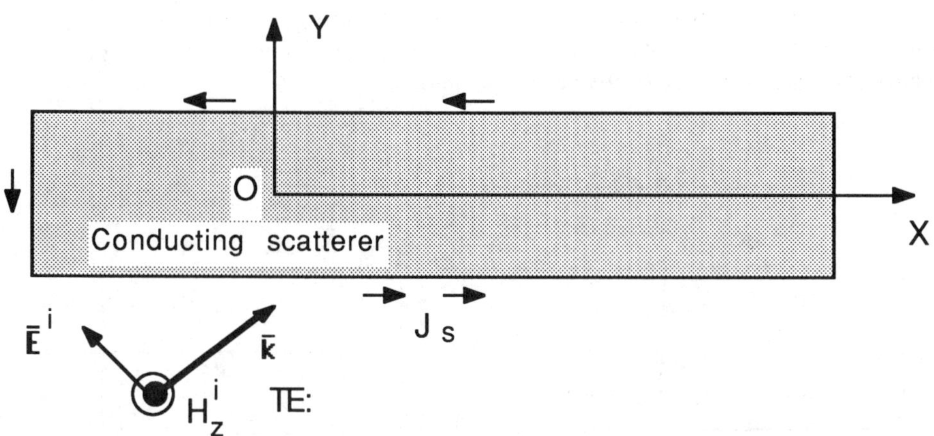

Figure 4.26a Electrically large perfectly conducting scatterer – TE excitation.

Figure 4.26a illustrates a two-dimensional conducting object. It is excited externally by a transverse electric polarized plane wave propagating in a direction normal to the z coordinate axis. As shown in Figure 4.26a, tangential-directed equivalent electric currents $\bar{J}_S(\bar{\rho}')$ reside on a virtual boundary conforming to the physical boundary of the scatterer.

Referring to Section 4.5, the EFIE for a two-dimensional, perfectly conducting scattering case is given by

$$E_S^i(\bar{\rho}) = \mathcal{L}_2[J_S(\bar{\rho}')] \qquad \bar{\rho} \text{ on } C \qquad (4.10.1a)$$

$$\mathcal{L}_2 = \frac{\omega\mu}{4} \int_C (\hat{s} \cdot \hat{s}')\, H_0^{(2)}(k|\bar{\rho} - \bar{\rho}'|)\, dL' + \frac{1}{4\omega\varepsilon} \frac{\partial}{\partial s} \int_C H_0^{(2)}(k|\bar{\rho} - \bar{\rho}'|)\, dL' \left[\frac{\partial}{\partial s'}\right] \tag{4.10.1b}$$

$\mathcal{L}_2$: EFIE integral operator for TE polarization.

In the direct method of moments solution, the tangential electric current J_s is expanded in terms of a suitable expansion function, such as pulse expansion functions, and tested on both sides by the weighting functions to reduce the integral equation (4.10.1a) to its equivalent matrix equation. As the electrical size of the scatterer increases, the direct solution technique puts impractical demands on the computer resources, in addition to various unsettled numerical accuracy and ill-conditioning problems of the large system matrix associated with electrically very large objects.

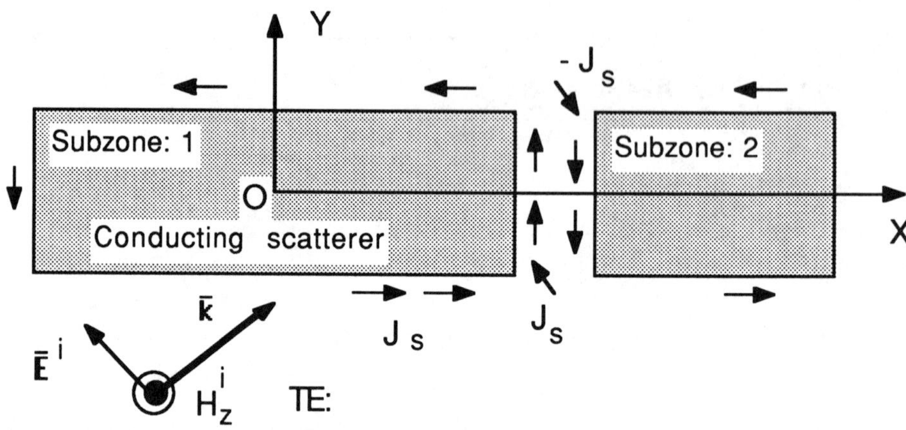

Figure 4.26b Spatial decomposition technique for an electrically large conducting scatterer – TE excitation.

To circumvent the high demand on computer resources and reduce the numerical difficulties, the formulation of the boundary value problem is modified, still retaining all the physics of electromagnetic scattering and interaction as depicted schematically in the Figures 4.26a and 4.26b. Figure 4.26a represents a rectangular scatterer geometry with unknown electric current distribution on a virtual boundary. An identical geometry is repeated in Figure 4.26b, but the virtual boundary contour is modified to define (in this case) two distinct spatial subzones. A key point to note is that at the virtual interface separating the two subzones, the tangential virtual currents on one side of the interface must be equal and opposite to the tangential virtual currents on the other side of the interface. In fact, the scatterer can be divided into an arbitrary number of N distinct spatial subzones in this manner (in Figure 4.26b, $N = 2$). Now, the usual electric field integral equation and method of moments technique is used to compute the electric currents along the enclosing boundary of one subzone. In effect, the subzone is treated as a distinct scatterer. It should be noted that a part of the subzone's boundary is the virtual interface separating it from the adjacent subzone. The excitation for this subzone consists of the original incident plane wave and additional excitation due to the electric currents residing on the boundaries of the remaining spatial subzones and radiating into the free space. Initially, the additional excitation due to the currents on the remaining spatial subzones is not known. However, these can be conveniently approximated to a zeroth order using, for example, the physical optics solution. Hence, for subzone 1, the modified EFIE takes the form

$$F_s^i(\bar{\rho}) = \mathcal{L}_2[J_{s1}(\bar{\rho}')] \qquad \text{for subzone 1, } \bar{\rho} \text{ on } C_1 \tag{4.10.2a}$$

$J_{s1}(\bar{\rho}')$: unknown electric current distribution for the subzone 1,

where the boundary contour C_1 encloses completely the spatial subzone 1, and the total excitation for spatial subzone 1 is given by plane wave excitation and additional excitation due to the remaining spatial subzones, $n = 2, 3, 4, \ldots, N$:

$$F_s^i(\bar{\rho}) = E_s^i(\bar{\rho}) - \sum_{n=2}^{N} \mathcal{L}_2[J_{sn}(\bar{\rho}'_n)] \tag{4.10.2b}$$

for subzone 1, and
$\bar{\rho}'_n$ on n^{th} subzone boundary C_n

As discussed in Section 3.11, the analysis is now shifted to the next adjacent subzone. The excitation for this subzone consists of the original plane wave and additional excitation due to the currents on boundaries of the remaining subzones, including the updated currents on the first subzone. In this manner, the step-by-step analysis approach can be sequentially implemented for rest of the subzones, effectively scanning the original scatterer subzone by subzone, always using the incident plane wave and the latest surface currents as the excitation for the subzone of interest. As can be seen in the SDT iterative process, the subzones are used rather than the complete structure, and the subzones are not completely related to individual blocks of the original full matrix

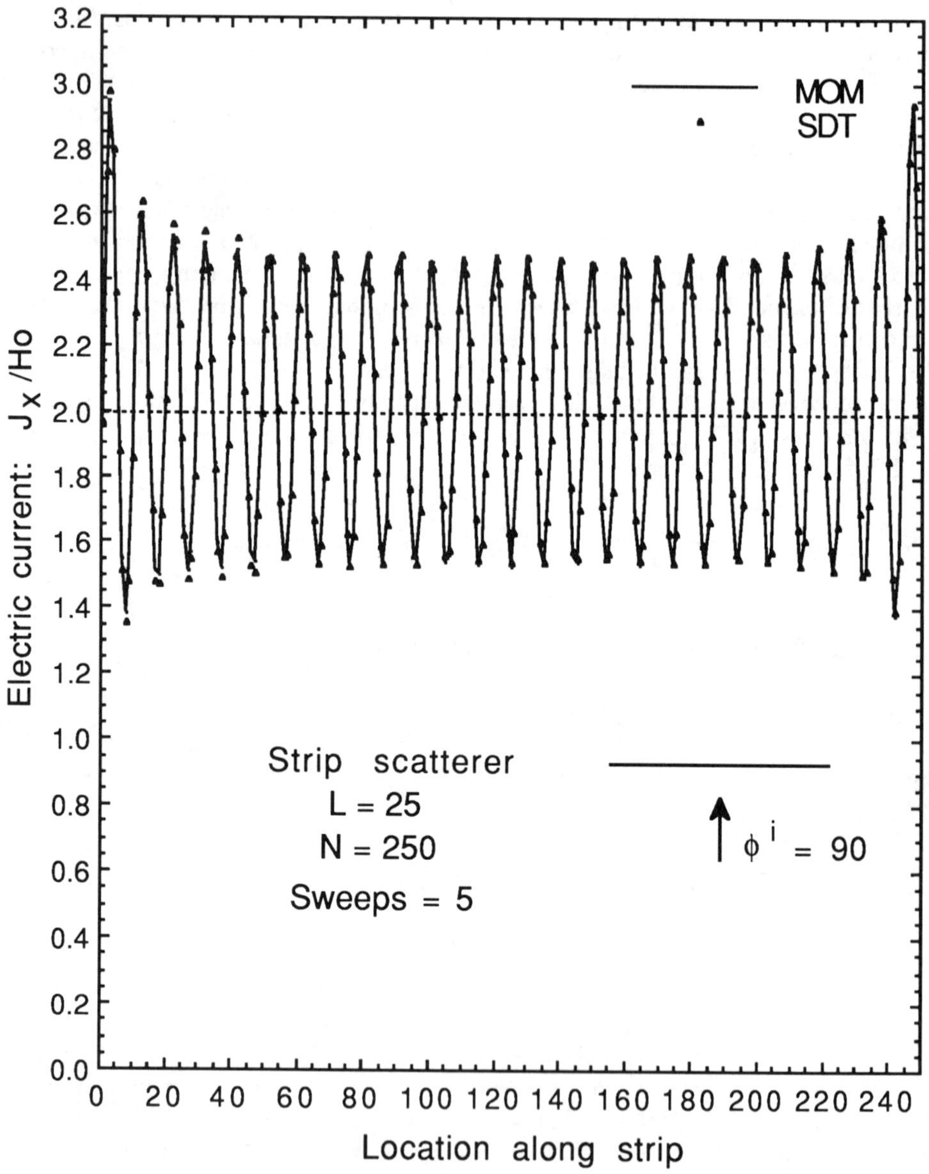

Figure 4.27 Magnitude of electric current distribution on an electrically large thin-strip conducting scatterer, normal incidence – TE excitation.

problem, rather the subzones are defined as distinct targets, having fully enclosing boundaries with additional virtual electric current unknowns (still maintaining the boundary conditions) introduced as needed to define the virtual interface consistently between the subzones. Since each subzone is electrically small, spanning few wavelengths, it is expected that the conditioning of the method of moments system matrix resulting for each subzone is acceptable for numerical processing with limited demands on computer resources.

To illustrate results of the spatial decomposition technique, the electromagnetic scattering by a perfectly conducting thin-strip is considered. The distribution of tangential electric current on the thin-strip scatterer can be directly obtained by applying the method of moments numerical technique using a pulse expansion set and also testing the EFIE expression (4.10.1a) by the same pulses. Figure 4.27 shows a plot of the magnitude of electric current on a thin-strip scatterer of total length, $L = 25\lambda$, excited with an angle of incidence $\phi^i = 90^o$. This result is obtained using a full matrix of size (250 x 250) with a current resolution of 10 pulse samples per wavelength.

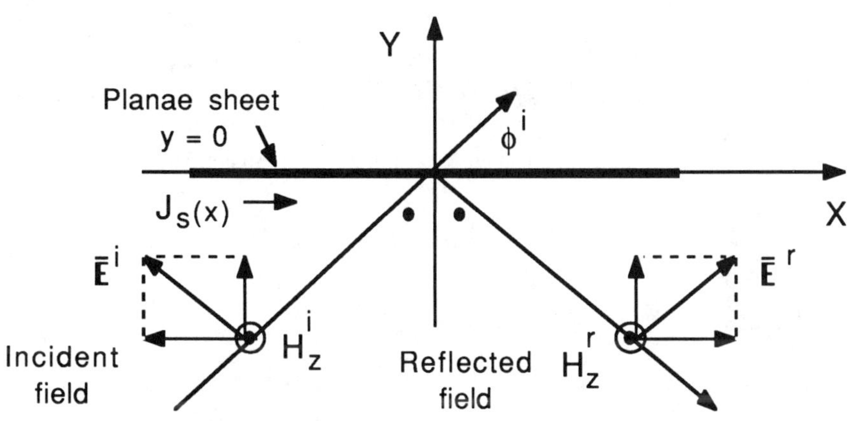

Figure 4.28 Calculation of PO currents on an infinitely large sheet.

To apply the spatial decomposition technique, the thin-strip scatterer is divided, for example, into five subzones, $N = 5$. Maintaining the same current resolution of 10 samples per wavelength, the matrix size is now only (50 x 50) for each subzone modeling. On the lefthand side of SDT expression (4.10.2b), the total excitation for the spatial subzone 1 is given by the plane wave excitation and additional excitation due to the electric currents in remaining spatial subzones, n = 2, 3, 4, and 5. But, the distribution of electric current in the remaining subzones, n = 2, 3, 4, and 5, are not known initially. However, they can be obtained approximately using the physical optics by just considering incident and reflected fields on an infinitely large planar conducting sheet. This is accomplished by considering the following boundary value problem.

The TE polarized incident electric and magnetic fields are given by

$$H_z^i(x,y) = \frac{E_0}{\eta} e^{-j(k_x x+k_y y)} \tag{4.10.3a}$$

$$\vec{E}^i(x,y) = E_0 [-\hat{x} \sin\phi^i + \hat{y} \cos\phi^i] e^{-j(k_x x+k_y y)} \tag{4.10.3b}$$

Using the boundary condition that the s tangential component of total electric field is equal to zero on the scatterer, referring to Figure 4.28, the TE polarized reflected electric and magnetic fields are given by

$$H_z^r(x,y) = \frac{E_0}{\eta} e^{-j(k_x x-k_y y)} \tag{4.10.4a}$$

$$\vec{E}^r(x,y) = E_0 [\hat{x} \sin\phi^i + \hat{y} \cos\phi^i] e^{-j(k_x x-k_y y)} \tag{4.10.4b}$$

The physical optics induced electric current on the infinite planar sheet is obtained by

$$\vec{J}_x(x',y=0) = -\hat{y}' \times \left[\vec{H}^i(x',y=0) + \vec{H}^r(x',y=0)\right] \tag{4.10.5a}$$

$$J_x(x') = -\frac{2E_0}{\eta} e^{-jkx\cos\phi^i} \tag{4.10.5b}$$

With the normal broadside excitation, $\phi^i = 90^o$, this PO expression yields an initial current distribution, which is a flat current with no standing wave oscillations. In fact, the goal of the SDT is to sequentially update this initial current for each subzone. After determining the currents for the subzone 1, the analysis is now shifted to adjacent subzone $n = 2$. The excitation for subzone 2 consists of the original plane wave plus additional excitation due to the approximate currents on boundaries of the remaining subzones, $n = 3$, 4, and 5 including the updated current on the first subzone. This step-by-step analysis approach is sequentially implemented in an iterative sweep for each subzone from one end of the scatterer to the other end. Once the first sweep is completed, a first-order approximate distribution of the electric current on the thin-strip scatterer is numerically obtained. It is noted here that no additional boundary conditions are enforced in the application of SDT. Better approximation of the distribution of magnitude and phase of the electric current on the thin-strip scatterer can be obtained by more iterative sweeps.

Figure 4.27 shows the distribution of magnitude of electric current calculated based on the SDT (with five spatial subzones, $N = 5$) on the thin-strip scatterer excited with an angle of incidence $\phi^i = 90^\circ$. The results shown are obtained for five sweeps with less than one percent error in the region separating two adjacent subzones. It should be noted that the electric currents are valid only for the specified angle of incidence, and the SDT iterative process should be repeated for other angles of incidence.

The distribution of the electric current can now be utilized for calculating either the near electric and magnetic field distributions or the bistatic radar cross section data. In fact, the number of successive iterative sweeps is determined based on the degree of convergence required of the electric current or the radar cross section data. Figure 4.29 shows a plot of the bistatic radar cross section obtained using SDT compared with the direct method of moments solution. In the angular range of $\phi = 30^\circ$ to 150°, the bistatic radar cross section converges in two sweeps with less than one percent error, but for the grazing angles more sweeps are required, and for the results shown five sweeps are utilized. Figure 4.30 shows a plot of the monostatic radar cross section obtained using SDT compared with respect to the direct MOM solution.

Details of the computational burden to obtain bistatic radar cross section of two thin, perfectly conducting strips of lengths 10λ and 25λ using the direct MOM technique and the SDT are presented in Table 4.10 for TE excitation. It can be inferred from the Table 4.10 that the spatial decomposition technique provides a 100:1 savings in computer storage and about 40:1 savings in run time over the direct MOM solution for the larger strip geometry. This comparative savings appear to increase with the size of the scatterer. It may be noted that, as the MOM matrix size increases by a factor of R, the direct MOM run time increases by R^3, whereas the SDT run time increases by R.

Table 4.10

Computer Resources: Direct MOM vs. SDT
(bistatic radar cross section of thin-strip)

Strip size	*Direct MOM matrix size*	*MOM run time (sec)*	*SDT matrix size*	*SDT(*) run time (sec)*
10λ	200	282	50	45
25λ	500	4210	50	109

on Sun 4.0 Workstation (*) run time for one SDT sweep which gives a first-order approximate result.

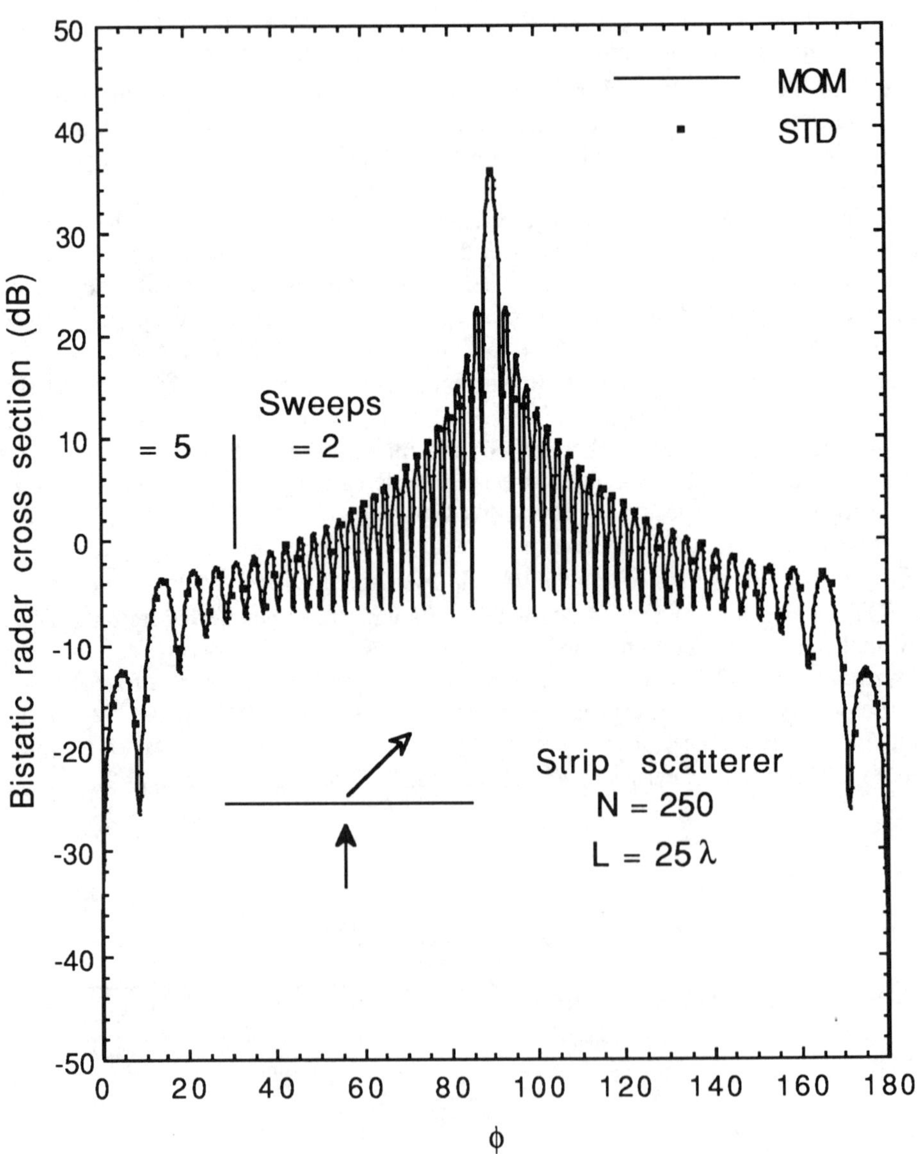

Figure 4.29 Bistatic radar cross section of the electrically large, perfectly conducting thin-strip scatterer - TE excitation.

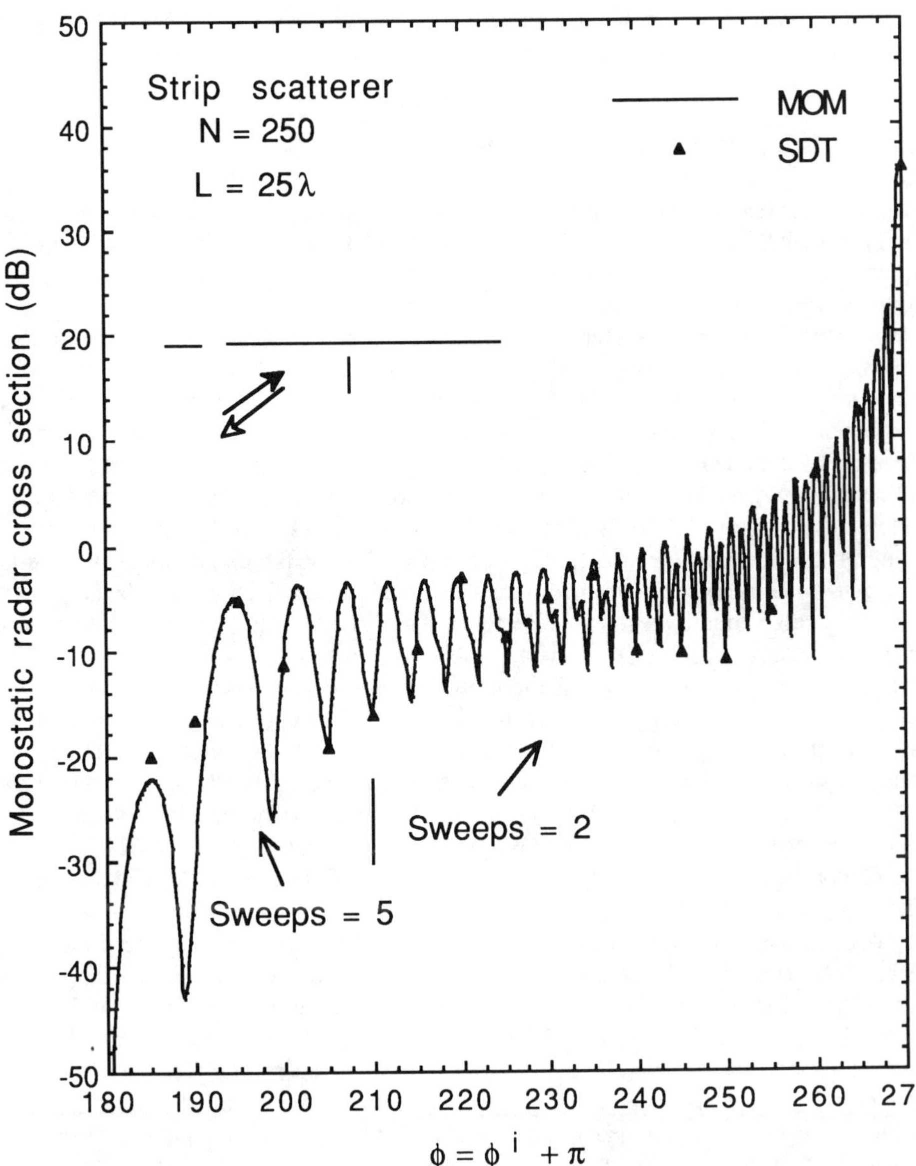

Figure 4.30 Monostatic radar cross section of the electrically large, perfectly conducting thin-strip scatterer - TE excitation.

Other geometric shapes can also be considered based on this SDT approach for TE excitation. Further, similar to the TM case, error analysis can also be repeated using the condition number of the subzone matrices, which are relatively well behaved.

4.11 ANALYSIS OF APERTURES

Apertures represent dual geometries compared to perfectly conducting objects. In Chapter 3, the basic theory is discussed for the TE case to show electromagnetic equivalence as well as duality. In the following sections, the earlier study on apertures is extended for the dual TM excitation.

The analysis of electromagnetic scattering and interaction by an arbitrary shaped, two-dimensional conducting object, either close or open type, has been studied for TE excitation in a systematic manner starting from classical Maxwell's equations. To set up an exterior scattering boundary value problem, based on either EFIE or MFIE, generally the perfectly conducting object is assumed to be located in a large free-space medium that is linear, homogeneous, and isotropic, having constant permittivity and permeability. Referring to Figures 3.40a and 3.40b, for either the closed-type or open-type of scatterer, the induced electric currents on the surface of the conducting object is an unknown quantity to be determined first. Once the induced electric currents are known, then Maxwell's equations can be directly utilized to obtain the scattered electric and magnetic fields in the region of interest.

Specifically, in the open-type scatterer as shown in Figure 3.40b, an aperture region separates an interior cavity region from the exterior unbounded region. The induced electric currents on the surface of the conducting object produce electric and magnetic fields in both interior and exterior regions, including appropriate fields in the aperture region. Even though the formulation of the boundary value problem based on the induced electric current is rigorous, from the numerical analysis point of view the shape and size of the aperture, and the loaded cavity behind it, play a critical role in the process of determining the field distributions.

As discussed in Section 3.12, it is also possible to introduce the concept of electromagnetic equivalence and the duality principle to come up with an *alternative formulation* for the boundary value problem based on the distribution of fields in the aperture region. This is accomplished using the boundary conditions that the total tangential electric and magnetic field distributions in the aperture region are continuous. Initially, the distribution of electric and magnetic fields in the aperture region are unknown quantities. A boundary value problem is set up in terms of either a differential or an integral equation for the unknown aperture electric field distribution or the corresponding equivalent magnetic current distribution. As pointed out earlier, this alternative approach basically leads to the well-known aperture theory, in turn leading to the dual integral equations. In fact, the purpose of this section is to explore and apply familiar integral equations to boundary value problems involving infinitely large, perfectly conducting planar sheet with a perforated aperture or hole in it. Further, the relevance of the integral equation for the study of conducting object with aperture is also discussed for TM excitation.

4.11.1 APERTURE EQUIVALENCES

A detailed discussion on the general aperture equivalences and modeling of the aperture region fields in terms of the magnetic current distribution is presented in Section 3.12.1. Referring to Figure 3.41, let us consider an infinitely large, perfectly conducting screen perforated with an arbitrary shaped aperture or hole located in the *yz* plane separating media 1 and 2, having different permittivity, permeability, and conductivity properties. There are electric and magnetic field distributions in both media 1 and 2, including at the aperture interface separating the two media. The electromagnetic boundary conditions are such that the tangential total electric and magnetic fields in the aperture region are continuous. For the $x < 0$ region covering medium 1, let

ε_1 : permittivity of the left half-space medium 1;

μ_1 : permeability of the left half-space medium 1;

$\sigma_1 = 0$: conductivity of the left half-space medium 1;

$(\bar{E}^i, \bar{H}^i)$: electric and magnetic incident fields in the left half-space medium 1;

$(\bar{E}_1, \bar{H}_1)$: electric and magnetic total fields in the left half-space medium 1.

For the $x > 0$ region covering medium 2, let

ε_2 : permittivity of the right half-space medium 2;

μ_2 : permeability of the right half-space medium 2;

σ_2 : conductivity of the right half-space medium 2;

$(\bar{E}_2, \bar{H}_2)$: electric and magnetic total fields in the right half-space medium 2;

and for the $x = 0$ region, where the conducting screen and the aperture are located, let

$\sigma \Rightarrow \infty$: conductivity of the perfectly conducting screen located at $x = 0$;

$\bar{E}_t^a$: electric field distribution in the aperture region;

$\bar{H}_t^a$: magnetic field distribution in the aperture region.

Left Half-Space – Medium 1 Equivalence

To obtain the electric and magnetic field distributions in medium 1, an equivalent boundary value problem is now set up that is valid only for the $x < 0$ region. The aperture region is short-circuited or physically closed using a perfectly conducting planar sheet having an identical shape and size of the aperture, so that the medium 2 is completely isolated from the medium 1.

Referring to Figure 3.42b, to maintain the distribution of original electric and magnetic fields in medium 1, the effect of aperture radiation is taken into account in terms of equivalent magnetic currents given by

$$\bar{M}_S(\bar{r}') = - \bar{E}_t^a(\bar{r}=0-) \times \hat{x}' \tag{4.11.1}$$

The total fields in medium 1 are obtained by the sum of radiated field from the equivalent magnetic currents just in front of the short-circuited aperture region and the so-called short-circuited field consisting of the incident field plus the reflected field from the uniform perfectly conducting screen. Hence

$$\bar{E}_1(\bar{r}) = \bar{E}_1^{sc}(\bar{r}) + \bar{E}_1^{s}(\bar{r}) \tag{4.11.2a}$$

$$\bar{H}_1(\bar{r}) = \bar{H}_1^{sc}(\bar{r}) + \bar{H}_1^{s}(\bar{r}) \tag{4.11.2b}$$

where the superscript sc stands for the short-circuited fields and the superscript s corresponds to the radiated fields. The conducting screen is eliminated by introducing an image magnetic current distribution in the plane of aperture. Figure 3.43 shows the equivalent boundary value problem for calculating the radiated fields in medium 1. In terms of the dual potential functions

$\bar{F}_1(\bar{r},\omega)$: electric vector potential function at the field point

$$= F_{1x}(\bar{r},\omega)\hat{x} + F_{1y}(\bar{r},\omega)\hat{y} + F_{1z}(\bar{r},\omega)\hat{z} \tag{4.11.3}$$

$\Psi_1(\bar{r},\omega)$: magnetic scalar potential function at the field point;

and the electric and magnetic radiated fields are given by

$$\bar{E}_1^{s}(\bar{r},\omega) = - \frac{1}{\varepsilon'_1} \nabla \times \bar{F}_1(\bar{r},\omega) \tag{4.11.4a}$$

$$\bar{H}_1^{s}(\bar{r},\omega) = - j\omega\bar{F}_1(\bar{r},\omega) - \nabla\Psi_1(\bar{r},\omega) \tag{4.11.4b}$$

and

$$\nabla[\nabla \bullet \bar{F}_1(\bar{r},\omega)] = - j\omega\mu_1\varepsilon'_1 \nabla\Psi_1(\bar{r},\omega) \tag{4.11.5}$$

which is the Lorentz gauge condition. Further, the following vector *Helmholtz* partial differential equation is satisfied by the electric vector potential

$$\nabla^2\bar{F}_1(\bar{r},\omega) + k_1^2\bar{F}_1(\bar{r},\omega) = - \varepsilon'_1[-2\bar{M}_S(\bar{r},\omega)] \tag{4.11.6a}$$

where

ε'_1 : effective permittivity of medium 1

$$= \varepsilon_1\left(1 - j\frac{\sigma_1}{\omega\varepsilon_1}\right) \tag{4.11.6b}$$

k_1 : propagation constant of medium 1

$$= \omega(\mu_1\varepsilon'_1)^{1/2} \tag{4.11.6c}$$

The total electric vector potential at the field point is given by the surface integral over the complete magnetic current density in the source region S:

$$\bar{F}_1(\bar{r}) = \frac{\varepsilon'_1}{4\pi}\iint_S - 2\bar{M}_S(\bar{r}')\,\frac{e^{-jk_1R}}{R}\,ds(\bar{r}') \tag{4.11.7}$$

for $\bar{r}$ in $x < 0$ region

Right Half-Space – Medium 2 Equivalence

Similar expressions can be obtained for the medium 2 as well. To obtain the electric and magnetic field distributions in medium 2, an equivalent boundary value problem is now set up that is valid only for the $x > 0$ region. The aperture region is short-circuited or physically closed using a perfectly conducting planar sheet having an identical shape and size of the aperture, so that medium 1 is completely isolated from medium 2.

Referring to Figure 3.44b, to maintain the distribution of the original electric and magnetic fields in medium 2, the effect of aperture radiation is taken into account in terms of equivalent magnetic currents given by

$$\bar{M}_S(\bar{r}') = \bar{E}_t^a(\bar{r}=0+) \times \hat{n}' \tag{4.11.8}$$

The total fields in medium 2 are obtained directly as the radiated fields from the equivalent magnetic currents just in front of the short-circuited aperture region. It should be noted there is no incident excitation in medium 2. Hence

$$\bar{E}_2(\bar{r}) = \bar{E}_2^S(\bar{r}) \tag{4.11.9a}$$

$$\bar{H}_2(\bar{r}) = \bar{H}_2^S(\bar{r}) \tag{4.11.9b}$$

The conducting screen is eliminated by introducing an image magnetic current distribution in the plane of aperture. Figure 3.45 shows the equivalent boundary value problem for calculating the radiated fields in medium 2. In terms of the dual-potential functions,

$\bar{F}_2(\bar{r},\omega)$: electric vector potential function at the field point

$$= F_{2x}(\bar{r},\omega)\hat{x} + F_{2y}(\bar{r},\omega)\hat{y} + F_{2z}(\bar{r},\omega)\hat{z} \tag{4.11.10}$$

$\Psi_2(\bar{r},\omega)$: magnetic scalar potential function at the field point;

and the electric and magnetic radiated fields are given by

$$\bar{E}_2^S(\bar{r},\omega) = -\frac{1}{\varepsilon'_2} \nabla \times \bar{F}_2(\bar{r},\omega) \tag{4.11.11a}$$

$$\bar{H}_2^S(\bar{r},\omega) = -j\omega\bar{F}_2(\bar{r},\omega) - \nabla\Psi_2(\bar{r},\omega) \tag{4.11.11b}$$

and

$$\nabla[\nabla \bullet \bar{F}_2(\bar{r},\omega)] = -j\omega\mu_2\varepsilon'_2 \nabla\Psi_2(\bar{r},\omega) \tag{4.11.12}$$

which is again the Lorentz gauge condition. Further, the following vector *Helmholtz* partial differential equation is satisfied by the electric vector potential:

$$\nabla^2\bar{F}_2(\bar{r},\omega) + k_2^2\bar{F}_2(\bar{r},\omega) = -\varepsilon'_2[2\bar{M}_S(\bar{r},\omega)] \tag{4.11.13a}$$

ε'_2 : effective permittivity of medium 2

$$= \varepsilon_2\left(1 - j\frac{\sigma_2}{\omega\varepsilon_2}\right) \tag{4.11.13b}$$

k_2 : propagation constant of medium 2

$$= \omega(\mu_2 \varepsilon'_2)^{1/2} \tag{4.11.13c}$$

The total electric vector potential at the field point is given by the surface integral over the complete magnetic current density in the source region S:

$$\bar{F}_2(\bar{r}) = \frac{\varepsilon'_2}{4\pi} \iint_S 2\bar{M}_S(\bar{r}') \frac{e^{-jk_2R}}{R} ds(\bar{r}') \tag{4.11.14}$$

for $\bar{r}$ in $x > 0$ region.

4.12 TWO-DIMENSIONAL APERTURE

The preceding field expressions are applicable for the three-dimensional problem where a perfectly conducting screen with a perforated aperture separates two media, 1 and 2, having different medium properties. In the following, these field expressions are specialized to a two-dimensional problem. This can be accomplished by introducing the spectral domain transformation, as discussed in Section 3.3.

In the following, an analysis procedure is presented for the electromagnetic scattering and penetration by a two-dimensional aperture-conducting screen based on an integral equation formulation. Referring to Figure 4.31, an infinitely large, perfectly conducting screen with a uniform aperture or slot is located in the yz plane. The aperture is oriented with its axis coinciding with the z axis of the coordinate system. Along the axis, the aperture is infinitely long and has a uniform cross section. It is placed in the $x = 0$ plane, and it separates media 1 ($x < 0$) and 2 ($x > 0$). In medium 1, the aperture-screen is excited externally by a time-harmonic transverse magnetic (TM to z) polarized plane wave. As shown in the Figure 4.31, the TM polarized plane wave is incident at an arbitrary angle of incidence with its incident electric field polarized parallel and the corresponding incident magnetic field polarized perpendicular to the z coordinate axis. It should be clearly noted that there is no propagation of the incident plane wave parallel to the axis of the two-dimensional aperture. For this special case of the time-harmonic transverse magnetic *normal* excitation, there is only a transverse component of the induced magnetic currents on the surface $x = 0$ of aperture region. The induced magnetic currents produce only the axial scattered or radiated electric field and the corresponding transverse magnetic field distributions in the two media. The total electric and magnetic field distributions near the surface of the aperture are such that they satisfy appropriate electromagnetic boundary conditions. The distribution of induced magnetic currents on the aperture is still an unknown quantity and can be determined by enforcing the familiar boundary condition that the total tangential component of the magnetic field should be continuous in the aperture region, $x = 0$.

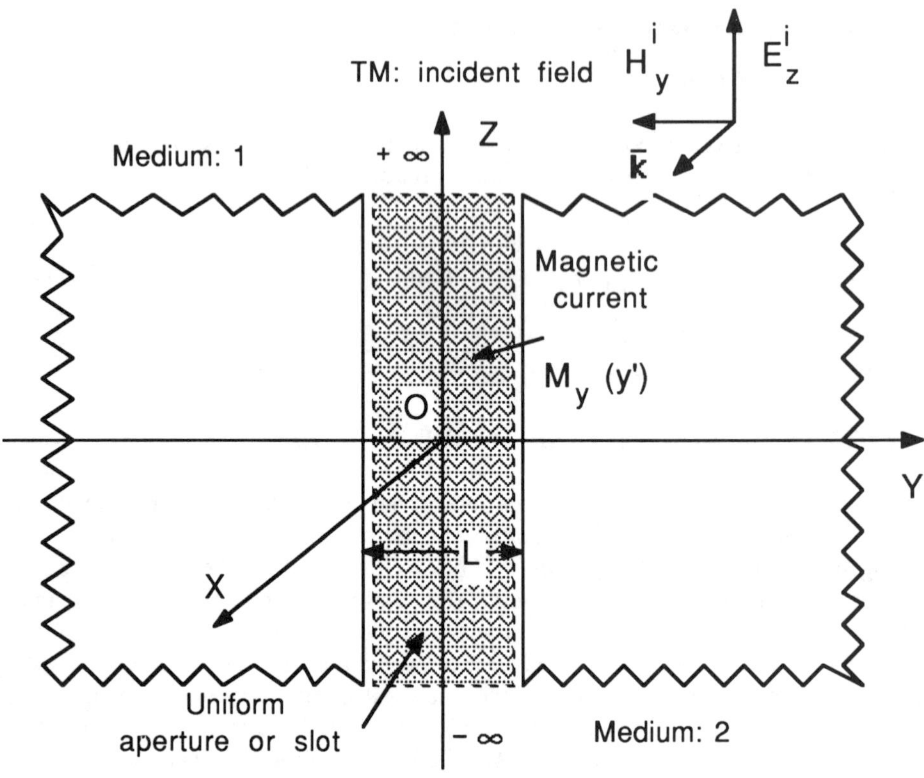

Figure 4.31 Geometry of a two-dimensional aperture-screen – TM excitation.

Similar to the classical approach discussed in Section 4.11 based on the aperture equivalences, the distribution of induced transverse magnetic currents in the aperture region is treated as unknown, and a boundary value integral equation is formulated. It is shown that the integral equation for the aperture distribution with TM excitation is similar to the integral equation of the perfectly conducting, thin-strip scatterer with TE excitation. The two cases form dual boundary value problems so that if the results for one case is known, then results for the other case can be repeated by duality.

Referring to Figure 4.31 and the discussion in Section 2.19 regarding plane wave fields in the frequency domain, the incident plane wave electric and magnetic fields can be written as follows. The plane wave incident electric field vector is polarized along the z axis, and the incident magnetic field vector is polarized in a plane perpendicular to the z axis. There is no propagation of the fields parallel to the z coordinate axis, indicating that the incident excitation fields are uniform and completely independent of the z coordinate variable. In fact, the corresponding field quantities, such as the induced magnetic currents, the scattered or radiated electric and magnetic field distributions in the two media, 1 and 2, are also independent of the *z* coordinate variable.

According to expressions (2.19.11a–b), the TM polarized incident electric and magnetic fields are given by

$$\bar{E}_0 = E_0\hat{z} \tag{4.12.1a}$$

$$\bar{H}_0 = H_{0x}\hat{x} + H_{0y}\hat{y} \tag{4.12.1b}$$

$$\bar{E}^i(\bar{\rho},\omega) = E_z^i(\bar{\rho},\omega)\hat{z} \tag{4.12.2a}$$

$$E_z^i(\bar{\rho},\omega) = E_0\, e^{-j\bar{k}\bullet\bar{\rho}} \tag{4.12.2b}$$

$$\bar{H}^i(\bar{\rho},\omega) = \bar{H}_0\, e^{-j\bar{k}\bullet\bar{\rho}} \tag{4.12.3}$$

$$\bar{k} = k_1\hat{k} \tag{4.12.4}$$

k_1 : propagation constant for homogeneous, isotropic lossless medium

$$= \omega(\mu_1\varepsilon_1)^{1/2} \tag{4.12.5}$$

$\hat{k}$: actual direction of the plane wave propagation;

ω : frequency of excitation, in radians per second;

ϕ^i : arbitrary angle of incidence of the TE plane wave field;

$k_z = 0$: propagation constant along the *z* coordinate direction.

As discussed earlier, the normally excited plane wave is uniform in the *z* coordinate direction, and hence, the axial induced magnetic currents and the radiated electric and magnetic fields in the two media are also uniform with no variation with respect to the *z* coordinate variable. Hence, the various field expressions obtained in the Sections 4.11 can be completely simplified by substituting either

$$\frac{\partial}{\partial z} \Rightarrow 0 \qquad \text{or} \qquad k_z = 0 \tag{4.12.6}$$

It is interesting to note that the spectrum of the spectral variable k_z varies from $-\infty$ to ∞, and the case of normal incident excitation yielding condition (4.12.6) is just a special case to isolate the dependence of various field quantities with respect to the z coordinate variable.

Table 4.11

Transverse magnetic fields for normal excitation in rectangular coordinates

$$\bar{E}^s_m(x,y,\omega) = E^s_{mz}(x,y,\omega)\hat{z} \tag{4.12.7a}$$

$$\bar{H}^s_m(x,y,\omega) = H^s_{mx}(x,y,\omega)\hat{x} + H^s_{my}(x,y,\omega)\hat{y} \tag{4.12.7b}$$

$$\bar{F}(x,y,\omega) = F_{my}(x,y,\omega)\hat{y} \tag{4.12.8}$$

$$E^s_{mz}(x,y,\omega) = -\frac{1}{\varepsilon'_m}\frac{\partial}{\partial x}F_{my}(\bar{r},\omega) \tag{4.12.9}$$

$$H^s_{mx}(x,y,\omega) = -\frac{j\omega}{k_m^2}\frac{\partial^2}{\partial x\partial y}F_{my}(\bar{r},\omega) \tag{4.12.10a}$$

$$H^s_{my}(x,y,\omega) = -j\omega F_{my}(\bar{r},\omega) - \frac{\partial}{\partial y}\Psi_m(\bar{r},\omega) \tag{4.12.10b}$$

for medium $m = 1, x < 0-$

for medium $m = 2, x > 0+$

Table 4.11 gives various scattered or radiated field components corresponding to the TM excitation. Referring to Figures 4.31 and expressions either (4.11.4a–b) or (4.11.11a–b), various field components in the media 1 and 2 can be obtained by substituting

$\bar{F}(x,y,\omega)$: electric vector potential at the field point

$$= F_{1y}(x,y,\omega)\hat{y} \qquad \text{for } x < 0- \tag{4.12.11a}$$

$$= F_{2y}(x,y,\omega)\hat{y} \qquad \text{for } x > 0+ \tag{4.12.11b}$$

where the y component of the electric vector potential satisfies the scalar Helmholtz differential equation based on relationship (4.11.6a) or (4.11.13a). The solutions for the electric vector potentials are given by the expressions (4.11.7) and (4.11.14). Based on condition (4.12.6), the incident fields, short-circuited fields, aperture radiated fields, and the aperture unknown magnetic current distribution are independent of the z coordinate variable and the field quantities are just determined in the xy transverse plane.

4.12.1 APERTURE INTEGRAL EQUATION – TM CASE

Following the aperture equivalences discussed in Sections 3.12 and 4.11, relevant expressions for the total electric and magnetic fields is written in terms of the unknown aperture magnetic current distribution for the two media, 1 and 2. By enforcing the boundary condition that the tangential component of the total magnetic field should be continuous in the aperture region, an integral equation can be deduced.

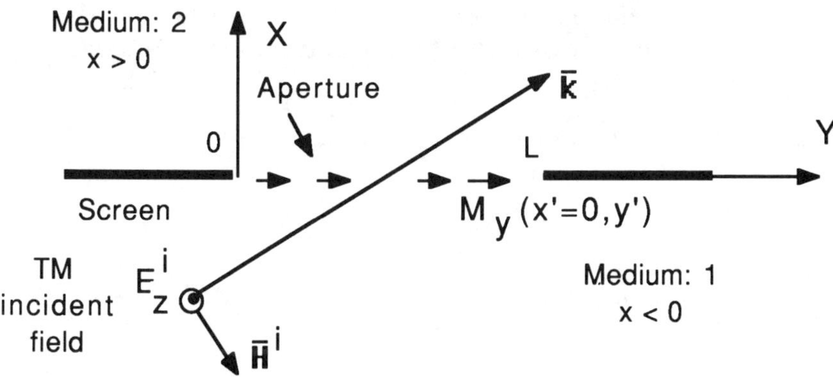

Figure 4.32 Geometry of the TM excited aperture.

Figure 4.32 shows the geometry of a TM excited aperture or slot. In fact, the geometry shown is derived directly from the Figure 4.31 with a transverse cut at the z = 0 plane. The total length of the aperture is L and it separates two media, 1 and 2. For medium 1 ($x < 0$), the effect of the aperture field distribution is taken into account in terms of the y component of equivalent magnetic current given by

$$\overline{M}_y = - \overline{E}_z^a(x=0-,y,z) \times \hat{n}' \tag{4.12.12}$$

The total magnetic field in medium 1 can be obtained as the sum of the radiated field from the equivalent magnetic current and the short-circuited field. Hence

$$\bar{H}_1(x,y) = \bar{H}_1^{SC}(x,y) - j\omega F_{1y}(x,y)\hat{y} - \nabla\Psi_1(x,y) \tag{4.12.13}$$

Referring to Figure 4.32,

$- 2M_y(x'=0-,y')$: magnetic current density distribution;

$\bar{r}'$: position vector corresponding to the source point $(x'=0, y', z')$

$$= y'\hat{y} + z'\hat{z} \tag{4.12.14a}$$

$\bar{r}$: position vector corresponding to the field point $(x, y, z=0)$ in $x < 0$ region, medium 1

$$= x\hat{x} + y\hat{y} \tag{4.12.14b}$$

The y component of electric vector potential at the field point in the xy plane is given by the surface integral over the complete magnetic current density in the aperture region:

$$F_{1y}(\bar{r}) = \frac{\varepsilon'_1}{4\pi}\iint_S - 2M_y(x'=0,y')\,\frac{e^{-jk_1|\bar{r}-\bar{r}'|}}{|\bar{r}-\bar{r}'|}\,dy'dz' \tag{4.12.15a}$$

$$F_{1y}(\bar{r}) = \frac{\varepsilon'_1}{4\pi}\int_0^L - 2M_y(y')\left[\int_{z'=-\infty}^{z'=\infty}\frac{e^{-jk_1R_1}}{R_1}\,dz'\right]dy' \tag{4.12.15b}$$

$$R_1 = \left[(x)^2+(y-y')^2+z'^2\right]^{1/2} \tag{4.12.15c}$$

The bracketed term in expression (4.12.15b) is the well-known integral representation for the Hankel function of zero order and second kind, Appendix B,

$$H_0^{(2)}(k_1R) = \frac{j}{\pi}\int_{z'=-\infty}^{z'=\infty}\frac{e^{-jk_1R_1}}{R_1}\,dz' \tag{4.12.16}$$

After substituting this relationship for the Hankel function into the potential integral representation (4.12.15b), the y component of vector electric potential for the two-dimensional case takes the form

$$F_{1y}(x,y) = \frac{\varepsilon'_1}{4j} \int_0^L -2M_y(y')H_0^{(2)}(k_1R)\,dy' \tag{4.12.17a}$$

Referring to the discussion in Section 2.28, the magnetic scalar potential can also be written in terms of the magnetic current distribution by replacing the magnetic charge in terms of the divergence of magnetic current distribution to obtain

$$\Psi_1(x,y) = \frac{1}{4\omega\mu_1} \int_0^L \frac{\partial}{\partial y'}[-2M_y(y')]H_0^{(2)}(k_1R)\,dy' \tag{4.12.17b}$$

$$R = \left[(x)^2+(y-y')^2\right]^{1/2} \qquad \text{for } (x, y) \text{ in medium 1} \tag{4.12.17c}$$

Similarly, these steps can be repeated for the medium 2 ($x > 0$). The effect of the aperture field distribution is taken into account in terms of the y component of equivalent magnetic current given by

$$\bar{M}_y = \bar{E}_z^a(x=0+,y,z) \times \hat{x}' \tag{4.12.18}$$

The total magnetic field in medium 2 can be obtained as the sum of the radiated field from the equivalent magnetic current. Hence

$$\bar{H}_2(x,y) = -j\omega F_{2y}(x,y)\hat{y} - \nabla\Psi_2(x,y) \tag{4.12.19}$$

Referring to Figure 4.32,

$2M_y(x'=0+,y')$: magnetic current density distribution;

$\bar{r}'$: position vector corresponding to the source point $(x'=0, y', z')$

$$= y'\hat{y} + z'\hat{z} \tag{4.12.20a}$$

$\bar{r}$: position vector corresponding to the field point $(x, y, z=0)$ in $x > 0$ region, medium 2

$$= x\hat{x} + y\hat{y} \tag{4.12.20b}$$

The y component of electric vector potential at the field point in the xy plane is given by the surface integral over the complete magnetic current density in the aperture region:

$$F_{2y}(\bar{r}) = \frac{\varepsilon'_2}{4\pi} \iint_S 2M_y(x'=0,y') \frac{e^{-jk_2|\bar{r}-\bar{r}'|}}{|\bar{r}-\bar{r}'|} dy'dz' \tag{4.12.21a}$$

$$F_{2y}(\bar{r}) = \frac{\varepsilon'_2}{4\pi} \int_0^L 2M_y(y') \left[\int_{z'=-\infty}^{z'=\infty} \frac{e^{-jk_2 R_2}}{R_2} dz' \right] dy' \tag{4.12.21b}$$

$$R_2 = \left[(x)^2 + (y-y')^2 + z'^2 \right]^{1/2} \tag{4.12.21c}$$

The bracketed term in expression (4.12.21b) is the well-known integral representation for the Hankel function of zero order and second kind, Appendix B, given by expression (4.12.16). After substituting relationship (4.12.16) for the Hankel function into the potential integral representation (4.12.21b), the y component of vector electric potential for the two-dimensional case takes the form

$$F_{2y}(x,y) = \frac{\varepsilon'_2}{4j} \int_0^L 2M_y(y') H_0^{(2)}(k_2 R)\, dy' \tag{4.12.22a}$$

and similarly, the magnetic scalar potential can be written as

$$\Psi_2(x,y) = \frac{1}{4\omega\mu_2} \int_0^L \frac{\partial}{\partial y'} \left[2M_y(y') \right] H_0^{(2)}(k_2 R)\, dy' \tag{4.12.22b}$$

$$R = \left[(x)^2 + (y-y')^2 \right]^{1/2} \qquad \text{for } (x, y) \text{ in medium 2} \tag{4.12.22c}$$

To obtain the integral equation for the aperture region, the y components of the total magnetic field expressions (4.12.13) and (4.12.19) are equated at the $x = 0$ plane where the two-dimensional aperture is located. Thus

$$\bar{H}_1(x=0,y) \bullet \hat{y} = \bar{H}_2(x=0,y) \bullet \hat{y} \tag{4.12.23a}$$

$$- j\omega F_{1y}(x=0,y) - \frac{\partial}{\partial y} \Psi_1(x=0,y) + j\omega F_{2y}(x=0,y) + \frac{\partial}{\partial y} \Psi_2(x=0,y) = - \bar{H}_1^{SC}(x=0,y) \bullet \hat{y} \tag{4.12.23b}$$

The short-circuited field on the perfectly conducting screen located on the x = 0 plane is given by negative of twice the incident magnetic field based on the boundary conditions. After substituting for the electric vector potentials and the magnetic scalar potentials for the two media, 1 and 2, the boundary condition (4.12.23b) yields the following integral equation for the TM-excited aperture in terms of the unknown y component of the magnetic current distribution:

$$\frac{\omega\varepsilon'_1}{4}\int_0^L M_y(y')H_0^{(2)}(k_1|y-y'|)\,dy'$$

$$+\frac{1}{4\omega\mu_1}\frac{\partial}{\partial y}\int_0^L \frac{\partial}{\partial y'}\left[M_y(y')\right]H_0^{(2)}(k_1|y-y'|)\,dy'$$

$$+\frac{\omega\varepsilon'_2}{4}\int_0^L M_y(y')H_0^{(2)}(k_2|y-y'|)\,dy'$$

$$+\frac{1}{4\omega\mu_2}\frac{\partial}{\partial y}\int_0^L \frac{\partial}{\partial y'}\left[M_y(y')\right]H_0^{(2)}(k_2|y-y'|)\,dy'$$

$$= H_y^i(x=0,y) \qquad (4.12.24a)$$

The integral equation (4.12.24a) can be solved based on the method of moments techniuqe. The unknown magnetic current distribution can be expanded in terms of a piecewise pulse expansion functions, and the integral equation is weighted on both sides using the same pulse functions to reduce it to a matrix form of equation. For the special case of medium 2 same as medium 1, having identical permittivity and permeability properties:

$$\varepsilon = \varepsilon_1 = \varepsilon_2$$

$$\mu = \mu_1 = \mu_2$$

$$k = k_1 = k_2$$

Integral equation (4.12.24a) for the TM-excited aperture reduces to the following form:

$$\frac{\omega\varepsilon}{4}\int_0^L 2M_y(y')H_0^{(2)}(k|y-y'|)\,dy'$$

$$+\frac{1}{4\omega\mu}\frac{\partial}{\partial y}\int_0^L \frac{\partial}{\partial y'}[2M_y(y')]H_0^{(2)}(k|y-y'|)\,dy'$$

$$= H_y^i(x=0,y) \qquad (4.12.24b)$$

This is in a dual form compared to the integral equation of the thin-strip conducting scatterer discussed in Section 4.8. Based on the concept of duality, it may be noted that, by replacing the permittivity by permeability, and incident magnetic field by incident electric field, the integral equation (4.12.24b) for the TM-excited aperture is identical to integral equation (4.8.23b) for the TE-excited thin-strip scatterer. As far as the unknown quantities are concerned, twice the aperture magnetic current distribution is to be treated like the net electric current distribution on the thin-strip scatterer.

The numerical computer algorithm developed for the thin-strip scatterer can be directly utilized in order to determine the aperture magnetic current distribution. The numerical results reported in section 4.8 for the thin-strip scatterer can be correspondingly interpreted for the TM excited two-dimensional aperture. Based on the duality principle, the scattered electric field and the scattered magnetic field distributions for the thin-strip scatterer are to be treated respectively as the radiated magnetic field and the radiated electric field distributions for the TM excited aperture with appropriate normalization.

Chapter 5

Two-Dimensional Homogeneous Dielectric Object: TM and TE Polarizations

In Chapters 3 and 4, a detailed analysis of the electromagnetic scattering and interaction by a two-dimensional, arbitrary shaped, perfectly conducting object was presented. The perfectly conducting two-dimensional geometries, in fact, form a class of special interaction structures useful for understanding electromagnetic interaction phenomena. With the external incident excitation, the total electric and magnetic field distributions basically exist near and outside the conducting object, and practically no electric and magnetic fields penetrate into the conducting object. Further extension of the scattering and interaction study is discussed in the following sections, which deal with the use of analytical and numerical methods to analyze electromagnetic boundary value problems involving forward- and backscattering from arbitrarily shaped, homogeneous penetrable objects. Both transverse magnetic and transverse electric incident polarizations are considered. Not only is the analysis discussed here applicable for determining important electromagnetic scattering cross-sectional data, it is also useful for studying field penetration and power absorption in a biological environment, scattering by a microstrip, and integrated circuits consisting of perfectly conducting and dielectric boundary regions. Even the study of complicated dielectric rod and optical waveguide structures can be undertaken based on the theory presented.

In this chapter, a detailed analysis of the electromagnetic scattering and interaction by an arbitrary shaped, two-dimensional, homogeneous dielectric object is studied in a systematic manner starting from classical Maxwell's equations. Figure 5.1 shows an arbitrary two-dimensional penetrable object. The geometry of homogeneous dielectric object is assumed to be uniform and infinite in length along the z coordinate axis. Further, the cross-sectional area of the scattering geometry is uniform and the same along its axis which is assumed to coincide with the z axis of coordinate system. As discussed earlier, such a two-dimensional object, which is uniform and infinite in length along its axis, is difficult to simulate. But, this special geometry forms quite a valuable boundary value problem for understanding various aspects of the electromagnetic analytical formulation and subsequent solution based on the numerical technique. Various physics of the electromagnetic scattering, penetration and interaction due to the geometry configuration of the scatterer, such as the straight and curved arbitrary edges, or the sharp wedge and smooth rounded corners, can be extracted conveniently from the two-dimensional study.

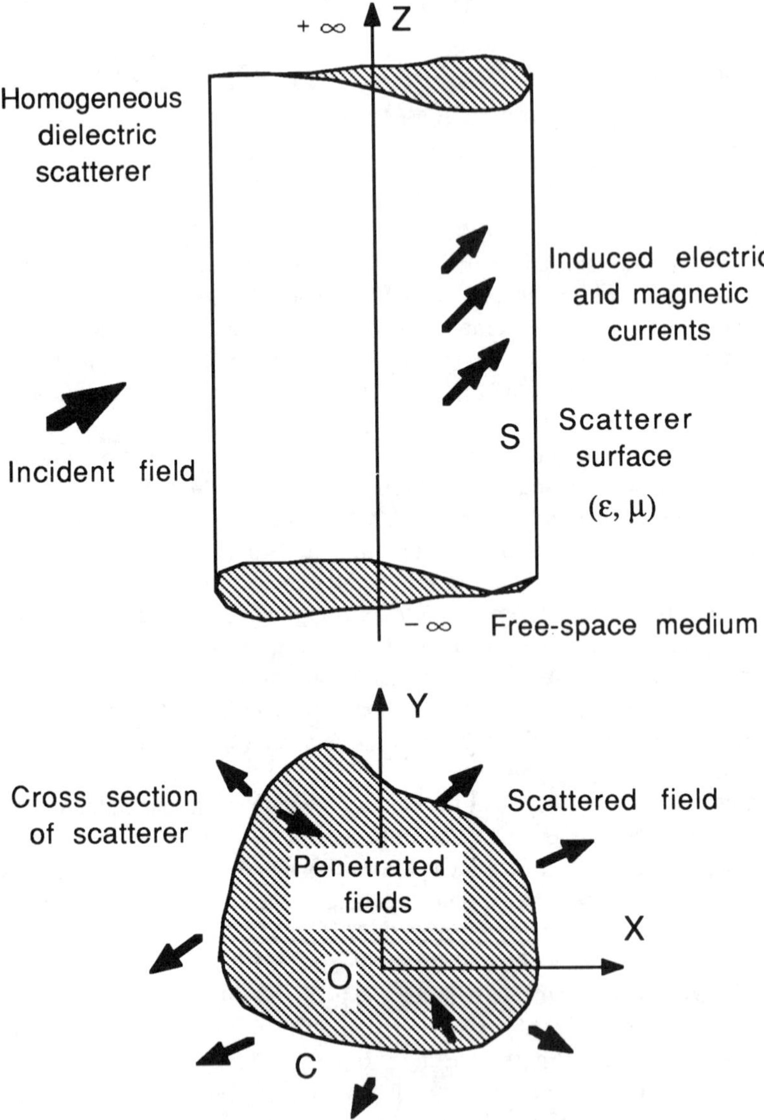

Figure 5.1 Two-dimensional homogeneous dielectric scatterer.

In the following study, the homogeneous dielectric object, Figure 5.1, is assumed to be located in a large free-space medium which is linear, homogeneous, and isotropic, having constant permittivity and permeability. The analysis procedure is applicable both for the lossless and lossy scatterers. The time-harmonic incident electric and magnetic fields are simulated and propagated in the free-space medium by their corresponding time-harmonic primary electric and charge sources, which are located at a large distance from the interacting object under study. For all practical modeling considerations, the incident electric and magnetic excitation fields are taken to be those field distributions existing in the free-space medium in absence of the interacting object. When the two-dimensional homogeneous dielectric object is introduced into the medium, the incident fields interact to induce secondary polarization currents and charges on and inside the penetrable object. These induced polarization currents produce and propagate the scattered electric and magnetic fields in all directions. In the free-space medium, the total fields consist of the sum of the incident and scattered fields. Also, inside the penetrable homogeneous dielectric object, there exists both penetrated electric and magnetic field distributions. Further, the complete electric and magnetic field distributions in the free-space medium and inside the penetrable object are such that they satisfy the familiar electromagnetic boundary conditions. According to the electromagnetic boundary conditions, the total tangential electric field is continuous on the surface of the dielectric object. Similarly, the total tangential magnetic field is continuous on the surface of the dielectric object.

In most electromagnetic scattering, penetration, and interaction studies involving a homogeneous penetrable object, the induced electric and magnetic field distributions on the surface of object are treated as unknown quantities. In fact, the unknown surface electric and magnetic field distributions can be converted into the corresponding unknown equivalent magnetic current and electric current distributions. For the unknown induced surface current distributions, a coupled boundary value problem is set up either in terms of a differential equation or an integral equation or an integro-differential equation subject to the appropriate surface boundary conditions. As discussed earlier, the coupled boundary value equations can be directly solved using analytical method based on an appropriate eigenfunction expansion for a canonical type of scatterer, such as a circular dielectric cylinder. But, for an arbitrary shaped, two-dimensional, homogeneous dielectric object, the boundary value equations can be conveniently solved using the method of moments numerical technique which is suitable for either the differential equations or integral equations. Once the induced surface electric currents and magnetic currents are known, then Maxwell's equations can be directly utilized to obtain the scattered electric and magnetic fields in the two regions of interest. The relevant electromagnetic scattering and penetration properties, such as near-field and far-field distributions can be obtained.

The formulation to analyze homogeneous, isotropic dielectric scatterer is also extended to the study of electromagnetic scattering by a two-dimensional, homogeneous, anisotropic dielectric or magnetic material scatterer.

5.1 GENERAL FIELD EQUATIONS

An electromagnetic coupled boundary value problem can formulated in terms of the unknown surface electric and magnetic field distributions. To accomplish a systematic and rigorous formulation of the coupled electromagnetic boundary value problem, classical Maxwell's equations with both electric current and magnetic current sources are utilized. A complete summary of *Maxwell's equations in the frequency domain* is given in Tables 2.5 and 2.6 with the electric current sources. Similarly, a complete summary of *Maxwell's equations in the frequency domain* is given in Tables 2.7 and 2.8 with the magnetic current sources. The electric and magnetic field distributions in a given region consist of the sum of two partial field distributions generated by the electric and magnetic current sources.

Referring to Figure 5.2, let us consider a large unbounded region which is a linear, homogeneous, and isotropic lossless medium. Let the electric current and charge sources be confined to a small source region, given by

$\rho_{ve}(\bar{r},\omega)$: volume electric charge density, in coulombs per cubic meter;

$\bar{J}_V(\bar{r},\omega)$: volume electric current density, in amperes per square meter.

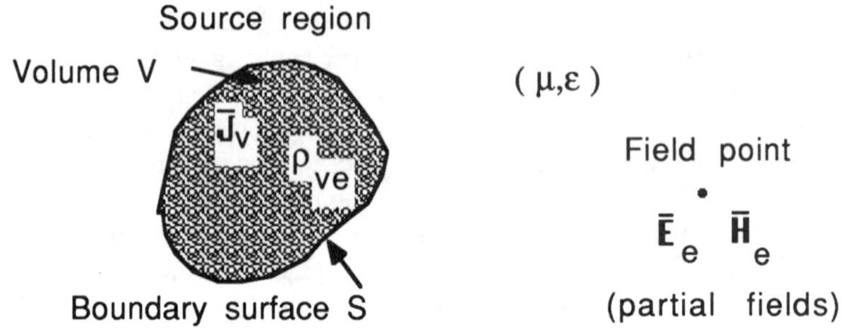

Figure 5.2 Electric current and charge sources in an unbounded medium.

The electric current and charge sources produce the electric and magnetic field distributions both inside and outside the source region. At any field point in the medium, the partial electromagnetic field quantities with the electric type of sources are given by

$\bar{E}_e(\bar{r},\omega)$: electric field distribution, in volts per meter;

$\bar{D}_e(\bar{r},\omega)$: electric flux density distribution, coulombs per square meter;

$\bar{H}_e(\bar{r},\omega)$: magnetic field distribution, in amperes per meter;

$\bar{B}_e(\bar{r},\omega)$: magnetic flux density distribution, in webers per square meter.

For all observation points within the source region, referring to Table 2.5, the partial electric and magnetic fields satisfy the following Maxwell's equations in differential form:

$$\nabla \times \bar{E}_e(\bar{r},\omega) = -j\omega\bar{B}_e(\bar{r},\omega) \tag{5.1.1a}$$

$$\nabla \times \bar{H}_e(\bar{r},\omega) = \bar{J}_V(\bar{r},\omega) + j\omega\bar{D}_e(\bar{r},\omega) \tag{5.1.1b}$$

$\bar{r}$ inside the source region

For an unbounded lossless region assumed to be linear, homogeneous, and isotropic, the various constitutive relationships are given by

$$\bar{D}_e(\bar{r},\omega) = \varepsilon\bar{E}_e(\bar{r},\omega) \tag{5.1.2a}$$

$$\bar{B}_e(\bar{r},\omega) = \mu\bar{H}_e(\bar{r},\omega) \tag{5.1.2b}$$

ε : permittivity of the medium, in farads per meter

μ : permeability of the medium, in henrys per meter

Constitutive relationships (5.1.2a–b) are now substituted into the frequency-dependent Maxwell's equations (5.1.1a–b) valid for the source region. It is noted again that the permeability and permittivity parameters of the medium are constants and depend on neither the frequency parameter variable nor the spatial coordinate variables. Hence, the electric and magnetic fields in the source region satisfy the following frequency-dependent Maxwell's equations

$$\nabla \times \bar{E}_e(\bar{r},\omega) = -j\omega\mu\bar{H}_e(\bar{r},\omega) \tag{5.1.3a}$$

$$\nabla \times \bar{H}_e(\bar{r},\omega) = j\omega\varepsilon\bar{E}_e(\bar{r},\omega) + \bar{J}_V(\bar{r},\omega) \tag{5.1.3b}$$

For the electric and magnetic fields, their corresponding divergence relationships can be derived by utilizing Maxwell's equations (5.1.1a–b), given by

$$\nabla \cdot \varepsilon\bar{E}_e(\bar{r},\omega) = \rho_{Ve}(\bar{r},\omega) \tag{5.1.3c}$$

$$\nabla \bullet \mu \bar{H}_e(\bar{r},\omega) = 0 \tag{5.1.3d}$$

where the electric current and charge sources are related through the continuity equation

$$\nabla \bullet \bar{J}_v(\bar{r},\omega) = - j\omega\rho_{ve}(\bar{r},\omega) \tag{5.1.3e}$$

For a lossy medium, this set of equations still can be utilized by redefining the medium constants as the corresponding effective permittivity and effective permeability. The set of equations defined for the source region, the expressions (5.1.3a–b), are the coupled vector equations. They can be reduced to the scalar form of the coupled equations by assuming a rectangular coordinate system. Using the rectangular coordinate system *(x, y, z)*, at any field point in the medium, the electromagnetic field equations are given by Table 3.1.

Referring to the discussion in Section 2.21, the complete solution for the partial electric and magnetic field quantities can be expressed in terms of two arbitrary potential functions given by a vector magnetic potential and a scalar electric potential. Let

$\bar{A}(\bar{r},\omega)$: magnetic vector potential at the field point

$$= \bar{A}_\tau(\bar{r},\omega) + A_z(\bar{r},\omega)\hat{z} \tag{5.1.4}$$

$\bar{A}_\tau(\bar{r},\omega)$: transverse component of the magnetic vector potential;

$\bar{A}_z(\bar{r},\omega)$: axial component of the magnetic vector potential;

$\Phi(\bar{r},\omega)$: electric scalar potential at the field point.

Then, the expressions for magnetic and electric field distributions are given by

$$\bar{H}_e(\bar{r},\omega) = \frac{1}{\mu} \nabla \times \bar{A}(\bar{r},\omega) \tag{5.1.5}$$

$$\bar{E}_e(\bar{r},\omega) = - j\omega\bar{A}(\bar{r},\omega) - \nabla\Phi(\bar{r},\omega) \tag{5.1.6}$$

where the two potentials satisfy the Lorentz gauge condition

$$\nabla[\nabla \bullet \bar{A}(\bar{r},\omega)] = - j\omega\mu\varepsilon\nabla\Phi(\bar{r},\omega) \tag{5.1.7a}$$

$$\Phi(\bar{r},\omega) = \frac{j\omega}{k^2}\nabla \bullet \bar{A}(\bar{r},\omega) \tag{5.1.7b}$$

By enforcing the gauge condition, the following vector and scalar *Helmholtz* partial differential equations are obtained for the magnetic vector potential and the electric scalar potential:

$$\nabla^2 \bar{A}(\bar{r},\omega) + k^2 \bar{A}(\bar{r},\omega) = -\mu \bar{J}_v(\bar{r},\omega) \tag{5.1.8a}$$

$$\nabla^2 \Phi(\bar{r},\omega) + k^2 \Phi(\bar{r},\omega) = -\frac{\rho_{ve}(\bar{r},\omega)}{\varepsilon} \tag{5.1.8b}$$

$$k^2 = \omega^2 \mu\varepsilon \tag{5.1.8c}$$

Section 2.22 gives a method to solve for the Helmholtz differential equations. Let

$\bar{r}'$: position vector corresponding to the source point (x', y', z')

$$= x'\hat{x} + y'\hat{y} + z'\hat{z} \tag{5.1.9a}$$

$\bar{r}$: position vector corresponding to the field point (x, y ,z)

$$= x\hat{x} + y\hat{y} + z\hat{z} \tag{5.1.9b}$$

R : distance of the line joining the source point to the field point

$$= |\bar{r} - \bar{r}'| \tag{5.1.9c}$$

The magnetic vector potential and the electric scalar potential at the field point $P(x, y, z)$ are given by the corresponding volume integral over the complete electric current and electric charge source distributions in volume V:

$$\bar{A}(\bar{r}) = \frac{\mu}{4\pi} \iiint_V \bar{J}_v(\bar{r}') \frac{e^{-jkR}}{R}\, dv(\bar{r}') \tag{5.1.10}$$

$$\Phi(\bar{r}) = \frac{1}{4\pi\varepsilon} \iiint_V \rho_{ve}(\bar{r}') \frac{e^{-jkR}}{R}\, dv(\bar{r}') \tag{5.1.11}$$

If other types of source distributions, such as the surface currents and charges and the line currents and charges are present, the two potential integrals (5.1.10) and (5.1.11) are correspondingly modified with respect to the surface and line integrals. The components of the electric and magnetic field distributions based on the two arbitrary vector and scalar potentials are given in Table 3.3 in terms of the rectangular coordinate system.

The preceding analysis procedure is repeated below for the magnetic type of sources. Referring to Figure 5.3, let us consider a large unbounded region which is a linear, homogeneous, and isotropic lossless medium. Let the magnetic current and charge sources be confined to a small source region, given by

$\rho_{vm}(\bar{r},\omega)$: volume magnetic charge density, webers per cubic meter;

$\bar{M}_v(\bar{r},\omega)$: volume magnetic current density, in volts per square meter.

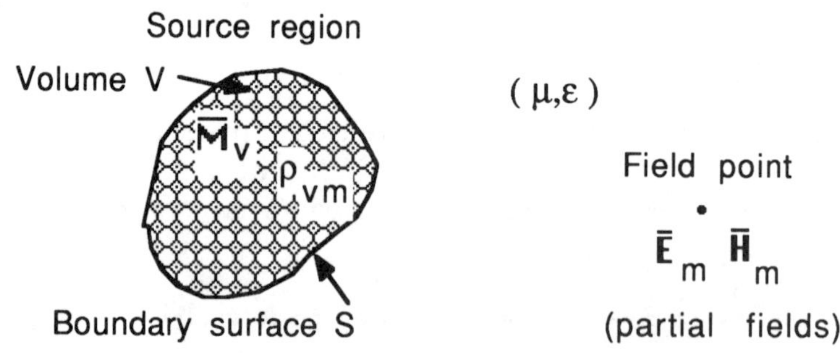

Figure 5.3 Magnetic current and charge sources in an unbounded medium.

The magnetic current and charge sources produce the electric and magnetic field distributions both inside and outside the source region. At any field point in the medium, the partial electromagnetic field quantities with magnetic type of sources are given by

$\bar{E}_m(\bar{r},\omega)$: electric field distribution, in volts per meter;

$\bar{D}_m(\bar{r},\omega)$: electric flux density distribution, coulombs per square meter;

$\bar{H}_m(\bar{r},\omega)$: magnetic field distribution, in amperes per meter;

$\bar{B}_m(\bar{r},\omega)$: magnetic flux density distribution, in webers per square meter.

For all observation points within the source region, referring to Table 2.7, the partial electric and magnetic fields satisfy the following Maxwell's equations:

$$\nabla \times \bar{E}_m(\bar{r},\omega) = - j\omega\bar{B}_m(\bar{r},\omega) - \bar{M}_v(\bar{r},\omega) \tag{5.1.12a}$$

$$\nabla \times \bar{H}_m(\bar{r},\omega) = j\omega\bar{D}_m(\bar{r},\omega) \tag{5.1.12b}$$

$\bar{r}$ inside the source region

For an unbounded lossless region assumed to be linear, homogeneous, and isotropic, the various constitutive relationships are given by

$$\bar{D}_m(\bar{r},\omega) = \varepsilon \bar{E}_m(\bar{r},\omega) \tag{5.1.13a}$$

$$\bar{B}_m(\bar{r},\omega) = \mu \bar{H}_m(\bar{r},\omega) \tag{5.1.13b}$$

ε : permittivity of the medium, in farads per meter

μ : permeability of the medium, in henrys per meter

Constitutive relationships (5.1.13a–b) are now substituted into frequency-dependent Maxwell's equations (5.1.12a–b) for the source region. Then, the electric and magnetic fields in the source region satisfy the following frequency-dependent Maxwell's equations

$$\nabla \times \bar{E}_m(\bar{r},\omega) = -j\omega\mu \bar{H}_m(\bar{r},\omega) - \bar{M}_v(\bar{r},\omega) \tag{5.1.14a}$$

$$\nabla \times \bar{H}_m(\bar{r},\omega) = j\omega\varepsilon \bar{E}_m(\bar{r},\omega) \tag{5.1.14b}$$

For the electric and magnetic fields, their corresponding divergence relationships can be derived by utilizing the Maxwell's equations (5.1.12a–b), given by

$$\nabla \bullet \varepsilon \bar{E}_m(\bar{r},\omega) = 0 \tag{5.1.14c}$$

$$\nabla \bullet \mu \bar{H}_m(\bar{r},\omega) = \rho_{vm}(\bar{r},\omega) \tag{5.1.14d}$$

where the magnetic current and the corresponding charge sources are related through the continuity equation

$$\nabla \bullet \bar{M}_v(\bar{r},\omega) = -j\omega\rho_{vm}(\bar{r},\omega) \tag{5.1.14e}$$

The set of equations defined previouly for the source region, expressions (5.1.14a–b), are the coupled vector equations. Referring to the discussion in Section 2.27, the complete solution for electric and magnetic field quantities can also be expressed in terms of two arbitrary dual potential functions, given by a vector electric potential and a scalar magnetic potential.

Following the discussion presented in Section 2.27, the partial electric field and magnetic field distributions can be represented in terms of the following arbitrary dual potential functions:

$\bar{F}(\bar{r},\omega)$: electric vector potential at the field point

$$= \bar{F}_\tau(\bar{r},\omega) + F_z(\bar{r},\omega)\hat{z} \tag{5.1.15}$$

$\bar{F}_\tau(\bar{r},\omega)$: transverse component of the electric vector potential;

$\bar{F}_z(\bar{r},\omega)$: axial component of the electric vector potential;

$\Psi(\bar{r},\omega)$: magnetic scalar potential at the field point.

Then, the expressions for the electric field and the magnetic field distributions are given by

$$\bar{E}_m(\bar{r},\omega) = -\frac{1}{\varepsilon}\nabla \times \bar{F}(\bar{r},\omega) \tag{5.1.16}$$

$$\bar{H}_m(\bar{r},\omega) = -j\omega\bar{F}(\bar{r},\omega) - \nabla\Psi(\bar{r},\omega) \tag{5.1.17}$$

where the two potentials satisfy the Lorentz gauge condition

$$\nabla[\nabla \bullet \bar{F}(\bar{r},\omega)] = -j\omega\mu\varepsilon\nabla\Psi(\bar{r},\omega) \tag{5.1.18a}$$

$$\Psi(\bar{r},\omega) = \frac{j\omega}{k^2}\nabla \bullet \bar{F}(\bar{r},\omega) \tag{5.1.18b}$$

By enforcing the gauge condition, the following vector and scalar *Helmholtz* partial differential equations are obtained for the electric vector potential and the magnetic scalar potential:

$$\nabla^2\bar{F}(\bar{r},\omega) + k^2\bar{F}(\bar{r},\omega) = -\varepsilon\bar{M}_v(\bar{r},\omega) \tag{5.1.19a}$$

$$\nabla^2\Psi(\bar{r},\omega) + k^2\Psi(\bar{r},\omega) = -\frac{\rho_{vm}(\bar{r},\omega)}{\mu} \tag{5.1.19b}$$

$$k^2 = \omega^2\mu\varepsilon \tag{5.1.19c}$$

Following the discussion in Section 2.27, the solution for Helmholtz differential equations can be written. Let

$\bar{r}'$: position vector corresponding to the source point (x', y', z');

$\bar{r}$: position vector corresponding to the field point (x, y, z);
R : distance of the line joining the source point to the field point

$$= |\bar{r} - \bar{r}'| \tag{5.1.20}$$

The electric vector potential and the magnetic scalar potential at the field point $P(x, y, z)$ are given by the corresponding volume integral over the complete magnetic current and charge source distributions in volume V:

$$\bar{F}(\bar{r}) = \frac{\varepsilon}{4\pi} \iiint_V \bar{M}_V(\bar{r}') \frac{e^{-jkR}}{R} \, dv(\bar{r}') \tag{5.1.21}$$

$$\Psi(\bar{r}) = \frac{1}{4\pi\mu} \iiint_V \rho_{vm}(\bar{r}') \frac{e^{-jkR}}{R} \, dv(\bar{r}') \tag{5.1.22}$$

As mentioned earlier, if other types of source distributions, such as the surface currents and charges and the line currents and charges are present, then dual potential integrals (5.1.21) and (5.1.22) are to be modified with respect to the surface and line integrals. With these representations for the partial electric and magnetic fields in terms of the two sets of dual vector and scalar potentials, it is now possible to write the composite electric and magnetic field expressions as the sum of two partial solutions defined in expressions (5.1.5), (5.1.6) and (5.1.16), (5.1.17). Hence

$$\bar{E}(\bar{r}) = \bar{E}_e(\bar{r}) + \bar{E}_m(\bar{r}) \tag{5.1.23}$$

$$\bar{H}(\bar{r}) = \bar{H}_e(\bar{r}) + \bar{H}_m(\bar{r}) \tag{5.1.24}$$

5.2 FIELD EQUATIONS IN THE SPECTRAL DOMAIN

The electromagnetic field expressions, derived in the previous section, are in terms of the electromagnetic potential functions suitable for the analysis of general three-dimensional material boundary value problems. To get insight into the electromagnetic scattering, penetration, and interaction problem involving two-dimensional homogeneous dielectric objects, the three-dimensional fields are now reduced to the specific two-dimensional fields corresponding to the transverse magnetic and the transverse electric polarizations. These two polarizations were studied in detail in the previous two chapters for perfectly conducting scatterers. As pointed out, it is possible to simplify the analysis by transforming all the sources and their field quantities from the z spatial coordinate domain to the corresponding spectral transform domain. This process basically converts the three-dimensional sources and their field quantities to the corresponding two-dimensional sources and their field quantities. As discussed in Chapters 3 and 4, this is similar to defining an equivalent Fourier integral transform to convert all the sources, potentials, and fields with (x, y, z) dependence to the corresponding (x, y, k_z) dependence, where k_z is the spectral transform parameter. To get back the original field quantities with (x, y, z) dependence, the inverse integral transform is performed for the spectral domain field quantities. The concept of the spectral transform domain is briefly discussed in Section 3.3.

Let $F(z)$ represent an arbitrary function in the z coordinate domain. Then, function $F(z)$ can be transformed into the spectral domain function $F(k_z)$ based on the following direct spectral integral transform:

$$\mathcal{S}[F(z)] = F(k_z) = \frac{1}{\sqrt{2\pi}} \int_{-\infty}^{\infty} F(z)\, e^{jk_z z} dz \tag{5.2.1}$$

where k_z is the spectral domain variable and is similar to a propagation phase constant along the z coordinate axis, in radians per meter. The function $F(k_z)$ is always a complex function referred to as a *phasor* quantity in the two-dimensional coordinate domain having both magnitude and phase variation as a function of the spectral parameter variable k_z in the range varying from $-\infty$ to ∞.

The original function $F(z)$ can be recovered by performing the inverse integral transform:

$$\mathcal{S}^{-1}[F(k_z)] = F(z) = \frac{1}{\sqrt{2\pi}} \int_{-\infty}^{\infty} F(k_z)\, e^{-jk_z z} dk_z \tag{5.2.2}$$

Based on these integral transforms, all the relevant electromagnetic equations are converted into the spectral domain. It can be noted that the basic transformation from the z coordinate domain to the spectral transform domain involves defining the various electromagnetic sources and the corresponding electric and magnetic fields to vary as

$$\bar{E}(x,y,z) = \bar{E}(x,y,k_z)\, e^{-jk_z z} \tag{5.2.3a}$$

$$\bar{H}(x,y,z) = \bar{H}(x,y,k_z)\, e^{-jk_z z} \tag{5.2.3b}$$

Hence, the various electromagnetic equations can be transformed using the following operators given by

$$\frac{\partial}{\partial z} \leftrightarrow -jk_z \tag{5.2.4a}$$

$$\frac{\partial^2}{\partial z^2} \leftrightarrow -k_z^2 \tag{5.2.4b}$$

$$\int \partial z \leftrightarrow \frac{1}{-jk_z} \tag{5.2.4c}$$

5.2.1 TM AND TE POLARIZATIONS

Classical Maxwell's equations for the source-free region are simplified in the following by expressing the vector electric and magnetic fields in terms of their axial and transverse components. Referring to Figure 5.4, the vector electric and magnetic fields are resolved into two distinct components, namely the axial electric and magnetic field components, which are parallel to the z coordinate axis and the transverse components, parallel to a plane perpendicular to the z coordinate axis.

Thus, the transverse components of electric and magnetic fields can be expressed in the following form:

$$\bar{E}_\tau(\bar{r},\omega) = \bar{E}_\tau(x,y,\omega)\, e^{-jk_z z} \tag{5.2.5a}$$

$$= \hat{\tau} E_\tau(x,y,\omega)\, e^{-jk_z z} \tag{5.2.5b}$$

$$\bar{H}_\tau(\bar{r},\omega) = \bar{H}_\tau(x,y,\omega)\, e^{-jk_z z} \tag{5.2.5c}$$

and similarly, the axial components of electric field and magnetic field can be expressed in the following form:

$$\bar{E}_z(\bar{r},\omega) = \bar{E}_z(x,y,\omega)\, e^{-jk_z z} \tag{5.2.5d}$$

$$= \hat{z} E_z(x,y,\omega)\, e^{-jk_z z} \tag{5.2.5e}$$

$$\bar{H}_z(\bar{r},\omega) = \bar{H}_z(x,y,\omega)\, e^{-jk_z z} \tag{5.2.5f}$$

$$= \hat{z} H_z(x,y,\omega)\, e^{-jk_z z} \tag{5.2.5g}$$

where

$\hat{\tau}$: unit vector in the transverse plane;
$\hat{z}$: unit vector in the axial z coordinate direction;
k_z : propagation constant in the z coordinate direction.

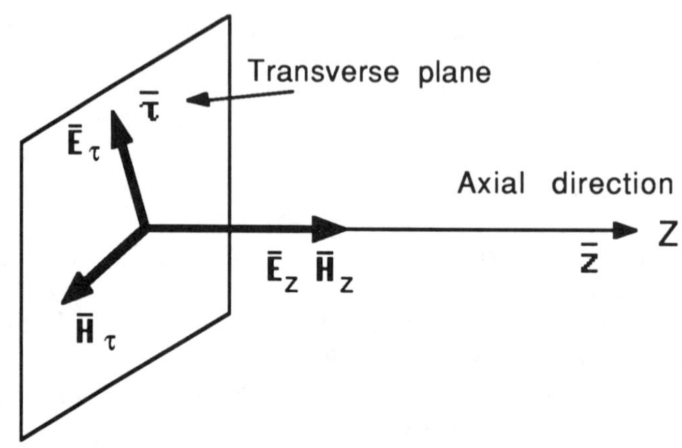

Figure 5.4 Components of general electric and magnetic fields.

These field representations are now substituted into the two coupled Maxwell's equations to determine the relationships between the axial and the transverse components of the electric and magnetic fields. For the source-free region, using Faraday's law expressions (5.1.3a) and (5.1.14a),

$$\nabla \times \bar{E}(\bar{r},\omega) = - j\omega\mu\bar{H}(\bar{r},\omega) \tag{5.2.6a}$$

$$\left[\frac{\partial}{\partial x}\hat{x} + \frac{\partial}{\partial y}\hat{y} + \frac{\partial}{\partial z}\hat{z}\right] \times \bar{E}(\bar{r},\omega) = - j\omega\mu\bar{H}(\bar{r},\omega) \tag{5.2.6b}$$

$$\left[\nabla_\tau + \frac{\partial}{\partial z}\hat{z}\right] \times \bar{E}(\bar{r},\omega) = - j\omega\mu\bar{H}(\bar{r},\omega) \tag{5.2.6c}$$

where the transverse operator in the rectangular coordinate system is given by

$$\nabla_\tau = \frac{\partial}{\partial x}\hat{x} + \frac{\partial}{\partial y}\hat{y} \tag{5.2.6d}$$

The electric and magnetic fields in expression (5.2.6c) are now replaced in terms of the axial and transverse components of the electric and magnetic field distributions defined in expressions (5.2.5a) to (5.2.5g):

$$\left[\nabla_\tau + \frac{\partial}{\partial z}\hat{z}\right] \times [\bar{E}_\tau(x,y,\omega) + \bar{E}_z(x,y,\omega)]\, e^{-jk_z z}$$
$$= -j\omega\mu[\bar{H}_\tau(x,y,\omega) + \bar{H}_z(x,y,\omega)]\, e^{-jk_z z} \qquad (5.2.6e)$$

$$\nabla_\tau \times \bar{E}_\tau(x,y,\omega) + (-jk_z)\hat{z} \times \bar{E}_\tau(x,y,\omega) - \hat{z} \times \nabla_\tau E_z(x,y,\omega)$$
$$= -j\omega\mu\bar{H}_\tau(x,y,\omega) - j\omega\mu\bar{H}_z(x,y,\omega) \qquad (5.2.6f)$$

The axial and transverse components are now equated in the preceding expression to obtain the following field relationships:

$$\nabla_\tau \times \bar{E}_\tau(x,y,\omega) = -j\omega\mu\bar{H}_z(x,y,\omega) \qquad (5.2.7a)$$

$$jk_z\hat{z} \times \bar{E}_\tau(x,y,\omega) + \hat{z} \times \nabla_\tau E_z(x,y,\omega) = j\omega\mu\bar{H}_\tau(x,y,\omega) \qquad (5.2.7b)$$

Similarly, using Ampere's law expressions given by (5.1.3b) and (5.1.14b),

$$\nabla \times \bar{H}(\bar{r},\omega) = j\omega\varepsilon\bar{E}(\bar{r},\omega) \qquad (5.2.8a)$$

$$\left[\nabla_\tau + \frac{\partial}{\partial z}\hat{z}\right] \times \bar{H}(\bar{r},\omega) = j\omega\varepsilon\bar{E}(\bar{r},\omega) \qquad (5.2.8b)$$

The electric and magnetic fields in expression (5.2.8b) are now replaced in terms of the axial and transverse components of the electric and magnetic field distributions defined in expressions (5.2.5a) to (5.2.5g):

$$\left[\nabla_\tau + \frac{\partial}{\partial z}\hat{z}\right] \times [\bar{H}_\tau(x,y,\omega) + \bar{H}_z(x,y,\omega)]\, e^{-jk_z z}$$
$$= j\omega\varepsilon[\bar{E}_\tau(x,y,\omega) + \bar{E}_z(x,y,\omega)]\, e^{-jk_z z} \qquad (5.2.8c)$$

$$\nabla_\tau \times \bar{H}_\tau(x,y,\omega) + (-jk_z)\hat{z} \times \bar{H}_\tau(x,y,\omega) - \hat{z} \times \nabla_\tau H_z(x,y,\omega)$$
$$= j\omega\varepsilon\bar{E}_\tau(x,y,\omega) + j\omega\varepsilon\bar{E}_z(x,y,\omega) \qquad (5.2.8d)$$

The axial and transverse components are then equated in the preceding expression to obtain the following field relationships:

$$\nabla_\tau \times \bar{H}_\tau(x,y,\omega) = j\omega\varepsilon\bar{E}_z(x,y,\omega) \tag{5.2.9a}$$

$$jk_z\hat{z} \times \bar{H}_\tau(x,y,\omega) + \hat{z} \times \nabla_\tau H_z(x,y,\omega) = -j\omega\varepsilon\bar{E}_\tau(x,y,\omega) \tag{5.2.9b}$$

Further, Gauss's law divergence expressions for electric and magnetic fields can be simplified. Referring to expressions (5.1.3c) and (5.1.14c) for source-free region,

$$\nabla \bullet \bar{E}(\bar{r},\omega) = 0 \tag{5.2.10a}$$

After substituting for the electric field in terms of the axial and transverse components, expression (5.2.10a) reduces to

$$\left[\nabla_\tau + \frac{\partial}{\partial z}\hat{z}\right] \bullet [\bar{E}_\tau(x,y,\omega) + \bar{E}_z(x,y,\omega)]\, e^{-jk_z z} = 0 \tag{5.2.10b}$$

$$\nabla_\tau \bullet \bar{E}_\tau(x,y,\omega) = jk_z E_z(x,y,\omega) \tag{5.2.11}$$

Similarly, the divergence of the magnetic field as defined in expressions (5.1.3d) and (5.1.14d) is repeated for the source-free region,

$$\nabla \bullet \bar{H}(\bar{r},\omega) = 0 \tag{5.2.12a}$$

After substituting for the magnetic field in terms of the axial and transverse components, expression (5.2.12a) reduces to

$$\left[\nabla_\tau + \frac{\partial}{\partial z}\hat{z}\right] \bullet [\bar{H}_\tau(x,y,\omega) + \bar{H}_z(x,y,\omega)]\, e^{-jk_z z} = 0 \tag{5.2.12b}$$

$$\nabla_\tau \bullet \bar{H}_\tau(x,y,\omega) = jk_z H_z(x,y,\omega) \tag{5.2.13}$$

Hence, the set of coupled field expressions given by (5.2.7a,b), (5.2.9a,b), (5.2.11), and (5.2.13) are to be solved together to obtain the complete field distributions in a given material region subject to the appropriate electromagnetic boundary conditions. Table 5.1 gives a complete summary of these derived field equations connecting the axial and the transverse electric and magnetic field components. These are intricate coupled field relationships and, in general, are quite complicated to solve.

Table 5.1

General field expressions in terms of axial and transverse components

---

$$E_z \neq 0 \text{ and } H_z \neq 0$$

Faraday's law:

$$\nabla_\tau \times \bar{E}_\tau(x,y,\omega) = -j\omega\mu\bar{H}_z(x,y,\omega) \qquad (5.2.14a)$$

$$jk_z\hat{z} \times \bar{E}_\tau(x,y,\omega) + \hat{z} \times \nabla_\tau E_z(x,y,\omega) = j\omega\mu\bar{H}_\tau(x,y,\omega) \qquad (5.2.14b)$$

Ampere's law:

$$\nabla_\tau \times \bar{H}_\tau(x,y,\omega) = j\omega\varepsilon\bar{E}_z(x,y,\omega) \qquad (5.2.14c)$$

$$jk_z\hat{z} \times \bar{H}_\tau(x,y,\omega) + \hat{z} \times \nabla_\tau H_z(x,y,\omega) = -j\omega\varepsilon\bar{E}_\tau(x,y,\omega) \qquad (5.2.14d)$$

Gauss's law:

$$\nabla_\tau \bullet \bar{E}_\tau(x,y,\omega) = jk_z E_z(x,y,\omega) \qquad (5.2.14e)$$

$$\nabla_\tau \bullet \bar{H}_\tau(x,y,\omega) = jk_z H_z(x,y,\omega) \qquad (5.2.14f)$$

In order to obtain an insight into the two-dimensional electromagnetic boundary value problem involving homogeneous dielectric scatterers, specific excitations yielding either the transverse magnetic (TM) polarization or the transverse electric (TE) polarization of the electric and magnetic fields are considered. These polarizations can be conveniently identified based on a specific type of either an axial component of the electric field or an axial component of the magnetic field distribution which can exist in a specific boundary value problem depending upon the polarization of the incident excitation. Referring back to the earlier discussion in Chapter 3, for the case of transverse (to z coordinate axis) magnetic polarization, there is only the axial component of electric field distribution and the axial component of magnetic field distribution zero as depicted in Figure 5.5a.

The general coupled field relationships obtained in the Table 5.1 are now specialized in Table 5.2 for the case of transverse magnetic (TM) polarization. Further, it is assumed that the external field excitations are selected in such way that the various electric and magnetic field distributions are completely independent of the axial z coordinate variable.

Detailed analysis procedure is taken up in a later section to analyze homogeneous dielectric scatterers for TM excitation. In fact, the analysis for the TE excitation can be shown to be dual of the TM case.

The normal transverse magnetic excitation case is basically derived from

$$\frac{\partial}{\partial z} = 0 \qquad \text{or} \qquad k_z = 0 \tag{5.2.15}$$

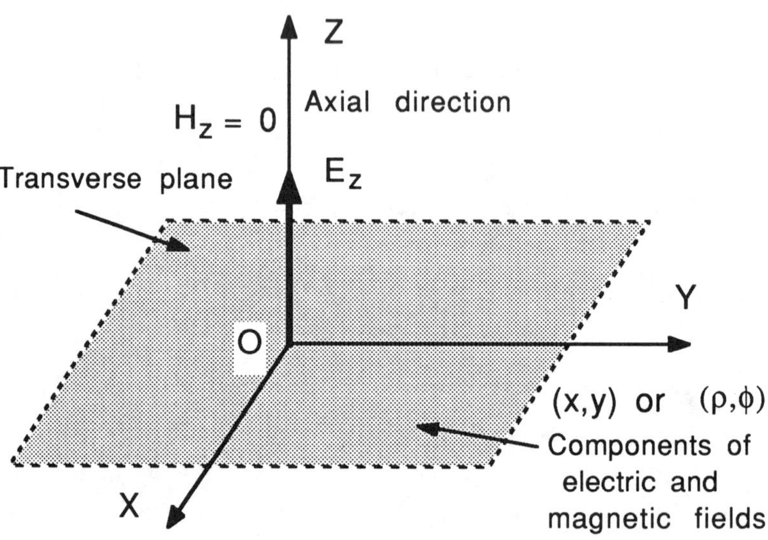

Figure 5.5a Transverse magnetic fields.

Table 5.2
General field expressions for the TM case

$E_z \neq 0$ and $H_z = 0$

$$\hat{z} \times \nabla_\tau E_z(x,y,\omega) = j\omega\mu \bar{H}_\tau(x,y,\omega) \tag{5.2.16a}$$

$$\nabla_\tau \times \bar{H}_\tau(x,y,\omega) = j\omega\varepsilon \bar{E}_z(x,y,\omega) \tag{5.2.16b}$$

$$\nabla_\tau \bullet \bar{H}_\tau(x,y,\omega) = 0 \tag{5.2.16c}$$

Referring to Table 5.2 and Figure 5.5a, for the special transverse magnetic polarization, only an axial component of the electric field distribution is parallel to the z coordinate axis and a transverse component of the magnetic field distribution is parallel to the transverse plane. Suppose, a generalized cylindrical coordinate system given by the orthogonal coordinate variables (n, s, z) is selected corresponding to the two-dimensional arbitrary cross sectional geometry, Figure 5.1. Then, the orthogonal coordinate variables (n, s) are in a plane parallel to the transverse plane and the coordinate variables (s, z) are oriented tangential to the cylindrical surface as discussed previously in Figure 4.6. Further, the transverse component of the magnetic field, which is parallel to the transverse plane, can be resolved into two distinct orthogonal components along the coordinate variables.

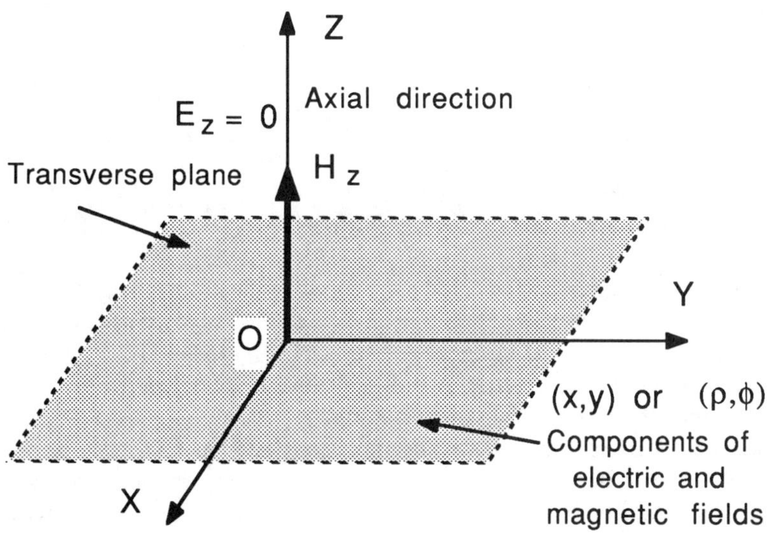

Figure 5.5b Transverse electric fields.

Similar expressions are presented in Table 5.3 for the case of transverse electric excitation. Referring back to the discussions in Chapter 4, for transverse (to z coordinate axis) electric polarization, there is only an axial component of magnetic field distribution, and the axial component of electric field distribution is zero as depicted in Figure 5.5b. Again, the external field excitations are selected in such way that the various electric and magnetic field distributions are completely independent of the z coordinate variable.

As can be seen from Table 5.3 for the special transverse electric polarization, only an axial component of the magnetic field distribution is parallel to the z coordinate axis and a transverse component of the electric field distribution is parallel to the transverse plane. The transverse component of the electric field parallel to transverse plane, can be resolved into two distinct orthogonal components along the coordinate variables *(n, s)*.

Table 5.3
General field expressions for the TE case

$$E_z = 0 \text{ and } H_z \neq 0$$

$$\nabla_\tau \times \bar{E}_\tau(x,y,\omega) = -j\omega\mu\bar{H}_z(x,y,\omega) \tag{5.2.17a}$$

$$\hat{z} \times \nabla_\tau H_z(x,y,\omega) = -j\omega\varepsilon\bar{E}_\tau(x,y,\omega) \tag{5.2.17b}$$

$$\nabla_\tau \bullet \bar{E}_\tau(x,y,\omega) = 0 \tag{5.2.17c}$$

5.3 FORMULATION OF INTEGRAL EQUATIONS

A detailed analysis procedure is presented for the case of electromagnetic scattering and penetration by the two-dimensional homogeneous dielectric object based on a rigorous boundary value *combined field integral equation* (CFIE) formulation. Referring to Figure 5.6a, a homogeneous dielectric, two-dimensional lossy object is oriented with its axis coinciding with the *z* axis of cylindrical coordinate system.

Along the axis, the dielectric object is infinitely long and has uniform arbitrary cross section. It is placed in a linear, homogeneous, and isotropic lossless medium and excited externally by a time-harmonic transverse magnetic (TM to *z*) polarized plane wave. The TM-polarized plane wave is incident on the dielectric object at an arbitrary angle of incidence with the incident electric field polarized parallel and the corresponding incident magnetic field polarized perpendicular to the *z* coordinate axis. It should be noted that there is no propagation of the incident plane wave parallel to the axis of the dielectric object. Various electric and magnetic fields, namely, the incident field, the scattered field, and the penetrated field quantities, are completely independent of the *z* coordinate variable. Hence, it is sufficient to analyze the electromagnetic scattering and penetration in the transverse plane $z = 0$, which is shown in Figure 5.6b.

For the special case of time-harmonic transverse magnetic *normal* excitation, there are only two distinct electromagnetic field components, Table 5.2, as discussed rigorously in the previous section. For the analysis of an arbitrary cross-sectional geometry, the generalized cylindrical coordinate system given by the orthogonal coordinate variables *(n, s, z)* is selected.

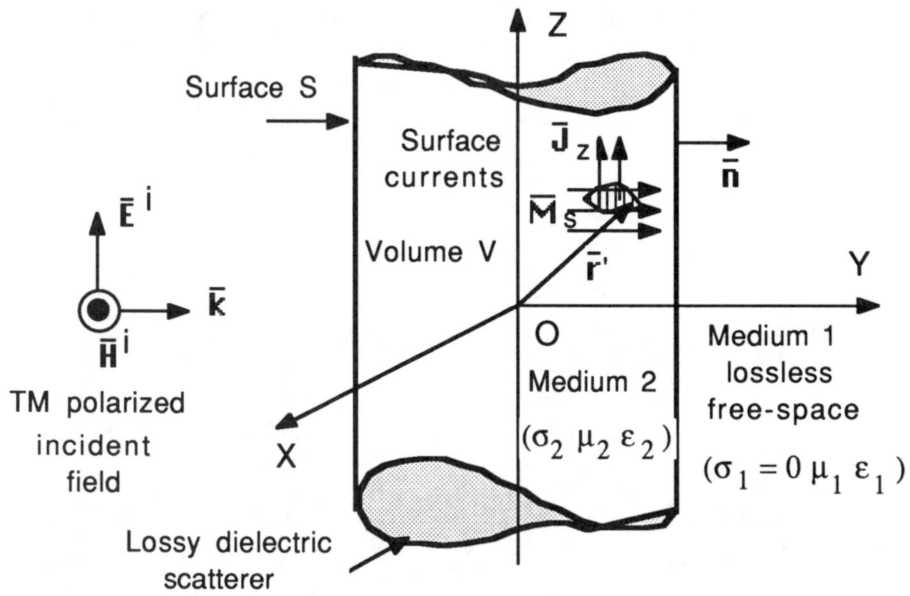

Figure 5.6a Homogeneous dielectric two-dimensional scatterer – TM excitation.

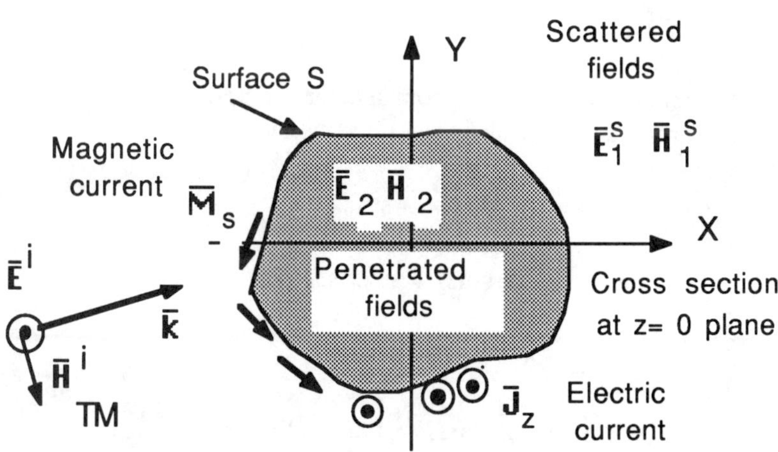

Figure 5.6b Cross sectional geometry – TM excitation.

Figure 5.7 shows various interrelationships between orthogonal unit vectors corresponding to the points on the surface of the arbitrary cross-sectional geometry. At the boundary surface separating the dielectric medium and the free-space medium, the coordinates variable *(n, s)* are in a plane parallel to the transverse plane and the coordinates variables *(s, z)* are oriented tangential to the cylindrical boundary surface as depicted in the Figure 5.7.

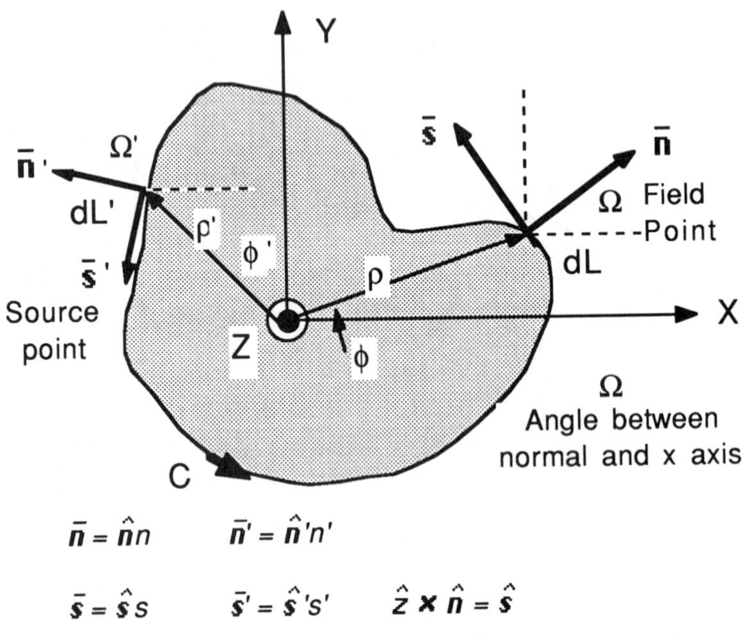

n : coordinate variable along normal to the surface
s : coordinate variable along tangent to the surface

$$\hat{n} = \hat{x}\cos\Omega + \hat{y}\sin\Omega$$

$$\hat{s} = -\hat{x}\sin\Omega + \hat{y}\cos\Omega$$

$$\hat{n}' = \hat{x}\cos\Omega' + \hat{y}\sin\Omega'$$

$$\hat{s}' = -\hat{x}\sin\Omega' + \hat{y}\cos\Omega'$$

Figure 5.7 Relationships between various unit vectors.

Thus, for the case of normal transverse magnetic excitation, there are only the following components of electric and magnetic field distributions including at the free-space and dielectric boundary surface:

$E_z(\rho,\phi)\hat{\boldsymbol{z}}$: total axial component of the electric field distribution;

$H_n(\rho,\phi)\hat{\boldsymbol{n}}$: total normal component of the magnetic field distribution;

$H_s(\rho,\phi)\hat{\boldsymbol{s}}$: total tangential component of the magnetic field distribution.

The total electric and magnetic field distributions are, in fact, unknown field quantities. An electromagnetic boundary value problem can be set up in terms of these unknown total surface electric and magnetic field distributions subject to the appropriate tangential boundary conditions. The external plane wave excitation produces scattered electric and magnetic field distributions in the surrounding free-space medium and also penetrated electric and magnetic field distributions inside the homogeneous dielectric object. Hence, in the free-space medium outside the dielectric object, the total field distribution consists of the sum of the incident field and scattered field. Within the dielectric scatterer region, the total field distribution consists of the penetrated field itself. Further, the total tangential electric and magnetic field distributions near the free-space/dielectric boundary are such that they satisfy appropriate electromagnetic boundary conditions as discussed in Section 2.16. The tangential components of the total electric field should be continuous, and similarly, the tangential components of the total magnetic field should be continuous at the boundary surface separating the two media.

Similar to the previous studies related to the perfectly conducting object, a classical approach is to treat the total tangential surface fields on the boundary surface of the homogeneous dielectric object as unknowns and set up corresponding boundary value coupled integral equations. In fact, the unknown total tangential surface magnetic field and surface electric field distributions can also be converted into the corresponding unknown equivalent electric current and equivalent magnetic current distributions. Rigorous boundary value coupled integral equations can be formulated in terms of the unknown equivalent currents. Even though the integral equations can be derived in number of different ways, discussion concerning the boundary value integral equations presented in this section is based upon the electromagnetic equivalence principle. Different types of integral equations, such as the electric field, the magnetic field, and the combined fields can be formulated. In the electric field type of integral equation formulation only the tangential electric field boundary condition is utilized, and similarly, in the magnetic field type of integral equation formulation only the tangential magnetic field boundary condition is utilized. Further, in the case of rigorous combined field integral equation formulation, both types of boundary conditions relating to the tangential electric field and the tangential magnetic field are enforced to derive the coupled set of boundary value equations.

Figures 5.6a and 5.6b show the geometry of a homogeneous, dielectric, lossy scatterer located in an isotropic lossless free-space medium. The scatterer has a volume V contained in medium 2 and is bounded by a surface S. Outside the volume of medium 2 is medium 1 representing the free-space region, and the externally excited TM to z polarized incident plane wave field is contained in it.

For the media 1 and 2, let

ε_1 : permittivity of free-space medium 1;

μ_1 : permeability of free-space medium 1;

$\sigma_1 = 0$: conductivity of free-space medium 1;

$(\bar{E}^i, \bar{H}^i)$: electric and magnetic incident fields in medium 1;

$(\bar{E}_1^S, \bar{H}_1^S)$: electric and magnetic scattered fields in medium 1;

$(\bar{E}_1, \bar{H}_1)$: electric and magnetic total fields in medium 1

$$= (\bar{E}^i, \bar{H}^i) + (\bar{E}_1^S, \bar{H}_1^S) \tag{5.3.1a}$$

and

ε_2 : permittivity of dielectric medium 2;

μ_2 : permeability of dielectric medium 2;

σ_2 : conductivity of dielectric medium 2;

$(\bar{E}_2^S, \bar{H}_2^S)$: electric and magnetic penetrated fields in medium 2;

$(\bar{E}_2, \bar{H}_2)$: electric and magnetic total fields in medium 2

$$= (\bar{E}_2^S, \bar{H}_2^S) \tag{5.3.1b}$$

Referring to the electromagnetic boundary conditions discussed in Section 2.16, the total electric and magnetic field distributions in the respective dielectric and free-space media are such that the following boundary conditions are satisfied on the boundary surface separating the two regions.

$$\bar{H}_1(\bar{r})\Big|_{\tan} = \bar{H}_2(\bar{r})\Big|_{\tan} \qquad \bar{r} \text{ on } S \tag{5.3.2}$$

$$\bar{E}_1(\bar{r})\Big|_{\tan} = \bar{E}_2(\bar{r})\Big|_{\tan} \qquad \bar{r} \text{ on } S \tag{5.3.3}$$

tan : tangential component of field

Referring to Figure 5.6b and the discussions in Section 2.19, the TM incident plane wave electric and magnetic fields can be written as

$$\bar{E}^i(\bar{\rho},\omega) = E_z^i(\bar{\rho},\omega)\hat{z} \tag{5.3.4a}$$

$$E_z^i(\bar{\rho},\omega) = E_0 e^{-j\bar{k}_1 \cdot \bar{\rho}} \tag{5.3.4b}$$

$$\bar{H}^i(\bar{\rho},\omega) = \bar{H}_0 e^{-j\bar{k}_1 \cdot \bar{\rho}} \tag{5.3.4c}$$

and the incident plane wave electric and magnetic fields are related through

$$\bar{H}_0 \times \hat{k} = \frac{\omega\varepsilon_1}{k_1}\bar{E}_0 \tag{5.3.5}$$

It is useful to introduce and work in terms of generalized cylindrical coordinate system, as shown in Figure 5.7, for the case of arbitrary cross-sectional scatterers. On the surface of the homogeneous, lossy dielectric scatterer, the three mutually orthogonal coordinates given by *(n, s, z)* satisfy the right hand system:

$$\hat{z} \times \hat{n} = \hat{s} \tag{5.3.6}$$

5.4 ELECTROMAGNETIC EQUIVALENCE

A direct analytical formulation of the electromagnetic scattering and penetration associated with the homogeneous, dielectric, lossy scatterer is accomplished based on the electromagnetic equivalence principle. In Figure 5.6b, the total electric and magnetic field distributions both are in the free-space medium and also within the dielectric material medium. Further, the electric and magnetic field distributions are such that they satisfy appropriate continuity of the total tangential surface field distributions on the boundary surface separating the dielectric medium and the free-space medium. In fact, in any electromagnetic boundary value problem, if the surface tangential electric and magnetic field distributions are known at the boundary surface separating the two media, then it is possible to calculate the scattered fields in the free-space medium and also the penetrated fields in the dielectric medium. Hence, the surface tangential electric and magnetic field quantities are initially treated as unknowns and, if required, can be expressed in terms of an unknown surface tangential magnetic current and an unknown surface tangential electric current distribution. It should be clearly noted that, for all practical purposes, the modeling of the boundary value problem based upon the surface unknown field distributions or the corresponding surface unknown electric and magnetic current distributions are the same.

In the analysis technique discussed in this section, the concept of unknown surface *electric currents and magnetic currents* has been extensively introduced to maintain consistency in the analysis procedure for the study of general shaped and complex material objects. The surface electric and magnetic current distributions, in fact, act as dependent sources for the scattered electromagnetic fields in a given region of space, and the fields so produced also satisfy the original boundary conditions at the boundary surface. In Section 5.1, a detailed analysis procedure is presented for the two types of electric and magnetic current dependent sources to calculate the scattered and penetrated fields using the electromagnetic potentials. The linearity and superposition principle is further invoked to obtain the electric and magnetic fields when both the electric current and the magnetic current sources are present simultaneously. The introduction of the electric and magnetic current sources is just an intermediate convenient step in the boundary value formulation. In fact, it is also feasible to introduce any arbitrary unknown electric and magnetic current sources at any predetermined boundary surface as long as the appropriate original boundary conditions are satisfied, and eventually the same original total fields in the free-space medium and penetrated fields in the dielectric material medium are simulated.

An electromagnetic equivalent problem follows that is suitable for the analysis of the homogeneous, lossy dielectric object. The dependent sources simulated in the equivalent problem are the *equivalent* or *virtual electric and magnetic current sources*, but are located in a linear, homogeneous, and isotropic medium. The electric and magnetic fields produced by the equivalent currents are identical to the electric and magnetic fields of the original boundary value problem and also satisfy the relevant dielectric surface boundary conditions.

Figure 5.8a shows the original homogeneous, lossy dielectric boundary value problem. Only the cross section of the geometry at the $z = 0$ plane is shown. The arbitrary shaped dielectric scattering geometry has a cross-sectional area S which is bounded by a contour C. For normal excitation, the incident, the scattered, and the penetrated electric and magnetic fields are completely independent of the z coordinate variable. Referring to Figure 5.8a, let

Original dielectric Scatterer – TM excitation

ε_1 : permittivity of free-space medium in region 1;

μ_1 : permeability of free-space medium in region 1;

$\sigma_1 = 0$: conductivity of free-space medium in region 1;

and

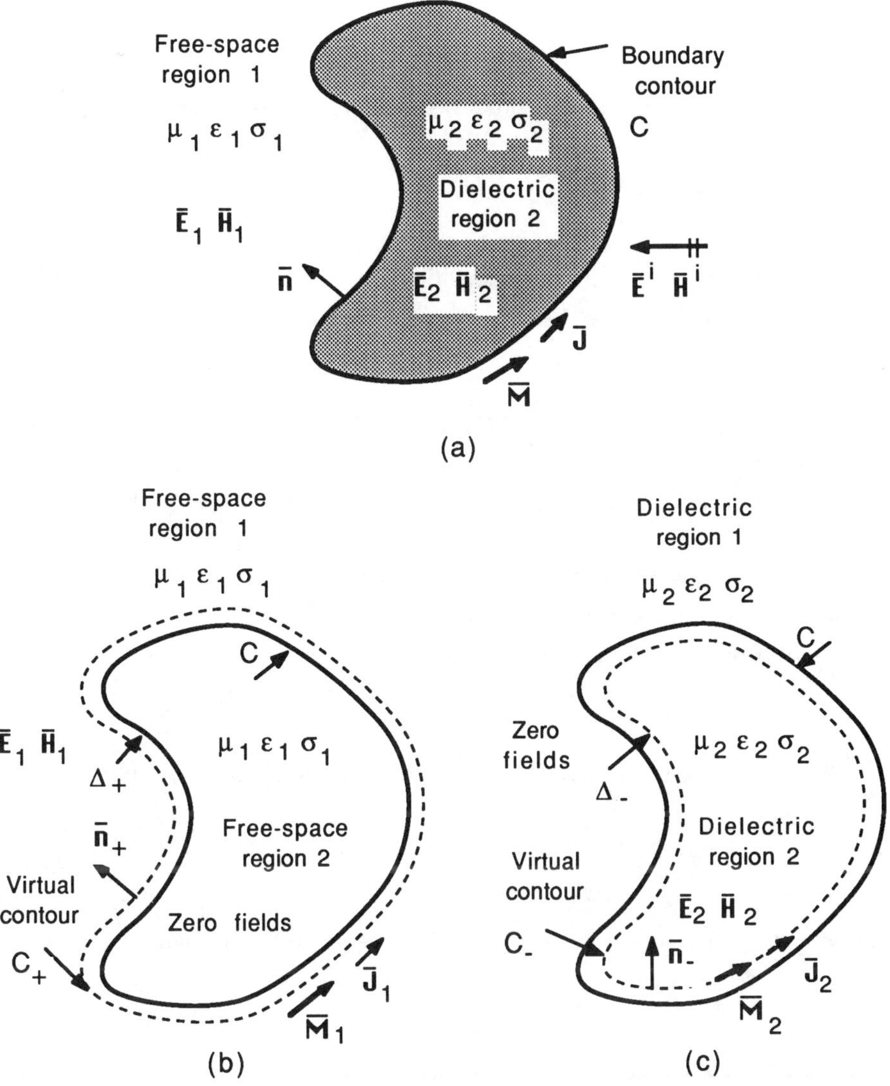

Figure 5.8 Electromagnetic equivalence for homogeneous, dielectric scatterer (a) original scatterer, (b) exterior equivalence, and (c) interior equivalence.

$(\vec{E}^i, \vec{H}^i)$: electric and magnetic incident fields in free-space medium

$$= (\hat{z}E_z^i, \vec{H}_\tau^i) \tag{5.4.1a}$$

$(\vec{E}_1^s, \vec{H}_1^s)$: electric and magnetic scattered fields in free-space medium

$$= (\hat{z}E_{1z}^s, \vec{H}_{1\tau}^s) \tag{5.4.1b}$$

$(\vec{E}_1, \vec{H}_1)$: electric and magnetic total fields in the free space medium

$$= (\vec{E}^i, \vec{H}^i) + (\vec{E}_1^s, \vec{H}_1^s) \tag{5.4.1c}$$

$$= (\hat{z}E_z^i, \vec{H}_\tau^i) + (\hat{z}E_{1z}^s, \vec{H}_{1\tau}^s) \tag{5.4.1d}$$

Similarly, for medium 2, containing the homogeneous, lossy dielectric scatterer

ε_2 : permittivity of dielectric medium in region 2
μ_2 : permeability of dielectric medium in region 2
σ_2 : conductivity of dielectric medium in region 2

and

$(\vec{E}_2^s, \vec{H}_2^s)$: electric and magnetic penetrated fields in dielectric medium

$$= (\hat{z}E_{2z}^s, \vec{H}_{2\tau}^s) \tag{5.4.2a}$$

$(\vec{E}_2, \vec{H}_2)$: electric and magnetic total fields in dielectric medium

$$= (\vec{E}_2^s, \vec{H}_2^s) \tag{5.4.2b}$$

$$= (\hat{z}E_{2z}^s, \vec{H}_{2\tau}^s) \tag{5.4.2c}$$

Referring to the general dielectric boundary conditions discussed in Section 2.16, on the contour C the tangential components of total electric field are continuous, and similarly, the tangential components of total magnetic field are continuous. Hence, the following boundary conditions are satisfied:

$$\bar{E}_1(\bar{\rho})\Big/_{\tan} = \bar{E}_2(\bar{\rho})\Big/_{\tan} \qquad \bar{\rho} \text{ on } C \tag{5.4.3a}$$

$$\bar{H}_1(\bar{\rho})\Big/_{\tan} = \bar{H}_2(\bar{\rho})\Big/_{\tan} \qquad \bar{\rho} \text{ on } C \tag{5.4.3b}$$

5.4.1 EXTERIOR EQUIVALENCE

The following procedure is adapted to set up an equivalent boundary value problem valid only for region 1. In Figure 5.8b, a dielectric scatterer geometry is shown that is identical in shape to the original dielectric scatterer. Virtual contour C_+ is drawn just outside the scatterer at a distance Δ_+ from the cross-sectional boundary to enclose the scatterer cross section completely. In a limit as $\Delta_+ \rightarrow 0$, virtual contour C_+ coincides with actual scatterer boundary C. The electric and magnetic fields outside the scatterer in region 1 are identical to the fields of the original problem. For region 1, let the various medium parameters and the electric and magnetic field distributions be same as the original problem. Then, for the case of TM excitation, for all field points located in region 1 on or outside virtual contour C_+

ε_1 : permittivity of free space medium, region 1;

μ_1 : permeability of free space medium, region 1;

$\sigma_1 = 0$: conductivity of free space medium, region 1;

and

$(\bar{E}^i, \bar{H}^i)$: electric and magnetic incident fields in free-space medium

$$= [\hat{z}E_z^i(\bar{\rho}), \bar{H}_\tau^i(\bar{\rho})] \tag{5.4.4a}$$

$(\bar{E}_1^S, \bar{H}_1^S)$: electric and magnetic scattered fields in free-space medium

$$= [\hat{z}E_{1z}^S(\bar{\rho}), \bar{H}_{1\tau}^S(\bar{\rho})] \tag{5.4.4b}$$

$(\bar{E}_1, \bar{H}_1)$: electric and magnetic total fields in free-space medium

$$= [\bar{E}^i(\bar{\rho}), \bar{H}^i(\bar{\rho})] + [\bar{E}_1^S(\bar{\rho}), \bar{H}_1^S(\bar{\rho})] \tag{5.4.4c}$$

$$= [\hat{z}E_z^i(\bar{\rho}), \bar{H}_\tau^i(\bar{\rho})] + [\hat{z}E_{1z}^S(\bar{\rho}), \bar{H}_{1\tau}^S(\bar{\rho})] \tag{5.4.4d}$$

In Figure 5.8b, on virtual contour C_+ certain tangential distribution of electric currents and magnetic currents are simulated. In a limit as virtual contour C_+ tends to original contour C, the simulated tangential electric currents and magnetic currents are such that they produce the same scattered electric and magnetic fields in region 1, expression (5.4.4b), identical to the original dielectric scatterer problem shown in Figure 5.8a. The simulated tangential electric currents and magnetic currents on virtual contour C_+ are generally referred to as the *equivalent or virtual electric currents and magnetic currents* which are valid only for the calculation of region 1 field distributions. As far as setting up the electromagnetic equivalent problem valid for region 1 total fields, it really does not matter what type of medium exists within virtual contour C_+, and similarly, it should be of no concern in the analysis regarding exact nature of the field distributions inside virtual contour C_+. Hence, for region 2 inside the virtual contour C_+, the total electric and magnetic fields are taken as zero (for convenience) to decouple the exterior boundary value problem from the interior. Hence, the region 2 homogeneous dielectric material medium can be completely removed from inside region of virtual contour C_+ and exactly replaced by a medium having properties identical to the region 1 free-space medium. Thus, in a limit as $\Delta_+ \rightarrow 0$, there are floating virtual or equivalent electric current and magnetic current distributions on contour C whose shape is identical to the original scatterer boundary, but located in a large linear, homogeneous, isotropic free-space medium. Referring to Figure 5.8b

ε_1 : permittivity of free-space medium, region 1 and region 2

μ_1 : permeability of free-space medium, region 1 and region 2

$\sigma_1 = 0$: conductivity of free-space medium, region 1 and region 2

In Figure 5.8b, the equivalent electric and magnetic currents on the virtual contour C_+ can be obtained based on the tangential component of the total surface magnetic field and the tangential component of the total surface electric field. Referring to Sections 2.16 and 2.25, expressions (2.16.2b) and (2.25.21), the equivalent surface electric and magnetic currents can be defined based on the following boundary conditions on contour C_+

$\bar{J}_{1s_{eq}}(\bar{\rho}')$: equivalent electric currents valid only for region 1 fields

$$= \hat{n}'_1 \times \bar{H}_1(\bar{\rho}') \qquad \bar{\rho}' \text{ on } C_+ \rightarrow C \tag{5.4.5a}$$

$\bar{M}_{1s_{eq}}(\bar{\rho}')$: equivalent magnetic currents valid only for region 1 fields

$$= -\hat{n}'_1 \times \bar{E}_1(\bar{\rho}') \qquad \bar{\rho}' \text{ on } C_+ \rightarrow C \tag{5.4.5b}$$

$\hat{n}'_1 = \hat{n}'$: outward normal to contour C

It should noted that the equivalent electric currents and magnetic currents, defined in expression (5.4.5a–b), are just based on the total surface field distributions of region 1 only, since for region 2 total fields are simulated as zero in the equivalent problem referred to region 1.

The complete mathematical development carried out in Section 5.1 based on the representation of scattered fields in terms of the electromagnetic potentials can now be utilized to write the region 1 general field distributions. For the case of equivalent electric current distribution, let

$\bar{A}_1(\bar{\rho},\omega)$: magnetic vector potential at the field point in region 1

$$= \bar{A}_{1\tau}(\bar{\rho},\omega) + A_{1z}(\bar{\rho},\omega)\hat{z} \tag{5.4.6}$$

$\bar{A}_{1\tau}(\bar{\rho},\omega)$: transverse component of the magnetic vector potential in region 1;

$\bar{A}_{1z}(\bar{\rho},\omega)$: axial component of the magnetic vector potential in region 1;

$\Phi_1(\bar{\rho},\omega)$: electric scalar potential at the field point in region 1.

Then the partial expressions for the magnetic field and the electric field distributions are given by

$$\bar{H}_{1e}(\bar{\rho},\omega) = \frac{1}{\mu_1} \nabla \times \bar{A}_1(\bar{\rho},\omega) \tag{5.4.7a}$$

$$\bar{E}_{1e}(\bar{\rho},\omega) = - j\omega\bar{A}_1(\bar{\rho},\omega) - \nabla\Phi_1(\bar{\rho},\omega) \tag{5.4.7b}$$

and the following vector and scalar *Helmholtz* partial differential equations are satisfied by the magnetic vector potential and the electric scalar potential:

$$\nabla^2\bar{A}_1(\bar{\rho},\omega) + k_1^2\bar{A}_1(\bar{\rho},\omega) = - \mu_1\bar{J}_{1s_{eq}}(\bar{\rho},\omega) \tag{5.4.8a}$$

$$\nabla^2\Phi_1(\bar{\rho},\omega) + k_1^2\Phi_1(\bar{\rho},\omega) = - \frac{\rho_{1se_{eq}}(\bar{\rho},\omega)}{\varepsilon_1} \tag{5.4.8b}$$

$$k_1 = \omega(\mu_1\varepsilon_1)^{1/2} \tag{5.4.8c}$$

Referring to the discussions in Section 5.1 and expressions (5.1.10) and (5.1.11), the solutions for the two potentials in the region 1 can be written. Let

$\bar{r}'$: position vector corresponding to the source point (x', y', z')

$$= x'\hat{x} + y'\hat{y} + z'\hat{z} \tag{5.4.9a}$$

$$= \bar{\rho}' + z'\hat{z} \tag{5.4.9b}$$

$\bar{r}_1$: position vector corresponding to the field point (x_1, y_1, z_1)

$$= x_1\hat{x} + y_1\hat{y} + z_1\hat{z} \tag{5.4.9c}$$

$$= \bar{\rho} + z_1\hat{z} \qquad \bar{\rho} = \bar{\rho}_1 \text{ on or outside } C_+ \tag{5.4.9d}$$

R_1 : distance of the line joining the source point to the field point

$$= |\bar{r}_1 - \bar{r}'| \tag{5.4.9e}$$

The magnetic vector potential and the electric scalar potential at the field point in region 1 are given by the corresponding surface integral over the complete equivalent current and charge source distributions on surface S:

$$\bar{A}_1(\bar{r}) = \frac{\mu_1}{4\pi} \iint_S \bar{J}_{1s_{eq}}(\bar{r}') \frac{e^{-jk_1R_1}}{R_1} ds(\bar{r}') \tag{5.4.10}$$

$$\Phi_1(\bar{r}) = \frac{1}{4\pi\varepsilon_1} \iint_S \rho_{1seeq}(\bar{r}') \frac{e^{-jk_1R_1}}{R_1} ds(\bar{r}') \tag{5.4.11}$$

In these two superposition integrals for the magnetic vector potential and the electric scalar potential, the integrands contain the three-dimensional free-space Green's function. The two superposition integrals can be easily reduced for further application to two-dimensional problems. As discussed earlier, the various sources and their field distributions are completely independent of the z coordinate variable. Hence, the two integrals can be simplified to yield corresponding superposition potential integrals containing just the two-dimensional free-space Green's function. The integral expression (5.4.10) can be rewritten as

$$\bar{A}_1(\bar{\rho}, z_1{=}0) = \frac{\mu_1}{4\pi} \int_C \bar{J}_{1s_{eq}}(\bar{\rho}') \left[\int_{z'=-\infty}^{z'=\infty} \frac{e^{-jk_1R_1}}{R_1} dz' \right] dL' \tag{5.4.12a}$$

$$R_1 = \left[(x_1 - x')^2 + (y_1 - y')^2 + (z')^2 \right]^{1/2} \tag{5.4.12b}$$

The bracketed term in expression (5.4.12a) is the well-known integral representation for the Hankel function of zero order and second kind, Appendix B, given by

$$H_0^{(2)}(k_1|\bar{\rho} - \bar{\rho}'|) = \frac{j}{\pi} \int_{z'=-\infty}^{z'=\infty} \frac{e^{-jk_1R_1}}{R_1}\, dz' \tag{5.4.12c}$$

After substituting this relationship for the Hankel function into potential integral representations (5.4.10) and (5.4.11), the vector magnetic potential and the electric scalar potential for the two-dimensional case take the form

$$\bar{A}_1(\bar{\rho}) = \frac{\mu_1}{4j} \int_C \bar{J}_{1eq}(\bar{\rho}') H_0^{(2)}(k_1|\bar{\rho} - \bar{\rho}'|)\, dL(\bar{\rho}') \tag{5.4.13a}$$

$$\Phi_1(\bar{\rho}) = \frac{1}{4j\varepsilon_1} \int_C \rho_{1e_{eq}}(\bar{\rho}') H_0^{(2)}(k_1|\bar{\rho} - \bar{\rho}'|)\, dL(\bar{\rho}') \tag{5.4.13b}$$

Similarly, the dual expressions derived in Section 5.1 can be utilized for the case of equivalent magnetic current and charge distributions. Let

$\bar{F}_1(\bar{\rho},\omega)$: electric vector potential at the field point in region 1

$$= \bar{F}_{1\tau}(\bar{\rho},\omega) + F_{1z}(\bar{\rho},\omega)\hat{z} \tag{5.4.14}$$

$\bar{F}_{1\tau}(\bar{\rho},\omega)$: transverse component of the electric vector potential in region 1;

$\bar{F}_{1z}(\bar{\rho},\omega)$: axial component of the electric vector potential in region 1;

$\Psi_1(\bar{\rho},\omega)$: magnetic scalar potential function at the field point in region 1.

Then, the partial expressions for the electric field and magnetic field distributions are given by

$$\bar{E}_{1m}(\bar{\rho},\omega) = -\frac{1}{\varepsilon_1} \nabla \times \bar{F}_1(\bar{\rho},\omega) \tag{5.4.15a}$$

$$\bar{H}_{1m}(\bar{\rho},\omega) = -j\omega\bar{F}_1(\bar{\rho},\omega) - \nabla\Psi_1(\bar{\rho},\omega) \tag{5.4.15b}$$

and the following vector and scalar *Helmholtz* partial differential equations are satisfied by the electric vector potential and the magnetic scalar potential:

$$\nabla^2 \bar{F}_1(\bar{\rho},\omega) + k_1^2 \bar{F}_1(\bar{\rho},\omega) = -\varepsilon_1 \bar{M}_{1s_{eq}}(\bar{\rho},\omega) \tag{5.4.16}$$

$$\nabla^2 \Psi_1(\bar{\rho},\omega) + k_1^2 \Psi_1(\bar{\rho},\omega) = -\frac{\rho_{1sm_{eq}}(\bar{\rho},\omega)}{\mu_1} \tag{5.4.17}$$

The electric vector potential and the magnetic scalar potential at the field point in region 1 are given by the corresponding surface integral over the complete source distributions on surface *S*:

$$\bar{F}_1(\bar{r}) = \frac{\varepsilon_1}{4\pi} \iint_S \bar{M}_{1s_{eq}}(\bar{r}') \frac{e^{-jk_1 R_1}}{R_1} \, ds(\bar{r}') \tag{5.4.18}$$

$$\Psi_1(\bar{r}) = \frac{1}{4\pi\mu_1} \iint_S \rho_{1sm_{eq}}(\bar{r}') \frac{e^{-jk_1 R_1}}{R_1} \, ds(\bar{r}') \tag{5.4.19}$$

Again, in these two superposition integrals for the electric and magnetic potentials, the integrands contain the three-dimensional free-space Green's function. Referring to the simplification in expression (5.4.12), integral expression (5.4.18) can be rewritten as

$$\bar{F}_1(\bar{\rho}, z_1=0) = \frac{\varepsilon_1}{4\pi} \int_C \bar{M}_{1s_{eq}}(\bar{\rho}') \left[\int_{z'=-\infty}^{z'=\infty} \frac{e^{-jk_1 R_1}}{R_1} dz' \right] dL' \tag{5.4.20a}$$

$$R_1 = \left[(x_1 - x')^2 + (y_1 - y')^2 + (z')^2 \right]^{1/2} \tag{5.4.20b}$$

The bracketed term in expression (5.4.20a) is the well-known integral representation for the Hankel function of zero order and second kind, which is given in expression (5.4.12c). After substituting expression (5.4.12c) into representations (5.4.18) and (5.4.19), for the two-dimensional case the vector electric potential and the magnetic scalar potential take the following form:

$$\bar{F}_1(\bar{\rho}) = \frac{\varepsilon_1}{4j} \int_C \bar{M}_{1_{eq}}(\bar{\rho}') H_0^{(2)}(k_1 |\bar{\rho} - \bar{\rho}'|) \, dL(\bar{\rho}') \tag{5.4.21a}$$

$$\Psi_1(\bar{\rho}) = \frac{1}{4j\mu_1} \int_C \rho_{1m_{eq}}(\bar{\rho}') H_0^{(2)}(k_1 |\bar{\rho} - \bar{\rho}'|) \, dL(\bar{\rho}') \tag{5.4.21b}$$

It is now possible to write the composite total electric field and the total magnetic field distributions using the scattered field expressions defined in (5.4.7a–b) and (5.4.15a–b), for $\bar{\rho}$ on or outside C_+

$$\bar{E}_1(\bar{\rho}) = \bar{E}^i(\bar{\rho}) + \bar{E}_{1e}(\bar{\rho}) + \bar{E}_{1m}(\bar{\rho}) \tag{5.4.22}$$

$$\bar{H}_1(\bar{\rho}) = \bar{H}^i(\bar{\rho}) + \bar{H}_{1e}(\bar{\rho}) + \bar{H}_{1m}(\bar{\rho}) \tag{5.4.23}$$

and in terms of the vector and scalar potentials, in the region 1, the total electric and magnetic field distributions are given by

$$\bar{E}_1(\bar{\rho}) = \bar{E}^i(\bar{\rho}) - j\omega\bar{A}_1(\bar{\rho},\omega) - \nabla_\tau\Phi_1(\bar{\rho},\omega) - \frac{1}{\varepsilon_1}\nabla_\tau \times \bar{F}_1(\bar{\rho},\omega) \tag{5.4.24}$$

$$\bar{H}_1(\bar{\rho}) = \bar{H}^i(\bar{\rho}) + \frac{1}{\mu_1}\nabla_\tau \times \bar{A}_1(\bar{\rho},\omega) - j\omega\bar{F}_1(\bar{\rho},\omega) - \nabla_\tau\Psi_1(\bar{\rho},\omega) \tag{5.4.25}$$

$\bar{\rho}$ on or outside C_+

5.4.2 INTERIOR EQUIVALENCE

A similar procedure is adapted to set up an equivalent boundary value problem valid for region 2. The procedure for setting up an equivalent problem for the interior lossy dielectric region is similar to the one discussed for the free-space region, but the equivalent electric and magnetic current sources are valid only for the region 2 penetrated fields. In Figure 5.8c, dielectric scatterer geometry is shown that is identical in shape to the original dielectric scatterer. Virtual contour C_- is drawn just inside the scatterer at a distance Δ_- from the cross-sectional boundary to enclose the scatterer cross section completely. In a limit as $\Delta_- \to 0$, virtual contour C_- coincides with actual scatterer boundary C. The electric and magnetic fields inside penetrable region 2 are identical to the fields of the original problem.

Then, for TM excitation, for all field points located in region 2 on or inside virtual contour C_-

ε_2: permittivity of dielectric medium, region 2

μ_2: permeability of dielectric medium, region 2

σ_2: conductivity of dielectric medium, region 2

Referring to Section 2.18, the effect of electrical conductivity or lossy nature of the homogeneous dielectric scatterer can be easily accounted for by redefining the permittivity parameter in terms of an effective permittivity as in expression (2.18.5d). Let

ε'_2 : effective permittivity of dielectric medium, region 2

$$= \varepsilon_2 \left[1 - j\frac{\sigma_2}{\omega\varepsilon_2}\right] \tag{5.4.26}$$

and

$(\vec{E}^S_2, \vec{H}^S_2)$: electric and magnetic penetrated fields in dielectric medium;

$(\bar{E}_2, \bar{H}_2)$: electric and magnetic total fields in dielectric medium

$$= [\vec{E}^S_2(\bar{\rho}), \vec{H}^S_2(\bar{\rho})] \tag{5.4.27a}$$

$$= [\hat{z}E^S_{2z}(\bar{\rho}), \vec{H}^S_{2\tau}(\bar{\rho})] \tag{5.4.27b}$$

In Figure 5.8c, on virtual contour C_- certain tangential distributions of the electric and magnetic currents are simulated. In a limit as virtual contour C_- tends to the original contour C, the simulated tangential electric currents and magnetic currents are such that they produce the same penetrated total electric and magnetic fields in region 2, expression (5.4.27b), which are identical to the original dielectric scatterer problem shown in Figure 5.8a. The simulated tangential electric currents and magnetic currents on virtual contour C_- are again referred to as the *equivalent or virtual electric currents and magnetic currents* valid only for the calculation of region 2 field distributions. As far as setting up the electromagnetic equivalent problem valid for the region 2 total fields, it really does not matter what type of medium exists outside virtual contour C_-, and similarly, it should be of no concern in the analysis regarding the exact nature of the field distributions outside of virtual contour C_-. Hence, for region 1 outside virtual contour C_-, the total electric and magnetic fields are taken as zero (for convenience) to decouple the interior boundary value problem from the exterior. Hence, the region 1 free-space medium can be completely removed and exactly replaced by a medium having properties identical to the region 2 homogeneous, lossy dielectric medium. Thus, in a limit as $\Delta_- \rightarrow 0$, there are floating virtual or equivalent electric current and magnetic current distributions on contour C whose shape is identical to the original scatterer boundary, but located in a linear, homogeneous, isotropic, lossy dielectric medium.

Referring to Figure 5.8c, let

ε_2 : permittivity of dielectric medium, region 2 and region 1;

μ_2 : permeability of dielectric medium, region 2 and region 1;
σ_2 : conductivity of dielectric medium, region 2 and region 1;
ε'_2 : effective permittivity of dielectric medium, region 2 and 1

$$= \varepsilon_2 \left[1 - j\frac{\sigma_2}{\omega\varepsilon_2}\right] \quad (5.4.28)$$

In Figure 5.8c, the equivalent electric and magnetic currents on virtual contour C_- can be obtained based on the tangential component of the total surface magnetic field and the tangential component of the total surface electric field. Referring to Sections 2.16 and 2.25, expressions (2.16.2b) and (2.25.21), the equivalent surface electric and magnetic currents can be defined based on the following boundary conditions on contour C_-

$\bar{J}_{2s_{eq}}(\bar{\rho}')$: equivalent electric currents valid only for region 2 fields

$$= \hat{n}'_2 \times \bar{H}_2(\bar{\rho}') \qquad \bar{\rho}' \text{ on } C_- \rightarrow C \quad (5.4.29a)$$

$\bar{M}_{2s_{eq}}(\bar{\rho}')$: equivalent magnetic currents valid only for region 2 fields

$$= - \hat{n}'_2 \times \bar{E}_2(\bar{\rho}') \qquad \bar{\rho}' \text{ on } C_- \rightarrow C \quad (5.4.29b)$$

$\hat{n}'_2 = - \hat{n}'$: inward normal to contour C

It should noted that the equivalent electric and magnetic currents are based on the total surface field distributions of region 2 only, since the total fields for region 1 are simulated as zero in the equivalent problem referred to region 2. The complete mathematical development carried out in Section 5.1 based on the representation of fields in terms of electromagnetic potential can now be utilized to write the region 2 general field distributions. For the equivalent electric current distribution, let

$\bar{A}_2(\bar{\rho},\omega)$: magnetic vector potential at the field point in region 2

$$= \bar{A}_{2\tau}(\bar{\rho},\omega) + A_{2z}(\bar{\rho},\omega)\hat{z} \quad (5.4.30)$$

$\bar{A}_{2\tau}(\bar{\rho},\omega)$: transverse component of the magnetic vector potential in region 2;
$\bar{A}_{2z}(\bar{\rho},\omega)$: axial component of the magnetic vector potential in region 2;
$\Phi_2(\bar{\rho},\omega)$: electric scalar potential at the field point in region 2.

Then the partial expressions for the magnetic field and the electric field distributions are given by

$$\bar{H}_{2e}(\bar{\rho},\omega) = \frac{1}{\mu_2} \nabla \times \bar{A}_2(\bar{\rho},\omega) \tag{5.4.31a}$$

$$\bar{E}_{2e}(\bar{\rho},\omega) = - j\omega\bar{A}_2(\bar{\rho},\omega) - \nabla\Phi_2(\bar{\rho},\omega) \tag{5.4.31b}$$

and the following vector and scalar *Helmholtz* partial differential equations are satisfied by the magnetic vector potential and the electric scalar potential:

$$\nabla^2\bar{A}_2(\bar{\rho},\omega) + k_2^2\bar{A}_2(\bar{\rho},\omega) = - \mu_2\bar{J}_{2s_{eq}}(\bar{\rho},\omega) \tag{5.4.32a}$$

$$\nabla^2\Phi_2(\bar{\rho},\omega) + k_2^2\Phi_2(\bar{\rho},\omega) = - \frac{\rho_{2se_{eq}}(\bar{\rho},\omega)}{\varepsilon'_2} \tag{5.4.32b}$$

$$k_2 = \omega(\mu_2\varepsilon'_2)^{1/2} \tag{5.4.32c}$$

Referring to the discussions in Section 5.1 and expressions (5.1.10) and (5.1.11), the solutions for the two potentials in region 2 can be written. Let

$\bar{r}'$: position vector corresponding to the source point (x', y', z')

$$= x'\hat{x} + y'\hat{y} + z'\hat{z} \tag{5.4.33a}$$

$$= \bar{\rho}' + z'\hat{z} \tag{5.4.33b}$$

$\bar{r}_2$: position vector corresponding to the field point (x_2, y_2, z_2)

$$= x_2\hat{x} + y_2\hat{y} + z_2\hat{z} \tag{5.4.33c}$$

$$= \bar{\rho} + z_2\hat{z} \qquad \bar{\rho} = \bar{\rho}_2 \text{ on or inside } C_- \tag{5.4.33d}$$

R_2: distance of the line joining the source point to the field point

$$= |\bar{r}_2 - \bar{r}'| \tag{5.4.33e}$$

The magnetic vector potential and the electric scalar potential at the field point in region 2 are given by the corresponding surface integral over the complete equivalent current and charge source distributions on surface S:

$$\bar{A}_2(\bar{r}) = \frac{\mu_2}{4\pi} \iint_S \bar{J}_{2s_{eq}}(\bar{r}') \frac{e^{-jk_2R_2}}{R_2} ds(\bar{r}') \tag{5.4.34}$$

$$\Phi_2(\bar{r}) = \frac{1}{4\pi\varepsilon'_2}\iint_S \rho_{2seeq}(\bar{r}')\frac{e^{-jk_2R_2}}{R_2}\,ds(\bar{r}') \tag{5.4.35}$$

In these superposition integrals for the magnetic vector potential and the electric scalar potential, the integrands contain the Green's function for the three-dimensional homogeneous, lossy dielectric region. As discussed earlier, the various sources and their field distributions are independent of the z coordinate variable. Hence, these two integrals can be simplified to yield corresponding superposition potential integrals containing just the two-dimensional, homogeneous, lossy dielectric region Green's function. The integral expression (5.4.34) can be rewritten as

$$\bar{A}_2(\bar{\rho}, z_2=0) = \frac{\mu_2}{4\pi}\int_C \bar{J}_{2s_{eq}}(\bar{\rho}')\Big[\int_{z'=-\infty}^{z'=\infty}\frac{e^{-jk_2R_2}}{R_2}dz'\Big]dL' \tag{5.4.36a}$$

$$R_2 = \big[(x_2-x')^2 + (y_2-y')^2 + (z')^2\big]^{1/2} \tag{5.4.36b}$$

The bracketed term in expression (5.4.36a) is the well-known integral representation for the Hankel function of zero order and second kind, Appendix B, given by

$$H_0^{(2)}(k_2|\bar{\rho} - \bar{\rho}'|) = \frac{j}{\pi}\int_{z'=-\infty}^{z'=\infty}\frac{e^{-jk_2R_2}}{R_2}\,dz' \tag{5.4.36c}$$

After substituting the relationship for the Hankel function into potential integral representations (5.4.34) and (5.4.35), the vector magnetic potential and the electric scalar potential for the two-dimensional case take the form

$$\bar{A}_2(\bar{\rho}) = \frac{\mu_2}{4j}\int_C \bar{J}_{2eq}(\bar{\rho}')H_0^{(2)}(k_2|\bar{\rho} - \bar{\rho}'|)\,dL(\bar{\rho}') \tag{5.4.37a}$$

$$\Phi_2(\bar{\rho}) = \frac{1}{4j\varepsilon'_2}\int_C \rho_{2e_{eq}}(\bar{\rho}')H_0^{(2)}(k_2|\bar{\rho} - \bar{\rho}'|)\,dL(\bar{\rho}') \tag{5.4.37b}$$

Similarly, the dual expressions derived in Section 5.1 can be utilized for the case of equivalent magnetic current distribution. Let

$\bar{F}_2(\bar{\rho},\omega)$: electric vector potential at the field point in region 2

$$= \bar{F}_{2\tau}(\bar{\rho},\omega) + F_{2z}(\bar{\rho},\omega)\hat{z} \quad (5.4.38)$$

$\bar{F}_{2\tau}(\bar{\rho},\omega)$: transverse component of the electric vector potential in region 2;

$\bar{F}_{2z}(\bar{\rho},\omega)$: axial component of the electric vector potential in region 2;

$\Psi_2(\bar{\rho},\omega)$: magnetic scalar potential at the field point in region 2.

Then the partial expressions for the electric field and the magnetic field are given by

$$\bar{E}_{2m}(\bar{\rho},\omega) = -\frac{1}{\varepsilon'_2}\nabla \times \bar{F}_2(\bar{\rho},\omega) \quad (5.4.39a)$$

$$\bar{H}_{2m}(\bar{\rho},\omega) = -j\omega\bar{F}_2(\bar{\rho},\omega) - \nabla\Psi_2(\bar{\rho},\omega) \quad (5.4.39b)$$

and the following vector and scalar *Helmholtz* partial differential equations are satisfied by the electric vector potential and the magnetic scalar potential:

$$\nabla^2\bar{F}_2(\bar{\rho},\omega) + k_2^2\bar{F}_2(\bar{\rho},\omega) = -\varepsilon'_2\bar{M}_{2s_{eq}}(\bar{\rho},\omega) \quad (5.4.40a)$$

$$\nabla^2\Psi_2(\bar{\rho},\omega) + k_2^2\Psi_2(\bar{\rho},\omega) = -\frac{\rho_{2sm_{eq}}(\bar{\rho},\omega)}{\mu_2} \quad (5.4.40b)$$

The electric vector potential and the magnetic scalar potential at the field point in region 2 is given by the corresponding surface integral over the complete source distributions on surface S:

$$\bar{F}_2(\bar{r}) = \frac{\varepsilon'_2}{4\pi}\iint_S \bar{M}_{2s_{eq}}(\bar{r}')\frac{e^{-jk_2R_2}}{R_2}\,ds(\bar{r}') \quad (5.4.41)$$

$$\Psi_2(\bar{r}) = \frac{1}{4\pi\mu_2}\iint_S \rho_{2sm_{eq}}(\bar{r}')\frac{e^{-jk_2R_2}}{R_2}\,ds(\bar{r}') \quad (5.4.42)$$

Again, in these two superposition integrals for the electric vector potential and the scalar magnetic potentials, the integrands contain the Green's function for the three-dimensional homogeneous, lossy dielectric region. Referring to the simplification in expression (5.4.36), integral expression (5.4.41) can be rewritten as

$$\bar{F}_2(\bar{\rho},z_2=0) = \frac{\varepsilon'_2}{4\pi}\int_C \bar{M}_{2s_{eq}}(\bar{\rho}')\left[\int_{z'=-\infty}^{z'=\infty} \frac{e^{-jk_2R_2}}{R_2}dz'\right]dL' \qquad (5.4.43a)$$

$$R_2 = \left[(x_2-x')^2 + (y_2-y')^2 + (z')^2\right]^{1/2} \qquad (5.4.43b)$$

The bracketed term in expression (5.4.43a) is the integral representation for the Hankel function of zero order and second kind, given in expression (5.4.36c). After substituting expression (5.4.36c) into representations (5.4.41) and (5.4.42), the electric vector potential and the magnetic scalar potential take the following form:

$$\bar{F}_2(\bar{\rho}) = \frac{\varepsilon'_2}{4j}\int_C \bar{M}_{2eq}(\bar{\rho}')H_0^{(2)}(k_2|\bar{\rho} - \bar{\rho}'|)\, dL(\bar{\rho}') \qquad (5.4.44a)$$

$$\Psi_2(\bar{\rho}) = \frac{1}{4j\mu_2}\int_C \rho_{2m_{eq}}(\bar{\rho}')H_0^{(2)}(k_2|\bar{\rho} - \bar{\rho}'|)\, dL(\bar{\rho}') \qquad (5.4.44b)$$

It is now possible to write the composite total electric field and the total magnetic field distributions using the penetrated field expressions defined in (5.4.31a), (5.4.31b) and (5.4.39a), (5.4.39b), for $\bar{\rho}$ on or inside C_-

$$\bar{E}_2(\bar{\rho}) = \bar{E}_{2e}(\bar{\rho}) + \bar{E}_{2m}(\bar{\rho}) \qquad (5.4.45)$$

$$\bar{H}_2(\bar{\rho}) = \bar{H}_{2e}(\bar{\rho}) + \bar{H}_{2m}(\bar{\rho}) \qquad (5.4.46)$$

and in terms of the vector and scalar potentials, in region 2, the total electric and magnetic field distributions are given by

$$\bar{E}_2(\bar{\rho}) = -j\omega\bar{A}_2(\bar{\rho},\omega) - \nabla_\tau\Phi_2(\bar{\rho},\omega) - \frac{1}{\varepsilon'_2}\nabla_\tau\times\bar{F}_2(\bar{\rho},\omega) \qquad (5.4.47)$$

$$\bar{H}_2(\bar{\rho}) = \frac{1}{\mu_2}\nabla_\tau\times\bar{A}_2(\bar{\rho},\omega) - j\omega\bar{F}_2(\bar{\rho},\omega) - \nabla_\tau\Psi_2(\bar{\rho},\omega) \qquad (5.4.48)$$

$\bar{\rho}$ on or inside C_-

Thus, for the exterior free-space region 1 and the interior homogeneous, lossy dielectric region 2, the complete expressions for the total electric and magnetic field distributions are rigorously obtained based on the exterior and the interior electromagnetic equivalences. It should be noted that the equivalent electric and magnetic current sources are still unknown and to be determined by enforcing the appropriate boundary conditions on boundary contour C which basically yield a set of coupled combined field integral equations. The combined field integral equations are to be solved for the equivalent electric and magnetic current distributions. This formulation, in fact, is ideal for treating homogeneous material objects. Extension of this analysis to treat layered objects is discussed in the next chapter.

5.5 COMBINED FIELD INTEGRAL EQUATIONS

A detailed analysis procedure based on the electromagnetic equivalence was presented in the previous section for the study of electromagnetic scattering, penetration, and interaction by a homogeneous lossy dielectric object. By invoking the electromagnetic equivalence principle, general expressions for the total electric field and magnetic field distributions are given in terms of the unknown equivalent electric and magnetic surface currents. It is now possible to formulate a set of combined field coupled integral equations using the general dielectric boundary conditions discussed in Section 2.16.

On contour C separating the two media consisting of the free-space region 1 and the lossy dielectric region 2, referring to Figure 5.6, the tangential components of the total electric field and the total magnetic field are continuous. Hence,

$$\bar{E}_1(\bar{\rho})\Big|_{\tan} = \bar{E}_2(\bar{\rho})\Big|_{\tan} \qquad \bar{\rho} \text{ on } C \tag{5.5.1a}$$

$$\bar{H}_1(\bar{\rho})\Big|_{\tan} = \bar{H}_2(\bar{\rho})\Big|_{\tan} \qquad \bar{\rho} \text{ on } C \tag{5.5.1b}$$

In terms of the various vector and scalar potentials referred to the free space region 1, the total electric and magnetic field distributions, referring to the expressions (5.4.22) to (5.4.25), are given by

$$\bar{E}_1(\bar{\rho}) = \bar{E}^i(\bar{\rho}) - j\omega\bar{A}_1(\bar{\rho},\omega) - \nabla_\tau\Phi_1(\bar{\rho},\omega) - \frac{1}{\varepsilon_1}\nabla_\tau \times \bar{F}_1(\bar{\rho},\omega) \tag{5.5.2a}$$

$$\bar{H}_1(\bar{\rho}) = \bar{H}^i(\bar{\rho}) + \frac{1}{\mu_1}\nabla_\tau \times \bar{A}_1(\bar{\rho},\omega) - j\omega\bar{F}_1(\bar{\rho},\omega) - \nabla_\tau\Psi_1(\bar{\rho},\omega) \tag{5.5.2b}$$

$\bar{\rho}$ on or outside C_+

where using the expressions (5.4.13a–b) and (5.4.21a–b), for the region 1 the vector and scalar potentials are given by

$$\bar{A}_1(\bar{\rho}) = \frac{\mu_1}{4j}\int_C \bar{J}_{1eq}(\bar{\rho}')H_0^{(2)}(k_1|\bar{\rho} - \bar{\rho}'|)\; dL(\bar{\rho}') \tag{5.5.3a}$$

$$\Phi_1(\bar{\rho}) = \frac{1}{4\omega\varepsilon_1}\int_C [\nabla'_\tau \bullet \bar{J}_{1eq}(\bar{\rho}')]H_0^{(2)}(k_1|\bar{\rho} - \bar{\rho}'|)\; dL(\bar{\rho}') \tag{5.5.3b}$$

$$\bar{F}_1(\bar{\rho}) = \frac{\varepsilon_1}{4j}\int_C \bar{M}_{1eq}(\bar{\rho}')H_0^{(2)}(k_1|\bar{\rho} - \bar{\rho}'|)\; dL(\bar{\rho}') \tag{5.5.3c}$$

$$\Psi_1(\bar{\rho}) = \frac{1}{4\omega\mu_1}\int_C [\nabla'_\tau \bullet \bar{M}_{1eq}(\bar{\rho}')]H_0^{(2)}(k_1|\bar{\rho} - \bar{\rho}'|)\; dL(\bar{\rho}') \tag{5.5.3d}$$

$$k_1 = \omega(\mu_1\varepsilon_1)^{1/2} \tag{5.5.3e}$$

$\bar{\rho}$ on or outside C_+

It should be noted that the electric charge term and the magnetic charge term in the scalar potential integrals, expressions (5.5.3b) and (5.5.3d), are replaced in terms of the corresponding divergence of the electric current and divergence of the magnetic current by using the respective current-charge continuity equation stated in expressions (5.1.3e) and (5.1.14e).

Similarly, referring to expressions (5.4.45) to (5.4.48), the total electric field and magnetic field distributions in terms of the vector and scalar potentials, as referred to lossy dielectric region 2, are given by

$$\bar{E}_2(\bar{\rho}) = -j\omega\bar{A}_2(\bar{\rho},\omega) - \nabla_\tau\Phi_2(\bar{\rho},\omega) - \frac{1}{\varepsilon'_2}\nabla_\tau \times \bar{F}_2(\bar{\rho},\omega) \tag{5.5.4a}$$

$$\bar{H}_2(\bar{\rho}) = \frac{1}{\mu_2}\nabla_\tau \times \bar{A}_2(\bar{\rho},\omega) - j\omega\bar{F}_2(\bar{\rho},\omega) - \nabla_\tau\Psi_2(\bar{\rho},\omega) \tag{5.5.4b}$$

$\bar{\rho}$ on or inside C_-

where using the expressions (5.4.37 a,b) and (5.4.44 a,b), for the region 2 the vector and scalar potentials are given by

$$\bar{A}_2(\bar{\rho}) = \frac{\mu_2}{4j}\int_C \bar{J}_{2eq}(\bar{\rho}')H_0^{(2)}(k_2|\bar{\rho} - \bar{\rho}'|)\, dL(\bar{\rho}') \tag{5.5.5a}$$

$$\Phi_2(\bar{\rho}) = \frac{1}{4\omega\varepsilon'_2}\int_C [\nabla'_\tau \bullet \bar{J}_{2eq}(\bar{\rho}')]H_0^{(2)}(k_2|\bar{\rho} - \bar{\rho}'|)\, dL(\bar{\rho}') \tag{5.5.5b}$$

$$\bar{F}_2(\bar{\rho}) = \frac{\varepsilon'_2}{4j}\int_C \bar{M}_{2eq}(\bar{\rho}')H_0^{(2)}(k_2|\bar{\rho} - \bar{\rho}'|)\, dL(\bar{\rho}') \tag{5.5.5c}$$

$$\Psi_2(\bar{\rho}) = \frac{1}{4\omega\mu_2}\int_C [\nabla'_\tau \bullet \bar{M}_{2eq}(\bar{\rho}')]H_0^{(2)}(k_2|\bar{\rho} - \bar{\rho}'|)\, dL(\bar{\rho}') \tag{5.5.5d}$$

$$k_2 = \omega(\mu_2\varepsilon'_2)^{1/2} \tag{5.5.5e}$$

$\bar{\rho}$ on or inside C_-

It should be noted that the equivalent electric and magnetic current sources valid for free-space region 1 and lossy dielectric region 2 are still unknown. They can be determined by enforcing the appropriate boundary conditions on boundary contour C, which basically yield a set of coupled combined field integral equations. On substituting expressions (5.5.2a–b) and (5.5.4a–b) into the two boundary conditions (5.5.1a–b):

$$\bar{E}^i(\bar{\rho})\Big|_{tan} = \Big[j\omega\bar{A}_1(\bar{\rho},\omega) + \nabla_\tau\Phi_1(\bar{\rho},\omega) + \frac{1}{\varepsilon_1}\nabla_\tau \times \bar{F}_1(\bar{\rho},\omega) - j\omega\bar{A}_2(\bar{\rho},\omega) - \nabla_\tau\Phi_2(\bar{\rho},\omega) - \frac{1}{\varepsilon'_2}\nabla_\tau \times \bar{F}_2(\bar{\rho},\omega)\Big]\Big|_{tan} \tag{5.5.6a}$$

$$\bar{H}^i(\bar{\rho})\Big|_{tan} = \Big[-\frac{1}{\mu_1}\nabla_\tau \times \bar{A}_1(\bar{\rho},\omega) + j\omega\bar{F}_1(\bar{\rho},\omega) + \nabla_\tau\Psi_1(\bar{\rho},\omega) + \frac{1}{\mu_2}\nabla_\tau \times \bar{A}_2(\bar{\rho},\omega) - j\omega\bar{F}_2(\bar{\rho},\omega) - \nabla_\tau\Psi_2(\bar{\rho},\omega)\Big]\Big|_{tan} \tag{5.5.6b}$$

$\bar{\rho}$ on C

In Figure 5.8b, the equivalent electric currents and magnetic currents on virtual contour $C_+ \rightarrow C$ is defined based on the tangential components of the total surface magnetic field and the total surface electric field. Referring to the earlier discussion, expressions (5.4.5a) and (5.4.5b), the equivalent surface electric currents and magnetic currents can be obtained using the following boundary conditions on contour C:

$\bar{J}_{1_{eq}}(\bar{\rho}')$: equivalent electric currents valid only for region 1 fields

$$= \hat{n}' \times \bar{H}_1(\bar{\rho}') \qquad \bar{\rho}' \text{ on } C \tag{5.5.7a}$$

$$= \bar{J}_{eq}(\bar{\rho}') \tag{5.5.7b}$$

$\bar{H}_1(\bar{\rho}')$: total surface magnetic field distribution on contour C;

$\hat{n}'$: outward normal to contour C;

and similarly,

$\bar{M}_{1_{eq}}(\bar{\rho}')$: equivalent magnetic currents valid only for region 1 fields

$$= - \hat{n}' \times \bar{E}_1(\bar{\rho}') \qquad \bar{\rho}' \text{ on } C \tag{5.5.7c}$$

$$= \bar{M}_{eq}(\bar{\rho}') \tag{5.5.7d}$$

$\bar{E}_1(\bar{\rho}')$: total surface electric field distribution on contour C.

Again, referring to Figure 5.8c, the equivalent electric currents and magnetic currents on virtual contour $C_- \rightarrow C$ is defined based on the tangential components of the total surface magnetic field and the total surface electric field. Referring to the earlier discussion, expressions (5.4.29a) and (5.4.29b), the equivalent surface electric currents and magnetic currents can be obtained using the following boundary conditions on contour C:

$\bar{J}_{2_{eq}}(\bar{\rho}')$: equivalent electric currents valid only for region 2 fields

$$= - \hat{n}' \times \bar{H}_2(\bar{\rho}') \qquad \bar{\rho}' \text{ on } C \tag{5.5.8a}$$

using the surface boundary condition (5.5.1b)

$$= - \hat{n}' \times \bar{H}_1(\bar{\rho}') \qquad \bar{\rho}' \text{ on } C \tag{5.5.8b}$$

$$= - \bar{J}_{eq}(\bar{\rho}') \tag{5.5.8c}$$

$\bar{H}_2(\bar{\rho}')$: total surface magnetic field distribution on contour C;
$\hat{n} = -\hat{n}_2$: outward normal to contour C;

and similarly,

$\bar{M}_{2eq}(\bar{\rho}')$: equivalent magnetic currents valid only for region 2 fields

$$= \hat{n}' \times \bar{E}_2(\bar{\rho}') \qquad \bar{\rho}' \text{ on } C \tag{5.5.8d}$$

using the surface boundary condition (5.5.1a)

$$= \hat{n}' \times \bar{E}_1(\bar{\rho}') \qquad \bar{\rho}' \text{ on } C \tag{5.5.8e}$$

$$= -\bar{M}_{eq}(\bar{\rho}') \tag{5.5.8f}$$

$\bar{E}_2(\bar{\rho}')$: total surface electric field distribution on contour C.

Hence, the coupled combined field integral equations derived in expressions (5.5.6a) and (5.5.6b) can be rewritten as

$$\bar{E}^i(\bar{\rho})\Big|_{\tan} = \Big\{j\omega\big[\bar{A}_1(\bar{\rho},\omega) - \bar{A}_2(\bar{\rho},\omega)\big] + \nabla_\tau\big[\Phi_1(\bar{\rho},\omega) - \Phi_2(\bar{\rho},\omega)\big]$$

$$+ \nabla_\tau \times \Big[\frac{1}{\varepsilon_1}\bar{F}_1(\bar{\rho},\omega) - \frac{1}{\varepsilon'_2}\bar{F}_2(\bar{\rho},\omega)\Big]\Big\}\Big|_{\tan} \tag{5.5.9a}$$

$$\bar{H}^i(\bar{\rho})\Big|_{\tan} = \Big\{j\omega\big[\bar{F}_1(\bar{\rho},\omega) - \bar{F}_2(\bar{\rho},\omega)\big] + \nabla_\tau\big[\Psi_1(\bar{\rho},\omega) - \Psi_2(\bar{\rho},\omega)\big]$$

$$- \nabla_\tau \times \Big[\frac{1}{\mu_1}\bar{A}_1(\bar{\rho},\omega) - \frac{1}{\mu_2}\bar{A}_2(\bar{\rho},\omega)\Big]\Big\}\Big|_{\tan} \tag{5.5.9b}$$

$\bar{\rho}$ on C

where the vector and scalar potential integrals are given by the following expressions:

$$\bar{A}_1(\bar{\rho}) = \frac{\mu_1}{4j}\int_C \bar{J}_{eq}(\bar{\rho}')H_0^{(2)}(k_1|\bar{\rho} - \bar{\rho}'|)\, dL(\bar{\rho}') \tag{5.5.10a}$$

$$\Phi_1(\bar{\rho}) = \frac{1}{4\omega\varepsilon_1}\int_C [\nabla'_\tau \bullet \bar{J}_{eq}(\bar{\rho}')]H_0^{(2)}(k_1|\bar{\rho} - \bar{\rho}'|)\, dL(\bar{\rho}') \tag{5.5.10b}$$

$$\bar{F}_1(\bar{\rho}) = \frac{\varepsilon_1}{4j}\int_C \bar{M}_{eq}(\bar{\rho}')H_0^{(2)}(k_1|\bar{\rho} - \bar{\rho}'|)\, dL(\bar{\rho}') \tag{5.5.10c}$$

$$\Psi_1(\bar{\rho}) = \frac{1}{4\omega\mu_1}\int_C [\nabla'_\tau \bullet \bar{M}_{eq}(\bar{\rho}')]H_0^{(2)}(k_1|\bar{\rho} - \bar{\rho}'|)\, dL(\bar{\rho}') \tag{5.5.10d}$$

$\bar{\rho}$ on C

and

$$\bar{A}_2(\bar{\rho}) = \frac{\mu_2}{4j}\int_C -\bar{J}_{eq}(\bar{\rho}')H_0^{(2)}(k_2|\bar{\rho} - \bar{\rho}'|)\, dL(\bar{\rho}') \tag{5.5.11a}$$

$$\Phi_2(\bar{\rho}) = \frac{1}{4\omega\varepsilon'_2}\int_C -[\nabla'_\tau \bullet \bar{J}_{eq}(\bar{\rho}')]H_0^{(2)}(k_2|\bar{\rho} - \bar{\rho}'|)\, dL(\bar{\rho}') \tag{5.5.11b}$$

$$\bar{F}_2(\bar{\rho}) = \frac{\varepsilon'_2}{4j}\int_C -\bar{M}_{eq}(\bar{\rho}')H_0^{(2)}(k_2|\bar{\rho} - \bar{\rho}'|)\, dL(\bar{\rho}') \tag{5.5.11c}$$

$$\Psi_2(\bar{\rho}) = \frac{1}{4\omega\mu_2}\int_C -[\nabla'_\tau \bullet \bar{M}_{eq}(\bar{\rho}')]H_0^{(2)}(k_2|\bar{\rho} - \bar{\rho}'|)\, dL(\bar{\rho}') \tag{5.5.11d}$$

$\bar{\rho}$ on C

In coupled integral equations (5.5.9a) and (5.5.9b), there are only two unknown distribution of the equivalent electric and magnetic currents. Depending on the type of polarization of incident excitation, either transverse magnetic or transverse electric, only specific components of the equivalent electric and magnetic currents are excited on boundary contour C.

5.5.1 CFIE – TM CASE

As discussed in Table 5.2, for time-harmonic transverse magnetic *normal* excitation, there are only two distinct electromagnetic field components. For the analysis of arbitrary cross-sectional geometry, the generalized cylindrical coordinate system given by the orthogonal coordinate variables (n, s, z) is applied. Figure 5.7 shows various interrelationships among orthogonal unit vectors corresponding to the points on contour C of arbitrary cross-sectional geometry. At the boundary contour separating the dielectric medium and the free-space medium, the coordinates variable (n, s) are in a plane parallel to the transverse plane and the coordinates variables (s, z) are oriented tangential to the cylindrical boundary. Further, in regions 1 and 2 including at boundary contour C that separates the dielectric medium and the free-space medium, there are only the axial component of electric field distribution parallel to the z coordinate axis and the transverse component of magnetic field distribution parallel to the transverse plane. The transverse component of the magnetic field at the boundary surface can be resolved into two distinct orthogonal components along the coordinate variables n and s. It should be noted again, for normal transverse magnetic excitation, the various electric and magnetic field distributions in the two regions and the corresponding equivalent magnetic and electric current sources are completely independent of the z coordinate variable. Thus

$$\frac{\partial}{\partial z} = 0 \tag{5.5.12}$$

and there are only the following components of the electric and magnetic field distributions:

$E_z(\rho,\phi)\hat{\boldsymbol{z}}$: total axial component of the electric field distribution;

$H_n(\rho,\phi)\hat{\boldsymbol{n}}$: total normal component of the magnetic field distribution;

$H_s(\rho,\phi)\hat{\boldsymbol{s}}$: total tangential component of the magnetic field distribution.

Based on these two-dimensional nonzero electric and magnetic field components, for the case of TM normal excitation, only the axial z component of the equivalent electric current distribution and the tangential s component of the equivalent magnetic current distribution are on virtual contour C as defined in the following. Referring to Figure 5.9,

$J_z(\rho',\phi')\hat{\boldsymbol{z}}'$: unknown axial component of electric current distribution;

$M_s(\rho',\phi')\hat{\boldsymbol{s}}'$: unknown tangential component of magnetic current distribution;

$$\bar{J}_{eq}(\bar{\rho}') = \hat{z}'J_z(\bar{\rho}') \tag{5.5.13a}$$

$$= \hat{n}' \times \hat{s}'H_s(\rho',\phi') \qquad \bar{\rho}' \text{ on } C \tag{5.5.13b}$$

and

$$\bar{M}_{eq}(\bar{\rho}') = \hat{s}'M_s(\bar{\rho}') \tag{5.5.14a}$$

$$= \hat{z}'E_z(\rho',\phi') \times \hat{n}' \qquad \bar{\rho}' \text{ on } C \tag{5.5.14b}$$

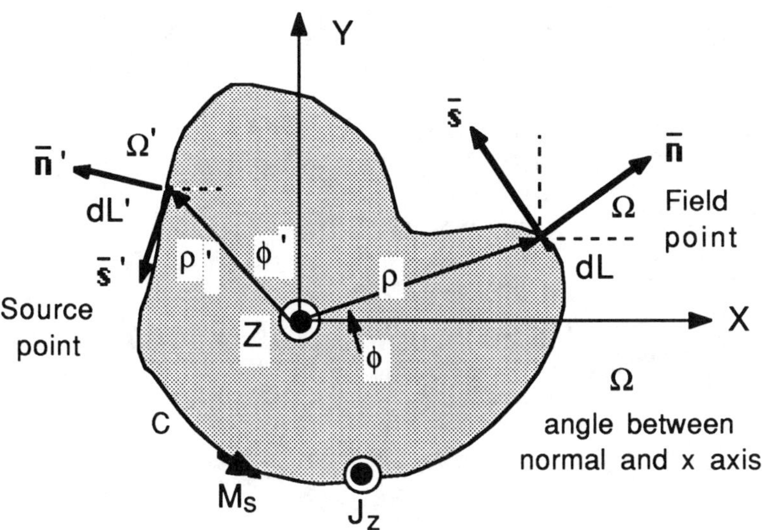

Figure 5.9 TM excitation – dielectric case.

Hence, for TM excitation, based on these two components of unknown electric and magnetic current distributions, the coupled combined field integral equations, (5.5.9a) and (5.5.9b), can be specialized. Taking the z component of integral equation (5.5.9a) and the s component of integral equation (5.5.9b), the following coupled integral equations are obtained:

$$\hat{z} \bullet \bar{E}^i(\bar{\rho}) = \hat{z} \bullet j\omega\left[\bar{A}_1(\bar{\rho},\omega) - \bar{A}_2(\bar{\rho},\omega)\right] + \hat{z} \bullet \nabla_\tau \times \left[\frac{1}{\varepsilon_1}\bar{F}_1(\bar{\rho},\omega) - \frac{1}{\varepsilon'_2}\bar{F}_2(\bar{\rho},\omega)\right] \quad (5.5.15a)$$

$$\hat{s} \bullet \bar{H}^i(\bar{\rho}) = \hat{s} \bullet j\omega\left[\bar{F}_1(\bar{\rho},\omega) - \bar{F}_2(\bar{\rho},\omega)\right] + \frac{\partial}{\partial s}\left[\Psi_1(\bar{\rho},\omega) - \Psi_2(\bar{\rho},\omega)\right] - \hat{s} \bullet \nabla_\tau \times \left[\frac{1}{\mu_1}\bar{A}_1(\bar{\rho},\omega) - \frac{1}{\mu_2}\bar{A}_2(\bar{\rho},\omega)\right] \quad (5.5.15b)$$

$\bar{\rho}$ on C

where the vector and scalar potential integrals are given by the following expressions:

$$\bar{A}_1(\bar{\rho}) = \frac{\mu_1}{4j}\int_C \hat{z}' J_z(\bar{\rho}') H_0^{(2)}(k_1|\bar{\rho} - \bar{\rho}'|)\, dL(\bar{\rho}') \tag{5.5.16a}$$

$$\bar{F}_1(\bar{\rho}) = \frac{\varepsilon_1}{4j}\int_C \hat{s}' M_s(\bar{\rho}') H_0^{(2)}(k_1|\bar{\rho} - \bar{\rho}'|)\, dL(\bar{\rho}') \tag{5.5.16b}$$

$$\Psi_1(\bar{\rho}) = \frac{1}{4\omega\mu_1}\int_C [\nabla'_\tau \bullet \hat{s}' M_s(\bar{\rho}')] H_0^{(2)}(k_1|\bar{\rho} - \bar{\rho}'|)\, dL(\bar{\rho}') \tag{5.5.16c}$$

and

$$\bar{A}_2(\bar{\rho}) = \frac{\mu_2}{4j}\int_C -\hat{z}' J_z(\bar{\rho}') H_0^{(2)}(k_2|\bar{\rho} - \bar{\rho}'|)\, dL(\bar{\rho}') \tag{5.5.17a}$$

$$\bar{F}_2(\bar{\rho}) = \frac{\varepsilon'_2}{4j}\int_C -\hat{s}' M_s(\bar{\rho}') H_0^{(2)}(k_2|\bar{\rho} - \bar{\rho}'|)\, dL(\bar{\rho}') \tag{5.5.17b}$$

$$\Psi_2(\bar{\rho}) = \frac{1}{4\omega\mu_2}\int_C -[\nabla'_\tau \bullet \hat{s}' M_s(\bar{\rho}')] H_0^{(2)}(k_2|\bar{\rho} - \bar{\rho}'|)\, dL(\bar{\rho}') \tag{5.5.17c}$$

$\bar{\rho}$ on C

After substituting these potential expressions into integral equation (5.5.15a), the following simplification is obtained:

$$\begin{aligned} E_z^i(\bar{\rho}) = {} & \frac{\omega\mu_1}{4}\int_C J_z(\bar{\rho}') H_0^{(2)}(k_1|\bar{\rho} - \bar{\rho}'|)\, dL(\bar{\rho}') \\ & + \frac{\omega\mu_2}{4}\int_C J_z(\bar{\rho}') H_0^{(2)}(k_2|\bar{\rho} - \bar{\rho}'|)\, dL(\bar{\rho}') \\ & + \hat{z} \bullet \nabla_\tau \times \frac{1}{4j}\int_C \hat{s}' M_s(\bar{\rho}') H_0^{(2)}(k_1|\bar{\rho} - \bar{\rho}'|)\, dL(\bar{\rho}') \\ & + \hat{z} \bullet \nabla_\tau \times \frac{1}{4j}\int_C \hat{s}' M_s(\bar{\rho}') H_0^{(2)}(k_2|\bar{\rho} - \bar{\rho}'|)\, dL(\bar{\rho}') \end{aligned} \tag{5.5.18a}$$

$\bar{\rho}$ on C

Similarly, after substituting these potential expressions into integral equation (5.5.15b), the following simplification is obtained:

$$
\begin{aligned}
H_s^i(\bar{\rho}) = & \frac{\omega\varepsilon_1}{4} \int_C (\hat{s} \cdot \hat{s}')M_s(\bar{\rho}')H_0^{(2)}(k_1|\bar{\rho} - \bar{\rho}'|)\, dL(\bar{\rho}') \\
& + \frac{\omega\varepsilon'_2}{4} \int_C (\hat{s} \cdot \hat{s}')M_s(\bar{\rho}')H_0^{(2)}(k_2|\bar{\rho} - \bar{\rho}'|)\, dL(\bar{\rho}') \\
& + \frac{1}{4\omega\mu_1}\frac{\partial}{\partial s} \int_C [\nabla'_\tau \cdot \hat{s}'M_s(\bar{\rho}')]H_0^{(2)}(k_1|\bar{\rho} - \bar{\rho}'|)\, dL(\bar{\rho}') \\
& + \frac{1}{4\omega\mu_2}\frac{\partial}{\partial s} \int_C [\nabla'_\tau \cdot \hat{s}'M_s(\bar{\rho}')]H_0^{(2)}(k_2|\bar{\rho} - \bar{\rho}'|)\, dL(\bar{\rho}') \\
& - \hat{s} \cdot \nabla_\tau \times \frac{1}{4j} \int_C \hat{z}'J_z(\bar{\rho}')H_0^{(2)}(k_1|\bar{\rho} - \bar{\rho}'|)\, dL(\bar{\rho}') \\
& - \hat{s} \cdot \nabla_\tau \times \frac{1}{4j} \int_C \hat{z}'J_z(\bar{\rho}')H_0^{(2)}(k_2|\bar{\rho} - \bar{\rho}'|)\, dL(\bar{\rho}') \qquad (5.5.18b)
\end{aligned}
$$

$\bar{\rho}$ on C

In these integral equations, the integral terms containing the divergence and curl operations can be further simplified. In expression (5.5.18b), the divergence term can be simplified as follows:

$$\nabla'_\tau \cdot \hat{s}'M_s(\bar{\rho}') = \left[\hat{n}'\frac{\partial}{\partial n'} + \hat{s}'\frac{\partial}{\partial s'}\right] \cdot \hat{s}'M_s(\bar{\rho}') \qquad (5.5.19a)$$

$$= \frac{\partial}{\partial s'} M_s(\bar{\rho}') \qquad (5.5.19b)$$

Further, referring to CFIE expressions (5.5.18a–b), the partial terms containing curl of the electric and magnetic vector potential integrals should be carefully analyzed. In these potential integrals, the kernel function appearing within the integrand of integral terms is singular. The curl operation cannot simply be taken inside, one has to consider the principal value of integral, specifically, when the field observation point coincides with the source segment in a limit. Complete details of the procedure for separating the principal value of integral is discussed in Section 4.5 based on the MFIE for conducting scatterers, expressions (4.5.10) to (4.5.12). In fact, similar analytical treatment can be adapted for separating the principal value even in the preceding coupled combined field integral equations.

Referring to the integral equation given by expression (5.5.18b) and the magnetic field boundary condition (5.5.1b), the partial s component of the scattered magnetic field in region 1 (based on the external equivalence) due to the axial z-directed electric currents is given by

$$\hat{s}H^{S}_{1e}(\bar{\rho}) = \frac{1}{\mu_1}\hat{s} \bullet \nabla_\tau \times \bar{A}_1(\bar{\rho}) \tag{5.5.20a}$$

$$= \frac{1}{4j}\hat{s} \bullet \nabla_\tau \times \int_C \hat{z}' J_z(\bar{\rho}')H_0^{(2)}(k_1|\bar{\rho} - \bar{\rho}'|)\, dL(\bar{\rho}') \tag{5.5.20b}$$

$$= \frac{1}{4j}\hat{s} \bullet \nabla_\tau \times \int_{C_0} \hat{z}' J_z(\bar{\rho}')H_0^{(2)}(k_1|\bar{\rho} - \bar{\rho}'|)\, dL(\bar{\rho}')$$

$$+ \frac{1}{4j}\hat{s} \bullet \nabla_\tau \times \fint_C \hat{z}' J_z(\bar{\rho}')H_0^{(2)}(k_1|\bar{\rho} - \bar{\rho}'|)\, dL(\bar{\rho}') \tag{5.5.20c}$$

$\bar{\rho}$ in region 1, approaching C from the outside

In expressions (5.5.20c), the integral term over the contour C_o represents the value of the integral when the field observation point approaches in a limit the source segment covered by contour C_o. In fact, it represents the principal value of integral that basically is the scattered partial magnetic field very close to the scatterer carrying the corresponding z-directed sheet of electric current. Hence, the principal value integral over contour C_o in expression (5.5.20c), as the field observation point in region 1 approaches the source segment, is equal to $j_z/2$, assuming the outward normal to contour C, as shown in Figure 5.9. Expression (5.5.20c) can be rewritten as

$$\hat{s}H^{S}_{1e}(\bar{\rho}) = \frac{j_z}{2}$$

$$+ \frac{1}{4j}\hat{s} \bullet \nabla_\tau \times \fint_C \hat{z}' J_z(\bar{\rho}')H_0^{(2)}(k_1|\bar{\rho} - \bar{\rho}'|)\, dL(\bar{\rho}') \tag{5.5.20d}$$

and similarly, for region 2,

$$\hat{s}H_{2e}^{S}(\bar{\rho}) = \frac{1}{\mu_2}\hat{s} \bullet \nabla_\tau \times \bar{A}_2(\bar{\rho}) \tag{5.5.21a}$$

$$= \frac{1}{4j}\hat{s} \bullet \nabla_\tau \times \int_C -\hat{z}'J_z(\bar{\rho}')H_0^{(2)}(k_2|\bar{\rho}-\bar{\rho}'|)\, dL(\bar{\rho}') \tag{5.5.21b}$$

$$= \frac{1}{4j}\hat{s} \bullet \nabla_\tau \times \int_{C_0} -\hat{z}'J_z(\bar{\rho}')H_0^{(2)}(k_2|\bar{\rho}-\bar{\rho}'|)\, dL(\bar{\rho}')$$

$$+ \frac{1}{4j}\hat{s} \bullet \nabla_\tau \times \fint_C -\hat{z}'J_z(\bar{\rho}')H_0^{(2)}(k_2|\bar{\rho}-\bar{\rho}'|)\, dL(\bar{\rho}') \tag{5.5.21c}$$

$\bar{\rho}$ in region 2, approaching C from the inside

Again, the principal value integral over the contour C_o in expression (5.5.21c), as the field observation point in the region 2 approaches the source segment, is equal to $-\jmath_z/2$ assuming inward normal to contour C. Retaining the same reference outward normal, expression (5.5.21c) can be rewritten as

$$\hat{s}H_{2e}^{S}(\bar{\rho}) = \frac{\jmath_z}{2}$$

$$- \frac{1}{4j}\hat{s} \bullet \nabla_\tau \times \fint_C \hat{z}'J_z(\bar{\rho}')H_0^{(2)}(k_1|\bar{\rho}-\bar{\rho}'|)\, dL(\bar{\rho}') \tag{5.5.21d}$$

After substituting into boundary condition (5.5.1b), the two principal value contributions cancel out. The other integral terms are written with a bar on the integral sign to recognize the field observation point on remaining of contour C leaving out the source segment. The kernel function in the integrand is well behaved, and thus, the transverse curl operation can be easily taken inside the integral to yield a normal derivative term similar to expression (4.5.5c).

Thus, the integral expressions (5.5.20d) and (5.5.21d) can be further simplified by noting that the transverse differential operator is in terms of unprimed coordinates and the sources are located at primed coordinates.

Referring to expressions (5.5.20b) and (5.5.21b),

$$\nabla_\tau \times \left[\hat{z}' J_z(\bar{\rho}') H_0^{(2)}(k_{1,2}|\bar{\rho} - \bar{\rho}'|)\right]$$

$$= \nabla_\tau \times \left[\bar{J}_z(\bar{\rho}') H_0^{(2)}(k_{1,2}|\bar{\rho} - \bar{\rho}'|)\right] \tag{5.5.22a}$$

$$= -\left[\hat{z}' J_z(\bar{\rho}') \times \nabla_\tau H_0^{(2)}(k_{1,2}|\bar{\rho} - \bar{\rho}'|)\right] \tag{5.5.22b}$$

Similar analysis can be adapted to separate out the principal value and simplify the integral terms containing curl operation in expression (5.5.18a).

After substituting these simplifications, the CFIE for the TM normal excitation take the following form:

$$E_z^i(\bar{\rho}) = \frac{\omega\mu_1}{4} \int_C J_z(\bar{\rho}') H_0^{(2)}(k_1|\bar{\rho} - \bar{\rho}'|)\, dL(\bar{\rho}')$$

$$+ \frac{\omega\mu_2}{4} \int_C J_z(\bar{\rho}') H_0^{(2)}(k_2|\bar{\rho} - \bar{\rho}'|)\, dL(\bar{\rho}')$$

$$- \hat{z} \bullet \frac{1}{4j} ⨍_C \hat{s}' M_s(\bar{\rho}') \times \nabla_\tau H_0^{(2)}(k_1|\bar{\rho} - \bar{\rho}'|)\, dL(\bar{\rho}')$$

$$- \hat{z} \bullet \frac{1}{4j} ⨍_C \hat{s}' M_s(\bar{\rho}') \times \nabla_\tau H_0^{(2)}(k_2|\bar{\rho} - \bar{\rho}'|)\, dL(\bar{\rho}') \tag{5.5.23a}$$

$\bar{\rho}$ on C

Similarly, integral equation (5.5.18b) reduces to the following form after substituting the simplification obtained in (5.5.19b) and (5.5.22b):

$$H_s^i(\bar{\rho}) = \frac{\omega\varepsilon_1}{4} \int_C (\hat{s} \bullet \hat{s}') M_s(\bar{\rho}') H_0^{(2)}(k_1|\bar{\rho} - \bar{\rho}'|)\, dL(\bar{\rho}')$$

$$+ \frac{\omega\varepsilon'_2}{4} \int_C (\hat{s} \bullet \hat{s}') M_s(\bar{\rho}') H_0^{(2)}(k_2|\bar{\rho} - \bar{\rho}'|)\, dL(\bar{\rho}')$$

$$+ \frac{1}{4\omega\mu_1} \frac{\partial}{\partial s} \int_C \left[\frac{\partial}{\partial s'} M_s(\bar{\rho}')\right] H_0^{(2)}(k_1|\bar{\rho} - \bar{\rho}'|)\, dL(\bar{\rho}')$$

$$+ \frac{1}{4\omega\mu_2}\frac{\partial}{\partial s}\int_C \left[\frac{\partial}{\partial s'} M_S(\bar{\rho}')\right] H_0^{(2)}(k_2|\bar{\rho}-\bar{\rho}'|)\, dL(\bar{\rho}')$$

$$+ \hat{s}\bullet\frac{1}{4j}\int_C \hat{z}' J_Z(\bar{\rho}')\times\nabla_\tau H_0^{(2)}(k_1|\bar{\rho}-\bar{\rho}'|)\, dL(\bar{\rho}')$$

$$+ \hat{s}\bullet\frac{1}{4j}\int_C \hat{z}' J_Z(\bar{\rho}')\times\nabla_\tau H_0^{(2)}(k_2|\bar{\rho}-\bar{\rho}'|)\, dL(\bar{\rho}') \tag{5.5.23b}$$

$\bar{\rho}$ on C

Referring to Figure 5.7, the combined field coupled integral equations just obtained, expressions (5.5.23a–b), can be further simplified. The various unit vectors on boundary contour C are now replaced in terms of the angle the normal makes with the x coordinate axis. In integral equation (5.5.23a), the integrand in the third integral term can be rewritten as

$$\frac{1}{4j}\,\hat{z}\bullet\hat{s}' M_S(\bar{\rho}')\times\nabla_\tau H_0^{(2)}(k_1|\bar{\rho}-\bar{\rho}'|) \tag{5.5.24a}$$

$$= \frac{1}{4j} M_S(\bar{\rho}')\,\hat{z}\bullet(-\,\hat{x}\sin\Omega' + \hat{y}\cos\Omega')$$
$$\times\left[\hat{x}\frac{\partial}{\partial x} + \hat{y}\frac{\partial}{\partial y}\right] H_0^{(2)}(k_1|\bar{\rho}-\bar{\rho}'|)$$

$$= \frac{1}{4j} M_S(\bar{\rho}')\left[-\cos\Omega'\frac{\partial}{\partial x} - \sin\Omega'\frac{\partial}{\partial y}\right] H_0^{(2)}(k_1|\bar{\rho}-\bar{\rho}'|) \tag{5.5.24b}$$

$$= -\frac{k_1}{4j} M_S(\bar{\rho}')\left[\frac{x-x'}{R}\cos\Omega' + \frac{y-y'}{R}\sin\Omega'\right] H_0^{(2)'}(k_1 R) \tag{5.5.24c}$$

and

$$R = |\bar{\rho}-\bar{\rho}'|$$

$$= \left[(x-x')^2 + (y-y')^2\right]^{1/2} \tag{5.5.24d}$$

Ω': angle the normal at the source point makes with the x coordinate axis.

After substituting these simplifications, the coupled CFIE (5.5.23a–b) for the *TM normal excitation* take the following form

$$E_z^i(\bar{\rho}) = \frac{\omega\mu_1}{4} \int_C J_z(\bar{\rho}')H_0^{(2)}(k_1R)\, dL(\bar{\rho}')$$

$$+ \frac{\omega\mu_2}{4} \int_C J_z(\bar{\rho}')H_0^{(2)}(k_2R)\, dL(\bar{\rho}')$$

$$+ \frac{k_1}{4j} \int_C M_s(\bar{\rho}')\Big[\frac{x-x'}{R}\cos\Omega' + \frac{y-y'}{R}\sin\Omega'\Big] H_0^{(2)'}(k_1R)\, dL(\bar{\rho}')$$

$$+ \frac{k_2}{4j} \int_C M_s(\bar{\rho}')\Big[\frac{x-x'}{R}\cos\Omega' + \frac{y-y'}{R}\sin\Omega'\Big] H_0^{(2)'}(k_2R)\, dL(\bar{\rho}')$$

$$\bar{\rho} \text{ on } C \tag{5.5.25a}$$

and

$$H_s^i(\bar{\rho}) = \frac{\omega\varepsilon_1}{4} \int_C \cos(\Omega - \Omega')M_s(\bar{\rho}')H_0^{(2)}(k_1R)\, dL(\bar{\rho}')$$

$$+ \frac{\omega\varepsilon'_2}{4} \int_C \cos(\Omega - \Omega')M_s(\bar{\rho}')H_0^{(2)}(k_2R)\, dL(\bar{\rho}')$$

$$+ \frac{1}{4\omega\mu_1}\frac{\partial}{\partial s} \int_C \Big[\frac{\partial}{\partial s'} M_s(\bar{\rho}')\Big]H_0^{(2)}(k_1R)\, dL(\bar{\rho}')$$

$$+ \frac{1}{4\omega\mu_2}\frac{\partial}{\partial s} \int_C \Big[\frac{\partial}{\partial s'} M_s(\bar{\rho}')\Big]H_0^{(2)}(k_2R)\, dL(\bar{\rho}')$$

$$+ \frac{k_1}{4j} \int_C J_z(\bar{\rho}')\Big[\frac{x-x'}{R}\cos\Omega + \frac{y-y'}{R}\sin\Omega\Big] H_0^{(2)'}(k_1R)\, dL(\bar{\rho}')$$

$$+ \frac{k_2}{4j} \int_C J_z(\bar{\rho}')\Big[\frac{x-x'}{R}\cos\Omega + \frac{y-y'}{R}\sin\Omega\Big] H_0^{(2)'}(k_2R)\, dL(\bar{\rho}')$$

$$\bar{\rho} \text{ on } C \tag{5.5.25b}$$

Similar to the earlier compact representations of EFIE and MFIE for a perfectly conducting object, for the case of dielectric object coupled CFIE expressions (5.5.25a–b) for the *TM normal excitation* take the form:

$$E_z^i(\bar{\rho}) = \mathcal{L}_{11}[J_z(\bar{\rho}')] + \mathcal{L}_{12}[M_s(\bar{\rho}')] \tag{5.5.26a}$$

$\bar{\rho}$ on C

$$H_s^i(\bar{\rho}) = \mathcal{L}_{21}[J_z(\bar{\rho}')] + \mathcal{L}_{22}[M_s(\bar{\rho}')] \tag{5.5.26b}$$

where the CFIE partial integral operators are given by

$$\mathcal{L}_{11} = \int_C \left[\frac{\omega\mu_1}{4} H_0^{(2)}(k_1R) + \frac{\omega\mu_2}{4} H_0^{(2)}(k_2R)\right] dL' \tag{5.5.27a}$$

$$\mathcal{L}_{12} = \oint_C \left[\frac{x-x'}{R}\cos\Omega' + \frac{y-y'}{R}\sin\Omega'\right]$$

$$\left[\frac{k_1}{4j} H_0^{(2)'}(k_1R) + \frac{k_2}{4j} H_0^{(2)'}(k_2R)\right] dL' \tag{5.5.27b}$$

$$\mathcal{L}_{21} = \oint_C \left[\frac{x-x'}{R}\cos\Omega + \frac{y-y'}{R}\sin\Omega\right]$$

$$\left[\frac{k_1}{4j} H_0^{(2)'}(k_1R) + \frac{k_2}{4j} H_0^{(2)'}(k_2R)\right] dL' \tag{5.5.27c}$$

$$\mathcal{L}_{22} = \int_C \cos(\Omega - \Omega')\left[\frac{\omega\varepsilon_1}{4} H_0^{(2)}(k_1R) + \frac{\omega\varepsilon'_2}{4} H_0^{(2)}(k_2R)\right] dL'$$

$$+ \frac{\partial}{\partial s}\int_C dL' \left[\frac{1}{4\omega\mu_1} H_0^{(2)}(k_1R) + \frac{1}{4\omega\mu_2} H_0^{(2)}(k_2R)\right]\frac{\partial}{\partial s'} \tag{5.5.27d}$$

$\mathcal{L}_{11}$: self integral operator for electric currents;

$\mathcal{L}_{22}$: self integral operator for magnetic currents;

$\mathcal{L}_{12}$: mutual integral operator for magnetic currents;

$\mathcal{L}_{21}$: mutual integral operator for electric currents.

5.6 DIELECTRIC CIRCULAR CYLINDER – TM EXCITATION

The coupled combined field integral equations, expressions (5.5.26a–b), derived in the previous section can be directly solved to analyze the electromagnetic scattering, penetration, and interaction due to a homogeneous, lossy dielectric circular cylinder excited externally by a TM to z polarized plane wave. In the CFIE, there are two unknown equivalent surface distributions given by the axial z component of electric current and the tangential s component of magnetic current on boundary C. If these two current distributions are known, then the complete scattered and penetrated fields in free-space and dielectric regions can be calculated.

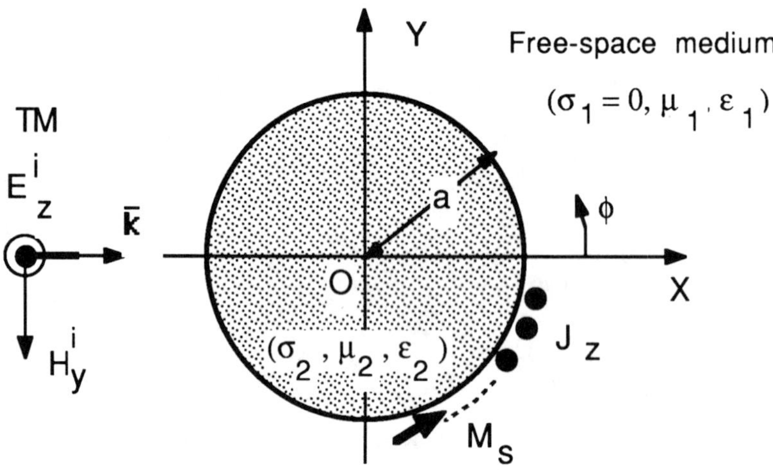

Figure 5.10 Homogeneous lossy dielectric circular scatterer – TM excitation.

Figure 5.10 shows the geometry of a two-dimensional, lossy dielectric circular cylinder having radius equal to a. The circular scatterer geometry has perfect symmetry with respect to the angular variable ϕ of the cylindrical coordinate system. The complete scattering and penetration analysis can be carried out by just assuming the incidence plane wave to be propagating along the x coordinate direction, which gives the angle of incidence to be zero. The induced axial electric current and tangential magnetic current along circular contour C varies with respect to the angular variable and, in fact, repeats itself for every 2π radians. Following the analysis in Section 3.8 for TM axial current excitation and, similarly, in Section 4.6 for TE tangential current excitation, it is possible to expand the distribution of unknown electric current and unknown magnetic current on the circular dielectric scatterer in terms of two independent infinite Fourier series expansions. The two Fourier series expansions can be substituted into the CFIE, and the current modal coefficients can be determined by invoking orthogonality condition. Instead of this procedure, an alternative indirect formulation discussed in Sections 3.8.3 and 4.6.1 is followed.

Referring to the discussions in Section 2.19, the TM-polarized time-harmonic incident plane wave is given by

$$E_z^i(\bar{\rho},\omega) = E_0\, e^{-j\bar{k}_1 \bullet \bar{\rho}} \tag{5.6.1a}$$

$$\bar{k}_1 = k_1 \hat{k} \tag{5.6.1b}$$

k_1 : propagation constant for homogeneous, isotropic free-space medium 1

$$= \omega(\mu_1 \varepsilon_1)^{1/2} \tag{5.6.1c}$$

ω : frequency of excitation, in radians per second;

ε_1 : permittivity of the free-space medium, region 1;

μ_1 : permeability of the free-space medium, region 1;

$\phi^i = 0$: angle of incidence of the TM plane wave field.

For free-space region 1, let

$(\bar{E}^i, \bar{H}^i)$: electric and magnetic incident fields in free-space medium 1;

$(\bar{E}_1^S, \bar{H}_1^S)$: electric and magnetic scattered fields in free-space medium;

$(\bar{E}_1, \bar{H}_1)$: electric and magnetic total fields in free-space medium 1

$$= (\bar{E}^i, \bar{H}^i) + (\bar{E}_1^S, \bar{H}_1^S) \tag{5.6.2}$$

As discussed in the previous sections, for the case of TM normal excitation, various field distributions are independent of z coordinate variable. In expression (5.6.2), there are only an axial component of electric field and an angular component of magnetic field distributions. Following the discussion in Section 3.8 concerning the plane wave transformation, the plane wave electric field propagating in the x coordinate direction can be written in terms of an infinite series of cylindrical modes:

$$E_z^i(\rho,\phi) = E_0 e^{-jk_1\rho\,\cos\phi} \tag{5.6.3a}$$

$$= E_0 \sum_{m=-\infty}^{\infty} j^{-m} J_m(k_1\rho)\, e^{jm\phi} \tag{5.6.3b}$$

Based on the electromagnetic modeling discussed for the CFIE, the total tangential electric and magnetic fields on the surface of dielectric scatterer can be viewed as a collection of tangential-directed and axial-directed magnetic current and electric current distributions. These line-source type distributions basically generate fields with cylindrical symmetry in the free-space region and also in the dielectric region. From expression (3.8.20a), the scattered electric field distribution in the free-space region can be expressed in terms of an infinite number of outgoing propagating cylindrical modes:

$$E^{s}_{1z}(\rho,\phi) = E_0 \sum_{m=-\infty}^{\infty} A_m H^{(2)}_m(k_1\rho)\, e^{jm\phi} \qquad \rho \geq a \tag{5.6.4}$$

A_m : unknown scattered modal coefficients for the free-space region.

Further, for the homogeneous, lossy dielectric scatterer region 2, let

ε_2 : permittivity of the dielectric medium, region 2;

μ_2 : permeability of the dielectric medium, region 2;

σ_2 : conductivity of the dielectric medium, region 2.

The electrical conductivity of the dielectric scatterer can be accounted by redefining the permittivity in terms of an effective permittivity as in expression (5.4.26):

ε'_2 : effective permittivity of the dielectric medium, region 2

$$= \varepsilon_2 \left[1 - j\frac{\sigma_2}{\omega\varepsilon_2}\right] \tag{5.6.5}$$

and

$(\bar{E}^{s}_2, \bar{H}^{s}_2)$: electric and magnetic penetrated fields in the dielectric medium 2;

$(\bar{E}_2, \bar{H}_2)$: electric and magnetic total fields in the dielectric medium 2

$$= (\bar{E}^{s}_2, \bar{H}^{s}_2) \tag{5.6.6}$$

Again for the TM normal excitation, various field distributions are independent of the *z* coordinate variable in the dielectric region, and there are only an axial component of electric field and an angular component of magnetic field distributions. The penetrated electric field distribution in the dielectric region can be expressed in terms of an infinite number of nonpropagating cylindrical modes:

$$E_{2z}^{s}(\rho,\phi) = E_0 \sum_{m=-\infty}^{\infty} B_m J_m(k_2\rho)\, e^{jm\phi} \qquad \rho \le a \tag{5.6.7}$$

B_m : unknown penetrated modal coefficients for the dielectric region;
k_2 : propagation constant for the homogeneous lossy dielectric region

$$= \omega(\mu_2 \varepsilon'_2)^{1/2} \tag{5.6.8}$$

Hence, based on these field representations, the total electric field distributions in the free-space medium and the dielectric medium are given by

$$E_{1z}(\rho,\phi)\Big|_{\text{total}} = E_{1z}^{i}(\rho,\phi) + E_{1z}^{s}(\rho,\phi) \tag{5.6.9a}$$

$$= E_0 \sum_{m=-\infty}^{\infty} \left[j^{-m} J_m(k_1\rho) + A_m H_m^{(2)}(k_1\rho)\right] e^{jm\phi} \tag{5.6.9b}$$

$$\rho \ge a$$

$$E_{2z}(\rho,\phi)\Big|_{\text{total}} = E_{2z}^{s}(\rho,\phi) \tag{5.6.9c}$$

$$= E_0 \sum_{m=-\infty}^{\infty} \left[B_m J_m(k_2\rho)\right] e^{jm\phi} \tag{5.6.9d}$$

$$\rho \le a$$

Similarly, the corresponding expressions for the tangential or angular component of the magnetic field distributions can be written. Referring to Section 3.8.3 and using the source-free Faraday's law stated in expression (3.1.5a)

$$- j\omega\mu \mathbf{H}^{s}(\rho,\phi,\omega) = \nabla \times \hat{z} E_z^{s}(\rho,\phi,\omega) \tag{5.6.10a}$$

and expressing the scattered electric and magnetic fields in terms of the cylindrical coordinate system,

$$- j\omega\mu\left[H_\rho^{s}\hat{\rho} + H_\phi^{s}\hat{\phi}\right] = \frac{1}{\rho}\frac{\partial}{\partial\phi}E_z^{s}\,\hat{\rho} - \frac{\partial}{\partial\rho}E_z^{s}\,\hat{\phi} \tag{5.6.10b}$$

Hence, the angular component of the total magnetic field can be expressed in terms of the incident and scattered electric fields in free-space region 1, given by

$$H_{1\phi}(\rho,\phi)\Big|_{\text{total}} = \frac{1}{j\omega\mu_1}\frac{\partial}{\partial\rho}\left[E^i_{1z}(\rho,\phi) + E^s_{1z}(\rho,\phi)\right] \tag{5.6.11a}$$

$$= \frac{1}{j\omega\mu_1}\frac{\partial}{\partial\rho}E_0 \sum_{m=-\infty}^{\infty}\left[j^{-m}J_m(k_1\rho) + A_m H^{(2)}_m(k_1\rho)\right] e^{jm\phi} \tag{5.6.11b}$$

$$= -\frac{jk_1}{\omega\mu_1}E_0 \sum_{m=-\infty}^{\infty}\left[j^{-m}J'_m(k_1\rho) + A_m H^{(2)'}_m(k_1\rho)\right] e^{jm\phi} \tag{5.6.11c}$$

$$\rho \geq a$$

Similarly, the angular component of the total magnetic field can be expressed in terms of the penetrated electric field in dielectric region 2, given by

$$H_{2\phi}(\rho,\phi)\Big|_{\text{total}} = \frac{1}{j\omega\mu_2}\frac{\partial}{\partial\rho}\left[E^s_{2z}(\rho,\phi)\right] \tag{5.6.12a}$$

$$= \frac{1}{j\omega\mu_2}\frac{\partial}{\partial\rho}E_0 \sum_{m=-\infty}^{\infty}\left[B_m J_m(k_2\rho)\right] e^{jm\phi} \tag{5.6.12b}$$

$$= -\frac{jk_2}{\omega\mu_2}E_0 \sum_{m=-\infty}^{\infty}\left[B_m J'_m(k_2\rho)\right] e^{jm\phi} \tag{5.6.12c}$$

$$\rho \leq a$$

In these total field expressions, the two sets of modal coefficients A_m and B_m are still unknown. They can be determined by enforcing electromagnetic boundary conditions (5.5.1a) and (5.5.1b) on the surface of the dielectric scatterer. The axial components of total electric field should be continuous, and similarly, the angular components of total magnetic field should be continuous on the dielectric boundary contour. Hence

$$E_{1z}(\rho,\phi) = E_{2z}(\rho,\phi) \tag{5.6.13a}$$

$$\rho = a \text{ on } C$$

$$H_{1\phi}(\rho,\phi) = H_{2\phi}(\rho,\phi) \tag{5.6.13b}$$

After substituting total electric field expressions (5.6.9b) and (5.6.9d) and the total magnetic field expressions (5.6.11c) and (5.6.12c) into the above boundary conditions,

$$\sum_{m=-\infty}^{\infty} \left[j^{-m} J_m(k_1 a) + A_m H_m^{(2)}(k_1 a) \right] e^{jm\phi}$$

$$= \sum_{m=-\infty}^{\infty} \left[B_m J_m(k_2 a) \right] e^{jm\phi} \tag{5.6.14a}$$

$$\frac{k_1}{\mu_1} \sum_{m=-\infty}^{\infty} \left[j^{-m} J_m'(k_1 a) + A_m H_m^{(2)'}(k_1 a) \right] e^{jm\phi}$$

$$= \frac{k_2}{\mu_2} \sum_{m=-\infty}^{\infty} \left[B_m J_m'(k_2 a) \right] e^{jm\phi} \tag{5.6.14b}$$

Now the cylindrical modes orthogonality condition is enforced on the preceding two boundary expressions. Referring to the discussion in Section 3.7, a set of independent weighting terms are selected to test these two expressions on both sides, where each weighting term forms an entire domain harmonic function in the domain of scatterer contour C. To determine the scattered cylindrical mode coefficients, relationships (5.6.14a) and (5.6.14b) are multiplied on both sides by the entire domain weighting function $exp(-j\ell\phi)$ and integrated between the limits 0 to 2π with respect to ϕ. Thus, for all integer values $\ell = -\infty$ to ∞, the two boundary condition expressions simplify to the following form:

$$j^{-m} J_m(k_1 a) + A_m H_m^{(2)}(k_1 a) = B_m J_m(k_2 a) \tag{5.6.15a}$$

$$\frac{k_1}{\mu_1} \left[j^{-m} J_m'(k_1 a) + A_m H_m^{(2)'}(k_1 a) \right]$$

$$= \frac{k_2}{\mu_2} \left[B_m J_m'(k_2 a) \right] \tag{5.6.15b}$$

After rearranging, the two unknown modal coefficients are now obtained by solving the matrix equation:

$$\begin{bmatrix} H_m^{(2)}(k_1a) & -J_m(k_2a) \\ \frac{k_1}{\mu_1} H_m^{(2)'}(k_1a) & -\frac{k_2}{\mu_2} J_m'(k_2a) \end{bmatrix} \begin{bmatrix} A_m \\ B_m \end{bmatrix} = \begin{bmatrix} -j^{-m} J_m(k_1a) \\ -\frac{k_1}{\mu_1} j^{-m} J_m'(k_1a) \end{bmatrix} \qquad (5.6.16)$$

Hence, for each mode m, modal matrix equation (5.6.16) is first generated and the two unknown modal coefficients, A_m and B_m, are calculated. Once the modal coefficients are known, the axial components of the scattered and penetrated electric fields can be calculated based on expressions (5.6.4) and (5.6.7). Similarly, the angular components of the scattered and penetrated magnetic fields can be calculated based on expressions (5.6.11a) and (5.6.12a). Following the procedure explained in Section 3.8, for the free-space region 1, the far-field distribution and the corresponding radar cross section can be calculated by taking the large argument approximation of the Hankel function in expression (5.6.4).

Figures 5.11 and 5.12 show the distribution of the induced surface electric current and the induced surface magnetic current on boundary C of a lossless dielectric scatterer. In fact, these two current distributions are conveniently obtained by first calculating the total surface tangential field distributions on the dielectric scatterer and then converting them according to expressions (5.5.13) and (5.5.14). The conductivity is taken as zero and the relative permittivity and relative permeability are selected as equal to 4 and 2 respectively, for the lossless dielectric circular scatterer. The frequency of excitation is selected as 300 MHz, so that the free-space wavelength is equal to 1 meter. The free-space propagation constant k_1 is equal 2π, and the electrical radius of the circular dielectric scatterer, as referred to the free-space medium, is $k_1a = 5$.

Figure 5.11 shows the real and imaginary parts of the equivalent electric current distribution, and similarly, Figure 5.12 shows the real and imaginary parts of the equivalent magnetic current distribution. The electric current is normalized with respect to the incident magnetic field, and the magnetic current is normalized with respect to the incident electric field. The two current distributions on the lossless dielectric circular cylinder are plotted as a function of angular variable ϕ with the TM incident electric field assumed to be 1 volt per meter. Only the surface current distributions in the angular range of 180° to 360° are shown and, in fact, the results are symmetrical in the remaining angular range. Specifically the two distributions of induced electric and magnetic currents are symmetrical with respect to the direction of incident angle. In

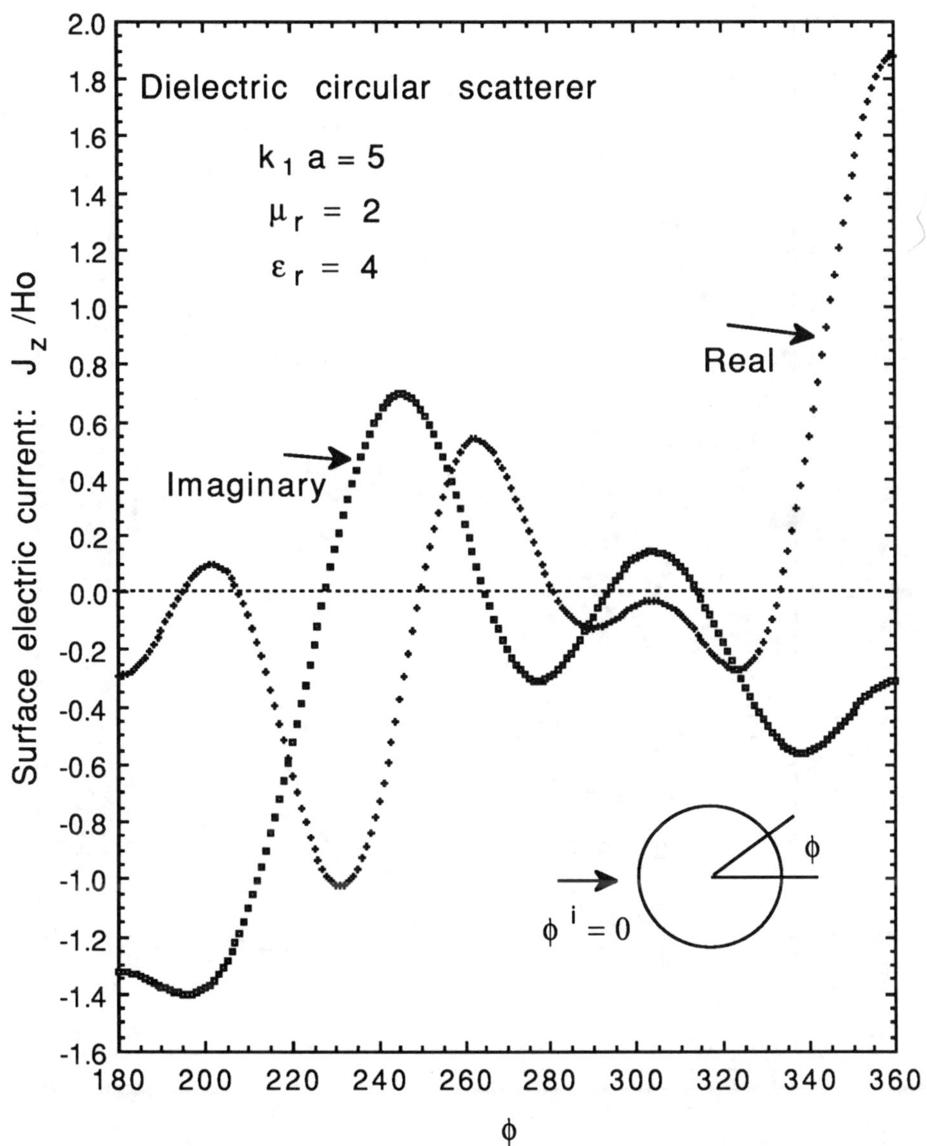

Figure 5.11 Equivalent electric current on a lossless circular dielectric scatterer – TM excitation.

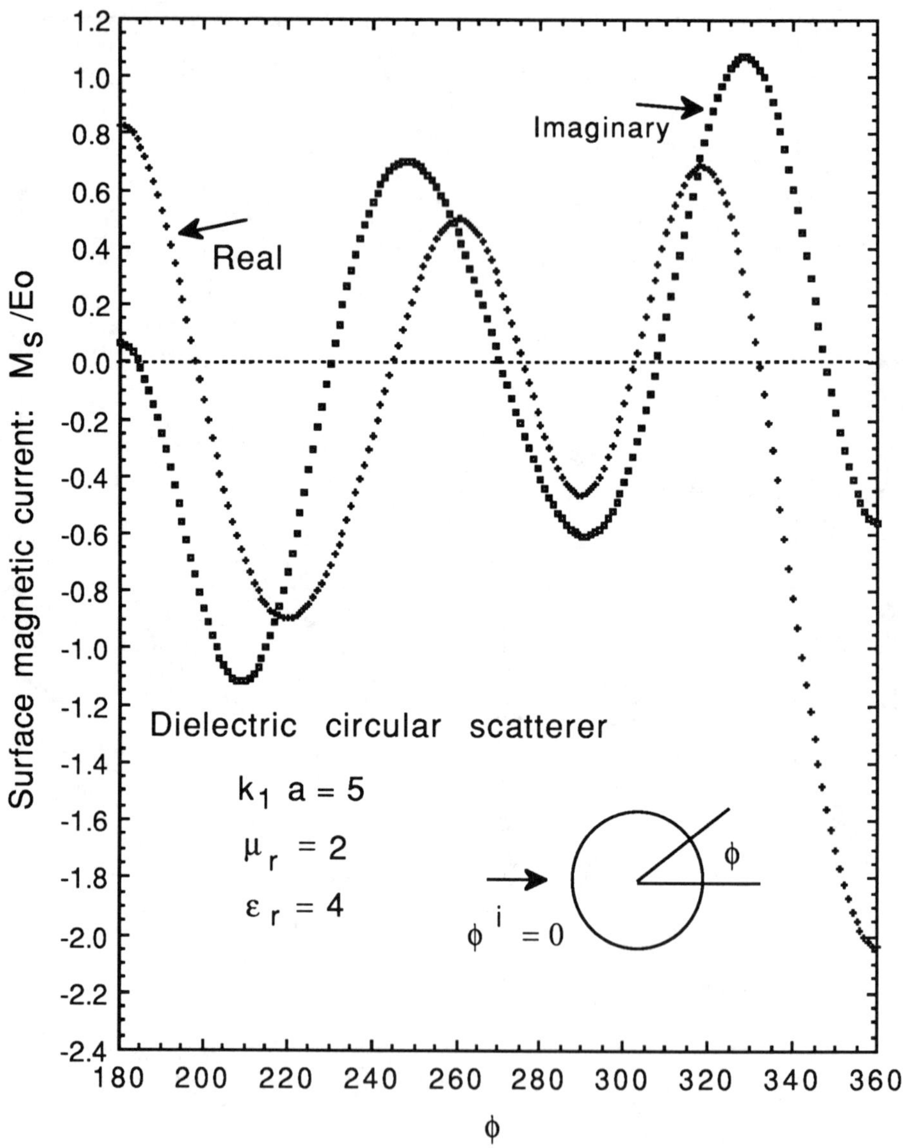

Figure 5.12 Equivalent magnetic current on a lossless circular dielectric scatterer – TM excitation.

Figures 5.11 and 5.12, the region around angle $\phi = 180^o$ forms the illuminated side and the region around angle $\phi = 0^o$ forms the backside of the circular cylindrical scatterer. There are both exterior scattering and interior penetration of the electric and magnetic fields from the illuminated side toward the backside of dielectric circular scatterer.

5.6.1 SCATTERED FIELD DISTRIBUTION

The z component of the scattered electric field distribution in the free-space medium and on the surface of the homogeneous, lossy dielectric circular scatterer is given by general expression (5.6.4). It should be noted that the first and the second kind Bessel and Hankel functions of the complex argument and integer order are well documented in the literature. By substituting field observation point (ρ, ϕ) and performing the summation over N cylindrical modes, the scattered electric field distribution can be calculated. As pointed out earlier, in the numerical calculation should include enough number of modes to obtain convergence of the infinite series. Generally the number of modes required for summation is just more than the electrical distance of the observation point. Thus for n positive, the upper limit N is taken to be $|k\rho|$ plus an additional five to ten modes, depending upon the numerical accuracy required. For very large distances from the scatterer, numerical generation of the Bessel and Hankel functions is quite slow, and thus, it is preferable to use their large argument approximation. Referring to expressions (5.6.4) and (5.6.7), the scattered and the penetrated electric fields at any general field point are repeated:

$$E_{1z}^{s}(\rho,\phi) = E_0 \sum_{m=-\infty}^{\infty} A_m H_m^{(2)}(k_1\rho)\, e^{jm\phi} \qquad \rho \geq a \tag{5.6.17}$$

$$E_{2z}^{s}(\rho,\phi) = E_0 \sum_{m=-\infty}^{\infty} B_m J_m(k_2\rho)\, e^{jm\phi} \qquad \rho \leq a \tag{5.6.18}$$

where the field modal coefficient A_m and B_m are obtained by solving modal matrix equation (5.6.16). Further, based on expression (5.6.10b), the angular and radial components of magnetic field distribution can be calculated in terms of the axial component of the corresponding electric field:

$$- j\omega\mu[H_\rho^s \hat{\rho} + H_\phi^s \hat{\phi}] = \frac{1}{\rho}\frac{\partial}{\partial\phi} E_z^s \hat{\rho} - \frac{\partial}{\partial\rho} E_z^s \hat{\phi} \tag{5.6.19}$$

Hence, the angular and radial components of the scattered and the penetrated magnetic field in terms of the corresponding the electric field in the free-space region and in the dielectric region are given by

$$H^s_{1\phi}(\rho,\phi) = -\frac{jk_1}{\omega\mu_1}E_0\sum_{m=-\infty}^{\infty} A_m H^{(2)'}_m(k_1\rho)\, e^{jm\phi} \tag{5.6.20a}$$

$$H^s_{1\rho}(\rho,\phi) = \frac{j}{\omega\mu_1}\frac{1}{\rho}E_0\sum_{m=-\infty}^{\infty} jmA_m H^{(2)}_m(k_1\rho)\, e^{jm\phi} \tag{5.6.20b}$$

$$\rho \geq a$$

$$H^s_{2\phi}(\rho,\phi) = -\frac{jk_2}{\omega\mu_2}E_0\sum_{m=-\infty}^{\infty} B_m J'_m(k_2\rho)\, e^{jm\phi} \tag{5.6.21a}$$

$$H^s_{2\rho}(\rho,\phi) = \frac{j}{\omega\mu_2}\frac{1}{\rho}E_0\sum_{m=-\infty}^{\infty} jmB_m J_m(k_2\rho)\, e^{jm\phi} \tag{5.6.21b}$$

$$\rho \leq a$$

To obtain the total field distributions, the corresponding incident field terms are added to the preceding scattered and penetrated field expressions. Figure 5.13 shows the distribution of z component of the total electric field in the neighborhood of the dielectric circular scatterer, $k_1 a = 5$, calculated based on expressions (5.6.17) and (5.6.18). Both the penetrated electric field and the near total electric field are displayed in Figure 5.13 for about a tenth of a free-space wavelength outside and a tenth of a dielectric region wavelength inside the lossless dielectric scatterer. As can be seen, the distribution of the z component of total electric field is continuous across the boundary surface. The numerical data shown are obtained for fixed angles $\phi = 0$, 30, 60, 90, 120, 150, and 180, and the radial variable ρ is varied from a point on the surface of scatterer in small increments. Using the expressions (5.6.20a) and (5.6.21a), Figure 5.14 shows the distribution of the angular component of the total magnetic field, which also exhibits continuous behavior at the boundary surface. Similarly, using expressions (5.6.20b) and (5.6.21b), Figure 5.15 shows the distribution of radial component of the total magnetic field. The radial component of magnetic field is clearly discontinuous at the boundary surface. It can be easily verified by multiplying with the respective region

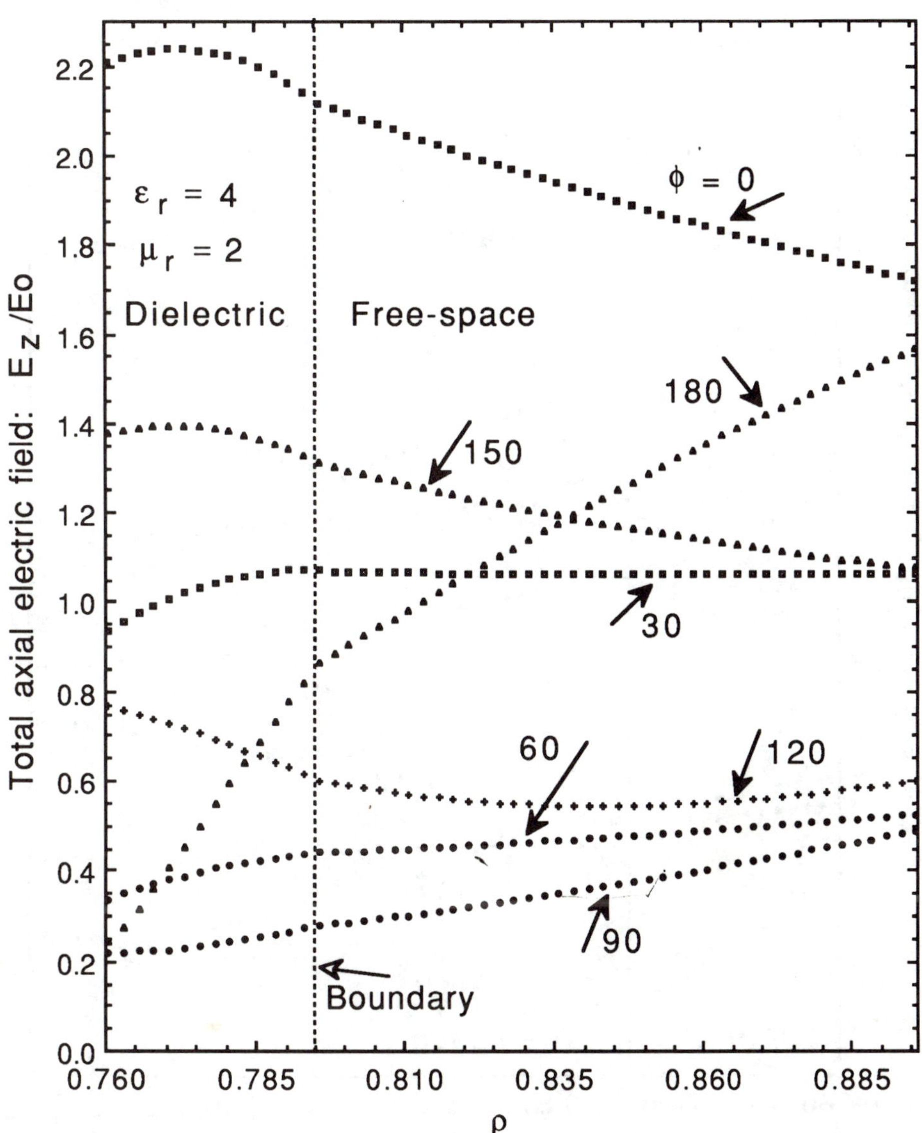

Figure 5.13 Distribution of the axial component of the total electric field near a circular dielectric scatterer – TM excitation.

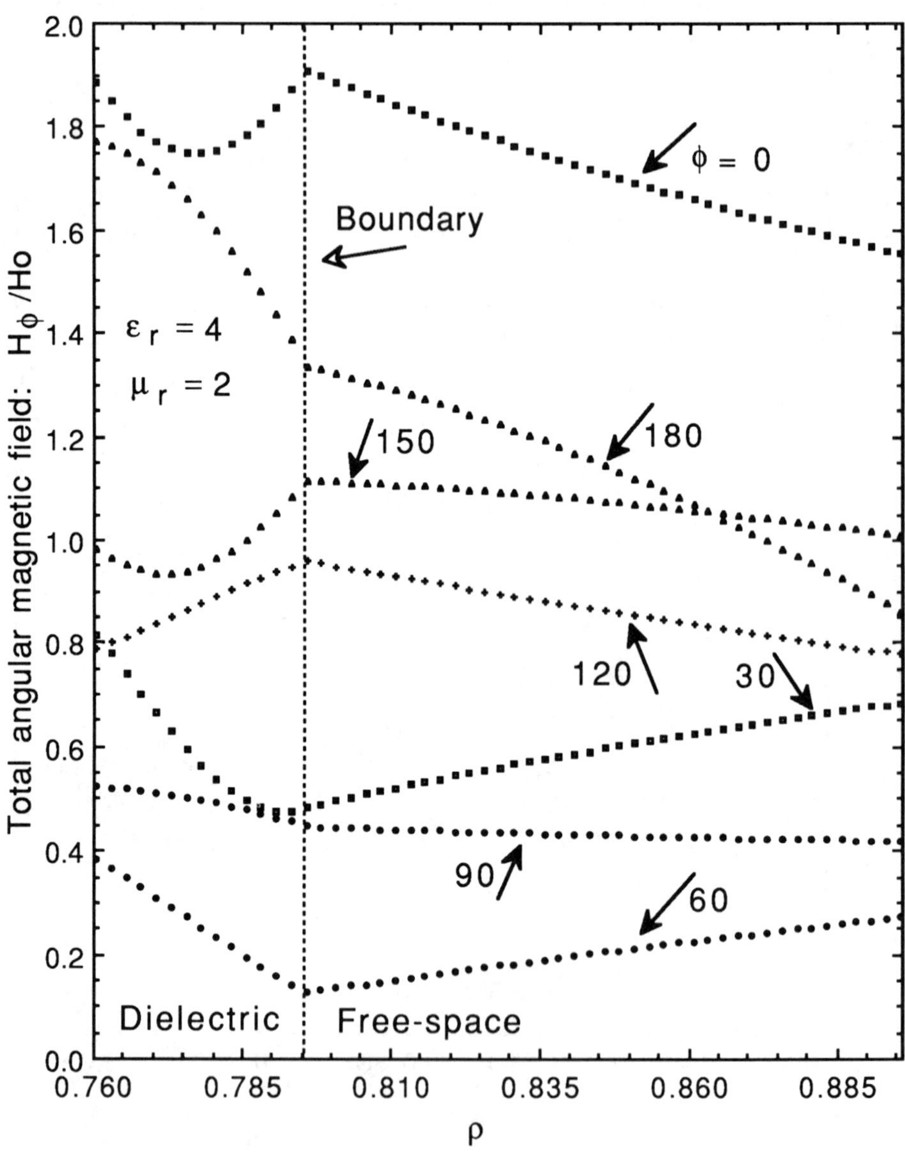

Figure 5.14 Distribution of the angular component of the total magnetic field near a circular dielectric scatterer – TM excitation.

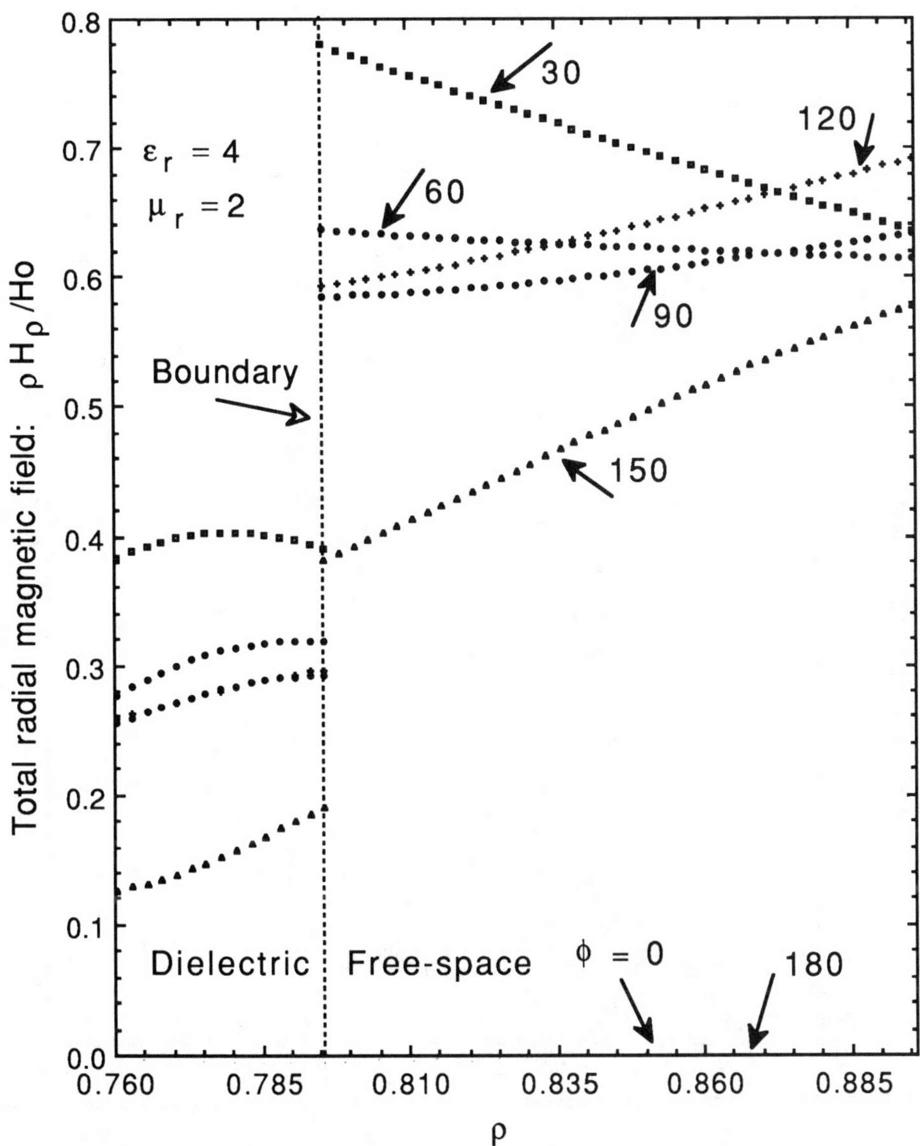

Figure 5.15 Distribution of the radial component of the total magnetic field near a circular dielectric scatterer – TM excitation.

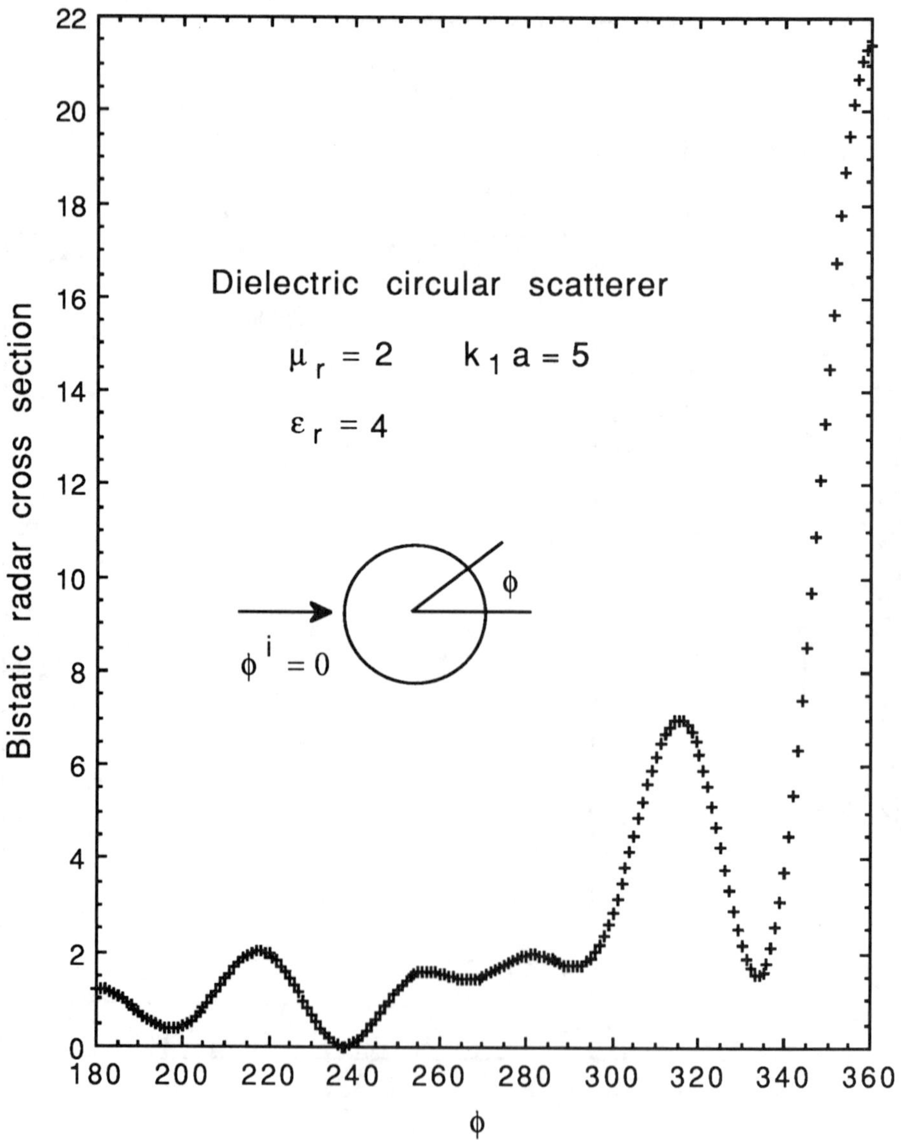

Figure 5.16 Bistatic radar cross section of the circular dielectric scatterer – TM excitation.

permeability parameter that the distribution of the radial component of the total magnetic flux density is continuous across the boundary surface.

Far Field and Radar Cross Section

The z component of the scattered field distribution in the far-field region is obtained by substituting for the argument $|k_1\rho| \to \infty$. In the large argument approximation, Appendix B, the Hankel function can be written as

$$H_m^{(2)}(k_1\rho) \sim \sqrt{\frac{2j}{\pi k_1\rho}}\, j^m\, e^{-jk_1\rho} \tag{5.6.22}$$

After substituting the large argument approximation given in expression (5.6.22) into infinite series solution (5.6.17), the z component of the scattered field in the far-field region reduces to

$$E_{1z}^s(\rho,\phi,\omega) \sim E_0 \sqrt{\frac{2j}{\pi k_1\rho}}\, e^{-jk_1\rho} \sum_{m=-\infty}^{\infty} A_m j^m e^{jm\phi} \tag{5.6.23}$$

and the magnitude of the far-field pattern is given by

$$\left| \frac{E_{1z}^s(\phi,\omega)}{E_z^i(\phi,\omega)} \right| = \sqrt{\frac{2}{\pi k_1\rho}} \left| \sum_{m=-\infty}^{\infty} A_m j^m e^{jm\phi} \right|_{k_1\rho \to \infty} \tag{5.6.24}$$

Using the expression (3.8.18), the bistatic or the monostatic radar cross section of the homogeneous dielectric scatterer can be calculated. The bistatic radar cross section for a lossless dielectric scatterer with $k_1 a = 5$ is shown in Figure 5.16. For the scatterer relative permeability is 2 and relative permittivity is 4.

5.7 ARBITRARY CROSS SECTION – CFIE – TM EXCITATION

The electromagnetic scattering, penetration, and interaction by a homogeneous, lossy dielectric circular cylinder is discussed in the previous section based on a rigorous analytical approach. But, the circular dielectric scatterer happens to be a special scattering geometry wherein the scattered and the penetrated electric field and the corresponding magnetic field distributions in the free-space and dielectric regions can be conveniently expanded in terms of two separate sets of eigenfunction series expansions. In a two-dimensional, homogeneous, lossy dielectric scatterer with an arbitrary cross section having arbitrary edges and corners, the analysis technique is quite involved. In the following, an approximate analysis procedure is discussed in detail to obtain a numerical solution for the electromagnetic scattering and penetration by solving the coupled CFIE (5.5.26a–b). This is, in fact, accomplished through the method of moments technique described in Section 3.7.

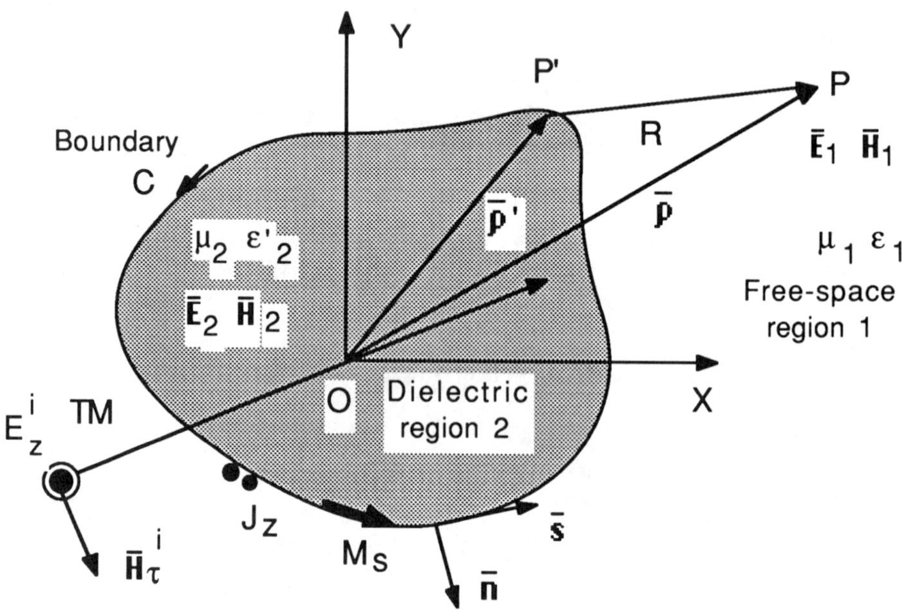

Figure 5.17 Geometry of an arbitrary shaped, homogeneous, lossy dielectric scatterer – TM excitation.

Let us consider the geometry of a two-dimensional, homogeneous, lossy dielectric scatterer having an arbitrary cross section and placed in a linear, homogeneous, and isotropic lossless medium, shown in Figure 5.17. The scatterer is uniform along its axis having the same arbitrary cross section. It is assumed that the z coordinate axis of cylindrical coordinate system (ρ, ϕ, z) or generalized cylindrical coordinate system (n, s, z) coincides with the axis of the scatterer. The external incident plane wave field is transverse magnetic polarized with respect to the z coordinate axis of the scatterer, and propagates with its direction of propagation normal to the z axis. The angle ϕ^i represents the incident angle that the direction of propagation makes with the x coordinate axis. As discussed earlier, due to the normal excitation, various field quantities are independent of the z coordinate variation.

Referring to Figure 5.17 and expression (5.6.1), the TM to z polarized incident plane wave electric field is given by

$$E_z^i(\bar{\rho},\omega) = E_0\, e^{-j\bar{k}_1 \cdot \bar{\rho}} \tag{5.7.1}$$

ϕ^i : arbitrary angle of incidence of the TM plane wave field.

Hence, for a plane wave propagating normal to the axis of the scatterer,

$$E_z^i(\rho,\phi) = E_0\, e^{-jk_1\rho\,\cos(\phi - \phi^i)} \tag{5.7.2}$$

E_0 : amplitude of the incident plane wave electric field, in volts per meter.

The corresponding transverse component of incident magnetic field in the transverse plane is given by

$$\bar{H}_\tau^i(\rho,\phi) = [\sin\phi^i\hat{x} - \cos\phi^i\hat{y}]\, H_0\, e^{-jk_1\rho\,\cos(\phi - \phi^i)} \tag{5.7.3}$$

Referring to Figure 5.7, in terms of the generalized cylindrical coordinate system *(n, s, z)*, the normal and tangential unit vectors corresponding to field points on the surface of the dielectric boundary can be written as

$$\hat{n} = \hat{x}\cos\Omega + \hat{y}\sin\Omega \tag{5.7.4a}$$

$$\hat{s} = -\hat{x}\sin\Omega + \hat{y}\cos\Omega \tag{5.7.4b}$$

Ω : angle the normal at field point makes with the x coordinate axis.

The component of incident magnetic field tangential to the dielectric boundary is obtained by

$$H_s^i(\rho,\phi) = \hat{s} \cdot \bar{H}_\tau^i(\rho,\phi)$$

$$= - H_0 \cos(\Omega - \phi^i)\, e^{-jk_1\rho \cos(\phi - \phi^i)} \tag{5.7.5a}$$

H_0 : amplitude of incident plane wave magnetic field, in amperes per meter

$$= \frac{E_0}{\eta_1} \tag{5.7.5b}$$

η_1 : intrinsic impedance of the homogeneous, isotropic free-space medium.

In Section 5.5, a complete and systematic derivation of the coupled CFIE, expressions (5.5.25a–b), for the *TM normal excitation* is presented. In these two coupled equations, the axial component of the electric current and the tangential component of the magnetic current distributions on boundary contour C are unknown and are to be solved. The two coupled CFIE integral equations are repeated the following:

$$E_z^i(\bar{\rho}) = \int_C J_z(\bar{\rho}')G_{11}(\bar{\rho}, \bar{\rho}')\, dL' + ⨍_C M_s(\bar{\rho}')G_{12}(\bar{\rho}, \bar{\rho}')\, dL' \tag{5.7.6a}$$

$$H_s^i(\bar{\rho}) = ⨍_C J_z(\bar{\rho}')G_{21}(\bar{\rho}, \bar{\rho}')\, dL' + \int_C M_s(\bar{\rho}')\, G_{22}(\bar{\rho}, \bar{\rho}')dL'$$

$$+ \frac{\partial}{\partial s} \int_C \left[\frac{\partial}{\partial s'} M_s(\bar{\rho}')\right] G'_{22}(\bar{\rho}, \bar{\rho}')\, dL' \tag{5.7.6b}$$

These integral equations are valid for the field point $\bar{\rho}$ on C. The various composite self and mutual kernel functions in the integrands of the coupled combined field integral equations are given by

$$G_{11}(\bar{\rho}, \bar{\rho}') = \frac{\omega\mu_1}{4} H_0^{(2)}(k_1R) + \frac{\omega\mu_2}{4} H_0^{(2)}(k_2R) \tag{5.7.7a}$$

$$G_{12}(\bar{\rho}, \bar{\rho}') = \frac{k_1}{4j}\left[\frac{x-x'}{R}\cos\Omega' + \frac{y-y'}{R}\sin\Omega'\right] H_0^{(2)'}(k_1R)$$

$$+ \frac{k_2}{4j}\left[\frac{x-x'}{R}\cos\Omega' + \frac{y-y'}{R}\sin\Omega'\right] H_0^{(2)'}(k_2R) \tag{5.7.7b}$$

$$G_{21}(\bar{\rho}, \bar{\rho}') = \frac{k_1}{4j}\Big[\frac{x-x'}{R}\cos\Omega + \frac{y-y'}{R}\sin\Omega\Big] H_0^{(2)'}(k_1R)$$

$$+ \frac{k_2}{4j}\Big[\frac{x-x'}{R}\cos\Omega + \frac{y-y'}{R}\sin\Omega\Big] H_0^{(2)'}(k_2R) \qquad (5.7.7c)$$

$$G_{22}(\bar{\rho}, \bar{\rho}') = \frac{\omega\varepsilon_1}{4}\cos(\Omega - \Omega')H_0^{(2)}(k_1R)$$

$$+ \frac{\omega\varepsilon'_2}{4}\cos(\Omega - \Omega')H_0^{(2)}(k_2R) \qquad (5.7.7d)$$

$$G'_{22}(\bar{\rho}, \bar{\rho}') = \frac{1}{4\omega\mu_1}H_0^{(2)}(k_1R) + \frac{1}{4\omega\mu_2}H_0^{(2)}(k_2R) \qquad (5.7.7e)$$

where

$$R = |\bar{\rho} - \bar{\rho}'|$$

$$= [(x - x')^2 + (y - y')^2]^{1/2} \qquad (5.7.7f)$$

To solve the coupled combined field integral equations (5.7.6a–b), the method of moments numerical technique, as described in Section 3.7, is applied. This is accomplished by a suitable selection of a set of independent, piecewise current expansion functions and a corresponding set of independent, piecewise weighting functions.

A number of modeling considerations should be understood before a suitable choice can be made to represent approximately the unknown equivalent electric current and the unknown magnetic current distributions in terms of a set of current expansion functions, and similarly, on the suitable choice of a set of weighting functions to enforce equality on both sides of the CFIE.

5.7.1 EXPANSION FUNCTIONS

On contour C, two types of unknown current distributions are to be determined. If one has knowledge of the types of distribution of the electric and the magnetic currents, then it is easier to make a choice on the approximate representation by suitable sampling of the distribution functions. Depending on the electrical size of the dielectric scatterer and its arbitrary cross section, having arbitrary edges and corners, the electric and the magnetic currents may have a singularity distribution or a well-behaved continuous distribution with nulls or even rapid phase changes. Similar to the modeling of conducting scatterers excited by either TM or TE polarization, a number of considerations should be given as to the choice and arrangement of the expansion functions. The modeling of the axial electric current distribution is similar to the modeling of TM currents, and the modeling of the tangential magnetic current distribution is similar to the modeling of TE currents. In fact, the distribution of the axial electric current is discontinuous near sharp corners. But, the distribution of the tangential magnetic current is always continuous along the closed contour, even for the scatterer with arbitrary edges and sharp corners.

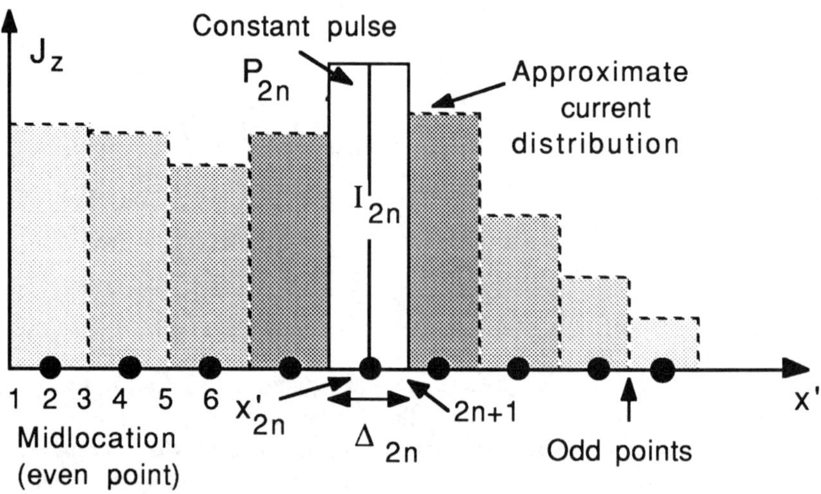

Figure 5.18 Piecewise pulse expansion functions for an axial current.

Referring to Section 3.9 concerning modeling of TM-type current distribution, contour C is first sampled at a number of spatial locations, consisting of odd and even points as shown in Figure 5.18. The various pulse functions are defined between two consecutive odd points. The piecewise pulse expansion function P_{2n} is located such that the even point x'_{2n} represents its midlocation and is defined between the two consecutive odd points given by

$$x'_{2n+1} = x'_{2n} + \frac{\Delta_{2n}}{2} \tag{5.7.8a}$$

$$x'_{2n-1} = x'_{2n} - \frac{\Delta_{2n}}{2} \tag{5.7.8b}$$

where the length of the pulse segment is given by

$$\Delta_{2n} = x'_{2n+1} - x'_{2n-1} \tag{5.7.8c}$$

It is assumed that the distribution of the electric current is constant within a pulse expansion function and corresponds to its distribution at the mid (even) point. Referring to Figure 5.18, the unknown complex current coefficients I_{2n} is constant within the piecewise pulse P_{2n} and represents approximately the magnitude as well as the phase of the unknown current distribution within that specific pulse region. Then, the unknown electric current distribution can be approximated in a piecewise sense and written in terms of the piecewise pulse or constant expansion terms as

$$J_z(x') \approx \sum_{n=1}^{N} I_{2n} P_{2n}(x') \tag{5.7.9a}$$

and the piecewise pulse function

$$P_{2n}(x') = 1 \qquad \text{for } x'_{2n-1} \leq x' \leq x'_{2n+1}$$

$$= 0 \qquad \text{otherwise} \tag{5.7.9b}$$

Similarly, referring to Figure 5.19, the distribution of the unknown magnetic current can be sampled at discrete spatial points and represented in terms of a set of independent pulse expansion functions. The various pulse functions are defined between two consecutive even points. The piecewise pulse function P_{2n-1} is located at the odd point x'_{2n-1} and is defined between the two consecutive even points given by

$$x'_{2n} = x'_{2n-1} + \frac{\Delta_{2n}}{2} \tag{5.7.10a}$$

$$x'_{2n-2} = x'_{2n-1} - \frac{\Delta_{2n-2}}{2} \tag{5.7.10b}$$

For the dielectric contour, the first point is same as the last sampling point. It is should be noted that a pulse current expansion term spans two halves of adjacent straightline segments. Suppose contour C of the scatterer is not smooth, but has sharp corners, then the unknown current expansion pulse is arranged such that half the pulse spans along one side of corner and the remaining half spans the other side of corner. In Figure 5.19, the unknown complex current coefficients I_{2n-1} is constant within the piecewise pulse P_{2n-1} and represents approximately the magnitude as well as the phase of unknown current distribution within the specific pulse region. Thus

$$M_s(x') \approx \sum_{n=1}^{N} I_{2n-1} P_{2n-1}(x') \tag{5.7.11a}$$

and the piecewise pulse function

$$P_{2n-1}(x') = 1 \qquad \text{for } x'_{2n-2} \leq x' \leq x'_{2n}$$

$$= 0 \qquad \text{otherwise} \tag{5.7.11b}$$

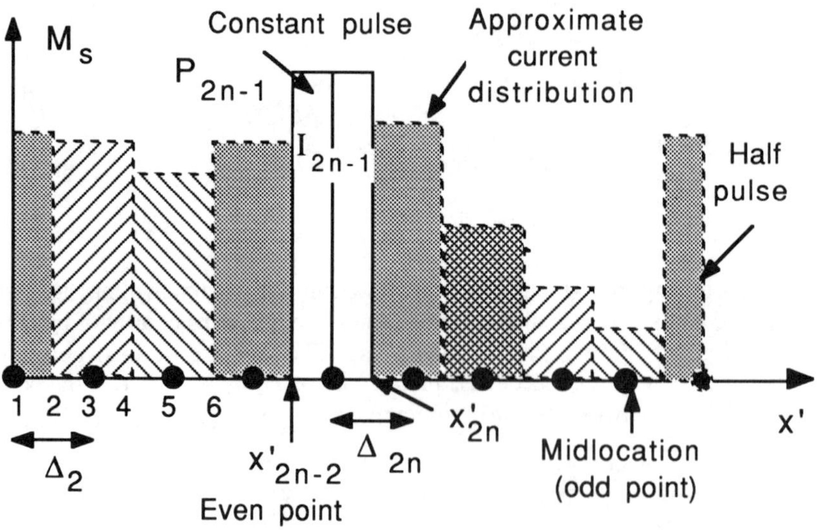

Figure 5.19 Piecewise pulse expansion functions for a tangential current.

Thus, the two expansion functions are *staggered by half a pulse* by locating the unknown electric currents at the even points and the unknown magnetic currents at the odd points. This arrangement of the current expansion pulses is quite useful in treating numerically the continuity as well as the derivative of the tangential magnetic current distribution. Further, smooth and sharp corners can be modeled in a straightforward way.

As discussed earlier in the method of moments numerical procedure, Section 3.7, equality of the coupled integral equations on both sides should be enforced at every point in the domain of validity of the integral equations. This is accomplished by testing through the scalar product in the domain of validity, as defined in expression (3.7.1), on both sides of the integral equations with respect to a convenient weighting function. In the numerical analysis presented, the piecewise pulse expansion terms discussed earlier are directly adapted as weighting functions. In fact, the weighting functions with the corresponding variable m are arranged to coincide with same locations as the current expansion terms.

5.8 REDUCTION TO A PARTITIONED MATRIX EQUATION

The coupled integral equations (5.7.6a–b) are reduced to their equivalent functional form by first testing on both sides using the pulse type of weighting functions. If contour C of the dielectric scatterer is completely smooth with no geometrical discontinuities, then it is possible to pick arbitrary sampling regions along contour C, and the pulse expansion functions developed in Figures 5.18 and 5.19 can be adapted directly. On the other hand, if contour C is not smooth, then the weighting functions are to be properly set up on contour C with respect to the various piecewise sampling regions shown in Figures 5.20a and 5.20b.

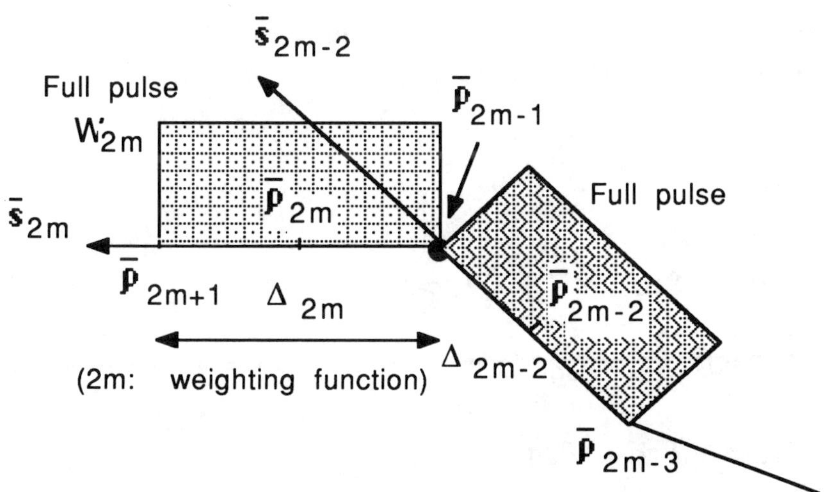

Figure 5.20a Positioning of weighting function W_{2m} near discontinuities.

In Figure 5.20a, the weighting functions W_{2m}, which are located at the even points $\bar{\rho}_{2m}$, $m = 1, 2, 3, \ldots, M$ are defined as follows within the limits:

$$\bar{\rho}_{2m+1} = \bar{\rho}_{2m} + \frac{\Delta_{2m}}{2} \hat{s}_{2m} \tag{5.8.1a}$$

$$\bar{\rho}_{2m-1} = \bar{\rho}_{2m} - \frac{\Delta_{2m}}{2} \hat{s}_{2m} \tag{5.8.1b}$$

$$\Delta_{2m} = |\bar{\rho}_{2m+1} - \bar{\rho}_{2m-1}| \tag{5.8.1c}$$

where

$\hat{s}_{2m}$: tangential unit vector along the length of pulse segment Δ_{2m}.

The weighting function in terms of piecewise pulse terms is given by

$$W_{2m}(\bar{\rho}) = 1 \qquad \text{for } \bar{\rho}_{2m-1} \le \bar{\rho} \le \bar{\rho}_{2m+1}$$

$$= 0 \qquad \text{otherwise} \tag{5.8.2}$$

$$m = 1, 2, 3, \ldots, M$$

Similarly, in Figure 5.20b, the weighting functions W_{2q-1} located at the odd points $\bar{\rho}_{2q-1}$, $q = 1, 2, 3, \ldots, Q$ are defined as follows within the limits:

$$\bar{\rho}_{2q} = \bar{\rho}_{2q-1} + \frac{\Delta_{2q}}{2} \hat{s}_{2q} \tag{5.8.3a}$$

$$\bar{\rho}_{2q-2} = \bar{\rho}_{2q-1} - \frac{\Delta_{2q-2}}{2} \hat{s}_{2q-2} \tag{5.8.3b}$$

$$\Delta_{2q} = |\bar{\rho}_{2q+1} - \bar{\rho}_{2q-1}| \tag{5.8.3c}$$

$$\Delta_{2q-2} = |\bar{\rho}_{2q-1} - \bar{\rho}_{2q-3}| \tag{5.8.3d}$$

where

$\hat{s}_{2q-2}$: tangential unit vector along the length of pulse segment Δ_{2q-2};
$\hat{s}_{2q}$: tangential unit vector along the length of pulse segment Δ_{2q}.

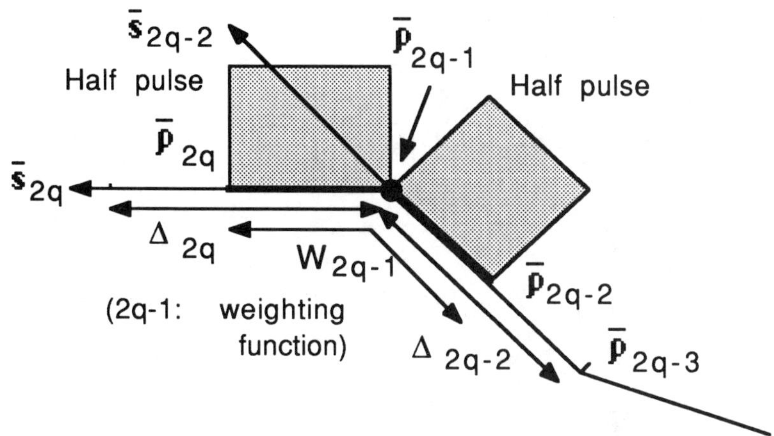

Figure 5.20b Positioning of weighting function W_{2q-1} near discontinuities.

Correspondingly, the weighting function in terms of piecewise pulse terms is given by

$$W_{2q-1}(\bar{\rho}) = 1 \qquad \text{for } \bar{\rho}_{2q-2} \le \bar{\rho} \le \bar{\rho}_{2q}$$

$$= 0 \qquad \text{otherwise} \tag{5.8.4}$$

$$q = 1, 2, 3, \ldots, Q$$

It should be noted at this stage, contour C of scatterer is not smooth, but has sharp corners. Thus, care has been taken in properly positioning the location of the weighting functions and the expansion functions with respect to the geometrical discontinuity. From a numerical modeling point of view, the specific weighting function close to the wedge-type corner is always properly arranged on both sides of the corner so that it encloses the geometrical discontinuity point. This, in fact, is completely similar to the numerical treatment for the case of transverse electric current modeling discussed in Section 4.8. Referring to Figures 5.20a and 5.20b, all even sampling points are arranged half a cell away from the corner point of the geometry, and all odd sampling points are arranged to coincide with the corner point of the geometry and spanned by the weighting function on either side by half a cell.

The pulse weighting functions, expressions (5.8.2) and (5.8.4), are utilized in the following to test the coupled combined field integral equations.

Pulse Testing of CFIE CFIE integral expression (5.7.6a) is now multiplied on both sides by weighting function (5.8.2) and integrated along scatterer contour C to yield the following functional relationship:

$$\int_{\bar{\rho}_{2m-1}}^{\bar{\rho}_{2m+1}} W_{2m}(\bar{\rho}) E_z^i(\bar{\rho})\, dL(\bar{\rho})$$

$$= \int_{\bar{\rho}_{2m-1}}^{\bar{\rho}_{2m+1}} W_{2m}(\bar{\rho}) \left[\int_C J_z(\bar{\rho}') G_{11}(\bar{\rho}, \bar{\rho}') dL' \right] dL(\bar{\rho})$$

$$+ \int_{\bar{\rho}_{2m-1}}^{\bar{\rho}_{2m+1}} W_{2m}(\bar{\rho}) \left[\fint_C M_s(\bar{\rho}') G_{12}(\bar{\rho}, \bar{\rho}') dL' \right] dL(\bar{\rho})$$

$$\bar{\rho} \text{ on } C,\ m = 1, 2, 3, \ldots, M \qquad (5.8.5)$$

Numerical integration can be implemented to calculate these weighted integrals. If the integrands are smooth varying functions, it can be assumed that the integrands on the lefthand and righthand sides of expression (5.8.5) do not vary over the limits of integration interval defined by the piecewise weighting pulse function. Thus, suitable approximations can be introduced by calculating the integrand at the point $\bar{\rho}_{2m}$ and multiplying it by the corresponding segment length of the weighting function:

$$\Delta_{2m} E_z^i(\bar{\rho}) = \Delta_{2m} \left[\int_C J_z(\bar{\rho}') G_{11}(\bar{\rho}_{2m}, \bar{\rho}') dL' \right]$$

$$+ \Delta_{2m} \left[\fint_C M_s(\bar{\rho}') G_{12}(\bar{\rho}_{2m}, \bar{\rho}') dL' \right]$$

$$m = 1, 2, 3, \ldots, M \qquad (5.8.6a)$$

After substituting the various kernel functions defined in expressions (5.7.7a) and (5.7.7b), the weighted equation simplifies to the following form:

$$\Delta_{2m} E_0\, e^{-jk_1 \rho_{2m} \cos(\phi_{2m} - \phi^i)}$$

$$= \Delta_{2m} \int_C J_z(\bar{\rho}') \left[\frac{\omega\mu_1}{4} H_0^{(2)}(k_1|\bar{\rho}_{2m} - \bar{\rho}'|) + \frac{\omega\mu_2}{4} H_0^{(2)}(k_2|\bar{\rho}_{2m} - \bar{\rho}'|) \right] dL'$$

$$+ \Delta_{2m} \fint_C M_s(\bar{\rho}') \left[\frac{x_{2m} - x'}{|\bar{\rho}_{2m} - \bar{\rho}'|} \cos\Omega' + \frac{y_{2m} - y'}{|\bar{\rho}_{2m} - \bar{\rho}'|} \sin\Omega' \right]$$

$$\left[\frac{k_1}{4j} H_0^{(2)'}(k_1|\bar{\rho}_{2m} - \bar{\rho}'|) + \frac{k_2}{4j} H_0^{(2)'}(k_2|\bar{\rho}_{2m} - \bar{\rho}'|) \right] dL'$$

$$m = 1, 2, 3, \ldots, M \qquad (5.8.6b)$$

Further, CFIE integral expression (5.7.6b) is now multiplied on both sides by weighting function (5.8.4) and integrated along scatterer contour C to yield the following functional relationship:

$$\int_{\bar{\rho}_{2q-2}}^{\bar{\rho}_{2q}} W_{2q-1}(\bar{\rho}) H_s^i(\bar{\rho})\, dL(\bar{\rho})$$

$$= \int_{\bar{\rho}_{2q-2}}^{\bar{\rho}_{2q}} W_{2q-1}(\bar{\rho}) \left[\fint_C J_z(\bar{\rho}') G_{21}(\bar{\rho}, \bar{\rho}') dL' \right] dL(\bar{\rho})$$

$$+ \int_{\bar{\rho}_{2q-2}}^{\bar{\rho}_{2q}} W_{2q-1}(\bar{\rho}) \left[\int_C M_s(\bar{\rho}') G_{22}(\bar{\rho}, \bar{\rho}') dL' \right] dL(\bar{\rho})$$

$$+ \int_{\bar{\rho}_{2q-2}}^{\bar{\rho}_{2q}} W_{2q-1}(\bar{\rho}) \left\{ \frac{\partial}{\partial s} \int_C \left[\frac{\partial}{\partial s'} M_s(\bar{\rho}') \right] G'_{22}(\bar{\rho}, \bar{\rho}') dL' \right\} dL(\bar{\rho})$$

$$\bar{\rho} \text{ on } C,\ q = 1, 2, 3, \ldots, Q \qquad (5.8.7a)$$

Further, these integrals can be split into two integrals, corresponding to the two half straightline segments of the weighting pulse term, and expression (5.8.7a) takes the form

$$\left[\int_{\bar{\rho}_{2q-2}}^{\bar{\rho}_{2q-1}} + \int_{\bar{\rho}_{2q-1}}^{\bar{\rho}_{2q}}\right] H_s^i(\bar{\rho})\, dL(\bar{\rho})$$

$$= \left[\int_{\bar{\rho}_{2q-2}}^{\bar{\rho}_{2q-1}} + \int_{\bar{\rho}_{2q-1}}^{\bar{\rho}_{2q}}\right]\left[⨍_C J_z(\bar{\rho}')G_{21}(\bar{\rho}, \bar{\rho}')dL'\right] dL(\bar{\rho})$$
$$+ \left[\int_{\bar{\rho}_{2q-2}}^{\bar{\rho}_{2q-1}} + \int_{\bar{\rho}_{2q-1}}^{\bar{\rho}_{2q}}\right]\left[\int_C M_s(\bar{\rho}')G_{22}(\bar{\rho}, \bar{\rho}')dL'\right] dL(\bar{\rho})$$
$$+ \left[\int_{\bar{\rho}_{2q-2}}^{\bar{\rho}_{2q-1}} + \int_{\bar{\rho}_{2q-1}}^{\bar{\rho}_{2q}}\right]\left\{\frac{\partial}{\partial s}\int_C \left[\frac{\partial}{\partial s'} M_s(\bar{\rho}')\right]G'_{22}(\bar{\rho}, \bar{\rho}')dL'\right\} dL(\bar{\rho})$$

$$\bar{\rho} \text{ on } C,\; q = 1, 2, 3, \ldots, Q \qquad (5.8.7b)$$

Hence, after substituting the kernel terms (5.7.7c), (5.7.7d), and (5.7.7e) into the preceding tested equation

$$- \frac{\Delta_{2q-2}}{2} H_0 \cos(\Omega_{2q-2} - \phi^i)\, e^{-jk_1\rho_{2q-1}\cos(\phi_{2q-1} - \phi^i)}$$
$$- \frac{\Delta_{2q}}{2} H_0 \cos(\Omega_{2q} - \phi^i)\, e^{-jk_1\rho_{2q-1}\cos(\phi_{2q-1} - \phi^i)}$$

$$= \frac{\Delta_{2q-2}}{2} ⨍_C J_z(\bar{\rho}')\left[\frac{x_{2q-1}-x'}{|\bar{\rho}_{2q-1}-\bar{\rho}'|}\cos\Omega_{2q-2} + \frac{y_{2q-1}-y'}{|\bar{\rho}_{2q-1}-\bar{\rho}'|}\cos\Omega_{2q-2}\right]$$
$$\left[\frac{k_1}{4j}H_0^{(2)'}(k_1|\bar{\rho}_{2q-1}-\bar{\rho}'|) + \frac{k_2}{4j}H_0^{(2)'}(k_2|\bar{\rho}_{2q-1}-\bar{\rho}'|)\right] dL'$$

$$+\frac{\Delta_{2q}}{2}\int_C J_z(\bar{\rho}')\left[\frac{x_{2q-1}-x'}{|\bar{\rho}_{2q-1}-\bar{\rho}'|}\cos\Omega_{2q}+\frac{y_{2q-1}-y'}{|\bar{\rho}_{2q-1}-\bar{\rho}'|}\sin\Omega_{2q}\right]$$

$$\left[\frac{k_1}{4j}H_0^{(2)'}(k_1|\bar{\rho}_{2q-1}-\bar{\rho}'|)+\frac{k_2}{4j}H_0^{(2)'}(k_2|\bar{\rho}_{2q-1}-\bar{\rho}'|)\right]dL'$$

$$+\frac{\Delta_{2q-2}}{2}\int_C M_s(\bar{\rho}')\cos(\Omega_{2q-2}-\Omega')$$

$$\left[\frac{\omega\varepsilon_1}{4}H_0^{(2)}(k_1|\bar{\rho}_{2q-1}-\bar{\rho}'|)+\frac{\omega\varepsilon'_2}{4}H_0^{(2)}(k_2|\bar{\rho}_{2q-1}-\bar{\rho}'|)\right]dL'$$

$$+\frac{\Delta_{2q}}{2}\int_C M_s(\bar{\rho}')\cos(\Omega_{2q}-\Omega')$$

$$\left[\frac{\omega\varepsilon_1}{4}H_0^{(2)}(k_1|\bar{\rho}_{2q-1}-\bar{\rho}'|)+\frac{\omega\varepsilon'_2}{4}H_0^{(2)}(k_2|\bar{\rho}_{2q-1}-\bar{\rho}'|)\right]dL'$$

$$-\int_C \frac{\partial}{\partial s'}M_s(\bar{\rho}')\left[\frac{1}{4\omega\mu_1}H_0^{(2)}(k_1|\bar{\rho}_{2q-2}-\bar{\rho}'|)\right.$$

$$\left.+\frac{1}{4\omega\mu_2}H_0^{(2)}(k_2|\bar{\rho}_{2q-2}-\bar{\rho}'|)\right]dL'$$

$$+\int_C \frac{\partial}{\partial s'}M_s(\bar{\rho}')\left[\frac{1}{4\omega\mu_1}H_0^{(2)}(k_1|\bar{\rho}_{2q}-\bar{\rho}'|)\right.$$

$$\left.+\frac{1}{4\omega\mu_2}H_0^{(2)}(k_2|\bar{\rho}_{2q}-\bar{\rho}'|)\right]dL'$$

$$q = 1, 2, 3, \ldots, Q \qquad (5.8.7c)$$

In expression (5.8.7b), numerical integration can be implemented to calculate the weighted integrals. Instead, if the integrands are smooth varying functions, it can be assumed that the integrands on the lefthand and righthand sides of expression (5.8.7b) do not vary over the limits of integration interval defined by the small range of piecewise weighting pulse function. Thus, in expression (5.8.7c), a suitable approximation has been implemented by calculating the integrand at the point $\bar{\rho}_{2q-1}$ and multiplying it by the corresponding half-segment length of the weighting function.

Further, following the procedure discussed in Section 4.8, specifically expressions (4.8.3) and (4.8.4) for evaluating the gradient of the scalar potential term, the last integral term on the righthand side of expression (5.8.7b) can be recognized to have a similar form. Hence, it has been simplified by replacing the spatial derivative term by its finite difference approximation to yield the numerically approximate weighted equation, expression (5.8.7c). Thus, the functional form of expressions (5.8.6b) and (5.8.7c) form a set of coupled weighted equations.

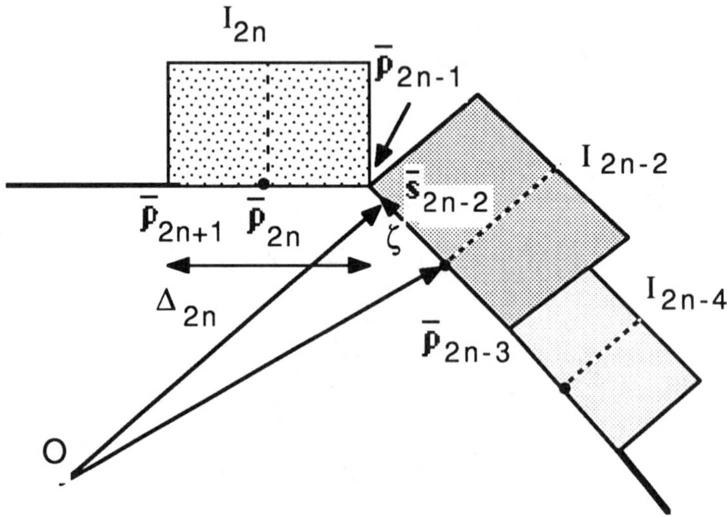

Figure 5.21a Positioning of electric current expansion functions near discontinuities.

Electric and Magnetic Current Expansions Referring to Figures 5.21a and 5.21b, the unknown electric current $J_z(\bar{\rho}')$ and the unknown magnetic current $M_S(\bar{\rho}')$ distributions are now expanded in terms of two sets of piecewise linearly independent pulse expansion functions. Similar to the discussion on the weighting functions, if contour C of the scatterer is completely smooth with no geometrical discontinuities, then it is possible to pick arbitrary sampling regions for the unknown currents along contour C. On the other hand, if contour C of the scatterer is not smooth, but has sharp corners, as indicated in Figures 5.21a and 5.21b, care should to be taken in properly positioning the unknown current expansion functions with respect to the geometrical discontinuity. From the numerical modeling point of view, the axial electric current expansion functions are always arranged to be half a cell away from the geometrical corner so that it is not enclosed. Similarly, the tangential magnetic current expansion functions close to the wedge-type corner are always arranged on both sides of the corner so that they enclose

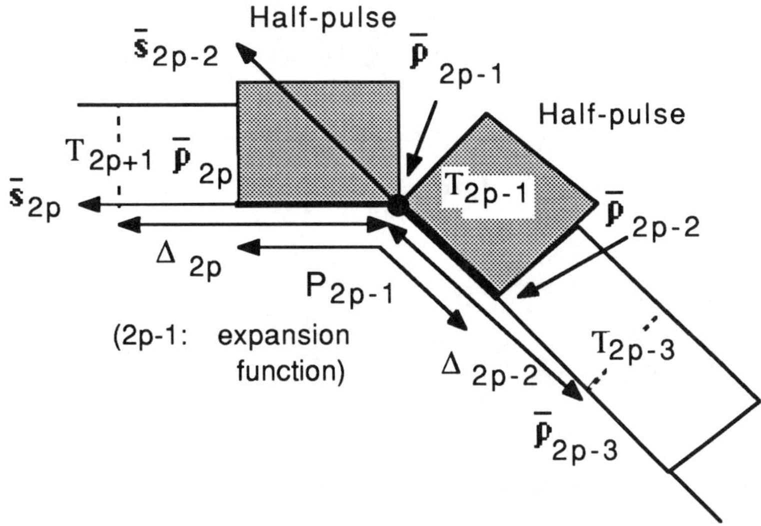

Figure 5.21b Positioning of magnetic current expansion functions near discontinuities.

completely the geometrical discontinuity point. Such an arrangement also provides continuity of the tangential magnetic current at the corner. Thus, the corner of the geometry is spanned by the magnetic current expansion function on either side by half a cell.

Referring to Figure 5.21a, the unknown electric current distribution is represented in terms of the piecewise expansion pulse P_{2n}, which is located such that it spans the even points $\bar{\rho}_{2n}$, n = 1, 2, 3, . . . , N and defined between two consecutive odd points. Hence, the electric current is expanded in terms of a set of piecewise linearly independent pulse expansion functions, given by

$$J_z(\bar{\rho}') \approx \sum_{n=1}^{N} I_{2n} P_{2n}(\bar{\rho}') \tag{5.8.8a}$$

$$P_{2n}(\bar{\rho}') = 1 \qquad \text{for } \bar{\rho}_{2n-1} \leq \bar{\rho}' \leq \bar{\rho}_{2n+1}$$

$$= 0 \qquad \text{otherwise} \tag{5.8.8b}$$

$$n = 1, 2, 3, \quad . \quad . \quad . \quad , N$$

I_{2n} : unknown electric current coefficients.

With respect to the *primed* coordinates, the electric current pulse expansions are located between the odd points:

$$\bar{\rho}_{2n+1} = \bar{\rho}_{2n} + \frac{\Delta_{2n}}{2}\hat{s}_{2n} \tag{5.8.8c}$$

$$\bar{\rho}_{2n-1} = \bar{\rho}_{2n} - \frac{\Delta_{2n}}{2}\hat{s}_{2n} \tag{5.8.8d}$$

$$\Delta_{2n} = |\bar{\rho}_{2n+1} - \bar{\rho}_{2n-1}| \tag{5.8.8e}$$

where

$\hat{s}_{2n}$: tangential unit vector along the length of pulse segment Δ_{2n}.

Similarly, the unknown magnetic current distribution is expanded in terms of piecewise expansion pulse P_{2p-1} which is located such that it spans odd points $\bar{\rho}_{2p-1}$, where $p = 1, 2, 3, \ . \ . \ . \ , P$ and is defined between two consecutive even points:

$$\bar{\rho}_{2p} = \bar{\rho}_{2p-1} + \frac{\Delta_{2p}}{2}\hat{s}_{2p} \tag{5.8.9a}$$

$$\bar{\rho}_{2p-2} = \bar{\rho}_{2p-1} - \frac{\Delta_{2p-2}}{2}\hat{s}_{2p-2} \tag{5.8.9b}$$

$$\Delta_{2p} = |\bar{\rho}_{2p+1} - \bar{\rho}_{2p-1}| \tag{5.8.9c}$$

$$\Delta_{2p-2} = |\bar{\rho}_{2p-1} - \bar{\rho}_{2p-3}| \tag{5.8.9d}$$

where

$\hat{s}_{2p-2}$: tangential unit vector along the length of pulse segment Δ_{2p-2};

$\hat{s}_{2p}$: tangential unit vector along the length of pulse segment Δ_{2p}.

Then, the tangential-directed magnetic current distribution in terms of the piecewise pulse expansion terms is given by

$$M_S(\bar{\rho}') \approx \sum_{p=1}^{P} T_{2p-1} P_{2p-1}(\bar{\rho}') \qquad (5.8.10a)$$

$$P_{2p-1}(\bar{\rho}') = 1 \qquad \text{for } \bar{\rho}_{2p-2} \le \bar{\rho}' \le \bar{\rho}_{2p}$$

$$= 0 \qquad \text{otherwise} \qquad (5.8.10b)$$

$$p = 1, 2, 3, \ . \ . \ . \ , P$$

T_{2p-1} : unknown magnetic current coefficients.

The two current expansion functions (5.8.8a) and (5.8.10a) are now substituted into weighted equations (5.8.6b) and (5.8.7c) to obtain the following partitioned matrix equations. Hence

$$\Delta_{2m} E_0 \, e^{-jk_1 \rho_{2m} \cos(\phi_{2m} - \phi^i)}$$

$$= \Delta_{2m} \int_C \sum_{n=1}^{N} I_{2n} P_{2n}(\bar{\rho}') \Big[\frac{\omega\mu_1}{4} H_0^{(2)}(k_1 |\bar{\rho}_{2m} - \bar{\rho}'|)$$

$$+ \frac{\omega\mu_2}{4} H_0^{(2)}(k_2 |\bar{\rho}_{2m} - \bar{\rho}'|) \Big] dL'$$

$$+ \Delta_{2m} \oint_C \sum_{p=1}^{P} T_{2p-1} P_{2p-1}(\bar{\rho}') \Big[\frac{x_{2m} - x'}{|\bar{\rho}_{2m} - \bar{\rho}'|} \cos\Omega' + \frac{y_{2m} - y'}{|\bar{\rho}_{2m} - \bar{\rho}'|} \sin\Omega' \Big]$$

$$\Big[\frac{k_1}{4j} H_0^{(2)'}(k_1 |\bar{\rho}_{2m} - \bar{\rho}'|) + \frac{k_2}{4j} H_0^{(2)'}(k_2 |\bar{\rho}_{2m} - \bar{\rho}'|) \Big] dL'$$

$$m = 1, 2, 3, \ . \ . \ . \ , M \qquad (5.8.11a)$$

The preceding partial matrix equation can be rewritten as

$$\Delta_{2m} E_0 e^{-jk_1 \rho_{2m} \cos(\phi_{2m} - \phi^i)}$$

$$= \Delta_{2m} \sum_{n=1}^{N} I_{2n} \int_{\bar{\rho}_{2n-1}}^{\bar{\rho}_{2n+1}} \Big[\frac{\omega\mu_1}{4} H_0^{(2)}(k_1|\bar{\rho}_{2m} - \bar{\rho}'|)$$

$$+ \frac{\omega\mu_2}{4} H_0^{(2)}(k_2|\bar{\rho}_{2m} - \bar{\rho}'|) \Big] dL'$$

$$+ \Delta_{2m} \sum_{p=1}^{P} T_{2p-1} \Big[\int_{\bar{\rho}_{2p-2}}^{\bar{\rho}_{2p-1}} + \int_{\bar{\rho}_{2p-1}}^{\bar{\rho}_{2p}} \Big] \Big[\frac{x_{2m} - x'}{|\bar{\rho}_{2m} - \bar{\rho}'|} \cos\Omega' + \frac{y_{2m} - y'}{|\bar{\rho}_{2m} - \bar{\rho}'|} \sin\Omega' \Big]$$

$$\Big[\frac{k_1}{4j} H_0^{(2)'}(k_1|\bar{\rho}_{2m} - \bar{\rho}'|) + \frac{k_2}{4j} H_0^{(2)'}(k_2|\bar{\rho}_{2m} - \bar{\rho}'|) \Big] dL'$$

$$m = 1, 2, 3, \quad . \ . \ . \ . \ , M \qquad (5.8.11b)$$

In functional equation (5.8.7c), the electric current, magnetic current and its derivative are the unknown distributions. Further, the derivative of the magnetic current distribution, which represents the magnetic charge distribution, can also be expanded in terms of a set of charge expansion functions by suitably introducing a finite difference approximation based on the magnetic current expansion functions. The numerical charge expansion functions are shown in Figure 5.22.

Hence, referring to the discussion in Section 4.8, the derivative of magnetic current can be expanded as follows:

$$\frac{\partial}{\partial s'} M_s(\bar{\rho}') \approx \sum_{p=1}^{P} \frac{T_{2p+1} - T_{2p-1}}{\Delta_{2p}} P'_{2p}(\bar{\rho}') \qquad (5.8.12a)$$

$$P'_{2p}(\bar{\rho}') = 1 \qquad \text{for } \bar{\rho}_{2p+1} \le \bar{\rho}' \le \bar{\rho}_{2p-1}$$

$$= 0 \qquad \text{otherwise} \qquad (5.8.12b)$$

$$p = 1, 2, 3, \quad . \ . \ . \ . \ , P$$

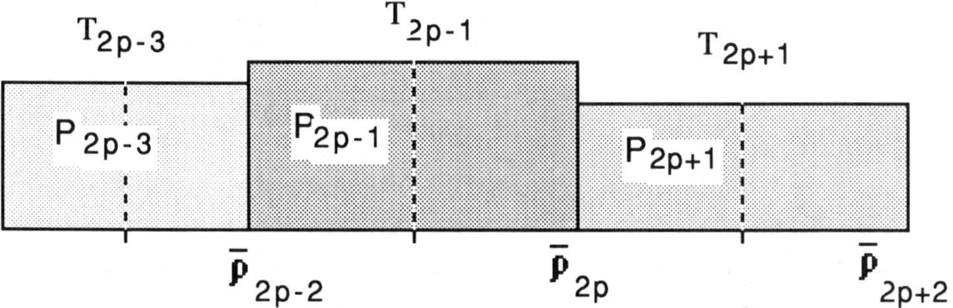

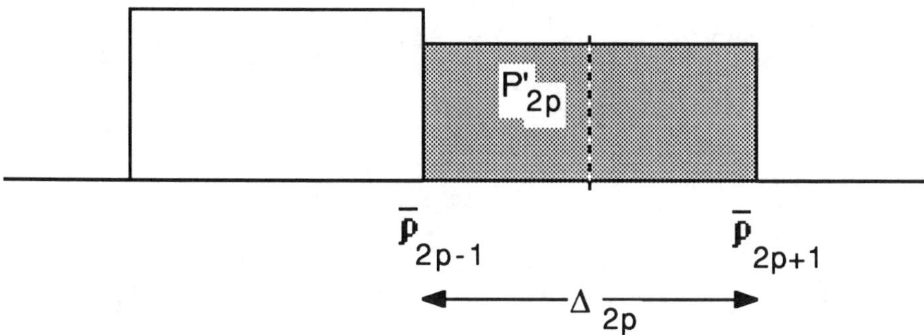

Figure 5.22 Arrangement of magnetic current and magnetic charge expansion pulses.

Hence, after substituting the electric current, magnetic current and magnetic charge expansion functions into functional expression (5.8.7c), the following partitioned matrix equation is obtained:

$$- \frac{\Delta_{2q-2}}{2} H_0 \cos(\Omega_{2q-2} - \phi^i)\, e^{-jk_1\rho_{2q-1}\cos(\phi_{2q-1} - \phi^i)}$$

$$- \frac{\Delta_{2q}}{2} H_0 \cos(\Omega_{2q} - \phi^i)\, e^{-jk_1\rho_{2q-1}\cos(\phi_{2q-1} - \phi^i)}$$

$$= \frac{\Delta_{2q-2}}{2} \int_C \sum_{n=1}^{N} I_{2n} P_{2n}(\bar{\rho}') \left[\frac{x_{2q-1} - x'}{|\bar{\rho}_{2q-1} - \bar{\rho}'|} \cos\Omega_{2q-2} + \frac{y_{2q-1} - y'}{|\bar{\rho}_{2q-1} - \bar{\rho}'|} \sin\Omega_{2q-2} \right]$$

$$\left[\frac{k_1}{4j} H_0^{(2)'}(k_1|\bar{\rho}_{2q-1} - \bar{\rho}'|) + \frac{k_2}{4j} H_0^{(2)'}(k_2|\bar{\rho}_{2q-1} - \bar{\rho}'|) \right] dL'$$

$$+ \frac{\Delta_{2q}}{2} \int_C \sum_{n=1}^{N} I_{2n} P_{2n}(\bar{\rho}') \left[\frac{x_{2q-1} - x'}{|\bar{\rho}_{2q-1} - \bar{\rho}'|} \cos\Omega_{2q} + \frac{y_{2q-1} - y'}{|\bar{\rho}_{2q-1} - \bar{\rho}'|} \sin\Omega_{2q} \right]$$

$$\left[\frac{k_1}{4j} H_0^{(2)'}(k_1|\bar{\rho}_{2q-1} - \bar{\rho}'|) + \frac{k_2}{4j} H_0^{(2)'}(k_2|\bar{\rho}_{2q-1} - \bar{\rho}'|) \right] dL'$$

$$+ \frac{\Delta_{2q-2}}{2} \int_C \sum_{p=1}^{P} T_{2p-1} P_{2p-1}(\bar{\rho}') \cos(\Omega_{2q-2} - \Omega')$$

$$\left[\frac{\omega\varepsilon_1}{4} H_0^{(2)}(k_1|\bar{\rho}_{2q-1} - \bar{\rho}'|) + \frac{\omega\varepsilon'_2}{4} H_0^{(2)}(k_2|\bar{\rho}_{2q-1} - \bar{\rho}'|) \right] dL'$$

$$+ \frac{\Delta_{2q}}{2} \int_C \sum_{p=1}^{P} T_{2p-1} P_{2p-1}(\bar{\rho}') \cos(\Omega_{2q} - \Omega')$$

$$\left[\frac{\omega\varepsilon_1}{4} H_0^{(2)}(k_1|\bar{\rho}_{2q-1} - \bar{\rho}'|) + \frac{\omega\varepsilon'_2}{4} H_0^{(2)}(k_2|\bar{\rho}_{2q-1} - \bar{\rho}'|) \right] dL'$$

$$- \int_C \sum_{p=1}^{P} \frac{T_{2p+1} - T_{2p-1}}{\Delta_{2p}} P'_{2p}(\bar{\rho}')$$

$$\left[\frac{1}{4\omega\mu_1} H_0^{(2)}(k_1|\bar{\rho}_{2q-2} - \bar{\rho}'|) + \frac{1}{4\omega\mu_2} H_0^{(2)}(k_2|\bar{\rho}_{2q-2} - \bar{\rho}'|) \right] dL'$$

$$+ \int_C \sum_{p=1}^{P} \frac{T_{2p+1} - T_{2p-1}}{\Delta_{2p}} P'_{2p}(\bar{\rho}')$$

$$\left[\frac{1}{4\omega\mu_1} H_0^{(2)}(k_1|\bar{\rho}_{2q} - \bar{\rho}'|) + \frac{1}{4\omega\mu_2} H_0^{(2)}(k_2|\bar{\rho}_{2q} - \bar{\rho}'|)\right] dL'$$

$$q = 1, 2, 3, \ldots, Q \qquad (5.8.13a)$$

Hence, partial matrix equation (5.8.13a) can be rewritten as

$$- \frac{\Delta_{2q-2}}{2} H_0 \cos(\Omega_{2q-2} - \phi^i) e^{-jk_1\rho_{2q-1}\cos(\phi_{2q-1} - \phi^i)}$$

$$- \frac{\Delta_{2q}}{2} H_0 \cos(\Omega_{2q} - \phi^i) e^{-jk_1\rho_{2q-1}\cos(\phi_{2q-1} - \phi^i)}$$

$$= \frac{\Delta_{2q-2}}{2} \sum_{n=1}^{N} I_{2n} \int_{\bar{\rho}_{2n-1}}^{\bar{\rho}_{2n+1}} \left[\frac{x_{2q-1} - x'}{|\bar{\rho}_{2q-1} - \bar{\rho}'|} \cos\Omega_{2q-2} + \frac{y_{2q-1} - y'}{|\bar{\rho}_{2q-1} - \bar{\rho}'|} \sin\Omega_{2q-2}\right]$$

$$\left[\frac{k_1}{4j} H_0^{(2)'}(k_1|\bar{\rho}_{2q-1} - \bar{\rho}'|) + \frac{k_2}{4j} H_0^{(2)'}(k_2|\bar{\rho}_{2q-1} - \bar{\rho}'|)\right] dL'$$

$$+ \frac{\Delta_{2q}}{2} \sum_{n=1}^{N} I_{2n} \int_{\bar{\rho}_{2n-1}}^{\bar{\rho}_{2n+1}} \left[\frac{x_{2q-1} - x'}{|\bar{\rho}_{2q-1} - \bar{\rho}'|} \cos\Omega_{2q} + \frac{y_{2q-1} - y'}{|\bar{\rho}_{2q-1} - \bar{\rho}'|} \sin\Omega_{2q}\right]$$

$$\left[\frac{k_1}{4j} H_0^{(2)'}(k_1|\bar{\rho}_{2q-1} - \bar{\rho}'|) + \frac{k_2}{4j} H_0^{(2)'}(k_2|\bar{\rho}_{2q-1} - \bar{\rho}'|)\right] dL'$$

$$+ \frac{\Delta_{2q-2}}{2} \sum_{p=1}^{P} T_{2p-1} \left[\int_{\bar{\rho}_{2p-2}}^{\bar{\rho}_{2p-1}} + \int_{\bar{\rho}_{2p-1}}^{\bar{\rho}_{2p}} \right] \cos(\Omega_{2q-2} - \Omega')$$

$$\left[\frac{\omega\varepsilon_1}{4} H_0^{(2)}(k_1|\bar{\rho}_{2q-1} - \bar{\rho}'|) + \frac{\omega\varepsilon'_2}{4} H_0^{(2)}(k_2|\bar{\rho}_{2q-1} - \bar{\rho}'|) \right] dL'$$

$$+ \frac{\Delta_{2q}}{2} \sum_{p=1}^{P} T_{2p-1} \left[\int_{\bar{\rho}_{2p-2}}^{\bar{\rho}_{2p-1}} + \int_{\bar{\rho}_{2p-1}}^{\bar{\rho}_{2p}} \right] \cos(\Omega_{2q} - \Omega')$$

$$\left[\frac{\omega\varepsilon_1}{4} H_0^{(2)}(k_1|\bar{\rho}_{2q-1} - \bar{\rho}'|) + \frac{\omega\varepsilon'_2}{4} H_0^{(2)}(k_2|\bar{\rho}_{2q-1} - \bar{\rho}'|) \right] dL'$$

$$- \sum_{p=1}^{P} \frac{T_{2p-1}}{\Delta_{2p-2}} \int_{\bar{\rho}_{2p-3}}^{\bar{\rho}_{2p-1}} \left[\frac{1}{4\omega\mu_1} H_0^{(2)}(k_1|\bar{\rho}_{2q-2} - \bar{\rho}'|) \right.$$

$$\left. + \frac{1}{4\omega\mu_2} H_0^{(2)}(k_2|\bar{\rho}_{2q-2} - \bar{\rho}'|) \right] dL'$$

$$+ \sum_{p=1}^{P} \frac{T_{2p-1}}{\Delta_{2p}} \int_{\bar{\rho}_{2p-1}}^{\bar{\rho}_{2p+1}} \left[\frac{1}{4\omega\mu_1} H_0^{(2)}(k_1|\bar{\rho}_{2q-2} - \bar{\rho}'|) \right.$$

$$\left. + \frac{1}{4\omega\mu_2} H_0^{(2)}(k_2|\bar{\rho}_{2q-2} - \bar{\rho}'|) \right] dL'$$

$$+ \sum_{p=1}^{P} \frac{T_{2p-1}}{\Delta_{2p-2}} \int_{\bar{\rho}_{2p-3}}^{\bar{\rho}_{2p-1}} \left[\frac{1}{4\omega\mu_1} H_0^{(2)}(k_1|\bar{\rho}_{2q} - \bar{\rho}'|) \right.$$

$$\left. + \frac{1}{4\omega\mu_2} H_0^{(2)}(k_2|\bar{\rho}_{2q} - \bar{\rho}'|) \right] dL'$$

$$- \sum_{p=1}^{P} \frac{T_{2p-1}}{\Delta_{2p}} \int_{\bar{\rho}_{2p-1}}^{\bar{\rho}_{2p+1}} \Big[\frac{1}{4\omega\mu_1} H_0^{(2)}(k_1|\bar{\rho}_{2q} - \bar{\rho}'|)$$

$$+ \frac{1}{4\omega\mu_2} H_0^{(2)}(k_2|\bar{\rho}_{2q} - \bar{\rho}'|) \Big] \, dL'$$

$$q = 1, 2, 3, \; . \; . \; . \; , Q \qquad (5.8.13b)$$

where in the above partial matrix equation

Ω_{2q-2} : angle the normal to the straightline field segment Δ_{2q-2} makes with the x coordinate axis;

Ω_{2q} : angle the normal to the straight line field segment Δ_{2q} makes with the x coordinate axis.

The two expressions (5.8.11b) and (5.8.13b) form a set of $M + Q$ linear simultaneous algebraic equations for the $N + P$ unknown electric and magnetic current coefficients. A direct numerical solution can be obtained by choosing the number of equations equal to the number of unknowns. Expressions (5.8.11b) and (5.8.13b) can also be rewritten in terms of a compact generalized partitioned matrix equation:

$$\begin{bmatrix} Z_{mn}^{JJ} & Z_{mp}^{JM} \\ \\ Y_{qn}^{MJ} & Y_{qp}^{MM} \end{bmatrix} \begin{bmatrix} I_n \\ \\ T_p \end{bmatrix} = \begin{bmatrix} E_m \\ \\ H_q \end{bmatrix} \qquad (5.8.14)$$

where in partial matrix equation (5.8.11b) and (5.8.13b),

Z_{mn}^{JJ} : generalized self partitioned impedance matrix;

Z_{mp}^{JM}: generalized mutual partitioned impedance matrix;

Y_{qn}^{MJ}: generalized mutual partitioned admittance matrix;

Y_{qp}^{MM}: generalized self partitioned admittance matrix.

The elements of the generalized matrix are list in the following:

Z_{mn}^{JJ}: generalized self partitioned impedance matrix

$$= \Delta_{2m} \frac{\omega\mu_1}{4} \int_{-\Delta_{2n}/2}^{\Delta_{2n}/2} H_0^{(2)}(k_1 R_1)\, d\zeta$$

$$+ \Delta_{2m} \frac{\omega\mu_2}{4} \int_{-\Delta_{2n}/2}^{\Delta_{2n}/2} H_0^{(2)}(k_2 R_1)\, d\zeta \tag{5.8.15a}$$

$$R_1 = \left|\bar{\rho}_{2m} - (\bar{\rho}_{2n} + \hat{s}_{2n}\zeta)\right| \tag{5.8.15b}$$

$$= \left\{[(x_{2m} - x_{2n}) - \cos\alpha_{2n}\zeta]^2 + [(y_{2m} - y_{2n}) - \cos\beta_{2n}\zeta]^2\right\}^{1/2} \tag{5.8.15c}$$

$$\hat{s}_{2n} = \hat{x}\cos\alpha_{2n} + \hat{y}\cos\beta_{2n} \tag{5.8.15d}$$

$m = 1, 2, 3, \ldots, M$

$n = 1, 2, 3, \ldots, N$

Z_{mp}^{JM}: generalized mutual partitioned impedance matrix

$$= \Delta_{2m} \int_{-\Delta_{2p-2}/2}^{0} \Big\{ [(x_{2m} - x_{2p-1}) - \cos\alpha_{2p-2}\zeta] \frac{\cos\Omega_{2p-2}}{R_2}$$

$$+ [(y_{2m} - y_{2p-1}) - \cos\beta_{2p-2}\zeta] \frac{\sin\Omega_{2p-2}}{R_2} \Big\}$$

$$\Big[\frac{k_1}{4j} H_0^{(2)'}(k_1 R_2) + \frac{k_2}{4j} H_0^{(2)'}(k_2 R_2) \Big] d\zeta$$

$$+ \Delta_{2m} \int_{0}^{\Delta_{2p}/2} \Big\{ [(x_{2m} - x_{2p-1}) - \cos\alpha_{2p}\zeta] \frac{\cos\Omega_{2p}}{R_3}$$

$$+ [(y_{2m} - y_{2p-1}) - \cos\beta_{2p}\zeta] \frac{\sin\Omega_{2p}}{R_3} \Big\}$$

$$\Big[\frac{k_1}{4j} H_0^{(2)'}(k_1 R_3) + \frac{k_2}{4j} H_0^{(2)'}(k_2 R_3) \Big] d\zeta \quad (5.8.16a)$$

$$R_2 = \left| \bar{\boldsymbol{\rho}}_{2m} - (\bar{\boldsymbol{\rho}}_{2p-1} + \hat{\boldsymbol{s}}_{2p-2}\zeta) \right| \quad (5.8.16b)$$

$$= \Big\{ [(x_{2m} - x_{2p-1}) - \cos\alpha_{2p-2}\zeta]^2$$

$$+ [(y_{2m} - y_{2p-1}) - \cos\beta_{2p-2}\zeta]^2 \Big\}^{1/2} \quad (5.8.16c)$$

$$\hat{\boldsymbol{s}}_{2p-2} = \hat{\boldsymbol{x}} \cos\alpha_{2p-2} + \hat{\boldsymbol{y}} \cos\beta_{2p-2} \quad (5.8.16d)$$

$$R_3 = \left| \bar{\boldsymbol{\rho}}_{2m} - (\bar{\boldsymbol{\rho}}_{2p-1} + \hat{\boldsymbol{s}}_{2p}\zeta) \right| \quad (5.8.16e)$$

$$= \Big\{ [(x_{2m} - x_{2p-1}) - \cos\alpha_{2p}\zeta]^2$$

$$+ [(y_{2m} - y_{2p-1}) - \cos\beta_{2p}\zeta]^2 \Big\}^{1/2} \quad (5.8.16f)$$

$$\hat{\boldsymbol{s}}_{2p} = \hat{\boldsymbol{x}} \cos\alpha_{2p} + \hat{\boldsymbol{y}} \cos\beta_{2p} \quad (5.8.16g)$$

$m = 1, 2, 3, \ldots, M$

$p = 1, 2, 3, \ldots, P$

Referring to partial matrix equation (5.8.13b), the various admittance matrix elements are given by

Y_{qn}^{MJ} : generalized mutual partitioned admittance matrix

$$= \frac{\Delta_{2q-2}}{2} \int_{-\Delta_{2n}/2}^{\Delta_{2n}/2} \Big\{ [(x_{2q-1} - x_{2n}) - \cos\alpha_{2n}\zeta] \frac{\cos\Omega_{2q-2}}{R_4} + [(y_{2q-1} - y_{2n}) - \cos\beta_{2n}\zeta] \frac{\sin\Omega_{2q-2}}{R_4} \Big\} \Big[\frac{k_1}{4j} H_0^{(2)'}(k_1 R_4) + \frac{k_2}{4j} H_0^{(2)'}(k_2 R_4) \Big] d\zeta$$

$$+ \frac{\Delta_{2q}}{2} \int_{-\Delta_{2n}/2}^{\Delta_{2n}/2} \Big\{ [(x_{2q-1} - x_{2n}) - \cos\alpha_{2n}\zeta] \frac{\cos\Omega_{2q}}{R_4} + [(y_{2q-1} - y_{2n}) - \cos\beta_{2n}\zeta] \frac{\sin\Omega_{2q}}{R_4} \Big\} \Big[\frac{k_1}{4j} H_0^{(2)'}(k_1 R_4) + \frac{k_2}{4j} H_0^{(2)'}(k_2 R_4) \Big] d\zeta \tag{5.8.17a}$$

$$R_4 = \left| \bar{\rho}_{2q-1} - (\bar{\rho}_{2n} + \hat{s}_{2n}\zeta) \right| \tag{5.8.17b}$$

$$= \Big\{ [(x_{2q-1} - x_{2n}) - \cos\alpha_{2n}\zeta]^2 + [(y_{2q-1} - y_{2n}) - \cos\beta_{2n}\zeta]^2 \Big\}^{1/2} \tag{5.8.17c}$$

$$\hat{s}_{2n} = \hat{x} \cos\alpha_{2n} + \hat{y} \cos\beta_{2n} \tag{5.8.17d}$$

$q = 1, 2, 3, \; . \; . \; . \; , Q$

$n = 1, 2, 3, \; . \; . \; . \; , N$

and

Y_{qp}^{MM} : generalized self partitioned admittance matrix

$$+\frac{\Delta_{2q-2}}{2}\int_{-\Delta_{2p-2}/2}^{0}\cos(\Omega_{2q-2}-\Omega'_{2p-2})\left[\frac{\omega\varepsilon_1}{4}H_0^{(2)}(k_1R_5)+\frac{\omega\varepsilon'_2}{4}H_0^{(2)}(k_2R_5)\right]d\zeta$$

$$+\frac{\Delta_{2q-2}}{2}\int_{0}^{\Delta_{2p}/2}\cos(\Omega_{2q-2}-\Omega'_{2p})\left[\frac{\omega\varepsilon_1}{4}H_0^{(2)}(k_1R_6)+\frac{\omega\varepsilon'_2}{4}H_0^{(2)}(k_2R_6)\right]d\zeta$$

$$+\frac{\Delta_{2q}}{2}\int_{-\Delta_{2p-2}/2}^{0}\cos(\Omega_{2q}-\Omega'_{2p-2})\left[\frac{\omega\varepsilon_1}{4}H_0^{(2)}(k_1R_5)+\frac{\omega\varepsilon'_2}{4}H_0^{(2)}(k_2R_5)\right]d\zeta$$

$$+\frac{\Delta_{2q}}{2}\int_{0}^{\Delta_{2p}/2}\cos(\Omega_{2q}-\Omega'_{2p})\left[\frac{\omega\varepsilon_1}{4}H_0^{(2)}(k_1R_6)+\frac{\omega\varepsilon'_2}{4}H_0^{(2)}(k_2R_6)\right]d\zeta$$

$$-\frac{1}{\Delta_{2p-2}}\int_{-\Delta_{2p-2}/2}^{\Delta_{2p-2}/2}\left[\frac{1}{4\omega\mu_1}H_0^{(2)}(k_1R_7)+\frac{1}{4\omega\mu_2}H_0^{(2)}(k_2R_7)\right]d\zeta$$

$$+\frac{1}{\Delta_{2p}}\int_{-\Delta_{2p}/2}^{\Delta_{2p}/2}\left[\frac{1}{4\omega\mu_1}H_0^{(2)}(k_1R_8)+\frac{1}{4\omega\mu_2}H_0^{(2)}(k_2R_8)\right]d\zeta$$

$$+ \frac{1}{\Delta_{2p-2}} \int_{-\Delta_{2p-2}/2}^{\Delta_{2p-2}/2} \left[\frac{1}{4\omega\mu_1} H_0^{(2)}(k_1 R_9) + \frac{1}{4\omega\mu_2} H_0^{(2)}(k_2 R_9) \right] d\zeta$$

$$- \frac{1}{\Delta_{2p}} \int_{-\Delta_{2p}/2}^{\Delta_{2p}/2} \left[\frac{1}{4\omega\mu_1} H_0^{(2)}(k_1 R_{10}) + \frac{1}{4\omega\mu_2} H_0^{(2)}(k_2 R_{10}) \right] d\zeta \quad (5.8.18a)$$

$$R_5 = \left| \bar{\rho}_{2q-1} - (\bar{\rho}_{2p-1} + \hat{s}_{2p-2}\zeta) \right| \quad (5.8.18b)$$

$$= \left\{ [(x_{2q-1} - x_{2p-1}) - \cos\alpha_{2p-2}\zeta]^2 + [(y_{2q-1} - y_{2p-1}) - \cos\beta_{2p-2}\zeta]^2 \right\}^{1/2} \quad (5.8.18c)$$

$$R_6 = \left| \bar{\rho}_{2q-1} - (\bar{\rho}_{2p-1} + \hat{s}_{2p}\zeta) \right| \quad (5.8.18d)$$

$$= \left\{ [(x_{2q-1} - x_{2p-1}) - \cos\alpha_{2p}\zeta]^2 + [(y_{2q-1} - y_{2p-1}) - \cos\beta_{2p}\zeta]^2 \right\}^{1/2} \quad (5.8.18e)$$

and

$$R_7 = \left| \bar{\rho}_{2q-2} - (\bar{\rho}_{2p-2} + \hat{s}_{2p-2}\zeta) \right| \quad (5.8.18f)$$

$$= \left\{ [(x_{2q-2} - x_{2p-2}) - \cos\alpha_{2p-2}\zeta]^2 + [(y_{2q-2} - y_{2p-2}) - \cos\beta_{2p-2}\zeta]^2 \right\}^{1/2} \quad (5.8.18g)$$

$$R_8 = \left| \bar{\rho}_{2q-2} - (\bar{\rho}_{2p} + \hat{s}_{2p}\zeta) \right| \quad (5.8.18h)$$

$$= \left\{ [(x_{2q-2} - x_{2p}) - \cos\alpha_{2p}\zeta]^2 + [(y_{2q-2} - y_{2p}) - \cos\beta_{2p}\zeta]^2 \right\}^{1/2} \quad (5.8.18i)$$

$$R_9 = \left| \bar{\rho}_{2q} - (\bar{\rho}_{2p-2} + \hat{s}_{2p-2}\zeta) \right| \quad (5.8.18j)$$

$$= \{[(x_{2q} - x_{2p-2}) - \cos\alpha_{2p-2}\zeta]^2 + [(y_{2q} - y_{2p-2}) - \cos\beta_{2p-2}\zeta]^2\}^{1/2} \tag{5.8.18k}$$

$$R_{10} = |\bar{\rho}_{2q} - (\bar{\rho}_{2p} + \hat{s}_{2p}\zeta)| \tag{5.8.18ℓ}$$

$$= \{[(x_{2q} - x_{2p}) - \cos\alpha_{2p}\zeta]^2 + [(y_{2q} - y_{2p}) - \cos\beta_{2p}\zeta]^2\}^{1/2} \tag{5.8.18m}$$

$$q = 1, 2, 3, \ldots, Q$$

$$p = 1, 2, 3, \ldots, P$$

Further, referring to expressions (5.8.11b) and (5.8.13b), partitioned matrix equation (5.8.14) has two excitation terms for the TM-polarized plane wave propagating with an angle of incidence ϕ^i and they are given by

$$E_m = \Delta_{2m} E_0 \, e^{-jk_1\rho_{2m}\cos(\phi_{2m} - \phi^i)} \tag{5.8.19a}$$

and

$$H_q = -\frac{\Delta_{2q-2}}{2} H_0 \cos(\Omega_{2q-2} - \phi^i)\, e^{-jk_1\rho_{2q-1}\cos(\phi_{2q-1} - \phi^i)} - \frac{\Delta_{2q}}{2} H_0 \cos(\Omega_{2q} - \phi^i)\, e^{-jk_1\rho_{2q-1}\cos(\phi_{2q-1} - \phi^i)} \tag{5.8.19b}$$

The unknown electric and magnetic current distributions are obtained by inverting generalized partitioned matrix equation (5.8.14).

5.9 NUMERICAL RESULTS – TM CASE

Based on the method of moments numerical solution procedure for the CFIE discussed in the previous section, a number of canonical two-dimensional, homogeneous, dielectric scattering geometries are analyzed to determine their electromagnetic scattering and penetration properties. A computer algorithm can now be developed to solve the coupled partitioned matrix equation obtained in expression (5.8.14). To validate the numerical algorithm, the electromagnetic scattering by a two-dimensional, lossless dielectric circular cylinder located in a linear, homogeneous, and isotropic lossless medium is considered. The frequency of excitation is selected as 300 MHz, and the corresponding free-space wavelength is 1 meter and the free-space propagation constant $k_1 = 2\pi$. The radius of the circular cylinder is chosen to be $a = 0.1$ meter and its permittivity $\varepsilon_2 = 2\varepsilon_0$. It is excited by an external monochromatic plane wave propagating at an angle of incidence $\phi^i = 0$. It should be noted for the TM polarization, the incident electric field is polarized parallel to z axis and the corresponding incident magnetic field is polarized parallel to the $-y$ axis.

Figure 5.23a shows magnitude distribution of the equivalent electric current and the equivalent magnetic current along the surface of dielectric circular cylinder. The electric current is normalized with respect to the incident magnetic field, and the magnetic current is normalized with respect to the incident electric field. Similarly, Figure 5.23b shows phase distribution of the equivalent electric and the magnetic currents along the surface of dielectric circular cylinder. The surface currents are plotted for the angular range of $\phi = 180^o$ to 360° and the distributions are symmetrical in the remaining angular range with respect to the direction of incidence. Along the surface of the circular cylinder, the electric and the magnetic currents are sampled uniformly and located at staggered points as discussed in the numerical solution. The numerical data shown in Figures 5.23a and 5.23b are obtained with pulse expansion functions $N = 40$ for the electric current and $P = 40$ for the magnetic current, so that the total size of the partitioned matrix equation is $N + P = 80$. The numerical data is also rigorously validated based on the eigenfunction series solution discussed in Section 5.6. The series solution is truncated by using number of modes equal to 15 in the numerical calculation. Excellent validation is depicted in the Figures 5.23a and 5.23b.

To further validate the computer algorithm, the electromagnetic scattering by an air dielectric square cylinder is considered, Figures 5.24a–c. This case study directly provides an assessment on the numerical simulation of the incident electric and magnetic plane wave fields in terms of the unknown electric and magnetic current distributions. Each sidelength of the air dielectric square cylinder is selected to be $L = 0.25$ meter and has a permittivity $\varepsilon_2 = \varepsilon_0$. Further, each sidelength of the air dielectric square cylinder is divided into twenty four equal segments so that the total number of unknown electric current pulses is selected as $N = 96$. Similarly, the total number of unknown magnetic current pulses is selected as $P = 96$. As discussed earlier, the various pulse expansion terms are staggered and arranged along the square cylinder with the even points representing the midlocation of the electric current pulses and the corresponding odd points representing the midlocation of the magnetic current pulses, as shown in the Figure 5.24c. The frequency of excitation is selected as 300 MHz, and the corresponding free-space wavelength is $\lambda = 1$ meter and the free-space propagation constant $k_1 = 2\pi$. The air dielectric scatterer is excited at an angle of incidence $\phi^i = 0^o$.

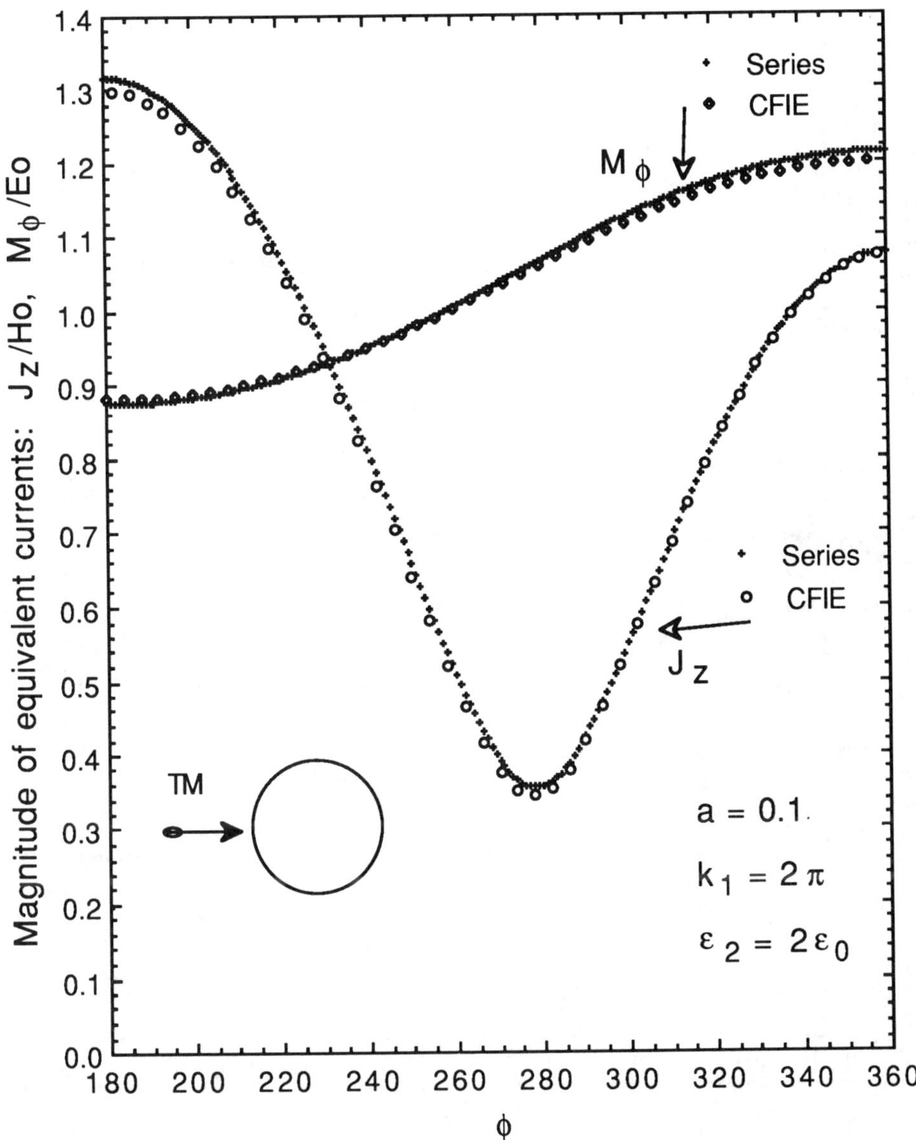

Figure 5.23a Magnitude distribution of the equivalent electric and magnetic currents on a circular dielectric scatterer – TM excitation.

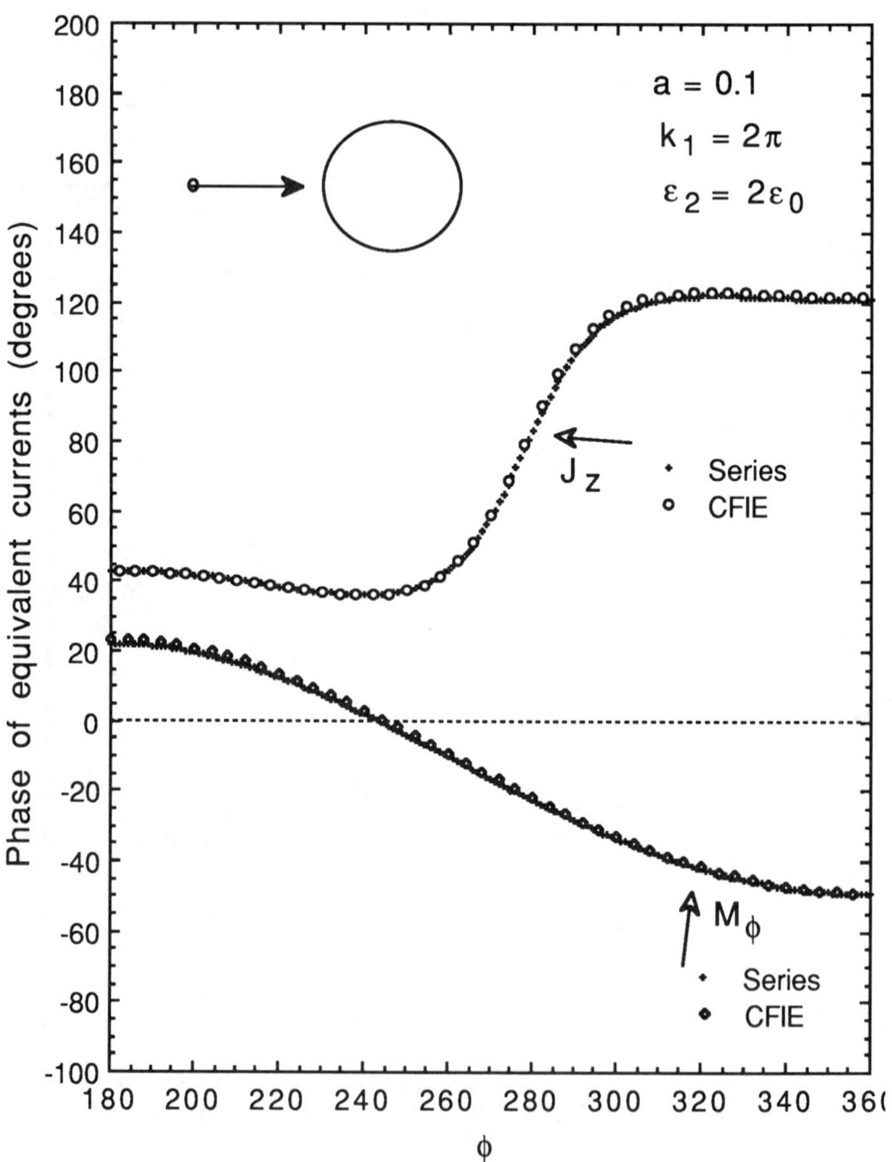

Figure 5.23b Phase distribution of the equivalent electric and magnetic currents on a circular dielectric scatterer – TM excitation.

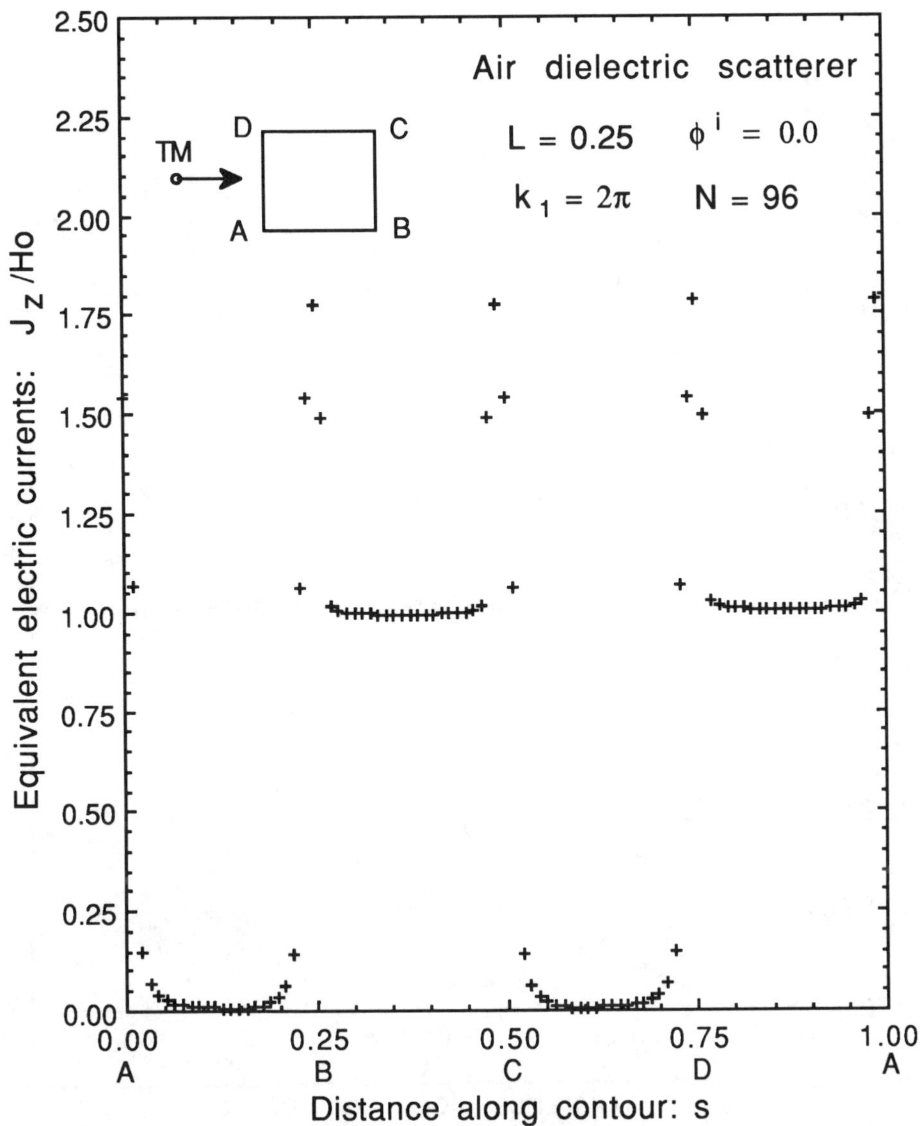

Figure 5.24a Magnitude distribution of the equivalent electric current on a square air dielectric scatterer – TM excitation.

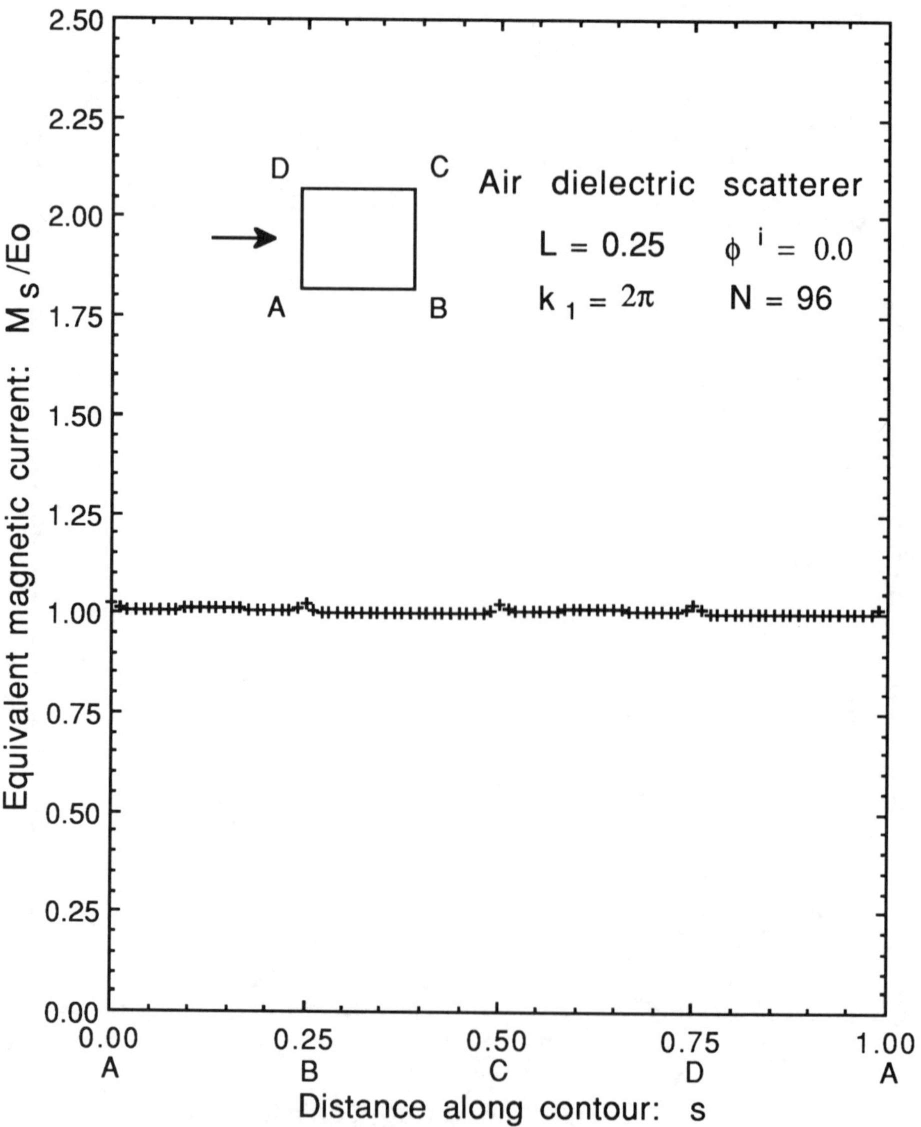

Figure 5.24b Magnitude distribution of the equivalent magnetic current on a square air dielectric scatterer – TM excitation.

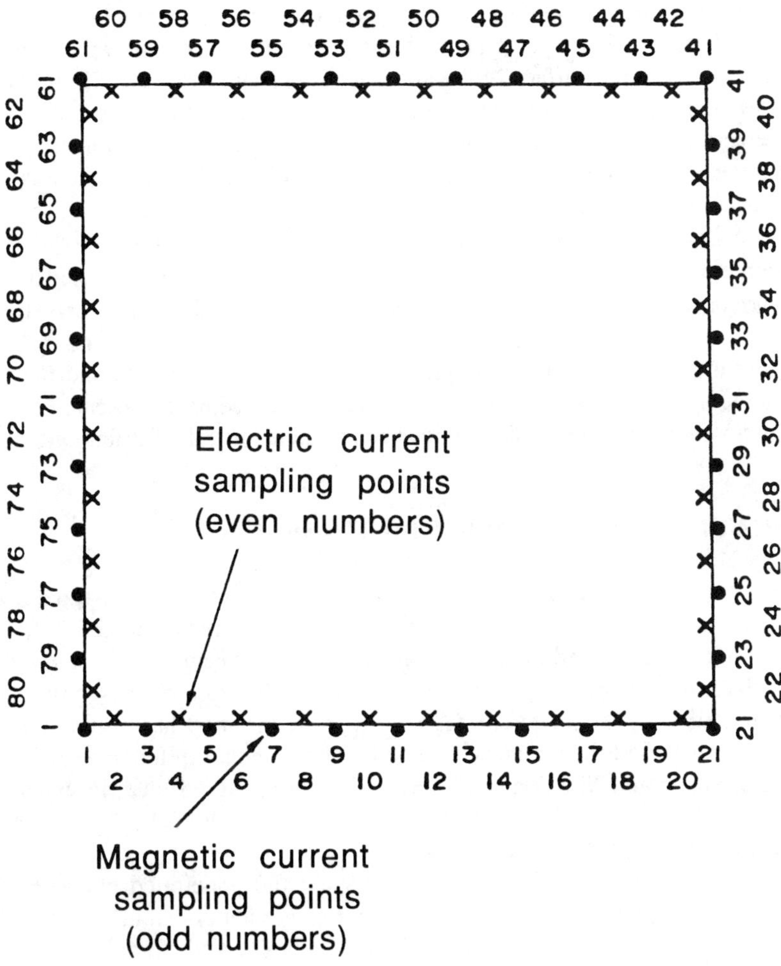

Figure 5.24c Arrangement of pulse expansion functions on the surface of a square dielectric scatterer – TM excitation.

Again for TM polarization, the incident electric field is polarized parallel to the z axis and the corresponding incident magnetic field is polarized parallel to the $-y$ axis. Since the scatterer is the air dielectric geometry, in practice there are no scattered electric and magnetic fields in the two regions. The distribution of the near total fields completely depend on the magnitude and phase of the incident plane wave excitation. Figure 5.24a shows magnitude distribution of the equivalent electric current or the $-y$ directed incident magnetic field along the surface of the air dielectric square cylinder. Similarly, Figure 5.24b shows magnitude distribution of the equivalent magnetic current or the z directed incident electric field along the surface of the air dielectric square cylinder. As can be seen from the numerical results, excellent simulation of the plane wave field is obtained along the surface of the scatterer.

Figures 5.25a and 5.25b show similar results of the distribution of equivalent electric current and equivalent magnetic current with dielectric permittivity selected as $\varepsilon_2 = 2\varepsilon_0$ for a homogeneous, lossless dielectric square scatterer. Additional numerical data are reported in later sections. The results of the equivalent electric current and the magnetic current distributions can now be utilized to calculate the near-field and the far-field distributions. In the following, a detailed numerical procedure is presented concerning the calculation of the near-field distributions and also the far-field distributions.

5.10 SCATTERED FIELD DISTRIBUTIONS

The pulse expansion coefficients of the equivalent electric and magnetic current distributions, as obtained from the solution of matrix expression (5.8.14), are now substituted into the scattered field expressions derived in Sections 5.4.1 and 5.4.2. For TM polarization, the near scattered electric field distribution has only the axial component, and the near scattered magnetic field distribution has only the transverse component. In terms of the various vector and scalar potentials referred to the free-space region 1, expressions (5.5.2) and (5.5.3) yield the scattered electric and magnetic distributions. The corresponding vector and the scalar potentials for region 1 are given by expressions (5.4.13a–b) and (5.4.21a–b).

Similarly, expressions (5.5.4) and (5.5.5) yield the penetrated electric field and magnetic field distributions in terms of the vector and scalar potentials as referred to the lossy dielectric region 2. Expressions (5.4.37a–b) and (5.4.44a–b) give the corresponding vector and the scalar potentials for region 2. It should be noted that the equivalent currents referred to regions 1 and 2 are directly interrelated. In fact, referring to expressions (5.5.7) and (5.5.8), they are equal and have opposite sign.

To calculate the near-field component in a given direction, first the location of the field point is selected, and then a unit vector is defined along which the components of the electric and magnetic fields are to be calculated. Hence, the components of scattered electric and magnetic fields in region 1 are given by:

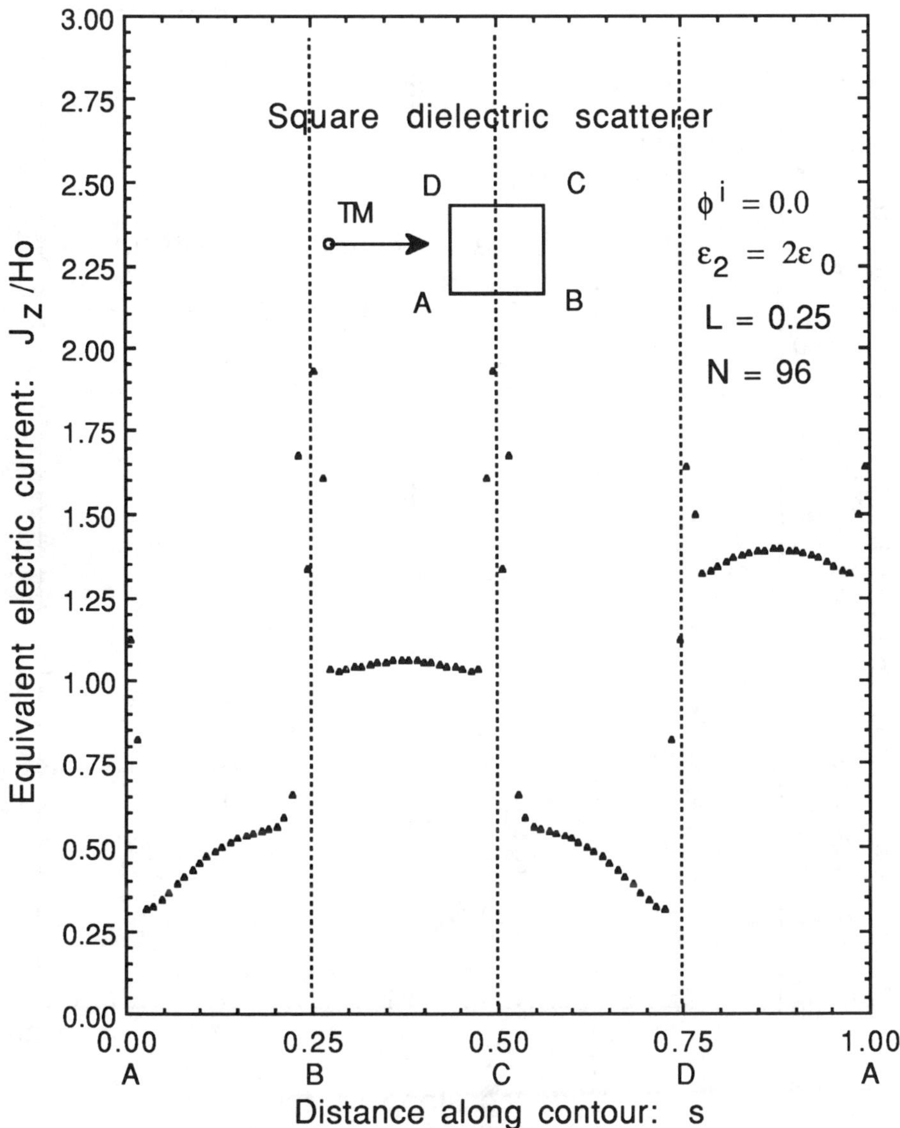

Figure 5.25a Magnitude distribution of the equivalent electric current on a square dielectric scatterer – TM excitation.

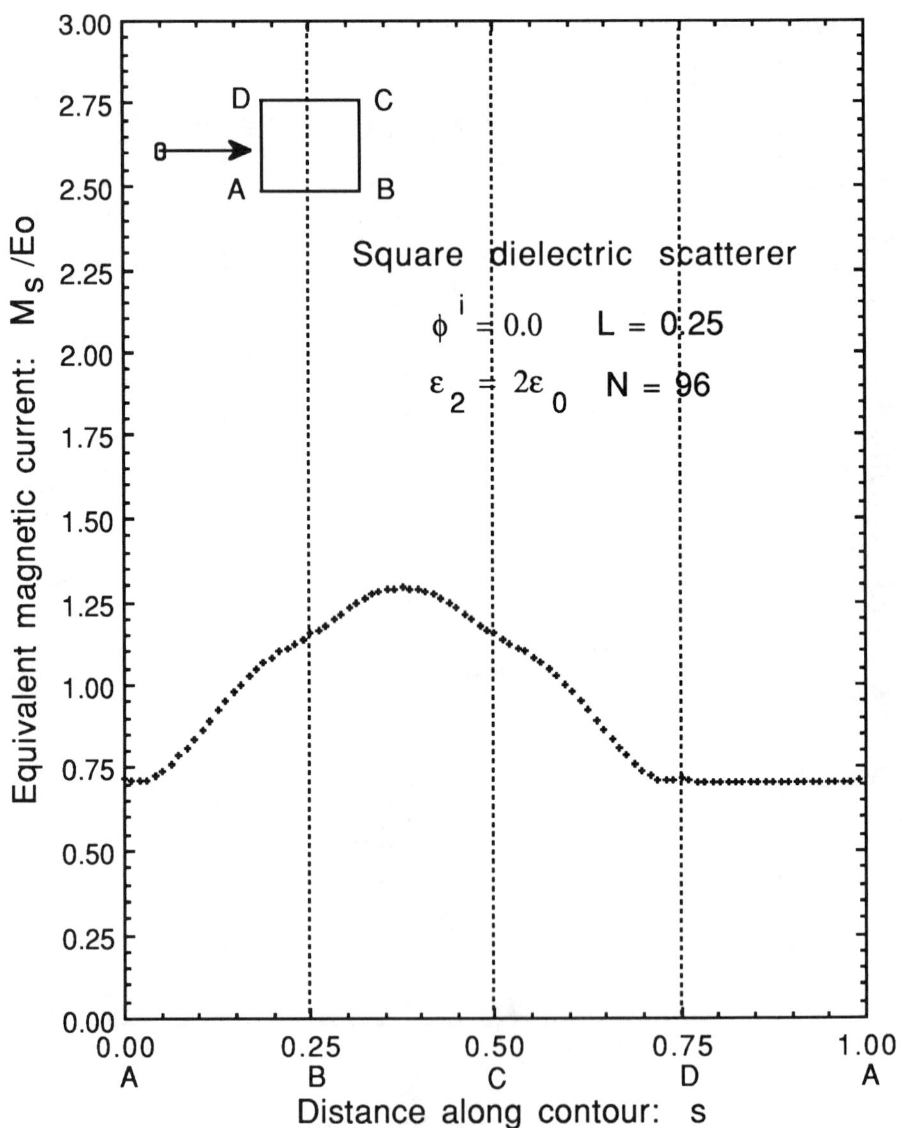

Figure 5.25b Magnitude distribution of the equivalent magnetic current on a square dielectric scatterer – TM excitation.

$$E_{1z}^{s}(\bar{\rho}) = -\sum_{n=1}^{N} I_{2n} \int_{\bar{\rho}_{2n-1}}^{\bar{\rho}_{2n+1}} \frac{\omega\mu_1}{4} H_0^{(2)}(k_1R)\, dL'$$

$$-\sum_{p=1}^{P} T_{2p-1}\Big[\int_{\bar{\rho}_{2p-2}}^{\bar{\rho}_{2p-1}} + \int_{\bar{\rho}_{2p-1}}^{\bar{\rho}_{2p}}\Big]\Big[\frac{x-x'}{R}\cos\Omega' + \frac{y-y'}{R}\sin\Omega'\Big]$$

$$\Big[\frac{k_1}{4j} H_0^{(2)'}(k_1R)\Big]\, dL' \qquad (5.10.1a)$$

and

$$H_{1s}^{s}(\bar{\rho}) = -\sum_{n=1}^{N} I_{2n} \int_{\bar{\rho}_{2n-1}}^{\bar{\rho}_{2n+1}} \Big[\frac{x-x'}{R}\cos\phi + \frac{y-y'}{R}\sin\phi\Big]$$

$$\Big[\frac{k_1}{4j} H_0^{(2)'}(k_1R)\Big] dL'$$

$$-\sum_{p=1}^{P} T_{2p-1}\Big[\int_{\bar{\rho}_{2p-2}}^{\bar{\rho}_{2p-1}} + \int_{\bar{\rho}_{2p-1}}^{\bar{\rho}_{2p}}\Big]\cos(\phi-\Omega')\Big[\frac{\omega\varepsilon_1}{4} H_0^{(2)}(k_1R)\Big]\, dL'$$

$$+\frac{1}{\Delta}\sum_{p=1}^{P}\frac{T_{2p-1}}{\Delta_{2p-2}} \int_{\bar{\rho}_{2p-3}}^{\bar{\rho}_{2p-1}} \frac{1}{4\omega\mu_1} H_0^{(2)}(k_1R_-)\, dL'$$

$$-\frac{1}{\Delta}\sum_{p=1}^{P}\frac{T_{2p-1}}{\Delta_{2p}} \int_{\bar{\rho}_{2p-1}}^{\bar{\rho}_{2p+1}} \frac{1}{4\omega\mu_1} H_0^{(2)}(k_1R_-)\, dL'$$

$$- \frac{1}{\Delta} \sum_{p=1}^{P} \frac{T_{2p-1}}{\Delta_{2p-2}} \int_{\bar{\rho}_{2p-3}}^{\bar{\rho}_{2p-1}} \frac{1}{4\omega\mu_1} H_0^{(2)}(k_1 R_+)\, dL'$$

$$+ \frac{1}{\Delta} \sum_{p=1}^{P} \frac{T_{2p-1}}{\Delta_{2p}} \int_{\bar{\rho}_{2p-1}}^{\bar{\rho}_{2p+1}} \frac{1}{4\omega\mu_1} H_0^{(2)}(k_1 R_+)\, dL' \tag{5.10.1b}$$

where corresponding to the selected field point

$$R_+ = |\bar{\rho}_+ - \bar{\rho}| \tag{5.10.1c}$$

$$R_- = |\bar{\rho}_- - \bar{\rho}| \tag{5.10.1d}$$

$$\bar{\rho}_+ = \bar{\rho} + \hat{s}\frac{\Delta}{2} \tag{5.10.1e}$$

$$\bar{\rho}_- = \bar{\rho} - \hat{s}\frac{\Delta}{2} \tag{5.10.1f}$$

5.10.1 FAR-FIELD DISTRIBUTION

The electric and magnetic field distributions in the far-field region can be obtained by using the large argument approximation for the free-space Green's function. For TM polarization, in the far-field region, there are only the z component of the scattered electric field distribution and the corresponding ϕ component of the scattered magnetic field distribution. It should be clearly noted, in the far-field region, the scattered electric and magnetic fields behave as pure *cylindrical waves*, propagating in the radial direction, and further, they are directly related through the medium intrinsic impedance. Hence, in expressions (5.4.24) and (5.4.25), only the magnetic vector potential and the electric vector potential terms are retained, which basically are the dominant terms in the far-field calculation. The other remaining terms corresponding to the gradient of magnetic scalar potential and curl of the vector potentials do not contribute toward the far-field calculation. The z component of the electric field in the far-field region can be calculated by the following procedure. Due to the equivalent electric current, the partial z component of the electric field in the far-field region is given by

$$E^{S}_{1z}(\bar{\rho})\Big|_{\text{due to } J_z} = -j\omega\hat{z}\bullet\bar{A}_1(\bar{\rho},\omega) \tag{5.10.2a}$$

and due to the equivalent magnetic current, the partial z component of the electric field in the far-field region is given by

$$E^{S}_{1z}(\bar{\rho})\Big|_{\text{due to } M_s} = -\eta_1 H^{S}_{1\phi}(\bar{\rho},\omega) \tag{5.10.2b}$$

and

$$H^{S}_{1\phi}(\bar{\rho}) = -j\omega\hat{\phi}\bullet\bar{F}_1(\bar{\rho},\omega) \tag{5.10.2c}$$

After summing expressions (5.10.2a) and (5.10.2b) and substituting the integral expressions (5.5.10a) and (5.5.10c) for the vector potentials, the z component of the total electric field in the far-field region takes the following form:

$$E^{S}_{1z}(\rho,\phi) = -\frac{\omega\mu_1}{4}\int_C J_z(\bar{\rho}')H_0^{(2)}(k_1|\bar{\rho}-\bar{\rho}'|)\,dL'$$

$$+\eta_1\frac{\omega\varepsilon_1}{4}\int_C (\hat{\phi}\bullet\hat{s}')M_s(\bar{\rho}')H_0^{(2)}(k_1|\bar{\rho}-\bar{\rho}'|)\,dL' \tag{5.10.3}$$

In the large argument approximation, $|k_1\rho| \to \infty$, the Hankel function can be written as

$$H_0^{(2)}(k_1\rho) \sim \sqrt{\frac{2j}{\pi k_1\rho}}\,e^{-jk_1\rho} \tag{5.10.4}$$

Substituting for the equivalent electric current (5.8.8a) and equivalent magnetic current (5.8.10a) distributions, the scattered electric field distribution in the far-field region reduces to

$$E^{S}_{1z}(\rho,\phi) = \sum_{n=1}^{N} I_{2n}\Big[-\frac{k_1\eta_1}{4}H_0^{(2)}(k_1|\bar{\rho}-\bar{\rho}_{2n}|)\Big]\Delta_{2n}$$

$$+\sum_{p=1}^{P} T_{2p-1}\Big[\frac{k_1}{4}\cos(\phi-\Omega_{2p-2})\,H_0^{(2)}(k_1|\bar{\rho}-\bar{\rho}_{2p-1}|)\Big]\frac{\Delta_{2p-2}}{2}$$

$$+ \sum_{p=1}^{P} T_{2p-1}\left[\frac{k_1}{4} \cos(\phi - \Omega_{2p})\, H_0^{(2)}(k_1|\bar{\rho} - \bar{\rho}_{2p-1}|)\right]\frac{\Delta_{2p}}{2} \tag{5.10.5a}$$

In this expression, the Hankel function is now replaced by its large argument approximation defined in (5.10.4):

$$E_{1z}^{s}(\rho,\phi) \sim \sum_{n=1}^{N} I_{2n}\left[-\frac{k_1\eta_1}{4}\sqrt{\frac{2j}{\pi(k_1|\bar{\rho}-\bar{\rho}_{2n}|)}}\; e^{-j(k_1|\bar{\rho} - \bar{\rho}_{2n}|)}\right]\Delta_{2n}$$

$$+ \sum_{p=1}^{P} T_{2p-1}\left[\frac{k_1}{4} \cos(\phi - \Omega_{2p-2})\right.$$

$$\left.\sqrt{\frac{2j}{\pi(k_1|\bar{\rho}-\bar{\rho}_{2p-1}|)}}\; e^{-j(k_1|\bar{\rho} - \bar{\rho}_{2p-1}|)}\right]\frac{\Delta_{2p-2}}{2}$$

$$+ \sum_{p=1}^{P} T_{2p-1}\left[\frac{k_1}{4} \cos(\phi - \Omega_{2p})\right.$$

$$\left.\sqrt{\frac{2j}{\pi(k_1|\bar{\rho}-\bar{\rho}_{2p-1}|)}}\; e^{-j(k_1|\bar{\rho} - \bar{\rho}_{2p-1}|)}\right]\frac{\Delta_{2p}}{2} \tag{5.10.5b}$$

Referring to Figures 3.23 and 4.17, this expression can be easily simplified. In the far-field region, $|k_1\rho| \rightarrow \infty$, the magnitude terms in the denominator of expression (5.10.5b) can be written as

$$|\bar{\rho} - \bar{\rho}_{2n}| \approx \rho \tag{5.10.6a}$$

$$|\bar{\rho} - \bar{\rho}_{2p-1}| \approx \rho \tag{5.10.6b}$$

and the exponential terms contributing to the phase distribution of the currents can be approximated as

$$|\bar{\rho} - \bar{\rho}_{2n}| \approx \rho - \rho_{2n}\cos(\phi - \phi_{2n}) \tag{5.10.6c}$$

$$|\bar{\rho} - \bar{\rho}_{2p-1}| \approx \rho - \rho_{2p-1} \cos(\phi - \phi_{2p-1}) \tag{5.10.6d}$$

After substituting these far-field approximations, expressions (5.10.6a–d), the scattered electric field distribution in the far-field region reduces to the following form:

$$\begin{aligned} E^{s}_{1z}(\rho,\phi) \sim \; & \sum_{n=1}^{N} I_{2n}\Big[k_1\eta_1 \mathcal{K} e^{jk_1\bar{\rho}_{2n}\cos(\phi-\phi_{2n})}\Big]\Delta_{2n} \\ & - \sum_{p=1}^{P} T_{2p-1}\Big[k_1\mathcal{K}\cos(\phi - \Omega_{2p-2}) \\ & \qquad e^{jk_1\bar{\rho}_{2p-1}\cos(\phi-\phi_{2p-1})}\Big]\frac{\Delta_{2p-2}}{2} \\ & - \sum_{p=1}^{P} T_{2p-1}\Big[k_1\mathcal{K}\cos(\phi - \Omega_{2p}) \\ & \qquad e^{jk_1\bar{\rho}_{2p-1}\cos(\phi-\phi_{2p-1})}\Big]\frac{\Delta_{2p}}{2} \end{aligned} \tag{5.10.7a}$$

where

$$\mathcal{K} = \frac{1}{\sqrt{8\pi k_1\rho}}\, e^{-jk_1\rho}\, e^{-j3\pi/4} \tag{5.10.7b}$$

Further, the bistatic radar cross section of the lossy, homogeneous dielectric scatterer having arbitrary edges and corners can now be calculated using expression (3.8.18), given by

$$\text{RCS}(\phi) = \lim_{\rho\to\infty} 2\pi\rho \left| \frac{E^{s}_{1z}(\phi,\omega)}{E^{i}_{z}(\phi,\omega)} \right|^2 \tag{5.10.8}$$

Figure 5.26 shows a plot of the bistatic radar cross section of the lossless, homogeneous dielectric circular scatterer having radius $a = 0.1$, calculated based the expressions (5.10.7) and (5.10.8). This bistatic RCS data is numerically generated by using the TM equivalent electric current and magnetic current results presented in Figures 5.23a and 5.23b. As can be seen, the bistatic data exhibits symmetrical distribution with respect to the direction of excitation about which the circular dielectric scatterer has the geometrical symmetry. Since the bistatic RCS distribution is proportional to the

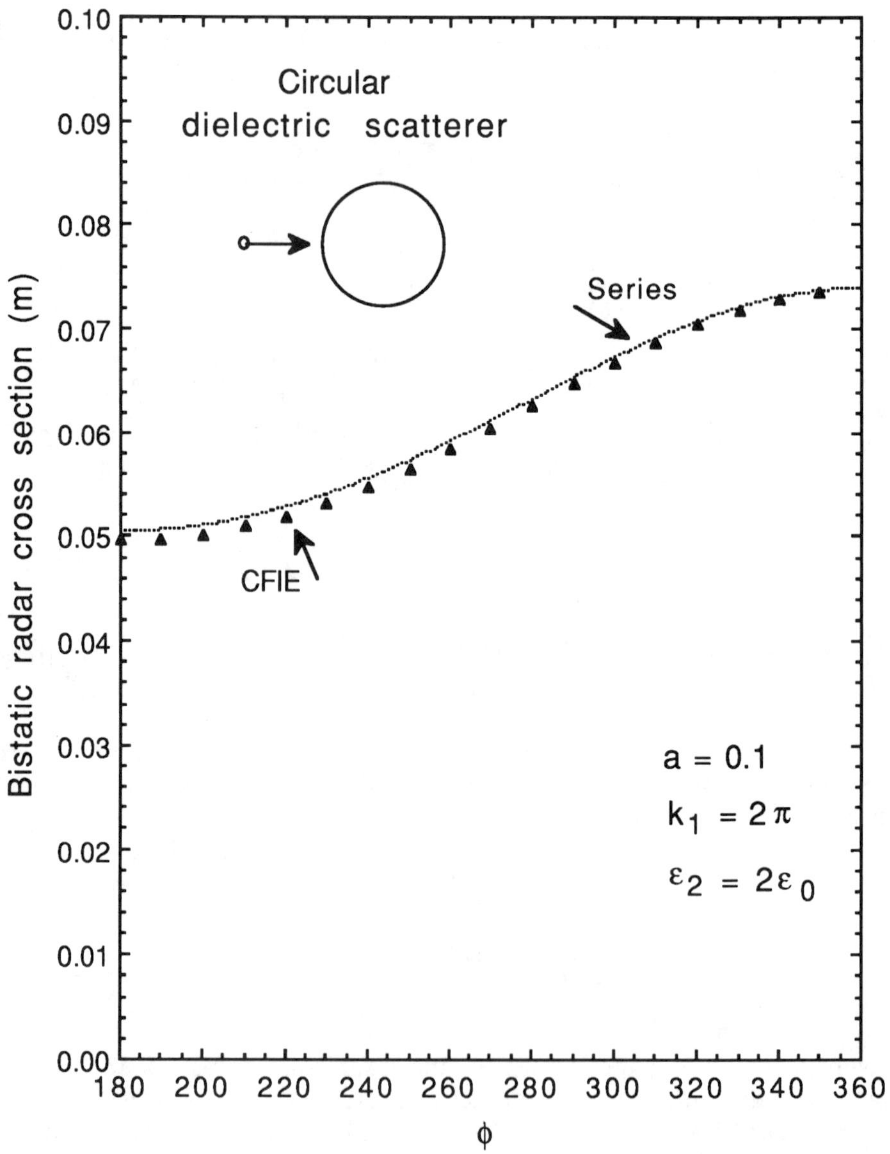

Figure 5.26 Bistatic radar cross section of a circular dielectric scatterer – TM excitation.

scattering power, the far-field scattering pattern is directly obtained by taking square root of the RCS distribution. The RCS data is also compared with respect to the series solution given by expressions (5.6.23) and (5.6.24).

Similarly, Figure 5.27 shows a plot of the bistatic radar cross section of the homogeneous dielectric square scatterer with sidelength $L = 0.25$ for the incident plane wave excitation, $\phi^i = 0^o$. Again, the bistatic data exhibits symmetrical distribution with respect to the direction of excitation about which the square scatterer also has the geometrical symmetry. In many practical applications, the monostatic or backscattering data is quite important which depends on number of parameters, such as the angle of incidence, the electrical size, and shape of the geometry. To produce the monostatic RCS distribution, referring to expression (5.8.14), the inverse of the generalized impedance matrix can be stored separately and multiplied by the excitation term recalculated for every angle of incidence to obtain the corresponding electric and magnetic current distributions. Figure 5.28a shows a plot of the monostatic radar cross section for incident angles ranging from $\phi^i = 90^o$ to 135° for the case of a rectangular, homogeneous dielectric scatterer. Similarly, Figure 5.28b shows a plot of the bistatic radar cross section for incident angle $\phi^i = 90^o$ for the same rectangular dielectric scatterer.

5.11 CFIE – TE CASE

The study of electromagnetic scattering by a two-dimensional, homogeneous dielectric scatterer with transverse electric excitation is similar to the study of transverse magnetic excitation, because the TE case forms the *dual* of the TM case. The complete analysis for the TM case as discussed in earlier sections, including the numerical technique, is also applicable to the TE case with proper dual interchange of the field quantities, the equivalent currents, and the medium parameters.

For the special case of time-harmonic transverse electric *normal* excitation, there are only two distinct electromagnetic field components, as given in Table 5.3. In regions 1 and 2 including at boundary contour C, which separates the dielectric medium and the free-space medium, only the axial component of the magnetic field distribution is parallel to the z coordinate axis, and the transverse component of the electric field distribution is parallel to the transverse plane. The transverse component of the electric field at the boundary surface can be resolved into two distinct orthogonal components along coordinate variables n and s. Further, the electric and magnetic field distributions in the two regions and the corresponding equivalent electric and magnetic current sources are completely independent of the z coordinate variable. Thus

$$\frac{\partial}{\partial z} = 0 \tag{5.11.1}$$

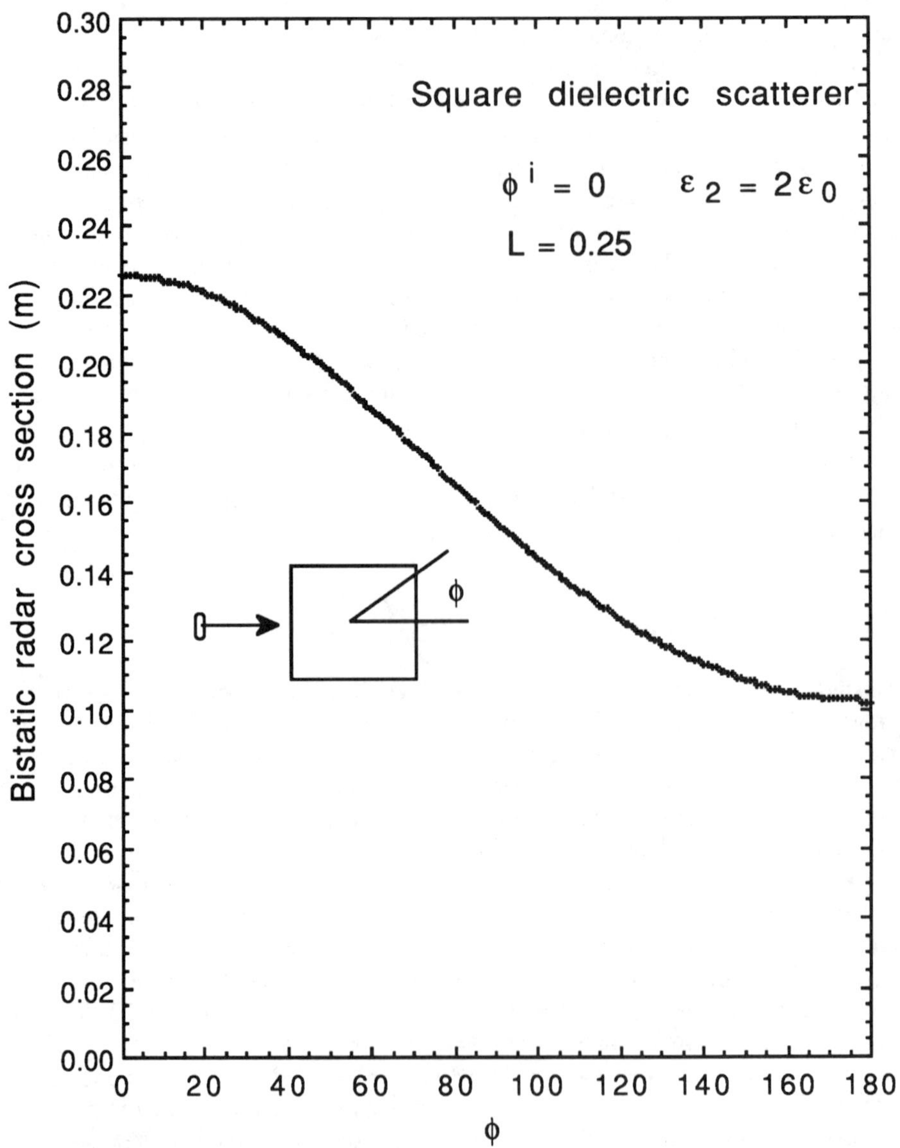

Figure 5.27 Bistatic radar cross section of a square dielectric scatterer – TM excitation.

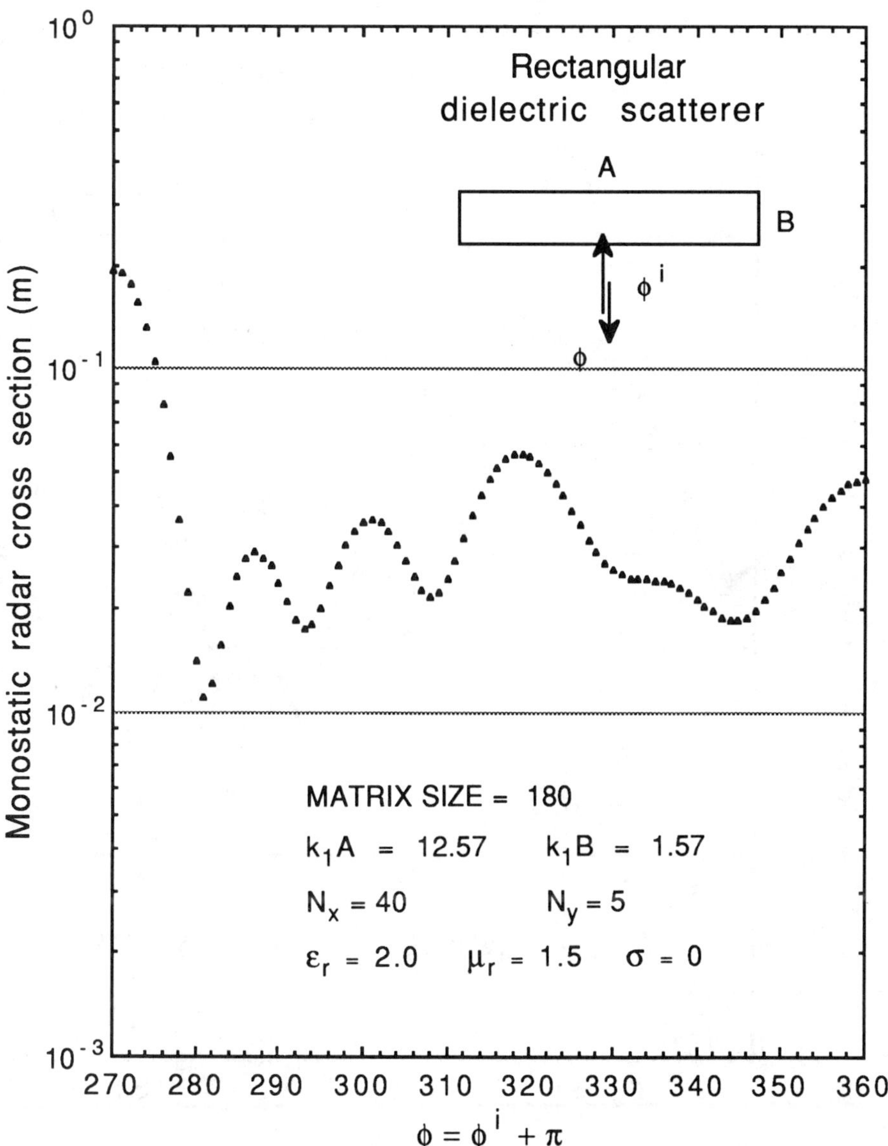

Figure 5.28a Monostatic radar cross section of a rectangular dielectric scatterer – TM excitation.

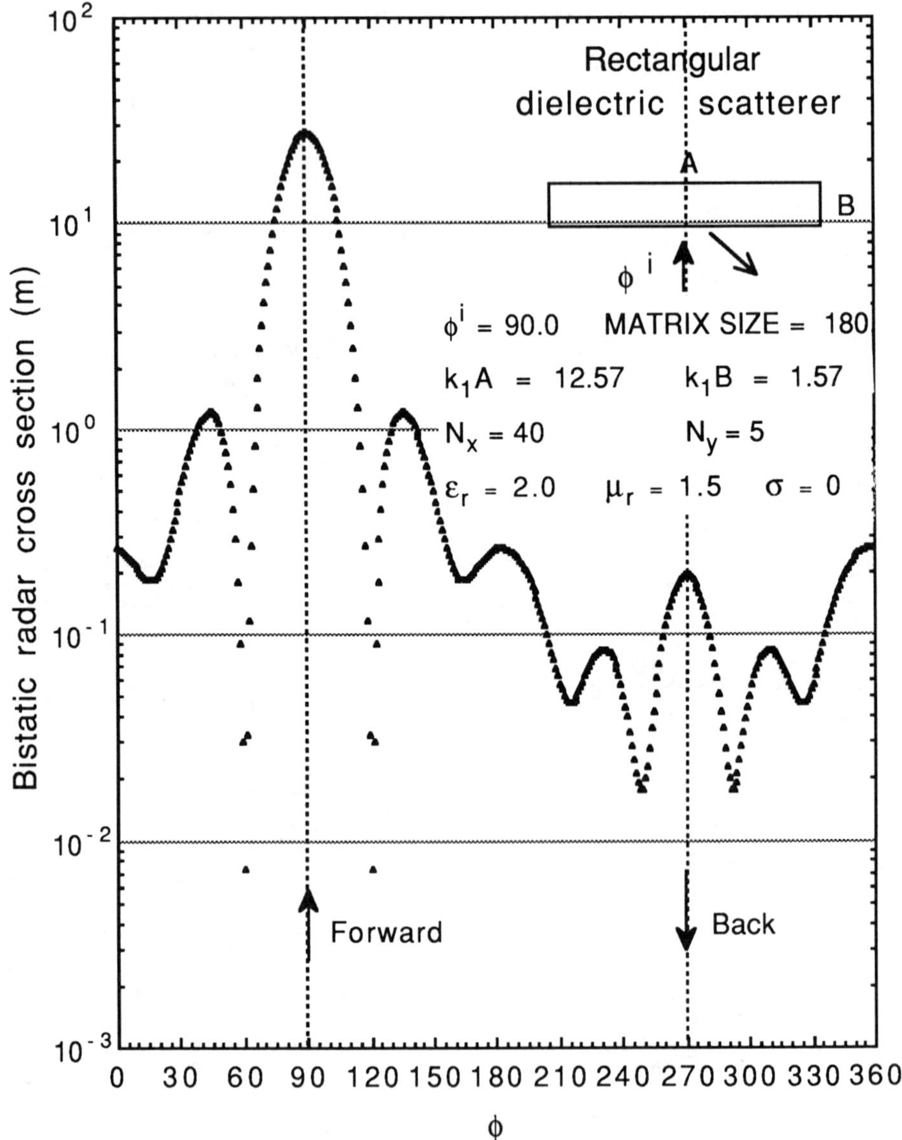

Figure 5.28b Bistatic radar cross section of a rectangular dielectric scatterer – TM normal excitation.

For TE excitation, there are only the following components of the magnetic and electric field distributions

$H_z(\rho,\phi)\hat{z}$: total axial component of the magnetic field distribution;

$E_n(\rho,\phi)\hat{n}$: total normal component of the electric field distribution;

$E_s(\rho,\phi)\hat{s}$: total tangential component of the electric field distribution.

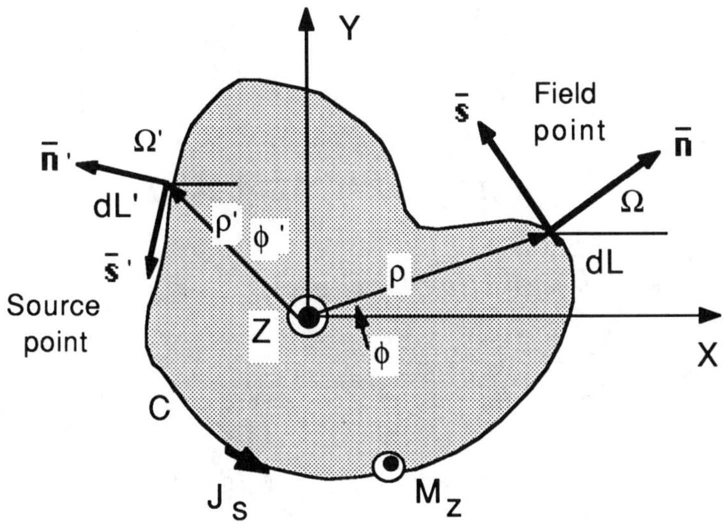

Figure 5.29 TE excitation – dielectric case.

Based on these nonzero magnetic and electric field components, for TE normal excitation, only the axial z component of the equivalent magnetic current distribution and the tangential s component of the equivalent electric current distribution are on virtual contour C as defined in the following. Referring to the Figure 5.29,

$M_z(\rho',\phi')\hat{z}'$: unknown axial component of magnetic current distribution;

$J_s(\rho',\phi')\hat{s}'$: unknown tangential component of electric current distribution.

$$\bar{M}_{eq}(\bar{\rho}') = \hat{z}'M_z(\bar{\rho}') \tag{5.11.2a}$$

$$= \hat{s}'E_s(\rho',\phi') \times \hat{n}' \qquad \bar{\rho}' \text{ on } C \tag{5.11.2b}$$

and

$$\bar{J}_{eq}(\bar{\rho}') = \hat{s}'J_s(\bar{\rho}') \tag{5.11.3a}$$

$$= \hat{n}' \times \hat{z}'H_z(\rho',\phi') \qquad \bar{\rho}' \text{ on } C \tag{5.11.3b}$$

Hence, for TE excitation, based on these components of the unknown magnetic and electric current distributions, the coupled combined field integral equations derived earlier in expressions (5.5.9a) and (5.5.9b) can be specialized. Taking the z component of integral equation (5.5.9b) and the s component of integral equation (5.5.9a), the following coupled integral equations are obtained:

$$\hat{z} \bullet \bar{H}^i(\bar{\rho}) = \hat{z} \bullet j\omega\left[\bar{F}_1(\bar{\rho},\omega) - \bar{F}_2(\bar{\rho},\omega)\right]$$

$$- \hat{z} \bullet \nabla_\tau \times \left[\frac{1}{\mu_1}\bar{A}_1(\bar{\rho},\omega) - \frac{1}{\mu_2}\bar{A}_2(\bar{\rho},\omega)\right] \tag{5.11.4a}$$

$$\hat{s} \bullet \bar{E}^i(\bar{\rho}) = \hat{s} \bullet j\omega\left[\bar{A}_1(\bar{\rho},\omega) - \bar{A}_2(\bar{\rho},\omega)\right]$$

$$+ \frac{\partial}{\partial s}\left[\Phi_1(\bar{\rho},\omega) - \Phi_2(\bar{\rho},\omega)\right]$$

$$+ \hat{s} \bullet \nabla_\tau \times \left[\frac{1}{\varepsilon_1}\bar{F}_1(\bar{\rho},\omega) - \frac{1}{\varepsilon'_2}\bar{F}_2(\bar{\rho},\omega)\right] \tag{5.11.4b}$$

$\bar{\rho}$ on C

where the vector and scalar potential integrals are given by the following expressions:

$$\bar{F}_1(\bar{\rho}) = \frac{\varepsilon_1}{4j}\int_C \hat{z}'M_z(\bar{\rho}')H_0^{(2)}(k_1|\bar{\rho} - \bar{\rho}'|)\, dL(\bar{\rho}') \tag{5.11.5a}$$

$$\bar{A}_1(\bar{\rho}) = \frac{\mu_1}{4j}\int_C \hat{s}'J_s(\bar{\rho}')H_0^{(2)}(k_1|\bar{\rho} - \bar{\rho}'|)\, dL(\bar{\rho}') \tag{5.11.5b}$$

$$\Phi_1(\bar{\rho}) = \frac{1}{4\omega\varepsilon_1}\int_C [\nabla'_\tau \bullet \hat{s}'J_s(\bar{\rho}')]H_0^{(2)}(k_1|\bar{\rho} - \bar{\rho}'|)\, dL(\bar{\rho}') \tag{5.11.5c}$$

$\bar{\rho}$ on C

and

$$\bar{F}_2(\bar{\rho}) = \frac{\varepsilon'_2}{4j} \int_C - \hat{z}' M_z(\bar{\rho}') H_0^{(2)}(k_2|\bar{\rho} - \bar{\rho}'|) \, dL(\bar{\rho}') \tag{5.11.6a}$$

$$\bar{A}_2(\bar{\rho}) = \frac{\mu_2}{4j} \int_C - \hat{s}' J_s(\bar{\rho}') H_0^{(2)}(k_2|\bar{\rho} - \bar{\rho}'|) \, dL(\bar{\rho}') \tag{5.11.6b}$$

$$\Phi_2(\bar{\rho}) = \frac{1}{4\omega\varepsilon'_2} \int_C - [\nabla'_\tau \bullet \hat{s}' J_s(\bar{\rho}')] H_0^{(2)}(k_2|\bar{\rho} - \bar{\rho}'|) \, dL(\bar{\rho}') \tag{5.11.6c}$$

$\bar{\rho}$ on C

After substituting these potential expressions into integral equation (5.11.4a), the following simplification is obtained:

$$H_z^i(\bar{\rho}) = \frac{\omega\varepsilon_1}{4} \int_C M_z(\bar{\rho}') H_0^{(2)}(k_1|\bar{\rho} - \bar{\rho}'|) \, dL(\bar{\rho}')$$

$$+ \frac{\omega\varepsilon'_2}{4} \int_C M_z(\bar{\rho}') H_0^{(2)}(k_2|\bar{\rho} - \bar{\rho}'|) \, dL(\bar{\rho}')$$

$$- \hat{z} \bullet \nabla_\tau \times \frac{1}{4j} \int_C \hat{s}' J_s(\bar{\rho}') H_0^{(2)}(k_1|\bar{\rho} - \bar{\rho}'|) \, dL(\bar{\rho}')$$

$$- \hat{z} \bullet \nabla_\tau \times \frac{1}{4j} \int_C \hat{s}' J_s(\bar{\rho}') H_0^{(2)}(k_2|\bar{\rho} - \bar{\rho}'|) \, dL(\bar{\rho}') \tag{5.11.7a}$$

$\bar{\rho}$ on C

Similarly, after substituting these potential expressions into integral equation (5.11.4b), the following simplification is obtained:

$$E_s^i(\bar{\rho}) = \frac{\omega\mu_1}{4} \int_C (\hat{s} \bullet \hat{s}') J_s(\bar{\rho}') H_0^{(2)}(k_1|\bar{\rho} - \bar{\rho}'|) \, dL(\bar{\rho}')$$

$$+ \frac{\omega\mu_2}{4} \int_C (\hat{s} \bullet \hat{s}') J_s(\bar{\rho}') H_0^{(2)}(k_2|\bar{\rho} - \bar{\rho}'|) \, dL(\bar{\rho}')$$

$$+ \frac{1}{4\omega\varepsilon_1}\frac{\partial}{\partial s}\int_C [\nabla'_\tau \bullet \hat{s}' J_s(\bar{\rho}')] H_0^{(2)}(k_1|\bar{\rho} - \bar{\rho}'|)\, dL(\bar{\rho}')$$

$$+ \frac{1}{4\omega\varepsilon'_2}\frac{\partial}{\partial s}\int_C [\nabla'_\tau \bullet \hat{s}' J_s(\bar{\rho}')] H_0^{(2)}(k_2|\bar{\rho} - \bar{\rho}'|)\, dL(\bar{\rho}')$$

$$+ \hat{s} \bullet \nabla_\tau \times \frac{1}{4j}\int_C \hat{z}' M_z(\bar{\rho}') H_0^{(2)}(k_1|\bar{\rho} - \bar{\rho}'|)\, dL(\bar{\rho}')$$

$$+ \hat{s} \bullet \nabla_\tau \times \frac{1}{4j}\int_C \hat{z}' M_z(\bar{\rho}') H_0^{(2)}(k_2|\bar{\rho} - \bar{\rho}'|)\, dL(\bar{\rho}') \qquad (5.11.7b)$$

$\bar{\rho}$ on C

In these coupled integral equations, the integral terms containing the divergence and the curl operations can be further simplified as in expressions (5.5.19) and (5.5.20). Hence, in expression (5.11.7b), the divergence term can be simplified as

$$\nabla'_\tau \bullet \hat{s}' J_s(\bar{\rho}') = \frac{\partial}{\partial s'} J_s(\bar{\rho}') \qquad (5.11.8)$$

Further, referring to the integral equation given by expression (5.11.7b) and electric field boundary condition (5.5.1a), the partial s component of the scattered electric field in the two regions (based on the external and the internal equivalences) due to the axial z-directed magnetic currents is given by

$$-\hat{s} E^S_{1m}(\bar{\rho}) = \frac{1}{\varepsilon_1}\hat{s} \bullet \nabla_\tau \times \bar{F}_1(\bar{\rho}) \qquad (5.11.9a)$$

$$= \frac{1}{4j}\hat{s} \bullet \nabla_\tau \times \int_C \hat{z}' M_z(\bar{\rho}') H_0^{(2)}(k_1|\bar{\rho} - \bar{\rho}'|)\, dL(\bar{\rho}') \qquad (5.11.9b)$$

$$= \frac{1}{4j}\hat{s} \bullet \nabla_\tau \times \int_{C_o} \hat{z}' M_z(\bar{\rho}') H_0^{(2)}(k_1|\bar{\rho} - \bar{\rho}'|)\, dL(\bar{\rho}')$$

$$+ \frac{1}{4j}\hat{s} \bullet \nabla_\tau \times \fint_C \hat{z}' M_z(\bar{\rho}') H_0^{(2)}(k_1|\bar{\rho} - \bar{\rho}'|)\, dL(\bar{\rho}') \qquad (5.11.9c)$$

$\bar{\rho}$ in region 1, approaching C from the outside

and

$$\hat{s}E^S_{2m}(\bar{\rho}) = -\frac{1}{\varepsilon'_2}\hat{s} \bullet \nabla_\tau \times \bar{F}_2(\bar{\rho}) \tag{5.11.10a}$$

$$= \frac{1}{4j}\hat{s} \bullet \nabla_\tau \times \int_C \hat{z}' M_z(\bar{\rho}') H_0^{(2)}(k_2|\bar{\rho} - \bar{\rho}'|)\, dL(\bar{\rho}') \tag{5.11.10b}$$

$$= \frac{1}{4j}\hat{s} \bullet \nabla_\tau \times \int_{C_0} \hat{z}' M_z(\bar{\rho}') H_0^{(2)}(k_2|\bar{\rho} - \bar{\rho}'|)\, dL(\bar{\rho}')$$
$$+ \frac{1}{4j}\hat{s} \bullet \nabla_\tau \times \int\!\!\!\!-_C \hat{z}' M_z(\bar{\rho}') H_0^{(2)}(k_2|\bar{\rho} - \bar{\rho}'|)\, dL(\bar{\rho}') \tag{5.11.10c}$$

$\bar{\rho}$ in region 2, approaching C from the inside

In expressions (5.11.9c) and (5.11.10c), the integral terms over the contour C_0 represent the value of the corresponding integral when the field observation point approaches in a limit the source segment covered by contour C_0. In fact, they represent the respective principal value of the integrals, which basically are the scattered partial electric fields very close to the scatterer carrying the corresponding sheet of magnetic current. Hence, the principal value integral over contour C_0 in expression (5.11.9c) as the field observation point in region 1 approaches the source segment is equal to - $m_z/2$ assuming the outward normal to contour C, as shown in Figure 5.29. Similarly, the principal value integral over contour C_0 in expression (5.11.10c) as the field observation point in region 2 approaches the source segment is equal to $m_z/2$ assuming the same outward normal to contour C. Hence, the two principal value contributions cancel out.

The second integral terms are written with a bar on the integral sign to recognize the field observation point on the remainder of contour C leaving out the source segment. The kernel in the integrand of the second integral is well behaved, and thus, the transverse curl operation can be easily taken inside the integral to yield a normal derivative term similar to expression (4.5.11c).

Hence, the coupled CFIE for TE normal excitation take the following form

$$H^i_z(\bar{\rho}) = \frac{\omega\varepsilon_1}{4}\int_C M_z(\bar{\rho}') H_0^{(2)}(k_1|\bar{\rho} - \bar{\rho}'|)\, dL(\bar{\rho}')$$
$$+ \frac{\omega\varepsilon'_2}{4}\int_C M_z(\bar{\rho}') H_0^{(2)}(k_2|\bar{\rho} - \bar{\rho}'|)\, dL(\bar{\rho}')$$

$$+ \hat{z} \bullet \frac{1}{4j} \oint_C \hat{s}' J_S(\bar{\rho}') \times \nabla_\tau H_0^{(2)}(k_1|\bar{\rho} - \bar{\rho}'|) \, dL(\bar{\rho}')$$

$$+ \hat{z} \bullet \frac{1}{4j} \oint_C \hat{s}' J_S(\bar{\rho}') \times \nabla_\tau H_0^{(2)}(k_2|\bar{\rho} - \bar{\rho}'|) \, dL(\bar{\rho}') \qquad (5.11.11a)$$

$$\bar{\rho} \text{ on } C$$

and

$$E_S^i(\bar{\rho}) = \frac{\omega\mu_1}{4} \int_C (\hat{s} \bullet \hat{s}') J_S(\bar{\rho}') H_0^{(2)}(k_1|\bar{\rho} - \bar{\rho}'|) \, dL(\bar{\rho}')$$

$$+ \frac{\omega\mu_2}{4} \int_C (\hat{s} \bullet \hat{s}') J_S(\bar{\rho}') H_0^{(2)}(k_2|\bar{\rho} - \bar{\rho}'|) \, dL(\bar{\rho}')$$

$$+ \frac{1}{4\omega\varepsilon_1} \frac{\partial}{\partial s} \int_C [\frac{\partial}{\partial s'} J_S(\bar{\rho}')] H_0^{(2)}(k_1|\bar{\rho} - \bar{\rho}'|) \, dL(\bar{\rho}')$$

$$+ \frac{1}{4\omega\varepsilon'_2} \frac{\partial}{\partial s} \int_C [\frac{\partial}{\partial s'} J_S(\bar{\rho}')] H_0^{(2)}(k_2|\bar{\rho} - \bar{\rho}'|) \, dL(\bar{\rho}')$$

$$- \hat{s} \bullet \frac{1}{4j} \oint_C \hat{z}' M_Z(\bar{\rho}') \times \nabla_\tau H_0^{(2)}(k_1|\bar{\rho} - \bar{\rho}'|) \, dL(\bar{\rho}')$$

$$- \hat{s} \bullet \frac{1}{4j} \oint_C \hat{z}' M_Z(\bar{\rho}') \times \nabla_\tau H_0^{(2)}(k_2|\bar{\rho} - \bar{\rho}'|) \, dL(\bar{\rho}') \qquad (5.11.11b)$$

$$\bar{\rho} \text{ on } C$$

The combined field coupled integral equations just obtained, expressions (5.11.11a–b), can be further simplified. The various unit vectors on boundary contour C are now replaced in terms of the angle the normal makes with the x coordinate axis. In the integral equation (5.11.11a), the third integral term can be rewritten as

$$\frac{1}{4j} \hat{z} \bullet \hat{s}' J_S(\bar{\rho}') \times \nabla_\tau H_0^{(2)}(k_1|\bar{\rho} - \bar{\rho}'|) \qquad (5.11.12a)$$

$$= \frac{1}{4j} J_S(\bar{\rho}') \, \hat{z} \bullet (- \hat{x} \sin\Omega' + \hat{y} \cos\Omega')$$

$$\times [\hat{x} \frac{\partial}{\partial x} + \hat{y} \frac{\partial}{\partial y}] H_0^{(2)}(k_1|\bar{\rho} - \bar{\rho}'|)$$

$$= \frac{1}{4j} J_s(\bar{\rho}')[-\cos\Omega' \frac{\partial}{\partial x} - \sin\Omega' \frac{\partial}{\partial y}] H_0^{(2)}(k_1|\bar{\rho} - \bar{\rho}'|) \quad (5.11.12b)$$

$$= -\frac{k_1}{4j} J_s(\bar{\rho}')[\frac{x-x'}{R}\cos\Omega' + \frac{y-y'}{R}\sin\Omega'] H_0^{(2)'}(k_1 R) \quad (5.11.12c)$$

and

$$R = |\bar{\rho} - \bar{\rho}'|$$
$$= [(x - x')^2 + (y - y')^2]^{1/2} \quad (5.11.12d)$$

Ω': angle the normal at the source point makes with x coordinate axis.

After substituting these simplifications, the coupled CFIE (5.11.11a–b) for the *TE normal excitation* take the following form:

$$H_z^i(\bar{\rho}) = \frac{\omega\varepsilon_1}{4} \int_C M_z(\bar{\rho}') H_0^{(2)}(k_1 R)\, dL(\bar{\rho}')$$
$$+ \frac{\omega\varepsilon'_2}{4} \int_C M_z(\bar{\rho}') H_0^{(2)}(k_2 R)\, dL(\bar{\rho}')$$
$$- \frac{k_1}{4j} \int_C J_s(\bar{\rho}')[\frac{x-x'}{R}\cos\Omega' + \frac{y-y'}{R}\sin\Omega'] H_0^{(2)'}(k_1 R)\, dL(\bar{\rho}')$$
$$- \frac{k_2}{4j} \int_C J_s(\bar{\rho}')[\frac{x-x'}{R}\cos\Omega' + \frac{y-y'}{R}\sin\Omega'] H_0^{(2)'}(k_2 R)\, dL(\bar{\rho}')$$
$$\bar{\rho} \text{ on } C \quad (5.11.13a)$$

and

$$E_s^i(\bar{\rho}) = \frac{\omega\mu_1}{4} \int_C \cos(\Omega - \Omega') J_s(\bar{\rho}') H_0^{(2)}(k_1 R)\, dL(\bar{\rho}')$$
$$+ \frac{\omega\mu_2}{4} \int_C \cos(\Omega - \Omega') J_s(\bar{\rho}') H_0^{(2)}(k_2 R)\, dL(\bar{\rho}')$$
$$+ \frac{1}{4\omega\varepsilon_1} \frac{\partial}{\partial s} \int_C [\frac{\partial}{\partial s'} J_s(\bar{\rho}')] H_0^{(2)}(k_1 R)\, dL(\bar{\rho}')$$

$$+ \frac{1}{4\omega\varepsilon'_2}\frac{\partial}{\partial s}\int_C \left[\frac{\partial}{\partial s'} J_s(\bar{\rho}')\right] H_0^{(2)}(k_2R)\, dL(\bar{\rho}')$$

$$- \frac{k_1}{4j}\oint_C M_z(\bar{\rho}')\left[\frac{x-x'}{R}\cos\Omega + \frac{y-y'}{R}\sin\Omega\right] H_0^{(2)'}(k_1R)\, dL(\bar{\rho}')$$

$$- \frac{k_2}{4j}\oint_C M_z(\bar{\rho}')\left[\frac{x-x'}{R}\cos\Omega + \frac{y-y'}{R}\sin\Omega\right] H_0^{(2)'}(k_2R)\, dL(\bar{\rho}')$$

$$\bar{\rho} \text{ on } C \qquad (5.11.13b)$$

Similar to the earlier compact representation of CFIE, expressions (5.5.26) for the TM-dielectric object, the combined field integral equations (5.11.13a–b) for the *TE normal excitation* take the following operator form:

$$H_z^i(\bar{\rho}) = \mathcal{K}_{11}[M_z(\bar{\rho}')] + \mathcal{K}_{12}[J_s(\bar{\rho}')] \qquad (5.11.14a)$$

$$\bar{\rho} \text{ on } C$$

$$E_s^i(\bar{\rho}) = \mathcal{K}_{21}[M_z(\bar{\rho}')] + \mathcal{K}_{22}[J_s(\bar{\rho}')] \qquad (5.11.14b)$$

where the CFIE partial integral operators are given by

$$\mathcal{K}_{11} = \int_C \left[\frac{\omega\varepsilon_1}{4} H_0^{(2)}(k_1R) + \frac{\omega\varepsilon'_2}{4} H_0^{(2)}(k_2R)\right] dL' \qquad (5.11.15a)$$

$$\mathcal{K}_{12} = -\oint_C \left[\frac{x-x'}{R}\cos\Omega' + \frac{y-y'}{R}\sin\Omega'\right]$$

$$\left[\frac{k_1}{4j} H_0^{(2)'}(k_1R) + \frac{k_2}{4j} H_0^{(2)'}(k_2R)\right] dL' \qquad (5.11.15b)$$

$$\mathcal{K}_{21} = -\oint_C \left[\frac{x-x'}{R}\cos\Omega + \frac{y-y'}{R}\sin\Omega\right]$$

$$\left[\frac{k_1}{4j} H_0^{(2)'}(k_1R) + \frac{k_2}{4j} H_0^{(2)'}(k_2R)\right] dL' \qquad (5.11.15c)$$

$$\mathcal{K}_{22} = \int_C \cos(\Omega - \Omega')\Big[\frac{\omega\mu_1}{4} H_0^{(2)}(k_1R) + \frac{\omega\mu_2}{4} H_0^{(2)}(k_2R)\Big]\, dL'$$

$$+ \frac{\partial}{\partial s} \int_C dL'\Big[\frac{1}{4\omega\varepsilon_1} H_0^{(2)}(k_1R) + \frac{1}{4\omega\varepsilon'_2} H_0^{(2)}(k_2R)\Big] \frac{\partial}{\partial s'} \quad (5.11.15d)$$

where

$\mathcal{K}_{11}$: self integral operator for magnetic currents;

$\mathcal{K}_{22}$: self integral operator for electric currents;

$\mathcal{K}_{12}$: mutual integral operator for electric currents;

$\mathcal{K}_{21}$: mutual integral operator for magnetic currents.

5.12 DIELECTRIC CIRCULAR CYLINDER – TE EXCITATION

In the CFIE, expression (5.11.14), two unknown equivalent surface distributions are given by the axial z component of the magnetic current and the tangential s component of the electric current on the boundary. If these two current distributions are known, then the complete scattered and penetrated fields in the free-space and the dielectric regions can be calculated.

Figure 5.30 shows the geometry of a two-dimensional, lossy dielectric circular cylinder having radius equal to a. The circular scatterer geometry has a perfect symmetry with respect to the angular variable ϕ of the cylindrical coordinate system. Hence, the complete scattering and penetration analysis can be carried out by just assuming the incidence plane wave to be propagating along the x coordinate direction, which gives the angle of incidence equal to zero. The induced axial magnetic current and the tangential electric currents along the circular contour C varies with respect to the angular variable and, in fact, repeats itself every 2π radians. Following the analysis carried over in Section 5.6 for TM excitation, it is possible to treat the TE excitation as a dual case and, thus, interchange various currents, fields, and mediam parameters.

Referring to the discussions in Section 2.19, the TE-polarized time-harmonic incident plane wave is given by

$$H_z^i(\bar{\rho},\omega) = H_0 e^{-j\bar{k}_1 \cdot \bar{\rho}} \quad (5.12.1a)$$

$$\bar{k}_1 = k_1 \hat{k} \quad (5.12.1b)$$

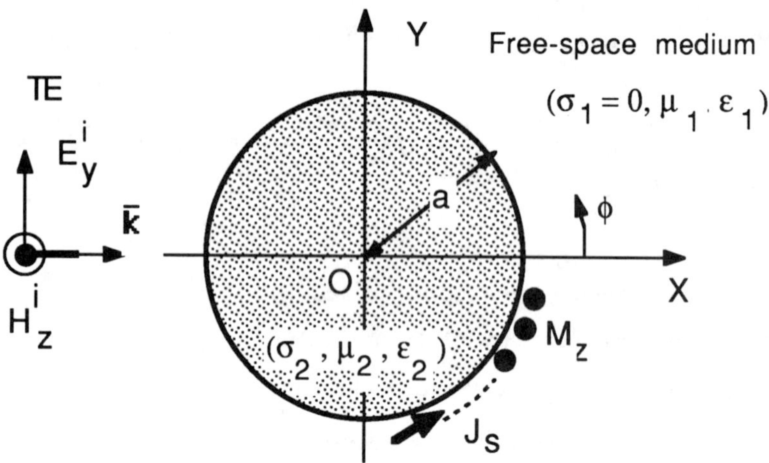

Figure 5.30 Homogeneous lossy dielectric circular scatterer – TE excitation.

k_1 : propagation constant for homogeneous, isotropic, free-space medium 1

$$= \omega(\mu_1 \varepsilon_1)^{1/2} \tag{5.12.1c}$$

ω : frequency of excitation, in radians per second;

ε_1 : permittivity of free-space medium, region 1;

μ_1 : permeability of free-space medium, region 1;

$\phi^i = 0$: angle of incidence of the TE plane wave field.

For the free space-region 1, let

$(\bar{E}^i, \bar{H}^i)$: electric and magnetic incident fields in free-space medium 1;

$(\bar{E}_1^S, \bar{H}_1^S)$: electric and magnetic scattered fields in free-space medium 1;

$(\bar{E}_1, \bar{H}_1)$: electric and magnetic total fields in free-space medium 1

$$= (\bar{E}^i, \bar{H}^i) + (\bar{E}_1^S, \bar{H}_1^S) \tag{5.12.2}$$

As discussed in previous sections, for TE normal excitation, various field distributions are independent of the z coordinate variable. In expression (5.12.2), there are only the axial component of magnetic field and the angular component of electric field distributions. Following the discussion in Section 3.8 concerning the plane wave transformation, the plane wave magnetic field propagating in the x coordinate direction can be written in terms of an infinite series of cylindrical wave modes:

$$H_z^i(\rho,\phi) = H_0\, e^{-jk_1\rho\cos\phi} \tag{5.12.3a}$$

$$= H_0 \sum_{m=-\infty}^{\infty} j^{-m} J_m(k_1\rho)\, e^{jm\phi} \tag{5.12.3b}$$

The scattered magnetic field distribution in the free-space region can be expressed in terms of an infinite number of outgoing propagating cylindrical modes

$$H_{1z}^s(\rho,\phi) = H_0 \sum_{m=-\infty}^{\infty} A_m H_m^{(2)}(k_1\rho)\, e^{jm\phi} \qquad \rho \geq a \tag{5.12.4}$$

A_m : unknown scattered modal coefficients for the free-space region.

Further, for the homogeneous, lossy dielectric scatterer region 2, let

ε_2 : permittivity of dielectric medium, region 2;
μ_2 : permeability of dielectric medium, region 2;
σ_2 : conductivity of dielectric medium, region 2.

The electrical conductivity of the dielectric scatterer can be accounted for by redefining the permittivity in terms of an effective permittivity as in expression (5.4.26)

ε'_2 : effective permittivity of dielectric medium, region 2

$$= \varepsilon_2 \left[1 - j\frac{\sigma_2}{\omega\varepsilon_2}\right] \tag{5.12.5}$$

and

$(\bar{E}_2^s, \bar{H}_2^s)$: electric and magnetic penetrated fields in dielectric medium 2;

$(\bar{E}_2, \bar{H}_2)$: electric and magnetic total fields in dielectric medium 2

$$= (\bar{E}_2^s, \bar{H}_2^s) \tag{5.12.6}$$

Again for TE normal excitation, the penetrated magnetic field in the dielectric region can be expressed in terms of infinite number of nonpropagating cylindrical modes:

$$H^s_{2z}(\rho,\phi) = H_0 \sum_{m=-\infty}^{\infty} B_m J_m(k_2\rho)\, e^{jm\phi} \qquad \rho \le a \tag{5.12.7}$$

B_m : unknown penetrated modal coefficients for the dielectric region;

k_2 : propagation constant for homogeneous, lossy dielectric region

$$= \omega(\mu_2 \varepsilon'_2)^{1/2} \tag{5.12.8}$$

Hence, based on these field representations, the total magnetic field distributions in the free-space medium and the dielectric medium are given by

$$H_{1z}(\rho,\phi)\Big|_{\text{total}} = H^i_{1z}(\rho,\phi) + H^s_{1z}(\rho,\phi) \tag{5.12.9a}$$

$$= H_0 \sum_{m=-\infty}^{\infty} \left[j^{-m} J_m(k_1\rho) + A_m H^{(2)}_m(k_1\rho)\right] e^{jm\phi} \tag{5.12.9b}$$

$$\rho \ge a$$

$$H_{2z}(\rho,\phi)\Big|_{\text{total}} = H^s_{2z}(\rho,\phi) \tag{5.12.9c}$$

$$= H_0 \sum_{m=-\infty}^{\infty} \left[B_m J_m(k_2\rho)\right] e^{jm\phi} \tag{5.12.9d}$$

$$\rho \le a$$

Similarly, the corresponding expressions for the tangential or angular component of the electric field distributions can be written. Referring to Section 3.8.3 and using the source-free Ampere's law stated in expression (3.1.5b):

$$j\omega\varepsilon \bar{E}^s(\rho,\phi,\omega) = \nabla \times \hat{z} H^s_z(\rho,\phi,\omega) \tag{5.12.10a}$$

and expressing the scattered electric and magnetic fields in terms of the cylindrical coordinate system

$$j\omega\varepsilon[E^s_\rho \hat{\rho} + E^s_\phi \hat{\phi}] = \frac{1}{\rho}\frac{\partial}{\partial\phi} H^s_z \hat{\rho} - \frac{\partial}{\partial\rho} H^s_z \hat{\phi} \tag{5.12.10b}$$

Hence, the angular component of the total electric field can be expressed in terms of the incident and scattered magnetic fields in free-space region 1, given by

$$E_{1\phi}(\rho,\phi)\Big|_{\text{total}} = -\frac{1}{j\omega\varepsilon_1}\frac{\partial}{\partial\rho}\left[H^i_{1z}(\rho,\phi) + H^s_{1z}(\rho,\phi)\right] \tag{5.12.11a}$$

$$= -\frac{1}{j\omega\mu_1}\frac{\partial}{\partial\rho}H_0\sum_{m=-\infty}^{\infty}\left[j^{-m}J_m(k_1\rho) + A_m H^{(2)}_m(k_1\rho)\right]e^{jm\phi} \tag{5.12.11b}$$

$$= \frac{jk_1}{\omega\varepsilon_1}H_0\sum_{m=-\infty}^{\infty}\left[j^{-m}J'_m(k_1\rho) + A_m H^{(2)'}_m(k_1\rho)\right]e^{jm\phi} \tag{5.12.11c}$$

$$\rho \geq a$$

Similarly, the angular component of the total electric field can be expressed in terms of the penetrated magnetic field in dielectric region 2, given by

$$E_{2\phi}(\rho,\phi)\Big|_{\text{total}} = -\frac{1}{j\omega\varepsilon'_2}\frac{\partial}{\partial\rho}\left[H^s_{2z}(\rho,\phi)\right] \tag{5.12.12a}$$

$$= -\frac{1}{j\omega\varepsilon'_2}\frac{\partial}{\partial\rho}H_0\sum_{m=-\infty}^{\infty}\left[B_m J_m(k_2\rho)\right]e^{jm\phi} \tag{5.12.12b}$$

$$= \frac{jk_2}{\omega\varepsilon'_2}H_0\sum_{m=-\infty}^{\infty}\left[B_m J'_m(k_2\rho)\right]e^{jm\phi} \tag{5.12.12c}$$

$$\rho \leq a$$

In these total field expressions, the two sets of modal coefficients, A_m and B_m, are still unknown. They can be determined by enforcing electromagnetic boundary conditions (5.5.1a) and (5.5.1b) on the surface of dielectric scatterer. According to the two boundary conditions, the axial component of total magnetic field should be continuous, and similarly, the angular component of total electric field should be continuous on the dielectric boundary contour. Hence

$$H_{1z}(\rho,\phi) = H_{2z}(\rho,\phi) \tag{5.12.13a}$$

$$\rho = a \text{ on } C$$

$$E_{1\phi}(\rho,\phi) = E_{2\phi}(\rho,\phi) \tag{5.12.13b}$$

After substituting the total magnetic field expressions, (5.12.9b) and (5.12.9d), and the total electric field expressions, (5.12.11c) and (5.12.12c), into the preceding boundary conditions,

$$\sum_{m=-\infty}^{\infty} [j^{-m} J_m(k_1 a) + A_m H_m^{(2)}(k_1 a)]\, e^{jm\phi}$$

$$= \sum_{m=-\infty}^{\infty} [B_m J_m(k_2 a)]\, e^{jm\phi} \tag{5.12.14a}$$

$$\frac{k_1}{\varepsilon_1} \sum_{m=-\infty}^{\infty} [j^{-m} J_m'(k_1 a) + A_m H_m^{(2)'}(k_1 a)]\, e^{jm\phi}$$

$$= \frac{k_2}{\varepsilon'_2} \sum_{m=-\infty}^{\infty} [B_m J_m'(k_2 a)]\, e^{jm\phi} \tag{5.12.14b}$$

Now the cylindrical mode orthogonality condition is enforced on these two boundary expressions. Referring back to the discussion in Section 3.7, a set of independent weighting terms are selected to test the two expressions on both sides, where each weighting term forms an entire domain harmonic function in the domain of scatterer contour C. To determine the scattered cylindrical mode coefficients, relationships (5.12.14a) and (5.12.14b) are multiplied on both sides by the entire domain weighting function $exp(-j\ell\phi)$ and integrated between the limits 0 to 2π with respect to ϕ.

Thus, for all integer values $\ell = -\infty$ to ∞, the two boundary condition expressions simplify to the following form:

$$j^{-m} J_m(k_1 a) + A_m H_m^{(2)}(k_1 a) = B_m J_m(k_2 a) \tag{5.12.15a}$$

$$\frac{k_1}{\varepsilon_1} [j^{-m} J_m'(k_1 a) + A_m H_m^{(2)'}(k_1 a)]$$

$$= \frac{k_2}{\varepsilon'_2} [B_m J_m'(k_2 a)] \tag{5.12.15b}$$

After rearranging, the following modal matrix equation is obtained based on the preceding two coupled boundary condition expressions. The two unknown modal coefficients are now obtained by solving the matrix equation:

$$\begin{bmatrix} H_m^{(2)}(k_1 a) & -J_m(k_2 a) \\ \frac{k_1}{\varepsilon_1} H_m^{(2)'}(k_1 a) & -\frac{k_2}{\varepsilon'_2} J_m'(k_2 a) \end{bmatrix} \begin{bmatrix} A_m \\ B_m \end{bmatrix} = \begin{bmatrix} -j^{-m} J_m(k_1 a) \\ -\frac{k_1}{\varepsilon_1} j^{-m} J_m'(k_1 a) \end{bmatrix}$$

(5.12.16)

Once the modal coefficients are known, the axial components of the magnetic field can be calculated based on expressions (5.12.4) and (5.12.7). Similarly, the angular components of the electric field can be calculated based on expressions (5.12.11c) and (5.12.12c). Following the procedure explained in Section 3.8, for free-space region 1, the far-field distribution and the corresponding radar cross section can be calculated by taking the large argument approximation of the Hankel function in expression (5.12.4).

Figures 5.31 and 5.32 show the distribution of the induced surface magnetic current in volts per meter and the induced surface electric current in amperes per meter on the boundary of a lossless circular dielectric scatterer for TE normal excitation. In fact, these two current distributions are conveniently obtained by first calculating the total surface tangential field distributions on the dielectric scatterer and then, converting them into the corresponding equivalent currents based on expressions (5.11.2) and (5.11.3). The conductivity is taken as zero and the relative permittivity and relative permeability are selected as equal to 4 and 2 respectively, for the lossless dielectric circular scatterer. The frequency of excitation is selected as 300 MHz, so that the free-space wavelength is equal to 1 meter. The free-space propagation constant k_1 is equal 2π, and the electrical radius of the circular dielectric scatterer, as referred to free-space medium, is $k_1 a = 5$. The magnetic current is normalized with respect to the incident electric field, and the electric current is normalized with respect to the incident magnetic field. The two current distributions on the lossless dielectric circular cylinder are plotted as a function of angular variable ϕ with the TE incident magnetic field assumed to be 1 ampere per meter. Only the surface current distributions in the angular range of 180° to 360° are shown and, in fact, are symmetrical in the remaining angular range. Specifically the two distributions of induced electric and magnetic currents are symmetrical with respect to the direction of incident angle.

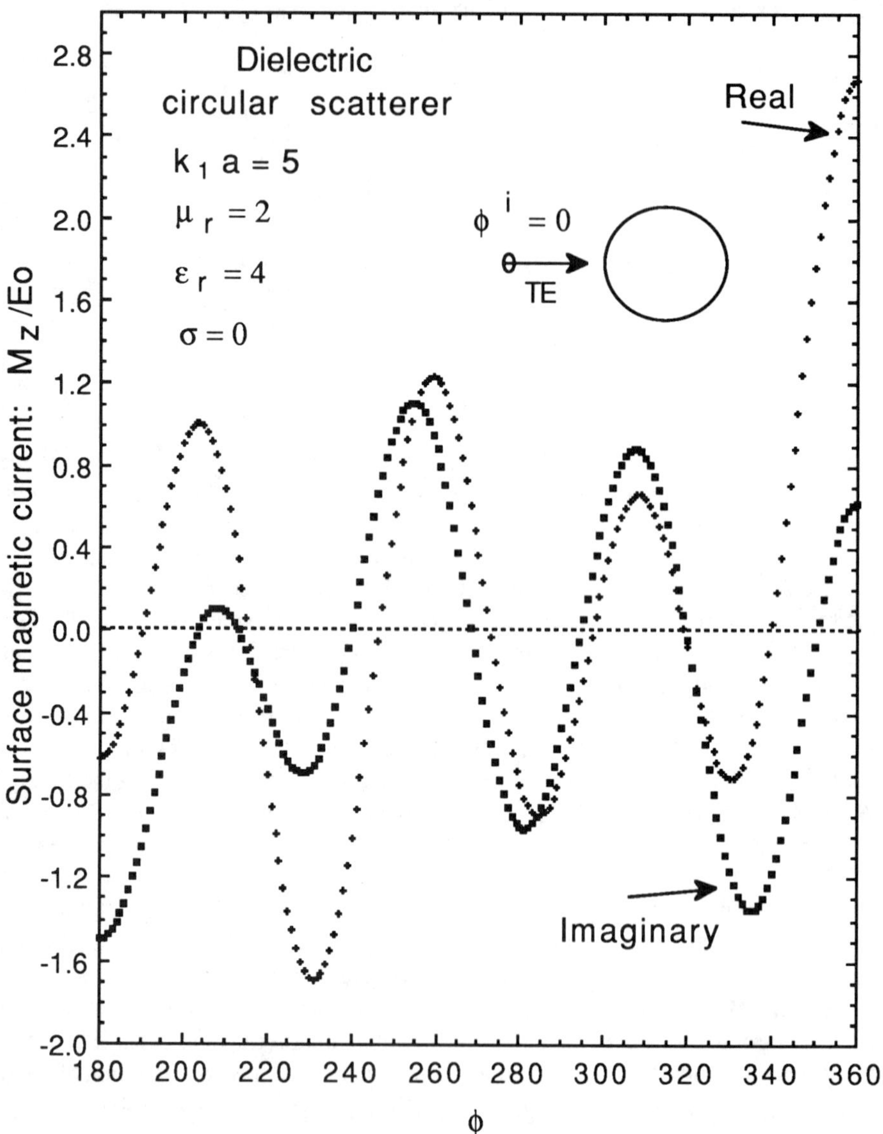

Figure 5.31 Equivalent magnetic current on a lossless circular dielectric scatterer – TE excitation.

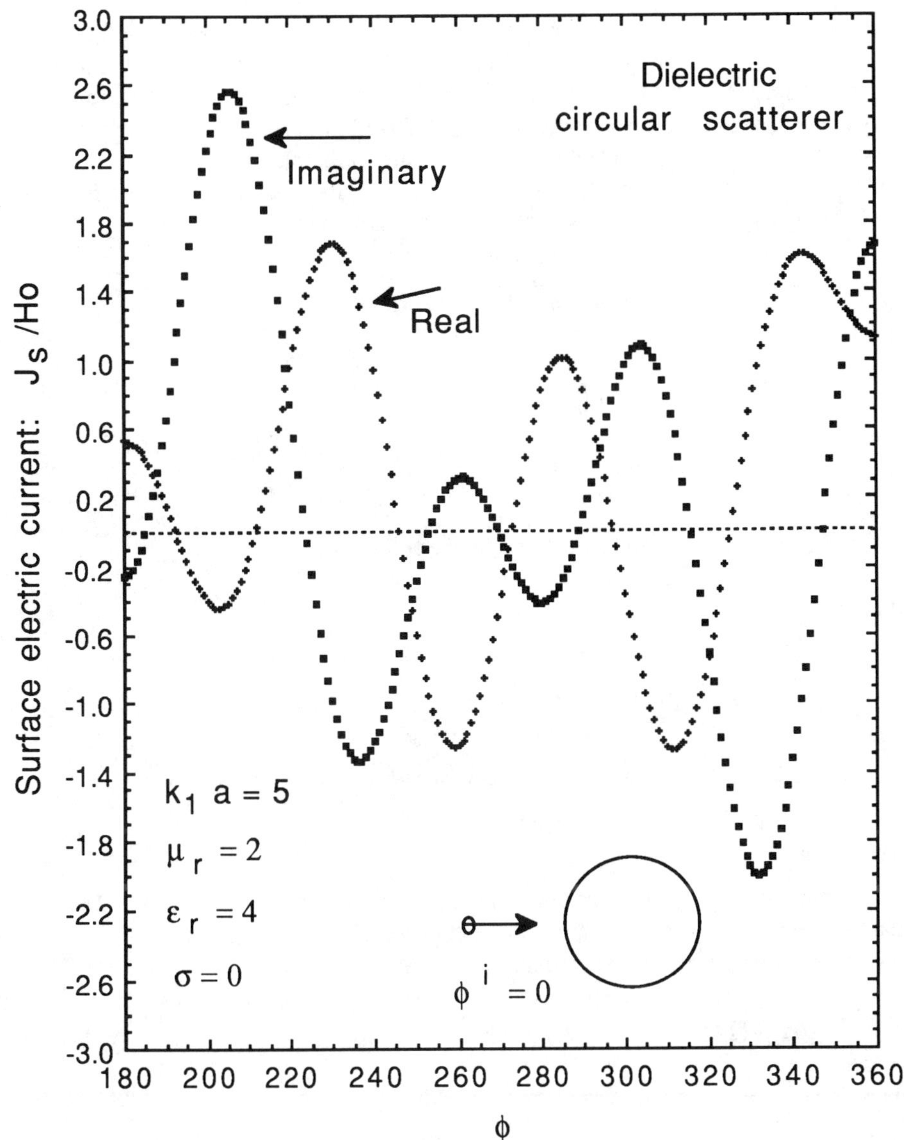

Figure 5.32 Equivalent electric current on a lossless circular dielectric scatterer – TE excitation.

5.12.1 DUALITY OF THE TM AND TE CASES

The numerical studies of the electromagnetic scattering and penetration for TM and TE excitations basically form a dual set. This should be evident by inspecting the various field expressions and the corresponding numerical results. In Section 5.6, a homogeneous, dielectric circular scatterer is analyzed in detail corresponding to the TM excitation. According to the electromagnetic duality principle, these expressions can also be utilized for solutions corresponding to the TE excitation. This is accomplished by interchanging the following field variables and medium parameters

TM case ⇐	⇒ TE case
E_z	H_z
H_ϕ	E_ϕ
ε	μ
μ	ε

The surface angular currents are given by the following:

TM case:

$M_\phi(\rho'=a,\phi')\hat{\phi}'$: angular component of magnetic current distribution

$$= \hat{z}' E_z(\rho'=a,\phi') \times \hat{\rho}' \tag{5.12.17a}$$

TE case:

$J_\phi(\rho'=a,\phi')\hat{\phi}'$: angular component of electric current distribution

$$= \hat{\rho}' \times \hat{z}' H_z(\rho'=a,\phi') \tag{5.12.17b}$$

Thus, numerically the angular component of the magnetic current for the TM case is equal and opposite to the angular component of the electric current for the TE case with appropriate incident field normalization:

$$M_\phi/E_0 \text{ (TM)} \equiv -J_\phi/H_0 \text{ (TE)} \tag{5.12.18}$$

The surface axial currents are given by the following:

TM case:

$J_z(\rho'=a,\phi')\hat{z}'$: axial component of electric current distribution

$$= \hat{\rho}' \times \hat{\phi}' H_\phi(\rho'=a,\phi') \tag{5.12.19a}$$

TE case:

$$M_z(\rho'=a,\phi')\hat{\phi}': \text{ axial component of magnetic current distribution}$$
$$= \hat{\phi}'E_\phi(\rho'=a,\phi') \times \hat{\rho}' \quad (5.12.19b)$$

Following the righthand rule, the angular component of the magnetic field in the TM case and the angular component of the electric field in the TE case are in opposite directions. Thus, numerically the axial component of the electric current for the TM case is equal to the axial component of the magnetic current for the TE case with appropriate incident field normalization:

$$J_z/H_0 \text{ (TM)} \equiv M_z/E_0 \text{ (TE)} \quad (5.12.20)$$

As discussed earlier, for TM excitation, Figures 5.11 and 5.12 show the distribution of the induced surface axial electric current and the induced surface angular magnetic current on the boundary of a lossless dielectric scatterer having the relative permittivity and relative permeability equal to 4 and 2, respectively. Based on the duality, expressions (5.12.18) and (5.12.20), the same results are obtained for the lossless dielectric circular scatterer with TE excitation. The results presented in Figures 5.11 and 5.12 can be interpreted as the distribution of the induced surface axial magnetic current and the induced surface angular electric current (with change in sign) on the boundary of a lossless dielectric scatterer having the relative permittivity and relative permeability equal to 2 and 4, respectively. It should be noted that the medium parameters are interchanged. In fact, this can be verified based on the TE analysis discussed earlier.

Similarly, for TE excitation, Figures 5.31 and 5.32 show the distribution of the induced surface axial magnetic current and the induced surface angular electric current on the boundary of a lossless dielectric scatterer having the relative permittivity and relative permeability equal to 4 and 2, respectively. Based on the duality, expressions (5.12.18) and (5.12.20), the same results are obtained for the lossless dielectric circular scatterer with TM excitation. The results presented in Figures 5.31 and 5.32 can be interpreted as the distribution of the induced surface axial electric current and the induced surface angular magnetic current (with change in sign) on the boundary of lossless dielectric scatterer having the relative permittivity and relative permeability equal to 2 and 4, respectively. Again, it should be noted that the medium parameters are interchanged. In fact, this can be conveniently verified based on the TM analysis discussed in Section 5.6.

5.12.2 SCATTERED FIELD DISTRIBUTIONS

The z component of the scattered magnetic field distribution in the free-space medium is given by general expression (5.12.4). By substituting field observation point (ρ,ϕ), the scattered magnetic field distribution can be calculated. In the numerical calculation, enough number of modes should be included to obtain convergence of the series. Generally, the number of modes required for summation is just more than the electrical distance of the observation point. Thus, for n positive, the upper limit N is taken to be $|k\rho|$ plus additional five to ten modes, depending on the numerical accuracy required.

The scattered magnetic field at any general point is given by expression (5.12.4), and is repeated here:

$$H^s_{1z}(\rho,\phi) = H_0 \sum_{m=-\infty}^{\infty} A_m H_m^{(2)}(k_1\rho)\, e^{jm\phi} \qquad \rho \ge a \tag{5.12.21a}$$

$$H^s_{2z}(\rho,\phi) = H_0 \sum_{m=-\infty}^{\infty} B_m J_m(k_2\rho)\, e^{jm\phi} \qquad \rho \le a \tag{5.12.21b}$$

where the scattered and the penetrated field modal coefficients, A_m and B_m, are obtained by solving modal matrix equation (5.12.16). Further, based on Ampere's law, expression (5.12.10b), the angular and radial components of the electric field distribution can be calculated in terms of the axial components of the magnetic field:

$$E^s_{1\phi}(\rho,\phi) = \frac{jk_1}{\omega\varepsilon_1} H_0 \sum_{m=-\infty}^{\infty} A_m H_m^{(2)'}(k_1\rho)\, e^{jm\phi} \tag{5.12.22a}$$

$$E^s_{1\rho}(\rho,\phi) = -\frac{j}{\omega\varepsilon_1}\frac{1}{\rho} H_0 \sum_{m=-\infty}^{\infty} jmA_m H_m^{(2)}(k_1\rho)\, e^{jm\phi} \tag{5.12.22b}$$

$$\rho \ge a$$

$$E^s_{2\phi}(\rho,\phi) = \frac{jk_2}{\omega\varepsilon_2} H_0 \sum_{m=-\infty}^{\infty} B_m H_m^{(2)'}(k_2\rho)\, e^{jm\phi} \tag{5.12.23a}$$

$$E^s_{2\rho}(\rho,\phi) = -\frac{j}{\omega\varepsilon_2}\frac{1}{\rho} H_0 \sum_{m=-\infty}^{\infty} jmB_m H_m^{(2)}(k_2\rho)\, e^{jm\phi} \tag{5.12.23b}$$

$$\rho \le a$$

To obtain the total field distributions, the corresponding incident field terms are added to the preceding scattered field expressions. In fact, duality can be easily invoked even for the near total fields. Hence, for the TE excitation, Figure 5.13 shows distribution of the z component of the total magnetic field in the neighborhood of dielectric circular scatterer having permittivity and permeability equal to 2 and 4, respectively. Both the penetrated magnetic field and the near total magnetic field are displayed in the Figure for about a tenth of the free-space wavelength outside and a tenth of the dielectric region wavelength inside the lossless dielectric scatterer. As can be seen, the distribution of the z component of total magnetic field is continuous across the boundary surface. The numerical data shown are obtained for fixed angles $\phi = 0$, 30, 60, 90, 120, 150, and 180, and the radial variable ρ is varied from a point near surface in the small increments. Similarly, based on the duality, Figure 5.14 shows distribution of the angular component of the total electric field, which also exhibits continuous behavior at the boundary surface. Further, Figure 5.15 shows distribution of the radial component of the total electric field. This distribution is clearly discontinuous at the boundary surface. By multiplying with the respective region permittivity parameter, the distribution of the radial component of electric flux density is continuous across the boundary surface.

Far-Field and Radar Cross Section

The z component of the scattered field distribution in the far-field region is obtained by substituting for the argument $|k_1\rho| \rightarrow \infty$. In the large argument approximation, Appendix B, the Hankel function can be written as

$$H_m^{(2)}(k_1\rho) \sim \sqrt{\frac{2j}{\pi k_1\rho}}\, j^m\, e^{-jk_1\rho} \tag{5.12.24a}$$

After substituting the large argument approximation given in expression (5.12.24a) into infinite series solution (5.12.21a), the scattered field in the far-field region reduces to

$$H_{1z}^s(\rho,\phi,\omega) \sim H_0 \sqrt{\frac{2j}{\pi k_1\rho}}\, e^{-jk_1\rho} \sum_{m=-\infty}^{\infty} A_m j^m\, e^{jm\phi} \tag{5.12.24b}$$

and the magnitude of the far-field pattern is given by, for $k_1\rho \rightarrow \infty$,

$$\left|\frac{H_{1z}^s(\phi,\omega)}{H_z^i(\phi,\omega)}\right| = \sqrt{\frac{2}{\pi k_1\rho}} \left|\sum_{m=-\infty}^{\infty} A_m j^m\, e^{jm\phi}\right| \tag{5.6.25}$$

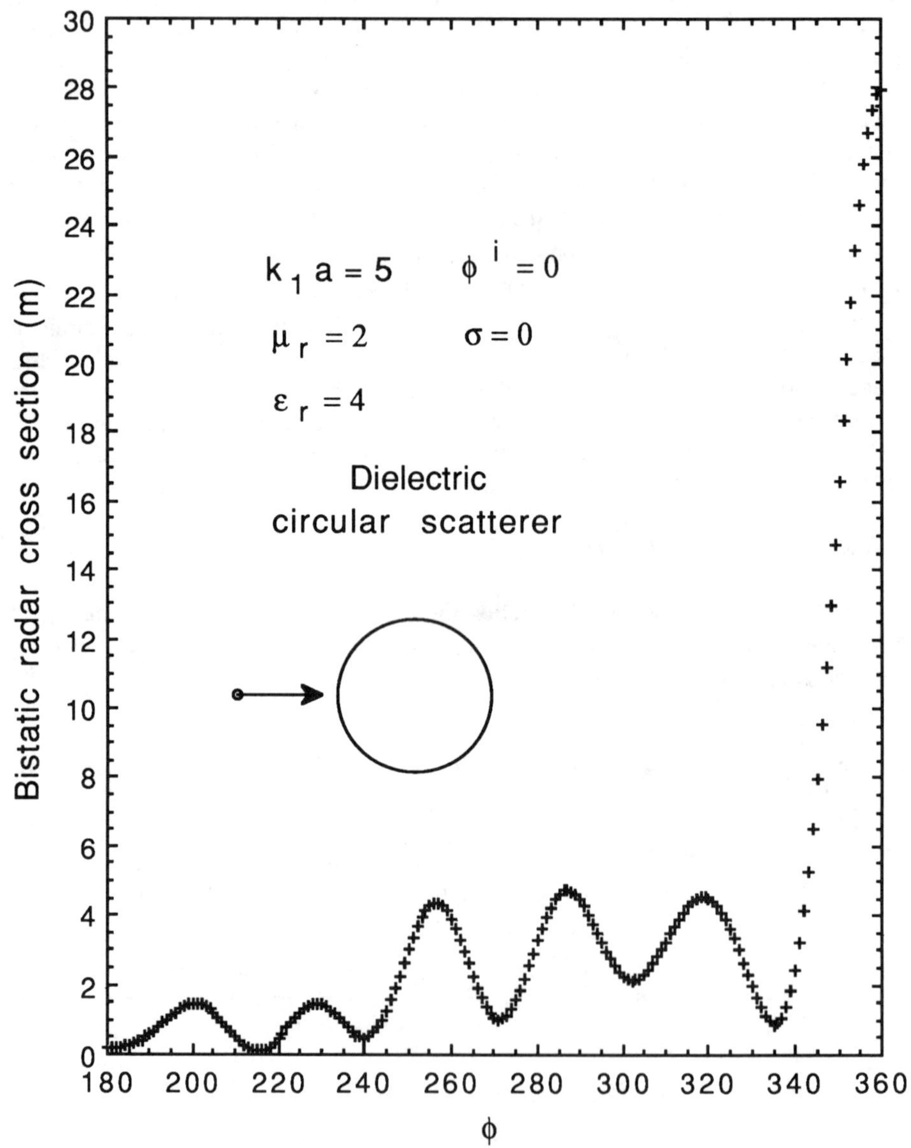

Figure 5.33 Bistatic radar cross section of the circular dielectric scatterer – TE excitation.

Using expression (3.8.18), the bistatic or the monostatic radar cross section of the homogeneous dielectric scatterer can be calculated. The bistatic radar cross section for a lossless dielectric scatterer with $k_1a = 5$, having relative permittivity 4 and relative permeability 2 is shown in Figure 5.16 for TM normal excitation. This result is also valid for TE excitation for the same size circular scatterer having relative permittivity 2 and relative permeability 4. Further, Figure 5.33 shows the bistatic radar cross section of the same lossless dielectric scatterer $k_1a = 5$ and having relative permittivity 4 and relative permeability 2 for TE normal excitation.

5.13 ARBITRARY CROSS SECTION - CFIE - TE EXCITATION

The electromagnetic scattering and penetration by a homogeneous, lossy dielectric circular cylinder is discussed in the previous section based on a rigorous analytical approach. In a two-dimensional, homogeneous, lossy dielectric scatterer with an arbitrary cross section having arbitrary edges and corners, the analysis technique is quite involved. In Section 5.7, a detailed numerical analysis was discussed for the case of TM excitation. The same numerical analysis can be also adopted for the TE excitation. This requires an appropriate modification to treat surface currents, or the duality principle discussed earlier can be invoked to derive the numerical results. In the following, the approximate analysis procedure is summarized to obtain a numerical solution for the electromagnetic scattering and penetration by solving CFIE (5.11.14a–b).

Let us consider the geometry of a twodimensional, homogeneous, lossy dielectric scatterer having an arbitrary cross section and placed in a linear, homogeneous, and isotropic lossless medium, shown in Figure 5.34. The scatterer is uniform along its axis having the same arbitrary cross section. The z coordinate axis of the cylindrical coordinate system (ρ, ϕ, z) or the generalized cylindrical coordinate system (n, s, z) coincide with the axis of the scatterer. The external incident plane wave field is assumed to be transverse electric polarized with respect to the z coordinate axis of the scatterer and propagates with its direction of propagation normal to the z axis. The angle ϕ^i represents the incident angle the direction of propagation makes with the x coordinate axis. As discussed earlier, due to the normal excitation, the various field quantities are independent of the z coordinate variation. Referring to Figure 5.34, the TE to z polarized incident plane wave field is given by

$$H_z^i(\bar{\rho},\omega) = H_0\, e^{-j\bar{k}_1 \cdot \bar{\rho}} \tag{5.13.1}$$

ϕ^i : arbitrary angle of incidence of the TE plane wave field

Hence, for the case of a plane wave propagating normal to the axis of the scatterer

$$H_z^i(\rho,\phi) = H_0\, e^{-jk_1\rho\, \cos(\phi - \phi^i)} \tag{5.13.2}$$

H_0 : amplitude of the incident plane wave magnetic field, in amperes per meter.

The corresponding transverse component of incident electric field in the transverse plane is given by

$$\bar{E}_\tau^i(\rho,\phi) = [\, - \sin\phi^i \hat{x} + \cos\phi^i \hat{y}]\, H_0\, e^{-jk_1\rho\, \cos(\phi - \phi^i)} \tag{5.13.3}$$

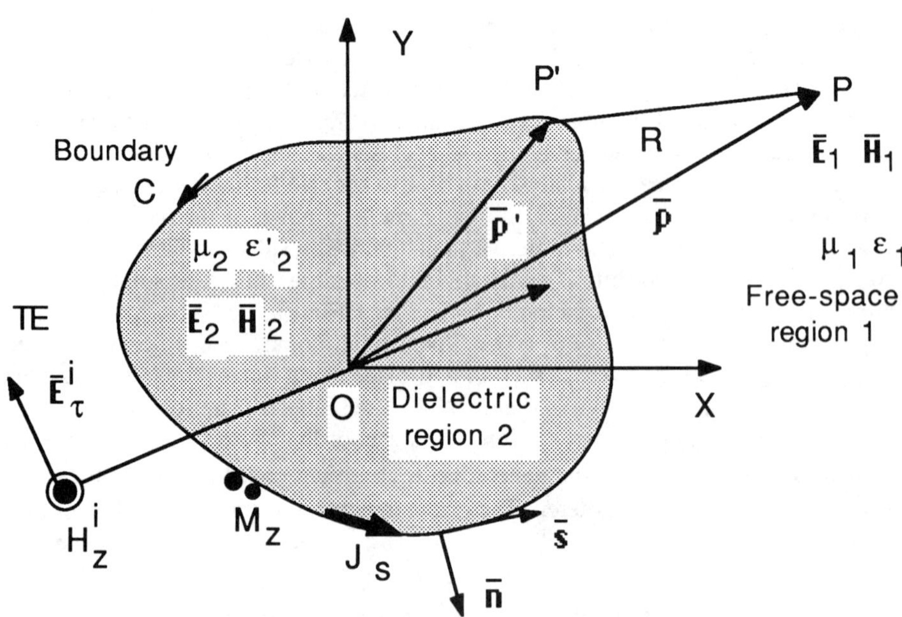

Figure 5.34 Geometry of arbitrary shaped, homogeneous, lossy dielectric scatterer, – TE excitation.

Referring to Figure 5.34, in terms of the generalized cylindrical coordinate system *(n, s, z)*, the normal and the tangential unit vectors corresponding to the field points on the surface of the dielectric boundary can be written as

$$\hat{n} = \hat{x}\, \cos\Omega + \hat{y}\, \sin\Omega \tag{5.13.4a}$$

$$\hat{s} = -\hat{x}\sin\Omega + \hat{y}\cos\Omega \tag{5.13.4b}$$

Ω : angle the normal at field point makes with the x coordinate axis.

The component of the incident electric field tangential to the dielectric boundary is obtained by taking scalar product of expression (5.13.3) with respect to the tangential unit vector defined in expression (5.13.4b)

$$E_s^i(\rho,\phi) = \hat{s} \bullet \vec{E}_\tau^i(\rho,\phi)$$

$$= E_0 \cos(\Omega - \phi^i)\, e^{-jk_1\rho\cos(\phi - \phi^i)} \tag{5.13.5a}$$

E_0 : amplitude of incident plane wave electric field, in volts per meter

$$= \eta_1 H_0 \tag{5.13.5b}$$

η_1 : intrinsic impedance of homogeneous, isotropic free-space medium.

In Section 5.11, a complete and systematic derivation of coupled CFIE (5.11.14a–b) for the *TE normal excitation* was presented. In these two coupled equations, the axial component of the magnetic current and the tangential component of the electric current distributions on boundary contour C are unknown and to be solved.

The two coupled CFIE for TE case are repeated here:

$$H_z^i(\bar{\rho}) = \int_C M_z(\bar{\rho}')K_{11}(\bar{\rho}, \bar{\rho}')\, dL' + \fint_C -J_s(\bar{\rho}')K_{12}(\bar{\rho}, \bar{\rho}')\, dL' \tag{5.13.6a}$$

$$-E_s^i(\bar{\rho}) = \fint_C M_z(\bar{\rho}')K_{21}(\bar{\rho}, \bar{\rho}')\, dL' + \int_C -J_s(\bar{\rho}')K_{22}(\bar{\rho}, \bar{\rho}')\, dL'$$

$$+ \frac{\partial}{\partial s}\int_C \left[-\frac{\partial}{\partial s'}J_s(\bar{\rho}')\right]K'_{22}(\bar{\rho}, \bar{\rho}')\, dL' \tag{5.13.6b}$$

It is interesting to compare this set of coupled equations with the CFIE for TM case given by expressions (5.7.6a–b). They form dual sets of integral equations. By comparing the two sets of equations, the TE-CFIE (5.13.6a–b) can be obtained directly from the TM-

CFIE (5.7.6a–b) by replacing the axial component of the TM incident electric field by the axial component of the TE incident magnetic field and by replacing the tangential component of the TM incident magnetic field by the tangential component of the TE incident electric field with a change in sign. Similarly, the unknown distributions are replaced by the axial component of the TE magnetic current for the axial component of the TM electric current; and the tangential component of the TE electric current (with change in sign) for the tangential component of the TM magnetic current. The duality is summarized in the following:

Duality for Homogeneous, Lossy Dielectric Scatterers

TM case $\Leftarrow$	$\Rightarrow$ TE case
E_z^i	H_z^i
H_s^i	$-E_s^i$
M_s	$-J_s$
J_z	M_z
ε'	μ
μ	ε'

Integral equations (5.13.6a–b) are valid for the field point $\bar{\rho}$ on C. The various composite self and mutual kernel functions in the integrands of the coupled combined field integral equations are given by the following expressions.

Again, duality can be verified by comparing with respect to expressions (5.7.7a–e). The medium permeability is to be replaced by permittivity, and similarly, the medium permittivity is to be replaced by the permeability.

$$K_{11}(\bar{\rho}, \bar{\rho}') = \frac{\omega\varepsilon_1}{4} H_0^{(2)}(k_1R) + \frac{\omega\varepsilon'_2}{4} H_0^{(2)}(k_2R) \tag{5.13.7a}$$

$$K_{12}(\bar{\rho}, \bar{\rho}') = \frac{k_1}{4j}\left[\frac{x-x'}{R}\cos\Omega' + \frac{y-y'}{R}\sin\Omega'\right] H_0^{(2)'}(k_1R)$$

$$+ \frac{k_2}{4j}\left[\frac{x-x'}{R}\cos\Omega' + \frac{y-y'}{R}\sin\Omega'\right] H_0^{(2)'}(k_2R) \tag{5.13.7b}$$

$$K_{21}(\bar{\rho}, \bar{\rho}') = \frac{k_1}{4j}\left[\frac{x-x'}{R}\cos\Omega + \frac{y-y'}{R}\sin\Omega\right] H_0^{(2)'}(k_1R)$$

$$+ \frac{k_2}{4j}\left[\frac{x-x'}{R}\cos\Omega + \frac{y-y'}{R}\sin\Omega\right] H_0^{(2)'}(k_2R) \qquad (5.13.7c)$$

$$K_{22}(\bar{\rho}, \bar{\rho}') = \frac{\omega\mu_1}{4}\cos(\Omega - \Omega')H_0^{(2)}(k_1R)$$

$$+ \frac{\omega\mu_2}{4}\cos(\Omega - \Omega')H_0^{(2)}(k_2R) \qquad (5.13.7d)$$

$$K'_{22}(\bar{\rho}, \bar{\rho}') = \frac{1}{4\omega\varepsilon_1} H_0^{(2)}(k_1R) + \frac{1}{4\omega\varepsilon'_2} H_0^{(2)}(k_2R) \qquad (5.13.7e)$$

where

$$R = |\bar{\rho} - \bar{\rho}'|$$
$$= [(x - x')^2 + (y - y')^2]^{1/2} \qquad (5.13.7f)$$

5.14 PARTITIONED MATRIX EQUATION

To solve coupled combined field integral equations (5.13.6a–b), the method of moments numerical technique can applied. This is accomplished by a suitable selection of a set of independent, piecewise current expansion functions and a corresponding set of independent, piecewise weighting functions. Complete details of the weighting and expansion functions are discussed in Section 5.7 for TM case. Based on the duality discussed, the same weighting and expansion functions are utilized to solve the CFIE for TE excitation. Referring to expression (5.8.14) and applying duality, the following partitioned matrix equation is obtained:

$$\begin{bmatrix} Y_{mn}^{MM} & Y_{mp}^{MJ} \\ Z_{qn}^{JM} & Z_{qp}^{JJ} \end{bmatrix} \begin{bmatrix} I_n \\ T_p \end{bmatrix} = \begin{bmatrix} H_m \\ E_q \end{bmatrix} \qquad (5.14.1)$$

where referring to partial matrix equation (5.8.11b) and (5.8.13b):

Y_{mn}^{MM} : generalized self partitioned admittance matrix;

Y_{mp}^{MJ} : generalized mutual partitioned admittance matrix;

Z_{qn}^{JM} : generalized mutual partitioned impedance matrix;

Z_{qp}^{JJ} : generalized self partitioned impedance matrix.

The elements of the generalized matrix are listed in the following:

Y_{mn}^{MM} : generalized self partitioned admittance matrix

$$= \Delta_{2m} \frac{\omega\varepsilon_1}{4} \int_{-\Delta_{2n}/2}^{\Delta_{2n}/2} H_0^{(2)}(k_1 R_1)\, d\zeta$$

$$+ \Delta_{2m} \frac{\omega\varepsilon'_2}{4} \int_{-\Delta_{2n}/2}^{\Delta_{2n}/2} H_0^{(2)}(k_2 R_1)\, d\zeta \qquad (5.14.2a)$$

$$R_1 = \left| \bar{\rho}_{2m} - (\bar{\rho}_{2n} + \hat{s}_{2n}\zeta) \right| \qquad (5.14.2b)$$

$$= \left\{ [(x_{2m} - x_{2n}) - \cos\alpha_{2n}\zeta]^2 + [(y_{2m} - y_{2n}) - \cos\beta_{2n}\zeta]^2 \right\}^{1/2} \qquad (5.14.2c)$$

$$\hat{s}_{2n} = \hat{x} \cos\alpha_{2n} + \hat{y} \cos\beta_{2n} \qquad (5.14.2d)$$

$m = 1, 2, 3, \quad . \ . \ . \ . \ , M$

$n = 1, 2, 3, \quad . \ . \ . \ . \ , N$

Y_{mp}^{MJ} : generalized mutual partitioned admittance matrix

$$= \Delta_{2m} \int_{-\Delta_{2p-2}/2}^{0} \Big\{ [(x_{2m} - x_{2p-1}) - \cos\alpha_{2p-2}\zeta] \frac{\cos\Omega_{2p-2}}{R_2} + [(y_{2m} - y_{2p-1}) - \cos\beta_{2p-2}\zeta] \frac{\sin\Omega_{2p-2}}{R_2} \Big\} \Big[\frac{k_1}{4j} H_0^{(2)'}(k_1R_2) + \frac{k_2}{4j} H_0^{(2)'}(k_2R_2) \Big] d\zeta$$

$$+ \Delta_{2m} \int_{0}^{\Delta_{2p}/2} \Big\{ [(x_{2m} - x_{2p-1}) - \cos\alpha_{2p}\zeta] \frac{\cos\Omega_{2p}}{R_3} + [(y_{2m} - y_{2p-1}) - \cos\beta_{2p}\zeta] \frac{\sin\Omega_{2p}}{R_3} \Big\} \Big[\frac{k_1}{4j} H_0^{(2)'}(k_1R_3) + \frac{k_2}{4j} H_0^{(2)'}(k_2R_3) \Big] d\zeta \quad (5.14.3a)$$

$$R_2 = \left| \bar{\rho}_{2m} - (\bar{\rho}_{2p-1} + \hat{s}_{2p-2}\zeta) \right| \quad (5.14.3b)$$

$$= \Big\{ [(x_{2m} - x_{2p-1}) - \cos\alpha_{2p-2}\zeta]^2 + [(y_{2m} - y_{2p-1}) - \cos\beta_{2p-2}\zeta]^2 \Big\}^{1/2} \quad (5.14.3c)$$

$$\hat{s}_{2p-2} = \hat{x} \cos\alpha_{2p-2} + \hat{y} \cos\beta_{2p-2} \quad (5.14.3d)$$

$$R_3 = \left| \bar{\rho}_{2m} - (\bar{\rho}_{2p-1} + \hat{s}_{2p}\zeta) \right| \quad (5.14.3e)$$

$$= \Big\{ [(x_{2m} - x_{2p-1}) - \cos\alpha_{2p}\zeta]^2 + [(y_{2m} - y_{2p-1}) - \cos\beta_{2p}\zeta]^2 \Big\}^{1/2} \quad (5.14.3f)$$

$$\hat{s}_{2p} = \hat{x} \cos\alpha_{2p} + \hat{y} \cos\beta_{2p} \quad (5.14.3g)$$

$m = 1, 2, 3, \ldots, M$

$p = 1, 2, 3, \ldots, P$

Z^{JM}_{qn} : generalized mutual partitioned impedance matrix

$$= \frac{\Delta_{2q-2}}{2} \int_{-\Delta_{2n}/2}^{\Delta_{2n}/2} \Big\{ [(x_{2q-1} - x_{2n}) - \cos\alpha_{2n}\zeta] \frac{\cos \Omega_{2q-2}}{R_4} + [(y_{2q-1} - y_{2n}) - \cos\beta_{2n}\zeta] \frac{\sin \Omega_{2q-2}}{R_4} \Big\} \Big[\frac{k_1}{4j} H_0^{(2)'}(k_1R_4) + \frac{k_2}{4j} H_0^{(2)'}(k_2R_4) \Big] d\zeta$$

$$+ \frac{\Delta_{2q}}{2} \int_{-\Delta_{2n}/2}^{\Delta_{2n}/2} \Big\{ [(x_{2q-1} - x_{2n}) - \cos\alpha_{2n}\zeta] \frac{\cos \Omega_{2q}}{R_4} + [(y_{2q-1} - y_{2n}) - \cos\beta_{2n}\zeta] \frac{\sin \Omega_{2q}}{R_4} \Big\} \Big[\frac{k_1}{4j} H_0^{(2)'}(k_1R_4) + \frac{k_2}{4j} H_0^{(2)'}(k_2R_4) \Big] d\zeta \tag{5.14.4}$$

$q = 1, 2, 3, \ldots, Q$

$n = 1, 2, 3, \ldots, N$

and

Z^{JJ}_{qp} : generalized self partitioned impedance matrix

$$= \frac{\Delta_{2q-2}}{2} \int_{-\Delta_{2p-2}/2}^{0} \cos(\Omega_{2q-2} - \Omega'_{2p-2}) \Big[\frac{\omega\varepsilon_1}{4} H_0^{(2)}(k_1R_5) + \frac{\omega\varepsilon'_2}{4} H_0^{(2)}(k_2R_5) \Big] d\zeta$$

$$+ \frac{\Delta_{2q-2}}{2} \int_{0}^{\Delta_{2p}/2} \cos(\Omega_{2q-2} - \Omega'_{2p})$$

$$\left[\frac{\omega\varepsilon_1}{4} H_0^{(2)}(k_1R_6) + \frac{\omega\varepsilon'_2}{4} H_0^{(2)}(k_2R_6)\right] d\zeta$$

$$+ \frac{\Delta_{2q}}{2} \int_{-\Delta_{2p-2}/2}^{0} \cos(\Omega_{2q} - \Omega'_{2p-2})$$

$$\left[\frac{\omega\varepsilon_1}{4} H_0^{(2)}(k_1R_5) + \frac{\omega\varepsilon'_2}{4} H_0^{(2)}(k_2R_5)\right] d\zeta$$

$$+ \frac{\Delta_{2q}}{2} \int_{0}^{\Delta_{2p}/2} \cos(\Omega_{2q} - \Omega'_{2p})$$

$$\left[\frac{\omega\varepsilon_1}{4} H_0^{(2)}(k_1R_6) + \frac{\omega\varepsilon'_2}{4} H_0^{(2)}(k_2R_6)\right] d\zeta$$

$$- \frac{1}{\Delta_{2p-2}} \int_{-\Delta_{2p-2}/2}^{\Delta_{2p-2}/2} \left[\frac{1}{4\omega\mu_1} H_0^{(2)}(k_1R_7) + \frac{1}{4\omega\mu_2} H_0^{(2)}(k_2R_7)\right] d\zeta$$

$$+ \frac{1}{\Delta_{2p}} \int_{-\Delta_{2p}/2}^{\Delta_{2p}/2} \left[\frac{1}{4\omega\mu_1} H_0^{(2)}(k_1R_8) + \frac{1}{4\omega\mu_2} H_0^{(2)}(k_2R_8)\right] d\zeta$$

$$+ \frac{1}{\Delta_{2p-2}} \int_{-\Delta_{2p-2}/2}^{\Delta_{2p-2}/2} \left[\frac{1}{4\omega\mu_1} H_0^{(2)}(k_1R_9) + \frac{1}{4\omega\mu_2} H_0^{(2)}(k_2R_9)\right] d\zeta$$

$$- \frac{1}{\Delta_{2p}} \int_{-\Delta_{2p}/2}^{\Delta_{2p}/2} \left[\frac{1}{4\omega\mu_1} H_0^{(2)}(k_1 R_{10}) + \frac{1}{4\omega\mu_2} H_0^{(2)}(k_2 R_{10}) \right] d\zeta \qquad (5.14.5)$$

$$q = 1, 2, 3, \ . \ . \ . \ , Q$$

$$p = 1, 2, 3, \ . \ . \ . \ , P$$

Further, the two excitation terms for the TE polarized plane wave propagating with an angle of incidence ϕ^i are given by

$$H_m = \Delta_{2m} H_0 \, e^{-jk_1 \rho_{2m} \cos(\phi_{2m} - \phi^i)} \qquad (5.14.6a)$$

and

$$E_q = \frac{\Delta_{2q-2}}{2} E_0 \cos(\Omega_{2q-2} - \phi^i) \, e^{-jk_1 \rho_{2q-1} \cos(\phi_{2q-1} - \phi^i)}$$

$$+ \frac{\Delta_{2q}}{2} E_0 \cos(\Omega_{2q} - \phi^i) \, e^{-jk_1 \rho_{2q-1} \cos(\phi_{2q-1} - \phi^i)} \qquad (5.14.6b)$$

The unknown electric and magnetic current distributions are obtained by inverting the generalized partitioned matrix equation (5.14.1).

5.15 NUMERICAL RESULTS – TE CASE

Based on the numerical solution for the TE-CFIE discussed in the previous section, a number of canonical two-dimensional, homogeneous, dielectric scattering geometries are analyzed to determine their electromagnetic scattering and penetration properties. A rigorous computer algorithm can now be developed to solve the coupled partitioned matrix equation obtained in expression (5.14.1). It should be noted again that the computer algorithm based on matrix equation (5.8.14) for TM excitation can be directly utilized for TE excitation by invoking the duality principle.

The electromagnetic scattering by a dielectric square cylinder is considered in Figures 5.35a and 5.35b. Each sidelength of the dielectric square cylinder is selected to be L = 0.25 meter and has a permittivity $\varepsilon_2 = 2\varepsilon_0$. Further, each sidelength of the dielectric square cylinder is divided into twenty four equal segments so that the total number of

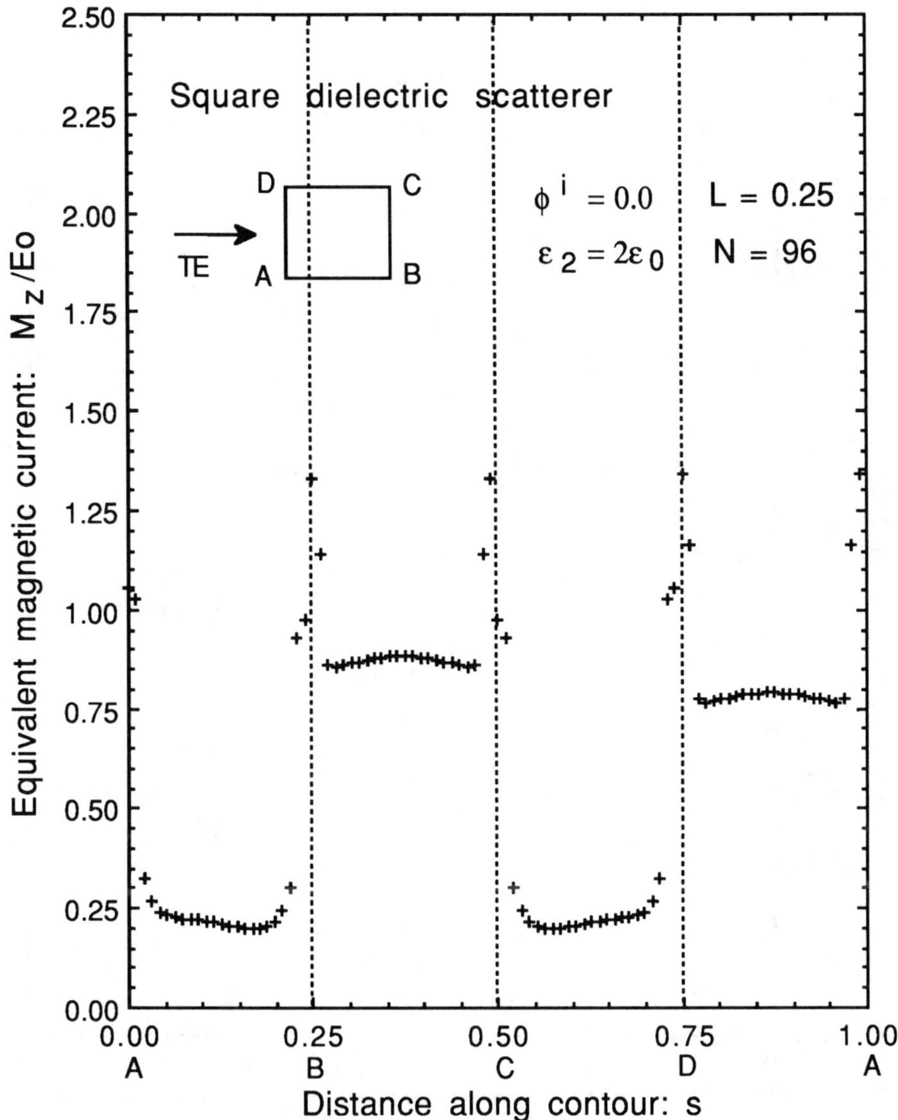

Figure 5.35a Magnitude distribution of the equivalent magnetic current on a square dielectric scatterer – TE excitation.

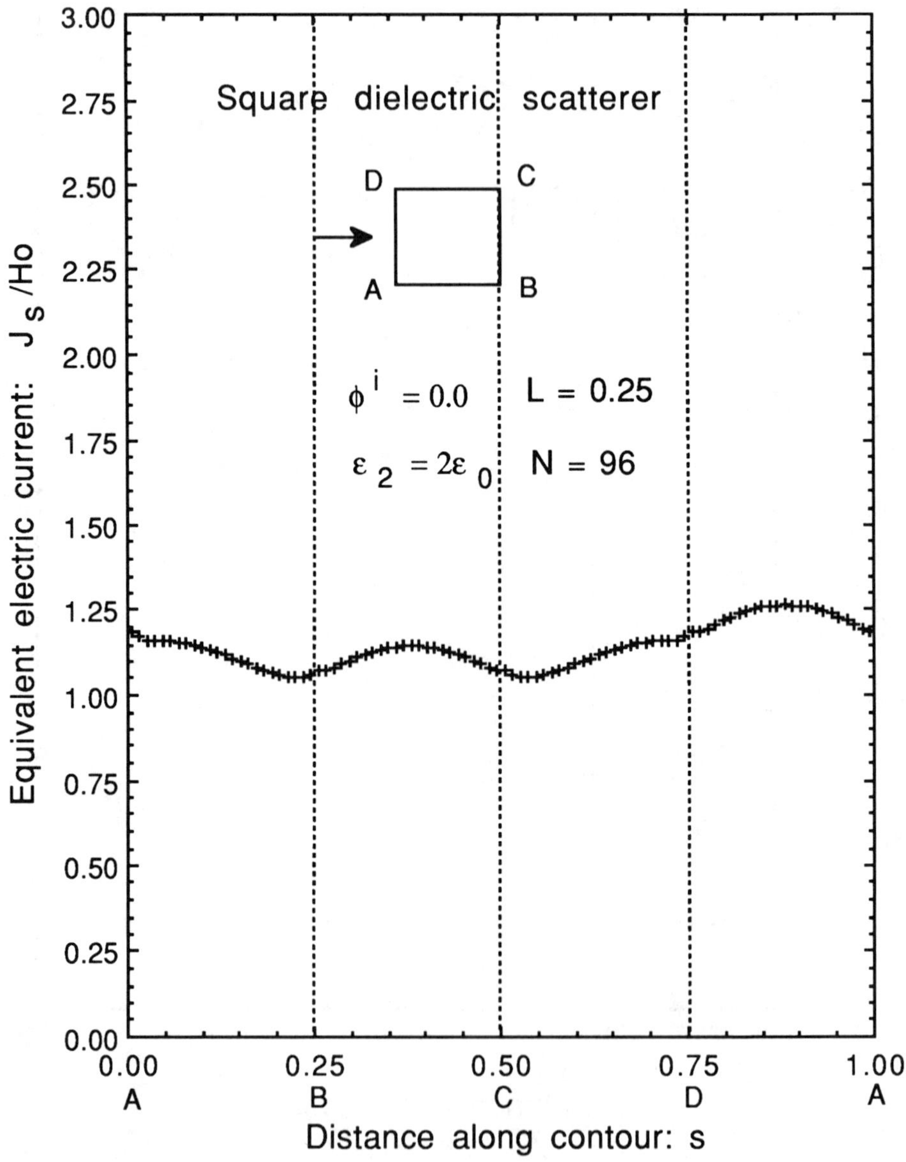

Figure 5.35b Magnitude distribution of the equivalent electric current on a square dielectric scatterer – TE excitation.

unknown magnetic current pulses is selected as $N = 96$. Similarly, the total number of unknown electric current pulses is selected as $P = 96$. As discussed earlier, the various pulse expansion terms are staggered and arranged along the square cylinder with the even points representing the midlocations of the magnetic current pulses and the corresponding odd points representing the midlocations of the electric current pulses, as in Figure 5.24c for the TM case. The frequency of excitation is selected as 300 MHz, and the corresponding free-space wavelength is 1 meter and the free-space propagation constant $k_1 = 2\pi$. The dielectric scatterer is excited at an angle of incidence $\phi^i = 0^o$. For the TE polarization, the incident magnetic field is polarized parallel to the z axis and the corresponding incident electric field is polarized parallel to the y axis. Figures 5.35a and 5.35b show the results of the distribution of equivalent magnetic and equivalent electric currents. Based on the duality, expressions (5.12.19c) and (5.12.20c), the same results are obtained for the lossless dielectric scatterer with TM excitation. The results presented in Figures 5.35a and 5.35b can be interpreted as the distribution of the induced axial electric current and the induced angular magnetic current (with a change in sign) on the boundary of a lossless dielectric scatterer having the relative permittivity and relative permeability equal to 1 and 2, respectively. Again, it should be noted that the medium parameters are interchanged. In fact, this can be conveniently verified based on the TM analysis discussed in Section 5.8. The results of the equivalent magnetic current and the electric current distributions can now be utilized to calculate the near-field and the far-field distributions. A brief numerical procedure is presented concerning the calculation of the far-field distribution.

5.15.1 FAR-FIELD DISTRIBUTION

The electric and magnetic field distributions in the far-field region can be obtained by using the large argument approximation for the free-space Green's function. For TE polarization, in the far-field region, there are only the z component of the scattered magnetic field distribution and the corresponding ϕ component of the scattered electric field distribution. It should be clearly noted, in the far-field region, the scattered electric and magnetic fields behave as pure *cylindrical waves* propagating in the radial direction, and further, they are directly related through the medium intrinsic impedance. Referring to Section 5.10 and invoking duality, the following expressions are obtained for TE far-field distributions.

Due to the equivalent magnetic current, the partial z component of the magnetic field in the far-field region is given by

$$H^S_{1z}(\bar{\rho}) \Big|_{\text{due to } M_z} = -j\omega\hat{z} \bullet \bar{F}_1(\bar{\rho},\omega) \tag{5.15.1}$$

and due to the equivalent electric current, the partial z component of the magnetic field in the far-field region is given by

$$H^{S}_{1z}(\bar{\rho})\Big|_{\text{due to } J_S} = \frac{E^{S}_{1\phi}(\bar{\rho},\omega)}{\eta_1} \tag{5.15.2a}$$

and

$$E^{S}_{1\phi}(\bar{\rho}) = -\,j\omega\hat{\phi}\bullet\bar{A}_1(\bar{\rho},\omega) \tag{5.15.2b}$$

On summing expressions (5.15.1) and (5.15.2a) and substituting the integral representation for the vector potentials, the z component of the magnetic field in the far-field region takes the following form:

$$H^{S}_{1z}(\rho,\phi) = -\frac{\omega\varepsilon_1}{4}\int_C M_z(\bar{\rho}')H_0^{(2)}(k_1|\bar{\rho}-\bar{\rho}'|)\,dL'$$

$$-\frac{\omega\varepsilon_1}{4\eta_1}\int_C (\hat{\phi}\bullet\hat{s}')J_S(\bar{\rho}')H_0^{(2)}(k_1|\bar{\rho}-\bar{\rho}'|)\,dL' \tag{5.15.3}$$

Substituting for the equivalent magnetic and electric current distributions in terms of pulse expansion functions, the scattered magnetic field distribution in the far-field region reduces to

$$H^{S}_{1z}(\rho,\phi) = \sum_{n=1}^{N} I_{2n}\Big[-\frac{k_1}{4\eta_1}H_0^{(2)}(k_1|\bar{\rho}-\bar{\rho}_{2n}|)\Big]\Delta_{2n}$$

$$-\sum_{p=1}^{P} T_{2p-1}\Big[\frac{k_1}{4}\cos(\phi-\Omega_{2p-2})\,H_0^{(2)}(k_1|\bar{\rho}-\bar{\rho}_{2p-1}|)\Big]\frac{\Delta_{2p-2}}{2}$$

$$-\sum_{p=1}^{P} T_{2p-1}\Big[\frac{k_1}{4}\cos(\phi-\Omega_{2p})\,H_0^{(2)}(k_1|\bar{\rho}-\bar{\rho}_{2p-1}|)\Big]\frac{\Delta_{2p}}{2} \tag{5.15.4}$$

In this expression, the Hankel function is now replaced by its large argument approximation defined in (5.15.4):

$$H_{1z}^{s}(\rho,\phi) \sim \sum_{n=1}^{N} I_{2n}\Big[- \frac{k_1}{4\eta_1}\sqrt{\frac{2j}{\pi(k_1|\bar{\rho}-\bar{\rho}_{2n}|)}}\; e^{-j(k_1|\bar{\rho} - \bar{\rho}_{2n}|)}\Big]\Delta_{2n}$$

$$- \sum_{p=1}^{P} T_{2p-1}\Big[\frac{k_1}{4}\cos(\phi - \Omega_{2p-2})$$

$$\sqrt{\frac{2j}{\pi(k_1|\bar{\rho}-\bar{\rho}_{2p-1}|)}}\; e^{-j(k_1|\bar{\rho} - \bar{\rho}_{2p-1}|)}\Big]\frac{\Delta_{2p-2}}{2}$$

$$- \sum_{p=1}^{P} T_{2p-1}\Big[\frac{k_1}{4}\cos(\phi - \Omega_{2p})$$

$$\sqrt{\frac{2j}{\pi(k_1|\bar{\rho}-\bar{\rho}_{2p-1}|)}}\; e^{-j(k_1|\bar{\rho} - \bar{\rho}_{2p-1}|)}\Big]\frac{\Delta_{2p}}{2} \qquad (5.15.5)$$

Referring to Figures 3.23 and 4.17, this expression can be easily simplified. In the far-field region, $|k_1\rho| \to \infty$, the magnitude terms in the denominator of the expression (5.15.5) can be written as

$$|\bar{\rho} - \bar{\rho}_{2n}| \approx \rho \qquad (5.15.6a)$$

$$|\bar{\rho} - \bar{\rho}_{2p-1}| \approx \rho \qquad (5.15.6b)$$

and the exponential terms contributing to the phase distribution of the currents can be approximated as

$$|\bar{\rho} - \bar{\rho}_{2n}| \approx \rho - \rho_{2n}\cos(\phi - \phi_{2n}) \qquad (5.15.6c)$$

$$|\bar{\rho} - \bar{\rho}_{2p-1}| \approx \rho - \rho_{2p-1}\cos(\phi - \phi_{2p-1}) \qquad (5.15.6d)$$

After substituting the above far-field approximations, expressions (5.15.6a–d), into the expression (5.15.5), the z component of the scattered magnetic field distribution in the far-field region reduces to the following form:

$$H^S_{1z}(\rho,\phi) \sim \sum_{n=1}^{N} I_{2n}\Big[\frac{k_1}{\eta_1}\mathcal{K} e^{jk_1\bar{\rho}_{2n}\cos(\phi-\phi_{2n})}\Big]\Delta_{2n}$$

$$- \sum_{p=1}^{P} T_{2p-1}\Big[k_1\mathcal{K}\cos(\phi - \Omega_{2p-2})$$

$$e^{jk_1\bar{\rho}_{2p-1}\cos(\phi-\phi_{2p-1})}\Big]\frac{\Delta_{2p-2}}{2}$$

$$- \sum_{p=1}^{P} T_{2p-1}\Big[k_1\mathcal{K}\cos(\phi - \Omega_{2p})$$

$$e^{jk_1\bar{\rho}_{2p-1}\cos(\phi-\phi_{2p-1})}\Big]\frac{\Delta_{2p}}{2} \tag{5.15.7a}$$

where

$$\mathcal{K} = \frac{1}{\sqrt{8\pi k_1\rho}}\, e^{-jk_1\rho}\, e^{-j3\pi/4} \tag{5.15.7b}$$

Further, the bistatic radar cross section of the lossy, homogeneous dielectric scatterer having arbitrary edges and corners can now be calculated using expression (3.8.18) given by

$$\mathrm{RCS}(\phi) = \lim_{\rho\to\infty} 2\pi\rho \left| \frac{H^S_{1z}(\phi,\omega)}{H^i_z(\phi,\omega)} \right|^2 \tag{5.15.8}$$

Figure 5.36 shows a plot of the bistatic radar cross section of the homogeneous dielectric square scatterer with sidelength $L = 0.25$ for the incident excitation, $\phi^i = 0^o$. The bistatic data exhibits symmetrical distribution with respect to the direction of excitation about which the square scatterer also has the geometrical symmetry. In many practical applications, the monostatic or backscattering data are quite important, which depends upon a number of parameters, such as the angle of incidence, the electrical size, and shape of the geometry. To produce the monostatic RCS distribution, referring to expression (5.14.1), the inverse of the generalized impedance matrix can be stored separately and multiplied by the excitation term recalculated for every angle of incidence to obtain the corresponding electric and magnetic current distributions. Figure 5.37a shows a plot of the monostatic radar cross section for incident angles ranging from ϕ^i =

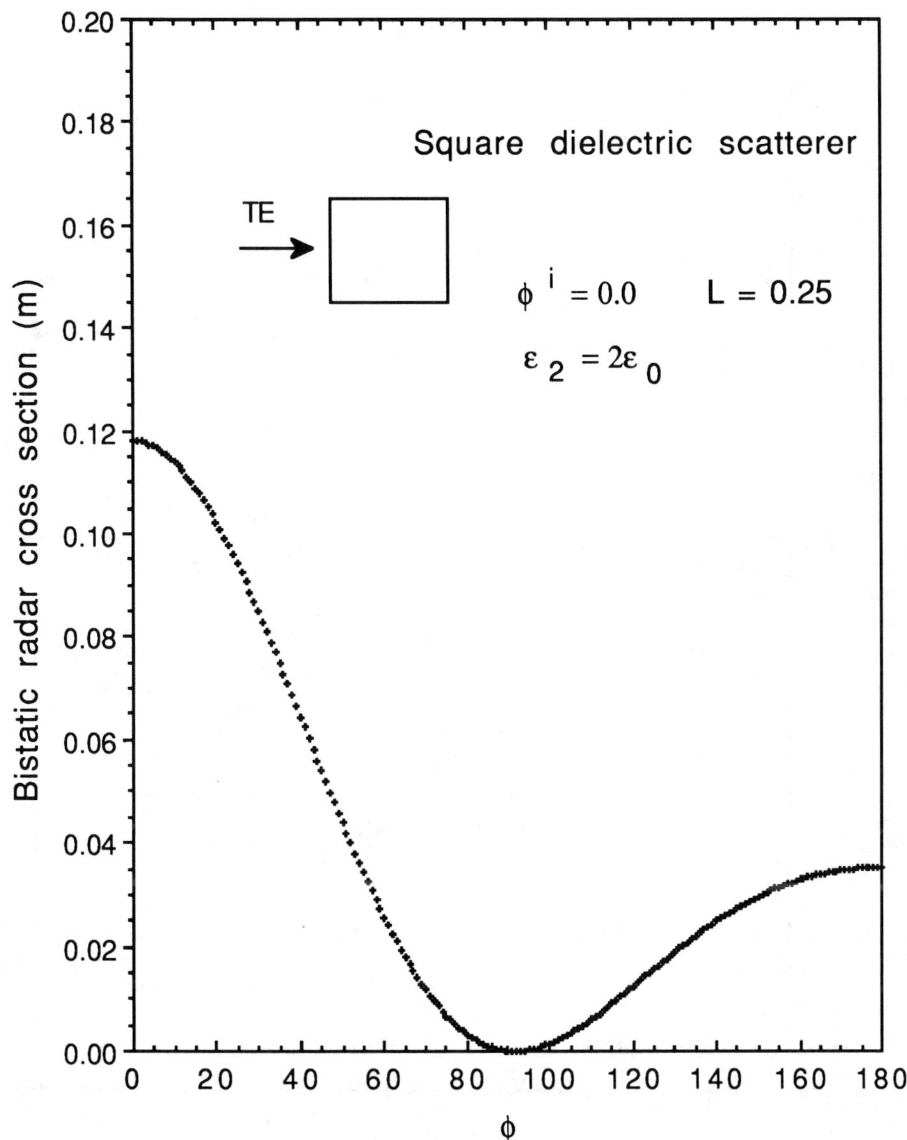

Figure 5.36 Bistatic radar cross section of a square dielectric scatterer – TE excitation.

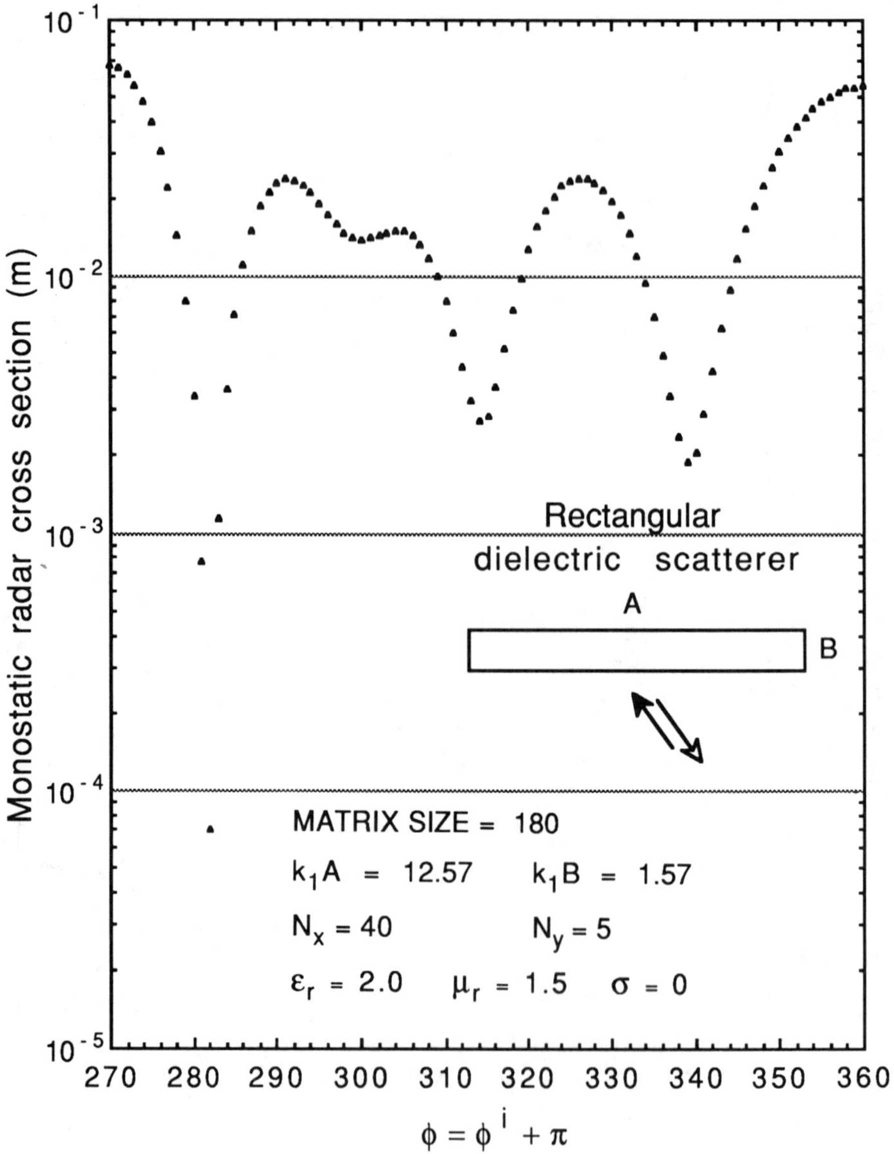

Figure 5.37a Monostatic radar cross section of a rectangular dielectric scatterer – TE excitation.

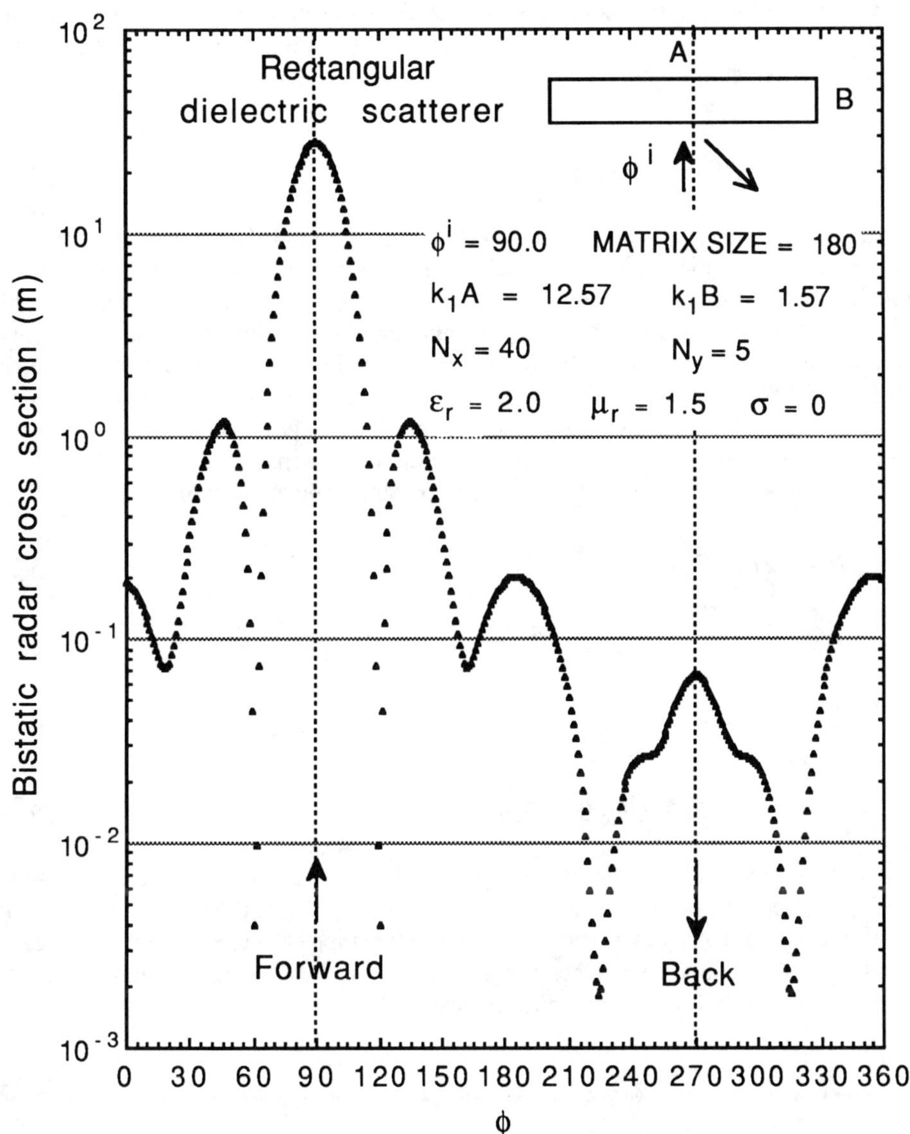

Figure 5.37b Bistatic radar cross section of a rectangular dielectric scatterer – TE excitation.

90° to 180° for the case of a rectangular dielectric scatterer. Similarly, Figure 5.37b shows a plot of the bistatic radar cross section for incident angle $\phi^i = 90^o$ for the same rectangular dielectric scatterer. These results can be easily checked out using the TM algorithm based on the duality concept.

5.16 ANISOTROPIC SCATTERER

A detailed study is discussed in Section 5.3 for analyzing the electromagnetic scattering and penetration by the two-dimensional, homogeneous, dielectric object based on the combined field integral equation formulation. In fact, the formulation can be easily extended even for analyzing an anisotropic material scatterer. The constitutive relationships (5.1.2a–b) are only special cases, applicable to the isotropic material medium. There are many special dielectric and magnetic materials, such as crystals, ferrites, and even composite materials, in which the field density in a given spatial direction can depend upon all the three coordinate directions. Let, the electric and magnetic fields, and their densities in the rectangular coordinate system are given by

$$\bar{D}(x,y,z) = D_x\hat{x} + D_y\hat{y} + D_z\hat{z} \tag{5.16.1a}$$

$$\bar{E}(x,y,z) = E_x\hat{x} + E_y\hat{y} + E_z\hat{z} \tag{5.16.1b}$$

$$\bar{B}(x,y,z) = B_x\hat{x} + B_y\hat{y} + B_z\hat{z} \tag{5.16.2a}$$

$$\bar{H}(x,y,z) = H_x\hat{x} + H_y\hat{y} + H_z\hat{z} \tag{5.16.2b}$$

In a general material medium, the electric flux density in a specific coordinate direction can depend upon the electric field along all the three coordinate directions, and similarly, the magnetic flux density in a specific coordinate direction can depend upon the magnetic field along all the three coordinate directions. Hence, the constitutive relationships for the homogeneous and anisotropic medium take the form:

$$\bar{D}(x,y,z) = \varepsilon_0 \bar{\bar{\varepsilon}} \bullet \bar{E}(x,y,z) \tag{5.16.3a}$$

$$\bar{B}(x,y,z) = \mu_0 \bar{\bar{\mu}} \bullet \bar{H}(x,y,z) \tag{5.16.3b}$$

where the elements of the permittivity and permeability parameters are referred to as *tensor elements* with

$\bar{\bar{\varepsilon}}$: tensor permittivity characteristics;

$\bar{\bar{\mu}}$: tensor permeability characteristics;

ε_{mn} : directional permittivity parameter;

μ_{mn} : directional permeability parameter

where $m = n$ corresponds to the self terms, and $m \neq n$ corresponds to the mutual terms of the tensor characteristics. The expressions (5.16.3a–b) can be rewritten in a (3 x 3) matrix form as

$$\begin{bmatrix} D_x \\ D_y \\ D_z \end{bmatrix} = \begin{bmatrix} \varepsilon_{xx} & \varepsilon_{xy} & \varepsilon_{xz} \\ \varepsilon_{yx} & \varepsilon_{yy} & \varepsilon_{yz} \\ \varepsilon_{zx} & \varepsilon_{zy} & \varepsilon_{zz} \end{bmatrix} \begin{bmatrix} \varepsilon_0 E_z \\ \varepsilon_0 E_y \\ \varepsilon_0 E_z \end{bmatrix} \qquad (5.16.4a)$$

and

$$\begin{bmatrix} B_x \\ B_y \\ B_z \end{bmatrix} = \begin{bmatrix} \mu_{xx} & \mu_{xy} & \mu_{xz} \\ \mu_{yx} & \mu_{yy} & \mu_{yz} \\ \mu_{zx} & \mu_{zy} & \mu_{zz} \end{bmatrix} \begin{bmatrix} \mu_0 H_z \\ \mu_0 H_y \\ \mu_0 H_z \end{bmatrix} \qquad (5.16.4b)$$

As can be seen from these expressions, the analysis of the electric and magnetic fields in an anisotropic material medium is quite intricate. But, the electromagnetic equivalences discussed earlier in Section 5.4 can be applied to write expressions for the electric and magnetic fields in the interior as well as the exterior regions of the anisotropic scatterer. However, an appropriate Green's function should be used for the anisotropic material region.

Figure 5.38a shows the original two-dimensional boundary value problem. Only the cross section of the geometry in the $z = 0$ plane is shown. The arbitrary shaped, anisotropic scattering geometry has a cross-sectional area S bounded by a contour C. For normal excitation, the incident, the scattered, and the penetrated electric and magnetic fields are completely independent of the z coordinate variable.

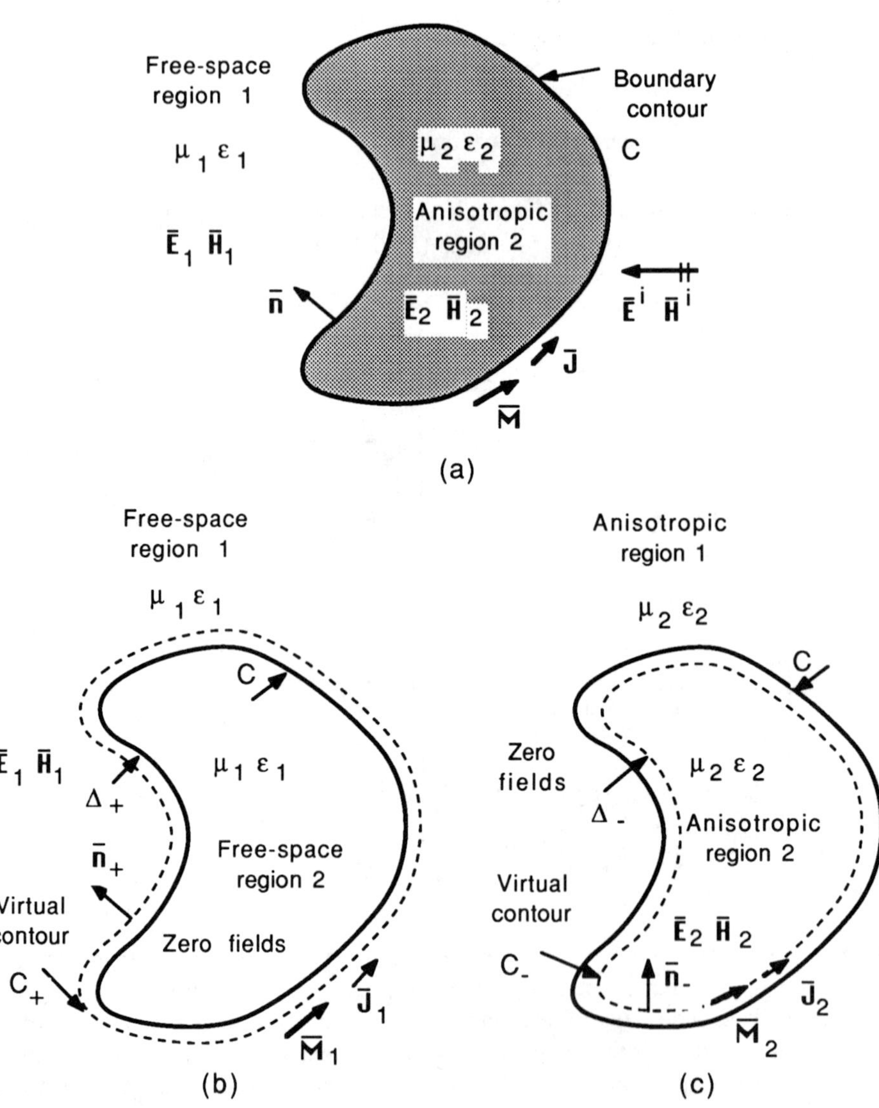

Figure 38 Electromagnetic equivalence for homogeneous, anisotropic scatterer
(a) original problem, (b) exterior equivalence, and
(c) interior equivalence.

Referring to Figures 5.8b and 5.38b, the analysis of the exterior equivalence is the same even for the anisotropic scatterer. Virtual contour C_+ is drawn just outside the scatterer at a distance Δ_+ from the cross-sectional boundary to enclose the scatterer cross section completely. In a limit as $\Delta_+ \to 0$, virtual contour C_+ coincides with actual scatterer boundary C. The electric and magnetic fields outside the scatterer in region 1 are identical to the fields of the original problem. Then, for TM excitation, for all field points located in region 1 on or outside virtual contour C_+,

ε_1 : permittivity of free-space medium, region 1;

μ_1 : permeability of free-space medium, region 1;

$\sigma_1 = 0$: conductivity of free-space medium, region 1;

$(\bar{E}^i, \bar{H}^i)$: electric and magnetic incident fields in free-space medium

$$= [\hat{z}E_z^i(\bar{\rho}), \bar{H}_\tau^i(\bar{\rho})] \tag{5.16.5a}$$

$(\bar{E}_1^S, \bar{H}_1^S)$: electric and magnetic scattered fields in free-space medium

$$= [\hat{z}E_{1z}^S(\bar{\rho}), \bar{H}_{1\tau}^S(\bar{\rho})] \tag{5.16.5b}$$

$(\bar{E}_1, \bar{H}_1)$: electric and magnetic total fields in free-space medium

$$= [\bar{E}^i(\bar{\rho}), \bar{H}^i(\bar{\rho})] + [\bar{E}_1^S(\bar{\rho}), \bar{H}_1^S(\bar{\rho})] \tag{5.16.5c}$$

$$= [\hat{z}E_z^i(\bar{\rho}), \bar{H}_\tau^i(\bar{\rho})] + [\hat{z}E_{1z}^S(\bar{\rho}), \bar{H}_{1\tau}^S(\bar{\rho})] \tag{5.16.5d}$$

Referring to Figure 5.38b, on virtual contour C_+ certain tangential distributions of electric and magnetic currents are simulated. In a limit as virtual contour C_+ tends to original contour C, the simulated tangential electric and magnetic currents are such that they produce the same scattered electric and magnetic fields in region 1, expression (5.4.4b), which are identical to the original anisotropic scatterer problem shown in Figure 5.38a. The simulated tangential electric and magnetic currents on virtual contour C_+ are the equivalent electric currents and magnetic currents valid only for the calculation of region 1 field distributions.

Referring to Sections 2.16 and 2.25, expressions (2.16.2b) and (2.25.21), the equivalent surface electric currents and magnetic currents can be defined based on the following boundary conditions on contour C_+:

$\bar{J}_{1eq}(\bar{\rho}')$: equivalent electric currents valid only for region 1 fields

$$= \hat{n}'_1 \times \bar{H}_1(\bar{\rho}') \qquad \bar{\rho}' \text{ on } C_+ \to C \tag{5.16.6a}$$

$\bar{M}_{1eq}(\bar{\rho}')$: equivalent magnetic currents valid only for region 1 fields

$$= - \hat{n}'_1 \times \bar{E}_1(\bar{\rho}') \qquad \bar{\rho}' \text{ on } C_+ \to C \tag{5.16.6b}$$

$\hat{n}'_1 = \hat{n}'$: outward normal to contour C.

The complete mathematical development carried out in Section 5.1 based on the representation of scattered fields in terms of the electromagnetic potentials can now be utilized to write the region 1 general field distributions. Thus, for the TM case, the expressions for the electric and magnetic field distributions are given by

$$\bar{E}_1(\bar{\rho}) = \bar{E}^i(\bar{\rho}) - j\omega\bar{A}_1(\bar{\rho},\omega) - \frac{1}{\varepsilon_1}\nabla_\tau \times \bar{F}_1(\bar{\rho},\omega) \tag{5.16.7}$$

$$\bar{H}_1(\bar{\rho}) = \bar{H}^i(\bar{\rho}) + \frac{1}{\mu_1}\nabla_\tau \times \bar{A}_1(\bar{\rho},\omega) - j\omega\bar{F}_1(\bar{\rho},\omega) - \nabla_\tau\Psi_1(\bar{\rho},\omega) \tag{5.16.8}$$

$\bar{\rho}$ on or outside C_+

where

$$\bar{A}_1(\bar{\rho}) = \frac{\mu_1}{4j}\int_C \bar{J}_{1eq}(\bar{\rho}')H_0^{(2)}(k_1|\bar{\rho} - \bar{\rho}'|)\, dL(\bar{\rho}') \tag{5.16.9}$$

$$\bar{F}_1(\bar{\rho}) = \frac{\varepsilon_1}{4j}\int_C \bar{M}_{1eq}(\bar{\rho}')H_0^{(2)}(k_1|\bar{\rho} - \bar{\rho}'|)\, dL(\bar{\rho}') \tag{5.16.10a}$$

$$\Psi_1(\bar{\rho}) = \frac{1}{4j\mu_1}\int_C \rho_{1meq}(\bar{\rho}')H_0^{(2)}(k_1|\bar{\rho} - \bar{\rho}'|)\, dL(\bar{\rho}') \tag{5.16.10b}$$

5.16.1 INTERIOR EQUIVALENCE

A similar procedure can be adapted to set up an equivalent boundary value problem valid for region 2. The procedure that follows for setting up of an equivalent problem for the interior anisotropic region is similar to the one discussed in Section 5.4.2, but the equivalent electric and magnetic current sources are valid only for the region 2 penetrated fields. Figure 5.38c shows the scatterer geometry is shown identical in shape to the original anisotropic scatterer. Virtual contour C_- is drawn just inside the scatterer at distance Δ_- from the cross-sectional boundary to enclose the scatterer cross section completely. In a limit as $\Delta_- \to 0$, virtual contour C_- coincides with actual scatterer boundary C. The electric and magnetic fields inside penetrable region 2 are identical to the fields of the original problem.

For TM excitation, for all field points located in region 2 on or inside the virtual contour C_-,

$\varepsilon_0 \bar{\bar{\varepsilon}}_2$: permittivity of anisotropic medium, region 2;

$\mu_0 \bar{\bar{\mu}}_2$: permeability of anisotropic medium, region 2.

$$\bar{D}_2(x,y,z) = \varepsilon_0 \bar{\bar{\varepsilon}}_2 \bullet \bar{E}_2(x,y,z) \tag{5.16.11a}$$

$$\bar{B}_2(x,y,z) = \mu_0 \bar{\bar{\mu}}_2 \bullet \bar{H}_2(x,y,z) \tag{5.16.11b}$$

where the permittivity and permeability *tensors* are given by

$$\bar{\bar{\varepsilon}}_2 = \begin{bmatrix} \varepsilon_{xx} & \varepsilon_{xy} & 0 \\ \varepsilon_{yx} & \varepsilon_{yy} & 0 \\ 0 & 0 & \varepsilon_{zz} \end{bmatrix} \tag{5.16.12a}$$

$$\bar{\bar{\mu}}_2 = \begin{bmatrix} \mu_{xx} & \mu_{xy} & 0 \\ \mu_{yx} & \mu_{yy} & 0 \\ 0 & 0 & \mu_{zz} \end{bmatrix} \tag{5.16.12b}$$

$(\bar{E}_2^S, \bar{H}_2^S)$: electric and magnetic penetrated fields in the anisotropic medium

$(\bar{E}_2, \bar{H}_2)$: electric and magnetic total fields in the anisotropic medium

$$= [\bar{E}_2^S(\bar{\rho}), \bar{H}_2^S(\bar{\rho})] \tag{5.16.13a}$$

$$= [\hat{z}E_{2z}^S(\bar{\rho}), \bar{H}_{2\tau}^S(\bar{\rho})] \tag{5.16.13b}$$

In Figure 5.38c, on virtual contour C_- certain tangential distributions of the electric and magnetic currents are simulated. In a limit as virtual contour C_- tends to original contour C, the simulated tangential electric and magnetic currents are such that they produce the same penetrated total electric and magnetic fields in region 2 identical to the original anisotropic scatterer problem shown in Figure 5.38a. As far as setting up the electromagnetic equivalent problem valid for the region 2 total fields, it really does not matter what type of medium exists outside virtual contour C_-, and similarly, it should be of no concern in the analysis regarding the exact nature of the field distributions outside of virtual contour C_-. Hence, for region 1 outside virtual contour C_-, the total electric and magnetic fields are taken as zero (for convenience) to decouple the interior boundary value problem from the exterior. Hence, the region 1 free-space medium can be completely removed and replaced by a medium having properties identical to the region 2 homogeneous, anisotropic medium. Thus, in a limit as $\Delta_- \rightarrow 0$, there are floating virtual or equivalent electric and magnetic current distributions on contour C whose shape is identical to the original scatterer boundary, but located in a linear, homogeneous, anisotropic medium.

Referring to Figure 5.38c, let

$\bar{J}_{2_{eq}}(\bar{\rho}')$: equivalent electric currents valid only for region 2 fields

$$= \hat{n}'_2 \times \bar{H}_2(\bar{\rho}') \qquad \bar{\rho}' \text{ on } C_- \rightarrow C \tag{5.16.14a}$$

$\bar{M}_{2_{eq}}(\bar{\rho}')$: equivalent magnetic currents valid only for region 2 fields

$$= - \hat{n}'_2 \times \bar{E}_2(\bar{\rho}') \qquad \bar{\rho}' \text{ on } C_- \rightarrow C \tag{5.16.14b}$$

$\hat{n}'_2 = - \hat{n}'$: inward normal to the contour C.

Referring to the development in Section 5.4.2, for the equivalent electric current distribution, let

$\bar{A}_2(\bar{\rho},\omega)$: magnetic vector potential at the field point in region 2

$$= A_{2z}(\bar{\rho},\omega)\hat{z} \tag{5.16.15}$$

Then, for TM excitation, the partial expressions for the magnetic field and the electric field distributions are given by

$$\bar{H}_{2e}(\bar{\rho},\omega) = \frac{1}{\mu_0} \bar{\bar{\mu}}_2^{-1} \bullet \nabla \times \bar{A}_2(\bar{\rho},\omega) \tag{5.16.16a}$$

$$\bar{E}_{2e}(\bar{\rho},\omega) = - j\omega\bar{A}_2(\bar{\rho},\omega) \tag{5.16.16b}$$

and the following scalar *Helmholtz* partial differential equation is satisfied by the magnetic vector potential:

$$\mathcal{L}^2 A_{2z}(\bar{\rho},\omega) + k_a^2 A_{2z}(\bar{\rho},\omega) = - \gamma\mu_0 J_{2z_{eq}}(\bar{\rho},\omega) \tag{5.16.17a}$$

$$k_0 = \omega(\mu_0\varepsilon_0)^{1/2} \tag{5.16.17b}$$

$$k_a^2 = k_0^2\,\gamma\varepsilon_{zz} \tag{5.16.17c}$$

$$\gamma = \mu_{xx}\,\mu_{yy} - \mu_{xy}\,\mu_{yx} \tag{5.16.17d}$$

and the differential operator for the TM case has the form

$$\mathcal{L}^2 = \mu_{xx}\frac{\partial^2}{\partial x^2} + \mu_{yy}\frac{\partial^2}{\partial y^2} + (\mu_{xy} + \mu_{yx})\frac{\partial^2}{\partial x\partial y} \tag{5.16.17e}$$

which, in fact, simplifies with the following restriction that the mutual tensor elements in most of the practical cases of two-dimensional, anisotropic scatterers have the relationship

$$\mu_{xy} = - \mu_{yx} \tag{5.16.18a}$$

$$\varepsilon_{xy} = - \varepsilon_{yx} \tag{5.16.18b}$$

Similarly, based on duality, for the equivalent magnetic current distribution, let

$\bar{F}_2(\bar{\rho},\omega)$: electric vector potential at the field point in region 2

$$= F_{2x}(\bar{\rho},\omega)\hat{x} + F_{2y}(\bar{\rho},\omega)\hat{y} \tag{5.16.19}$$

$\Psi_2(\bar{\rho},\omega)$: magnetic scalar potential at the field point in region 2

Then, for TM excitation, the partial expressions for the electric field and the magnetic field distributions are given by

$$\bar{E}_{2m}(\bar{\rho},\omega) = -\frac{1}{\varepsilon_0}\bar{\bar{\varepsilon}}_2^{-1}\bullet\nabla\times\bar{F}_2(\bar{\rho},\omega) \tag{5.16.20a}$$

$$\bar{H}_{2m}(\bar{\rho},\omega) = -j\omega\bar{F}_2(\bar{\rho},\omega) - \nabla\Psi_2(\bar{\rho},\omega) \tag{5.16.20b}$$

and the following scalar *Helmholtz* partial differential equations are satisfied by the modified electric vector potential and the magnetic scalar potential:

$$\mathcal{L}^2\Xi_{2x}(\bar{\rho},\omega) + k_a^2\Xi_{2x}(\bar{\rho},\omega) = -\gamma\varepsilon_0\varepsilon_{zz}M_{2x_{eq}}(\bar{\rho},\omega) \tag{5.16.21a}$$

$$\mathcal{L}^2\Xi_{2y}(\bar{\rho},\omega) + k_a^2\Xi_{2y}(\bar{\rho},\omega) = -\gamma\varepsilon_0\varepsilon_{zz}M_{2y_{eq}}(\bar{\rho},\omega) \tag{5.16.21a}$$

$$\mathcal{L}^2\psi_2(\bar{\rho},\omega) + k_a^2\psi_2(\bar{\rho},\omega) = -\frac{\rho_{2m_{eq}}(\bar{\rho},\omega)}{\mu_0} \tag{5.16.21c}$$

where the modified electric vector potential is related to the regular electric vector potential through the linear transformation

$$\Xi_{2x}(\bar{\rho},\omega)\hat{x} + \Xi_{2y}(\bar{\rho},\omega)\hat{y} = \bar{\bar{\mu}}_2\bullet\bar{F}_2(\bar{\rho},\omega) \tag{5.16.21d}$$

It should be noted that, to arrive at the consistent Helmholtz differential equations (5.16.21a–c), the electric vector potential and the magnetic scalar potential should satisfy the following Lorentz type gauge condition:

$$j\omega\nabla\bullet[\bar{\bar{\mu}}_2\bullet\bar{F}_2(\bar{\rho},\omega)] = k_a^2\psi_2(\bar{\rho},\omega) \tag{5.16.22}$$

It is now possible to write the composite total electric field and the total magnetic field distributions using the penetrated field expressions defined in (5.16.16a), (5.16.16b) and (5.16.20a), (5.16.20b), for $\bar{\rho}$ on or inside C_-,

$$\bar{E}_2(\bar{\rho}) = \bar{E}_{2e}(\bar{\rho}) + \bar{E}_{2m}(\bar{\rho}) \tag{5.16.23a}$$

$$\bar{H}_2(\bar{\rho}) = \bar{H}_{2e}(\bar{\rho}) + \bar{H}_{2m}(\bar{\rho}) \tag{5.16.23b}$$

Thus, for the TM case, the expressions for the electric and magnetic field distributions are given by

$$\bar{E}_2(\bar{\rho}) = -j\omega\bar{A}_2(\bar{\rho},\omega) - \frac{1}{\varepsilon_0}\bar{\bar{\varepsilon}}_2^{-1} \bullet \nabla \times \bar{F}_2(\bar{\rho},\omega) \tag{5.16.24}$$

$$\bar{H}_2(\bar{\rho}) = \frac{1}{\mu_0}\bar{\bar{\mu}}_2^{-1} \bullet \nabla \times \bar{A}_2(\bar{\rho},\omega) - j\omega\bar{F}_2(\bar{\rho},\omega) - \nabla_\tau\Psi_2(\bar{\rho},\omega) \tag{5.16.25}$$

$\bar{\rho}$ on or outside C_-.

5.16.2 POTENTIAL INTEGRALS – ANISOTROPIC MEDIUM

The expressions for the electric and magnetic fields obtained in (5.16.24) and (5.16.25) contain the vector and scalar potentials. The vector and the scalar potentials for the anisotropic case can be obtained by first solving the Helmholtz type differential equations (5.16.17a) and (5.16.21a–c). Introducing the parameter restriction (5.16.18a–b) and normalizing the coordinate variables by

$$\xi = \frac{x}{(\mu_{xx})^{1/2}} \tag{5.16.26a}$$

$$\zeta = \frac{y}{(\mu_{yy})^{1/2}} \tag{5.16.26b}$$

differential equation (5.16.17a) takes the standard form in terms of the transformed variables:

$$\nabla_\tau^2 A_{2z} + k_a^2 A_{2z} = -\gamma\mu_0 J_{2z_{eq}} \tag{5.16.27}$$

This expression yields the two-dimensional Green's function with the propagation constant k_a. Hence, the solution for the magnetic vector potential in terms of the regular coordinate variables can be written as

$$A_{2z}(\bar{\rho}) = \frac{c_1}{4j}\int_C J_{2_{eq}}(\bar{\rho}')H_0^{(2)}(k_a R_a)\, dL(\bar{\rho}') \tag{5.16.28a}$$

$$c_1 = \frac{\gamma\mu_0}{(\mu_{xx}\mu_{yy})^{1/2}} \tag{5.16.28b}$$

$$R_a = \left[\frac{(x-x')^2}{\mu_{xx}} + \frac{(y-y')^2}{\mu_{yy}}\right]^{1/2} \tag{5.16.28c}$$

Similarly, differential equations (5.16.21a–c) yield

$$\Xi_{2x}(\bar{\rho}) = \frac{c_2}{4j}\int_C M_{2x_{eq}}(\bar{\rho}')H_0^{(2)}(k_a R_a)\, dL(\bar{\rho}') \tag{5.16.29a}$$

$$\Xi_{2y}(\bar{\rho}) = \frac{c_2}{4j}\int_C M_{2y_{eq}}(\bar{\rho}')H_0^{(2)}(k_a R_a)\, dL(\bar{\rho}') \tag{5.16.29b}$$

$$c_2 = \frac{\gamma\varepsilon_0\varepsilon_{zz}}{(\mu_{xx}\mu_{yy})^{1/2}} \tag{5.16.29c}$$

and

$$\psi_2(\bar{\rho}) = \frac{c_3}{4j}\int_C \rho_{2m_{eq}}(\bar{\rho}')H_0^{(2)}(k_a R_a)\, dL(\bar{\rho}') \tag{5.16.30a}$$

$$c_3 = \frac{1}{\mu_0(\mu_{xx}\mu_{yy})^{1/2}} \tag{5.16.30b}$$

By substituting expressions (5.16.29a) and (5.16.29b) into the linear transformation (5.16.21d), the x and y components of the regular electric vector potentials are obtained.

5.16.3 CFIE – ANISOTROPIC CASE

Using relationship (5.16.21d), the components of the electric vector potential can be written. Further, the formulation of the CFIE can be accomplished by enforcing the continuity of the tangential components of the electric and magnetic fields at the surface of anisotropic scatterer.

Referring to Section 5.3, expressions (5.3.2) and (5.3.4) yield boundary conditions

$$\bar{H}_1(\bar{\rho})\Big|_{\tan} = \bar{H}_2(\bar{\rho})\Big|_{\tan} \qquad \bar{\rho} \text{ on } C \tag{5.16.31a}$$

$$\bar{E}_1(\bar{\rho})\Big|_{\tan} = \bar{E}_2(\bar{\rho})\Big|_{\tan} \qquad \bar{\rho} \text{ on } C \tag{5.16.31b}$$

For the TM case, the development of the numerical algorithm for analyzing an arbitrary shaped, anisotropic scatterer, in fact, is almost similar to the homogeneous dielectric scatterer discussed in the Section 5.8, by just changing the two-dimensional Green's function. Similarly, for the TE case, duality can be invoked to derive the CFIE integral equations.

Recently, extensive numerical results are reported in literature for anisotropic scatterers, such as the circular, elliptic and square scatterers. Results of the induced surface equivalent electric and magnetic currents, radar cross section are reported along with excellent numerical validation based on the finite-difference time-domain technique.

Chapter 6

Two-Dimensional Conducting and Dielectric Layered Objects: TM and TE Polarizations

In the earlier chapters, a detailed analysis of electromagnetic scattering and interaction by a two-dimensional, arbitrary shaped, perfectly conducting object and also a homogeneous, lossy, penetrable object are discussed. With external incident excitation, the total electric and magnetic field distributions basically exist near and outside the conducting object, and practically no fields penetrate it. In the case of a penetrable object, the total electric and magnetic field distributions exist both outside and inside the object. In fact, the electromagnetic field distributions such as the near fields, the far fields, and the penetrated electric and magnetic fields, completely depend upon the geometrical shape, electrical size, and material characteristics of the scatterer. Further extension of the scattering analysis to the case of layered material object is now considered.

In this chapter, the analysis of electromagnetic scattering and interaction by an arbitrary shaped, two-dimensional coated conducting and dielectric layered object is studied in a systematic manner. In fact, in many practical applications, combinations of both perfectly conducting and dielectric objects are widely utilized. For example, the electromagnetic properties of a perfectly conducting scatterer can be altered by installing absorbing materials consisting of either partially or fully coated layers of dielectric material on its surface. Another example is the case of a microstrip waveguide or microstrip antenna structure, wherein thin-strip conductors are etched on the dielectric material substrata, which in turn is supported by a thin conducting ground plane. Figure 6.1 shows a typical cross-sectional geometry of a canonical two-dimensional layered object. The individual layers are linear, isotropic, and homogeneous, having their own conductivity, permeability, and permittivity characteristics. In the figure, the geometry of the layered object is assumed to be uniform and infinite in length along the z coordinate axis. Further, the geometrical cross section of the layered object is uniform and identical along its axis, which is assumed to coincide with the z axis of the coordinate system. As discussed earlier, such a two-dimensional object, which is uniform and infinite in length along its axis, is quite a valuable boundary value problem for understanding various aspects of the electromagnetic analytical formulation and subsequent solution based on the numerical technique. The physics of the electromagnetic scattering, penetration, and interaction due to the constituent material properties and the geometry configuration of the layered scatterer, such as the straight and curved arbitrary edges or the sharp wedge and

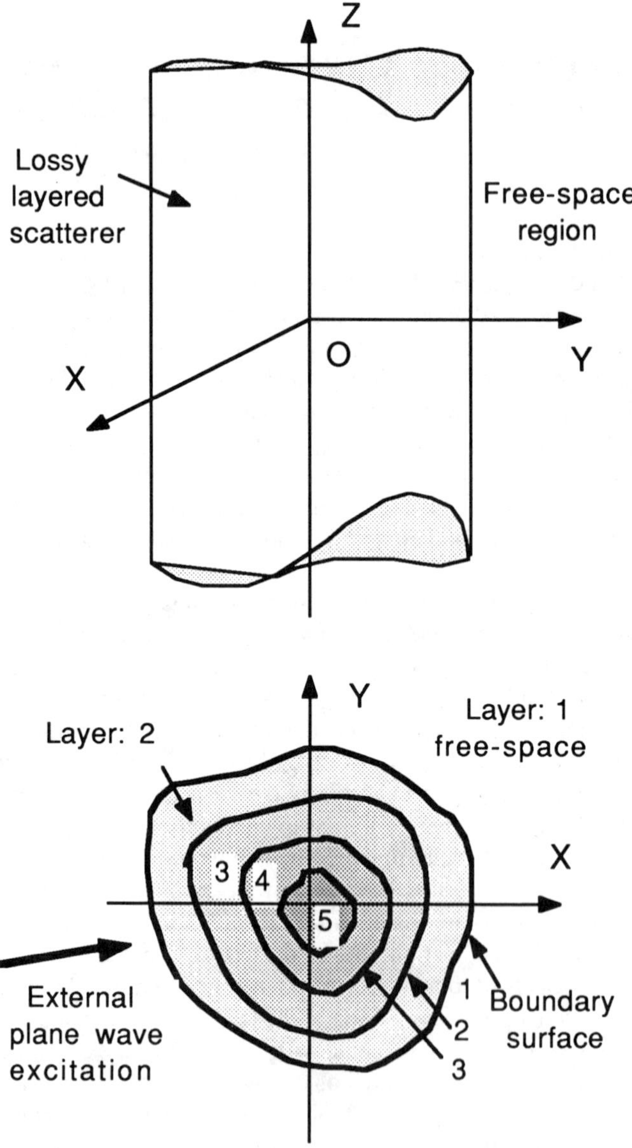

Figure 6.1 Geometry of a two-dimensional lossy layered scatterer.

smooth rounded corners, can be extracted conveniently from the two-dimensional study. The electric and magnetic material properties of the individual layer, the electrical thickness of the layer including dense and thin layer coating play important roles in various scattering and penetration studies.

In the following analysis, the layered object is assumed to be located in a large free-space medium that is linear, homogeneous, and isotropic, having constant permittivity and permeability. The systematic analysis procedure, in fact, is applicable both for partially and fully loaded lossless or lossy layered scatterers. The time-harmonic incident electric and magnetic fields are simulated and propagated in the free-space medium. For all practical modeling considerations, the incident excitation fields are taken to be those field distributions existing in the free-space medium in absence of the interacting object. When the two-dimensional, layered object is introduced into the medium, the incident fields interact to induce secondary polarization currents and charges on and inside the penetrable layered object. The induced polarization currents produce and propagate scattered electric and magnetic fields in all directions. In the free-space medium, the total fields now consist of the sum of incident and scattered fields. Also, inside each layer of the penetrable layered object are penetrated electric and magnetic field distributions. Further, the electric and magnetic field distributions in the free-space medium and inside the layered medium are such that they satisfy the familiar electromagnetic boundary conditions discussed in the earlier sections. According to the electromagnetic boundary conditions, the total tangential electric field is continuous on each boundary surface separating two adjacent layers, and similarly, the total tangential magnetic field is continuous on the boundary surface separating two adjacent layers of the layered object.

Following the previous modeling technique for a homogeneous dielectric object, the induced surface electric and magnetic field distributions on each boundary surface of the layered object are treated initially as unknown quantities. In fact, the unknown surface electric and magnetic field distributions can be converted into the corresponding unknown equivalent magnetic and electric current distributions. For the unknown induced surface current distributions on each boundary layer, a coupled boundary value problem is set up in terms of either a differential equation, an integral equation, or an integro-differential equation subject to the appropriate surface boundary conditions. As discussed earlier, the coupled boundary value equations can be directly solved using an analytical method based on an appropriate eigenfunction expansion for the canonical-type circular layered scatterer. But, for an arbitrary shaped two-dimensional, layered object, the boundary value equations can be conveniently solved using the method of moments numerical technique, which is suitable for the solution of differential equations or integral equations. Once the induced surface electric and magnetic currents are known on each boundary surface, then the penetrated electric and magnetic fields in the various layered regions of interest can be calculated. The relevant electromagnetic scattering and penetration properties, such as the penetrated field distribution in each layer, the near-field distribution, and the far-field distributions, can be further calculated.

6.1 FORMULATION OF INTEGRAL EQUATIONS

A detailed analysis procedure is presented for the case of electromagnetic scattering and penetration by a two-dimensional layered object based on a rigorous boundary value combined field integral equation formulation. Figure 6.2 shows the cross-sectional geometry of a lossy layered object. It is oriented with its axis coinciding with the z axis of a cylindrical coordinate system. Along the axis, the layered object is infinitely long and has a uniform arbitrary cross section. It is placed in a linear, homogeneous, and isotropic lossless medium excited externally by a time-harmonic transverse magnetic (TM to z) polarized plane wave. The TM-polarized plane wave is incident on the layered object at an arbitrary angle of incidence with its incident electric field polarized parallel and the incident magnetic field polarized perpendicular to the z coordinate axis. There is no propagation of the incident plane wave parallel to the axis of the two-dimensional object. Thus, the various electric and magnetic fields, namely, the incident field, the scattered field, and the penetrated field quantities are completely independent of the z coordinate variable.

For the special case of time-harmonic transverse magnetic *normal* excitation, there are only two distinct electromagnetic field components, as stated in Table 5.2. For the analysis of layered, arbitrary, cross-sectional geometry, the generalized cylindrical coordinate system given by the orthogonal coordinate variables (n, s, z) is selected. Figure 5.7 shows various interrelationships among orthogonal unit vectors corresponding to the points on a boundary surface. At the boundary surface separating two media, the coordinates variable (n, s) are in a plane parallel to the transverse plane, and the coordinates variables (s, z) are oriented tangential to the cylindrical boundary surface as depicted in Figure 5.7. In the various layered regions, there exist only the axial component of the electric field distribution parallel to the z coordinate axis and the transverse component of the magnetic field distribution parallel to the transverse plane. The transverse component of the magnetic field at the boundary surface separating two media can be resolved into two distinct orthogonal components along coordinate variables n and s. Referring to Figure 6.2, for the case of normal transverse magnetic excitation, on the mth boundary surface separating the m and $(m + 1)$ layers, there are only the following components of the electric and magnetic field distributions, for $m = 1, 2, 3, \ldots, M$:

$E_{mz}(\rho,\phi)\hat{z}$: total axial component of the electric field distribution on the mth boundary surface;

$H_{mn}(\rho,\phi)\hat{n}$: total normal component of the magnetic field distribution on the mth boundary surface;

$H_{ms}(\rho,\phi)\hat{s}$: total tangential component of the magnetic field distribution on the mth boundary surface;

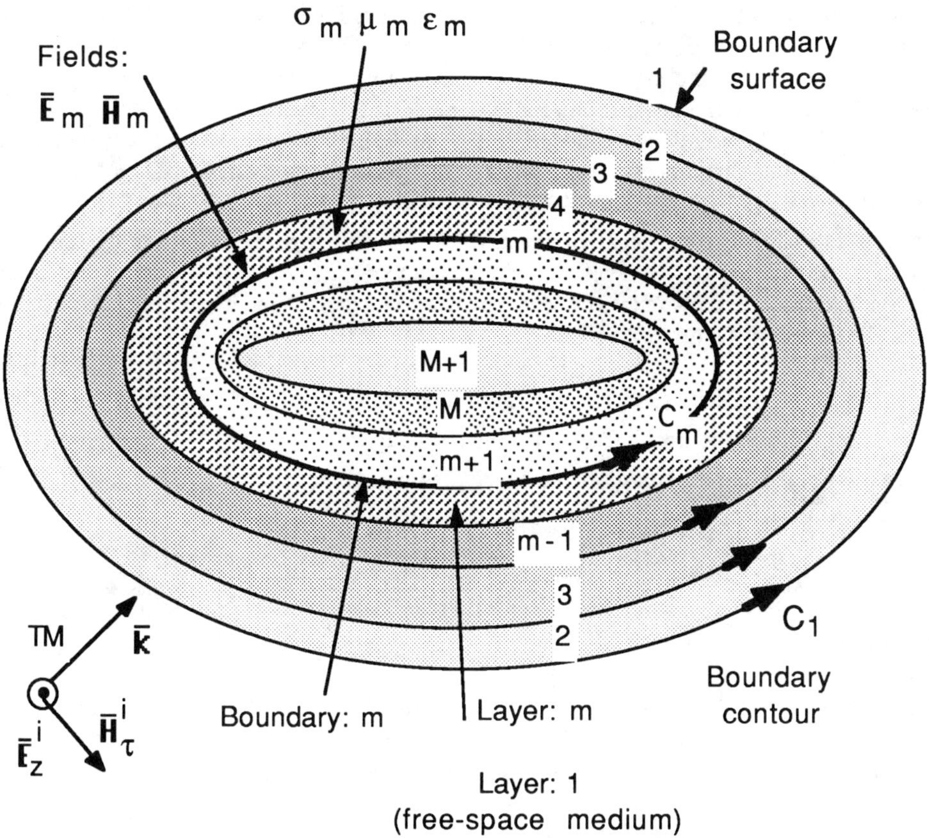

Figure 6.2 Cross-sectional geometry of a lossy, layered scatterer.

The total electric and magnetic field distributions on the mth boundary surface are, in fact, unknown field quantities. An electromagnetic boundary value problem is now set up in terms of these unknown total surface electric and magnetic field distributions subject to the appropriate tangential boundary conditions. The discussion concerning the boundary value formulation presented in this section is based upon the electromagnetic equivalence principle. Different types of integral equations, such as the electric field type, the magnetic field type and the combined fields type, can be formulated. In the electric field type of integral equation formulation only the tangential electric field boundary condition is utilized; and similarly, in the magnetic field type of integral equation formulation only the tangential magnetic field boundary condition is utilized.

Further, in the case of rigorous combined fields integral equation formulation, both types of boundary conditions relating to the tangential electric fields and the tangential magnetic fields are utilized to derive a coupled set of boundary value equations.

Figure 6.2 shows the geometry of a layered, lossy scatterer located in an isotropic, lossless, free-space medium. The total volume *V* of the layered scatterer consists of *M* arbitrary shaped, homogeneous layers or regions. The free-space region is treated as a layer or region so that there is a total of $(M + 1)$ layers or regions separated by *M* number of boundary surfaces. The mth arbitrary shaped boundary surface, $m = 1, 2, 3, \ldots M$, separates the two adjacent m^{th} and $(m + 1)^{th}$ homogeneous layers or regions. Outside the volume of the layered scatterer is medium 1, representing the free-space region, and the externally excited TM to *z* polarized incident plane wave field is contained in it. Referring to Figure 6.2, let

ε_1 : permittivity of free-space medium, layer 1;

μ_1 : permeability of free-space medium, layer 1;

$\sigma_1 = 0$: conductivity of free-space medium, layer 1;

$(\bar{E}^i, \bar{H}^i)$: electric and magnetic incident fields in layer 1;

$(\bar{E}_1^S, \bar{H}_1^S)$: electric and magnetic scattered fields in layer 1;

$(\bar{E}_1, \bar{H}_1)$: the electric and magnetic total fields in layer 1

$$= (\bar{E}^i, \bar{H}^i) + (\bar{E}_1^S, \bar{H}_1^S) \tag{6.1.1a}$$

Similarly, for each layer or region of the lossy layered scatterer,

ε_m : permittivity of layer m;

μ_m : permeability of layer m;

σ_m : conductivity of layer m;

$(\bar{E}_m^S, \bar{H}_m^S)$: electric and magnetic penetrated fields in layer m;

$(\bar{E}_m, \bar{H}_m)$: electric and magnetic total fields in layer m

$$= (\bar{E}_m^S, \bar{H}_m^S) \tag{6.1.1b}$$

Referring to the electromagnetic boundary conditions discussed in section 2.16, these electric and magnetic field distributions in the respective layers or regions are such that the following boundary conditions are satisfied. The continuity of the tangential total surface electric field, and the continuity of the tangential total surface magnetic field, always should be maintained at the boundary surface separating the two media, which yield the following boundary equations:

$$\bar{H}_m(\bar{r}_m)\Big|_{\tan} = \bar{H}_{m+1}(\bar{r}_m)\Big|_{\tan} \qquad \bar{r}_m \text{ on boundary } C_m \tag{6.1.2}$$

$$\bar{E}_m(\bar{r}_m)\Big|_{\tan} = \bar{E}_{m+1}(\bar{r}_m)\Big|_{\tan} \qquad \bar{r}_m \text{ on boundary } C_m \tag{6.1.3}$$

$$m = 1, 2, 3, \quad . \quad . \quad . \quad , M$$

Referring to Figure 6.2, the incident plane wave electric and magnetic fields can be written as

$$\bar{E}^i(\bar{\rho},\omega) = E_z^i(\bar{\rho},\omega)\hat{z} \tag{6.1.4a}$$

$$E_z^i(\bar{\rho},\omega) = E_0\, e^{-j\bar{k}_1\cdot\bar{\rho}} \tag{6.1.4b}$$

$$\bar{H}^i(\bar{\rho},\omega) = \bar{H}_0\, e^{-j\bar{k}_1\cdot\bar{\rho}} \tag{6.1.4c}$$

and the incident plane wave electric and magnetic fields are related through

$$\bar{H}_0 \times \hat{k} = \frac{\omega\varepsilon_1}{k_1}\bar{E}_0 \tag{6.1.5}$$

It is useful to work in terms of a generalized cylindrical coordinate system, as shown in Figure 5.7, for the case of an arbitrary cross-sectional scatterer. On the boundary surface separating two adjacent layers, let us consider three mutually orthogonal coordinates given by *(n, s, z)* such that the following righthand system is satisfied:

$$\hat{z}_m \times \hat{n}_m = \hat{s}_m \tag{6.1.6}$$

n: coordinate variable along normal to the surface;
s: coordinate variable along tangent to the surface;

Then, in terms of the rectangular coordinate system,

$$\hat{n}_m = \hat{x}\cos\Omega_m + \hat{y}\sin\Omega_m \tag{6.1.7a}$$

$$\hat{s}_m = -\hat{x}\sin\Omega_m + \hat{y}\cos\Omega_m \tag{6.1.7b}$$

Ω_m: angle the normal makes with the x coordinate axis.

6.2 ELECTROMAGNETIC EQUIVALENCE – LAYERED CASE

For the layered boundary value problem, if the surface tangential electric and magnetic field distributions are known on each boundary surface, then it is possible to calculate various field distributions in the layered regions. The analysis technique discussed in the following is similar to the equivalence principle discussed in section 5.4 for a homogeneous dielectric scatterer. The concept of surface *electric current and magnetic current* has been extensively introduced to maintain consistency of the analysis procedure for the study of general shaped and complex material objects.

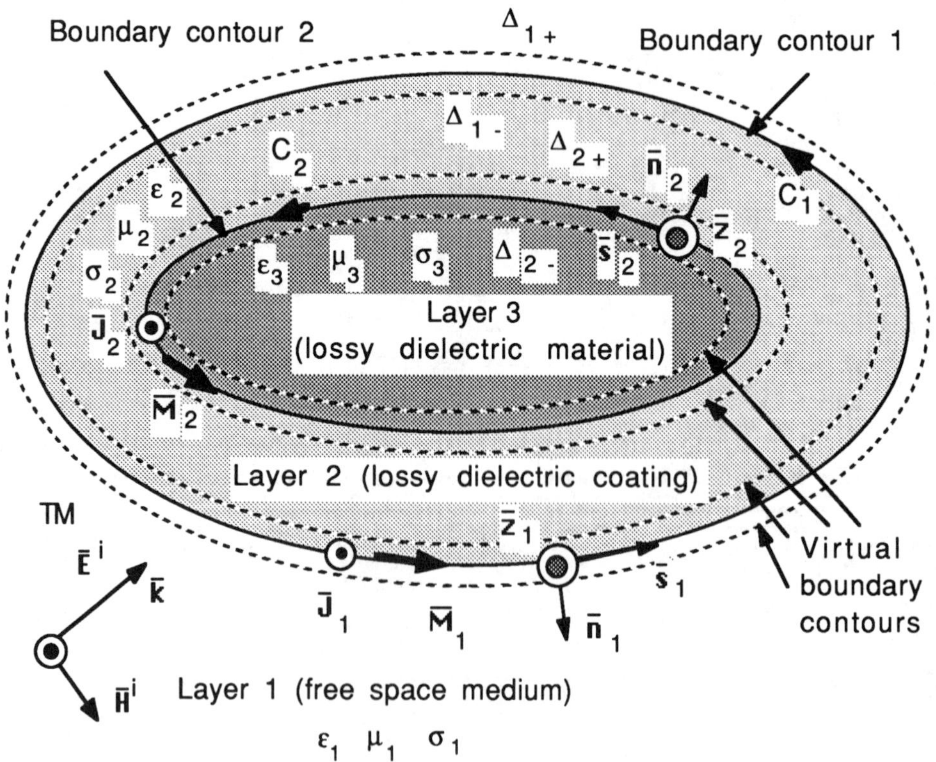

Figure 6.3a Original lossy, layered scatterer - TM excitation.

An electromagnetic equivalent problem is set up in the following that is suitable for the analysis of a layered object. The dependent sources simulated in the equivalent problem are the equivalent electric and magnetic current sources, but arranged in a linear, homogeneous, and isotropic medium. The electric and magnetic fields produced by the equivalent currents are identical to the electric and magnetic fields of the original boundary value problem and also satisfy the relevant boundary conditions.

To establish the analysis technique, Figure 6.3a shows the original boundary value problem of a homogeneous, lossy dielectric scatterer with *a single layer* of homogeneous lossy coating on it. Only the cross section of the geometry at the $z = 0$ plane is shown. The arbitrary shaped scattering geometry has a cross-sectional area whose layers are enclosed by boundary contours C_1 and C_2. For normal excitation, the incident, scattered and penetrated electric and magnetic fields are independent of the z coordinate variable. For the original layered scatterer with TM eccitation, referring to Figure 6.3a, let the medium parameters and the electric and magnetic fields for m^{th} layer, given by

ε_m : permittivity of layer m;

μ_m : permeability of layer m;

σ_m : conductivity of layer m;

$(\bar{E}^i, \bar{H}^i)$: electric and magnetic incident fields in free-space medium;

$(\bar{E}_1^s, \bar{H}_1^s)$: electric and magnetic scattered fields in free-space medium;

$(\bar{E}_1, \bar{H}_1)$: electric and magnetic total fields in free-space medium

$$= (\bar{E}^i, \bar{H}^i) + (\bar{E}_1^s, \bar{H}_1^s) \tag{6.2.1a}$$

$(\bar{E}_m, \bar{H}_m)$: electric and magnetic total fields in layer $m \geq 2$

$$= (\bar{E}_m^s, \bar{H}_m^s) \tag{6.2.1b}$$

6.2.1 EXTERIOR EQUIVALENCE – LAYER 1

The following procedure is adapted to set up an equivalent boundary value problem valid only for the layer 1 free-space medium. Figure 6.3b shows, again, the layered scatterer geometry which is similar to the original scatterer. Virtual contour C_{1+} is drawn just outside the scatterer at a distance Δ_{1+} from the cross-sectional boundary C_1 to enclose the scatterer cross section completely. In a limit as $\Delta_{1+} \to 0$, virtual contour C_{1+} coincides with actual scatterer boundary C_1. For layer 1, the various medium parameters and the electric and magnetic field distributions are same as in original scatterer problem. In Figure 6.3b, on virtual contour C_{1+} certain tangential distributions of the electric and magnetic currents are simulated. In a limit as virtual contour C_{1+} tends to original contour C_1, the simulated tangential electric and the magnetic currents are such that they produce the original electric and magnetic fields in the free-space medium. As far as setting up the electromagnetic equivalent problem valid for the region 1 fields, it really does not matter what type of medium exists within virtual contour C_{1+}, and similarly, it should be of no concern in the analysis regarding the exact nature of the field distributions inside virtual contour C_{1+}. Hence, for the region inside virtual contour C_{1+}, the total electric and magnetic fields are taken as zero (for convenience) to decouple the exterior boundary value problem from the interior. The inside region consisting of layers 2 and 3 is completely removed and replaced by a medium having properties identical to the region 1 free-space medium.

Thus, in a limit as $\Delta_{1+} \rightarrow 0$, there are floating equivalent electric and magnetic current distributions on contour C_1 whose shape is identical to the original scatterer boundary contour, but located in a large linear, homogeneous, isotropic free-space medium.

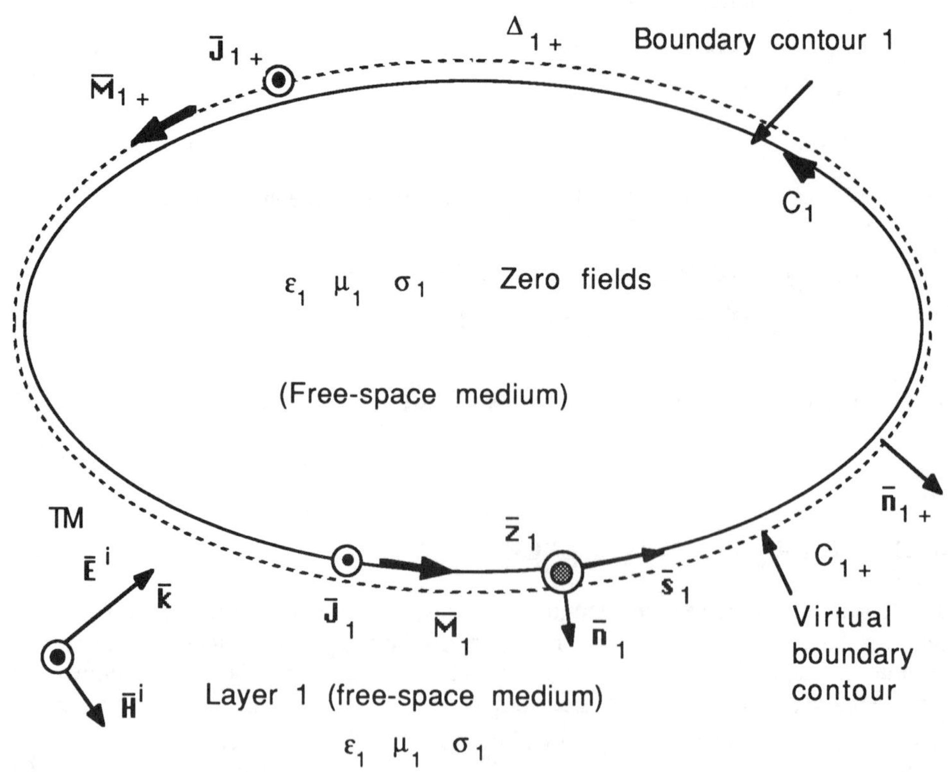

Figure 6.3b Exterior equivalence – Layer 1.

The equivalent surface electric and magnetic currents can be defined based on the following surface total fields on contour C_{1+}:

$\bar{J}_{1_{eq}}(\bar{\rho}')$: equivalent electric current valid only for the layer 1 fields;

$\bar{M}_{1_{eq}}(\bar{\rho}')$: equivalent magnetic current valid only for the layer 1 fields.

The composite electric and magnetic field distributions in terms of the vector and scalar potentials for layer 1 are given by

$$\bar{E}_1(\bar{\rho}) = \bar{E}^i(\bar{\rho}) - j\omega\bar{A}_1(\bar{\rho},\omega) - \nabla_\tau\Phi_1(\bar{\rho},\omega) - \frac{1}{\varepsilon_1}\nabla_\tau \times \bar{F}_1(\bar{\rho},\omega) \tag{6.2.2a}$$

$$\bar{H}_1(\bar{\rho}) = \bar{H}^i(\bar{\rho}) + \frac{1}{\mu_1}\nabla_\tau \times \bar{A}_1(\bar{\rho},\omega) - j\omega\bar{F}_1(\bar{\rho},\omega) - \nabla_\tau\Psi_1(\bar{\rho},\omega) \tag{6.2.2b}$$

$\bar{\rho}$ on or outside C_{1+}

The vector potentials and the scalar potentials at the field point in free-space layer 1 are given by the corresponding integrals over the complete equivalent current and charge source distributions:

$$\bar{A}_1(\bar{\rho}) = \frac{\mu_1}{4\,j}\int_{C_1} \bar{J}_{1_{eq}}(\bar{\rho}')H_0^{(2)}(k_1|\bar{\rho} - \bar{\rho}'|)\ dL(\bar{\rho}') \tag{6.2.3a}$$

$$\Phi_1(\bar{\rho}) = \frac{1}{4j\varepsilon_1}\int_{C_1} \rho_{1e_{eq}}(\bar{\rho}')H_0^{(2)}(k_1|\bar{\rho} - \bar{\rho}'|)\ dL(\bar{\rho}') \tag{6.2.3b}$$

$$\bar{F}_1(\bar{\rho}) = \frac{\varepsilon_1}{4\,j}\int_{C_1} \bar{M}_{1_{eq}}(\bar{\rho}')H_0^{(2)}(k_1|\bar{\rho} - \bar{\rho}'|)\ dL(\bar{\rho}') \tag{6.2.3c}$$

$$\Psi_1(\bar{\rho}) = \frac{1}{4j\mu_1}\int_{C_1} \rho_{1m_{eq}}(\bar{\rho}')H_0^{(2)}(k_1|\bar{\rho} - \bar{\rho}'|)\ dL(\bar{\rho}') \tag{6.2.3d}$$

and

$\bar{\rho}'$: position vector corresponding to the source point (x', y')

$$= x'\hat{x} + y'\hat{y} \qquad \bar{\rho}' \text{ on } C_{1+} \tag{6.2.4a}$$

$\bar{\rho}$: position vector corresponding to the field point (x, y)

$$= x\hat{x} + y\hat{y} \qquad \bar{\rho} \text{ on or outside } C_{1+} \tag{6.2.4b}$$

6.2.2 INTERIOR EQUIVALENCE - LAYER 2

The following procedure is adapted to set up an equivalent boundary value problem valid only for layer 2, which represents the single layer of a lossy coated region. Figure 6.3c again shows the layered scatterer geometry, which is similar to the original scatterer. Virtual contour C_{1-} is drawn just inside layer 2 at a distance Δ_{1-} from cross-sectional boundary C_1. In a limit as $\Delta_{1-} \to 0$, virtual contour C_{1-} coincides with actual scatterer boundary C_1. Similarly, a virtual contour C_{2+} is drawn just inside layer 2 at a distance Δ_{2+} from cross-sectional boundary C_2. In a limit as $\Delta_{2+} \to 0$, virtual contour C_{2+} coincides with actual scatterer boundary C_2.

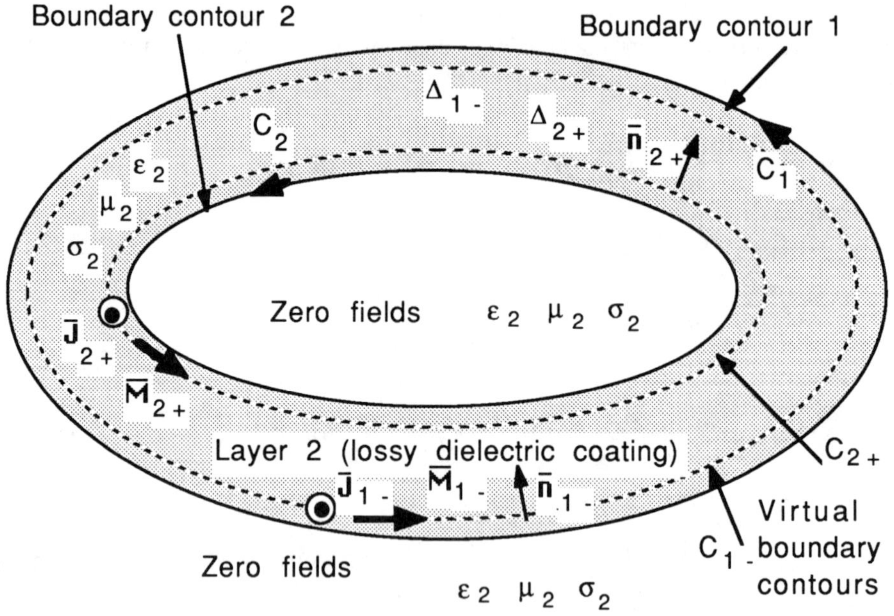

Figure 6.3c Interior equivalence – layer 2.

In Figure 6.3c, on virtual contour C_{1-} certain tangential distributions of electric and magnetic currents are simulated. In a limit as virtual contour C_{1-} tends to original contour C_1, the simulated tangential electric and magnetic currents are such that they contribute partially to the original electric and magnetic fields in layer 2. Similarly, on virtual contour C_{2+} certain tangential distributions of the electric and magnetic currents are simulated. In a limit as virtual contour C_{2+} tends to original contour C_2, the simulated tangential electric and the magnetic currents are such that they contribute to the remaining portion of the electric and magnetic fields in layer 2. As far as setting up the electromagnetic equivalent problem valid for the layer 2 fields, it really does not matter what type of medium exists within virtual contour C_{2+} and outside virtual contour C_{1-}. Similarly, it should be of no concern in the analysis regarding the exact nature of the field distributions inside virtual contour C_{2+} and also outside virtual contour C_{1-}. Hence, for the two regions the total electric and magnetic fields are taken as zero (for convenience) to decouple the layer 2 boundary value problem from the rest of the regions. The inside region consisting of layer 3 and the outside region consisting of layer 1, are completely removed and replaced by a medium having properties identical to the layer 2 dielectric-coated medium. Thus, in a limit as $\Delta_{1-} \to 0$ and $\Delta_{2+} \to 0$, there are floating equivalent electric and magnetic current distributions on contours C_1 and C_2 whose shapes are identical to the original scatterer boundary contours, but located in a linear, homogeneous, isotropic lossy medium. Referring to Figure 6.3c, for layer 2 and the complete inside and outside regions of layer 2,

ε_2 : permittivity of the medium;
μ_2 : permeability of the medium;
σ_2 : conductivity of the medium;
ε'_2 : effective permittivity of the medium

$$= \varepsilon_2 \left[1 - j\frac{\sigma_2}{\omega\varepsilon_2}\right] \tag{6.2.5}$$

The equivalent surface electric and magnetic currents can now be defined on contours C_{1-} and C_{2+} as

$\bar{J}_{1\text{-}eq}(\bar{\rho}')$: equivalent electric current valid only for the layer 2 fields

$$= -\bar{J}_{1eq}(\bar{\rho}') \tag{6.2.6a}$$

$\bar{M}_{1\text{-}eq}(\bar{\rho}')$: equivalent magnetic current valid only for the layer 2 fields

$$= -\bar{M}_{1eq}(\bar{\rho}') \tag{6.2.6b}$$

$\bar{J}_{2+eq}(\bar{\rho}')$: equivalent electric current valid only for the layer 2 fields

$$= \bar{J}_{2eq}(\bar{\rho}') \tag{6.2.6c}$$

$\bar{M}_{2+eq}(\bar{\rho}')$: equivalent magnetic current valid only for the layer 2 fields

$$= \bar{M}_{2eq}(\bar{\rho}') \tag{6.2.6d}$$

The composite electric and magnetic field distributions in terms of the vector and scalar potentials for layer 2 are given by

$$\bar{E}_2(\bar{\rho}) = - j\omega\bar{A}_2(\bar{\rho},\omega) - \nabla_\tau\Phi_2(\bar{\rho},\omega) - \frac{1}{\varepsilon'_2}\nabla_\tau \times \bar{F}_2(\bar{\rho},\omega) \tag{6.2.7a}$$

$$\bar{H}_2(\bar{\rho}) = \frac{1}{\mu_2}\nabla_\tau \times \bar{A}_2(\bar{\rho},\omega) - j\omega\bar{F}_2(\bar{\rho},\omega) - \nabla_\tau\Psi_2(\bar{\rho},\omega) \tag{6.2.7b}$$

$\bar{\rho}$ on or outside C_{2+} and on or inside C_{1-}

The vector potentials and scalar potentials at the field point in layer 2 are given by the corresponding integrals over the complete equivalent current and charge distributions:

$$\bar{A}_2(\bar{\rho}) = \frac{\mu_2}{4j}\int_{C_1} - \bar{J}_{1eq}(\bar{\rho}')H_0^{(2)}(k_2|\bar{\rho} - \bar{\rho}'|)\, dL(\bar{\rho}')$$

$$+ \frac{\mu_2}{4j}\int_{C_2} \bar{J}_{2eq}(\bar{\rho}')H_0^{(2)}(k_2|\bar{\rho} - \bar{\rho}'|)\, dL(\bar{\rho}') \tag{6.2.8a}$$

$$\Phi_2(\bar{\rho}) = \frac{1}{4j\varepsilon'_2}\int_{C_1} - \rho_{1e_{eq}}(\bar{\rho}')H_0^{(2)}(k_2|\bar{\rho} - \bar{\rho}'|)\, dL(\bar{\rho}')$$

$$+ \frac{1}{4j\varepsilon'_2}\int_{C_2} \rho_{2e_{eq}}(\bar{\rho}')H_0^{(2)}(k_2|\bar{\rho} - \bar{\rho}'|)\, dL(\bar{\rho}') \tag{6.2.8b}$$

$$\bar{F}_2(\bar{\rho}) = \frac{\varepsilon'_2}{4j}\int_{C_1} - \bar{M}_{1eq}(\bar{\rho}')H_0^{(2)}(k_2|\bar{\rho} - \bar{\rho}'|)\, dL(\bar{\rho}')$$

$$+ \frac{\varepsilon'_2}{4j} \int_{C_2} \bar{M}_{2eq}(\bar{\rho}')H_0^{(2)}(k_2|\bar{\rho} - \bar{\rho}'|)\, dL(\bar{\rho}') \tag{6.2.8c}$$

$$\Psi_2(\bar{\rho}) = \frac{1}{4j\mu_2} \int_{C_1} -\rho_{1m_{eq}}(\bar{\rho}')H_0^{(2)}(k_2|\bar{\rho} - \bar{\rho}'|)\, dL(\bar{\rho}')$$

$$+ \frac{1}{4j\mu_2} \int_{C_2} \rho_{2m_{eq}}(\bar{\rho}')H_0^{(2)}(k_2|\bar{\rho} - \bar{\rho}'|)\, dL(\bar{\rho}') \tag{6.2.8d}$$

6.2.3 INTERIOR EQUIVALENCE - LAYER 3

The following procedure is adapted to set up an equivalent boundary value problem valid only for layer 3, which represents the interior lossy dielectric region. Figure 6.3d once again shows the layered scatterer geometry, which is similar to the original scatterer. Virtual contour C_{2-} is drawn just inside layer 3 at a distance Δ_{2-} from the cross-sectional boundary C_2. In a limit as $\Delta_{2-} \to 0$, virtual contour C_{2-} coincides with actual scatterer boundary C_2.

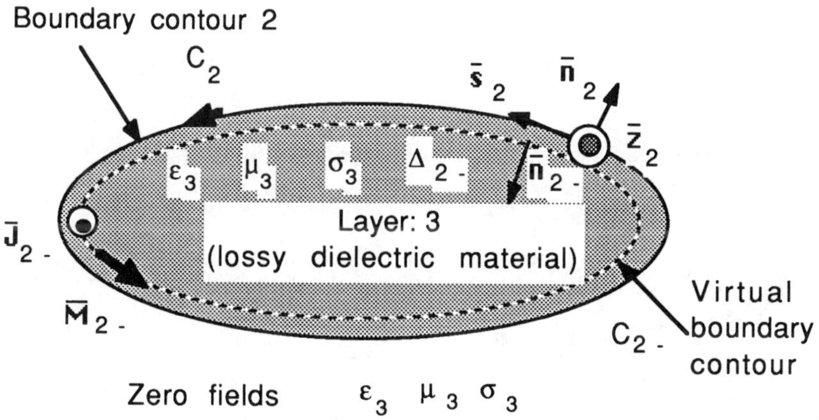

Figure 6.3d Interior equivalence – layer 3.

Referring to Figure 6.3d, on virtual contour C_{2-} certain tangential distributions of electric and magnetic currents are simulated. In a limit as virtual contour C_{2-} tends to original contour C_2, the simulated tangential electric and magnetic currents are such that they contribute to the electric and magnetic fields in layer 3. As far as setting up the electromagnetic equivalent problem valid for the layer 3 fields, it really does not matter what type of medium exists outside virtual contour C_{2-}. Similarly, it should be of no concern in the analysis regarding the exact nature of the field distributions outside virtual contour C_{2-}. Hence, for the outside region the total electric and magnetic fields are taken as zero (for convenience) to decouple the layer 3 boundary value problem from the rest of the regions. The outside region consisting of layers 1 and 2 are completely removed and replaced by a medium having properties identical to the layer 3 dielectric medium. Thus, in a limit as $\Delta_{2-} \to 0$, there are floating equivalent electric and magnetic current distributions on contour C_2 whose shape is identical to the original scatterer boundary contour, but located in a linear, homogeneous, isotropic lossy medium. Referring to Figure 6.3d, for layer 3 and the complete outside regions of layers 1 and 2,

ε_3 : permittivity of the medium;

μ_3 : permeability of the medium;

σ_3 : conductivity of the medium;

ε'_3 : effective permittivity of the medium

$$= \varepsilon_3 \left[1 - j\frac{\sigma_3}{\omega\varepsilon_3}\right] \tag{6.2.9}$$

The equivalent surface electric and magnetic currents can now be defined on the contour C_{2-} as

$\bar{J}_{2\text{-}eq}(\bar{\rho}')$: equivalent electric current valid only for the layer 2 fields

$$= -\bar{J}_{2eq}(\bar{\rho}') \tag{6.2.10a}$$

$\bar{M}_{2\text{-}eq}(\bar{\rho}')$: equivalent magnetic current valid only for the layer 2 fields

$$= -\bar{M}_{2eq}(\bar{\rho}') \tag{6.2.10b}$$

The composite electric and magnetic field distributions in terms of the vector and scalar potentials for layer 3 are given by

$$\bar{E}_3(\bar{\rho}) = -j\omega\bar{A}_3(\bar{\rho},\omega) - \nabla_\tau\Phi_3(\bar{\rho},\omega) - \frac{1}{\varepsilon'_3}\nabla_\tau \times \bar{F}_3(\bar{\rho},\omega) \tag{6.2.11a}$$

$$\bar{H}_3(\bar{\rho}) = \frac{1}{\mu_3} \nabla_\tau \times \bar{A}_3(\bar{\rho},\omega) - j\omega\bar{F}_3(\bar{\rho},\omega) - \nabla_\tau \Psi_3(\bar{\rho},\omega) \tag{6.2.11b}$$

$\bar{\rho}$ on or inside C_{2-}

The vector potentials and the scalar potentials at the field point in layer 3 are given by the corresponding integrals over the complete equivalent current and charge source distributions:

$$\bar{A}_3(\bar{\rho}) = \frac{\mu_3}{4\,j} \int_{C_2} -\bar{J}_{2eq}(\bar{\rho}') H_0^{(2)}(k_3|\bar{\rho} - \bar{\rho}'|)\, dL(\bar{\rho}') \tag{6.2.12a}$$

$$\Phi_3(\bar{\rho}) = \frac{1}{4j\varepsilon'_3} \int_{C_2} -\rho_{2e_{eq}}(\bar{\rho}') H_0^{(2)}(k_3|\bar{\rho} - \bar{\rho}'|)\, dL(\bar{\rho}') \tag{6.2.12b}$$

$$\bar{F}_3(\bar{\rho}) = \frac{\varepsilon'_3}{4\,j} \int_{C_2} -\bar{M}_{2eq}(\bar{\rho}') H_0^{(2)}(k_3|\bar{\rho} - \bar{\rho}'|)\, dL(\bar{\rho}') \tag{6.2.12c}$$

$$\Psi_3(\bar{\rho}) = \frac{1}{4j\mu_3} \int_{C_2} -\rho_{2m_{eq}}(\bar{\rho}') H_0^{(2)}(k_3|\bar{\rho} - \bar{\rho}'|)\, dL(\bar{\rho}') \tag{6.2.12d}$$

6.3 COMBINED FIELD INTEGRAL EQUATIONS – LAYERED CASE

An analysis procedure is presented in the previous section for the study of electromagnetic scattering, penetration, and interaction by a lossy, layered dielectric object. By invoking the electromagnetic equivalence principle, general expressions for the total electric and magnetic field distributions within each layer are obtained in terms of the unknown equivalent electric and magnetic surface currents on the appropriate boundary surfaces. It is now possible to set up combined field coupled integral equations using the general dielectric boundary conditions discussed in Section 2.16.

Let us consider the cross-sectional geometry of a layered, lossy dielectric scatterer shown in Figure 6.2. Layer 1 is the free-space medium, and the scatterer geometry is made up of M layers of homogeneous, lossy dielectric regions, so that in total $m = 1, 2, 3, \ldots, M + 1$ layers. These layers are separated by M closed boundary contours C_m, m = 1, 2, 3, . . . , M, on which are the corresponding unknown electric and magnetic current distributions. For example, the first boundary contour C_1 separates layer 1 (free-space medium) and layer 2. The Mth boundary contour C_M separates layer M and last layer $M+1$. In general, m^{th} boundary contour C_m separates layers m and $m+1$. It should be noted that layer m is formed by closed boundary contours C_{m-1} and C_m, and similarly, layer $m+1$ is formed by boundary contours C_m and C_{m+1}.

On boundary contour C_m, the tangential components of the total electric field are continuous, and similarly, the tangential components of the total magnetic field are continuous. Hence, the following boundary conditions are invoked:

$$\bar{E}_m(\bar{\rho})\Big|_{\tan} = \bar{E}_{m+1}(\bar{\rho})\Big|_{\tan} \qquad \bar{\rho} \text{ on } C_m \tag{6.3.1a}$$

$$\bar{H}_m(\bar{\rho})\Big|_{\tan} = \bar{H}_{m+1}(\bar{\rho})\Big|_{\tan} \qquad \bar{\rho} \text{ on } C_m \tag{6.3.1b}$$

In terms of the various vector and scalar potentials, the total electric and magnetic distributions in layer m, referring to Section 6.2.2, are given by

$$\bar{E}_m(\bar{\rho}) = \bar{E}^i(\bar{\rho})\delta(m-1) - j\omega\bar{A}_m(\bar{\rho}) - \nabla_\tau\Phi_m(\bar{\rho}) - \frac{1}{\varepsilon'_m}\nabla_\tau \times \bar{F}_m(\bar{\rho}) \tag{6.3.2a}$$

$$\bar{H}_m(\bar{\rho}) = \bar{H}^i(\bar{\rho})\delta(m-1) + \frac{1}{\mu_m}\nabla_\tau \times \bar{A}_m(\bar{\rho}) - j\omega\bar{F}_m(\bar{\rho}) - \nabla_\tau\Psi_m(\bar{\rho}) \tag{6.3.2b}$$

where

$$\delta(m-1) = 1 \qquad \text{for } m = 1$$

$$= 0 \qquad \text{for } m \neq 1 \tag{6.3.2c}$$

The vector potentials and the scalar potentials at the field point in layer m are given by the corresponding integrals over the complete equivalent current and charge source distributions:

$$\bar{A}_m(\bar{\rho}) = \frac{\mu_m}{4j}\int_{C_{m-1}} -\bar{J}_{m-1}(\bar{\rho}')H_0^{(2)}(k_m|\bar{\rho}-\bar{\rho}'|)\,dL'$$

$$+ \frac{\mu_m}{4j}\int_{C_m} \bar{J}_m(\bar{\rho}')H_0^{(2)}(k_m|\bar{\rho}-\bar{\rho}'|)\,dL' \tag{6.3.3a}$$

$$\Phi_m(\bar{\rho}) = \frac{1}{4\omega\varepsilon'_m} \int_{C_{m-1}} -[\nabla'_\tau \bullet \bar{J}_{m-1}(\bar{\rho}')] H_0^{(2)}(k_m|\bar{\rho} - \bar{\rho}'|)\, dL'$$

$$+ \frac{1}{4\omega\varepsilon'_m} \int_{C_m} [\nabla'_\tau \bullet \bar{J}_m(\bar{\rho}')] H_0^{(2)}(k_m|\bar{\rho} - \bar{\rho}'|)\, dL' \qquad (6.3.3b)$$

$$\bar{F}_m(\bar{\rho}) = \frac{\varepsilon'_m}{4j} \int_{C_{m-1}} -\bar{M}_{m-1}(\bar{\rho}') H_0^{(2)}(k_m|\bar{\rho} - \bar{\rho}'|)\, dL'$$

$$+ \frac{\varepsilon'_m}{4j} \int_{C_m} \bar{M}_m(\bar{\rho}') H_0^{(2)}(k_m|\bar{\rho} - \bar{\rho}'|)\, dL' \qquad (6.3.3c)$$

$$\Psi_m(\bar{\rho}) = \frac{1}{4\omega\mu_m} \int_{C_{m-1}} -[\nabla'_\tau \bullet \bar{M}_{m-1}(\bar{\rho}')] H_0^{(2)}(k_m|\bar{\rho} - \bar{\rho}'|)\, dL'$$

$$+ \frac{1}{4\omega\mu_m} \int_{C_m} [\nabla'_\tau \bullet \bar{M}_m(\bar{\rho}')] H_0^{(2)}(k_m|\bar{\rho} - \bar{\rho}'|)\, dL' \qquad (6.3.3d)$$

and

$$k_m^2 = \omega^2 \mu_m \varepsilon'_m \qquad (6.3.4)$$

$\bar{J}_{m-1}(\bar{\rho}')$: equivalent electric current on boundary contour C_{m-1};
$\bar{J}_m(\bar{\rho}')$: equivalent electric current on boundary contour C_m;
$\bar{M}_{m-1}(\bar{\rho}')$: equivalent magnetic current on boundary contour C_{m-1};
$\bar{M}_m(\bar{\rho}')$: equivalent magnetic current on boundary contour C_m.

It should be noted that the electric and the magnetic charge terms in the scalar potential integrals are replaced by the corresponding divergence of the electric current and divergence of the magnetic current using the current-charge continuity equations stated in expressions (5.1.3e) and (5.1.14e). Further, when considering the calculation of electric and magnetic fields in layer 1 (free-space medium), the equivalent electric and magnetic currents only on boundary contour C_1 are utilized, and the incident electric and magnetic fields are also included. Similarly, when considering the calculation of the electric and magnetic fields in layer M (interior dielectric region), the equivalent electric and magnetic currents only on boundary contour C_M are utilized. The equivalent electric and magnetic currents on the various boundary contours are still unknown. They can be determined by enforcing the appropriate boundary conditions the boundary contours C_m, $m = 1, 2, 3,$.

. . . , M, which basically yield a set of $2M$ coupled combined field integral equations for $2M$ unknowns. On substituting expressions (6.3.2a–b) into the two boundary conditions (6.3.1a–b) stated earlier, the following coupled combined field integral equations are obtained:

$$\Big[-j\omega\bar{A}_m(\bar{\rho}) - \nabla_\tau\Phi_m(\bar{\rho}) - \frac{1}{\varepsilon'_m}\nabla_\tau \times \bar{F}_m(\bar{\rho})$$
$$+ j\omega\bar{A}_{m+1}(\bar{\rho}) + \nabla_\tau\Phi_{m+1}(\bar{\rho}) + \frac{1}{\varepsilon'_{m+1}}\nabla_\tau \times \bar{F}_{m+1}(\bar{\rho})\Big]\Big/_{\tan}$$
$$= -\bar{E}^i(\bar{\rho})\delta(m-1)\Big/_{\tan} \tag{6.3.5}$$

$$\Big[-j\omega\bar{F}_m(\bar{\rho}) - \nabla_\tau\Psi_m(\bar{\rho}) + \frac{1}{\mu_m}\nabla_\tau \times \bar{A}_m(\bar{\rho})$$
$$+ j\omega\bar{F}_{m+1}(\bar{\rho}) + \nabla_\tau\Psi_{m+1}(\bar{\rho}) - \frac{1}{\mu_{m+1}}\nabla_\tau \times \bar{A}_{m+1}(\bar{\rho})\Big]\Big/_{\tan}$$
$$= -\bar{H}^i(\bar{\rho})\delta(m-1)\Big/_{\tan} \tag{6.3.6}$$

$\bar{\rho}$ on C_m, and m = 1, 2, 3, ... , M

In coupled integral equations (6.3.5) and (6.3.6), there are only two unknown distributions of the equivalent electric and magnetic currents on each boundary contour. Depending on the polarization of the excitation, transverse magnetic or transverse electric, only specific components of the equivalent electric and magnetic currents are excited on the boundary contours.

6.3.1 CFIE – TM CASE

For normal transverse magnetic excitation, the electric and magnetic field distributions in the various layers and the corresponding equivalent magnetic and electric currents are completely independent of the z coordinate variable. Thus

$$\frac{\partial}{\partial z} = 0 \tag{6.3.7}$$

and there are only the following components of the electric and magnetic field distributions:

$E_{mz}(\rho,\phi)\hat{z}$: total axial component of the electric field in layer m;

$H_{mn}(\rho,\phi)\hat{n}$: total normal component of the magnetic field in layer m;

$H_{ms}(\rho,\phi)\hat{s}$: total tangential component of the magnetic field in layer m.

For TM normal excitation, there are only the axial z component of the equivalent electric current distribution and the tangential s component of the equivalent magnetic current distribution on contour C_m, $m = 1, 2, 3, \ldots, M$:

$J_{mz}(\rho',\phi')\hat{z}'_m$: axial component of electric current on contour C_m

$$= \hat{n}'_m \times \hat{s}'_m H_{ms}(\rho',\phi') \qquad \bar{\rho}' \text{ on } C_m \tag{6.3.8}$$

$M_{ms}(\rho',\phi')\hat{s}'_m$: tangential component of magnetic current on contour C_m

$$= \hat{z}'_m E_{mz}(\rho',\phi') \times \hat{n}'_m \qquad \bar{\rho}' \text{ on } C_m \tag{6.3.9}$$

Hence, for TM excitation, taking the z component of integral equation (6.3.5) and s component of integral equation (6.3.6), the following coupled integral equations are obtained:

$$\begin{aligned} &-j\omega \hat{z}_m \cdot \bar{A}_m(\bar{\rho}_m) - \frac{1}{\varepsilon'_m}\hat{z}_m \cdot \nabla_\tau \times \bar{F}_m(\bar{\rho}_m) \\ &+ j\omega \hat{z}_m \cdot \bar{A}_{m+1}(\bar{\rho}_m) + \frac{1}{\varepsilon'_{m+1}}\hat{z}_m \cdot \nabla_\tau \times \bar{F}_{m+1}(\bar{\rho}_m) \\ &\qquad = -\ \hat{z}_m \cdot \bar{E}^i(\bar{\rho}_m)\delta(m-1) \end{aligned} \tag{6.3.10a}$$

$$\begin{aligned} &-j\omega \hat{s}_m \cdot \bar{F}_m(\bar{\rho}_m) - \hat{s}_m \cdot \nabla_\tau \Psi_m(\bar{\rho}_m) + \frac{1}{\mu_m}\hat{s}_m \cdot \nabla_\tau \times \bar{A}_m(\bar{\rho}_m) \\ &+ j\omega \hat{s}_m \cdot \bar{F}_{m+1}(\bar{\rho}_m) + \hat{s}_m \cdot \nabla_\tau \Psi_{m+1}(\bar{\rho}_m) \\ &- \frac{1}{\mu_{m+1}}\hat{s}_m \cdot \nabla_\tau \times \bar{A}_{m+1}(\bar{\rho}_m) \\ &\qquad = -\ \hat{s}_m \cdot \bar{H}^i(\bar{\rho}_m)\delta(m-1) \end{aligned} \tag{6.3.10b}$$

$\bar{\rho}_m$ on C_m, and $m = 1, 2, 3, \ldots, M$

where the vector and scalar potential integrals are given by the following expressions:

$$\bar{A}_m(\bar{\rho}_m) = \frac{\mu_m}{4j} \int_{C_{m-1}} -\hat{z}'_{m-1} J_{(m-1)z}(\bar{\rho}') H_0^{(2)}(k_m|\bar{\rho}_m - \bar{\rho}'|)\, dL'$$

$$+ \frac{\mu_m}{4j} \int_{C_m} \hat{z}'_m J_{mz}(\bar{\rho}') H_0^{(2)}(k_m|\bar{\rho}_m - \bar{\rho}'|)\, dL' \qquad (6.3.11a)$$

$$\bar{F}_m(\bar{\rho}_m) = \frac{\varepsilon'_m}{4j} \int_{C_{m-1}} -\hat{s}'_{m-1} M_{(m-1)s}(\bar{\rho}') H_0^{(2)}(k_m|\bar{\rho}_m - \bar{\rho}'|)\, dL'$$

$$+ \frac{\varepsilon'_m}{4j} \int_{C_m} \hat{s}'_m M_{ms}(\bar{\rho}') H_0^{(2)}(k_m|\bar{\rho}_m - \bar{\rho}'|)\, dL' \qquad (6.3.11b)$$

and

$$\Psi_m(\bar{\rho}_m) = \frac{1}{4\omega\mu_m} \int_{C_{m-1}} -\left[\frac{\partial}{\partial s'} M_{(m-1)s}(\bar{\rho}')\right] H_0^{(2)}(k_m|\bar{\rho}_m - \bar{\rho}'|)\, dL'$$

$$+ \frac{1}{4\omega\mu_m} \int_{C_m} \left[\frac{\partial}{\partial s'} M_{ms}(\bar{\rho}')\right] H_0^{(2)}(k_m|\bar{\rho}_m - \bar{\rho}'|)\, dL' \qquad (6.3.11c)$$

After substituting these potential expressions (6.3.11a–b) into integral equation (6.3.10a), the following simplification is obtained:

$$\frac{\omega\mu_m}{4} \int_{C_{m-1}} \hat{z}_m \bullet \hat{z}'_{m-1} J_{(m-1)z}(\bar{\rho}') H_0^{(2)}(k_m|\bar{\rho}_m - \bar{\rho}'|)\, dL'$$

$$- \frac{\omega\mu_m}{4} \int_{C_m} \hat{z}_m \bullet \hat{z}'_m J_{mz}(\bar{\rho}') H_0^{(2)}(k_m|\bar{\rho}_m - \bar{\rho}'|)\, dL'$$

$$- \frac{\omega\mu_{m+1}}{4} \int_{C_m} \hat{z}_m \bullet \hat{z}'_m J_{mz}(\bar{\rho}') H_0^{(2)}(k_{m+1}|\bar{\rho}_m - \bar{\rho}'|)\, dL'$$

$$+ \frac{\omega\mu_{m+1}}{4} \int_{C_{m+1}} \hat{z}_m \bullet \hat{z}'_{m+1} J_{(m+1)z}(\bar{\rho}') H_0^{(2)}(k_{m+1}|\bar{\rho}_m - \bar{\rho}'|)\, dL'$$

$$+ \frac{1}{4j}\hat{z}_m \bullet \nabla_\tau \times \int_{C_{m-1}} \hat{s}'_{m-1} M_{(m-1)s}(\bar{\rho}') H_0^{(2)}(k_m|\bar{\rho}_m - \bar{\rho}'|)\, dL'$$

$$- \frac{1}{4j}\hat{z}_m \bullet \nabla_\tau \times \int_{C_m} \hat{s}'_m M_{ms}(\bar{\rho}') H_0^{(2)}(k_m|\bar{\rho}_m - \bar{\rho}'|)\, dL'$$

$$- \frac{1}{4j}\hat{z}_m \bullet \nabla_\tau \times \int_{C_m} \hat{s}'_m M_{ms}(\bar{\rho}') H_0^{(2)}(k_{m+1}|\bar{\rho}_m - \bar{\rho}'|)\, dL'$$

$$+ \frac{1}{4j}\hat{z}_m \bullet \nabla_\tau \times \int_{C_{m+1}} \hat{s}'_{m+1} M_{(m+1)s}(\bar{\rho}') H_0^{(2)}(k_{m+1}|\bar{\rho}_m - \bar{\rho}'|)\, dL'$$

$$= - \hat{z}_m \bullet \bar{E}^i(\bar{\rho}_m)\delta(m-1) \tag{6.3.12a}$$

$\bar{\rho}_m$ on C_m, and $m = 1, 2, 3, \ldots, M$

Similarly, after substituting potential expressions (6.3.11a–c) into integral equation (6.3.10b), the following simplification is obtained:

$$\frac{\omega\varepsilon'_m}{4} \int_{C_{m-1}} \hat{s}_m \bullet \hat{s}'_{m-1} M_{(m-1)s}(\bar{\rho}') H_0^{(2)}(k_m|\bar{\rho}_m - \bar{\rho}'|)\, dL'$$

$$- \frac{\omega\varepsilon'_m}{4} \int_{C_m} \hat{s}_m \bullet \hat{s}'_m M_{ms}(\bar{\rho}') H_0^{(2)}(k_m|\bar{\rho}_m - \bar{\rho}'|)\, dL'$$

$$- \frac{\omega\varepsilon'_{m+1}}{4} \int_{C_m} \hat{s}_m \bullet \hat{s}'_m M_{ms}(\bar{\rho}') H_0^{(2)}(k_{m+1}|\bar{\rho}_m - \bar{\rho}'|)\, dL'$$

$$+ \frac{\omega\varepsilon'_{m+1}}{4} \int_{C_{m+1}} \hat{s}_m \bullet \hat{s}'_{m+1} M_{(m+1)s}(\bar{\rho}') H_0^{(2)}(k_{m+1}|\bar{\rho}_m - \bar{\rho}'|)\, dL'$$

$$+ \frac{1}{4\omega\mu_m}\frac{\partial}{\partial s} \int_{C_{m-1}} \left[\frac{\partial}{\partial s'} M_{(m-1)s}(\bar{\rho}')\right] H_0^{(2)}(k_m|\bar{\rho}_m - \bar{\rho}'|)\, dL'$$

$$- \frac{1}{4\omega\mu_m}\frac{\partial}{\partial s} \int_{C_m} \left[\frac{\partial}{\partial s'} M_{ms}(\bar{\rho}')\right] H_0^{(2)}(k_m|\bar{\rho}_m - \bar{\rho}'|)\, dL'$$

$$- \frac{1}{4\omega\mu_{m+1}} \frac{\partial}{\partial s} \int_{C_m} [\frac{\partial}{\partial s'} M_{ms}(\bar{\rho}')] H_0^{(2)}(k_{m+1}|\bar{\rho}_m - \bar{\rho}'|) \, dL'$$

$$+ \frac{1}{4\omega\mu_{m+1}} \frac{\partial}{\partial s} \int_{C_{m+1}} [\frac{\partial}{\partial s'} M_{(m+1)s}(\bar{\rho}')] H_0^{(2)}(k_{m+1}|\bar{\rho}_m - \bar{\rho}'|) \, dL'$$

$$- \frac{1}{4j} \hat{s}_m \bullet \nabla_\tau \times \int_{C_{m-1}} \hat{z}'_{m-1} J_{(m-1)z}(\bar{\rho}') H_0^{(2)}(k_m|\bar{\rho}_m - \bar{\rho}'|) \, dL'$$

$$+ \frac{1}{4j} \hat{s}_m \bullet \nabla_\tau \times \int_{C_m} \hat{z}'_m J_{mz}(\bar{\rho}') H_0^{(2)}(k_m|\bar{\rho}_m - \bar{\rho}'|) \, dL'$$

$$+ \frac{1}{4j} \hat{s}_m \bullet \nabla_\tau \times \int_{C_m} \hat{z}'_m J_{mz}(\bar{\rho}') H_0^{(2)}(k_{m+1}|\bar{\rho}_m - \bar{\rho}'|) \, dL'$$

$$- \frac{1}{4j} \hat{s}_m \bullet \nabla_\tau \times \int_{C_{m+1}} \hat{z}'_{m+1} J_{(m+1)z}(\bar{\rho}') H_0^{(2)}(k_{m+1}|\bar{\rho}_m - \bar{\rho}'|) \, dL'$$

$$= - \hat{s}_m \bullet \vec{H}^i(\bar{\rho}_m) \delta(m - 1) \qquad (6.3.12b)$$

$\bar{\rho}_m$ on C_m, and $m = 1, 2, 3, \ . \ . \ . \ , M$

Further, referring to CFIE integral equations (6.3.12a–b), the terms containing the curl of the electric and magnetic vector potential integrals should be carefully analyzed. Complete details of the procedure is discussed in Section 5.5, expressions (5.5.20) to (5.5.22), for separating the principal value in the integrals of the coupled combined field integral equations. A similar analysis procedure can be adapted to separate out the principal value and simplify the integral terms containing the curl operation in the combined field integral equations (6.3.12a–b). Hence, for TM normal excitation,

$$\frac{\omega\mu_m}{4} \int_{C_{m-1}} J_{(m-1)z}(\bar{\rho}') H_0^{(2)}(k_m R_m) \, dL'$$

$$- \frac{\omega\mu_m}{4} \int_{C_m} J_{mz}(\bar{\rho}') H_0^{(2)}(k_m R_m) \, dL'$$

$$- \frac{\omega\mu_{m+1}}{4} \int_{C_m} J_{mz}(\bar{\rho}') H_0^{(2)}(k_{m+1} R_m) \, dL'$$

$$+ \frac{\omega\mu_{m+1}}{4} \int_{C_{m+1}} J_{(m+1)z}(\bar{\rho}') H_0^{(2)}(k_{m+1}R_m)\, dL'$$

$$+ \frac{k_m}{4j} \int_{C_{m-1}} M_{(m-1)s}(\bar{\rho}') \Big[\frac{x_m - x'}{R_m} \cos\Omega'_{m-1}$$

$$+ \frac{y_m - y'}{R_m} \sin\Omega'_{m-1} \Big] H_0^{(2)\prime}(k_m R_m)\, dL'$$

$$- \frac{k_m}{4j} \int_{C_m} M_{ms}(\bar{\rho}') \Big[\frac{x_m - x'}{R_m} \cos\Omega'_m$$

$$+ \frac{y_m - y'}{R_m} \sin\Omega'_m \Big] H_0^{(2)\prime}(k_m R_m)\, dL'$$

$$- \frac{k_{m+1}}{4j} \int_{C_m} M_{ms}(\bar{\rho}') \Big[\frac{x_m - x'}{R_m} \cos\Omega'_m$$

$$+ \frac{y_m - y'}{R_m} \sin\Omega'_m \Big] H_0^{(2)\prime}(k_{m+1}R_m)\, dL'$$

$$+ \frac{k_{m+1}}{4j} \int_{C_{m+1}} M_{(m+1)s}(\bar{\rho}') \Big[\frac{x_{m+1} - x'}{R_{m+1}} \cos\Omega'_{m+1}$$

$$+ \frac{y_m - y'}{R_m} \sin\Omega'_{m-1} \Big] H_0^{(2)\prime}(k_{m+1}R_m)\, dL'$$

$$= - E_z^i(\bar{\rho}_m)\delta(m-1) \tag{6.3.13a}$$

$\bar{\rho}_m$ on C_m, and $m = 1, 2, 3, \ldots, M$

and

$$R_m = |\bar{\rho}_m - \bar{\rho}'|$$
$$= [(x_m - x')^2 + (y_m - y')^2]^{1/2} \tag{6.3.13b}$$

and the other integral equation (6.3.12b) takes the form:

$$\frac{\omega\varepsilon'_m}{4}\int_{C_{m-1}} \cos[\Omega_m - \Omega'_{m-1}]M_{(m-1)s}(\bar{\rho}')H_0^{(2)}(k_mR_m)\, dL'$$

$$-\frac{\omega\varepsilon'_m}{4}\int_{C_m} \cos[\Omega_m - \Omega'_m]J_{mz}(\bar{\rho}')H_0^{(2)}(k_mR_m)\, dL'$$

$$-\frac{\omega\varepsilon'_{m+1}}{4}\int_{C_m} \cos[\Omega_m - \Omega'_m]J_{mz}(\bar{\rho}')H_0^{(2)}(k_{m+1}R_m)\, dL'$$

$$+\frac{\omega\varepsilon'_{m+1}}{4}\int_{C_{m+1}} \cos[\Omega_m - \Omega'_{m+1}]J_{(m+1)z}(\bar{\rho}')H_0^{(2)}(k_{m+1}R_m)\, dL'$$

$$+\frac{1}{4\omega\mu_m}\frac{\partial}{\partial s}\int_{C_{m-1}} [\frac{\partial}{\partial s'} M_{(m-1)s}(\bar{\rho}')]H_0^{(2)}(k_mR_m)\, dL'$$

$$-\frac{1}{4\omega\mu_m}\frac{\partial}{\partial s}\int_{C_m} [\frac{\partial}{\partial s'} M_{ms}(\bar{\rho}')]H_0^{(2)}(k_mR_m)\, dL'$$

$$-\frac{1}{4\omega\mu_{m+1}}\frac{\partial}{\partial s}\int_{C_m} [\frac{\partial}{\partial s'} M_{ms}(\bar{\rho}')]H_0^{(2)}(k_{m+1}R_m)\, dL'$$

$$+\frac{1}{4\omega\mu_{m+1}}\frac{\partial}{\partial s}\int_{C_{m+1}} [\frac{\partial}{\partial s'} M_{(m+1)s}(\bar{\rho}')]H_0^{(2)}(k_{m+1}R_m)\, dL'$$

$$+\frac{k_m}{4j}\int_{C_{m-1}} J_{(m-1)z}(\bar{\rho}')[\frac{x_m-x'}{R_m}\cos\Omega_{m-1}$$

$$+\frac{y_m-y'}{R_m}\sin\Omega_{m-1}]H_0^{(2)'}(k_mR_m)\, dL'$$

$$-\frac{k_m}{4j}\int_{C_m} J_{mz}(\bar{\rho}')[\frac{x_m-x'}{R_m}\cos\Omega_m$$

$$+\frac{y_m-y'}{R_m}\sin\Omega_m]H_0^{(2)'}(k_mR_m)\, dL'$$

$$- \frac{k_{m+1}}{4j} \int_{C_m} J_{mz}(\bar{\rho}')\Big[\frac{x_m - x'}{R_m} \cos\Omega_m + \frac{y_m - y'}{R_m} \sin\Omega_m\Big] H_0^{(2)'}(k_{m+1}R_m)\, dL'$$

$$+ \frac{k_{m+1}}{4j} \int_{C_{m+1}} J_{(m+1)z}(\bar{\rho}')\Big[\frac{x_{m+1} - x'}{R_{m+1}} \cos\Omega_{m+1} + \frac{y_m - y'}{R_m} \sin\Omega_{m-1}\Big] H_0^{(2)'}(k_{m+1}R_m)\, dL'$$

$$= - \bar{H}_s^i(\bar{\rho}_m)\delta(m - 1) \tag{6.3.13c}$$

$\bar{\rho}_m$ on C_m, and $m = 1, 2, 3, \ldots, M$

6.4 LAYERED DIELECTRIC CIRCULAR CYLINDER

The layered circular scatterer happens to be a special geometry wherein fields can be expressed in terms of cylindrical wave modes. If required, the coupled combined field integral equations, expression (6.3.13a–c), derived in the previous section can be directly solved to analyze the electromagnetic scattering and penetration due to a layered, lossy dielectric circular cylinder excited externally by a TM-polarized plane wave. Figure 6.4 shows the cross-sectional geometry of a two-dimensional, lossy dielectric circular scatterer having radius a_2 and loaded with a single layer of lossy dielectric material having internal radius a_2 and external radius a_1. The layered scatterer has a perfect symmetry with respect to angular variable ϕ of the cylindrical coordinate system. Hence, the complete scattering and penetration analysis can be carried out by just assuming the incident plane wave to be propagating along the x coordinate direction, which gives the angle of incidence to be zero.

Referring to the discussions in Section 2.19, the TM-polarized time-harmonic incident plane wave is given by

$$E_z^i(\bar{\rho},\omega) = E_0\, e^{-j\bar{k}_1 \bullet \bar{\rho}} \tag{6.4.1}$$

$$\bar{k}_1 = k_1\hat{k} \tag{6.4.2}$$

The plane wave electric field propagating in the x coordinate direction can be written in terms of an infinite series of cylindrical wave modes:

$$E_z^i(\rho,\phi) = E_0 e^{-jk_1\rho\cos\phi} \tag{6.4.3a}$$

$$= E_0 \sum_{m=-\infty}^{\infty} j^{-m} J_m(k_1\rho)\, e^{jm\phi} \tag{6.4.3b}$$

Based on the electromagnetic modeling discussed for the derivation of combined field integral equations, the total tangential electric and magnetic fields on the boundary surfaces of the layered scatterer can be viewed as a collection of angular-directed magnetic current and axial-directed electric current distributions. These line source-type distributions basically generate fields with cylindrical symmetry.

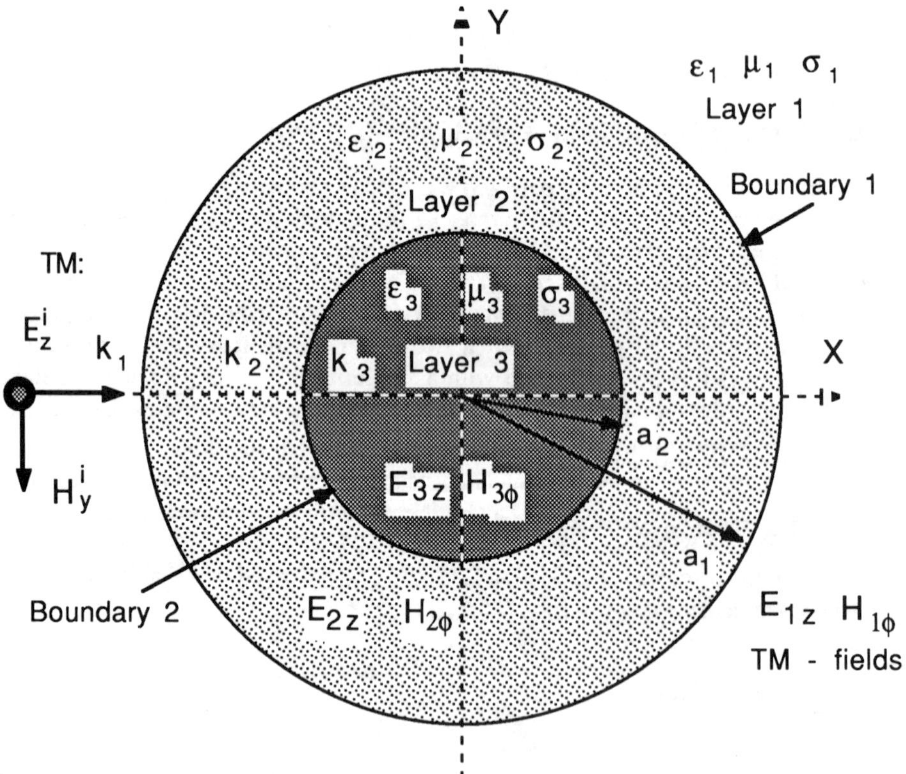

Figure 6.4 Geometry of a circular dielectric scatterer with a single layer of loading.

From expression (3.8.20a), the scattered electric field in the free-space region, layer 1, can be expressed in terms of an infinite number of outgoing propagating cylindrical wave modes:

$$E^{s}_{1z}(\rho,\phi) = E_0 \sum_{m=-\infty}^{\infty} C_{1m} H^{(2)}_{m}(k_1\rho)\, e^{jm\phi} \qquad \rho \geq a_1 \qquad (6.4.4)$$

and the total electric field in layer 1 is given by

$$E_{1z}(\rho,\phi) = E_0 \sum_{m=-\infty}^{\infty} j^{-m} J_m(k_1\rho)\, e^{jm\phi} + E_0 \sum_{m=-\infty}^{\infty} C_{1m} H^{(2)}_{m}(k_1\rho)\, e^{jm\phi} \qquad \rho \geq a_1 \qquad (6.4.5)$$

C_{1m} : unknown scattered modal coefficients for free-space, layer 1.

Further, for the homogeneous lossy dielectric scatterer region, layer 3, the total electric field distribution can be expressed in terms of an infinite number of nonpropagating cylindrical wave modes:

$$E_{3z}(\rho,\phi) = E_0 \sum_{m=-\infty}^{\infty} B_{3m} J_m(k_3\rho)\, e^{jm\phi} \qquad \rho \leq a_2 \qquad (6.4.6)$$

B_{3m} : unknown penetrated field modal coefficients for layer 3.

Further, for a single layer homogeneous, lossy dielectric region, layer 2, the total electric field distribution can be expressed in terms of an infinite number of both nonpropagating and propagating cylindrical wave modes

$$E_{2z}(\rho,\phi) = E_0 \sum_{m=-\infty}^{\infty} B_{2m} J_m(k_2\rho)\, e^{jm\phi} + E_0 \sum_{m=-\infty}^{\infty} C_{2m} H^{(2)}_{m}(k_2\rho)\, e^{jm\phi} \qquad a_2 \leq \rho \leq a_1 \qquad (6.4.7)$$

B_{2m}, C_{2m} : unknown modal coefficients for dielectric region, layer 2.

Similarly, the corresponding expressions for the angular component of the magnetic field distributions can be written. Referring to Section 3.8.3 and using the source-free Faraday's law stated in expression (3.1.5a), and expressing the electric and magnetic fields in terms of the cylindrical coordinate system,

$$-j\omega\mu[H_\rho\hat{\rho} + H_\phi\hat{\phi}] = \frac{1}{\rho}\frac{\partial}{\partial\phi}E_z\hat{\rho} - \frac{\partial}{\partial\rho}E_z\hat{\phi} \tag{6.4.8}$$

Hence, the angular component of the total magnetic field can be expressed in terms of the axial electric field in the respective layers, given by

$$H_{1\phi}(\rho,\phi) = -\frac{jk_1}{\omega\mu_1}E_0\sum_{m=-\infty}^{\infty}[j^{-m}J_m'(k_1\rho) + C_{1m}H_m^{(2)'}(k_1\rho)]\,e^{jm\phi} \qquad \rho \geq a_1 \tag{6.4.9}$$

$$H_{2\phi}(\rho,\phi) = -\frac{jk_2}{\omega\mu_2}E_0\sum_{m=-\infty}^{\infty}[B_{2m}J_m'(k_2\rho) + C_{2m}H_m^{(2)'}(k_2\rho)]\,e^{jm\phi} \qquad a_2 \leq \rho \leq a_1 \tag{6.4.10}$$

$$H_{3\phi}(\rho,\phi) = -\frac{jk_3}{\omega\mu_3}E_0\sum_{m=-\infty}^{\infty}[B_{3m}J_m'(k_3\rho)]\,e^{jm\phi} \qquad \rho \leq a_2 \tag{6.4.11}$$

In the above total field expressions, the set of modal coefficients C_{1m}, B_{2m}, C_{2m}, and B_{3m}, are still unknown. They can be determined by enforcing the electromagnetic boundary conditions (6.3.1a) and (6.3.1b) on the two circular boundaries 1 and 2. According to the two boundary conditions, the axial component of the total electric field should be continuous, and similarly, the angular component of the total magnetic field should be continuous on the dielectric boundary contours. Hence

$$E_{1z}(\rho,\phi) = E_{2z}(\rho,\phi) \tag{6.4.12a}$$

for $\rho = a_1$

$$H_{1\phi}(\rho,\phi) = H_{2\phi}(\rho,\phi) \tag{6.4.12b}$$

and

$$E_{2z}(\rho,\phi) = E_{3z}(\rho,\phi) \tag{6.4.12c}$$

for $\rho = a_2$

$$H_{2\phi}(\rho,\phi) = H_{3\phi}(\rho,\phi) \tag{6.4.12d}$$

Total electric field expressions (6.4.5) to (6.4.7) and total magnetic field expressions (6.4.19) to (6.4.11) are substituted into the preceding boundary conditions and invoking modal orthogonality, the following (4 x 4) modal matrix equation is obtained based on the coupled boundary condition expressions

$$[M][A] = [V] \tag{6.4.13}$$

where the matrix [*M*] has the following elements:

$$\begin{bmatrix} H_m^{(2)}(k_1a_1) & -J_m(k_2a_1) & -H_m^{(2)}(k_2a_1) & 0 \\ \frac{k_1}{\mu_1}H_m^{(2)'}(k_1a_1) & -\frac{k_2}{\mu_2}J_m'(k_2a_1) & -\frac{k_2}{\mu_2}H_m^{(2)'}(k_2a_1) & 0 \\ 0 & J_m(k_2a_2) & H_m^{(2)}(k_2a_2) & -J_m(k_3a_2) \\ 0 & \frac{k_2}{\mu_2}J_m'(k_2a_2) & \frac{k_2}{\mu_2}H_m^{(2)'}(k_2a_2) & -\frac{k_3}{\mu_3}J_m'(k_3a_2) \end{bmatrix} \tag{6.4.14a}$$

The excitation column vector [*V*] has the form

$$\begin{bmatrix} -j^{-m}J_m(k_1a_1) \\ -\frac{k_1}{\mu_1}j^{-m}J_m'(k_1a_1) \\ 0 \\ 0 \end{bmatrix} \tag{6.4.14b}$$

and the unknown modal coefficient column vector $[A]$ has the form

$$\begin{bmatrix} C_{1m} \\ B_{2m} \\ C_{2m} \\ B_{3m} \end{bmatrix} \qquad (6.4.14c)$$

Once the modal coefficients are known, the axial components of the electric field and the angular components of the magnetic field in the three regions can be calculated. Following the procedure explained in Section 3.8, in layer 1 (free-space medium), the far-field distribution and the corresponding radar cross section can be calculated by taking the large argument approximation of the Hankel function in expression (6.4.4).

Figures 6.5 and 6.6 show the distribution of the induced surface electric current and induced surface magnetic current on the outer boundary contour C_1 of a lossless single-layer loaded dielectric scatterer. In fact, these two current distributions are obtained by first calculating the total surface tangential field distributions on the boundary contour, and then converting them into the corresponding equivalent currents. The conductivity of the two layers is taken as zero. The relative permittivity and relative permeability of layer 3 are selected equal to 4 and 2, and similarly, the relative permittivity and relative permeability of layer 2 are selected equal to 2 and 1, respectively, for the circular layered scatterer. The frequency of excitation is selected as 300 MHz, so that the free-space wavelength is equal to 1 meter. The free-space propagation constant k_1 is equal 2π, and the electrical radii of the circular regions, as referred to the free-space medium, are given by $k_1 a_1 = 5$ and $k_1 a_2 = 3$. Figure 6.5 shows the real and imaginary parts of the equivalent electric current distribution, and similarly, Figure 6.6 shows the real and imaginary parts of the equivalent magnetic current distribution on boundary contour C_1 separating layers 1 and 2. The electric current is normalized with respect to the incident magnetic field, and the magnetic current is normalized with respect to the incident electric field. The two current distributions on the circular boundary are plotted as a function of angular variable ϕ with the TM incident electric field assumed to be 1 volt per meter. Only the surface current distributions in the angular range 180° to 360° are shown and, in fact, are symmetrical in the remaining angular range.

6.4.1 CIRCULAR MULTILAYER LOADING

As can be seen from matrix equation (6.4.14), for the case of two boundary contours separating three layers, there are four unknown modal coefficients. Further

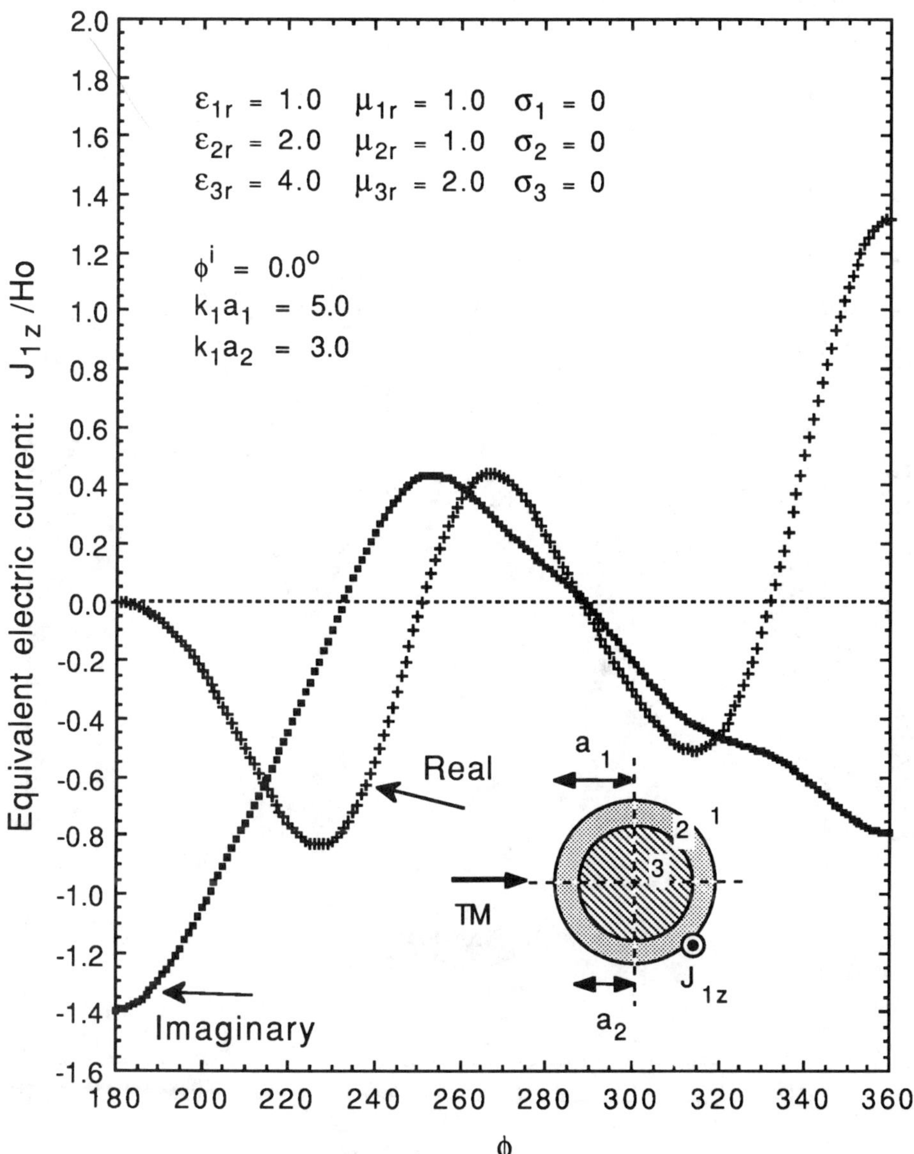

Figure 6.5 Equivalent surface electric current on the outer boundary of a circular dielectric scatterer with a single layer dielectric loading – TM excitation.

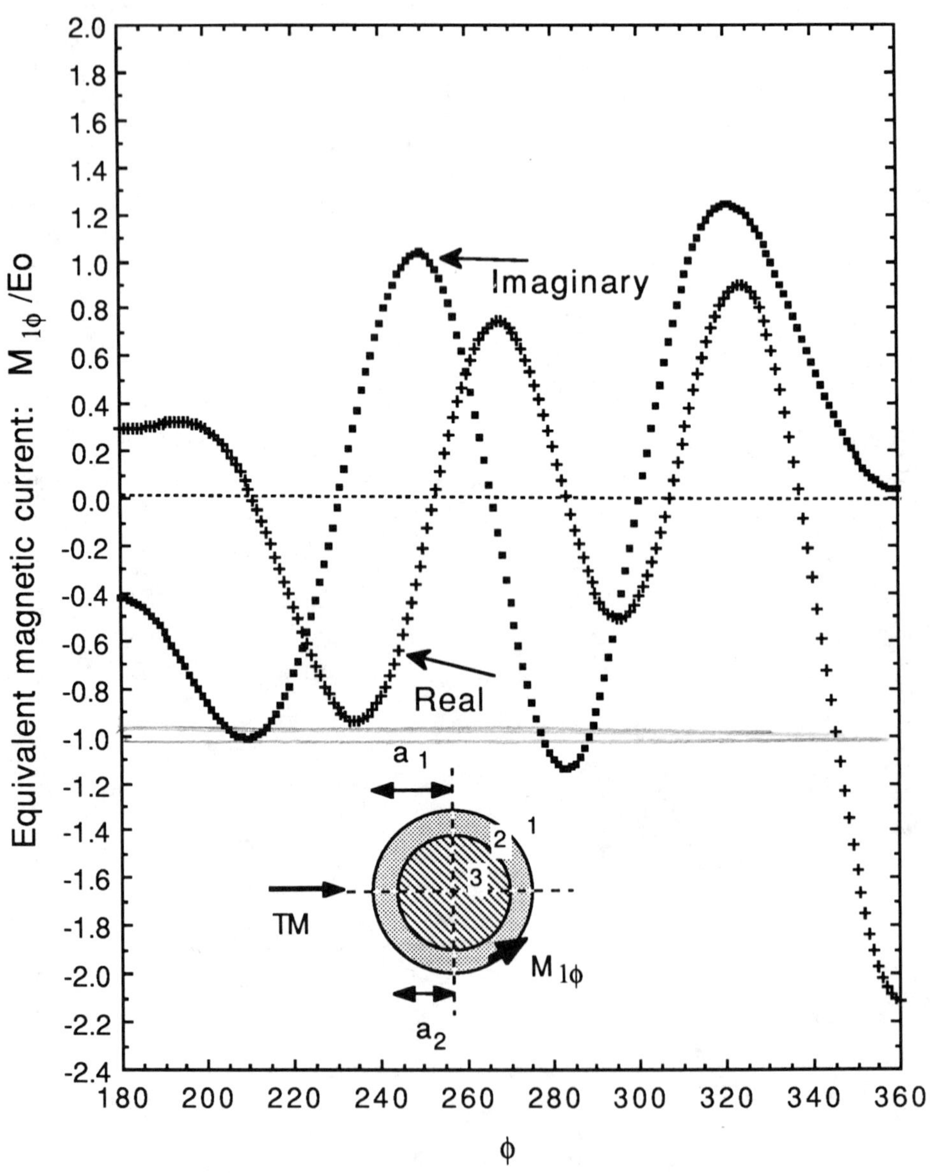

Figure 6.6 Equivalent surface magnetic current on the outer boundary of a circular dielectric scatterer with a single layer dielectric loading – TM excitation.

generalization can be easily accomplished for N boundary contours separating $(N+1)$ layers as depicted in Figure 6.7.

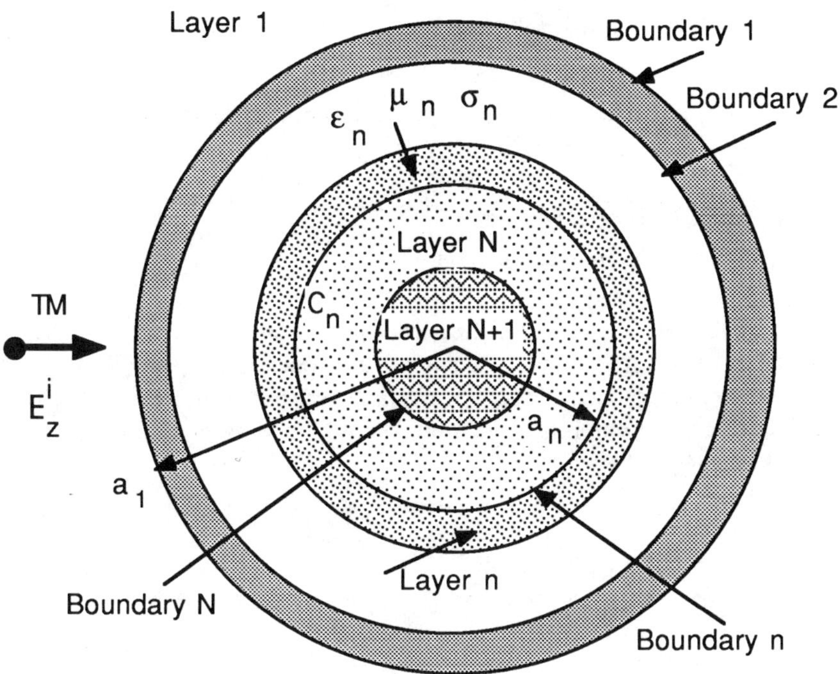

Figure 6.7 Circular dielectric scatterer with multilayer dielectric loading.

For TM normal excitation, referring to expression (6.4.5), for the homogeneous lossy dielectric layer n, the electric and magnetic field distributions can be written as

$$E_{nz}(\rho,\phi) = E_0 \sum_{m=-\infty}^{\infty} B_{nm} J_m(k_n\rho)\, e^{jm\phi}$$

$$+ E_0 \sum_{m=-\infty}^{\infty} C_{nm} H_m^{(2)}(k_n\rho)\, e^{jm\phi} \qquad a_n \le \rho \le a_{n-1} \tag{6.4.15}$$

and

$$H_{n\phi}(\rho,\phi) = -\frac{jk_n}{\omega\mu_n}E_0 \sum_{m=-\infty}^{\infty} [B_{nm}J_m'(k_n\rho)$$

$$+ C_{nm}H_m^{(2)'}(k_n\rho)]e^{jm\phi} \qquad a_n \leq \rho \leq a_{n-1} \tag{6.4.16}$$

B_{nm}, C_{nm} : unknown modal coefficients for the dielectric region, layer n,
n = 1, 2, 3, . . . , $N+1$;

a_n : radius of boundary contour C_n;
a_{n-1} : radius of boundary contour C_{n-1}.

Referring to the earlier analysis, for free-space layer 1 the modal coefficient $B_{1m} = j^{-m}$ and for the last dielectric layer $N + 1$ the modal coefficient $C_{(N+1)m} = 0$. In preceding total field expressions, the modal coefficients for the various layers are still unknown and can be determined by enforcing electromagnetic boundary conditions (6.3.1a) and (6.3.1b) on circular boundaries C_n, $n = 1, 2, 3, \ldots, N$. Hence

$$E_{nz}(\rho,\phi) = E_{(n+1)z}(\rho,\phi) \tag{6.4.17a}$$

$$H_{n\phi}(\rho,\phi) = H_{(n+1)\phi}(\rho,\phi) \tag{6.4.17b}$$

for $\rho = a_n$
$n = 1, 2, 3, \ldots, N.$

After substituting total electric field expression (6.4.15) and total magnetic field expression (6.4.16) into boundary conditions (6.4.17a–b), and enforcing the orthogonality condition for the cylindrical modes, a modal matrix equation of $2N \times 2N$ is obtained for the modal coefficients. In fact, the matrix equation is sparse and can be solved by back substitution as demonstrated later.

For example, with N = 5, the (10 x 10) sparse matrix equation is given by

$$[S][A] = [V] \tag{6.4.18a}$$

$[A]$: unknown modal coefficients column vector;
$[V]$: excitation column vector;

and the elements of the sparse matrix $[S]$ are given by

$$\begin{bmatrix} S_{11} & S_{12} & S_{13} & 0 & 0 & 0 & 0 & 0 & 0 & 0 \\ S_{21} & S_{22} & S_{23} & 0 & 0 & 0 & 0 & 0 & 0 & 0 \\ 0 & S_{32} & S_{33} & S_{33} & S_{35} & 0 & 0 & 0 & 0 & 0 \\ 0 & S_{42} & S_{43} & S_{44} & S_{45} & 0 & 0 & 0 & 0 & 0 \\ 0 & 0 & 0 & S_{54} & S_{55} & S_{56} & S_{57} & 0 & 0 & 0 \\ 0 & 0 & 0 & S_{64} & S_{65} & S_{66} & S_{67} & 0 & 0 & 0 \\ 0 & 0 & 0 & 0 & 0 & S_{76} & S_{77} & S_{78} & S_{79} & 0 \\ 0 & 0 & 0 & 0 & 0 & S_{86} & S_{87} & S_{88} & S_{89} & 0 \\ 0 & 0 & 0 & 0 & 0 & 0 & 0 & S_{98} & S_{99} & S_{910} \\ 0 & 0 & 0 & 0 & 0 & 0 & 0 & S_{108} & S_{109} & S_{1010} \end{bmatrix}$$

(6.4.18b)

Expression (6.4.18a) is solved to obtain the unknown modal coefficients column vector. It is not necessary to invert the matrix equation (6.4.18a). Since it is highly diagonalized, straightforward back substitution procedure can be adopted. This solution approach is discussed in a later section.

6.5 CIRCULAR CONDUCTOR WITH LAYERED DIELECTRIC

In the previous section, a detailed procedure was presented to analyze the electromagnetic scattering and penetration due to a layered, lossy dielectric circular cylinder excited externally by a TM-polarized plane wave. In fact, the field expressions are also applicable with minor modification of the boundary conditions for the special case of a perfectly conducting circular scatterer loaded externally with layers of lossy dielectric material. It should be noted that the total electric and magnetic fields inside the perfectly conducting region is zero. On the boundary of the perfectly conducting scatterer, the axial component of the total electric field is zero and the angular component of the total magnetic field gives directly the induced electric current distribution. Further, it should be noted that there is no magnetic current distribution on the boundary surface of the perfectly conducting scatterer.

Figure 6.8 shows the cross-sectional geometry of a two dimensional, perfectly conducting, circular scatterer having radius equal to a_2 and loaded with a single layer of lossy dielectric material having internal radius a_2 and external radius a_1.

As discussed in the previous sections, for TM normal excitation, the total electric fields are given by

$$E_{1z}(\rho,\phi) = E_0 \sum_{m=-\infty}^{\infty} j^{-m} J_m(k_1\rho)\, e^{jm\phi} + E_0 \sum_{m=-\infty}^{\infty} C_{1m} H_m^{(2)}(k_1\rho)\, e^{jm\phi} \qquad \rho \geq a_1 \tag{6.5.1}$$

$$E_{nz}(\rho,\phi) = E_0 \sum_{m=-\infty}^{\infty} B_{nm} J_m(k_n\rho)\, e^{jm\phi} + E_0 \sum_{m=-\infty}^{\infty} C_{nm} H_m^{(2)}(k_n\rho)\, e^{jm\phi} \qquad a_n \leq \rho \leq a_{n-1} \tag{6.5.2}$$

Similarly, the total magnetic fields are given by

$$H_{1\phi}(\rho,\phi) = -\frac{jk_1}{\omega\mu_1} E_0 \sum_{m=-\infty}^{\infty} \left[j^{-m} J_m'(k_1\rho) + C_{1m} H_m^{(2)'}(k_1\rho) \right] e^{jm\phi} \qquad \rho \geq a_1 \tag{6.5.3}$$

$$H_{n\phi}(\rho,\phi) = -\frac{jk_n}{\omega\mu_n} E_0 \sum_{m=-\infty}^{\infty} \left[B_{nm} J_m'(k_n\rho) + C_{nm} H_m^{(2)'}(k_n\rho) \right] e^{jm\phi} \qquad a_n \leq \rho \leq a_{n-1} \tag{6.5.4}$$

These field expressions are for N boundary contours separating N layers as depicted in Figure 6.9 and also applicable for the single-layer geometry with n = 2. In these total field expressions, the set of modal coefficients can be determined by enforcing the electromagnetic boundary conditions:

$$E_{nz}(\rho,\phi) = E_{(n+1)z}(\rho,\phi) \tag{6.5.5a}$$

for $\rho = a_n$

$$H_{n\phi}(\rho,\phi) = H_{(n+1)\phi}(\rho,\phi) \tag{6.5.5b}$$

and

$$E_{Nz}(\rho,\phi) = 0 \qquad \text{for } \rho = a_N \tag{6.5.5c}$$

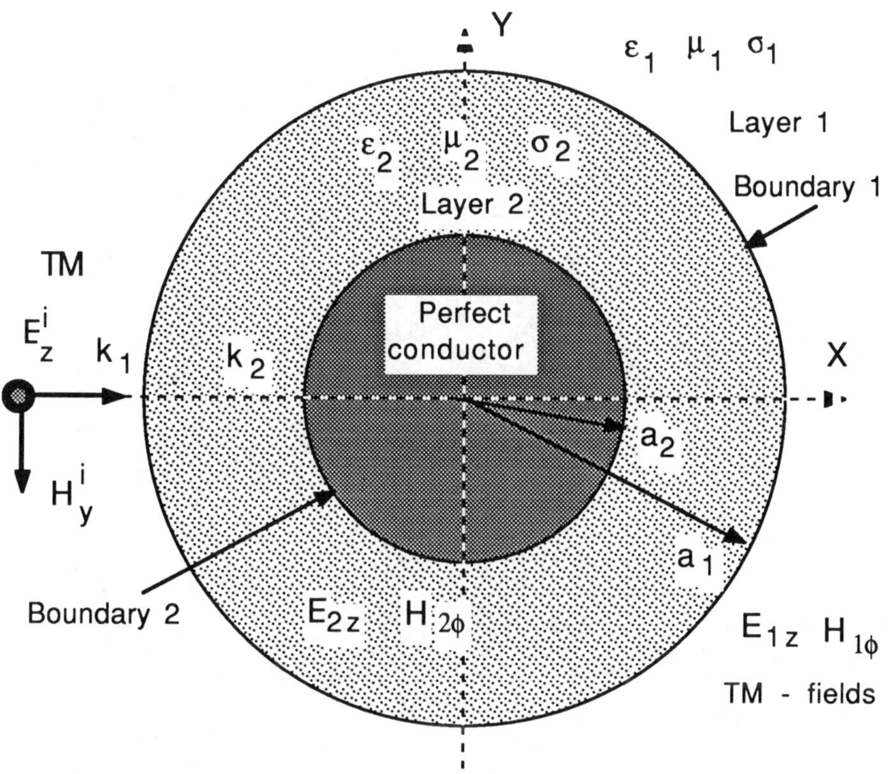

Figure 6.8 Geometry of a circular conducting scatterer with a single layer of loading.

The total electric field expressions (6.5.1) and (6.5.2), and the total magnetic field expressions (6.5.3) and (6.5.4) are substituted into the above boundary conditions. By invoking modal orthogonality, the expression (6.4.13) reduces to a (3 x 3) modal matrix equation by deleting the last row and column

$$[M][A] = [V] \tag{6.5.6}$$

where

$[A]$: unknown modal coefficients column vector;

$[V]$: excitation column vector;

and the matrix $[M]$ is of order (3 x 3) for the case of single-layer dielectric loading. It has the following elements:

$$\begin{bmatrix} H_m^{(2)}(k_1a_1) & -J_m(k_2a_1) & -H_m^{(2)}(k_2a_1) \\ \frac{k_1}{\mu_1}H_m^{(2)'}(k_1a_1) & -\frac{k_2}{\mu_2}J_m'(k_2a_1) & -\frac{k_2}{\mu_2}H_m^{(2)'}(k_2a_1) \\ 0 & J_m(k_2a_2) & H_m^{(2)}(k_2a_2) \end{bmatrix} \tag{6.5.7a}$$

The excitation column vector $[V]$ has the form

$$\begin{bmatrix} -j^{-m}J_m(k_1a_1) \\ -\frac{k_1}{\mu_1}j^{-m}J_m'(k_1a_1) \\ 0 \end{bmatrix} \tag{6.5.7b}$$

and the unknown modal coefficient column vector $[A]$ has the form

$$\begin{bmatrix} C_{1m} \\ B_{2m} \\ C_{2m} \end{bmatrix} \tag{6.5.7c}$$

Once the modal coefficients are known, the axial components of the electric field and the angular components of the magnetic field in the two regions can be calculated. Figures 6.10 and 6.11 show the distribution of induced surface electric current and induced surface magnetic current on outer boundary contour C_1 of a lossless, single-layer loaded perfectly conducting scatterer. The conductivity of the single-layer loading is taken as zero. The relative permittivity and relative permeability of layer 2 are selected equal to 4 and 2. The frequency of excitation is selected as 300 MHz, so that the free-space wavelength is equal to 1 meter. The free-space propagation constant k_1 is equal to 2π, and the electrical radii of the circular regions, as referred to the free-space medium, are given by $k_1a_1 = 5$ and $k_1a_2 = 3$. Figure 6.10 shows the real and imaginary parts of the equivalent electric current distribution, and similarly, Figure 6.11 shows the real and imaginary parts of the equivalent magnetic current distribution on boundary contour C_1 separating layers 1 and 2. The electric current is normalized with respect to the incident magnetic field, and the magnetic current is normalized with respect to the incident electric field. The two current distributions on the circular boundary are plotted as a function of angular variable ϕ with the TM incident electric field assumed to be 1 volt per meter.

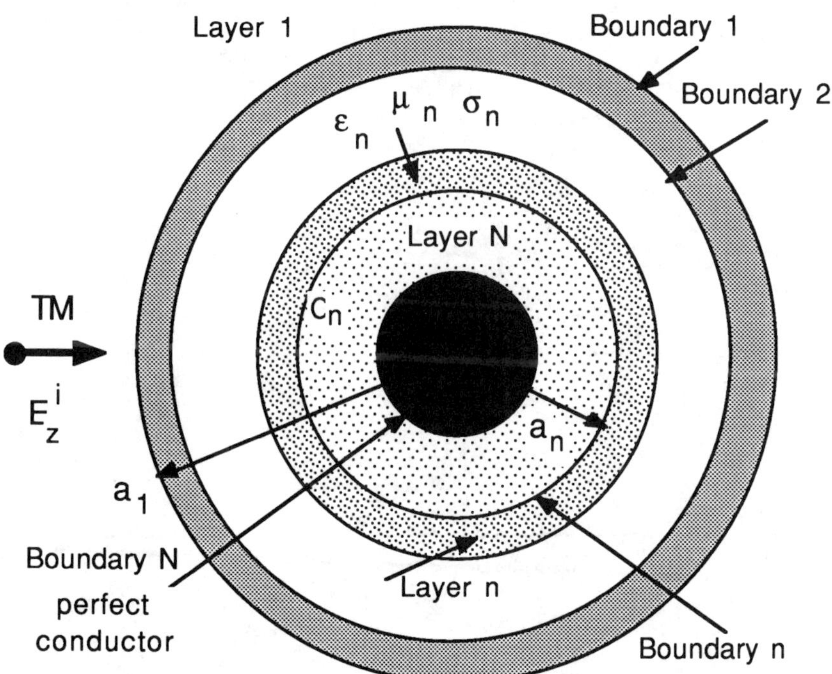

Figure 6.9 Geometry of a circular conducting scatterer with multilayer of loading.

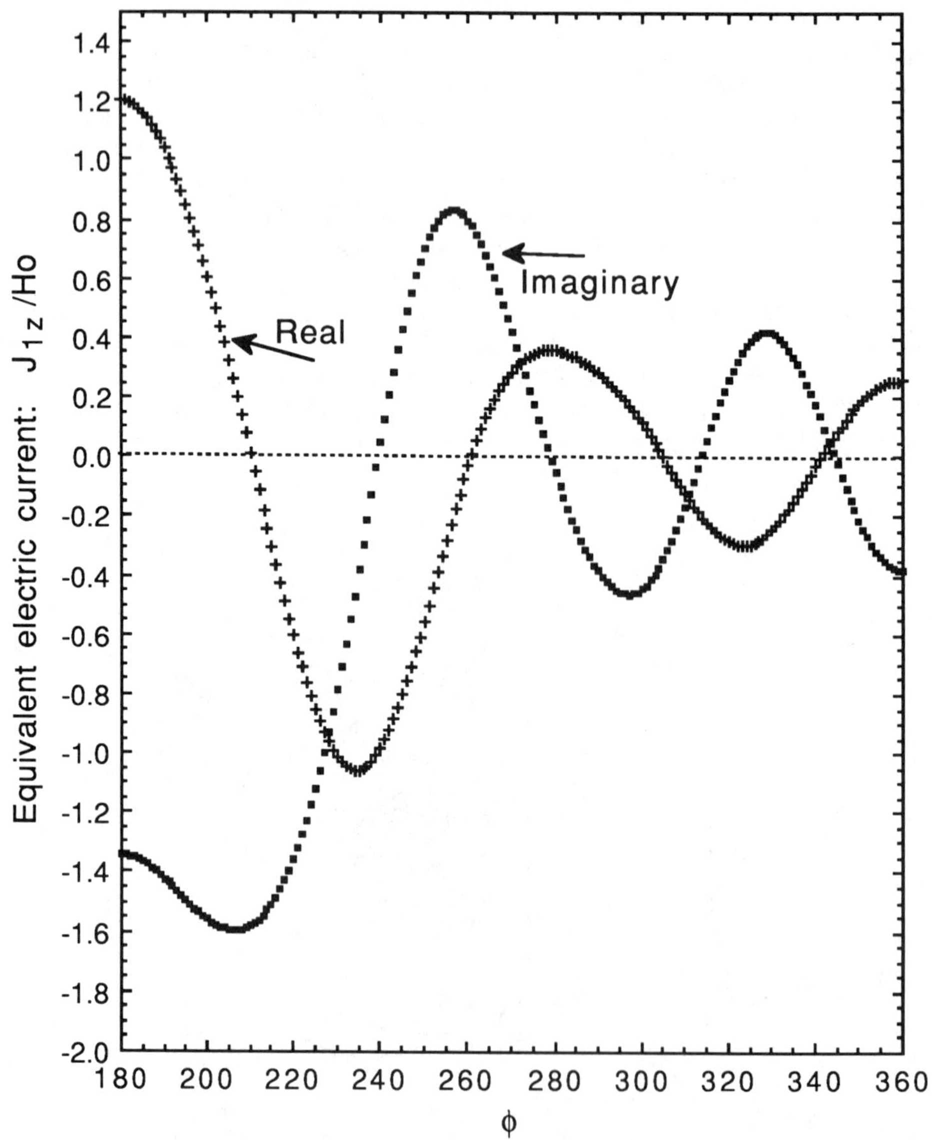

Figure 6.10 Equivalent surface electric current on the outer boundary of a circular conducting scatterer with a single layer dielectric loading – TM excitation.

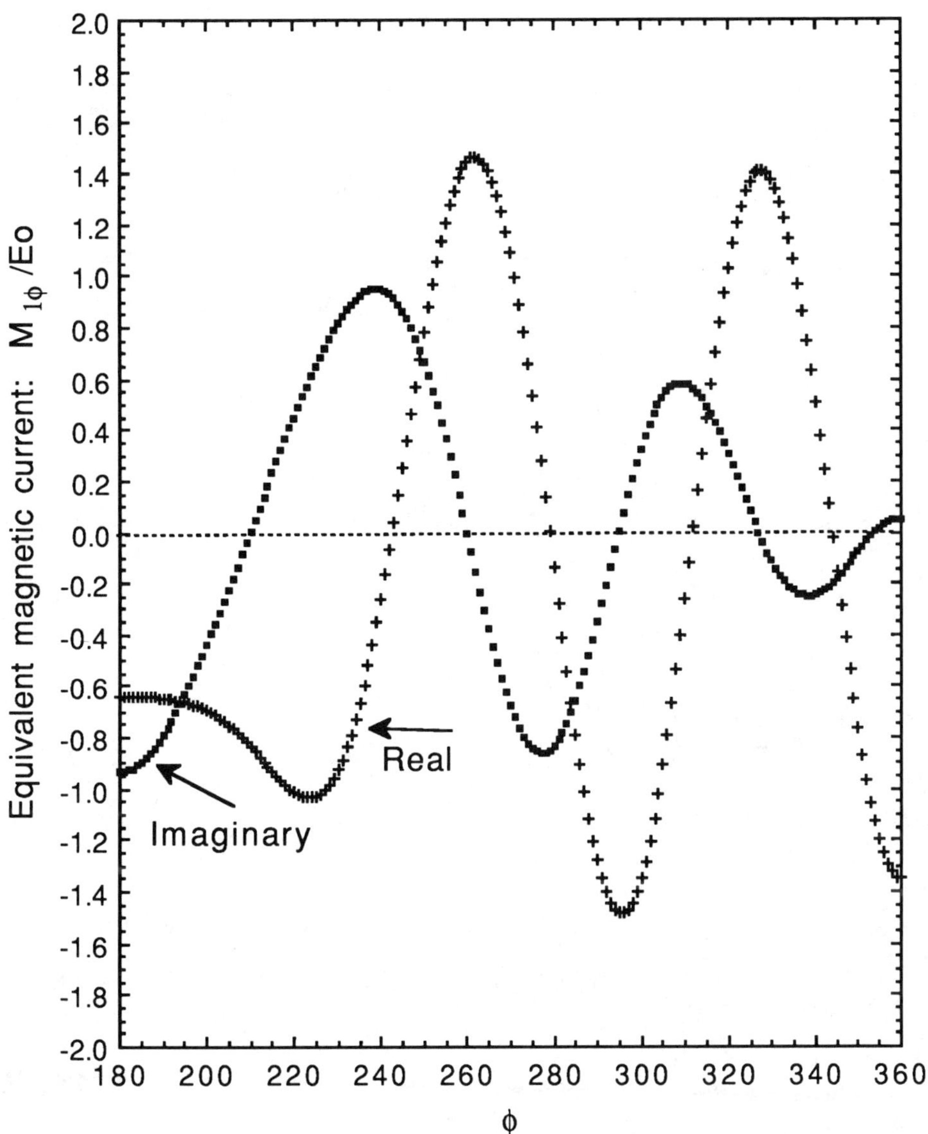

Figure 6.11 Equivalent surface magnetic current on the outer boundary of a circular conducting scatterer with a single layer dielectric loading – TM excitation.

6.5.1 SOLUTION BY BACK SUBSTITUTION

It is not always necessary to invert modal matrix equation (6.5.6) to determine the modal coefficients. In fact, for N layers matrix equation (6.5.6) is completely sparse, and the solution for the modal coefficients can be easily determined by back substitution. This is accomplished by redefining the expressions for the total axial component of the electric field and the total angular component of the magnetic field. Referring to the field expressions (6.5.2) and (6.5.4)

$$E_{nz}(\rho,\phi) = E_0 \sum_{m=-\infty}^{\infty} B_{nm}\Big[J_m(k_n\rho) + P_{nm}H_m^{(2)}(k_n\rho)\Big]\, e^{jm\phi} \qquad a_n \le \rho \le a_{n-1} \tag{6.5.8}$$

$$H_{n\phi}(\rho,\phi) = -\frac{jk_n}{\omega\mu_n} E_0 \sum_{m=-\infty}^{\infty} B_{nm}\Big[J_m'(k_n\rho) + P_{nm}H_m^{(2)'}(k_n\rho)\Big]\, e^{jm\phi} \qquad a_n \le \rho \le a_{n-1} \tag{6.5.9}$$

$$E_{(N+1)z}(\rho,\phi) = 0 \tag{6.5.10a}$$

$\rho \le a_N$, layer $N+1$

$$H_{(N+1)\phi}(\rho,\phi) = 0 \tag{6.5.10b}$$

where the modified modal coefficients

$$P_{nm} = \frac{C_{nm}}{B_{nm}} \qquad \text{for layers } n = 1, 2, 3, \; . \; . \; . \; , N \tag{6.5.11a}$$

$$B_{1m} = j^{-m} \tag{6.5.11b}$$

After substituting total electric field expressions (6.5.8) and total magnetic field expressions (6.5.9) into boundary conditions (6.5.5a–b) and enforcing the orthogonality condition for the cylindrical modes, the following relationships are obtained for the evaluation of each modal coefficient:

$$B_{nm}\left[J_m(k_n a_n) + P_{nm} H_m^{(2)}(k_n a_n)\right]$$

$$= B_{(n+1)m}\left[J_m[k_{(n+1)} a_n] + P_{(n+1)m} H_m^{(2)}[k_{(n+1)} a_n]\right] \tag{6.5.12a}$$

and

$$\frac{k_n}{\mu_n} B_{nm}\left[J_m'(k_n a_n) + P_{nm} H_m^{(2)'}(k_n a_n)\right]$$

$$= \frac{k_{(n+1)}}{\mu_{(n+1)}} B_{(n+1)m}\left[J_m'[k_{(n+1)} a_n] + P_{(n+1)m} H_m^{(2)'}[k_{(n+1)} a_n]\right] \tag{6.5.12b}$$

$$n = 1, 2, 3, \quad . \quad . \quad . \quad , N\text{-}1$$

Expression (6.5.12b) is divided by expression (6.5.12a) on both sides, and the resulting relationship can be cross multiplied and rearranged in the following form to express the mth modal coefficient of layer n in terms of the corresponding m^{th} modal coefficient of the adjacent layer $(n + 1)$ which are separated by the boundary contour C_n having the radius a_n

$$P_{nm} = \frac{-\dfrac{k_n}{\mu_n} J_m'(k_n a_n) + j\omega Y_{(n+1)m} J_m(k_n a_n)}{\dfrac{k_n}{\mu_n} H_m^{(2)'}(k_n a_n) - j\omega Y_{(n+1)m} H_m^{(2)}(k_n a_n)} \tag{6.5.13a}$$

$$Y_{(n+1)m} = \frac{k_{(n+1)}}{j\omega\mu_{(n+1)}} \frac{J_m'[k_{(n+1)} a_n] + P_{(n+1)m} H_m^{(2)'}[k_{(n+1)} a_n]}{J_m[k_{(n+1)} a_n] + P_{(n+1)m} H_m^{(2)}[k_{(n+1)} a_n]} \tag{6.5.13b}$$

Using expressions (6.5.13a–b), the following iterative steps are adapted for obtaining all the modal coefficients. The layers $(N + 1)$ and N are separated by the boundary contour C_N. The layer $(N + 1)$ is the perfectly conducting region. On the boundary contour C_N, the axial component of the total electric field is zero and the angular component of the total magnetic field is finite. Hence,

Step 1 For $n = N$, the radius $a_n = a_N$
The modal admittance $Y_{(N+1)m} \to \infty$. From expression (6.5.13a),

$$P_{Nm} = -\frac{J_m(k_N a_N)}{H_m^{(2)}(k_N a_N)} \tag{6.5.14}$$

Step 2 For $n = N\text{-}1$, the radius $a_n = a_{N-1}$
Substitute expression (6.5.14) for P_{Nm} into expression (6.5.13b) to calculate the modal admittance Y_{Nm}. Substitute modal admittance Y_{Nm} into expression (6.5.13a) to calculate the modal coefficient $P_{(N-1)m}$. Thus, all modal coefficients P_{nm} for $n = N, N\text{-}1, N\text{-}2, \ . \ . \ . \ , 1$, are iteratively calculated.

Step 3 Using expression (6.5.11b), the value of B_{1m} is known, for $n = 1$, the radius $a_n = a_1$. Choosing $n = 1$ in expression (6.5.12a), the modal coefficients B_{1m}, P_{1m}, P_{2m} are known, and modal coefficient B_{2m} is obtained. Choosing $n = 2$ in expression (6.5.12a), modal coefficients B_{2m}, P_{2m}, P_{3m} are known, and modal coefficient B_{3m} is obtained.

The back substitution is repeated for all the remaining values of $n = 3, 4, 5, \ . \ . \ . \ ,$ N and modal coefficients B_{nm} are calculated. Using expression (6.5.11a), modal coefficients C_{nm} are calculated.

Once the modal coefficients are known for the various layered regions, the distribution of the axial component of the electric field and the angular component of the magnetic fields can calculated based on expressions (6.5.8) and (6.5.9). The calculation steps discussed are quite useful in cases such as dense layers and layers having very large and very small permeability and permittivity characteristics.

6.5.2 NUMERICAL SOLUTION – LAYERED CASE

Based on the analysis procedure discussed in the previous sections, the near electric field and the near magnetic field distributions are obtained for a circular conducting cylinder loaded with concentric, multiple layers of dielectric loading. Figure 6.8 shows the geometry of a circular conducting scatterer with a single layer of dielectric loading. Layer 1 is the free-space medium, and layer 3 forms the perfectly conducting circular region. The layer 2 forms a single layer lossless dielectric loading on the circular conducting scatterer. Layer 2 parameters are selected as $\sigma_2 = 0$, $\varepsilon_{r2} = 4$, and $\mu_{r2} = 1$. The loaded scatterer is excited at a frequency of 300 MHz (free-space wavelength 1 meter) by an external TM-polarized plane wave propagating at an angle of incidence $\phi^i = 0$. The electrical radius of the circular conducting scatterer referred to the free-space medium ($k_1 = 2\pi$) is selected as $k_1 a_2 = 5$ and the thickness of layer 2 is selected as $a_1 - a_2 = 1/40$ meter.

Figure 6.12a shows the distribution of the z component of total electric field in layers 1 and 2 calculated based on expressions (6.5.8). Both the penetrated electric field in the layer 2 region and the near total electric field in the free-space region are displaced. As can be seen, the distribution of the z components of total electric field is continuous across the boundary surface. The numerical data shown are obtained for fixed angles ϕ = 0, 30, 60, 90, 120, 150, and 180, and the radial variable ρ is varied in small increments. Using expressions (6.5.9), Figure 6.12b shows the distribution of the angular components of the total magnetic field, which also exhibits continuous behavior at the boundary surface. Similarly, using expression (6.4.8), Figure 6.12c shows the distribution of the radial component of the total magnetic field.

Figures 6.13a–c show the results for a circular conducting scatterer with double layers of dielectric loading. Again, layer 1 is the free-space medium and layer 4 forms the perfectly conducting circular region. Layers 2 and 3 form the double layer of lossless dielectric loading on the circular conducting scatterer. Layer 2 parameters are selected as $\sigma_2 = 0$, $\varepsilon_{r2} = 1.5$ and $\mu_{r2} = 1.5$, and similarly, layer 3 parameters are selected as $\sigma_3 = 0$, $\varepsilon_{r3} = 2$ and $\mu_{r3} = 2$. The loaded scatterer is excited at a frequency of 300 MHz (free-space wavelength 1 meter) by an external TM-polarized plane wave propagating at an angle of incidence $\phi^i = 0$. The electrical radius of the circular conducting scatterer referred to the free-space medium ($k_1 = 2\pi$) is selected as $k_1 a_2 = 5$ and the widths of layers 2 and 3 are taken as $a_1 - a_2 = 1/10$ meter and $a_2 - a_3 = 1/10$ meter.

Figure 6.13a shows the distribution of the z component of the total electric field in three layers 1, 2, and 3 calculated based on the expressions (6.5.8). Both the penetrated electric field in the layers 2 and 3, and the near total electric field in the free-space region are displayed. As can be seen, the distribution of the z components of the total electric field is continuous across the boundary surfaces. The numerical data shown are obtained for fixed angles ϕ = 0, 30, 60, 90, 120, 150, and 180, and the radial variable ρ is varied in small increments. Similarly, Figure 6.13b shows the distribution of the angular components of the total magnetic field, and Figure 6.13c shows the distribution of the radial component of the total magnetic field.

To obtain the scattered field distributions, the corresponding incident field terms are subtracted from the total field expressions. The scattered far-field distribution is obtained by substituting the large argument approximation for the Hankel function, and the radar cross section can be calculated using the definition given by expression (5.10.21). Figure 6.14 shows the bistatic radar cross section for a circular conducting scatterer with five layers of dielectric loading. Case 1 corresponds to a nonpermeable dielectric loading

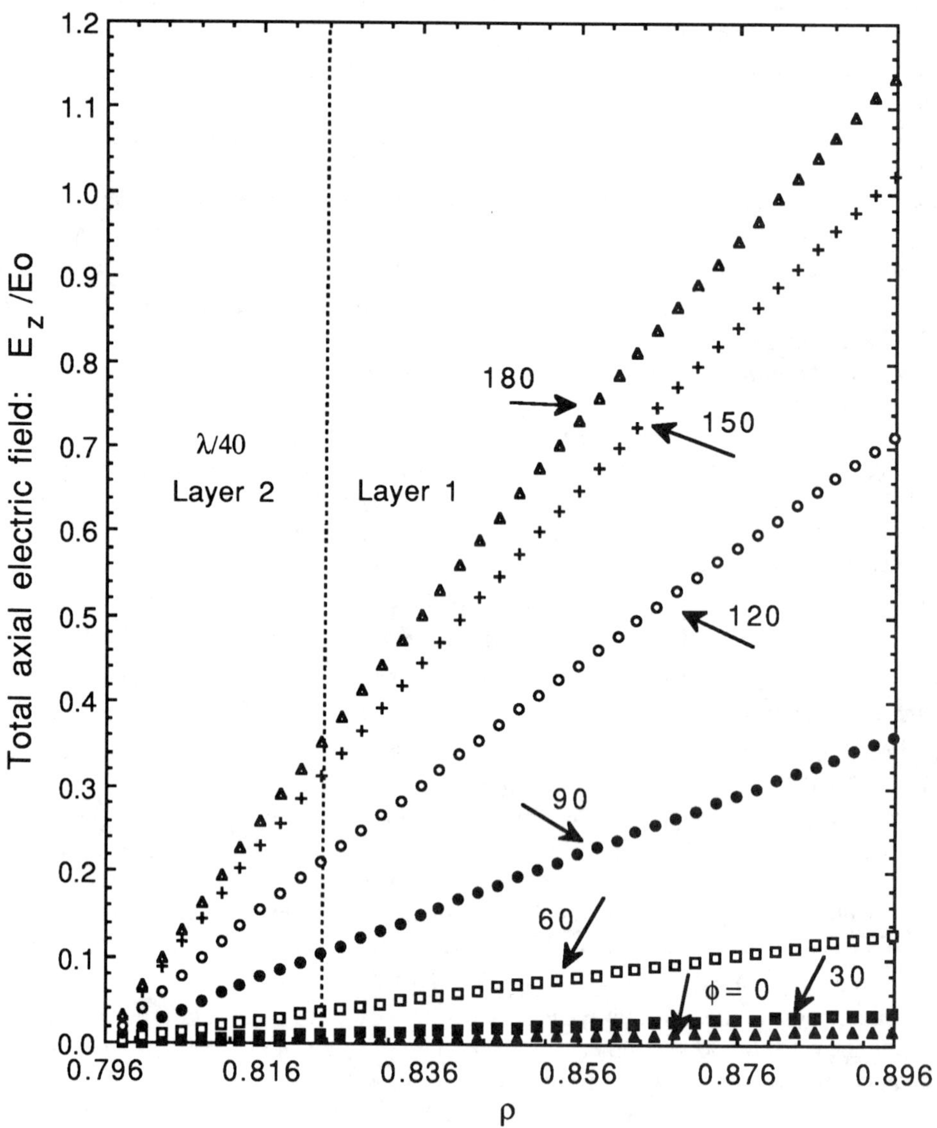

Figure 6.12a Distribution of the axial component of the total electric field near a circular conducting scatterer with a single layer of dielectric loading.

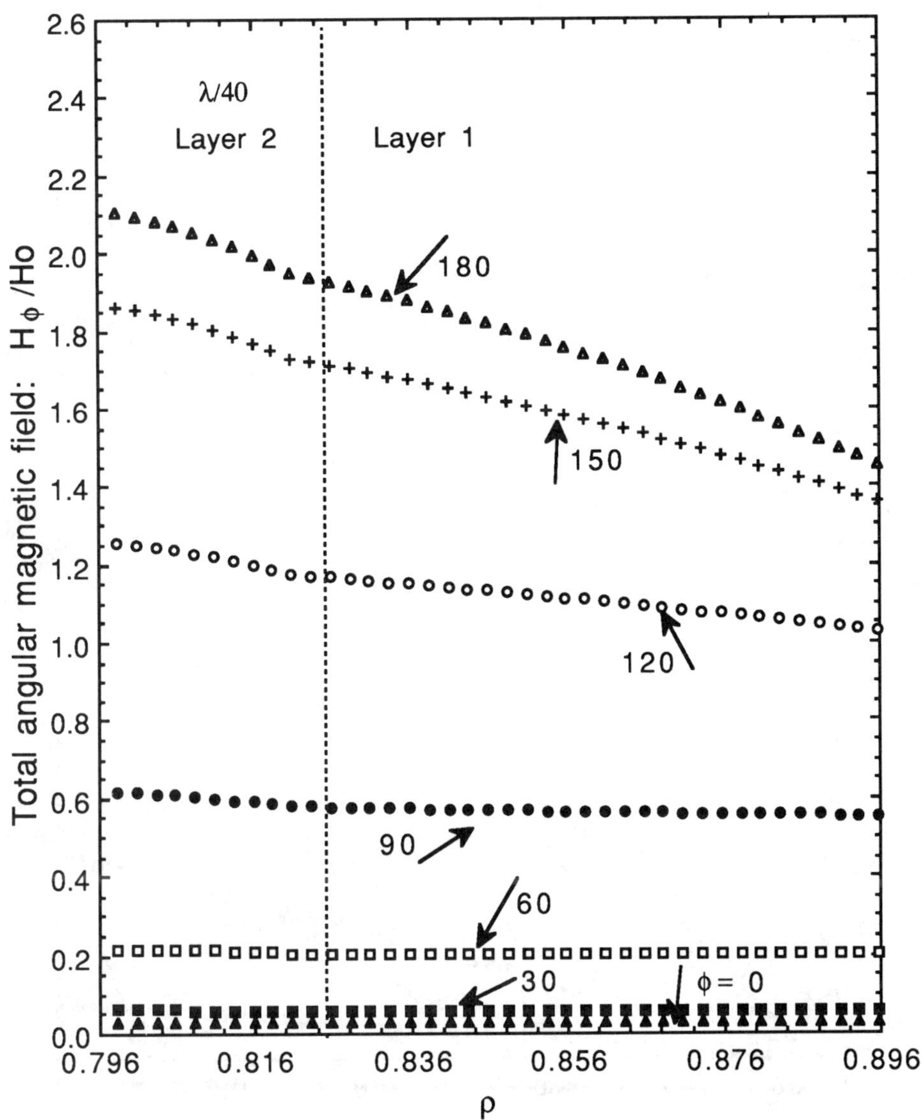

Figure 6.12b Distribution of the angular component of the total magnetic field near a circular conducting scatterer with a single layer of dielectric loading.

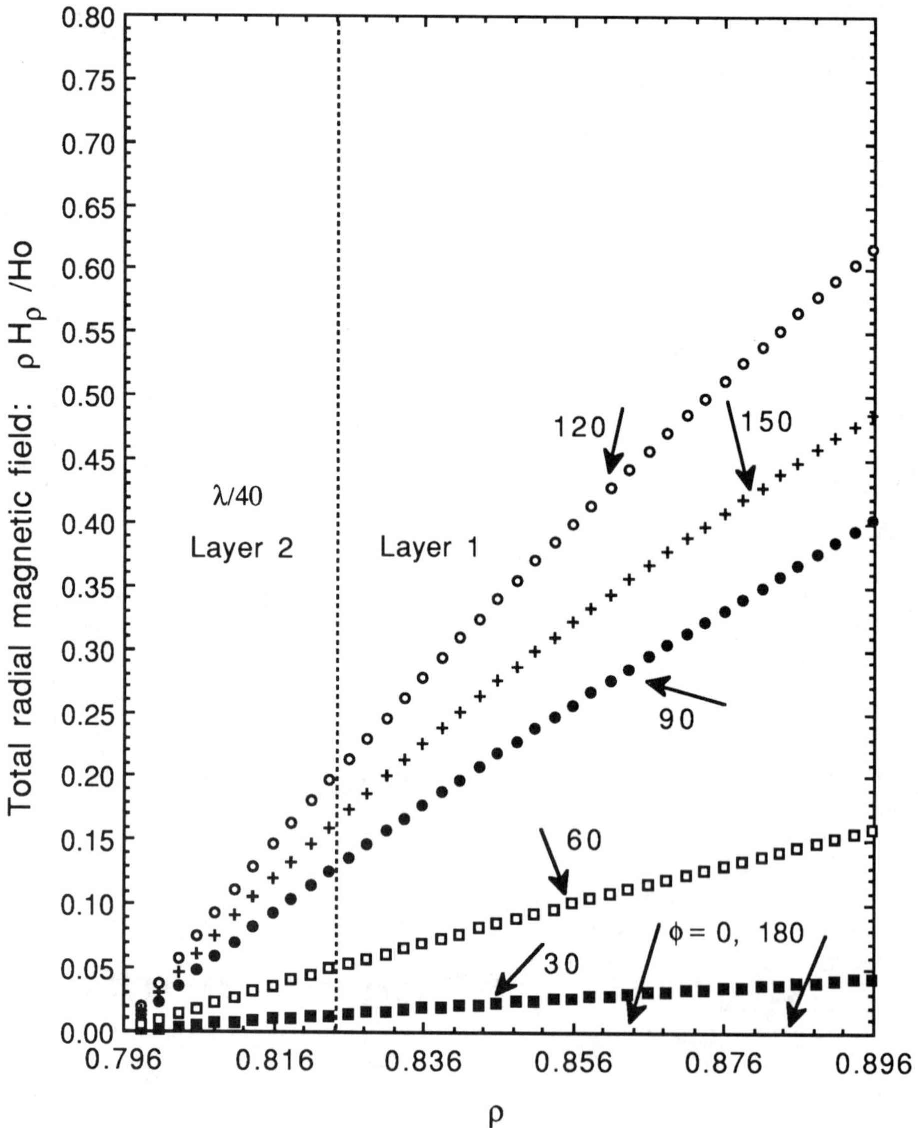

Figure 6.12c Distribution of the radial component of the total magnetic field near a circular conducting scatterer with a single layer of dielectric loading.

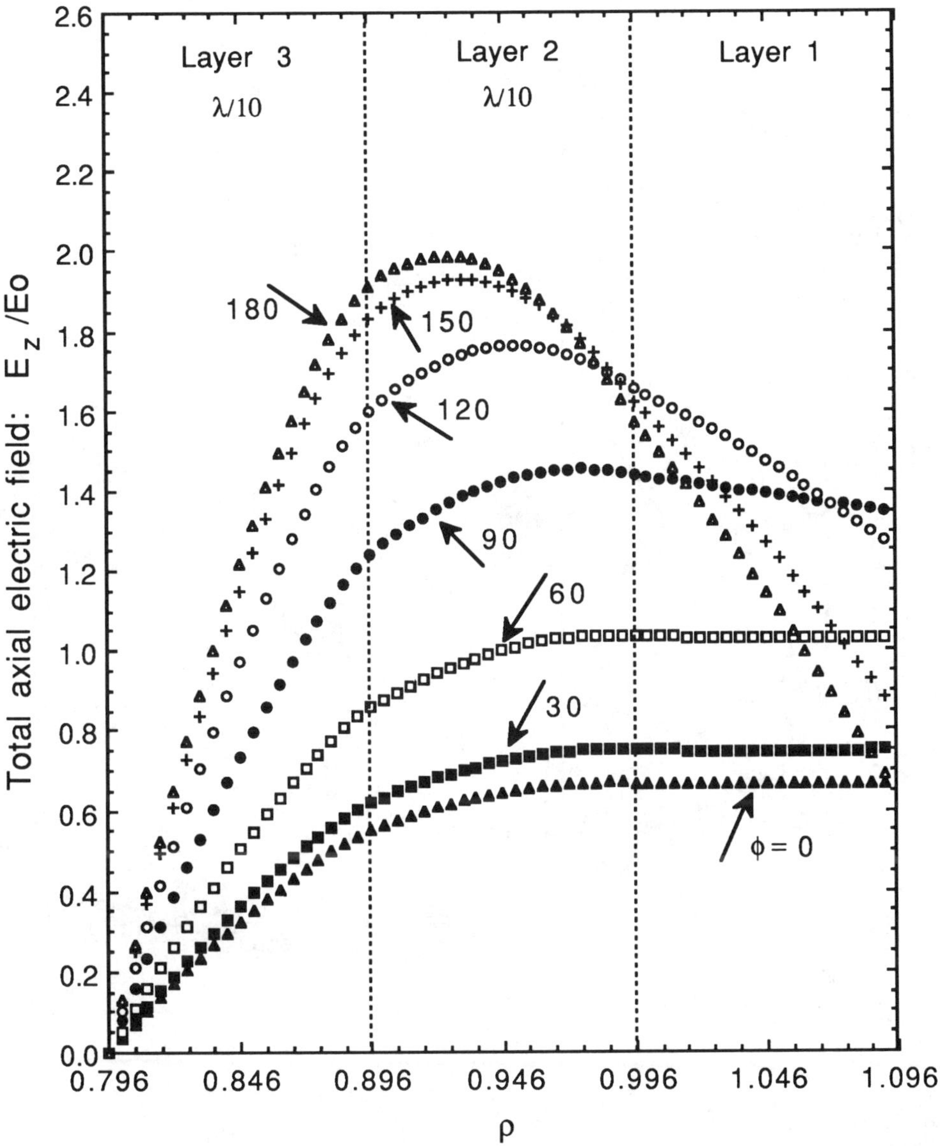

Figure 6.13a Distribution of the axial component of the total electric field near a circular conducting scatterer with double layers of dielectric loading.

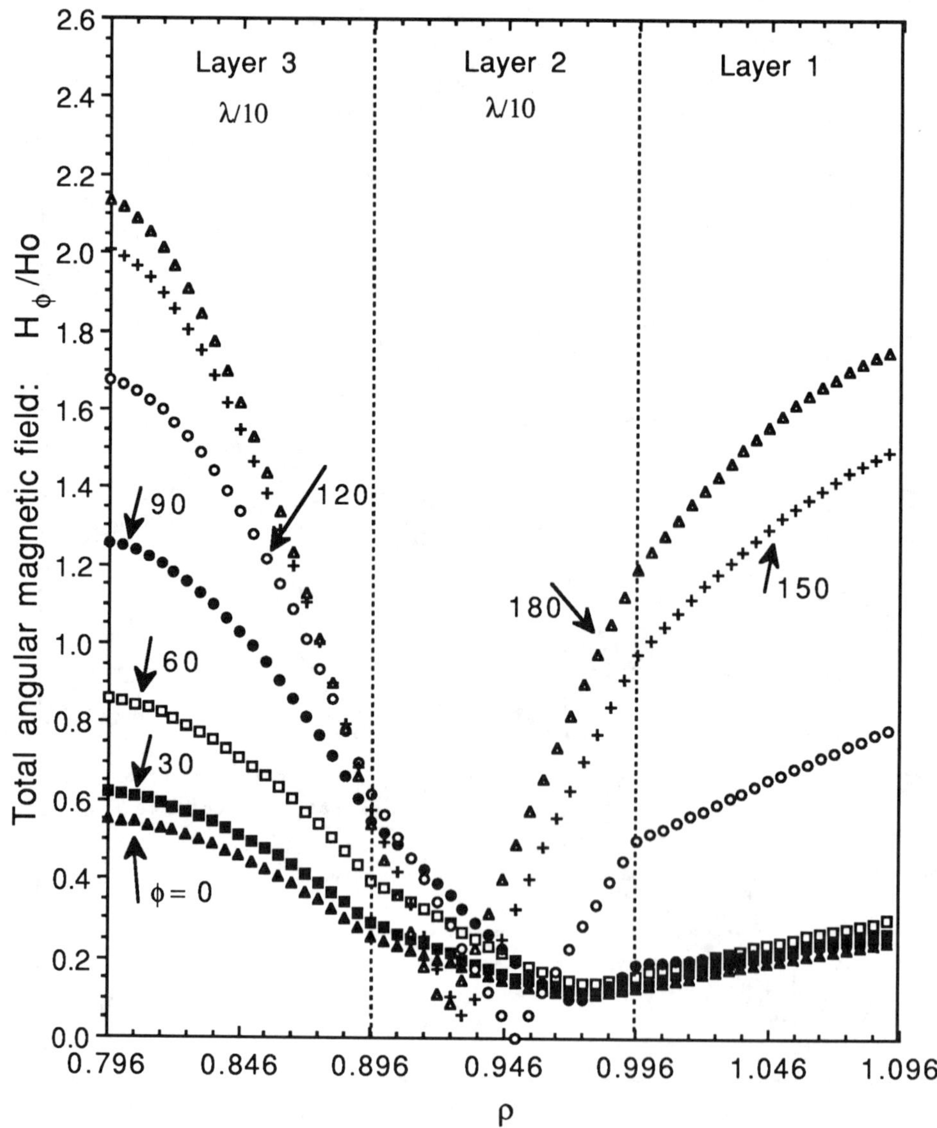

Figure 6.13b Distribution of the angular component of the total magnetic field near a circular conducting scatterer with double layers of dielectric loading.

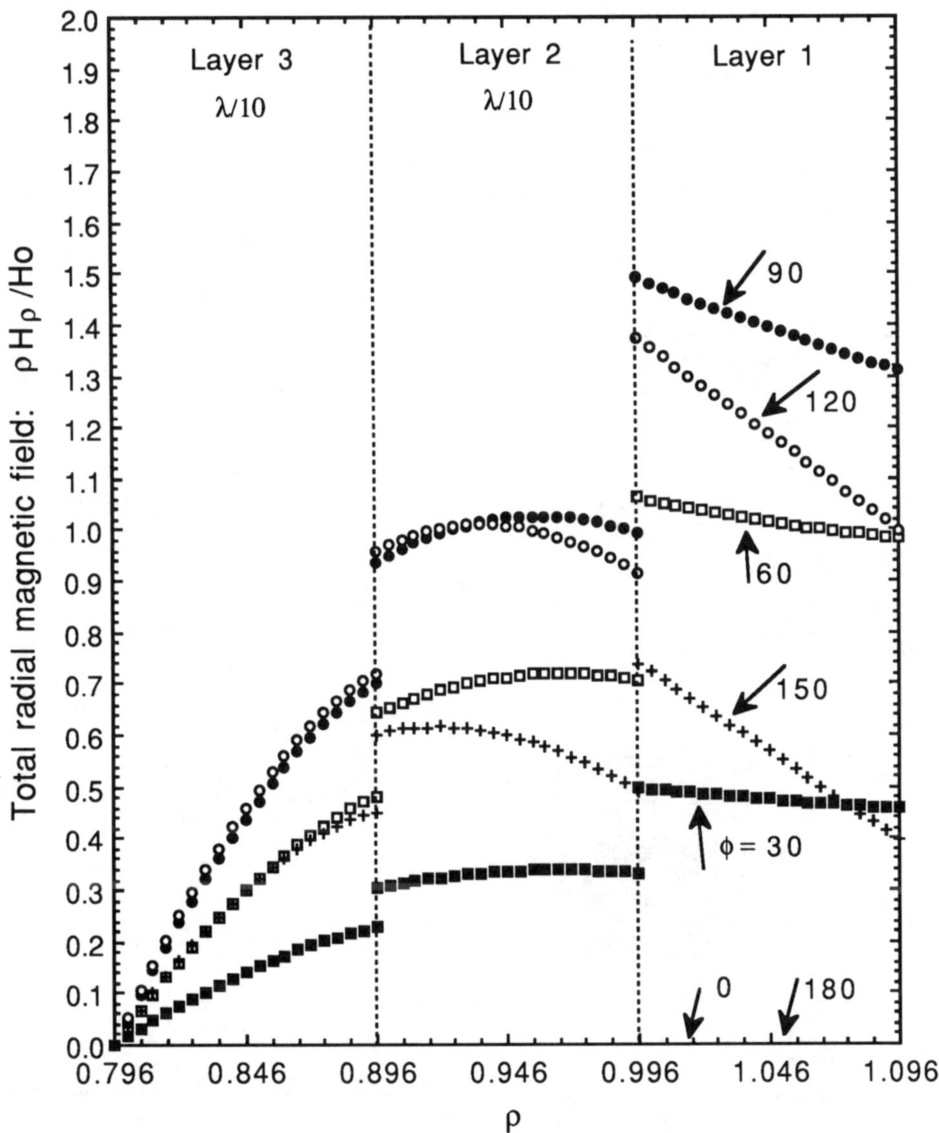

Figure 6.13c Distribution of the radial component of the total magnetic field near a circular conducting scatterer with double layers of dielectric loading.

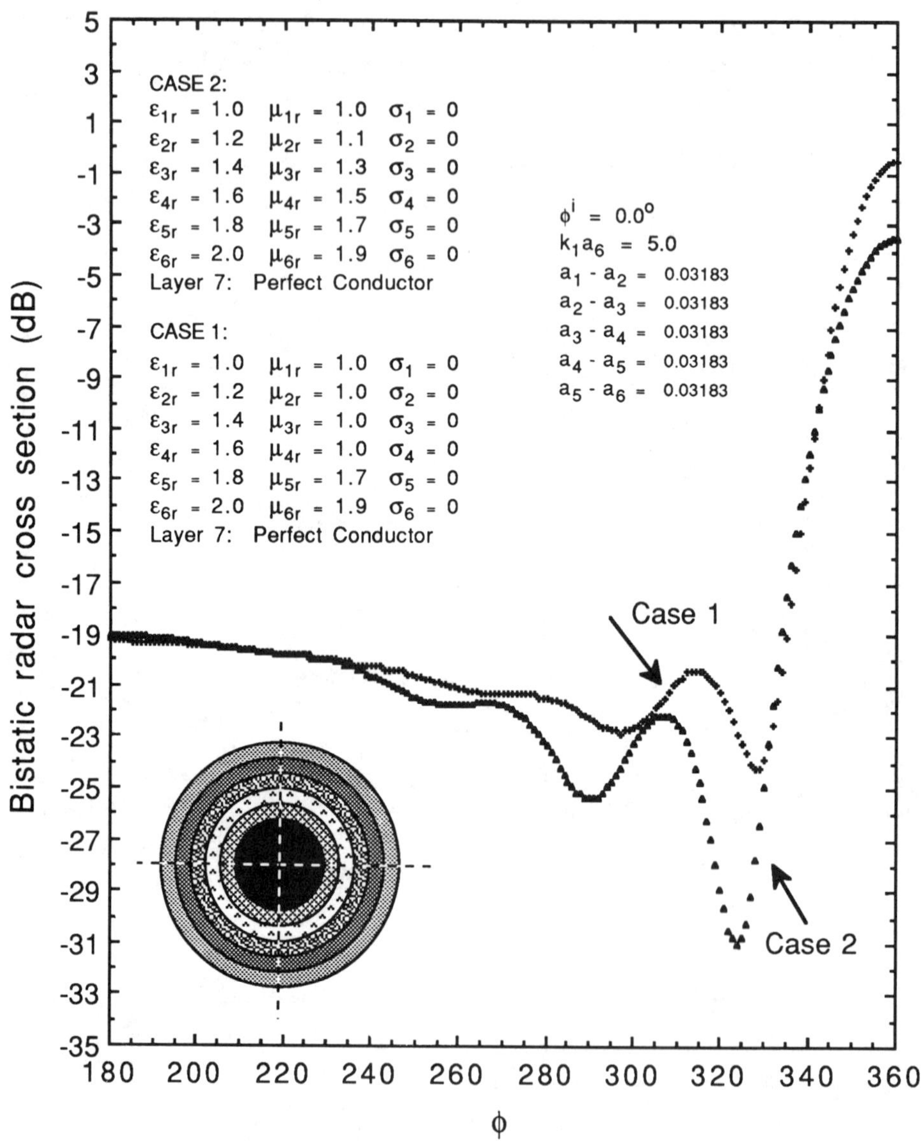

Figure 6.14 Bistatic radar cross section of a circular conducting scatterer with five layers of dielectric loading – TM excitation.

and case 2 corresponds to a general lossless dielectric loading having both permittivity and permeability variations.

6.6 LAYERED ARBITRARY CROSS SECTION - TM EXCITATION

The electromagnetic scattering and penetration by a conducting circular cylinder with concentric multiple layers of dielectric loading is discussed in the previous section based on a rigorous analytical approach. But, the circular loaded scatterer happens to be a special scattering geometry wherein the electric and magnetic fields can be conveniently expanded in terms of the eigenfunction series expansions. For a two-dimensional, loaded scatterer with an arbitrary cross section having arbitrary edges and corners, the analysis technique is quite involved.

In section 6.3, an analysis procedure was discussed in detail to formulate the CFIE (6.3.10a–b) for the multiple-layered case. In this section, a numerical analysis is briefly presented to solve the CFIE based on the method of moments technique described in section 3.7. In fact, many of the numerical steps closely follow the discussion given in section 5.7 as applied to a homogeneous dielectric scatterer. For a loaded scatterer, the numerical technique can extended to include the additional unknown current distributions along the boundary contours separating adjacent dielectric layers. For the single-layer loaded conducting scatterer case, on contour C_1, two unknown electric and magnetic current distributions exist, similarly on contour C_2, an unknown electric current distribution exists, and these three unknown distributions are to be determined based on the CFIE. If one has the knowledge of the types of distribution of electric and magnetic currents, then it is easier to choose an approximate representation by suitable sampling of the distribution functions. Depending on the electrical size of the dielectric scatterer and its arbitrary cross section having arbitrary edges and corners, the electric and magnetic currents may have a singularity distribution or a well-behaved continuous distribution with nulls or even rapid phase changes in the distribution. Similar to the modeling of conducting and dielectric scatterers excited by TM or TE polarization, a number of considerations should be given as to the choice and arrangement of the expansion functions. The modeling of the axial electric current distribution is similar to the modeling of the TM currents, and similarly, the modeling of the tangential magnetic current distribution is similar to the modeling of the TE currents. In fact, the distribution of the axial electric current is discontinuous near sharp corners. But, the distribution of the tangential magnetic current is always continuous along the closed contour, even for the scatterer with arbitrary edges and sharp corners.

The unknown distributions of the electric currents $\bar{\boldsymbol{J}}_{1z}$ on contour C_1 and $\bar{\boldsymbol{J}}_{2z}$ on contour C_2, and the unknown magnetic current $\overline{\boldsymbol{M}}_{1s}$ on contour C_1 are now expanded in terms of piecewise, linearly independent pulse expansion functions. If the contours of the loaded scatterer are completely smooth with no geometrical discontinuities, then it is possible to pick arbitrary sampling regions for the unknown currents along the contours.

On the other hand, if the contours of the loaded scatterer are not smooth, but have sharp corners, care should to be taken in properly positioning the unknown current expansion functions with respect to the geometrical discontinuity. From the numerical modeling point of view, the axial electric current expansion functions are always arranged to be half a cell away from the geometrical corner so that it is not enclosed. Similarly, the tangential magnetic current expansion functions close to the wedge-type corner are always arranged on both sides of the corner so that they enclose completely the geometrical discontinuity point. Such an arrangement also provides continuity of the tangential magnetic current at the corner. Thus, the corner of geometry is spanned by the magnetic current expansion function on either side by half a cell.

Hence, for the arbitrary shaped, single-layer dielectric loaded conducting scatterer a compact generalized partitioned matrix equation is obtained:

$$\begin{bmatrix} Z_{mn}^{J1J1} & Z_{mp}^{J1M1} & Z_{mu}^{J1J2} \\ Y_{qn}^{M1J1} & Y_{qp}^{M1M1} & Y_{qu}^{M1J2} \\ Z_{hn}^{J2J1} & Z_{hp}^{J2M1} & Z_{hu}^{J2J2} \end{bmatrix} \begin{bmatrix} I_n \\ M_p \\ I_u \end{bmatrix} = \begin{bmatrix} E_m \\ H_q \\ 0 \end{bmatrix} \tag{6.6.1}$$

The above partitioned matrix equation can be solved for the unknown expansion coefficients. For the case of multi-layered scatteter, the formulation leads to a highly sparse matrix equation which can be solved for by the method of back substitution.

Detailed computer algoritms can now be developed based on the numerical techniques presented in chapter 5 and in the earlier sections of this chapter.

6.7 NUMERICAL SOLUTION – ARBITRARY LAYERED CASE

Based on the numerical analysis discussed in the previous section, matrix equation (6.6.1) is now solved to obtain the equivalent surface electric and magnetic currents. First, the case of a circular conducting cylinder loaded with a concentric single layer of dielectric loading is analyzed. Figure 6.8 shows the geometry of a circular conducting scatterer with a single layer of dielectric loading. Layer 1 is the free-space medium, and layer 3 forms the perfectly conducting circular region. Layer 2 forms a single layer of lossless dielectric loading on the circular conducting scatterer. The layer 2 parameters are selected as $\sigma_2 = 0$, $\varepsilon_{r2} = 4$, and $\mu_{r2} = 2$. The loaded scatterer is excited at a frequency of 300 MHz (free-space wavelength 1 meter) by an external TM-polarized plane wave propagating at an angle of incidence $\phi^i = 0$. The electrical radii of the two boundary layers referred to the free-space medium ($k_1 = 2\pi$) are selected as $k_1 a_2 = 3$ and $k_1 a_1 = 5$. Figure 6.15 shows the distribution of the axial component of equivalent electric current on outer boundary C_1, and similarly, Figure 6.16 shows the distribution of the angular component of equivalent magnetic current on the same outer boundary C_1. On the outer boundary, the number of sampling pulses for the unknown electric current is chosen as $N_1 = 90$, and the corresponding number of sampling pulses for the unknown magnetic current is $N_1 = 90$. On the inner boundary, there is only the axial electric current distribution, and the number of sampling pulses is selected as $N_2 = 60$. Hence, the composite matrix size based on matrix equation (6.6.1) is 240 x 240. The numerical results obtained based on the matrix solution checks well with the series solution developed in the earlier sections. Only the results in the angular range of 180° to 360° are shown, and it should be noted that the distributions are symmetrical with respect to the direction of incident excitation.

Using the numerical results obtained in Figures 6.15 and 6.16, the bistatic radar cross section can be calculated based on the mathematical development given in expressions (5.10.15) to (5.10.21). Figure 6.17 shows the distribution of bistatic radar cross section for the single-layer circular dielectric loading on the concentric circular conducting scatterer. The result also checks closely with respect to the results of radar cross section obtained based on the series solution. Thus, the study of concentric loaded, circular geometry serves as an example to validate the numerical development scheme discussed in the previous section. In fact, it should be clearly noted that the numerical algorithm is valid for any arbitrary shape cross-sectional layered geometry with edges and corners.

To show the general applicability of the numerical development, a square conducting cylinder loaded with a uniform single layer of square dielectric loading is further analyzed. Again, layer 1 is the free-space medium, and layer 3 forms the perfectly conducting square region. Layer 2 forms a uniform, single layer of lossless dielectric loading on the square conducting scatterer. Two uniform and distinct single-layer loading cases are considered for comparison. The layer 2 parameters in Figure 6.18a correspond to the free-space medium, and in Figure 6.18b, the layer 2 parameters are selected as $\sigma_2 = 0$, $\varepsilon_{r2} = 2$, and $\mu_{r2} = 1$. The loaded square scatterer is excited at a frequency of 300 MHz (free-space wave length 1 meter) by an external TM-polarized plane wave propagating at an angle of incidence $\phi^i = 0$. The sides of the square cross-sectional geometry in terms of electrical widths for the two boundary layers referred to the free-space medium ($k_1 = 2\pi$) are selected as $k_1 s_2 = 2$ and $k_1 s_1 = 3$. For the first case, corresponding to the air dielectric loading, Figure 6.18a shows the distributions of the axial components of equivalent electric current on the two boundaries, C_1 and C_2, and also of the tangential component of equivalent magnetic current on outer boundary C_1. On the outer square boundary, the number of sampling pulses for the unknown

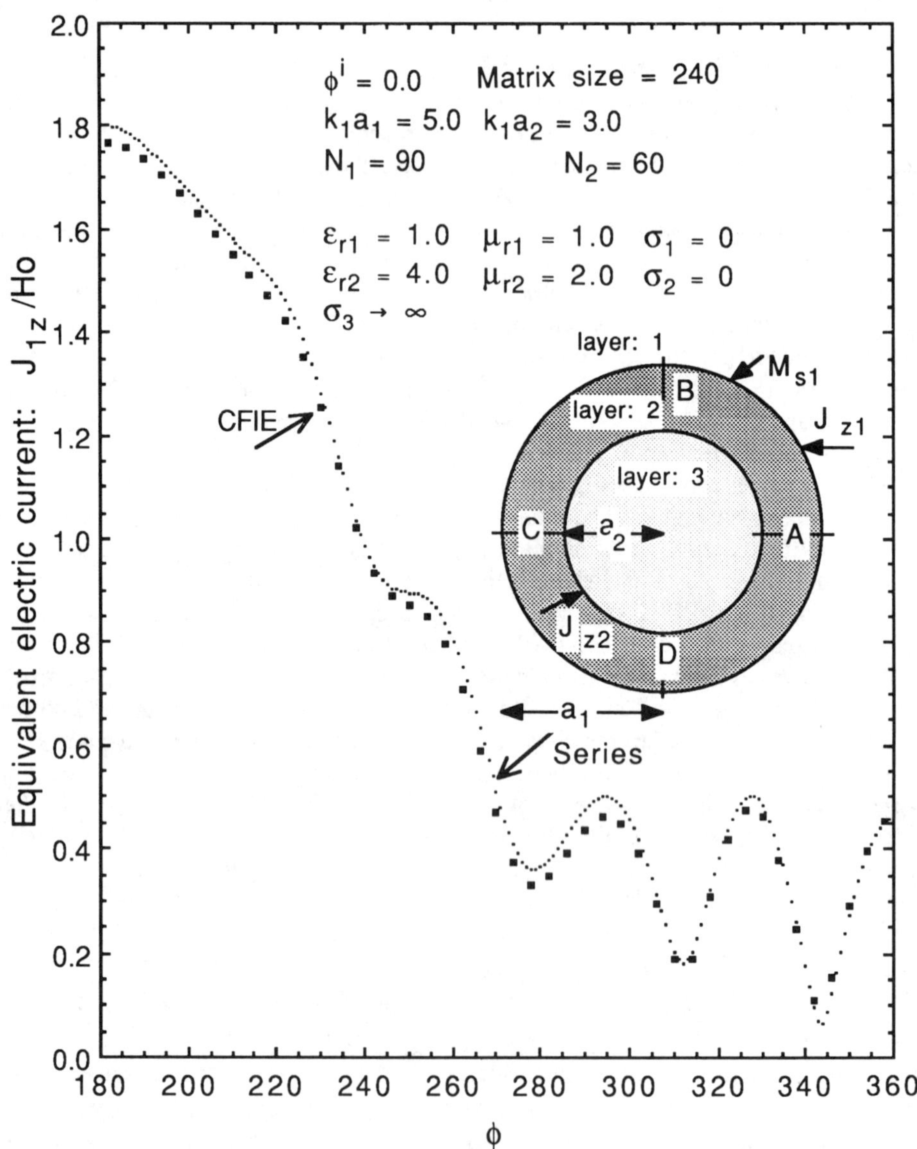

Figure 6.15 Distribution of equivalent electric current on the outer boundary of a circular conducting scatterer with a single layer of dielectric loading.

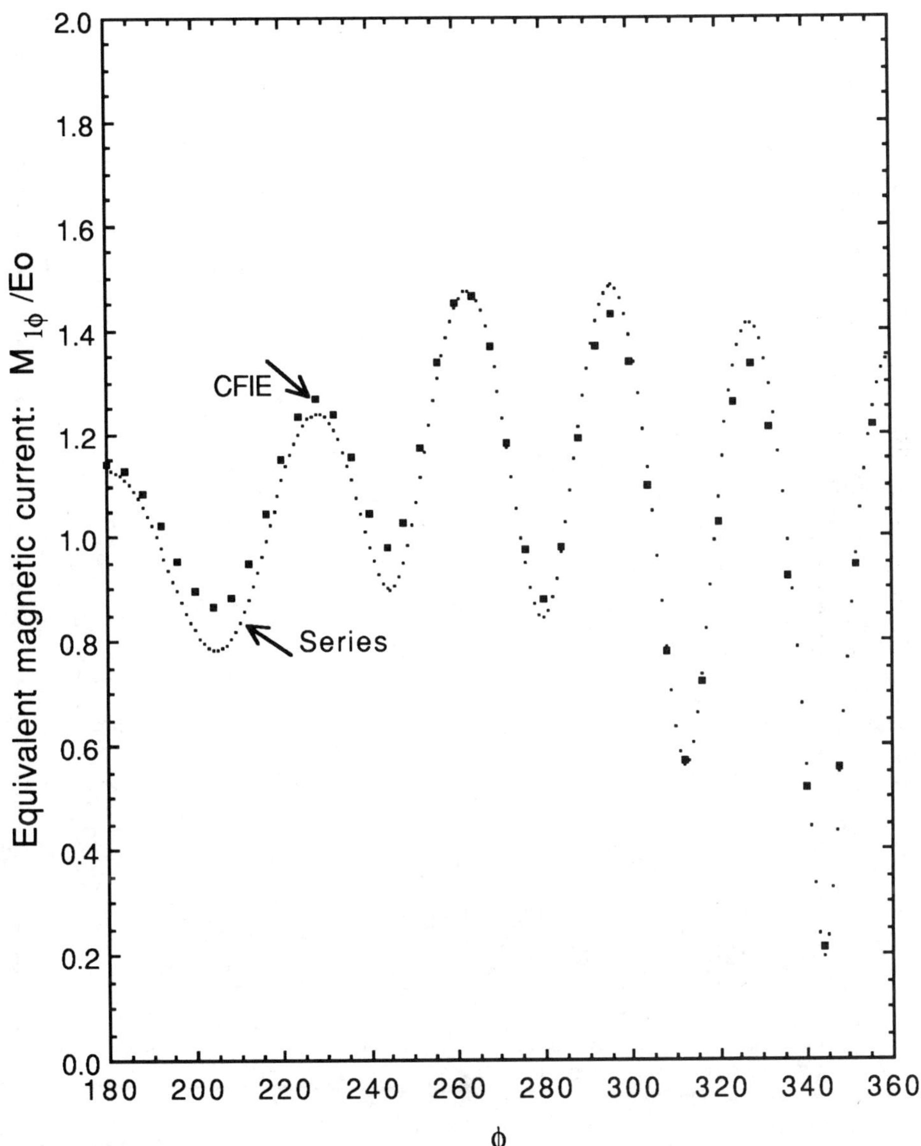

Figure 6.16 Distribution of equivalent magnetic current on the outer boundary of a circular conducting scatterer with a single layer of dielectric loading.

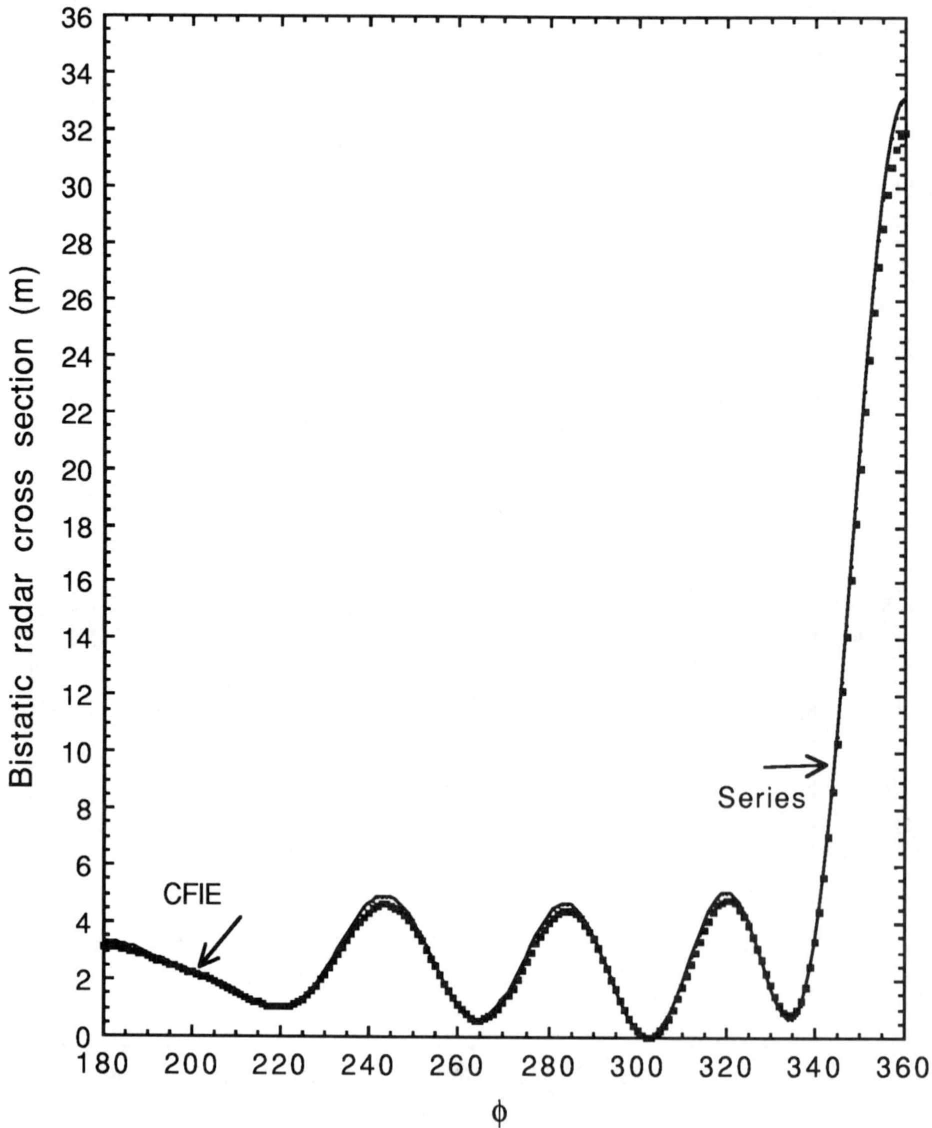

Figure 6.17 Bistatic radar cross section of a circular conducting scatterer with a single layer of dielectric loading – TM excitation.

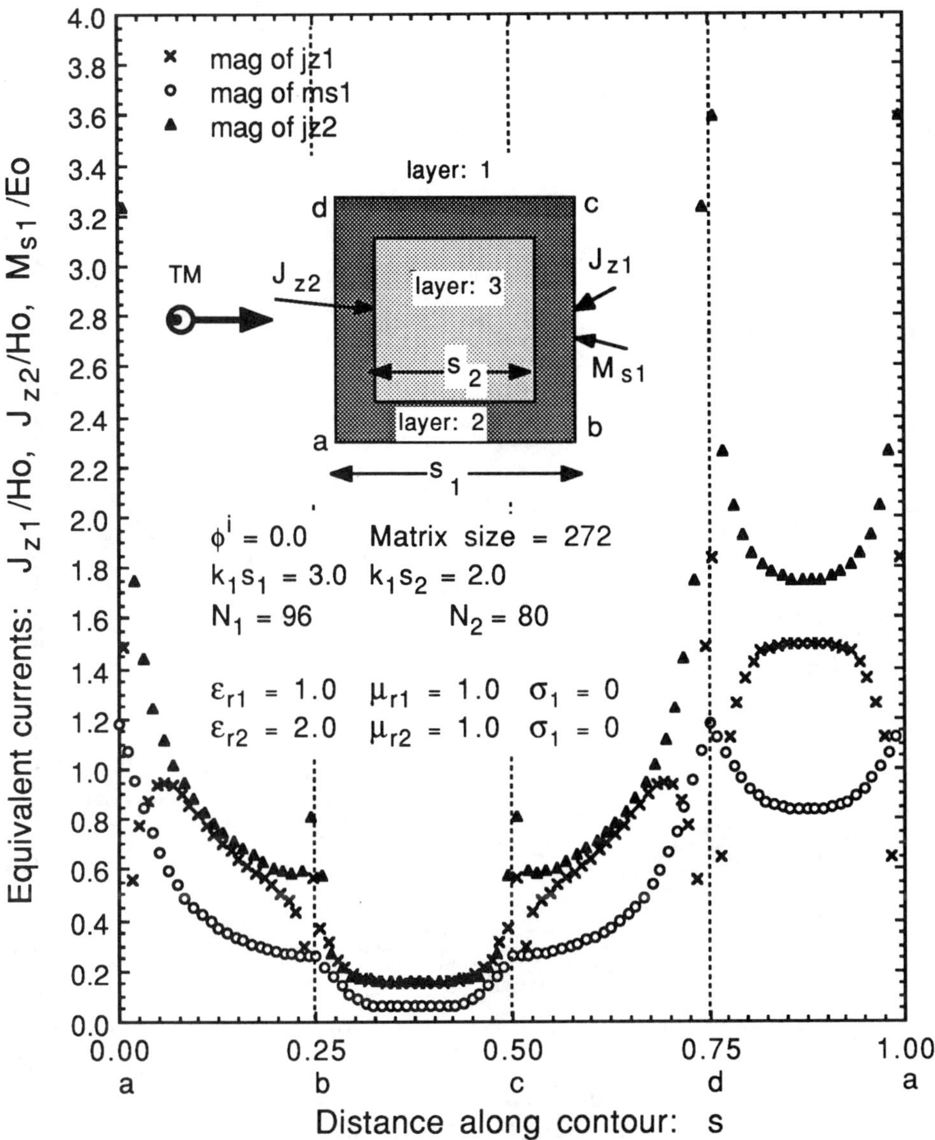

Figure 6.18a Distribution of equivalent currents on air and conducting boundaries of a square conducting scatterer with a single layer of dielectric loading.

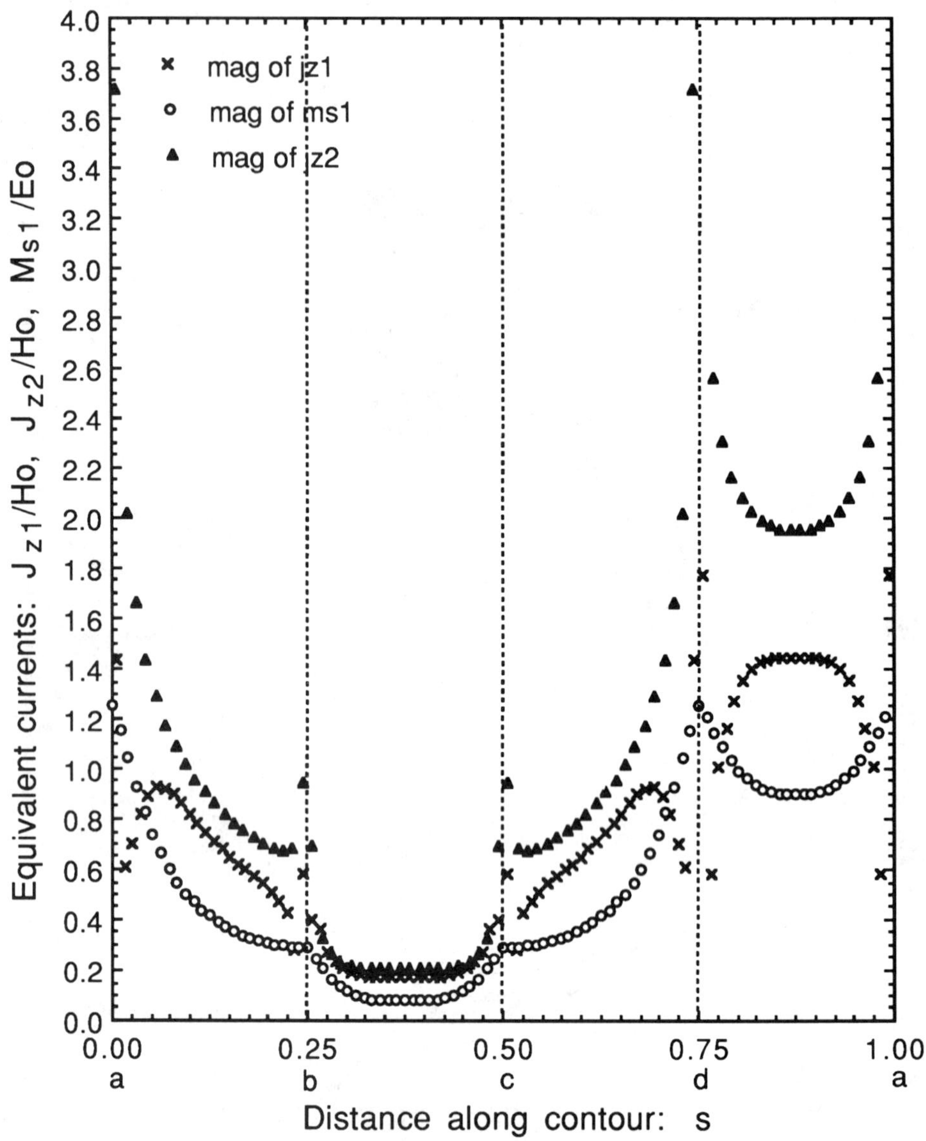

Figure 6.18b Equivalent currents on dielectric and conducting boundaries of a square conducting scatterer with a single layer of dielectric loading.

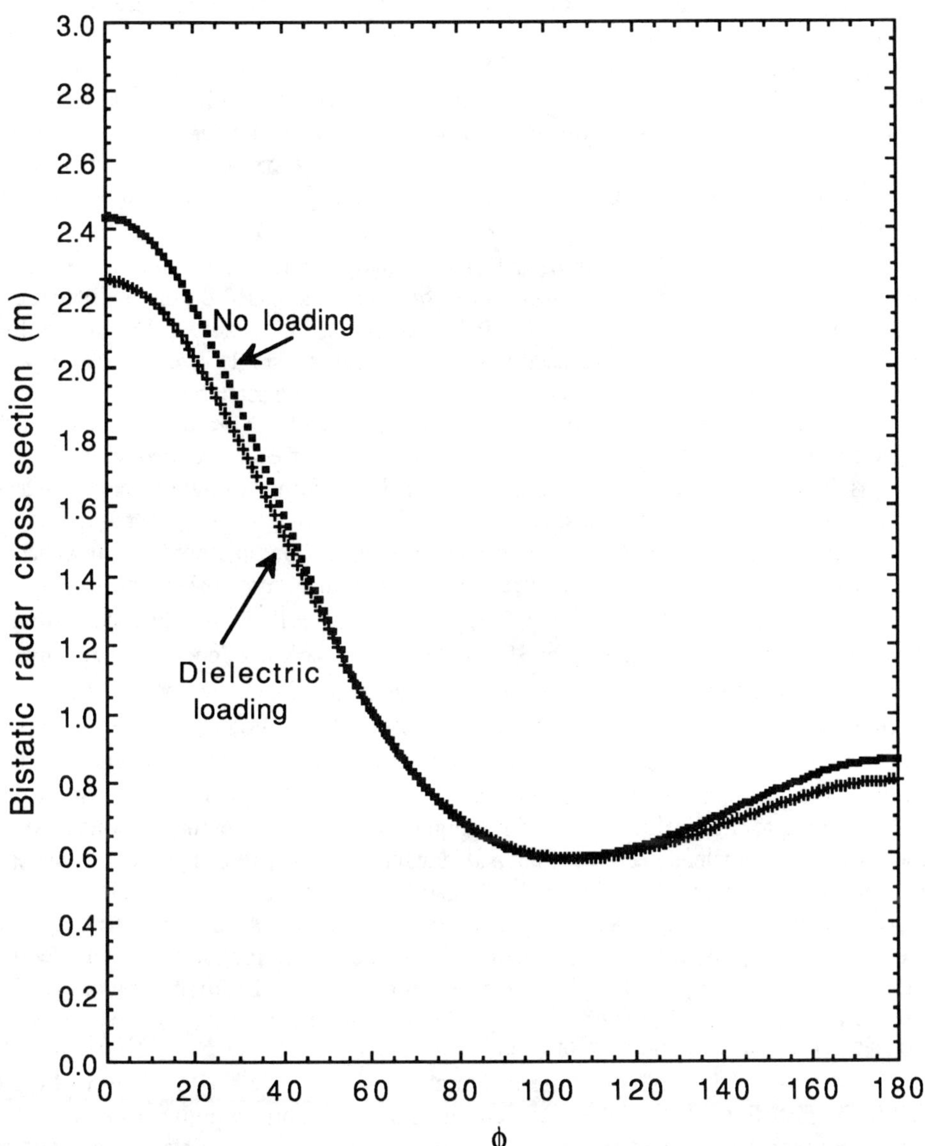

Figure 6.18c Bistatic radar cross section of a square conducting scatterer with a uniform single layer of dielectric loading – TM excitation.

electric current is chosen as $N_1 = 96$, and the corresponding number of sampling pulses for the unknown magnetic current is $N_1 = 96$. On the inner boundary, there is only the electric current, and the number of sampling pulses is selected as $N_2 = 80$. Hence, the composite matrix size for the numerical analysis is 272 x 272. Referring to Figure 6.18a, the distributions of the two axial electric currents clearly exhibit singular behavior at the right-angle corners, while the distribution of the tangential magnetic current shows continuity at the corners. In fact, the current distributions are symmetrical with respect to the direction of incident excitation. Similar results are shown in Figure 6.18b for the second case of single layer square dielectric loading with $\varepsilon_{r2} = 2$.

Using the numerical results of the electric and magnetic currents on the outer boundary, the bistatic radar cross section can be calculated based on the mathematical development given in expressions (5.10.15) to (5.10.21). Figure 6.18c shows the distribution of bistatic radar cross section for the two cases – single layer air loading and square dielectric loading on the symmetrical square conducting scatterer.

The case of a square conducting cylinder loaded with a single layer of circular dielectric loading is further presented. Again, layer 1 is the free-space medium and layer 3 forms the perfectly conducting square region. Layer 2 forms a single layer of lossless circular dielectric loading on the square conducting scatterer. Two distinct single-layer loading cases are considered for comparison. The layer 2 parameters in Figures 6.19a and 6.19b correspond to the free-space medium, and in Figures 6.19c and 6.19d, the layer 2 parameters are selected as $\sigma_2 = 0$, $\varepsilon_{r2} = 2$, and $\mu_{r2} = 1$. Again, the loaded square scatterer is excited at a frequency of 300 MHz (free-space wave length 1 meter) by an external TM-polarized plane wave propagating at an angle of incidence $\phi^i = 0$. Each side of the square cross sectional conducting geometry in terms of electrical widths for the inner boundary referred to the free-space medium ($k_1 = 2\pi$) is selected as $k_1 s_2 = 2$ and the radius of the single layer outer boundary is chosen as $k_1 a_1 = 3$. For the first case, corresponding to the air dielectric loading, Figure 6.19a shows the distributions of axial component of equivalent electric current and tangential component of equivalent magnetic current on outer circular boundary C_1, and also the Figure 6.19b shows the distribution of axial component of equivalent electric current on the inner square boundary C_2. On the outer circular boundary, the number of sampling pulses for the unknown electric current is chosen as $N_1 = 90$, and the corresponding number of sampling pulses for the unknown magnetic current is $N_1 = 90$. On the inner square boundary, there is only the electric current, and the number of sampling pulses is selected as $N_2 = 80$. Hence, the composite matrix size for the numerical analysis is 260 x 260. Referring to Figure 6.19b, the distribution of axial electric current clearly exhibits singular behavior at the square corners, and in fact, this result is identical to the case of a square conducting scatterer in a free space medium discussed in Chapter 3. The current distributions are symmetrical with respect to the direction of incident excitation. Similar results are shown in the Figures 6.19c and 6.19d for the second case of a single layer circular dielectric loading with $\varepsilon_{r2} = 2$.

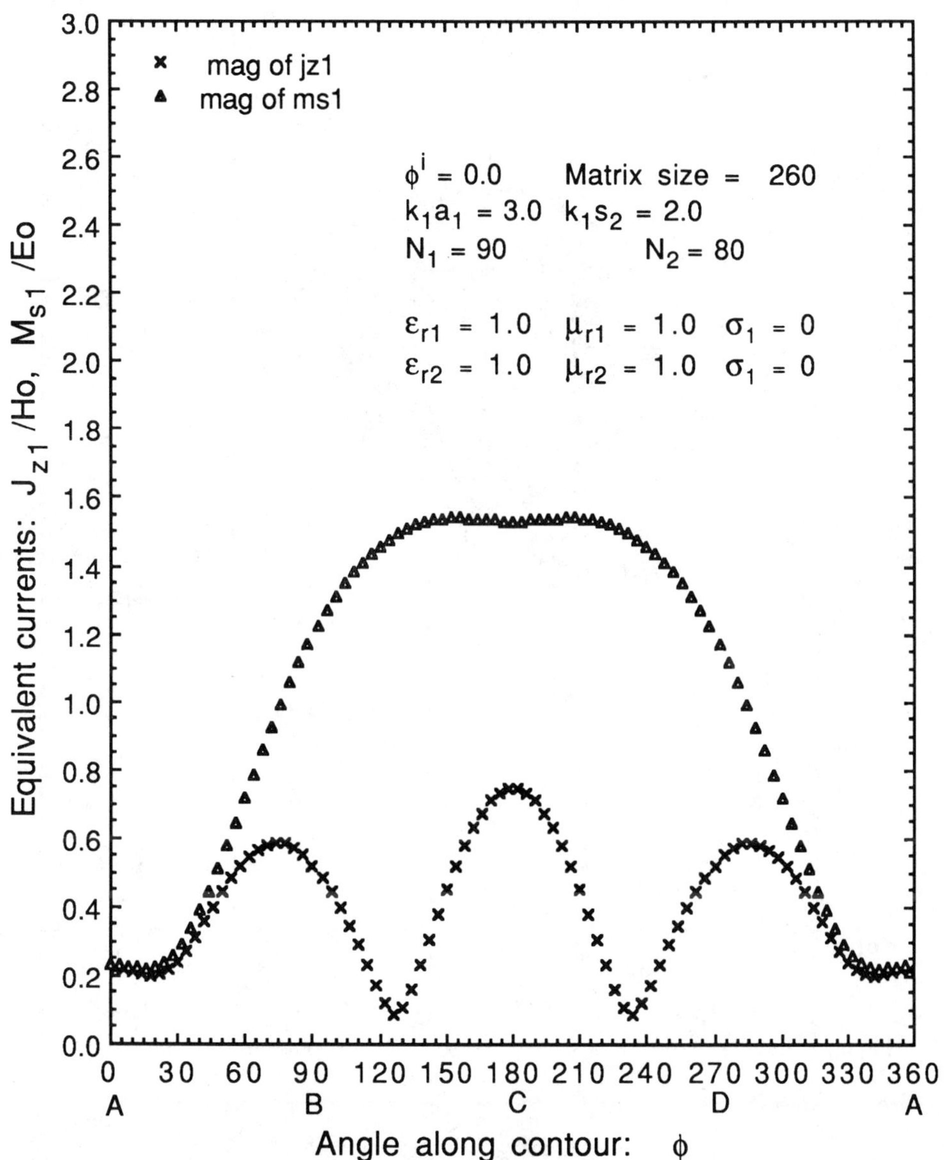

Figure 6.19a Equivalent currents on the outer air dielectric boundary for a square conducting scatterer with a circular single layer of air dielectric loading.

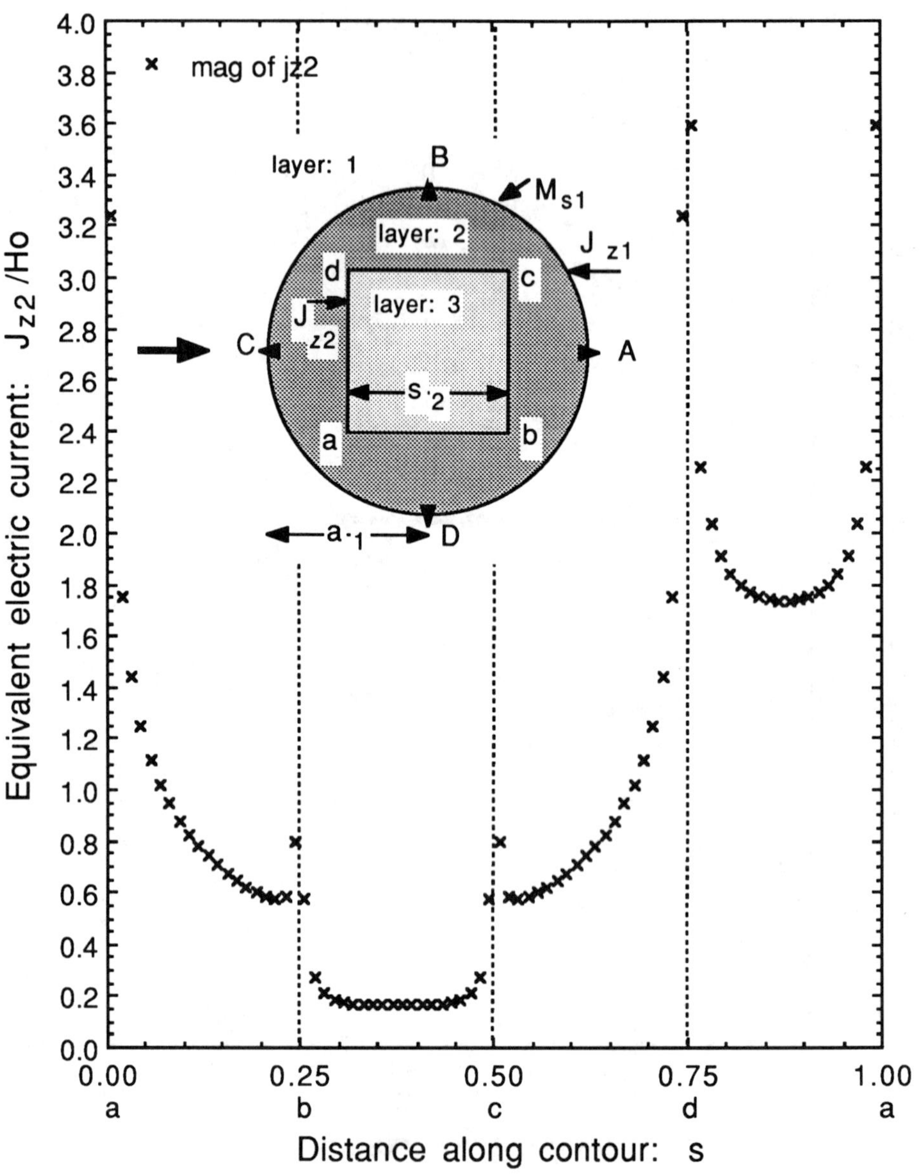

Figure 6.19b Equivalent electric current on the conducting boundary for a square conducting scatterer with a circular single layer of air dielectric loading.

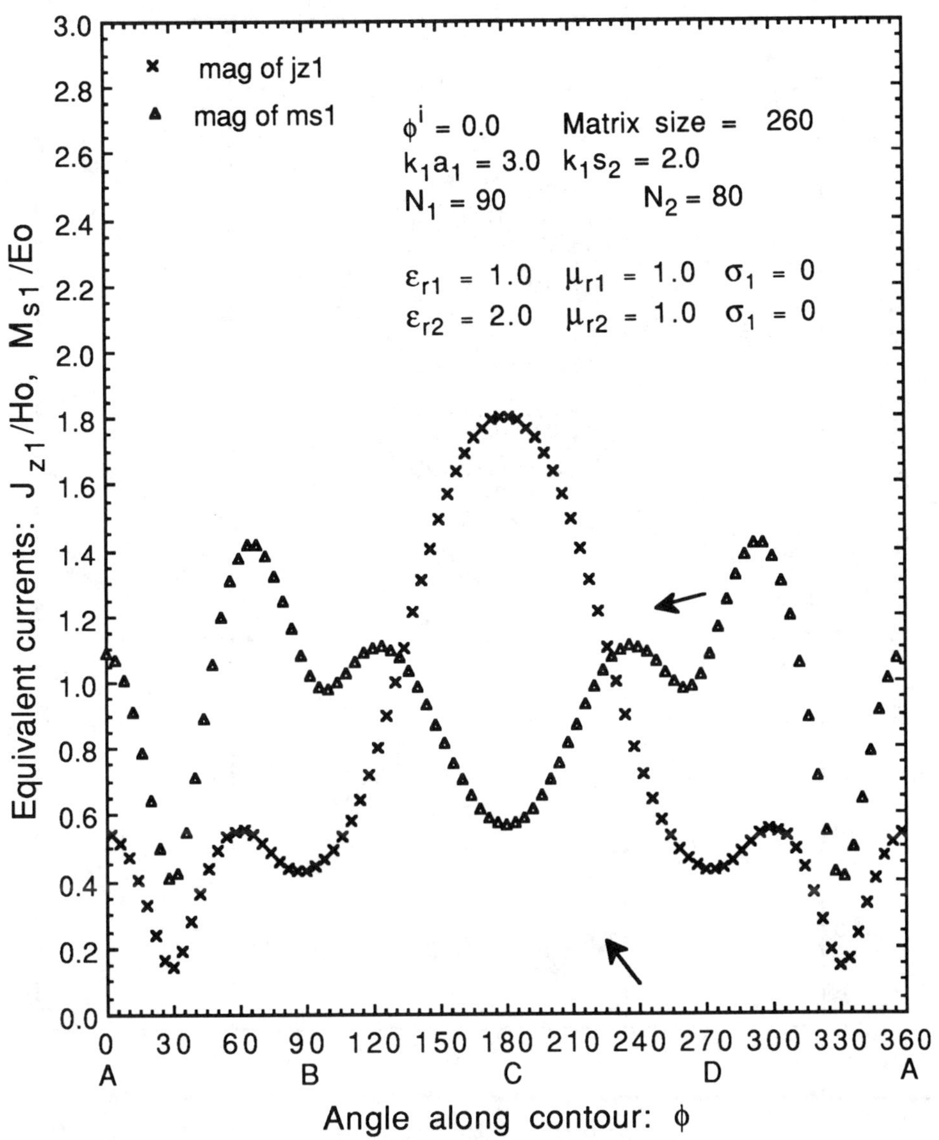

Figure 6.19c Equivalent currents on the outer dielectric boundary for a square conducting scatterer with a circular single layer of dielectric loading.

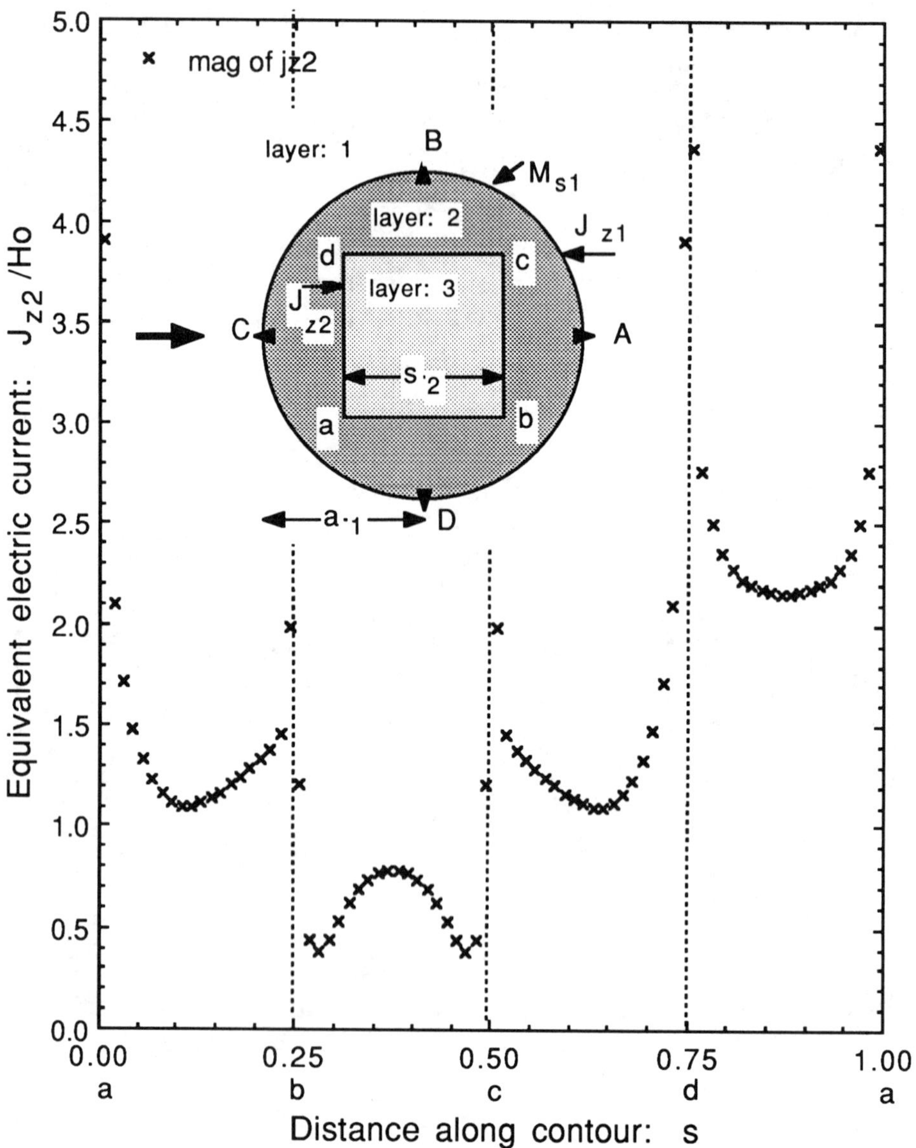

Figure 6.19d Equivalent electric current on the conducting boundary for a square conducting scatterer with a circular single layer of dielectric loading.

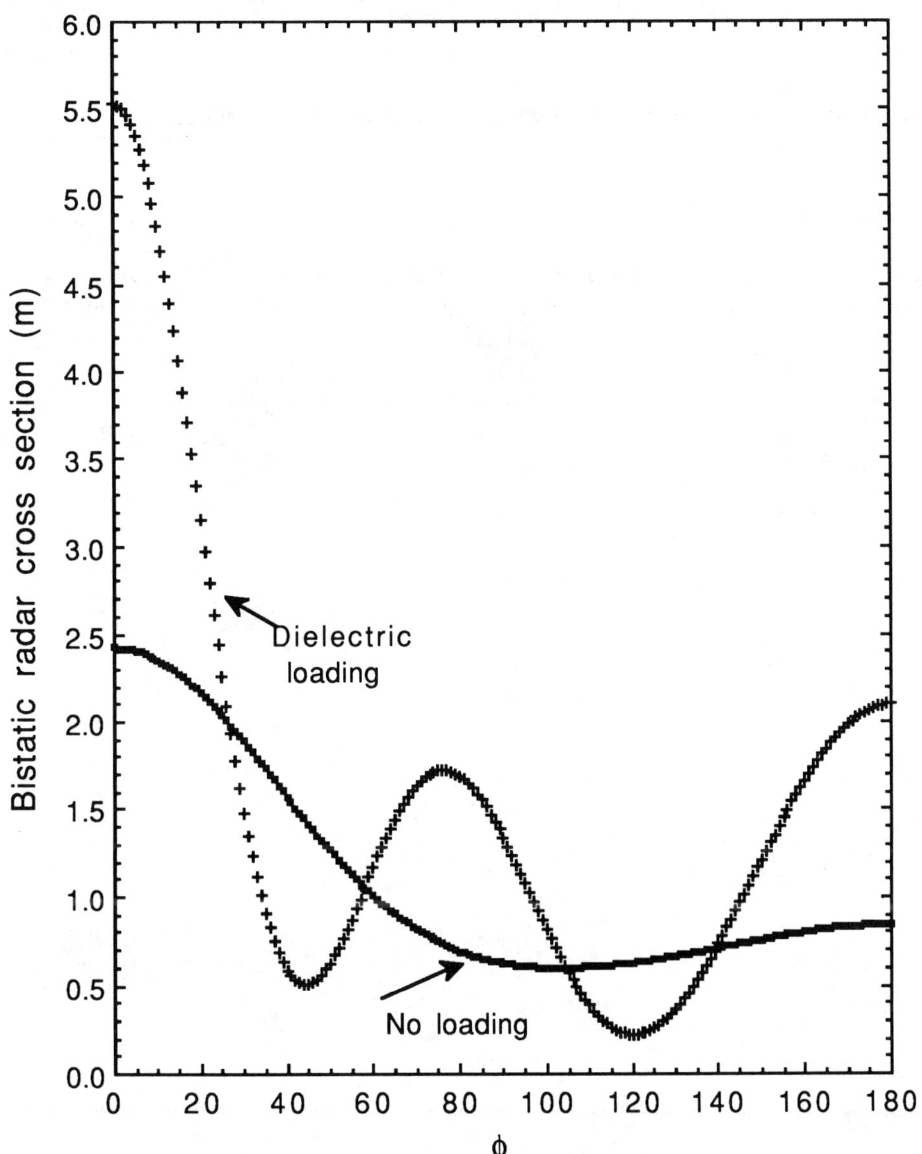

Figure 6.20 Bistatic radar cross section of a square conducting scatterer with a circular single layer of dielectric loading – TM excitation.

Using the numerical results of the electric and magnetic currents on the outer circular boundary, the bistatic radar cross section can be calculated based on the mathematical development given in expressions (5.10.15) to (5.10.21). Figure 6.20 shows the distribution of bistatic radar cross section for the two cases – single layer air loading and circular dielectric loading on the symmetrical square conducting scatterer.

6.7.1 CFIE - TE CASE

The study of layered conducting and dielectric scatterers with transverse electric excitation is, in fact, similar to the study of transverse magnetic excitation. The TE case is not discussed here, but can be easily obtained based on the analysis procedure discussed in Chapter 5. Various developments and numerical algorithms presented in the Chapters 5 and 6 can be easily applied to analyze TE scattering by the loaded conducting and dielectric scatterers. As discussed earlier, the role of the electric and magnetic currents, in fact, are reversed in the TE case.

Appendix A

Vector Identities

A.1 VECTOR OPERATIONS

Associative law	$\bar{A} + (\bar{B} + \bar{C}) = (\bar{A} + \bar{B}) + \bar{C}$	(A.1.1)
Commutative law	$\bar{A} + \bar{B} = \bar{B} + \bar{A}$	(A.1.2)
Commutative law	$m\bar{A} = \bar{A}m$	(A.1.3)
Associative law	$m(n\bar{A}) = (mn)\bar{A}$	(A.1.4)
Distributive law	$(m+n)\bar{A} = m\bar{A} + n\bar{A}$	(A.1.5)
Commutative law	$\bar{A} \bullet \bar{B} = \bar{B} \bullet \bar{A}$	(A.1.6)
	$\bar{A} \times \bar{B} = -\bar{B} \times \bar{A}$	(A.1.7)
Distributive law	$\bar{A} \bullet (\bar{B} + \bar{C}) = \bar{A} \bullet \bar{B} + \bar{A} \bullet \bar{C}$	(A.1.8)
	$\bar{A} \times (\bar{B} + \bar{C}) = \bar{A} \times \bar{B} + \bar{A} \times \bar{C}$	(A.1.9)
	$m(\bar{A} \bullet \bar{B}) = (m\bar{A}) \bullet \bar{B} = \bar{A} \bullet (m\bar{B})$	(A.1.10)
	$m(\bar{A} \times \bar{B}) = (m\bar{A}) \times \bar{B} = \bar{A} \times (m\bar{B})$	(A.1.11)
Scalar Product	$\bar{A} \bullet \bar{B} = A_x B_x + A_y B_y + A_z B_z$	(A.1.12)
Vector Product	$\bar{A} \times \bar{B} = \begin{matrix} \hat{x} & \hat{y} & \hat{z} \\ A_x & A_y & A_z \\ B_x & B_y & B_z \end{matrix}$	(A.1.13)

A.2 MIXED VECTOR OPERATIONS AND DERIVATIVES

$$\bar{A} \cdot (\bar{B} \times \bar{C}) = \bar{B} \cdot (\bar{C} \times \bar{A}) = \bar{C} \cdot (\bar{A} \times \bar{B}) \tag{A.2.1}$$

$$\bar{A} \times (\bar{B} \times \bar{C}) = \bar{B}\,(\bar{A} \cdot \bar{C}) - \bar{C}\,(\bar{A} \cdot \bar{B}) \tag{A.2.2}$$

$$(\bar{A} \times \bar{B}) \times \bar{C} = \bar{B}\,(\bar{A} \cdot \bar{C}) - \bar{A}\,(\bar{B} \cdot \bar{C}) \tag{A.2.3}$$

$$(\bar{A} \times \bar{B}) \cdot (\bar{C} \times \bar{D}) = (\bar{A} \cdot \bar{C})(\bar{B} \cdot \bar{D}) - (\bar{A} \cdot \bar{D})(\bar{B} \cdot \bar{C}) \tag{A.2.4}$$

$$(\bar{A} \times \bar{B}) \times (\bar{C} \times \bar{D}) = \bar{B}\,[\bar{A} \cdot (\bar{C} \times \bar{D})] - \bar{A}\,[\bar{B} \cdot (\bar{C} \times \bar{D})] \tag{A.2.5}$$

$$\frac{\partial(\bar{A}+\bar{B})}{\partial u} = \frac{\partial \bar{A}}{\partial u} + \frac{\partial \bar{B}}{\partial u} \tag{A.2.6}$$

$$\frac{\partial}{\partial u}(\bar{A} \cdot \bar{B}) = [\frac{\partial}{\partial u}\bar{A}] \cdot \bar{B} + \bar{A} \cdot [\frac{\partial}{\partial u}\bar{B}] \tag{A.2.7}$$

$$\frac{\partial}{\partial u}(\bar{A} \times \bar{B}) = [\frac{\partial}{\partial u}\bar{A}] \times \bar{B} + \bar{A} \times [\frac{\partial}{\partial u}\bar{B}] \tag{A.2.8}$$

$$\frac{\partial}{\partial u}(\bar{A}\Phi) = [\frac{\partial}{\partial u}\bar{A}]\Phi + \bar{A}[\frac{\partial}{\partial u}\Phi] \tag{A.2.9}$$

$$\frac{\partial^2 \bar{A}}{\partial x \partial y} = \frac{\partial^2 \bar{A}}{\partial y \partial x} \tag{A.2.10}$$

$$\frac{\partial}{\partial u}[\bar{A} \cdot (\bar{B} \times \bar{C})] = \frac{\partial}{\partial u}\bar{A} \cdot (\bar{B} \times \bar{C}) + \bar{A} \cdot (\frac{\partial}{\partial u}\bar{B} \times \bar{C}) + \bar{A} \cdot (\bar{B} \times \frac{\partial}{\partial u}\bar{C}) \tag{A.2.11}$$

A.3 DIFFERENTIAL OPERATOR

$$\nabla = \frac{\partial}{\partial x}\hat{x} + \frac{\partial}{\partial y}\hat{y} + \frac{\partial}{\partial z}\hat{z} \tag{A.3.1}$$

$$\nabla \bullet \nabla = \frac{\partial^2}{\partial x^2} + \frac{\partial^2}{\partial y^2} + \frac{\partial^2}{\partial z^2} \tag{A.3.2}$$

$$\nabla \bullet (\bar{B} + \bar{C}) = \nabla \bullet \bar{B} + \nabla \bullet \bar{C} \tag{A.3.3}$$

$$\nabla \times (\bar{B} + \bar{C}) = \nabla \times \bar{B} + \nabla \times \bar{C} \tag{A.3.4}$$

$$\nabla(\Phi + \Psi) = \nabla\Phi + \nabla\Psi \tag{A.3.5}$$

$$\nabla(\Phi\Psi) = \Psi\nabla\Phi + \Phi\nabla\Psi \tag{A.3.6}$$

$$\nabla \bullet (\Phi\bar{A}) = (\nabla\Phi) \bullet \bar{A} + \Phi(\nabla \bullet \bar{A}) \tag{A.3.7}$$

$$\nabla \times (\Phi\bar{A}) = (\nabla\Phi) \times \bar{A} + \Phi(\nabla \times \bar{A}) \tag{A.3.8}$$

$$\nabla \bullet (\bar{A} \times \bar{B}) = \bar{B} \bullet (\nabla \times \bar{A}) - \bar{A} \bullet (\nabla \times \bar{B}) \tag{A.3.9}$$

$$\nabla \times (\bar{A} \times \bar{B}) = (\bar{B} \bullet \nabla)\bar{A} - \bar{B}(\nabla \bullet \bar{A}) - (\bar{A} \bullet \nabla)\bar{B} + \bar{A}(\nabla \bullet \bar{B}) \tag{A.3.10}$$

$$\nabla(\bar{A} \bullet \bar{B}) = (\bar{B} \bullet \nabla)\bar{A} + (\bar{A} \bullet \nabla)\bar{B} + \bar{B} \times (\nabla \times \bar{A}) + \bar{A} \times (\nabla \times \bar{B}) \tag{A.3.11}$$

$$\nabla \times (\nabla\Phi) = 0 \tag{A.3.12}$$

$$\nabla \bullet (\nabla \times \bar{A}) = 0 \tag{A.3.13}$$

$$\nabla^2\bar{A} = \nabla^2 A_x\hat{x} + \nabla^2 A_y\hat{y} + \nabla^2 A_z\hat{z} \tag{A.3.14}$$

$$\nabla \times (\nabla \times \bar{A}) = \nabla(\nabla \bullet \bar{A}) - \nabla^2\bar{A} \tag{A.3.15}$$

$$\frac{d\Phi}{ds} = \nabla\Phi \bullet \frac{d\bar{r}}{ds} \tag{A.3.16}$$

A.4 TRANSFORMATION OF VECTORS

$\bar{A}(x,y,z) = A_x\hat{x} + A_y\hat{y} + A_z\hat{z}$ Rectangular coordinate system

$\bar{A}(\rho,\phi,z) = A_\rho\hat{\rho} + A_\phi\hat{\phi} + A_z\hat{z}$ Cylindrical coordinate system

$\bar{A}(r,\theta,\phi) = A_r\hat{r} + A_\theta\hat{\theta} + A_\phi\hat{\phi}$ Spherical coordinate system

Rectangular components in terms of cylindrical components

$$A_x = A_\rho \cos\phi - A_\phi \sin\phi \tag{A.4.1}$$
$$A_y = A_\rho \sin\phi + A_\phi \cos\phi \tag{A.4.2}$$
$$A_z = A_z \tag{A.4.3}$$

Rectangular components in terms of spherical components

$$A_x = A_r \sin\theta \cos\phi + A_\theta \cos\theta \cos\phi - A_\phi \sin\phi \tag{A.4.4}$$
$$A_y = A_r \sin\theta \sin\phi + A_\theta \cos\theta \sin\phi + A_\phi \cos\phi \tag{A.4.5}$$
$$A_z = A_r \cos\theta - A_\theta \sin\theta \tag{A.4.6}$$

Cylindrical components in terms of rectangular components

$$A_\rho = A_x \cos\phi + A_y \sin\phi \tag{A.4.7}$$
$$A_\phi = - A_x \sin\phi + A_y \cos\phi \tag{A.4.8}$$
$$A_z = A_z \tag{A.4.9}$$

Cylindrical components in terms of spherical components

$$A_\rho = A_r \sin\theta + A_\theta \cos\theta \tag{A.4.10}$$
$$A_z = A_r \cos\theta - A_\theta \sin\theta \tag{A.4.11}$$
$$A_\phi = A_\phi \tag{A.4.12}$$

Spherical components in terms of rectangular components

$$A_r = A_x \sin\theta \cos\phi + A_y \sin\theta \sin\phi + A_z \cos\theta \tag{A.4.13}$$
$$A_\theta = A_x \cos\theta \cos\phi + A_y \cos\theta \sin\phi - A_z \sin\theta \tag{A.4.14}$$
$$A_\phi = - A_x \sin\phi + A_y \cos\phi \tag{A.4.15}$$

Spherical components in terms of cylindrical components

$$A_r = A_\rho \sin\theta + A_z \cos\theta \qquad \text{(A.4.16)}$$

$$A_\theta = A_\rho \cos\theta - A_z \sin\theta \qquad \text{(A.4.17)}$$

$$A_\phi = A_\phi \qquad \text{(A.4.18)}$$

A.5 COORDINATE REPRESENTATIONS

Gradient of a scalar quantity, divergence and curl of a vector quantity, and Laplacian of a scalar quantity are given by:

GRADIENT OF A SCALAR

Rectangular $$\nabla\Phi(x,y,z) = \frac{\partial\Phi}{\partial x}\hat{x} + \frac{\partial\Phi}{\partial y}\hat{y} + \frac{\partial\Phi}{\partial z}\hat{z} \qquad \text{(A.5.1)}$$

Cylindrical $$\nabla\Phi(\rho,\phi,z) = \frac{\partial\Phi}{\partial\rho}\hat{\rho} + \frac{1}{\rho}\frac{\partial\Phi}{\partial\phi}\hat{\phi} + \frac{\partial\Phi}{\partial z}\hat{z} \qquad \text{(A.5.2)}$$

Spherical $$\nabla\Phi(r,\theta,\phi) = \frac{\partial\Phi}{\partial r}\hat{r} + \frac{1}{r}\frac{\partial\Phi}{\partial\theta}\hat{\theta} + \frac{1}{r\sin\theta}\frac{\partial\Phi}{\partial\phi}\hat{\phi} \qquad \text{(A.5.3)}$$

DIVERGENCE OF A VECTOR

Rectangular: $$\nabla\bullet\bar{A}(x,y,z) = \frac{\partial A_x}{\partial x} + \frac{\partial A_y}{\partial y} + \frac{\partial A_z}{\partial z} \qquad \text{(A.5.4)}$$

Cylindrical: $$\nabla\bullet\bar{A}(\rho,\phi,z) = \frac{1}{\rho}\frac{\partial}{\partial\rho}[\rho A_\rho] + \frac{1}{\rho}\frac{\partial A_\phi}{\partial\phi} + \frac{\partial A_z}{\partial z} \qquad \text{(A.5.5)}$$

Spherical: $$\nabla\bullet\bar{A}(r,\theta,\phi) = \frac{1}{r^2}\frac{\partial}{\partial r}[r^2 A_r] + \frac{1}{r\sin\theta}\frac{\partial}{\partial\theta}[A_\theta\sin\theta] + \frac{1}{r\sin\theta}\frac{\partial A_\phi}{\partial\phi} \qquad \text{(A.5.6)}$$

CURL OF A VECTOR

Rectangular $$\nabla \times \bar{A}(x,y,z) = \left\{\frac{\partial A_z}{\partial y} - \frac{\partial A_y}{\partial z}\right\}\hat{x} + \left\{\frac{\partial A_x}{\partial z} - \frac{\partial A_z}{\partial x}\right\}\hat{y} + \left\{\frac{\partial A_y}{\partial x} - \frac{\partial A_x}{\partial y}\right\}\hat{z} \quad \text{(A.5.7)}$$

Cylindrical $$\nabla \times \bar{A}(\rho,\phi,z) = \left\{\frac{1}{\rho}\frac{\partial A_z}{\partial \phi} - \frac{\partial A_\phi}{\partial z}\right\}\hat{\rho} + \left\{\frac{\partial A_\rho}{\partial z} - \frac{\partial A_z}{\partial \rho}\right\}\hat{\phi} + \frac{1}{\rho}\left\{\frac{\partial[\rho A_\phi]}{\partial \rho} - \frac{\partial A_\rho}{\partial \phi}\right\}\hat{z} \quad \text{(A.5.8)}$$

Spherical $$\nabla \times \bar{A}(r,\theta,\phi) = \frac{1}{r\sin\theta}\left\{\frac{\partial[A_\phi \sin\theta]}{\partial \theta} - \frac{\partial A_\theta}{\partial \phi}\right\}\hat{r} + \frac{1}{r}\left\{\frac{1}{\sin\theta}\frac{\partial A_r}{\partial \phi} - \frac{\partial[rA_\phi]}{\partial r}\right\}\hat{\theta} + \frac{1}{r}\left\{\frac{\partial[rA_\theta]}{\partial r} - \frac{\partial A_r}{\partial \theta}\right\}\hat{\phi} \quad \text{(A.5.9)}$$

LAPLACIAN OF A SCALAR

Rectangular $$\nabla^2\Phi(x,y,z) = \frac{\partial^2\Phi}{\partial x^2} + \frac{\partial^2\Phi}{\partial y^2} + \frac{\partial^2\Phi}{\partial z^2} \quad \text{(A.5.10)}$$

Cylindrical $$\nabla^2\Phi(\rho,\phi,z) = \frac{1}{\rho}\frac{\partial}{\partial \rho}\left\{\rho\frac{\partial \Phi}{\partial \rho}\right\} + \frac{1}{\rho^2}\frac{\partial^2\Phi}{\partial \phi^2} + \frac{\partial^2\Phi}{\partial z^2} \quad \text{(A.5.11)}$$

Spherical $$\nabla^2\Phi(r,\theta,\phi) = \frac{1}{r^2}\frac{\partial}{\partial r}\left\{r^2\frac{\partial \Phi}{\partial r}\right\} + \frac{1}{r^2\sin\theta}\frac{\partial}{\partial \theta}\left\{\sin\theta\frac{\partial \Phi}{\partial \theta}\right\} + \frac{1}{r^2\sin^2\theta}\frac{\partial^2\Phi}{\partial \phi^2} \quad \text{(A.5.12)}$$

A.6 INTEGRAL RELATIONSHIPS

Gauss-divergence theorem

$$\iint_S \bar{A} \bullet d\bar{s} = \iiint_V \nabla \bullet \bar{A}\, dv \tag{A.6.1}$$

Stoke's theorem

$$\oint_C \bar{A} \bullet d\bar{L} = \iint_S [\nabla \times \bar{A}] \bullet d\bar{s} \tag{A.6.2}$$

Appendix B

Bessel Functions

B.1 BASIC EQUATION

The Bessel differential equation is given by

$$z \frac{d}{dz}\left[z \frac{dB}{dz}\right] + [z^2 - n^2]\,B = 0 \qquad \text{(B.1.1)}$$

and for integer n, solutions are given by

$$B_n(z) \Rightarrow J_n(z),\ Y_n(z) \qquad \text{(B.1.2)}$$

z : argument of the Bessel function

$J_n(z)$: Bessel function of first kind and order n

$Y_n(z)$: Bessel function of second kind or Neumann function and order n

$H_n^{(1)}(z)$: Hankel function of first kind and order n

$$= J_n(z) + jY_n(z) \qquad \text{(B.1.3)}$$

$H_n^{(2)}(z)$: Hankel function of second kind and order n

$$= J_n(z) - jY_n(z) \qquad \text{(B.1.4)}$$

B.2 INTEGRAL REPRESENTATION

$$J_n(z) = \frac{j^n}{2\pi} \int_{\phi=0}^{2\pi} e^{-j(z\cos\phi + n\phi)}\, d\phi \qquad \text{(B.2.1)}$$

$$H_0^{(2)}(k\rho) = \frac{j}{\pi} \int_{\zeta=-\infty}^{\infty} \frac{e^{-jkR}}{R}\, d\zeta \tag{B.2.2a}$$

$$R = [\rho^2 + \zeta^2]^{1/2} \tag{B.2.2b}$$

B.3 SERIES REPRESENTATION

$$J_0(z) = 1 - (z/2)^2 + \frac{(z/2)^4}{1^2 \cdot 2^2} - \frac{(z/2)^6}{1^2 \cdot 2^2 \cdot 3^2} + \cdots \tag{B.3.1}$$

$$J_1(z) = (z/2) - \frac{(z/2)^3}{1^2 \cdot 2} + \frac{(z/2)^5}{1^2 \cdot 2^2 \cdot 3} - \cdots \tag{B.3.2}$$

$$J_{-n}(z) = (-1)^n J_n(z) \tag{B.3.3}$$

$$Y_0(z) = \frac{2}{\pi}\left(\ell n(\gamma z/2)\right) J_0(z) + \frac{2}{\pi}\frac{(z/2)^2}{(1!)^2} - \frac{2}{\pi}\frac{(z/2)^4}{(2!)^2}(1 + 1/2) + \frac{2}{\pi}\frac{(z/2)^6}{(3!)^2}(1 + 1/2 + 1/3) - \cdots \tag{B.3.4}$$

$$Y_1(z) = \frac{2}{\pi}\left(\ell n(\gamma z/2)\right) J_1(z) - \frac{2}{\pi z} - \frac{1}{\pi}\sum_{p=0}^{\infty} \frac{(-1)^p}{p!(p+1)!}\left(\frac{z}{2}\right)^{2p+1} \left\{2\left(1 + 1/2 + \cdots + 1/p\right) + 1/(p+1)\right\} \tag{B.3.5}$$

$$Y_{-n}(z) = (-1)^n Y_n(z) \tag{B.3.6}$$

$$\ell n\, \gamma = 0.577215665$$

The nth order Bessel and Neumann functions and their first order derivatives can be obtained using recurrence formulae

$$J_{n+1}(z) = - J_{n-1}(z) + \frac{2n}{z} J_n(z) \tag{B.3.7a}$$

$$J'_n(z) = \frac{1}{2} [J_{n-1}(z) - J_{n+1}(z)] \tag{B.3.7b}$$

$$Y_{n+1}(z) = - Y_{n-1}(z) + \frac{2n}{z} Y_n(z) \tag{B.3.8a}$$

$$Y'_n(z) = \frac{1}{2} [Y_{n-1}(z) - Y_{n+1}(z)] \tag{B.3.8b}$$

The generation of above Bessel functions can be verified using Wronskian

$$J_n(z)Y'_n(z) - J'_n(z)Y_n(z) = \frac{2}{\pi z} \tag{B.3.9}$$

With small argument, $z << 1$

$$J_0(z) \approx 1 \tag{B.3.10a}$$

$$Y_0(z) \approx \frac{2}{\pi} \ell n\left(\frac{\gamma z}{2}\right) \tag{B.3.10b}$$

$$J_1(z) \approx 0 \tag{B.3.11a}$$

$$Y_1(z) \approx \frac{-2}{\pi z} \tag{B.3.11b}$$

With large argument, $z \rightarrow \infty$

$$H_n^{(2)}(z) \sim \sqrt{\frac{2j}{\pi z}}\, j^n e^{-jz} \tag{B.3.12}$$

B.4 COORDINATE TRANSFORMATION

Based on coordinate shift, Hankel function can be written as

$$H_0^{(2)}(k|\bar{\boldsymbol{\rho}} - \bar{\boldsymbol{\rho}}'|) = \sum_{n=-\infty}^{\infty} J_n(k\rho')H_n^{(2)}(k\rho)e^{jn(\phi - \phi')} \tag{B.4.1}$$

for $\bar{\boldsymbol{\rho}} \geq \bar{\boldsymbol{\rho}}'$

$$H_0^{(2)}(k|\bar{\boldsymbol{\rho}} - \bar{\boldsymbol{\rho}}'|) = \sum_{n=-\infty}^{\infty} J_n(k\rho)H_n^{(2)}(k\rho')e^{jn(\phi - \phi')} \tag{B.4.2}$$

for $\bar{\boldsymbol{\rho}} \leq \bar{\boldsymbol{\rho}}'$

Bibliography

1. J.Bowman, T.B.A.Senior and P.L.E.Uslenghi, Electromagnetic and Acoustic Scattering by Simple Shapes, Amsterdam, North Holland, 1969.
2. L.B.Felsen, Ed., Transient Electromagnetic Fields, Springer-Verlag, New York, 1975.
3. R.C.Hansen, Ed., Geometric Theory of Diffraction - A Selected Reprint Volume, IEEE Press, New York, 1981.
4. R.G.Kouyoumijan, "Asymptotic High-Frequency Methods", Proceedings of the IEEE, Vol. 53, pp. 864-876, August 1965.
5. R.G.Kouyoumijan and P.H.Pathak, "A uniform geometrical theory of diffraction of an edge in a perfectly conducting surface," Proceedings of the IEEE, Vol. 62, No. 11, pp. 1448-1461, November 1974.
6. B.J.Strait, Ed., Applications of the Method of Moments to Electromagnetic Fields, The SCEEE Press, 1981.
7. R.F.Harrington, Field Computation by Moment Methods, MacMillan, New York, 1968.
8. K.Umashankar, "Numerical Analysis of Electromagnetic Wave Scattering and Interaction Based on Frequency-Domain Integral Equation and Method of Moments Techniques", Wave Motion, No.10, pp. 493-525, December 1988.
9. E.K.Miller, Ed., Computational Electromagnetics - A Selected Reprint Volume, IEEE Press, New York, 1992.
10. T.Itoh, Ed., Numerical Techniques for Microwaves and Millimeterwave Passive structures, John Wiley, New York, 1989.
11. C.L.Bennett, "Space Time Integral Equations Approach to the Large Body Scattering Problem", Final Technical Report, No. RADC-CR-73-70, AD 763794, May 1973.
12. K.Umashankar and A.Taflove, "Analytical Models for Electromagnetic Scattering", Final Technical Report, No. F19628-82-C-0140, RADC/ESD, Hanscom Airforce Base, Ma. June 1984.
13. K.Umashankar and A.Taflove, "A Novel Method to Analyze Electromagnetic Scattering of Complex Objects", IEEE Transactions on Electromagnetic Compatibility, Vol. 24, No. 4, pp. 397-405, November 1982.
14. A.Taflove and K.Umashankar, "Radar Cross Section General Three Dimensional Scatterers", IEEE Transactions on Electromagnetic Compatibility, Vol. 25, No. 4, pp. 433-440, November 1983.
15. A.Taflove and K.Umashankar, "Review of FD-TD Numerical Modeling of Electromagnetic Wave Scattering and Radar Cross Section", Proceedings of the IEEE, Special Issue on Radar Cross Section, Vol.77, No.5, pp.682-699, May 1989.

16. K.S.Kunz and K.M.Lee, "A Three-Dimensional Finite-Difference Solution of the External Response of an Aircraft to a Complex Transient EM Environment", IEEE Transactions on Electromagnetic Compatibility, Vol. 20, pp. 328-333, May 1978.

17. C.E.Baum, "On the Singularity Expansion Method for the Solution of Electromagnetic Interation Problems", Interaction Note 88, KAFB, December 1971.

18. K.K.Mei, "Unimoment Method of Solving Antenna and Scattering Problems", IEEE Transactions on Antennas and Propagation, Vol. 22, pp. 760-766, November 1974.

19. A.J.Poggio and E.K.Miller, "Integral Equation Solutions of Three-Dimensional Scattering Problems", in Computer Techniques for Electromagnetics, R.Mittra, Ed., Pergamon, 1973.

20. S.A.Schelkunoff, "Field Equivalence Theorems", Comm. Pure Appl. Math., Vol. 4, pp. 43-59, June 1951.

21. D.E.Livesay and K.M.Chen, "Electromagnetic Fields Induced inside Arbitrarily Shaped Biological Bodies", IEEE Transactions on Microwave Theory and Techniques, Vol. 22, No. 12, pp. 1273-1280, December 1974.

22. D.T.Borup and O.P.Gandhi, "Fast-Fourier Transform Method for Calculation of SAR Distributions in Finely Discretized Inhomogeneous Models of Biological Bodies", IEEE Transactions on Microwave Theory and Techniques, Vol. 32, No. 4, pp. 355-360, April 1984.

23. T.K.Sarkar, K.R.Siarkiewicz and R.F.Stratton, "Survey of Numerical Methods for Solution of Large Systems of Linear Equations for Electromagnetic Field Problems", IEEE Transactions on Antennas and propagation, Vol. 29, pp. 847-856, 1981.

24. S.Govind, "Numerical Computation of Electromagnetic Scattering by Inhomogeneous Penetrable Bodies", Ph.D. Dissertation, University of MIssissippi, 1978.

25. M.G.Andreasen, "Scattering from Bodies of Revolution", IEEE Transactions on Antennas and Propagation, Vol. 13, pp. 303-310, March 1965.

26. A.W.Glisson, "On the Development of Numerical Techniques for Treating Arbitrary-Shaped Surfaces", Ph.D. Dissertation, University of Mississippi, 1978.

27. P.L.Huddleston, L.N.Medgyesi-Mitschang, and J.M.Putnum, "Combined Field Integral Equation for Scattering by Dielectrically Coated Conducting Bodies", IEEE Transactions on Antennas and propagation, Vol. 34, No. 4, pp. 510-520, April 1986.

28. S.M.Rao, "Electromagnetic Scattering and Radiation of Arbitrarily Shaped Surfaces by Triangular Patch Modeling", Ph.D. Dissertation, University of Mississippi, August 1980.

29. K.Umashankar, A.Taflove and S.M.Rao, "Electromagnetic Scattering by Arbitrary Shaped Three Dimensional Homogeneous Lossy Dielectric Objects", IEEE Transactions on Antennas and Propagation, Vol. 34, No. 6, pp. 758-766, June 1986.

30. R.Mittra and C.A.Klein, "Stability and Convergence of Method of Moments Solutions", in Numerical and Asymptotic Techniques in Electromagnetics, R.Mittra, Ed., Springer-Verlag, pp. 129-133, New York, 1975.

31. B.Beker and K.Umashankar, "Analysis of Electromagnetic Scattering by Arbitrary Shaped Anisotropic Objects Using Combined Field Surface Integral Equation Formulation", Journal of Electromagnetics, Vol 9, No. 1, pp. 215-230, January 1989.

32. B.Beker, K.Umashankar and A.Taflove, "Numerical Analysis and Validation of the Combined Field Surface Integral Equations for Electromagnetic Scattering by Arbitrary Shaped Two-Dimensional Anisotropic Objects", IEEE Transactions on Antennas and Propagation, Vol.37, pp. 1573-1581, December 1989.

33. T.J.Kim and G.A.Thiele, "A Hybrid Diffraction Technique - General Theory and Applications", IEEE Transactions on Antennas and Propagation, Vol. 30, pp. 888-897, 1982.

34. K.Umashankar, S.Nimmagadda and A.Taflove, "Numerical Analysis of Electromagnetic Scattering by Electrically Large Objects Using Spatial Decomposition Technique", IEEE Transactions on Antennas and propagation, August 1992.

35. G.A.Kriegsmann, A.Taflove and K.Umashankar, "A New Formulation of Electromagnetic Wave Scattering Using an On-Surface Radiation Boundary Condition", IEEE Transactions on Antennas and Propagation, AP-35, pp. 153-161, 1987.

36. C.M.Butler and K.Umashankar, "Electromagnetic Excitation of a Wire Through an Aperture Perforated Conducting Screen", IEEE Transactions on Antennas and Propagation, Vol.24, No.4, pp. 456-462, July 1976.

37. C.M.Butler and K.Umashankar, "Electromagnetic Penetration Through an Aperture in an Infinite Planer Screen Separating Two Half Spaces of Different Electromagnetic Properties", Radio Science, Vol.11, No.7, pp. 611-619, July 1976.

38. M.Abramwitz and I.A.Stegun, Handbook of Mathematical Functions, Dover Publications, New York 1965.

39. S.A.Schelkunoff, Electromagnetic Waves, D.Van Nostrand Company, Inc., New York, 1964.

40. J.A. Stratton, Electromagnetic Theory, McGraw-Hill Book Company, Inc., New York, 1941.

41. S.Ramo, J.R.Whinnery and T.Van Duzer, Fields and Waves in Communication Electronics, 2nd Edition, John Wiley & Sons, New York, 1984.

42. C.A.Balanis, Advanced Engineering Electromagnetics, John Wiley & Sons, New York, 1989.

43. K.Umashankar, Introduction to Engineering Electromagnetic Fields, World Scientific Publishing Company, New Jersey, 1989.

44. H.P.Hsu, Vector Analysis, Simon and Schuster, New York, 1969.

45. M.P. Do Carmo, Differential Geometry of Curves and Surfaces, Prentice-Hall, Inc., New Jersey, 1976.

Index

The Artech House Antenna Library

Helmut E. Schrank, *Series Editor*

Advanced Technology in Satellite Communication Antennas: Electrical and Mechanical Design, Takashi Kitsuregawa

Analysis, Design, and Measurement of Small and Low-Profile Antennas, K. Hirasawa

Analysis of Wire Antennas and Scatterers: Software and User's Manual, A. R. Djordjevic, M. B. Bazdar, G. M. Bazdar, G. M. Vitosevic, T. K. Sarkar, and R. F. Harrington

Analysis Methods for Electromagnetic Wave Problems, E. Yamashita, editor

Antenna Measurement Techniques, Gary E. Evans

Antenna-Based Signal Processing Techniques for Radar Systems, Alfonso Farina

CAD for Linear and Planar Antenna Arrays of Various Radiating Elements: Software and User's Manual, Aleksandar Nesic and Miodrag Mikavica

Designer Notes for Microwave Antennas, Richard C. Johnson

Fixed and Mobile Terminal Antennas, A. Kumar

Generalized Multipole Technique for Computational Electromagnetics, Cristian Hafner

Integral Equation Methods for Electromagnetics, N. Morita, N. Kumagai, and J. Mautz

Four-Armed Spiral Antennas, Robert G. Corzine and Joseph A. Mosko

Interference Suppression Techniques for Microwave Antennas and Transmitters, Ernest Freeman

Introduction to Electromagnetic Wave Propagation, Paul Rohan

Introduction to the Uniform Geometrical Theory of Diffraction, D.A. McNamara

Microwave Cavity Antennas, A. Kumar and H. D. Hristov

Millimeter-Wave Microstrip and Printed Circuit Antennas, Prakash Bhartia

Modern Methods of Reflector Antenna Analysis and Design, Craig Scott

Moment Methods in Antennas and Scattering, Robert C. Hansen, editor

Monopole Elements on Circular Ground Planes, M.M. Weiner *et al.*

Near-Field Antenna Measurements, D. Slater

Polariztion in Electromagnetic Systems, Warren Stutzman

Practical Phased-Array Antenna Systems, Eli Brookner *et al.*

Practical Simulation of Radar Antennas and Radomes, Herbert L. Hirsch and Douglas C. Grove

Reflector and Lens Antennas: Software User's Manual and Example Book, Carlyle J. Sletten, editor

Shipboard Antennas, Second Edition, Preston Law

Small-Aperture Radio Direction-Finding, Herndon Jenkins

Spectral Domain Method in Electromagnetics, Craig Scott

Two-Dimensional Multiple Multipole Analysis Software and User's Manual, Christian Hafner